The International Encyclopedia of Aviation

The International Encyclopedia of
Aviation

General Editor
David Mondey
**Assistant Compiler
Jane's All The World's Aircraft**

**Foreword by Juan Trippe
Founder, Pan American
World Airways**

HAMLYN

First published in Great Britain in 1977 by
Octopus Books Limited

© 1977, 1988 Octopus Books Limited

This edition first published in 1988 by
The Hamlyn Publishing Group
an imprint of the Octopus Publishing Group plc
Michelin House
81 Fulham Road
London SW3 6RB

Second impression 1989

ISBN 0 600 56080 5

Produced by Mandarin Offset
Printed and bound in Hong Kong

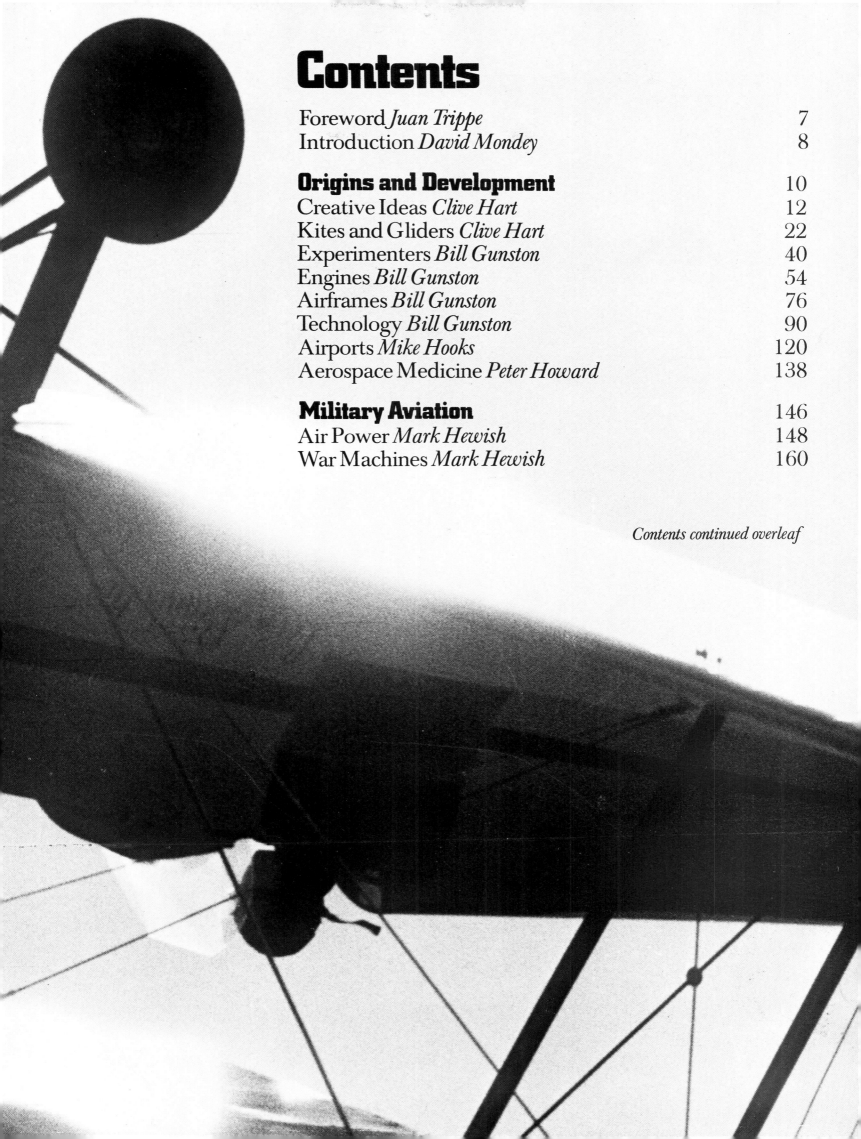

Contents

Contents continued overleaf

International Consultants

United States
John W Dennison, BS, MA
*US aviation historian currently employed
by the US Department of Defense*

United Kingdom
Kenneth Munson
*Assistant Compiler 'Jane's All the World's
Aircraft' and author and editor 'Pocket
Encyclopedia of World Aircraft'*

Australia
Henry R Rayner, MVO, BA
*Director of Public Relations, Defence
Department, Australia and author of a
history of the RAAF*

Canada
Neil A Macdougall, BSc, MA
*Contributing editor 'The Aeroplane' and
senior commercial pilot*

France
Jean Liron
*French aviation historian and contributing
editor 'Aviation Magazine International'*

Germany
Peter Pletschacher
*Co-editor 'FlugRevue', former press officer
Dornier GmbH and winner Otto
Lilienthal Award*

New Zealand
David Henderson
*Editor and contributor New Zealand
aviation publications*

South Africa
Ray L Watts
*African aviation historian and a director of
the Aviation Society of Africa*

Foreword

Juan Trippe

The penchant of human beings for perceiving themselves in flight goes back in time far beyond the Wright Brothers' effort in 1903. Dreams of flight are recorded among the earliest of myths. Only in the twentieth century, however, have the dreams come true; or, to put the matter another way, we of the twentieth century are able to see and do what our predecessors could only dream about – are able to see and undertake what earlier generations could visualize but could not bring about.

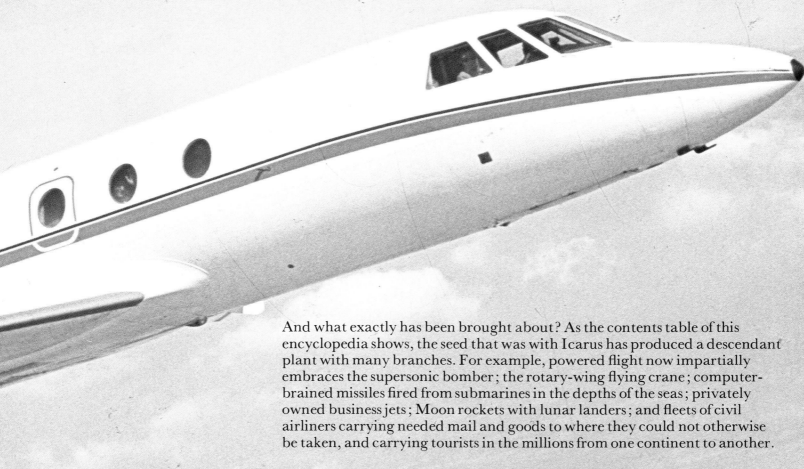

And what exactly has been brought about? As the contents table of this encyclopedia shows, the seed that was with Icarus has produced a descendant plant with many branches. For example, powered flight now impartially embraces the supersonic bomber; the rotary-wing flying crane; computer-brained missiles fired from submarines in the depths of the seas; privately owned business jets; Moon rockets with lunar landers; and fleets of civil airliners carrying needed mail and goods to where they could not otherwise be taken, and carrying tourists in the millions from one continent to another.

Those of us who are primarily concerned with civil aviation should be among the first to pay tribute to the accomplishments of military aviation, not only for its contribution to national security but for what military aviation consistently has done to advance the state of the art. However, we still harbor the hope and the conviction that it is the peaceful use of aviation that means the most, and will continue to mean the most, to the people of the world; that the low-fare airliner with its tourists is more significant than the bomber with its atom bombs; and that space flight, as epitomized by the American Space Shuttle programme, will be notable for its peaceful rather than its military application—as, for example, in capturing abundant solar energy for the use of all on Earth.

The record as set forth in this book is a good one, an impressive one. But as you read, remember that you are reading about the beginning, not the end.

Introduction

Almost invariably the most difficult section to write of any book is the introduction. When the aim has been, from the onset, to create an all-embracing volume to cover the entire history of aviation, the task is even more complicated.

This is no place for me to tell you anything about the excitement that is an integral part of the aviation scene. This you can read for yourself in the pages which follow. Rather would I explain how you can use this book to find the answer to a particular question, or to expand your knowledge of some branch of aviation. This is not to suggest that you should not browse, or read through an entire section: but in so doing you may find a subject mentioned briefly and seek the short-cut to more detailed information.

In broad outlines, you can guide your reading by reference to the contents page which gives, in some detail, the various sections of the book and their contributors. You may be

surprised to find that the conventional chronological format has not been followed. This was intentional from the moment that the work was planned. Because we were concerned not only to tell the tale but also to explain how and why aviation evolved in the way it did we chose a subject format rather than a strictly historical one. Thus in the chapter on War Machines the major sub-sections are Reconnaissance Aircraft, Bombers, Fighters, etc. rather than World War I, World War II, etc. The approach followed allowed for both analysis and straightforward description. An additional plus to this arrangement is that the book is easy to use as a reference source.

As you read through a particular section you will find sub-headings which will help you to follow a special interest. Invariably, brief mention of indirectly related subjects will appear in most sections, but the detailed index at the end of the volume should enable you to pinpoint the fact you are seeking. And

ENDPAPERS: *Red Arrows rehearsing in BAe Hawks*
HALF-TITLE: *Focke-Wulf FW.200 Condor*
TITLE SPREAD: *Grumman F-14A Tomcat*
CONTENTS SPREAD: *Stampe aerobatic aeroplane*
FOREWORD SPREAD: *Dassault (Mystère) Falcon 20*
THIS SPREAD: *Britten-Norman Trislander*

within this work there are a number of detailed studies of such items as 'How an aeroplane is flown', and these can be found listed in the index under the heading of 'Special features'.

Of necessity, the aim to cover the history of aviation, rocketry and space flight in a single, manageable volume, has imposed the need to condense the available material – written and pictorial – to key developments. Obviously, we have been as objective as possible in our selection, but inevitably a certain amount of disagreement will arise over the omissions. This is only to be expected, and, in fact, is a sign of the dynamism of aviation.

Aviation knows no boundaries. Great contributions have come from all nations. In acknowledging the international nature of aviation we have been aided considerably by our respected international consultants who have used their specialized knowledge to help maintain an equitable balance.

One other matter relating to an international readership is the need to give specification and performance details in both Imperial and metric notations. The metric equivalents used in this work are those in general use throughout the world, and which have been accepted internationally by common usage in the world's aviation press, and by such specialized publications as *Jane's All the World's Aircraft*.

As General Editor of this encyclopedia I must thank, most sincerely, those who have contributed enthusiastically the words and pictures which are the building blocks of any book. There must also be another great thank-you to the production team and artists who have taken words and pictures to create this book.

I trust you will enjoy our combined efforts.

D.M.
Surbiton 1977

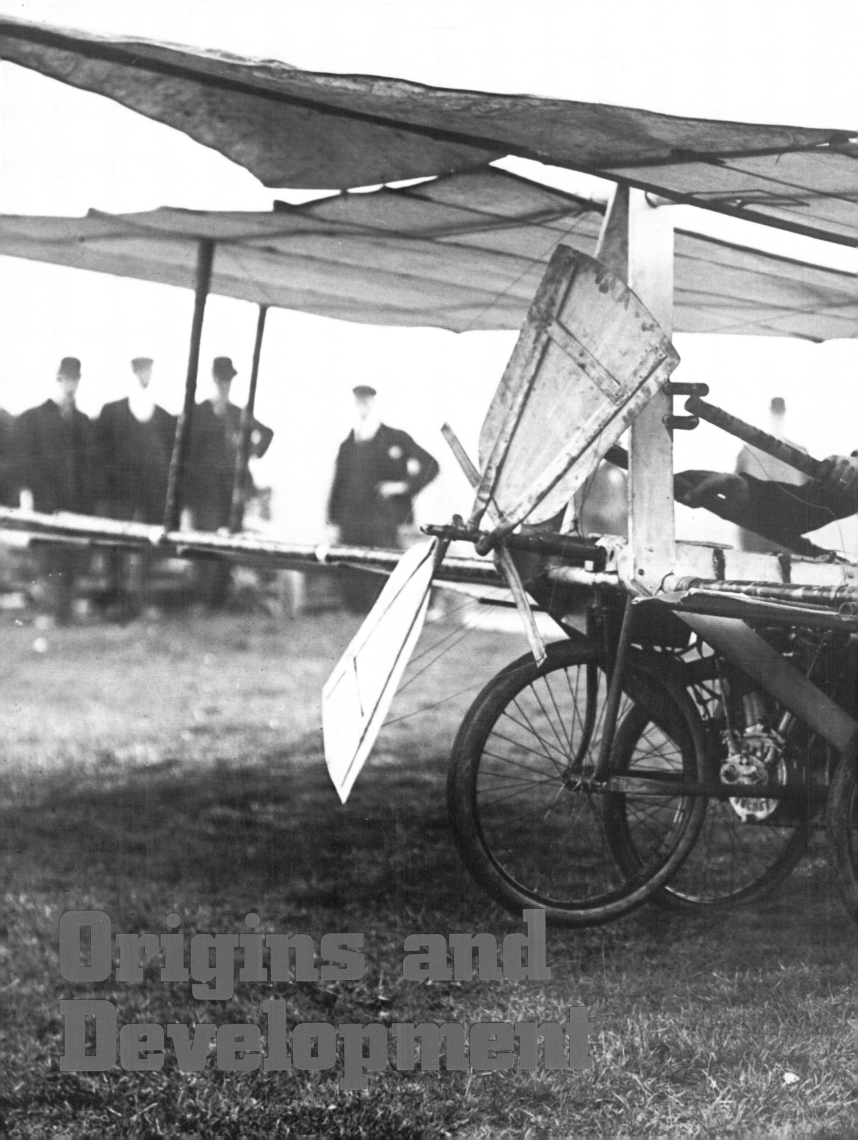

Origins and Development

Creative Ideas

BELOW: *Leonardo da Vinci's designs for a flying machine were based on the workings of a bird's wing. This drawing shows the pulleys intended to work the joints and the movable wingtip sections.*

BELOW: *Some of Leonardo's designs pushed ingenuity far beyond practicality. These wings were intended to be flapped by turning a handle which alternately wound and unwound cords wrapped round a drum.*

RIGHT: *For this ornithopter, drawn in about 1487, the downstroke of the wings was produced by kicking the legs backwards, while the upstroke, needing less strength, was performed by the hands and arms.*

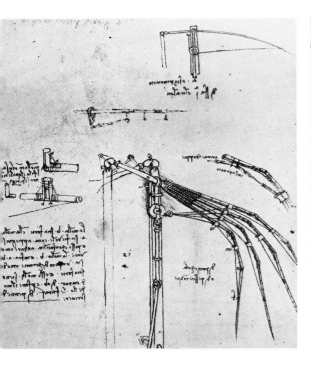

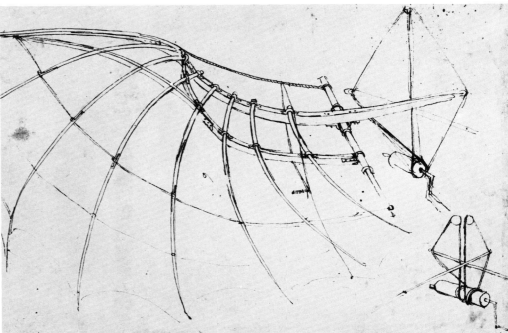

Although it is now natural to us to think of all forms of flight, except rocket propulsion, as dependent upon properties of the air, the essential role which air plays in providing a lifting force was not even vaguely understood until quite late in the history of man's attempts to fly.

Many early commentators, from at least as long ago as the days of Aristotle (4th century BC), thought it obvious that the flight of a bird, which aspiring aeronauts should attempt to imitate, was directly analogous to the action of swimming, which was in turn thought of as a process of pushing back against a resistance, as one pushes against the ground in walking. A bird, it was believed, flapped its wings downwards and backwards, pressing itself up and along against the air which lay behind.

Just as a swimmer is light in comparison with the water in which he is immersed, so it was assumed that a bird is built with sufficient inherent lightness to be able to sustain itself with little effort. In Aristotelian physics which, in a variety of forms, was generally dominant until the coming of the 'New Science' in the 17th century, both 'heaviness' and 'lightness' were believed to be positive qualities of the four 'elements' from which all things were made. The apparent density of an object was determined by the degree

PREVIOUS PAGES: *Early aviation experimenter Guillon gets ready to make an attempt on Epsom Downs on 11 April 1907 in his Guilly and Clouzy aeroplane.*

to which one or the other of these qualities dominated, and it seemed natural to early observers to think of birds as having a greater proportion of 'lightness' than man and other earthbound creatures.

The lightness, which was not just the absence of weight but a positive tendency to rise, was especially concentrated in the feathers, which explains why so many of those who, equipped with flimsy flying apparatus, threw themselves off towers and other eminences, so often chose feathers for their wings. As a bird can in fact fly neither because it is inherently light nor because it pushes down and back against the air (see p. 16), early concepts of flight suffered from a misunderstanding of two fundamental aspects of the physical laws associated with movement in a fluid medium: first, the principles of displacement and relative mass; second, the function of the flow of air over a curved surface to produce lift.

Mediaeval speculation

Although potential mediaeval fliers possessed neither an understanding of the essential principles nor the necessary technology, many thinkers engaged in active speculation about the possibility of human flight. Most famous of all, perhaps, was Friar Roger Bacon (c. 1220 to 1292), who wrote, 'It's possible to make Engines for flying, a man sitting in the midst whereof, by turning onely about an Instrument, which moves artificiall Wings made to beat the Aire, much after the fashion of a Birds flight'.

More impractical, but perhaps more ingenious and imaginative, was the idea proposed by Albert of Saxony (14th century), who suggested that as the element of 'fire' is lighter than air, just as air is lighter than water, so a ship filled with 'fire' might be made to float on the upper surface of the atmosphere. As the nature of elemental fire was a matter of controversy, Albert and his contemporaries were uncertain as to how hazardous such an undertaking might be. Any such ideas were nevertheless commonly held to be spiritually as well as physically dangerous, since, although flight would bring about the fulfilment of an age-old desire, it would also place man 'out of his element', above his station, and thus make him guilty of pride. The disastrous end to many practical attempts at flying served only to reinforce this powerful prejudice and slow the progress of aviation.

Leonardo da Vinci

In the early days many people engaged in dangerous flapping or gliding experiments unsupported by rational theory. Of the few who worked with intelligence and real seriousness, Leonardo da Vinci (1452 to 1519), was undoubtedly the most impressive. His ideas about bird flight were more sophisticated than those of most, and were based on the careful observations of a man blessed with the eye of a great painter. They were nonetheless totally mistaken, including not only the almost universal assumption of downward and backward wing movements, but also a

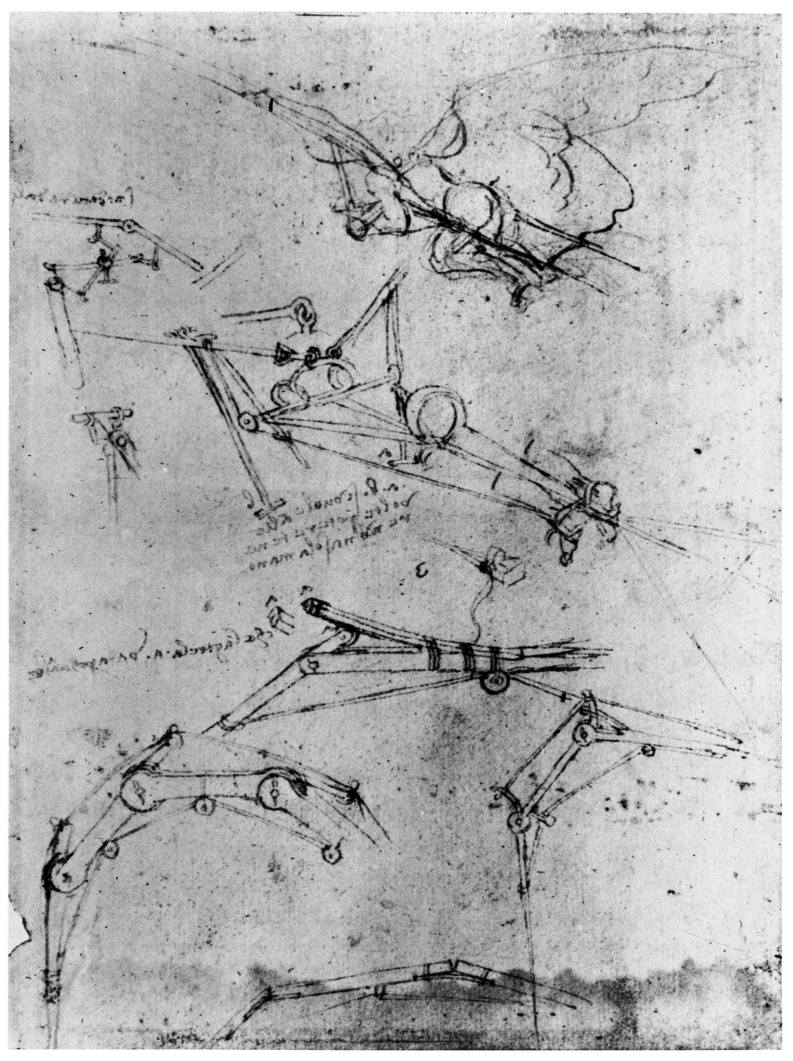

ABOVE: *The story of a flying monk, Kaspar Mohr, is depicted in this 17th century scene from a monastery in Schussenried (Germany). The artist has supplemented the usual arm movements by providing the flier with cords enabling him to add leg power to the downstroke.*
RIGHT: *When a brick is immersed in water its weight, as measured by the spring balance, is reduced by an amount equal to the weight of the water it displaces. The brick in air weighs 3kg (6·6lb) (top); when immersed the brick weighs 2·5kg (5·5lb) (bottom).*

belief that the bird 'squeezes' the air by a grasping motion of the wingtips, creating a series of dense air masses on which to support itself.

Basing his work on these ideas and on repeated observations of birds in soaring flight, Leonardo, who was a professional engineer as well as a painter, spent many years designing a variety of ornithopters, or flapping machines, some of which were relatively simple sets of artificial wings while others were full scale flying chariots. The plans were often meticulously thought out and included many fine points of engineering detail.

Leonardo believed that the ability of a bird to soar without flapping was dependent on the presence of a horizontal wind. If a bird raised the leading edge of its wings while facing upwind, so creating a positive 'angle of attack', it would, he thought, be raised by the wedgelike action of the air driving underneath. The inadequacy of contemporary ideas about gravity and relative motion misled Leonardo, rendering him incapable of grasping that (ignoring gusts and other changes of velocity) the relative motion of the air over the ground has no effect on the aerodynamics of flight. Although a bird's ground speed will of course be influenced by any wind which may be blowing, giving it far greater range downwind than upwind, its airspeed will be in no way affected, whatever its direction of flight.

Imagine a bird flying inside the cabin of a jumbo jet which is travelling fast over the ground. The bird may fly as it pleases, back and forth relative to the cabin; that the air inside the cabin is moving relative to the earth at speeds of hundreds of miles an hour does not affect the bird's motion relative to the air mass in which it is flying. A steady wind is moving relative to the earth in just the same way as the air inside the cabin, and affects the flight of the bird, or of any other flying object, just as little. Ground speed and airspeed are thus two very different frames of reference which need to be kept clearly distinguished. The earlier misconception about the effects of a steady wind on a flying machine or balloon was to form the basis of many futile designs from Renaissance times until almost the present day. Highly imaginative balloons and airships were often drawn with sails which, apart from the occasional effects of turbulence, would always have hung limply in the dead calm air bearing them along.

While Leonardo's hope that he might profit from the energy of the winds was totally unfounded, it resulted in some very sensitive analyses of the potentialities of gliding flight. Had he ever built a rigid glider, instead of devoting nearly all of his attention to flappers, he might well have made significant advances along the road which was later to be followed by Otto Lilienthal.

The nature of the air

The air is a mixture of gases, principally nitrogen (about 77 per cent) and oxygen (about 21 per cent). It surrounds the earth to a distance of several hundred miles, being most dense at sea level and thinning out rapidly with increase of height. At sea level the pressure which it exerts, in all directions, is sufficient to sustain a column of mercury (as used in barometers – literally 'pressure meters') to a height of about 760mm (29·92 in). Variations in climatic conditions cause this average pressure to fluctuate quite widely from day to day, and even from hour to hour, and as pressure readings are used to determine an aircraft's height above ground or sea level (on the 'altimeter'), pilots always need accurate and up to date information about variations in local barometric pressures.

Other than in special circumstances ('inversions'), there is a steady decrease in the temperature, as one rises through the lower regions of the atmosphere (the 'troposphere'), averaging about 2°C per 300m (1,000ft). After a few kilometres, the exact height varying with latitude, the temperature ceases to fall and remains fairly constant. Beyond that point, known as the tropopause, is the upper atmosphere, the stratosphere.

Apart from the effect on the altimeter of decreasing pressure, pilots need to take account of the change in performance of lifting surfaces, engine airspeed indicators and other instruments resulting from differences of both temperature and pressure. Winds at altitude are often stronger than those at sea level and may differ from them markedly in direction. Although steady winds have no effect on aircraft handling, they are of course a crucial factor in navigation, ground speed and therefore the aircraft's fuel consumption.

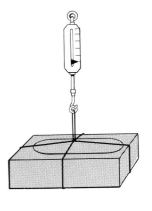

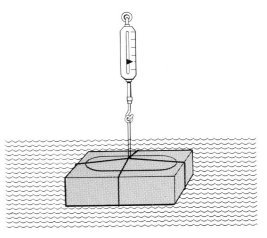

RIGHT: *This early 19th century engraving illustrates the means by which a balloon can be filled with a gas of a lower density than that of air to achieve aerostatic lift sufficient to carry passengers.*

Displacement

The ancient Greek scientist Archimedes (c. 287 to 212 BC) is credited with having been the first to understand that when an object is immersed in a fluid (e.g., water or air), its apparent weight is reduced by an amount equal to the weight of the fluid which it displaces. If a brick weighing, say, 3kg (6·6lb) occupies the same amount of space as ½kg (1·1lb) of water, it will appear to weigh only 2½kg (5·5lb) when totally immersed. The same principle applies even when the object to be immersed is initially lighter than the displaced fluid. A gas-filled balloon which weighs 200kg (441lb) and which occupies the same space as does 500kg (1,102lb) of air at sea level, will, when immersed in air, have its apparent weight reduced by 500kg (1,102lb). It will therefore weigh *minus* 300kg (661lb), which means the balloon will experience an upward lifting force of 300kg (661lb).

This kind of lift, the basic principle of aerostation (ballooning), has always been readily observable in the rising of hot air, which is lighter, as a result of expansion, than is cold air. But when smoke was seen to rise above fires, early speculators thought they were watching the effects of the inherent lightness of the 'element' of fire contained in the smoke. In the 15th century Giovanni da Fontana discussed the possibility of making a hot air balloon based on the rising of smoke, while even the Montgolfier brothers, who built the first man-carrying hot air balloon, tried to determine what kinds of smoke contained the highest concentration of 'lightness'.

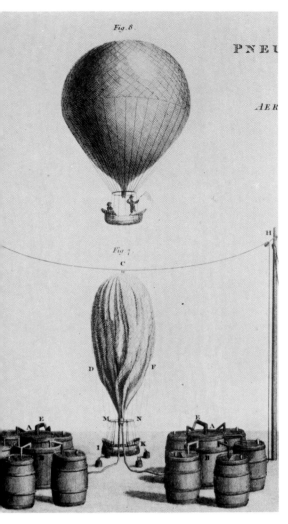

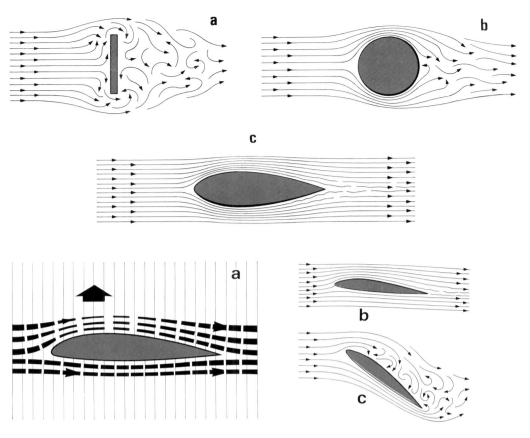

TOP: *The relative amounts of drag produced in a steady airstream by objects of the same cross-sectional area but with differing degrees of streamlining: (a) Flat plate (b) Ball (c) Streamlined shape.*

ABOVE: *(a) Cross-section of a typical aerofoil in an airstream. (b) Airflow over a wing in normal forward flight. (c) Airflow at the point of stall, where the wing is tilted at a high angle and lift rapidly decreases.*

Aerostatic lift

The same principle underlies the lift both of a hot air balloon and of a gas filled balloon. In both cases the envelope, being filled with a substance lighter than that which surrounds it, experiences an upward force equal to the difference between the weight of the displaced air and the weight of the gas or hot air in the balloon, plus the weight of the basket, etc. This difference in weight represents the maximum load which the balloon will then support in level flight.

As water is almost incompressible, its density remains virtually constant up to very considerable depths. By contrast the density of air, which is highly compressible, varies greatly with depth: at about 16km (10 miles) above the earth's surface it is only about one tenth of its value at sea level. Since a given volume of air is therefore lighter with increasing height, a point will be reached when the displaced air weighs the same as the whole balloon, after which the balloon will cease to rise. In most balloons the attainment of this height is delayed by the fact that the gas or hot air in the envelope, being subjected to less and less pressure, tends to make the envelope expand and displace more air, thus partially counteracting the effects of decreasing lift.

Aerodynamic lift

If an aerofoil such as an aircraft's wing is held at an appropriate angle of attack to an airstream, a difference is created in the total sums of the velocities of the parts of the air-stream flowing over the upper and lower surfaces. This difference, known as 'circulation', is the basis of aerodynamic lift. The sum of the velocities over the top of the aerofoil is greater than that over the bottom, and one of the basic physical laws of aerodynamics, 'Bernoulli's theorem', says in effect that, other things being equal, the faster a liquid or gas flows, the lower is its static pressure. This means that the downward pressure of the air on top of the wing is less than the upward pressure on the bottom, so that the wing experiences lift.

Up to the point at which the wing stalls, an increase in the angle of attack increases the 'circulation' and thus increases the total lift. For any given airspeed a pilot may therefore vary the lift by raising or lowering the aircraft's nose.

The circulation theory of lift was first enunciated by the Englishman F. W. Lanchester, in an address given in 1894 to the Birmingham Natural History Society and later published in his book *Aerodynamics* (1907). The idea was independently developed in Germany by Ludwig Prandtl.

Drag and vortices

The movement of air over an aircraft's wings (or of the aircraft through still air, which amounts to the same thing) results not only in lift, but also in the production of a number of different kinds of 'drag'. Apart from the retarding force created by friction between the airflow and the surface area of the aeroplane ('skin friction' on the so-called 'wetted area'), the main drag elements are 'profile drag' and 'induced drag'. Profile drag is a measure of the amount of energy consumed in disturbing the uniform horizontal flow, pushing it aside and to some degree breaking it up into turbulence. With an aircraft as with a motor car, profile drag may be reduced by 'streamlining', the effectiveness of which may be judged from comparing the drag created by a flat circular plate held at right angles to the airflow, with a sphere of the same diameter, and by a fully streamlined figure of the same maximum thickness.

While profile drag wastes energy, and while, consistent with other design considerations, it must be reduced as much as possible, it is less important than induced drag. When an ordinary wing is pushed through the air, the flow over the top and bottom surfaces is not exactly parallel with the direction of motion: on the bottom it is angled slightly outwards, towards the wing tips, while on the top it is angled slightly in, towards the fuselage. This is a result of the high pressure underneath trying to 'leak' around the wing tips towards the region of low pressure. When the two airflows meet at the trailing edge they are therefore moving at a slight angle to one another and form swirling vortices in the wake of the wing ('wake turbulence'). These normally roll up together to form a single powerful vortex trailing back from each wing tip, the direction of the spiral being such that when seen from behind, the lefthand vortex turns clockwise while the righthand vortex turns anti-

BELOW LEFT: *These diagrams illustrate the wide variations possible in the aspect ratios of different aircraft. The aspect ratio is the ratio of the span to the chord of an aerofoil: (a) a sailplane has a high aspect ratio – a large span and small chord; (b) a lightplane has a low aspect ratio – a small span and large chord.*

BELOW RIGHT: *Wing tip vortices being shed from a typical aircraft in flight. Such vortices can sometimes become visible at high altitudes, owing to the condensation of water vapour. The diagrams below, based on the work of the physicist F. W. Lanchester, explain in detail how these vortices are formed.*

the theory of lift was fully investigated in the late 19th century.

A bird's wing may be considered as performing two related but separate functions: first, it provides a horizontal lifting force, or 'thrust', which pulls the bird through the air and creates a horizontal airflow over the wings; second, its cambered aerofoil shape provides a vertical lifting force which sustains the bird's weight. For centuries it seemed desirable to most experimenters to imitate nature by aiming at a single unified system of lift and thrust through the construction of artificial flapping wings. Some work on ornithopters, by no means all of it negligible, continues to be done, but successful manned flight did not become a reality until the idea of separating the systems of lift and thrust were fully conceived. In a modern aeroplane the wing, which contributes most of the lift, remains essentially rigid. The thrust is provided by an engine driving a stream of air or hot gases backwards and thereby causing the aircraft to move with equal momentum in the opposite direction (compare the recoil from a gun).

Although the creation of a lightweight engine was necessary before the separation of lift and thrust could be successfully achieved, credit for the first important insight into the matter in connection with heavier-than-air flight must go to Sir George Cayley (1773 to 1857), rightly dubbed 'the Father of Aerial Navigation'. In 1799 Cayley had a silver disc engraved with a now famous design for a man-carrying aeroplane, equipped with rigid wings and flappers for propulsion. Three centuries earlier, Leonardo da Vinci went on one occasion at least part of the way towards abandoning his usual assumptions when he sketched a design for a hang glider with flappable wing tips, the inboard sections of the wings remaining rigid. It is possible that this machine might have brought some limited success, but there is no evidence that Leonardo ever tried it in practice, and he did not proceed with the idea.

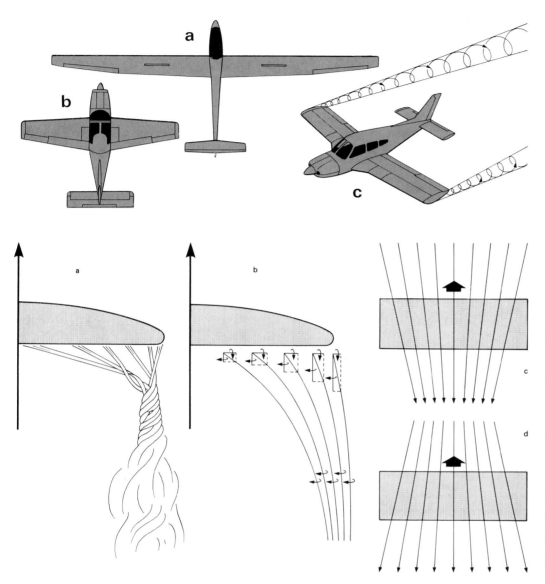

ABOVE: *Four diagrams from F. W. Lanchester's* Aerodynamics *(1907).* ON THE LEFT: *The production of trailing edge vortices and the way they gather together to form a single wingtip vortex.* CENTRE: *The components of airstream velocity which lead to the creation of the vortices.* RIGHT: *Angled airflow over the upper (above) and the lower (below) surfaces of a wing, resulting in vortices at the trailing edge.*

clockwise. The energy consumed in these vortices must of course be derived from the aircraft's engine.

The effect of speed on the two forms of drag differs. While profile drag increases with increasing speed, induced drag is greatest at low speeds and high angles of attack. The turbulence which results from induced vortices can be a serious hazard to other aircraft. Since an aeroplane flies most slowly when taking off and landing, it is then that the vortices are at their most powerful, and unless light aircraft remain several miles behind large, slow-moving jets,

there is a grave danger of their being thrown totally out of control.

Separation of lift and thrust

A bird flies by pushing its wings downward through the air so that the 'primary' or 'flight' feathers at the wing tips (or, in some birds, the wing tips as a whole) twist upwards at the trailing edge, while on the upstroke the trailing edges are twisted slightly down. Each downward movement of a wing is analogous to the movement of a propeller blade through the same arc, forward 'lift' being produced and used to pull the bird along. During the downstroke, the lift also has a vertical component which, together with the upward lift resulting from the forward motion of the whole wing, sustains the bird in the air. (The upstroke, which contributes further vertical lift, actually retards the bird's forward progress a little.) Although some aspects of these movements and forces were understood by G. A. Borelli as early as 1680, they were not properly analyzed until

Stability and flight control

Stability may be broadly defined as the tendency of an aircraft to remain in the condition of flight selected by the pilot: straight and level, turning, climbing, or descending. The degree of stability must be carefully designed, since if the aircraft is unstable it will be difficult or impossible to control, while if it is overendowed with stability it will be difficult to manoeuvre quickly or may require excessive effort (stick forces) from the pilot.

Since an aircraft is free to move in all three dimensions of space, it needs to be both stable and controllable about three axes: (a) a vertical axis perpendicular to the surface of the wings; (b) a longitudinal axis passing through the fuselage from front to back; (c) a horizontal axis passing through the wings approximately from tip to tip. The three kinds of movement about these axes are called respectively *yaw*, *roll* and *pitch*.

Automatic stability in yaw is effected by a vertical fin at the rear. If the nose of an aircraft is gusted to the left, say, the fin will be

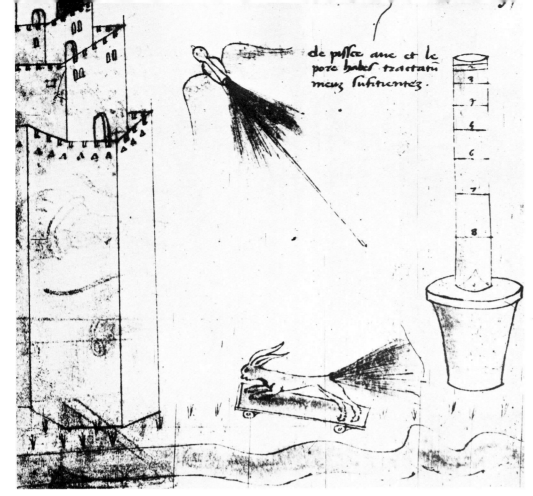

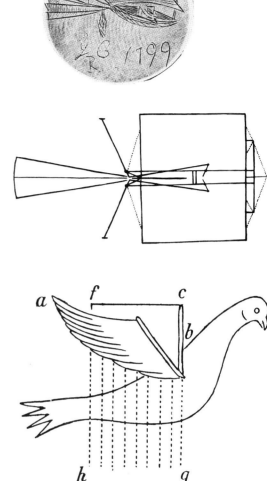

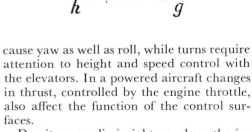

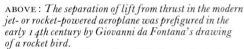

ABOVE: *The separation of lift from thrust in the modern jet- or rocket-powered aeroplane was prefigured in the early 14th century by Giovanni da Fontana's drawing of a rocket bird.*

ABOVE RIGHT: *The earliest of Sir George Cayley's aeronautical sketches was also one of the most prophetic. This engraving on a small silver disc, dated 1799, shows a design for a man-powered aircraft with a boat-shaped hull, curved fixed wing of low aspect ratio, adjustable cruciform tail unit and hand-operated flappers.*

CENTRE RIGHT: *Plan view of Cayley's first proposal for an aircraft.*

RIGHT: *In 1680 G. A. Borelli understood that a bird does not push backwards against the air, but that flapping causes the primary feathers of the wings to flex up and down about a line near the leading edge. In his great work,* De motu animalium, *Borelli proved conclusively that man was incapable of self-powered flight and thus opened new and productive paths of thought on flight.*

placed at an angle to the airflow, producing lift on its left side which will tend to pull the aircraft back into line with the airflow. A fixed horizontal tailplane confers stability in pitch by being so rigged that in normal flight its angle of attack is less than that of the wings and may often, in fact, be zero. As a consequence, any change in pitch angle of the whole aircraft causes a change in lift on the tailplane proportionally greater than that on the wings. This additional lift acts in the direction opposite to the original pitch change and so tends to return the aircraft to its normal flight path.

Stability in roll is achieved by setting the wings at a slight upward angle (*dihedral*), or by arranging the centre of gravity to be below the level of the wings (*pendulous stability*). The operation of the stabilizing forces in roll is slightly more complex than is the case with the other kinds of stability, and depends on the tendency of the aircraft to sideslip if rolled, or banked, sideways in level flight. If dihedral is used, the lift on the downgoing wing is increased by the slip, so introducing a correcting force which rights the aircraft, while pendulous stability is based on the tendency of the centre of gravity to swing back to the vertical as if it were momentarily suspended beneath a pressure point created by the lateral airflow in the slip. It should be noted that the initial ten-

dency to sideslip is essential to the function of both kinds of lateral stability.

In a conventional aircraft, control about the three axes is achieved by the use of a rudder (behind the fin), the elevators (behind the tailplane), and the ailerons (behind the wingtips). All three 'control surfaces' function by varying the total curvature and angle of attack of the aerofoil which they comprise in conjunction with the fixed surfaces ahead of them. Thus if the pilot lowers the elevators (by pushing forward on the stick or control column), the total effective curvature of the tailplane/elevator combination is increased, and, since the trailing edge is now lower than the leading edge, the angle of attack is also increased. The combined surface then acts to produce more lift, and the tail will rise. The rudder acts similarly, to move the rear of the aircraft from side to side (yaw), while the ailerons, which act simultaneously in opposite directions (one rising while the other is lowered), decrease the lift on one wing tip while increasing it on the other, so causing the aircraft to roll as a result of the asymmetrical forces on the wings.

The forces which operate in the three axes of movement are not independent, but interact in various ways, so that harmonized and coordinated use of the control surfaces is necessary. Use of rudder alone will result in roll as well as yaw, use of aileron alone will

cause yaw as well as roll, while turns require attention to height and speed control with the elevators. In a powered aircraft changes in thrust, controlled by the engine throttle, also affect the function of the control surfaces.

Despite sporadic insights, such as the incorporation by Cayley of a combined adjustable elevator and rudder into his model glider of 1804, or the foreshadowing of modern methods in a prophetic design by Alphonse Pénaud in 1876, the coordinated use of controls was slow to emerge. A full understanding of their interrelationship began only with the successful solution by the Wright brothers of the problems of 'adverse yaw'. An increase in the curvature and angle of attack of an aerofoil, such as occurs when a control surface is depressed, results not only in increased lift but also in increased drag, and the same occurs if the change in aerofoil configuration is brought about by the bending or twisting of a wing tip, as was done by the Wrights. While this increase in drag is not of great significance in the case of elevators and rudders, it was the cause of great difficulties in the control of roll until the Wrights found a cure in the coordinated use of rudder. In many modern aircraft sophisticated modifications of design detail reduce the adverse performance effects of aileron drag.

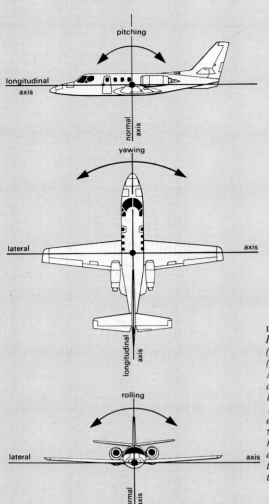

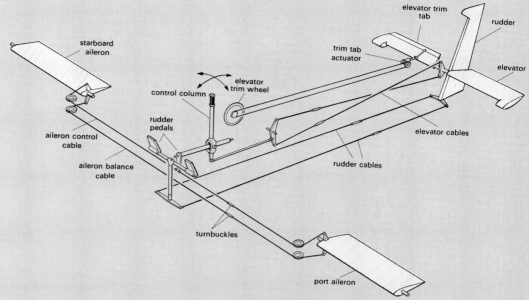

LEFT: *The three axes about which an aircraft moves. From top: the lateral (side-to-side) axis; the vertical (through top and belly) axis; and the longitudinal (nose-to-tail) axis.*
ABOVE: *The cable-operated control system of a conventional modern aircraft. This shows the location of levers, pulleys and cable turn-buckles used to adjust the system and attain the correct tension. Some aircraft have a different system: they use a single push-pull rod or tube.*
NEAR RIGHT ABOVE: *Stability in pitch is achieved by having the horizontal tailplane set at a smaller angle of attack than that of the wings. Changes in pitch angle then have a proportionally greater effect on the lift of the tailplane, tending to return the aircraft to level flight.*

NEAR RIGHT CENTRE: *Wings set at a dihedral angle give stability in roll. If, when out of balance, the aircraft slips sideways, the down-going wing produces proportionally greater lift, which counteracts the roll, and restores the aircraft to level flight.*
NEAR RIGHT BELOW: *Pendulous stability counteracts unwanted roll when the centre of gravity of the aircraft swings back to the level position beneath a lateral pressure centre created by the airstream in the direction of the sideslip.*
CENTRE RIGHT ABOVE: *The forces acting on an aircraft in balanced level flight, where the opposing forces are in equilibrium: lift balances weight, and thrust balances drag.*

Forces in flight

In straight and level flight the total aero-dynamic reaction of the wing in response to the airflow is a force which acts both upwards and backwards, the degree of backward slope of the force being determined by the 'lift: drag ratio'. By definition, the 'lift' is that part of the total force which acts at right angles to the airflow, irrespective of the aircraft's direction of travel relative to the earth, while the 'drag' is that part of the total force which acts at right angles to the lift, parallel to the airflow. Thus, if the aircraft is in a vertical dive, the lift will be parallel to the ground and the drag will act vertically upwards. More lift is produced at some points on the aerofoil surface than at others, the amounts varying with different angles of attack and airspeeds. For any given flight condition, however, a point may be found on the curve of the aerofoil through which all of the resultant forces may be said to act. That point is known as the 'centre of pressure'.

At a constant airspeed in straight and level flight, four forces act in equilibrium: the lift exactly balances the weight (if it is greater than the weight the aircraft will rise, while if it is less the aircraft will sink); the thrust, provided by the engine, exactly balances the drag (if it is greater than the drag the air-craft will accelerate, while if it is less the aircraft will slow down). Any changes in air-speed will also affect the velocity of the air moving over the upper and lower surfaces of the wings, increasing or decreasing the static pressure, and thus affecting the lift.

Forces in a balanced turn

In a balanced turn the aircraft is banked towards the inside of the circle. If a turn is made with insufficient bank a skid away from the centre of the intended circle will result, while too much bank will cause the aircraft to slip downwards towards the centre of the turn. The correct degree of bank depends on the radius of the turn and on the speed, the situation being analogous to the motion of a car around a curved road which is banked up on the outside of the turn. For any banked curve there is one precise speed at which the car will tend neither to skid outwards nor to slip inwards.

Since the aeroplane is banked in the turn, the lift, which is by definition at right angles to the airflow, is no longer vertical but is angled towards the inside of the circle, thus providing a horizontal force which in fact keeps the aeroplane turning. The upward component of the lift, being less than in straight and level flight, no longer balances the weight, and unless the pilot takes appro-priate action (by raising the nose a little, or by applying power, or both) the aircraft will sink and will begin to describe a 'spiral dive'. In a correctly banked turn, no matter how steep, the aircraft is in perfect balance: a pendulum will remain hanging parallel to the vertical axis of the fuselage and a glass of water will not spill.

Stalling

The lift produced by wings of conventional aerofoil shape may be continuously increased by increasing the angle of attack up to about 16°. Beyond that point the airflow over the upper surfaces is subject to forces too great to allow it to flow smoothly, with the result that it separates from the surface, becomes turbulent, and causes the lift to decay. At that point the wing is said to stall. Not only does the lift decrease beyond the stalling angle, often very rapidly, but the centre of pressure, which up to the stall has been moving continuously forward towards the leading edge of the wing, rapidly retreats in the direction of the trailing edge. Beyond the stall the aircraft therefore suddenly has insufficient lift and is nose heavy. If attempts are made to keep the nose up the aircraft will sink rapidly, and since the wings are stalled there will be a loss of control.

Small departures from symmetry, together with other factors such as the spiralling slip-stream from a propeller, may cause one wing to stall before the other, so that the aircraft falls sideways out of control in a sometimes alarming fashion. In a conventional aircraft, recovery consists in lowering the elevators, keeping straight with the rudder, and apply-ing power. The angle of attack is thereby reduced and the rapid build up of lift enables control to be regained.

The speed at which the stall occurs varies with the effective weight of the aircraft and with the loading on the wings (the total forces acting in the direction opposite to the lift). The loading is markedly increased by the centrifugal force of a tight turn or a pull out from a dive, and in these manoeuvres the

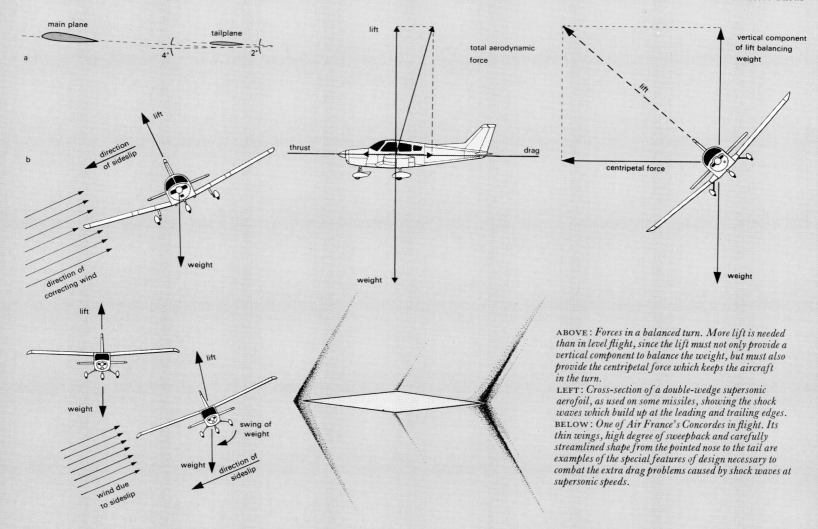

a

main plane   tailplane
                    4°   2°

b

lift

direction of sideslip

direction of correcting wind

weight

lift

thrust   drag

total aerodynamic force

weight

lift

vertical component of lift balancing weight

centripetal force

weight

lift

weight

lift

weight

swing of weight

weight   direction of sideslip

wind due to sideslip

ABOVE: *Forces in a balanced turn. More lift is needed than in level flight, since the lift must not only provide a vertical component to balance the weight, but must also provide the centripetal force which keeps the aircraft in the turn.*
LEFT: *Cross-section of a double-wedge supersonic aerofoil, as used on some missiles, showing the shock waves which build up at the leading and trailing edges.*
BELOW: *One of Air France's Concordes in flight. Its thin wings, high degree of sweepback and carefully streamlined shape from the pointed nose to the tail are examples of the special features of design necessary to combat the extra drag problems caused by shock waves at supersonic speeds.*

stalling angle may be reached at a speed very much higher than usual. A stall under such conditions is known as a 'high speed stall'.

Boundary layer control

Owing to the friction which exists between the airflow and the aerofoil, the layer of air close to the surface, the boundary layer, is subject to very powerful retarding forces which slow it up. Fast flowing layers of air therefore have to pass over slower layers beneath. If this flow pattern is everywhere smooth, the boundary layer is said to be laminar, while if the air is unable to travel smoothly the flow is said to be turbulent. The usual tendency is for the flow to be laminar near the leading edge and to grow turbulent at a point further back towards the trailing edge, known as the 'transition point'.

Boundary Layer Control (BLC) has become an important branch of applied aerodynamics, especially in connection with transonic and supersonic flight. A variety of devices has been invented to ensure that the flow remains laminar, although paradoxically it is sometimes beneficial to encourage a small amount of controlled turbulence very close to the wing, since this has the effect of preventing or delaying wholesale breakdown of the flow pattern ('separation').

Slats, slots and flaps all serve to control the boundary layer, as well as to alter the geometry and surface area of a wing, while the most readily observable devices for creating a small amount of boundary layer turbulence

are the vortex generators (small plates a centimeter or two in height), which may be seen in rows on the upper surfaces of some high speed wings.

Supersonic flight

At speeds approaching the speed of sound new factors, resulting from the compressibility of air, affect the function of an aerofoil. As an aircraft passes through the air, disturbances are continually generated, and in subsonic flight these are dissipated through the surrounding air in the form of waves, which escape from the region of the aircraft at the speed of sound. As that speed is approached, however, the disturbances ahead of the aircraft are unable to escape fast enough. A piling up of such disturbances,

known as a 'shock wave', results. (A similar shock wave, at very much lower speeds, may be seen trailing back on either side of a duck swimming across a still pond.)

Shock waves are generated by various parts of an aircraft travelling near, at or beyond the speed of sound, and in the case of a wing they appear, in the fully formed condition, at the leading and trailing edges. As they are a potential source of rapid increases in drag, they require special geometrical features in the design of aircraft intended for transonic or supersonic flight. These features include the use of thin wings, symmetrically shaped aerofoil sections, pointed leading and trailing edges, high degrees of sweepback, and the careful graduation of cross-sectional area of the entire aircraft from nose to tail.

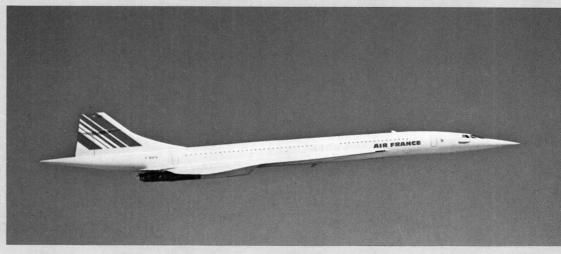

How an Aeroplane is Flown

Very early in the history of powered flight, the layout and function of the basic controls were standardized. The one significant variation among modern light aircraft is the designer's choice of either a 'stick' or a 'control column' for combined control of pitch and bank. The stick is free to move in any direction about a pivot on the floor between the pilot's legs, while the column moves only back and forth for control of pitch, control of roll being effected by turning a quadrant wheel or crossbar with handles. Many pilots, especially those who enjoy aerobatics, prefer the stick, which allows for easier harmonization of the controls. (A stick nevertheless has some minor disadvantages, such as making it harder to hold a map on one's knee.)

The primary controls consist of the stick or column, rudder pedals (or a turnable rudder bar), and the throttle, which is hand operated. As the throttle may be placed, according to the aircraft's design, on either side of the pilot, it is necessary to be able to fly ambidextrously. A light but positive grip is used on the stick, while the other hand operates the throttle, which is moved forward to increase power. Either hand may be needed, according to the particular aircraft, for flap levers, switches, and trimmers. If a variable pitch (or constant speed) propeller is fitted, its control lever is normally placed next to the throttle. Wheel brakes, which were uncommon in early aircraft, have now become standard; they are usually operated by toe (or sometimes heel) pedals associated with the rudder pedals.

The stick moves in the 'natural' sense. That is, the nose of the aircraft is lowered by pressing the stick forward, and raised by pressing it back. (Although these adjustments will often result in a descent or in a climb, respectively, they will not always do so, the relative movement in the vertical direction being dependent on other factors as well as attitude.) Movement of the stick left or right banks the aircraft left or right. Pilots sometimes disagree as to whether the rudder controls also work in a natural sense. Forward pressure with the right foot yaws the aircraft to the right, this movement being therefore opposite to the turning of a bicycle's handlebars. It may be more helpful to think of steering a skiff by hand held lines (right hand forward for a turn to the right), but in any case pilots rapidly grow accustomed to the required direction of movement.

What follows is a description of the bare essentials in the handling of a single-engined light aircraft flying one circuit at an aerodrome (i.e., take-off, return downwind, and landing). The aircraft is assumed to have a stick, left hand throttle, fixed pitch propeller rotating clockwise as seen by the pilot, and a fixed tricycle landing gear. Trimmers, flaps, brakes, and many other details are ignored. While substantial differences arise in the handling of other types, the conventional light aeroplane still provides a common basis of experience for the coordinated use of controls.

Once cleared for take-off and lined up on the runway, the pilot opens the throttle, pushing it smoothly forward, until the engine reaches take-off power (which may be the same as full power). The stick is held a little back, to relieve the pressure on the comparatively fragile nosewheel, and as the aircraft begins to move forward the airflow over the tail surfaces rapidly establishes rudder control. Engine torque and forces associated with the slipstream create a tendency to swing to the left, and this must be counteracted by pressure on the right rudder pedal. The control surfaces grow more responsive with increasing airspeed, and finer adjustments are needed to keep the aircraft running straight. When flying speed has been reached, progressive back pressure on the stick raises the nose, increasing the angle of attack of the wings which then provide sufficient lift for take-off. The stick is used to settle the aircraft into a climb at the best climbing speed: moving the stick forward increases

LEFT: *Cherokee. The Cherokee Archer II, made by the Piper Company, is typical of modern light touring aircraft designed for safety and comfort.*

ABOVE LEFT: *A Cessna cockpit, showing dual control columns, rudder pedals on the floor, flight instruments (left), engine instruments (right), throttle in the middle.*

ABOVE RIGHT: *Rallye. The range of low wing Rallye aircraft, made in France, are especially useful for their ability to operate from very short runways.*

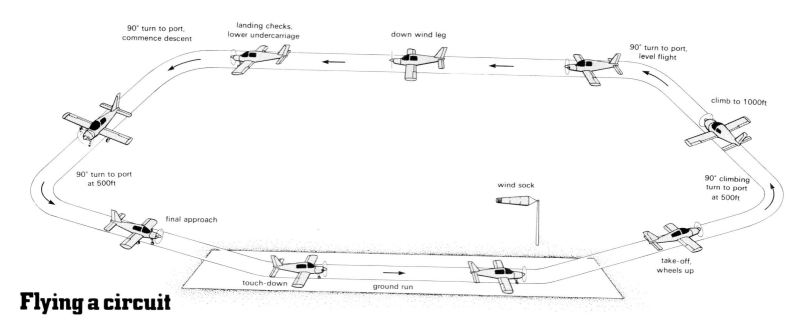

90° turn to port, commence descent

landing checks, lower undercarriage

down wind leg

90° turn to port, level flight

climb to 1000ft

90° turn to port at 500ft

90° climbing turn to port at 500ft

final approach

wind sock

touch-down

ground run

take-off, wheels up

Flying a circuit

the speed (by reducing the angle of attack, and hence the induced drag), while back pressure reduces speed. Angle of climb and rate of climb are controlled indirectly by maintaining a predetermined airspeed.

Once past the end of the runway and at a safe height (say about 150m, or 500ft), the pilot initiates a gentle turn (usually to the left), still climbing. This he does by gentle movement of the stick to the left, applying at the same time a little left rudder, which may need no more than a relaxation of the pressure which he has been maintaining with his right foot. To maintain the appropriate angle of bank once it has been established, the stick is returned to the central position, and may even need to be pressed a little to the right, to prevent the aircraft from overbanking.

With the turn completed, the aircraft is returned to the straight climb until 'circuit height' is reached (usually 300m, or 1,000ft). In order to level out, the pilot presses the stick gently forward until the aircraft is

in level flight. It now begins to accelerate and reaches circuit cruising speed in a few seconds. As that speed is approached, the pilot eases back the throttle to cruising power. With reduction of power, the tendency to yaw to the left decreases or may vanish altogether, and the pilot adjusts his use of rudder accordingly.

A level turn to the left is now made, to head downwind, parallel to the runway. This time the aircraft is banked a little more steeply and, in view of the lower power setting, a little more left rudder may be needed than in the climbing turn. The stick is gently moved both to the left and back, a slight increase in angle of attack, and a consequent slight loss of airspeed, being necessary in order to provide additional lift in the turn. If the turn were executed without backpressure on the stick, the nose would drop and the aircraft would lose height. A third turn at the end of the 'downwind leg' of the circuit is followed by the start of the descent, the pilot easing back the throttle and

using the elevators to maintain an appropriate airspeed. Once again the rudder pressure must be adjusted since at a very low power setting there may be a tendency to yaw to the right.

A final, descending turn establishes the aircraft on the descent path in line with the runway. In a descending turn the aircraft may resist the application of bank, so that a positive pressure of the stick to the left may prove necessary. Since there is more drag in the turn than in straight flight, the stick must be used to maintain airspeed.

On the final approach the aircraft's attitude, speed and angle of descent are controlled by coordinated use of stick and throttle. Once the runway threshold is reached, and when the aircraft is some 5m (16ft) or so above the ground, the stick is used to level off, power is progressively reduced, and the aircraft is kept flying for as long as possible by continuous very slight increases in angle of attack, produced by further

back pressure. When the moment of touchdown is reached, the aircraft should ideally be flying as slowly as possible, very near the stall, and therefore in a marked nose up attitude.

During the flight the pilot has been keeping a careful lookout for other traffic; he has been monitoring the flight and engine instruments, adjusting the aircraft's attitude in relation to the horizon, allowing for the effects of wind, and trying as much as possible to be at one with the machine. Man is not properly equipped for movement in three dimensions, and there is no doubt that flying by 'feel' alone can be extremely hazardous, as anyone knows who has tried to fly (accompanied by another pilot!) with his eyes closed. Coordination and a sense of balance are nevertheless essential, since an aeroplane is not pushed around the sky like an airborne motor car; to maintain balance it must be continuously and delicately manoeuvred in relation to its three axes of freedom .

Kites and Gliders

Although kites vary widely in their characteristics and design, they may generally be defined as aerodynes (flying machines that are heavier than air) tethered to the ground by a flexible line or lines, and capable of being sustained at a positive angle to the horizon by lift resulting from airspeed in a horizontal wind. (The airspeed may be augmented by 'groundpower', e.g. by running with the line or by attaching it to a moving vehicle.)

As with any flying machine, the kite's aerofoil surfaces produce both lift and drag, and the higher the lift:drag ratio the greater is the angle of flight. A diagram of forces may be drawn for a kite in steady flight analogous to that which represents an aeroplane flying straight and level. The aeroplane's weight is exactly balanced by its lift, and its drag by its thrust. In the case of the kite, the lift is balanced by the weight plus the downward vertical component of the 'thrust' (i.e., the pull of the flying line), while the drag is balanced by the horizontal component of the thrust. For the attainment of greatest height

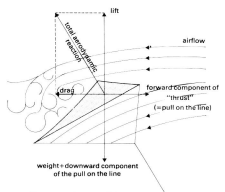

ABOVE: *Forces acting on a kite.*

and for the lifting of the heaviest loads, a kite should clearly be designed to have maximum overall lift combined with the best possible lift:drag ratio, while for towing drogues, vehicles, etc., a high ratio of drag is needed.

Although most kites consist of a symmetrical, rigid framework covered with a light fabric such as paper or cloth, several high performance kites developed in recent years have no rigid members at all, but use air pressure, controlled by shroud lines and flexible fabric keels, to maintain the correct aerodynamic shape. The absence of rigid members not only reduces the kite's weight, thereby increasing its angle of flight, but also enables the kite to assume a more aerodynamically satisfactory shape than is possible with most rigid designs.

In contrast to an aeroplane or glider in normal flight, most simple kites are usually stalled, the turbulence on the upper surface producing a great deal of drag. The best non-rigid kites are designed to fly at angles below the stall, and the cover of some simple kites, such as the Eddy bow, can be made to

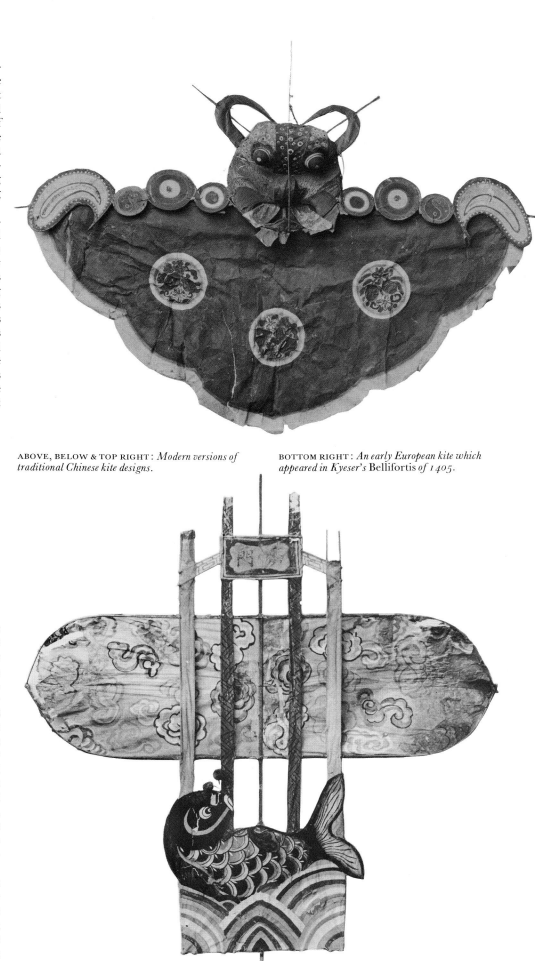

ABOVE, BELOW & TOP RIGHT: *Modern versions of traditional Chinese kite designs.*

BOTTOM RIGHT: *An early European kite which appeared in Kyeser's* Bellifortis *of 1405.*

assume a shape approximating to a curved aerofoil, which results in efficient, unstalled flight.

The early history of kites

Where and when kites began is not entirely clear. Although they were probably known in China several centuries before Christ, the earliest firm evidence of their having been used there dates from about the 2nd or 3rd

century BC. In various parts of the Pacific, kites rather different from those flown in China have been known for many centuries, and it is possible that they were independently invented. In any case, the kite is not only the earliest recorded form of aircraft but is also among the earliest of discrete technological objects having a continuous history.

Chinese kites and their derivatives in Japan, Korea and South-East Asia, fall into two broad categories. The first is a simple, rectangular design consisting of a framework of crossed sticks covered by paper or cloth and using a multileg bridle attached to various points on the surface. Such kites were used not only for creating spectacular effects at the kite flying festivals which used to be common in the east, but also for lifting heavy weights and even men. The second is a broad category of figure kites representing men, animals, dragons, insects, and inanimate objects. Such kites are usually smaller than the rectangular ones and have always been flown entirely for recreational purposes. It was not until about 1,000 years ago, however, that kites in China began to be associated with the play of children. Before then they appear to have been used exclusively as an adult form of amusement.

It is difficult to decide what may be the earliest references to kites in China. Their use by General Han Hsin (d 196 BC) to measure the distance between his troops and an enemy stronghold may be antedated by the inventions of Mo Tsu and Kungshu Phan (both contemporaries of Confucius), who are

credited with the construction of wooden birds which flew briefly before being destroyed. In one story Kungshu Phan is also said to have designed wooden man-lifting kites which were flown over a city in a state of siege.

However they originated, kites rapidly came to be used for religious, recreational and military purposes throughout most of Asia and Oceania. Their ceremonial func-

tion is an immediate reflexion of the powerful effect on the human psyche of the concept of rising upwards to the heavens, and of the godlike associations of flight in general. The western myths of Daedalus, Pegasus and the flights of angels have their equivalents in a fund of eastern flight myths, including a great many concerning kites. Polynesian myths are especially rich in stories of the personification of kites, or of the manifestation of the gods in kite form.

Early European kites

Many of the kites which today are common in the West had their origins in Asia, but it is possible that some of the earliest European kites were largely indigenous creations. The first clear evidence of a kite in Europe appears in a 14th-century book by Walter de Milemete, *De nobilitatibus sapientiis et prudentiis regum* ('Of the Nobility, Wisdom and Prudence of Kings') completed in 1326 to 1327. It is a roundtopped, plane surface kite with a long, sinuous tail of cloth. Three soldiers are using it to drop a bomb (interestingly fitted with fins) over the walls of an enemy castle. Although this illustration is not discussed in the text of the book, a very similar kite is fully explained in a German manuscript of about 1430, the author of which was clearly a skilled and practised kite maker and flier.

While these pennon-shaped kites may be a purely European invention, they may owe a part of their origin to a similar, more complex, but nonflying object: the military

ABOVE: *The medieval dragon kite occasionally turned up in the 17th century.*
LEFT: *Kyeser's* Bellifortis *often showed highly-decorated kites such as this grotesque three-leg 'bridle' type.*
BOTTOM LEFT: *Asian kites became popular in Europe after trade routes with the Far East were opened for sailing ships in the 17th century. This illustration, dated 1634, shows one performing in a firework display with jumping jacks tied to the kite's tail.*

standard known as the *draco* (dragon). The *draco*, which seems to have originated in the region of what is now Iran, was a banner consisting of a carved wooden head, with open mouth, placed on the top of a pole. Behind the head a cloth tube, or windsock, was allowed to billow out when the banner was held aloft by the standard bearer, or *draconarius*. In some cases the head seems to have been provided with a burning lamp to produce smoke from the dragon's mouth. These standards, used to frighten the enemy and encourage the home troops, were adopted by the Romans early in the 2nd century AD, and were familiar in Europe until the 16th century. They certainly influenced the design of the flat, pennon kites and may even have inspired them. In any case, the kites, like the *draco* standards, were used by adults rather than by children and were commonly associated with the noble art of warfare.

Later European kites

The pennon kites remained in use until late in the 16th century, when they began to be supplanted by smaller, neater and aerodynamically superior kites from the east. These were imported in increasing numbers by the sailors and merchants whose ships plied between Europe and the Indies. The ordinary lozenge, or 'kite shaped', kite reached us from Indonesia, together with the common variant which has a rounded top. Although the earliest known illustration of such kites in Europe shows it being flown by children in Middelburg, Holland (1618), the new kites, like the old ones, were commonly associated with adult activities. The first English illustration (1634), shows one being flown by a man taking part in a fireworks display. The tail of his kite is a string of 'jumping jacks', which are intended to astonish the audience by exploding noisily in the air.

By the second half of the 17th century kites had lost most of their interest for adults, but their popularity among children rapidly grew, as may be gathered from the increasing frequency with which they appear in illustrations of children at play. Not until the rapid rise of serious aerodynamic experimentation in the later 19th century did kites again catch more than the sporadic attention of European adults.

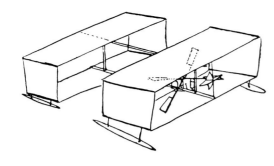

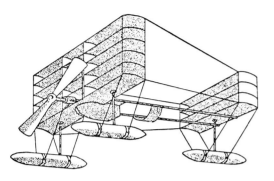

Lawrence Hargrave, working alone in far-off Australia, experimented widely with kites in pursuit of his goal of manned flight.
TOP: *His second design, in 1896, was for this man-carrying box-kite aeroplane.*
RIGHT: *Hargrave in the field.*
ABOVE: *Sidney Holland's design of 1902 was based on Hargrave's ideas. Made of tinplate and fitted with floats, it lacked a suitable engine.*
BELOW: *Hargrave's two-cell box kites.*

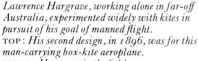

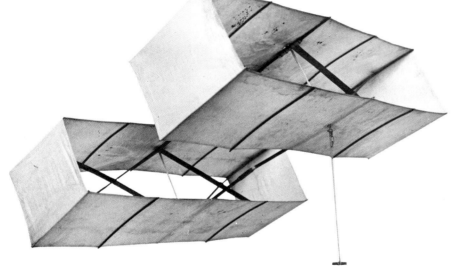

Lawrence Hargrave

Living in New South Wales, Australia, far from the main centres of aeronautical interest in Europe and America, Lawrence Hargrave (1850 to 1915) worked patiently and methodically at the problems of aeronautics during the decades just before and just after the turn of the century.

Although he was equipped with no more than a basic grounding in mathematics and the physical sciences, Hargrave was a good draftsman, a first class mechanic and a remarkably acute objective observer. In the 1880s and 1890s he undertook to explore the principles of flight by examining the behaviour of a series of small flying models powered by twisted rubber or by simple compressed air motors arranged to drive flapping wings or screw propellers. While the best of these

were capable of sustained flight over encouragingly long distances, Hargrave grew increasingly interested in how birds soar in flight without flapping their wings, a problem which he investigated by means of kites.

Hargrave's grasp of the principles of relative motion and fluid mechanics was not altogether sound. Although the theoretical arguments in his notebooks are more sophisticated than those of Leonardo da Vinci, he shared with him the mistaken belief that a body floating freely in a horizontally and uniformly flowing fluid medium can absorb energy from the flow. This error encouraged Hargrave in his work on so-called 'soaring kites', which he tried to make fly beyond the zenith (upwind of the flier). Part of the theory was sound, since Hargrave rightly believed, and was able to show, that a well made kite

may be made to absorb the energy of a gust and move momentarily upwind. His beautifully fashioned aerofoil wings gave him useful experience, but the general idea, being based on misapprehensions of physical laws, led to a dead end.

While the matter of soaring never entirely faded from Hargrave's mind, it sometimes took second place to his other main aerodynamic concern: the need for stability in the design of any practical flying machine. From this arose his invention of the now familiar box kite, which he always preferred to call a 'cellular kite'. The first of these, made early in 1893, contained a number of boxes, or cells, and thus resembled, as he said, 'two pieces of honeycomb put on the ends of a stick'. The multicellular kites soon gave way to the simpler design using only one cell, or box, at each end of the frame, and Hargrave achieved some fine results with a number of box kites whose surfaces were given the shape of curved aerofoils. In the first decade of the 20th century a great many European gliders and aeroplanes were powered derivatives of the Hargrave kite.

Among the most serious of Hargrave's mistakes was his excessive concern for stability. While he was commendably concerned to design a flying machine which would be safe to use, he developed the idea to the detriment of manoeuvrability, and it is significant that the stabilizing vertical surfaces of his box kites (sometimes called 'side curtains') disappeared from European aircraft after only a few years.

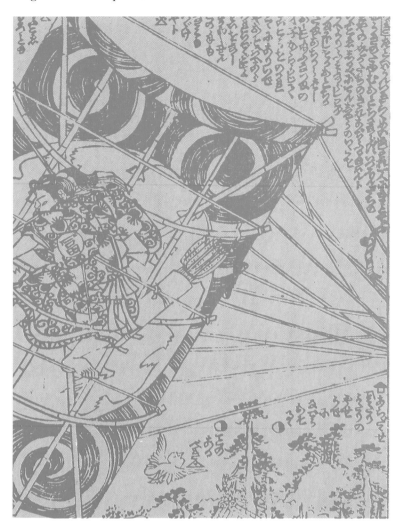

ABOVE: *Kite-flying in Japan is—even today—a very popular sport.*
LEFT: *No nation has ever been more devoted to the construction of large kites than Japan. Occasionally, rectangular, bowed models like this, ultimately of Chinese origin, were built large enough to carry a man, and there are many stories of warriors or robbers using them to soar off into the sky.*
TOP RIGHT: *Alexander Graham Bell was a prolific inventor who constructed several kites capable of lifting a person. This one, called* Siamese Twins, *was built in 1905.*
BELOW: *S. F. Cody (on the left) and his war kite of 1903.*

Alexander Graham Bell

Bell, an American, (1847–1922) took the idea of stability still further, with the result that his experimental kites, though often spectacular, were totally impractical as a basis for flying machines. Bell's enthusiasm and breadth of imagination partly compensated for a lack of serious scientific training in aeronautics. Having sufficient means to experiment on a fairly grand scale, he tried a wide variety of kites as a source of lift before deciding to set about refining the Hargrave box kite in order to make it stabler and stronger. As Bell soon found, a single Hargrave kite, which he called 'the high-water mark of progress in the nineteenth century', could not be scaled up to the size of a machine large enough to carry a man without encountering problems of weight and strength.

By early in the 20th century Bell's progressive developments of the box kite had led him to evolve an essentially new design, the tetrahedral kite. In its simplest form this consists of a regular four-sided figure, bounded on each side by an equilateral triangle and with a rigid frame member along each of the edges. By covering any two of the triangular sides, one arrives at a strong and simple structure whose large dihedral angle confers a good deal of stability. Any number of such cells may be connected together without changing the surface:weight ratio, and Bell accordingly began building multicellular kites capable of lifting a man.

The *Siamese Twins*, built in 1905, contained 1,829 cells, each about 25cm (10in)

on a side. Two years later Bell launched the *Cygnet*, consisting of 3,393 cells and provided with a central opening in which a man could lie. This was tested over Baddeck Bay, Nova Scotia, on 6 December 1907, with Lieutenant Thomas E. Selfridge of the US Army aboard. The kite rose to a height of 51m (168ft) and flew stably for 7 minutes, towed by a steamer. On landing in the water it was destroyed, but Selfridge escaped with a wetting. (Less than a year later Selfridge died in the first fatal crash in the history of powered flight.)

Although Bell experimented with the addition of engines and propellers to his kites, his unwieldy and fragile structures, dominated by an obsession with stability and safety, had already become little more than curiosities and soon faded from the scene. In recent years Bell tetrahedrals have nevertheless been enthusiastically adopted by sporting kite fliers.

The kite in war

Rectangular or round kites of enormous size, sometimes with multiple tails, were often built by the Chinese and Japanese, and very large kites, potentially capable of lifting a man, are sometimes seen in the east today. There is no doubt that the Chinese used them to carry soldiers aloft for observing enemy troops, and they may even have lifted men surreptitiously over enemy lines. In the West the kite was not seriously considered as a possible aid to warfare until the mid 19th century, when Admiral Sir Arthur Cochrane made some trials with kites about 4m (13ft)

long, to see if they might be useful in drawing torpedoes towards a target. Although Cochrane's trials were thought successful, little or nothing seems to have come of the idea. In the 1890s, however, there began an intense period of development of military kites which came to an end only with the adoption of the aeroplane as a fully practicable vehicle. Captain B. F. S. Baden-Powell, brother of the founder of the scouting movement and a keen member (later President) of the Aeronautical Society, began trials in 1894 with a view to developing a reliable system of man-lifters. After a year of experiments, during which he discovered that his first kite, about 11m (36ft) high, was unwieldy and was best replaced by trains of smaller ones, Baden-Powell patented a design which could easily lift observers to heights of 100m (328ft) or more. Although these 'levitor' or 'war' kites were never officially adopted, they provided some of the stimulus for Samuel F. Cody's subsequent development of a much stabler and safer system. For a short period in the 1900s the Cody kites formed a part of the British Army's official equipment.

Cody's manlifters, which were tried at sea as well as on land, were highly efficient winged box kites. A train of them was raised until sufficient lift had been created, and then a further kite, from which a man was suspended in a basket, was allowed to run up the line on a trolley. The man was equipped with a camera, a telescope, a firearm, a telephone for contact with the ground, and a system of lines and brakes for controlling his movement. Similar manlifting apparatus was also developed in other parts of the world, including Russia and France.

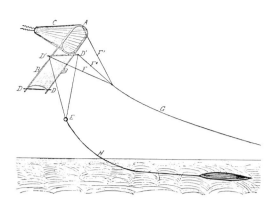

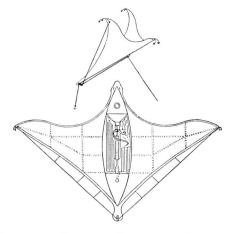

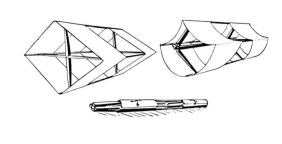

Although the adoption of the aeroplane as a military weapon soon rendered the man-lifters redundant, kites continued to be used in limited ways for military purposes up to the end of World War II. During World War I the German U-boats vastly increased their observational range by sending a man up in a basket suspended beneath a box kite which could later be folded for stowage and which was controlled, when aloft, by lines and pulleys operated by the observer. This simple version of the Cody system was soon replaced by a safer design which allowed the observer, now sitting on a keel-like structure within the kite itself, to convert the kite into a glider in case of a sudden emergency requiring the U-boat to dive.

During World War II the Germans again flew observation kites, this time using apparatus of a more highly developed kind. The

In the 19th century kites were taken seriously as a life-saving device for shipwrecks.
TOP LEFT: *Jobert's kite, from the man who also built twisted-rubber ornithopters.*
TOP CENTRE: *An ingenious but untried design, Chatfield's Storm Kite.*
TOP RIGHT: *From World War I, a German U-Boat's man-lifting observation box kite.*
BELOW AND RIGHT: *Using kites like these, made from leaves, kite-fishing has gone on for centuries in Indonesia and the South Pacific.*

kite was in effect a small unpowered autogiro with three rotating blades above a tubular steel fuselage. The pilot was provided with a control column and rudder bar. With both this and the earlier designs a sufficient wind to sustain the kites could always be maintained by the U-boat's forward movement over the surface.

The US Navy made use of kites for a time to carry barrage wires over convoys, and US trainee gunners were helped by the use of kites as targets. Paul E. Garber developed an especially successful target kite, consisting of a large bow kite with a rudder and twin control lines. A silhouette of an aeroplane was painted on the surface, and the kite could be manipulated very nimbly to simulate most aerobatic manoeuvres.

Survival equipment used by airmen having to ditch in the sea during World War II sometimes included a simple collapsible box kite which was used to raise a radio aerial and distress signal.

The kite in peace
Kite fishing Kite fishing is probably the oldest known peaceful application of kites. Until recent times it was widespread in parts of Indonesia, Melanesia and Polynesia, and may have sprung up independently among a number of Oceanic peoples. The technique varied a little from place to place, and a great many different kites were used, although most of them were made of large leaves sewn together. The kite was flown to a considerable height by a fisherman standing on the shore or paddling a canoe. In addition to the flying line, a second line hung from the kite into the water, ending either in a hook or in a lure which trailed along the surface. In some parts the lure was an ingenious loop about 10cm (4in) long, fashioned from the web of a special spider which was cultivated and 'milked' for the purpose. Long-snouted garfish would attack the loop, mistaking it for prey, and would become entangled. Both kite and line would then be hauled in, and the loop reconstituted for a further catch. Up to ten fish might be caught with a single loop. Kites enabled the fisherman both to work in dangerous or otherwise inaccessible places and to conceal his presence from the fish. Although the technique has now largely died out in Oceania, a number of Western fishermen have adopted the idea.

Meteorological kites Many people have heard of how Benjamin Franklin used a kite

in June 1752 to demonstrate the electrical nature of lightning. This, although one of the first uses of kites for meteorological purposes, was preceded three years earlier by the work of Alexander Wilson, a Scottish scientist who, with the help of a young man called Thomas Melvill, used a train of kites to raise thermometers to the height of the clouds. The thermometers were arranged to fall at predetermined intervals and were equipped with tassels to slow their rate of descent.

From the middle of the 18th century a number of investigators used kites to repeat Franklin's experiment or to gather other information about temperature, cloud height, etc., but it was not until the birth of meteorology as a strict science in the later part of the 19th century that kites were systematically used to study the weather. The British meteorologist E. D. Archibald used diamond-

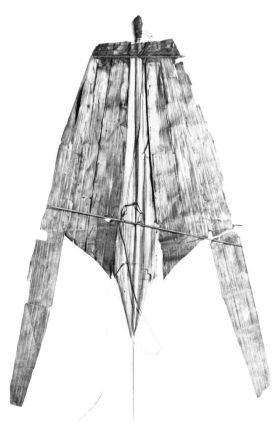

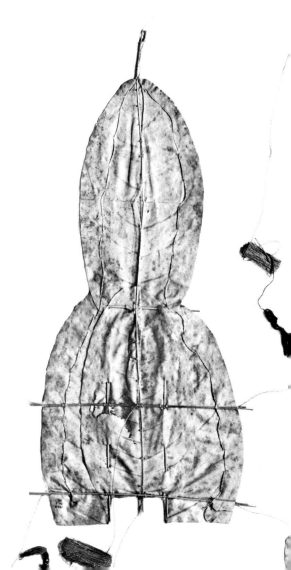

shaped kites in 1883 to measure wind velocities, and a year or two later the US Weather Bureau experimented with three-stick or 'barndoor' kites.

The invention in Australia, soon afterwards, of the Hargrave box kite provided a stronger and more reliable tool, and in the 1890s many weather stations both in America and in Europe used Hargrave kites to study conditions at heights of several kilometres. In July 1933 the last US Weather Bureau kite station was closed, after sounding balloons had rendered the kites obsolete. Not only did the launching and retrieval of trains of kites, sometimes pulling as much as 14·5km (9 miles) of wire, need much time and effort, but the growth of aviation had turned the previously harmless kites into a serious hazard.

Life saving As early as 1760 suggestions were offered for the use of kites to save lives in cases of shipwreck. As most shipwrecks occurred on a lee shore, kites flown from the ship could carry a line to potential rescuers. Many such devices, some of them steerable by multiple control lines, were patented by hopeful inventors.

Kite photography Although many aerial photographs were taken in the 19th century by balloonists, kites offered a cheaper and easier means of raising a camera. Lightweight shutter release systems were devised for taking photographs automatically from kites, often with excellent results. Kite photography is still occasionally practised.

RIGHT: *Thayer's life-saving kite system.*
BELOW: *Specially strengthened Hargrave box kites were used for two or three decades to gather weather information from heights of 8km (5 miles) or more. This one was used at the Blue Hill Observatory, Massachusetts, USA.*
BELOW RIGHT: *Many 18th and 19th century experimenters used kites to conduct electricity to the ground following Benjamin Franklin's example. The Frenchman de Romas showed that a pigeon insulated from the ground would not be hurt.*

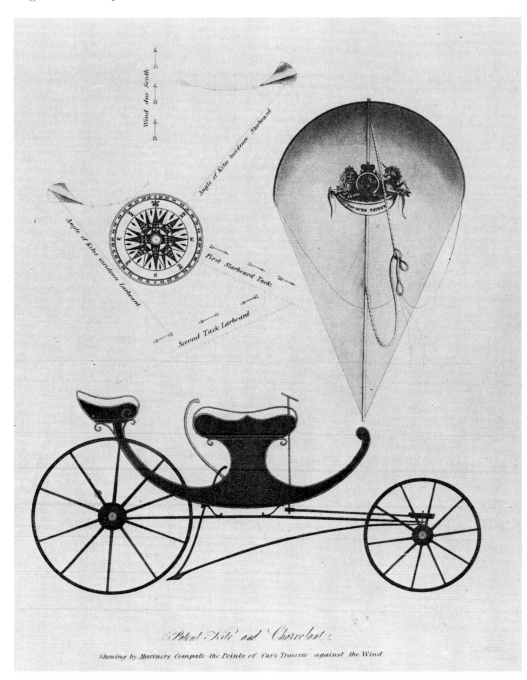

Leaflet and supply dropping It seems that the first leaflet raid was carried out in China in the 13th century. A besieged army raised leaflets on kites which were allowed to fall by cutting the strings at the appropriate moment. During the American Civil War, presidential amnesty notes were once distributed to Confederate sympathizers in this way. In 1870, during the Franco-Prussian War, French smugglers used kites to lift contraband over the walls of Paris.

Other uses Kites have always proved surprisingly versatile. Work on the bridge across Niagara Falls was begun by commissioning a boy to fly a line across on a kite. (He was paid $10.) In 1901 Marconi used a kite to raise the aerial in Newfoundland from which the first transatlantic transmission was sent. In the 19th century hunters occasionally used kites shaped like hawks which frightened game birds into staying in the undergrowth

LEFT: *One of the more novel aeronautical vehicles, which did work after a fashion, was this* Char Volant *or kite-carriage. Its designer, an Englishman called George Pocock, had merely attached a large round-topped kite to a lightweight horseless carriage and used the one to pull the other along. Surprisingly, the kite-carriage made a successful journey along the road from Bristol to Marlborough in 1827.*
BELOW LEFT: *Pocock en route, 1827.*
BELOW: *Hang gliding has benefited from Rogallo's research into flexible wings.*
RIGHT: *A parachutist of the US Army's display team flying a Parafoil.*

until the beaters arrived to put them up. Among the most exciting of recent suggestions is the idea of developing a light wind-powered generator which might be kept permanently in the fast jet streams by a kite, whose flying line would double as an electric cable. Naturally, careful thought would need to be given to the aviation hazard.

Traction

In ancient Polynesia kites seem to have been used to pull canoes, while in 18th century America Benjamin Franklin reports having allowed himself to be pulled across a pond by a simple paper kite. In about 1825 came one of the most celebrated instances of the use of kite for traction: George Pocock's invention of the *Char Volant*. Pocock, a Bristol school teacher, used large arch-topped kites to pull a specially designed lightweight carriage at speeds of up to 32 km/h (20mph). Lacking the modern hazards

of electric cables and telegraph poles, Pocock was able to cover long distances in the west of England. Since then kite traction has been provided principally for boats. If the kites are flown, as were Pocock's, by a system of control lines enabling their heading to be altered, the carriage or vessel can readily be manoeuvred. The forces act just as they do with a fixed sail, and tacking into wind is not difficult.

The most fruitful modern uses of kites in connection with vehicles did not, however, emerge until the perfection of the non-rigid kite. Various non-rigid designs have been tried since at least as early as the late 19th century, when Baden-Powell flew one. Although this was successful, he did not proceed with it, preferring instead to develop his hexagonal manlifters. While many of the earliest non-rigid kites were little more than asymmetrically rigged circular parachutes, the line being attached where the load is

normally carried, more sophisticated and more efficient shapes have emerged in recent years, most of them ultimately owing their existence to Francis Rogallo.

Francis Rogallo's flex wings

In 1945, after the end of World War II, Rogallo began serious research on the design of flexible wings, the aim being initially to develop a kite whose lift would approximate to that of an aeroplane wing, but whose shape would be determined by a combination of wind pressure and the tension on shroud lines. A stable non-rigid kite was evolved in 1948 and a patent application filed. The early Rogallo kites needed as many as 28 shroud lines, but as these were difficult to handle he simplified the design until as few as four lines were required. Since Rogallo's early experiments were not subsidized, he had to work independently, even installing in his house a large electric fan which created a draught through a doorway and so provided him with a primitive wind tunnel.

In the 1950s and '60s, a good deal of work was carried out on the application of the Flex wing kites, as they were called. Two American aircraft companies in particular, Ryan and North American, explored their potential as the basis of gliders and dirigible parachutes. Among other results of this research was the creation of the Fleep ('flying jeep'), a new form of shorthaul vehicle which can work from difficult terrain lifting quite heavy loads. For this and other applications the delta wing shape of the true Rogallo Flex wing was modified to include rigid members at the leading edges and at the centre, though these rigid members were sometimes no more than inflated cylinders of fabric. Apart from its use in such vehicles as the Fleep, the Rogallo configuration is seen in a number of ski kites and ground-powered manlifters, and it has grown still more familiar in the '70s with the growth of interest in delta-winged hang gliders.

The Jalbert Parafoil

The greatest single advance in kite design since Rogallo was made by Domina C. Jalbert, of the Jalbert Aerology Laboratory, Inc., in Boca Raton, Florida. The Jalbert Parafoil has undergone continuous development since its invention in the early 60s. Like a Rogallo Flex wing, it is entirely nonrigid, and is therefore light, efficient and almost indestructible. The Parafoil is essentially a high-lift wing of low aspect ratio, both upper and lower surfaces being formed of cloth or sheet plastic. The surfaces are separated by vertical ribs whose shape is that of a thick aerofoil section, and the leading edge is open so that wind pressure keeps the kite inflated to aerofoil shape. Ventral fins and shroud lines are used to stabilize it and to maintain the correct angle of attack. The Parafoil, which flies at angles below the stall, has an astonishingly high lift:drag ratio and has found numerous applications for both military and meteorological work. It is perhaps most commonly seen freed of its flying line and forming the basis of a steerable parachute, which makes possible controlled glides over long distances.

Mechanics' Magazine,

MUSEUM, REGISTER, JOURNAL, AND GAZETTE.

No. 1520.]   SATURDAY, SEPTEMBER 25, 1852.   [Price 3*d*., Stamped 4*d*.

Edited by J. C. Robertson, 166, Fleet-street.

SIR GEORGE CAYLEY'S GOVERNABLE PARACHUTES.

Fig. 2.

Fig. 1.

BELOW: *In 1804 Sir George Cayley made a prophetic model glider consisting of an arch-topped kite on a pole, with a cruciform tail unit the angle of which could be varied in any direction. At the front was a weight to adjust the centre of gravity.*
ABOVE: *Cayley's earlier designs were refined over the years, and by 1852 he was able to publish his outstanding design for a fixed-wing glider. He called this a 'governable parachute', and included a brilliant description of how to fly it. For this and his other wide-ranging theoretical works, Cayley is often regarded as the inventor of the aeroplane or the 'Father of Aviation'.*

ABOVE RIGHT: *A diagram to illustrate the balance of forces operating on a glider descending at the flattest glide angle and at constant speed.*
FACING PAGE
ABOVE: *Otto Lilienthal was a gifted gliding pioneer who lost his life following an uncontrolled stall in 1896. He is pictured here standing atop his artificial hill near Berlin in 1894. It is interesting to notice how 'bird-like' appeared the man whose work was based on nature's wings.*
BELOW RIGHT: *Perhaps this 2,000-year-old Egyptian wooden bird (or flying fish) is an ancient attempt to build a flying model.*

A glider is an aerodyne (heavier than air machine) which descends continuously through the air under the influence of gravity. It has no power unit, but produces lift over its wing surfaces as a result of its forward and downward motion relative to the air. An aeroplane whose engine is stopped or is fully throttled back is functioning as a glider.

When a glider is descending at a steady speed, the total aerodynamic force on the wing acts vertically upwards, exactly balancing the weight. The lift, by definition at right angles to the relative airflow (glide path), is angled forward, providing a forward component of force which exactly balances the backward horizontal component of the drag. All the forces are thus in equilibrium, and the glider accelerates neither forwards nor

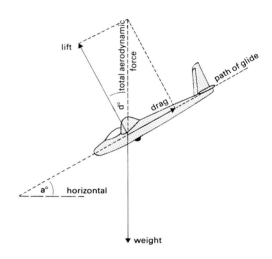

downwards. It is clear that the smaller the value of the drag, the flatter will be the angle of glide. The glide angle is, in fact, solely determined by the lift:drag ratio and, surprisingly though it may seem, is independent of the glider's weight. Other things being equal, increasing the glider's weight increases the speed of descent along the glide path but does not alter the angle of that path.

Gliders do not, of course, always descend continuously relative to the ground, but are capable of soaring to very considerable altitudes. Except occasionally when pulling up in certain manoeuvres and trading speed for height, a glider can rise only by flying in air which is rising faster than the glider's own inherent rate of sink. Glider pilots therefore seek out areas where the wind is deflected upwards by hills, or where differences of ground temperature have created *thermals* – columns of warm, rising air.

Ancestors of the glider

The first man-made gliders to have achieved

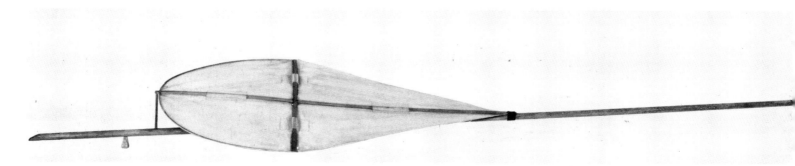

By 1809 he had built a full sized glider which was tested in short hops with a boy on board, and late in life, in 1849, he built a triplane glider which once or twice carried a boy for a few yards.

In 1852 Cayley designed an elegant and refined fixed wing monoplane, features of which may have been incorporated into a man-carrying machine (probably another triplane), built in 1853. In this glider Cayley's coachman was privileged to make the first recorded flight in an aircraft of approximately modern configuration. The coachman, doubtless unaware that he was making history, gave in his notice immediately afterwards, complaining that he was 'hired to drive and not to fly'.

Later 19th century gliding attempts

In 1804, the same year in which he made his first kite glider, Cayley built a 'whirling arm' for testing the lift on a flat surface at varying angles of attack. In 1808 he investigated the movement of the centre of pressure, and in 1809 described some of the properties of cambered aerofoils. Between that time and the end of the century the science of aerodynamics was gradually established. In 1864 Count Ferdinand d'Esterno published a book on bird flight which greatly increased the available information on soaring; the Englishman Francis Wenham, who delivered and had published, in 1866, a highly important paper on *Aerial Locomotion*, worked at the theory of aerofoils and investigated the properties of multiplane structures; in 1884 Horatio Phillips, another Englishman, published further valuable findings on aerofoils. Practical work was also continuing: Wenham tested multiplane gliders; Jean-Marie le Bris tried a bird-shaped glider in 1857, and another in 1868; in the 1880s John Montgomery made a series of rather unsuccessful trials in America. Despite the growth of theoretical knowledge, the experiments were often of a hit and miss nature, and it was not until the work of Otto Lilienthal that mature theoretical insight was combined with a high degree of practical aeronautical skill.

Otto Lilienthal

Lilienthal (1848–1896) was among the most gifted and remarkable of the pioneers immediately preceding the Wrights. A well-trained engineer, he had, from his earliest years, been devoted to the idea of manned flight, which he attempted to realize through a meticulous study of birds and through a series of controlled experiments. Like many before him, he was convinced that success could be achieved only by a close imitation of the flapping flight of birds, a belief which he

limited success may have been model birds. A number of accounts, going back 2,000 years and more, tell of the invention of ingenious flying objects, variously described as wooden birds or wooden kites. In the West the earliest mention of the construction of such a model by a real person is the story of the wooden dove said to have been made by Archytas of Tarentum, a Greek mathematician of the 4th century BC. Although this account speaks of the dove as having been propelled by the force of a spirit or gas contained within it, there seems little likelihood that any form of jet or rocket propulsion could have been devised at that time. If this story and one or two others like it have any basis in fact, the birds were probably lightweight wooden gliders carried some distance by the wind.

Tower jumpers

From the earliest times many foolish men – and also some skilled experimenters – have attempted to fly with the aid of wings strapped or tied to their arms and bodies. Although most of these attempts concern the use of unstable and totally ineffective flapping apparatus, it seems possible, even likely, that in a number of cases rigid wings were tried, and that a combination of luck and intuition led to the achievement of steep glides which usually ended in semi-controlled crashes.

Sir George Cayley

It was to need many centuries of fruitless experiment before the idea of flapping gave way to serious trials with fixed wings. Even Leonardo da Vinci rarely departed from the designs of ornithopters. The first successful full scale glider conceived and built after a rational consideration of aerodynamic problems was Cayley's man-carrying machine of 1853. As early as 1804 he had built a model glider using an arch-topped kite fixed to a pole, which served as combined wing and fuselage, with an adjustable cruciform tail unit and a sliding weight at the nose for altering the position of the centre of gravity.

expressed in a brilliant and highly influential book, *Der Vogelflug als Grundlage der Fliegekunst* ('Bird Flight as the Basis of Aviation'), published in 1889. Unlike many other ornithopterists, Lilienthal understood the general principles of flapping flight. In 1893 and 1895 he built two powered man-carrying ornithopters which, however inefficient, would have begun to provide controllable propulsion and lift. Although the first of these was tested as a glider, neither of them ever made powered flights. However misguided Lilienthal's concentration on flapping may have been, his observations of birds served him well, for he was a true airman, concerned to learn from experience in flight and not to theorize for too long before trying his ideas in practice.

Although, until his tragic and untimely death in 1896, Lilienthal saw the creation of a successful ornithopter as his ultimate aim, most of his experimenting was done with fixed-wing hang gliders from which he hoped to learn the secrets of stability and control. In the last years of his life he was moving towards the use of control surfaces, having tried wing tip steering airbrakes and a form of full-span leading edge flap, and had made notes about wing warping, rudder control and (in a limited form) elevator control. Before his death he had applied these ideas very little in practice, and all of his gliders were manoeuvred by changes of the position of the centre of gravity, brought about by swinging movements of the pilot's body.

From 1891 until his fatal crash 5 years later, he flew a total of 18 different types, the first 12 of which were monoplanes. For a time he was attracted to the possibilities of biplanes, which he found in some respects easier to manoeuvre, and Nos. 13, 14 and 15 were of biplane configuration. For his earliest flights Lilienthal launched himself from natural heights, but in 1893, after he had had a large hangar built, he began flying from its 10m (33ft) roof. After further trials from natural heights he finally made an artificial mound 15m (50ft) high, from which he could glide in any direction, according to the direction of the wind.

Lilienthal's gliders were strongly constructed and braced, with an engineer's attention to appropriate distributions of stress and loading. He conducted many careful investigations into the properties of aerofoil shapes and drew up tables of lifting forces under various conditions. For his last flight he used one of his No. 11 gliders, his 'Normal Flying Apparatus'. The accident which caused his death was the result not of a structural failure, but of the inadequacy of Lilienthal's control system to cope with sudden changes of attitude. When he was about 15m (50ft) high, a gust pitched the nose up beyond the stalling angle. The starboard wing stalled first, dropped sharply, and caused the glider to sideslip rapidly to the ground. Lilienthal had insufficient height in which to recover, and suffered a broken spine in the crash. He died the next day, 10 August 1896 in Berlin's Bergmann Clinic. However, his work continued to inspire subsequent aviation pioneers.

MAIN PICTURE: *Otto Lilienthal airborne in 1894. In the last five years of his life Lilienthal flew 18 different types of glider, 12 of which were monoplanes. He could fly whatever the wind direction, thanks to a 15m (50ft) conical mound he constructed in 1893 close to his hangar.*

BELOW: *Lilienthal gliding from his artificial hill in June, 1895. It is interesting to notice that he has raised his knees to adjust the monoplane's centre of gravity.*
BELOW RIGHT: *Lilienthal flying another of his biplane gliders in 1895.*

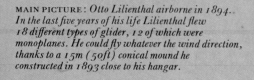

Percy Pilcher

The aeronautical work of the Scot, Percy Pilcher (1866–1899), was closely related to that of Lilienthal, whom Pilcher met and many of whose ideas and techniques he adopted. Although it had been an interest of long standing, Pilcher did not begin practical work on aviation until 1895, after he had spent some time in the navy and had trained as an engineer. While inspired by the successes of Lilienthal and indebted both to him and to Octave Chanute for much essential material, he was a man of independent temperament who wished to make his own contributions to aviation. Despite having received firm contrary advice from Lilienthal, he unaccountably insisted on first trying a glider which wholly lacked a tailplane. This, of course, he found to be uncontrollable in pitch, and he soon afterwards ruefully admitted his mistake. The *Bat*, as he called it, was equipped with a vertical fin and had monoplane wings with pronounced dihedral. Even after he had diagnosed the trouble and added horizontal stabilizing surfaces at the tail, the large vertical surface area, contributed by the combination of fin and dihedral, caused stability problems in gusty conditions at low airspeed, and in his later gliders he abandoned excessive dihedral in favour of a configuration essentially similar to that of Lilienthal's monoplanes.

The fourth of Pilcher's gliders, the *Hawk*, was the one with which he had most success, although ironically it was a structural failure in the *Hawk* which was to bring about his

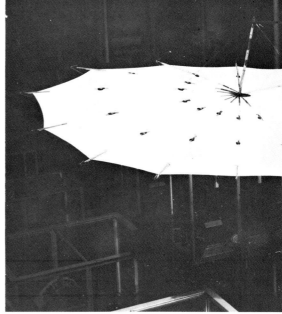

LEFT: *Around the turn of the century Octave Chanute built this glider which was to have a great influence on the Wright brothers.*
BOTTOM LEFT: *Three views of the influential biplane hang glider, which was flown by Chanute's young engineer-assistant A M Herring.*
BELOW: *Percy Pilcher's* Hawk, *showing the spring-loaded undercarriage and hinged tail of this successful – but fatal – hang glider.*
RIGHT: *When he died flying the* Hawk *in 1899, Pilcher left this powered glider design behind. Unfortunately, it was never tested.*
BOTTOM RIGHT: *One of Chanute's first multiplane gliders, which he began building, testing and refining in 1896.*

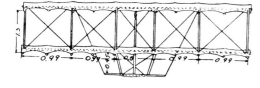

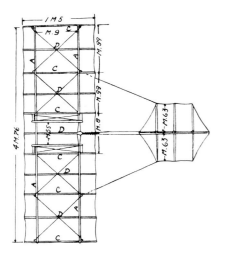

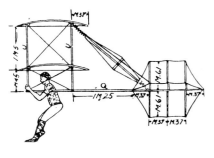

fatal accident. As with all of Lilienthal's and Pilcher's machines, the *Hawk* was a hang glider depending upon body movements to alter the position of the centre of gravity. It was equipped with a device which (with variations) he had adopted from Lilienthal: although the tail unit was not controllable by the pilot, it was hinged so that it could move freely upwards. Lilienthal's reason for incorporating this apparently odd arrangement was to ensure against a sharp pitch down of the nose (which he always feared) resulting from upward gusts at the tail. The *Hawk* was not only powerfully braced with the help of kingposts, but was also equipped with a sturdy spring loaded landing gear to absorb some of the energy of a heavy landing. When he was flying this glider in poor weather conditions on 30 September 1899, a bamboo rod snapped in the tailplane structure, causing the machine to fall out of control from a height of about 10m (33ft). Pilcher died of his injuries on 2 October 1899 without ever recovering consciousness.

Pilcher was an enthusiastic, imaginative and energetic young man, more ebullient than Lilienthal and inclined to want to proceed faster. At his death he was well advanced with the development of a powered glider, originally patented as a monoplane resembling the *Hawk*, but built as a multiplane. For this machine he even designed, built and bench-tested a lightweight motor. Unfortunately, his machine was never even tested as a glider.

Although it is just possible that Pilcher might, had he lived, have preceded the Wrights in the achievement of powered flight, he was surprisingly little interested in control surfaces, and was still a very long way from solving the problems of stability and control which the Wrights were soon to overcome. Pilcher was nevertheless outstanding in his concern to learn the feel of flying from experience in the air, and in this he showed a marked superiority to many of the enthusiastic 'powered hoppers' of the late 19th century.

Octave Chanute
The French-born American Octave Chanute (1832–1910) began practical aeronautical experiments only late in his life, but through his support and encouragement of the Wrights, the publication of his classic history *Progress in Flying Machines* (1894) and the success of some aspects of his own hang gliders, he exerted a powerful and beneficial influence on the subsequent development of aviation.

In 1896 Chanute began building and testing multiplane gliders which, because of his age, he wisely decided not to fly himself, employing instead the services of a young engineer, A. M. Herring. The successful outcome of these experiments was a biplane glider with cruciform tail unit which, while looking back to Cayley's triplane design and Lilienthal's biplanes, may also be seen as the immediate ancestor of the Wrights' machines. In 1900, the Wrights, who had already been experimenting for some time, wrote to

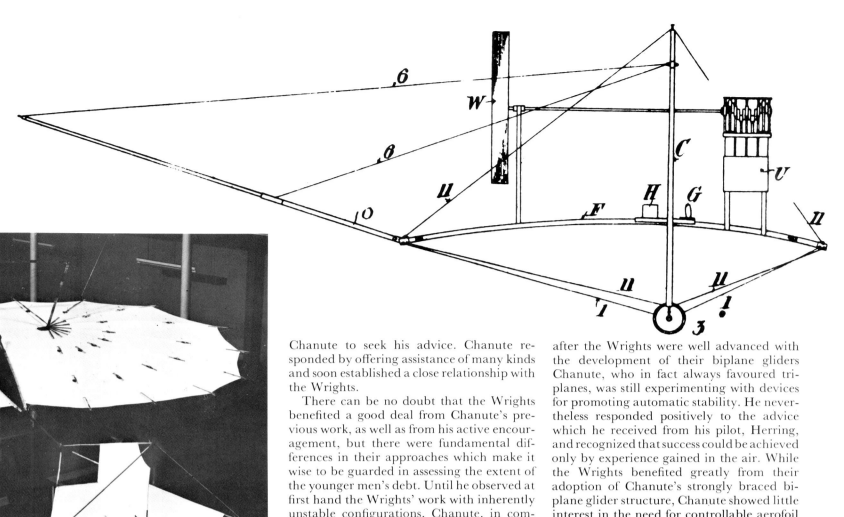

Chanute to seek his advice. Chanute responded by offering assistance of many kinds and soon established a close relationship with the Wrights.

There can be no doubt that the Wrights benefited a good deal from Chanute's previous work, as well as from his active encouragement, but there were fundamental differences in their approaches which make it wise to be guarded in assessing the extent of the younger men's debt. Until he observed at first hand the Wrights' work with inherently unstable configurations, Chanute, in common with many others, had given absolute priority to the idea of stability, and even

after the Wrights were well advanced with the development of their biplane gliders Chanute, who in fact always favoured triplanes, was still experimenting with devices for promoting automatic stability. He nevertheless responded positively to the advice which he received from his pilot, Herring, and recognized that success could be achieved only by experience gained in the air. While the Wrights benefited greatly from their adoption of Chanute's strongly braced biplane glider structure, Chanute showed little interest in the need for controllable aerofoil surfaces, and contributed little to a solution of the problems of flight control.

ABOVE: *Unlike the Wright machines, many early European gliders and powered aircraft owed a good deal to the Hargrave box kite. This float glider, constructed by Voisin and Archdeacon in 1905, shows clearly both Wright and Hargrave features. From Wright they borrowed biplane wings and forward elevator; from Hargrave a stabilizing tail unit.*

BELOW: *The Wright biplane evolved directly from their experiments with gliders.*

OPPOSITE PAGE

TOP: *The Wright brothers' patent design for a biplane kite, an original design which owed nothing to Hargrave. The kite was their first aircraft (built in August 1899). The special feature of the kite was the warped wing-tips which enabled the kite to bank or to right itself. It was fitted with a fixed horizontal tailplane and wings which could be moved in relation to one another in order to shift the centre of gravity.*

The Wright brothers

The Wrights worked towards ultimate success by undertaking a rationally ordered sequence of experiments, beginning in 1899 with the construction and testing of a biplane kite, the wings of which could be twisted, or 'warped'. Like Lilienthal, Wilbur Wright initially found bird flight the best source of understanding, though with the significant difference that he first concentrated not on propulsion or stability, but on control: 'My observations on the flight of buzzards led me to believe that they regain their lateral balance when partly overturned by a gust of wind, by a torsion of the tips of the wings.'

The biplane kite, about 1·5m (5ft) in span, was stabilized by a fixed tailplane and was so arranged that the two 'wings' could be moved in relation to each other to introduce 'stagger' and so shift the centre of gravity. The kite, which had no side-curtains and was not (as is sometimes asserted) in any way derived from Hargrave's box kites, provided useful information and, following their correspondence with Chanute in 1900, the Wrights progressed to tests with full-scale gliders. In the autumn of that year they flew their first glider over the sand dunes of Kitty Hawk, North Carolina, making a few short manned glides but testing it as a kite for the greater part of the time. The wing warping idea was incorporated, together with a forward elevator (characteristic of Wright aeroplanes for the next decade), but there was neither a rudder nor a fixed fin, the Wrights having not yet begun the development of

control surfaces for intentional manoeuvring.

It was during this series of experiments that the Wrights confirmed their suspicions about the disadvantages of excessive inherent stability. The three gliders which they flew in the later months of 1900, 1901 and 1902, were all built so as to be inherently unstable in roll. Unlike so many of the early aeroplanes and gliders which were to be flown in Europe in the first decade of the century, none of the Wright gliders was provided with dihedral or with vertical stabilizing surfaces.

The degree of camber, the wing loading and the rigging angle of the wings were all based on information which they gathered for themselves, with the aid of apparatus constructed at home, the Wrights having discovered that all previously published figures, even those of Lilienthal, were unreliable. It was with the use of these progressively developed gliders that they surmounted the basic control problems and were able to embark on the construction of a practical powered flying machine.

Their most salient success, and their greatest single contribution to the advance of aircraft control systems, was their correct diagnosis of a problem which they unexpectedly encountered when using the wing warping device. To their initial surprise, the upgoing wing, whose lift had been increased by a downward warp of the trailing edge at the tip, tended to be retarded, thus turning the glider against the bank, a condition which eventually led to the dropping of the raised wing as it slowed down and stalled in the inside of a tight circle. The trouble was caused by 'warp drag' ('aileron drag', or 'adverse yaw'), discussed in Chapter One. In an initial attempt to cure it, the Wrights equipped their gliders with a fixed fin at the rear, hoping thus to counteract the yaw, but it soon grew apparent that the adverse yaw could be avoided only by some positive use of rudder in the direction of bank. In September 1902, they modified their third glider so that the fins, now free to move as rudders, were coupled to the warping mechanism. The Wrights were then able not only to correct unintentional wing drops without sideslip, but were also able to initiate balanced turns. This glider, now fully controllable about all three axes, formed the basis of the *Flyer No. 1*, which was to make the celebrated first sustained and controlled powered flight from level ground on 17 December 1903.

The glider in war

Although their limited ability to manoeuvre makes gliders relatively defenceless, they have played a significant part in modern warfare. During World War II, in particular, large gliders were used as troop carriers. Towed by bombers or other large aircraft, they were released over suitable terrain such as open fields or beaches where they were usually abandoned.

A typical troop carrying glider, which saw much service in the Allied invasion of Europe, was the Airspeed Horsa, a wooden highwing cantilever monoplane with a span of 26.84m (88ft). The Horsa, which could accommodate up to 25 armed troops and

their equipment, had dual controls in the extreme nose position, giving the pilots excellent visibility. As it was essential that gliders of this type could land in very short distances, the Horsa was provided with spoilers on both the upper and the lower surfaces of the wings, and had large split flaps.

A rather bigger troop carrier designed in America was the Waco YCG–13, a tailwheel type which could lift several tons of freight and supplies, or 30 fully armed troops.

The Allied gliders looked small, however, in comparison with the Messerschmidt Gigant, used by the Germans. With a span of 55.2m (181ft) the Gigant could carry as many as 130 troops. Although it was used on various fronts during World War II, it was not entirely successful, being awkward to fly and needing a very powerful launching tug. The standard towing arrangement was a formation of three Bf 110s, but for a time the Germans experimented with an interesting compound tug aircraft, consisting of two twin engined Heinkel HE 111s joined side by side with a fifth engine at the point of junction. Later the Gigant was converted into a powered transport aircraft with six engines.

Gliders were sometimes used to carry supplies over very long distances. The Germans towed them as far as the Russian front, while on 4 July 1943, the British Transport Command set up a record by using a C–47 Dakota to tow a Waco CG–4A glider across the Atlantic in a total flying time of 28 hours. The glider carried vaccines, and spares for radios, aircraft, and land vehicles.

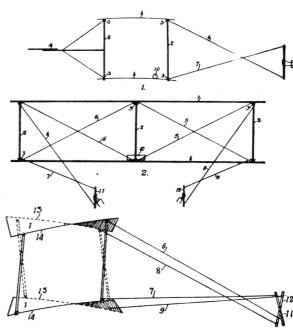

The glider as an auxiliary war weapon came into its own during World War II. Both the Allied and the Axis powers built giant gliders to carry troops and supplies. The two most notable achievements of glider troop carriers were the invasion of Crete by German airborne troops in May 1941 and the landing of Allied troops behind the Normandy beachheads on D-Day.
BELOW: *Deutsche Lastensgler GO 242.*
BOTTOM: *Deutsche Lastensgler GO 244B.*

Experimenters

By 1809 Sir George Cayley's extensive experiments and writings had laid a groundwork that made the confused ignorance of 1799 seem like a prehistoric era. He had discussed the fixed wing, its aerofoil section, its centre of pressure, the need for dihedral angle, the need for a fixed fin and tailplane for stability and the use of movable control surfaces such as a rudder and elevators. He had begun to use models for his research into wing lift and the basic problem of stability and control. Later he discussed converti-planes (having a wing which doubled as a powered lifting rotor), internal-combustion engines (never considered previously), jet propulsion for aircraft (with what we today call vectored thrust) and many other far-seeing ideas which, at the time, were incapable of being realized. He built a simple device which can be described as the first aeronautical research tool, which alone called for completely new thinking. It was a whirling arm, mounted on a vertical pillar rotated by a falling weight and thus having a constant drive torque. The arm was pivoted to the pillar like a see-saw; on one end was a section of wing and on the other a weight on a sliding horizontal scale. Cayley could thus measure the lift of different wing aerofoil sections, different planforms, different wing sizes (bearing in mind the constant effective thrust), and the effect of varying the angle of incidence. All this as long ago as 1804.

Cayley never doubted the eventual success of the aeroplane, nor that its achievement rested on the perfection of a suitable engine; in his words: 'It is only necessary to have a first mover, which will generate more power in a given time, in proportion to its weight, than the animal system of muscles.' He soon decided the steam engine was unlikely to offer sufficient power:weight ratio, and in a search for a higher power:weight

ABOVE: *Sir George Cayley (1773–1857) is known as 'The Father of the Aeroplane'. He was the first person to establish aerodynamics as a science, and to break away from bird flight with flapping wings and instead use separate lift and propulsion, thus producing the aeroplane.*
BELOW: *In 1809 it was reported that Jacob Degen of Vienna had flown with this muscle-powered ornithopter. In fact it had no hope of flying; what Degen did was hang it under a balloon and then make large jumps while pumping the 'wings' up and down.*
ABOVE RIGHT: *Henson's* Ariel, *an aerial steam carriage, was a futuristic drawing published in 1843. Though never built, it was not only a bold concept but also surprisingly well schemed, like a modern aeroplane. The chief omission was any form of flight control.*
RIGHT: *This model is a copy of the world's first successful flying model aeroplane, made by Félix Du Temple about 1857. Later this courageous naval officer actually made a 'hop' in a full-size machine, but as he had no proper engine could not remain airborne.*

ratio in the field of internal combustion experimented with a small engine running on gunpowder. In 1807 he published his specification for another completely new invention, the hot-air engine, but though he saw this through to industrial use, he never succeeded in his quest for a good aero engine. He therefore had to confine his full-scale flight research to gliders. His first such machine was almost a direct enlargement of models he had flown down the Yorkshire slopes around 1800 (which were probably the first stable and steerable glider models ever to fly). His first man-carrying glider appeared in 1809, and with 27·87m² (300sq ft) of wing often flew short distances with a man or boy, but usually with ballast.

In 1809 Cayley was spurred into publishing the first of a monumental three-part treatise *On Aerial Navigation*, by the report that Jacob Degen had flown in Vienna. The truth of the matter was that the Austrian had merely become the first to play at balloon-jumping. He had fixed himself under a balloon which almost lifted him off the ground and then 'flown' by flapping left and right clack-valve wings. A clack-valve is like those in the heart, which open automatically in one direction but shut tight in the other. The valves helped Degen lift his big wings on their upward strokes, and then shut as he pulled the wings down. Though it was good fun, and Degen certainly 'flew' after a fashion, he made no contribution to either ballooning or aeroplane flight.

Cayley explored various further configurations, including an unpublished tandem-wing layout in 1815, and his final machine was a triplane with passenger car suspended underneath. He was reluctant to use large braced surfaces, and this led him to restrict wing span and instead use more wings. His big 1849 triplane made several 'piloted'

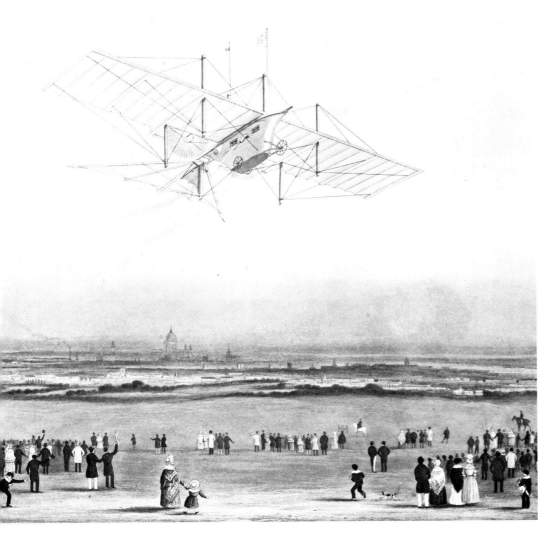

carriages or galloping coaches, and full-scale experiments, possibly with aircraft tethered in a high wind. There were two parts to the problem: how to design and construct an aeroplane; and how to fly it. It is remarkable that hardly any of the experimenters for the next 50 years tried to solve the second part, and most of them made no methodical attack on the first. It seems incredible that many experimenters simply built an aeroplane. Some then got aboard, while the majority got others to act as the first 'test pilots' and operate the controls. They appear to have taken it for granted that, if the the machine would fly, the 'aviator' would be able to fly it.

Cayley is not known ever to have flown his own gliders, and possibly the first man to have the confidence to fly his own machine (if we discount early parachutists) was L. C. Letur. His contraption was as much a parachute as a glider, and has some affinity with the sport parachutes of today, but though never intended for sustained flight it is important because the man flew it himself. He made several apparently stable glides from balloons in 1853 to 1854 before the wind dragged him over some trees, tipping him out and probably qualifying him as the first human to be killed in serious heavier-than-air flight.

This tragedy focused more attention on the problem. Michel Loup sketched a bird-like aeroplane which attempted to reconcile nature and engineering by fitting two propellers into bird-shaped fixed wings. In 1857

flights, the longest and last in 1853, but the occupants served only as ballast; the tail controls were locked and there was no lateral control, Cayley relying on the wing dihedral and the pendulum stability of the under-slung car. One of their many advanced features was the use of wheels with tensioned-wire spokes, yet another Cayley 'first' that much later became standard on bicycles and the first successful wheeled aeroplanes.

William Samuel Henson

William Samuel Henson, of Chard, Somerset, was more of a visionary, but fanciful impressions of his remarkable 'An Aerial Steam Carriage' were widely published in 1843, and firmly set people thinking for the first time of a machine along the lines of what we today recognize as an aeroplane. It had a practical wing (supposed to span 45·72m [150ft]) with sensible bracing, a cabin fuselage directly under the wing, pilot-controlled rudder and 'slab' tailplane, and twin screw propellers. Nor did Henson merely dream. He not only tried to fly at least one model, in 1847, but he built a working steam engine, and calculated that a refined version for his machine would give 30 horsepower for a weight of 272kg (600lb) with boiler and water. His friend John Stringfellow likewise consistently worked on lightweight steam engines, running excellent ones in 1844 and 1848. In the latter year he was busy trying to fly a steam-driven model even more advanced than Henson's but lacking a rear fin or rudder; it could not maintain height when flying free.

Experiments of the mid-19th century

By the mid-19th century there thus existed, thanks to Cayley, a bedrock of knowledge which took human heavier-than-air flight out of the realm of mythology (though it certainly did not make it popularly respectable, because anyone who dabbled in the subject was judged to be a crank). There were the prospects of suitable engines, and the stage was set for the experiments that would solve the basic problems of aviation, though this was to take until the next century. It is essential at this point to look afresh at the problem of flight and how it was attacked. It is probably self-evident today that flight could be achieved only after a methodical study, making the greatest possible use of models, whirling arms (such as Cayley had used), models held out of railway

Félix du Temple, a French naval officer, drew a more practical machine with single tractor propeller and simpler geometry. Probably in the following year he flew a model of this design which became the first powered heavier-than-air device ever to sustain itself in free flight. But in 1857 another French sailor, Jean-Marie Le Bris, had boldly attempted to fly a full-size glider. As a sea-captain Le Bris had intently studied the effortless flight of the albatross. When he retired he tried to build an albatross glider, with a span of 14·94m (49ft) and wing area of 20m² (215sq ft). He was one of those who thought the thing to do was simply get in and fly. His glider was potentially good, and it did make one short glide after launch downhill from a farm cart. Then it crashed, and Le Bris broke a leg. Undismayed, he even-

Experimental Aircraft, 1901-1908

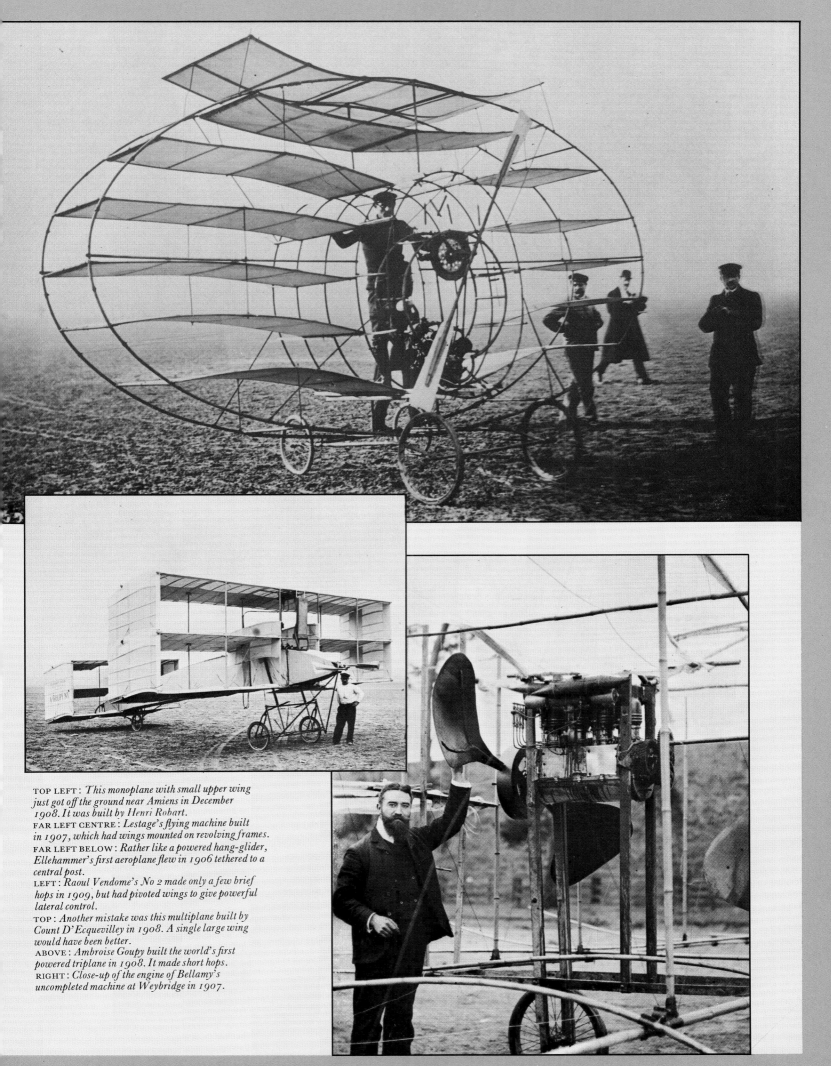

TOP LEFT: *This monoplane with small upper wing just got off the ground near Amiens in December 1908. It was built by Henri Robart.*

FAR LEFT CENTRE: *Lestage's flying machine built in 1907, which had wings mounted on revolving frames.*

FAR LEFT BELOW: *Rather like a powered hang-glider, Ellehammer's first aeroplane flew in 1906 tethered to a central post.*

LEFT: *Raoul Vendome's No 2 made only a few brief hops in 1909, but had pivoted wings to give powerful lateral control.*

TOP: *Another mistake was this multiplane built by Count D'Ecquevilley in 1908. A single large wing would have been better.*

ABOVE: *Ambroise Goupy built the world's first powered triplane in 1908. It made short hops.*

RIGHT: *Close-up of the engine of Bellamy's uncompleted machine at Weybridge in 1907.*

tually built a second glider, and in 1868 this, too, crashed, but with only ballast on board. The use of ballast highlights one of the basic faults of most of the early experimenters: they did not appreciate what Cayley had emphasized about the importance of control surfaces. Le Bris may have had a controllable horizontal tail, but he had no rudder and no wing controls. He was thus in no position to 'fly' the glider at all.

At about the same time, in 1858 to 1859, Frank H. Wenham in England was flying models with multiple superimposed wings. He had been told by Cayley that the leverage on large monoplane wings was 'terrific' and should be avoided (Cayley was criticizing Henson's Aerial Steam Carriage), so Wenham adopted the multi-wing layout. He was probably the first actually to fly multi-wing gliders, and his research not only confirmed Cayley's in showing the superiority of the cambered aerofoil, but also demonstrated that the centre of pressure was quite close to the leading edge. He therefore decided that most of the rear part of the wing was not worth having, which meant in effect that he built his wings with high aspect ratio. In 1866

thopter in which auxiliary 'ailerons' were fitted, but these were not for lateral control and, in fact, would have constituted a serious hazard had they been used. In 1870 Richard Harte patented a true aeroplane wing with ailerons; but even he failed to appreciate the need for lateral control, though he provided the means for it. His ailerons were deflected to counteract torque from the propeller, by imparting a roll in the opposite direction. As a second possibility he thought the ailerons could be used to help steer, by increasing drag at one wing-tip. The purpose for which ailerons are vitally needed never occurred to him, or to any other experimenters of the 19th century.

tractor propellers.

Of machines actually built, that of du Temple deserves special mention, because it made the first attempted flight by a powered man-carrying aeroplane. It was identical in configuration to du Temple's model and patent of 1857, apart from replacement of the boat-like fuselage by a light framework. It is difficult to see the reason for the swept-forward wings, which must have put the centre of pressure far ahead of the centre of gravity, besides making the machine unstable in pitch. But in 1874 it was ready to be tested, and its steam power plant had been run at full power with the aircraft tethered. Du Temple actually got it airborne after a run down a sloping ramp, but it flopped to the ground almost at once. This probably saved du Temple's life, because he had no proper control system.

An even less likely machine was the 'Aerial Steamer' tested at the Crystal Palace in 1875. It was built by Thomas Moy, who was a pioneer of the hydrofoil boat by proxy, because, to see how his wings worked, he tested them in 1861 under a boat, which was lifted 'quite out of the water'. His wing-supported

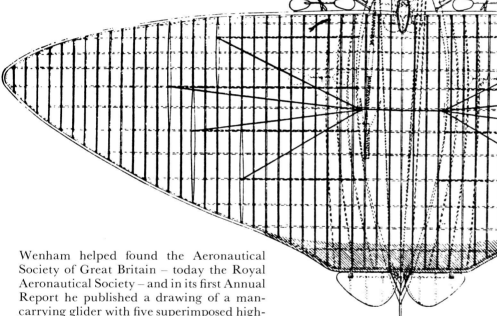

Wenham helped found the Aeronautical Society of Great Britain – today the Royal Aeronautical Society – and in its first Annual Report he published a drawing of a man-carrying glider with five superimposed high-aspect-ratio wings.

Two years later the infant Society held an aeronautical exhibition at London's Crystal Palace, the first such gathering of air-minded people in history. Prominent in the engraving on the front page of the *Illustrated Times* for 4 July 1868, was the triplane built by String-fellow, and this again was an attempt to follow Cayley's principles. Stringfellow used two propellers driven by an engine in the fin-like fuselage, but could not get the machine to maintain height. Like virtually all contemporary flying machines, it lacked a control system, but its world-wide publicity inspired almost every subsequent 19th-century aviator, and thus its influence was great.

Most designers of the 1860s had little idea of the kind of control system needed. Very few recognized the need for any movable surfaces other than a rudder (generally accepted, because they were needed on ships) and, less often, a movable horizontal surface, either in front or behind. In 1868 Matthew Boulton patented a highly impractical orni-

Experimenters of the late 19th century

In 1870–71 Alphonse Pénaud triggered off a craze that multiplied the number of people building successful model aircraft by at least 1,000. He did it by a simple model aeroplane which he called the 'Planophore', with a pusher propeller powered by twisted rubber. His was a beautifully simple cantilever monoplane with inherent stability along the lines laid down by Cayley, and though subsequent experimenters built some large monstrosities, this little model ensured that most flying models from 1871 onwards would actually fly. Pénaud's full-size aeroplane of 1876 remained a paper project only, but though it was virtually an all-wing design it had such advanced features as a glass-domed cockpit canopy and retractable landing gear. Its main advantage was in confirming the need for a rudder and elevators, both moved by the same control column, and

boat worked well, but unfortunately he had little idea how to construct an aeroplane, and when he tested it he was wise to stay on the ground and have the machine tethered on a circular track. It did manage to lift off, and was possibly the first aeroplane in history to do so entirely under its own power.

In 1879 Victor Tatin achieved complete success with a small model which again re-inforced the rightness of the 'aeroplane' shape with tractor screws, fixed wing and rear tail. He cheated, because instead of an on-board power source, the cylindrical fuse-lage housed a store of compressed air, which quickly ran down. But his model took off smoothly at 8m/sec (about 18mph) and gained much publicity. Less well known were the tandem-wing models of D. S. Brown of the early 1870s, which are thought to have been inspired by a proposal by Thomas Walker in 1831. Brown was a painstaking and scientific worker, and his scholarly report of 1874 certainly influenced Langley, whose work is described later. Another worker who was little-known in his day was the Russian, Alexander F. Mojhaiski who – probably after building models – constructed a large monoplane which, in 1884, was launched down a sloping ramp near St Petersburg,

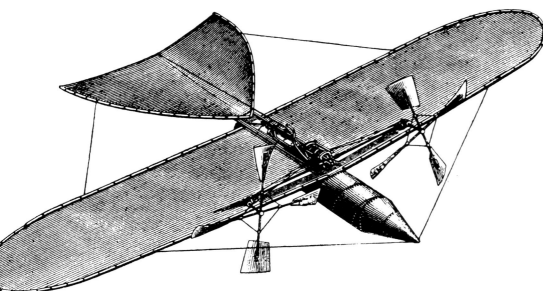

the 1870s, and in 1884 took out his first patent. In this he described the wing section as it has existed more or less ever since. His long studies showed him that a wing for the low speeds then envisaged should have separate upper and lower surfaces, appreciable thickness, a dipping leading edge (the 'Phillips entry') and an upper surface curved much more than the underside. He found, as Cayley had reported but many found hard to believe, that most of the lift was generated by the reduced pressure over the upper surface, especially over the front upper portion. He took out many further patents, and in 1893 successfully flew a large, unmanned aeroplane which lifted off a circular track at Harrow, although tethers restrained it. Sadly, this great researcher was convinced that the answer lay in many very narrow wings. His 1893 model did lift off, but he had no success with a later machine, built in 1907, which had 200 slim wings arranged in four groups, each looking like a Venetian blind. By that time he might have been expected to realize that this conviction was the one part of his great work that was faulty.

The outstanding work of the 1890s was the fabulous research of Lilienthal, outlined in Chapter 2. When he was killed, Otto Lilienthal had far more flight-time to his credit than every other heavier-than-air pilot in history combined, but he was unlike most experimenters in that his objective was to learn how to fly. He was toying with a powered flapping wing tip machine at the time of his death in August 1896, but powered flight was not his central ambition. On the other hand his great pupil, Englishman Percy S. Pilcher, was working towards powered flight and was well advanced with a lightweight oil engine when he, too, was killed gliding in 1899. The third important glider figure of the 1890s was the American Octave Chanute (Chapter 2), but he had little direct personal interest in engines.

piloted by I. N. Golubev. It had one tractor propeller and two pushers, the latter being located either in slots in the large rectangular wings, as in the model now on view in the Soviet Union, or behind the trailing edge according to a contemporary print; the latter shows a steam engine smoking furiously. This big machine took off, as did du Temple's, but sank again at once. This was fortunate, for it appears to have had no controls except small elevators.

Little appears to have happened in the 1880s, though by this time the number of experimenters was certainly above 100. Many of them found it hard to avoid thinking of birds, or even dragonflies, and one of the leading 'birdmen' was L. P. Mouillard. His book on bird flight, *L'Empire de l'Air* (1881), emphasized the importance to birds of sustained gliding. Mouillard had himself built many gliders from 1856, and he was helpful in once again ramming home the belief that, before one could hope to soar aloft in a powered aeroplane, its pilot had to learn to fly. Today we may find it difficult to comprehend that this belief was shared by few of the early experimenters. Hardly any made any attempt to 'walk before they could run' by building gliders or solving the basic problem of control. Today we can see that trying to take off in an improperly designed aeroplane, with no proper control system, would be almost suicidal; to do so without having any idea how to fly in the first place is simply nonsensical.

Yet this is just what happened at least 40 times between 1850 and 1910. Possibly the outstanding case came on 9 October 1890, at Armainvilliers, when Clément Ader tested his *Eole*. This flying machine contained a truly outstanding steam engine, which after a careful study in 1960, was calculated to have weighed only about 1·13kg (2·5lb) per horsepower (most contemporary engines weighed from 5·5 to 68kg ·[12 to 150lb] per horsepower). Unfortunately, almost

everything else about the *Eole* was wrong, and though it had great bat-like wings with a span of 14·94m (49ft), and a semi-enclosed cabin, the occupant had no way whatever of controlling the aircraft other than by moving a steam throttle valve. On the date above, Ader did become the first human ever to rise off level ground in a self-propelled flying machine, but he correctly called the flight 'tentative'. He was airborne for a distance estimated at 50m (164ft).

Far more significant progress was made by Horatio F. Phillips in England, but his work was less spectacular and less publicized. His outlook was the one needed; instead of rushing to build and try to fly a powered machine, he set to work with aerofoils and tested them day and night. Much of his research was done in wind tunnels, the obvious tool for an aircraft designer, although they appear never to have existed until one was built by Wenham, helped by John Browning, in 1871. This tunnel was immediately publicized through the Aeronautical Society, and Phillips used it and also built his own. He worked on wings throughout

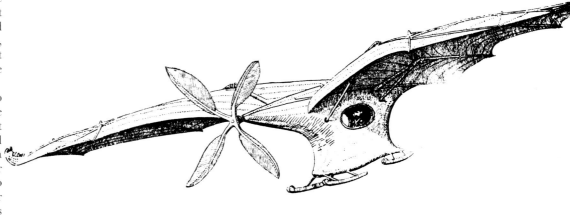

Hiram Maxim

The other great figure of the 1890s was wealthy American Hiram (later Sir Hiram) Maxim. Originally an electrical engineer, Maxim lived in England and invented among other things the machine gun that bears his name which led to the most famous gun of all in the history of armies and air forces, the Vickers. He had the resources to do things on the grand scale and, after spending a reputed £20,000, tested a gigantic 'flying machine' at Baldwyn's Park, Kent. He did much prior research, and in any case did not intend this machine to make a free flight. What impressed was not its design but its size: not for many years was the world to see another flying machine with a span of 31·70m (104ft), wing area of 371·6sq m (4,000sq ft) and weight of 3,629kg (8,000lb), complete with crew of 4! The two propellers, each 5·43m (17ft 10in) in diameter, were driven by a pair of amazingly advanced compound steam engines with very-high-pressure boilers and maximum output at 180 horse power each. Excluding the boilers and fuel, the specific weight was 0·67kg (1·55lb) per horsepower, which is little short of fantastic in comparison with other steam engines of 1894. Their combined power accelerated the monster biplane to an estimated 66·93km/h (42mph), at which it lifted from its rail track, broke one of the guard rails added to prevent free flight, and was safely brought to a halt. There seems every likelihood that, had full power been maintained, this vast 'flying test rig' would have climbed purposefully away.

It would almost certainly have crashed, because it had no controls other than fore and aft elevators.

In 1897 Ader began testing *Avion III* (his second machine was left unfinished), which was no advance on *Eole* apart from being even larger and having two engine/propeller units. Fortunately for Ader it never left the ground, but 9 years afterwards he claimed it had flown 300m (984ft), because by that time aeroplane flight was an accomplished fact and Ader wanted to be 'first'.

In 1899 a much more methodical experimenter visited England: Lawrence Hargrave. Working virtually isolated in Australia, he had by 1893 perfected the box kite, as related in Chapter 2, but only as a stepping-stone to the powered aeroplane. He, more than any other pioneer of flight, followed what might have been thought the logical and sensible path in perfecting a stable and controllable kite and then turning this into an aeroplane. Moreover, his work on engines was wholly sound and promising, concentrating on lightweight petrol engines and including the first rotary engine, with fixed crankshaft and rotating radial cylinders, ever to run (many credit Hargrave with inventing the rotary engine itself). Outstanding as was his engine work, Hargrave's greatest contribution was the box kite configuration, which combined light weight, structural elegance and rigidity, and acceptable flight stability. In the period 1905 to 1910 it was Hargrave's work in New South Wales in 1880 to 1899 that served as the basis for most of the world's aeroplanes.

Samuel Pierpont Langley

By the beginning of the 20th century aeronautic attention began to be focused on experiments in the United States. America's most eminent scientific institution, the Smithsonian, had as its Secretary an astronomer, Professor Samuel Pierpont Langley. He had become interested in heavier-than-air flight in 1886, and embarked on a sensible programme of powered models which differed from most in the scale of backing available in the form of facilities and test equipment. Langley decided to use the Brown tandem-wing layout, because the Englishman's experiments were so successful and documented in scholarly fashion. Langley called all his machines 'Aerodromes' and, apart from having no control system, were well designed and built. Features included highly cambered single-surface wings, usually of equal size, a fixed cruciform 'Penaud tail' and a central engine driving two pusher propellers. It took Langley ten years to produce a successful model, but in 1896 he made good flights with two models each weighing about 11·8kg (26lb) and with a span of about 1·83m (6ft). The best flights covered 914 and 1,280 m (3,000 and 4,200ft) in 45 seconds. Langley had devoted most of his time to the beautiful small steam engines that powered these models. He almost ignored the vital fact that he could not control their flight. He also chose to launch them by catapult over water at Quantico, Virginia, which with models may have been a good idea. They flew steadily, and this was an achievement in itself. Langley made no attempt to consider a machine that could be steered, and ignored any sort of landing gear.

In 1898 Langley had decided to let others carry on the good work; but war with Spain suddenly prompted the US Government to

see if flying machines might have practical use. They asked Langley to build a full-size man-carrying Aerodrome, and contracts were signed. This was extraordinary, and far more than just the first time anyone had considered buying a flying machine. It was the first time any official body had even taken the idea of aeroplane flight seriously, which was almost universally regarded as a subject for scorn and ridicule. The reasons for official interest lay in Langley's stature and eminence; the fact that he was a scientist rather than a crazy inventor; and the success of his models for if they flew, it was reasoned, so could a man-carrying machine.

Langley collaborated with the brilliant Charles M. Manley and Stephen M. Balzer, who created what was by far the most advanced and lightest engine for its power the world had ever seen (see p. 55). In 1901 he successfully flew a quarter-scale model of his man-carrying Aerodrome, and this was the first machine ever to fly with a petrol engine. Finally, on 7 October 1903, Manley climbed aboard the full-scale machine which, like the models, was to be launched by catapult from a houseboat. It plunged into the river Potomac. On 8 December it plunged into the river again, and that was the end of the Aerodromes.

It is almost beyond our modern comprehension why Langley, with so much going for him, should have ignored the whole question of flight control, and also chosen not to use a wheeled landing gear for land take-offs. Arguments have raged over what

ABOVE LEFT: *A model of the giant steam-driven machine of Sir Hiram Maxim, which became airborne in 1894 but accomplished little.*
CENTRE LEFT: *Ader's second machine,* Avion III *of 1897. It was even less successful than his* Eole *of 1890.*
BELOW LEFT: *Another view of Ader's* Avion III *with its bat-like wings folded.*
ABOVE: *S. P. Langley's important but wholly unsuccessful* Aerodrome *on its houseboat on the Potomac in 1903.*
RIGHT: *The Wright brothers succeeded because they were methodical. They tested aerofoils in the world's first wind tunnel, seen here.*

happened on his two launch attempts. Some say the machine suffered immediate structural failure, even before it left the catapult, and others insist it fouled the launching mechanism on each occasion. Certainly the splendid engine was not at fault, though the machine needed at least two instead of only one. In 1914, Glenn Curtiss took the same machine and made short hops with it over Lake Keika. Sensibly Curtiss fitted it with floats. For several years afterwards the Smithsonian made extravagant claims about the 1903 Aerodrome; what they did not disclose, but must have known, was that before flying it Curtiss had secretly and dishonestly altered it in major particulars, in order to try to diminish a patent claim of priority by the Wright brothers.

Somewhat similar to Langley's models, and often reported to have been based on them, the world's first flying boat was tested by Austrian Wilhelm Kress on the Tullnerbach reservoir in October 1901. It had twin waterproofed hulls, three wings in tandem

and a good tail with fin, rudder, tailplane and elevators. Two pusher propellers were driven by a heavy Daimler petrol engine said to weigh 13·61kg (30lb) per horsepower. Sadly, the machine capsized and sank when Kress swerved to avoid an obstruction, but both Chanute and Wilbur Wright thought it capable of flight. A much more dangerous European design was that of Karl Jatho of Hanover, which made short hops in 1903 – said to be of 18m (59ft) on 18 August and 60m (197ft) in November. This contrivance was just a kite with a 9hp engine and pusher propeller; Jatho had no control whatever.

The Wright Brothers

We come now to the great Wright brothers, whose extremely important research with gliders was outlined in Chapter 2. Their success where so many had failed rested upon a methodical approach, which, instead of deciding what to build and then trying to fly it, began with first principles, painstaking research with a wind tunnel and other tools, and eventually led to the brothers learning how to fly. It is significant that, to avoid a nose dive such as that in which Lilienthal and many others since have met their deaths, they adopted a pilot-controlled forward elevator or foreplane. This enabled the pilot, lying prone on the lower wing, to keep watching the angle of the elevator and match this with the elevator's position with respect to the horizon. Today canard (tail-first) aircraft are often the best arrangement, but at the turn of the century the front-elevator layout led the Wrights into a blind alley and could have cost either his life.

In particular, the adoption of an inherently unstable configuration was an unnecessary error, because by 1910 it was appreciated that full control could be achieved whilst still having an inherently stable aeroplane that would fly 'hands off' – which would have been suicidal in a Wright Flyer. Yet another strange quirk of the Wrights was their reluctance to take off normally, with wheels (which would have made it much easier to handle their Flyers on the ground); instead they stuck to skids resting on a trolley running on a prepared track, to cumbersome handling trucks, one fixed under each lower wing and removed when the Flyer was on its runway, and to catapult take-off with a cable pulled by a falling heavy weight.

None of these criticisms in any way detracts from the Wrights' achievement, which makes theirs by a very wide margin indeed the greatest name in the entire history of flight. But it reflects the sad situation in which they found themselves after 1905, when, unable to settle agreements for the sale of their machines, beset by spies and would-be copyists and with the world generally unaware or sceptical of their achievements, they stayed on the ground for more than two and a half years. Design choices that seemed prudent in 1902 to 1903 had by 1908 become outmoded, yet the painstaking and dogged brothers were very reluctant to make major changes, which might have helped them, to a refined and proven technique.

Powered flight

By 1903 Wilbur and Orville were the only experienced pilots in the world. They knew more about flying even than Lilienthal, because unlike the German they were using a control system built into the Flyer. This system comprised warping wings, operated by a hip cradle swung to right or left by the pilot lying on it; biplane elevators in front driven by a fore/aft lever which cunningly not only rocked the surfaces bodily but changed their camber from convex to concave to give maximum up or down thrust; and twin rear rudders linked to the warp cradle to give a smoothly banked turn.

During their gliding in 1901 to 1902 the Wrights had discovered that warping could have an effect opposite to that intended, the asymmetric drag at the wing tips swinging the machine in yaw and causing the wing it was desired to raise going down still further. The fixed fin made things worse by accentuating the yaw and speeding up the high wing. Long and potentially hazardous experiments in 1902 gradually led to the linked warping and rudder control, with the aid of which the Wrights became the first aviators in history to make turns, or, indeed, to control an aeroplane in flight at all.

Most people know how, searching for wind at bleak Kitty Hawk, Wilbur won the toss and attempted to fly into a 32 to 35km/h (20 to 22mph) wind on Monday 14 December 1903. The original powered Flyer had a span of 12·29m (40ft 4in), a length of 6·43m (21ft 1in), a flying weight (i.e. fully loaded) of about 338kg (746lb), and a water-cooled engine with four cylinders lying on their side and developing about 13 horsepower. No catapult was used with this Flyer. Wilbur simply opened the throttle at about 10.35am and accelerated down the launching rail with the skids resting on a free trolley. He became airborne, the nose rose, Wilbur over-corrected and the Flyer ploughed into the sand. It was maddening, but Wilbur was completely unhurt and the somewhat bent Flyer was repaired and made airworthy in 48 hours. Some reports claim Wilbur tried to climb too

ABOVE: *Orville Wright piloting the No 3 glider in the autumn of 1902. Until 1902 the Wrights had had complete faith in existing aeronautical data, but it made their second glider a failure. With No 3 they based the design entirely on their own research, and it flew well, making more than 950 successful glides on the bleak sandhills at Kitty Hawk, North Carolina.*
BELOW: *The original powered Wright Flyer at Kill Devil Hill, Kitty Hawk, on 14 December 1903. It was on this day that the first attempt to fly under power was made. Wilbur won the toss and took off, soon after this picture was taken, but ploughed into the sand, causing slight damage. After the Flyer had been repaired it was brought out on Thursday 17 December and made four good flights, the first in the hands of Orville.*
RIGHT: *A model of an early biplane built in 1908 by Gabriel Voisin. It was based loosely on the designs of the Australian pioneer Lawrence Hargrave drawn up for his box kites, as was an early machine constructed by Santos-Dumont. The vertical surfaces between the main planes (of which there were six in the Santos-Dumont machine) were reduced to four in this design in order to increase the lateral stability.*

Dayton, well pleased with their success.

In 1904 the brothers did all their flying at the world's first aerodrome, on Huffman Prairie, a 90-acre pasture at Simms Station about 13km (8 miles) east of Dayton. Flyer *No 2* was similar to the first, but its elevators and rudders were raised well clear of the ground and it had a new upright engine of 15–16hp. By November 1904 flights had become routine, with accelerated towed take-off to make them relatively independent of ground conditions. Durations ranged up to over five minutes, and most important of all was the ability to make sure turns and fly a circuit. Twice in 1904 the Wrights invited the Press, but the engine refused to start on both occasions; the newspapermen never came back, and completely disbelieved the tales of the few locals. However, recent investigation suggests that the 'failures' were deliberately staged to convince the Press that the Wrights were unable to fly and therefore keep the trials private. Subsequently the Wrights built Flyer *No 3*, generally regarded as the

ferably with a simple glider, and that a successful flying machine had to have some means of control. Instead, the European experimenters were satisfied to start by building their definitive powered aircraft, believing that they could be driven like a car. The only exceptions were French Captain Ferdinand Ferber who, in 1902 to 1904, flew copies of the Wright gliders, and the young French Voisin brothers Charles and, much more important, Gabriel. The latter was commissioned by Ernest Archdeacon to build a glider that could be towed by a motor boat on the Seine. Helped by another youth, Louis Blériot, Gabriel Voisin first built a Hargrave-style box kite landplane which was towed into the air by a car in March 1905, without a pilot. This brief test cannot have offered Voisin much information.

He changed the design for Archdeacon's float-glider, adding a large biplane box-kite tail which contributed to the lift. Control in pitch was provided by a foreplane, as in the Wright machines, but there were no other

steeply and stalled. Whatever the cause, he was content to claim a 'flight' lasting just $3\frac{1}{2}$ seconds, and this is not considered to rank as a flight at all. On the Wednesday the Wrights were confident of success on the morrow, and sent word to the staff at the weather and life-saving station four miles away. Next morning, 17 December, was blustery and bitterly cold, with a north wind gusting at about 40 km/h (25mph) and ice on the puddles from recent rains. Five men arrived from the weather and lifesaving station, and a man and a boy came from other localities. The Wrights rigged up a camera on a tripod to photograph the first take-off, and assigned one of the visitors to work it at the appropriate moment. It was Orville's turn, and at about 10.35am he accelerated along the rail. The camera was operated, and it shows not only the Flyer climbing away on man's first flight, but also the muddy footsteps left by Wilbur as he ran alongside at the start, holding the bottom wing. About 12 seconds were covered at about 13km/h (8mph) ground speed, or 48km/h (30mph) airspeed, equivalent to a ground distance of 36·58m (120ft) and an air distance of over 152·4m (500ft).

Man had flown at last! Orville had wobbled and undulated, as was to be expected in such conditions, but there was not the slightest doubt the Flyer could fly. After man-handling the machine back to the rail and making an inspection it was flown again: Wilbur covered 53·34m (175ft) ground distance (213 m [700ft] through the wind) at higher speed in 11 seconds. Then Orville flew for 15 seconds, covering 61m (200ft) (air distance 259m [850ft]). Finally, at about noon, Wilbur kept going for no less than 59 seconds, covering 259·7m (852ft) of ground (an air distance of about 792m [2,600ft]). Unfortunately the elevator was damaged in the final landing. The brothers were gaining experience, and had petrol enough for a flight of 18 minutes. They had intended after lunch to fly all the way to the weather station, and there is no reason to doubt that one of them could have done it. But with a repair needed and winter set in they went back to

world's first completely practical aeroplane. Though it flew only 49 times, compared with 105 flights by *No 2*, it logged much more flight-time – a total of 3hr 5min, with flights lasting up to 38 minutes and covering 39km (24·25 miles). Seven Flyer *No 4* models were built, one being shipped to France in 1908, where Wilbur amazed the Europeans, who were not only unable to fly under control, but had not even appreciated that control was needed. By this time passengers were being carried, flights were extending to over 2hr duration and, by the end of 1908, the brothers had logged 36hr 20min in the air, or roughly six times the total flight-time of all other aviators combined.

European experiments, 1900–1914

During the first six years of the 20th century (1901 to 1907) European experimenters made extraordinarily poor progress. With Lilienthal and Pilcher dead, there seemed to be nobody left with the basic understanding that an aviator had first to learn to fly, pre-

controls. Fortunately the glider was basically stable – possibly a matter of luck – and it flew quite well with Gabriel Voisin at the controls on 8 June 1905. A little later a slightly different float-glider, built jointly by Voisin and Blériot, showed that supposed inherent stability was no substitute for a control system. On its first take-off, on 18 July 1905, it tilted and slid stright into the Seine, Voisin narrowly escaping being drowned.

Gabriel Voisin was not deterred, but before he achieved real success the first European powered flight had been made by the glamorous little Brazilian, Alberto Santos-Dumont, who had already become the rage of Paris by his exploits in airships. In early 1906 he was building his first heavier-than-air machine. At first glance it looks quite normal, with the engine and propeller at one end, close to the biplane wings, and the biplane tail at the other; it is only when we see which way the pilot is facing that it is seen to be an odd freak, with the tail far in front and the wings right at the back.

Experimental Aircraft, 1908-1914

ABOVE LEFT: *50hp Gnome rotary engine of Spencer's biplane 1911.*
ABOVE: *This Farman III was built for Roger Sommer, who flew it at the Reims meeting.*
LEFT: *Cody seated in his much-modified biplane, at the Lanark meeting in August 1910.*
BELOW: *This Voisin was fitted with a 37mm Hotchkiss cannon as early as 1910.*
BOTTOM: *T. O. M. Sopwith flew his Howard Wright biplane from Kent to Belgium in 1910.*

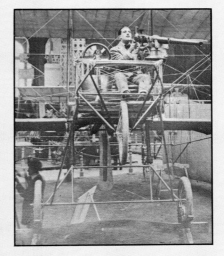

TOP FAR LEFT: *After flying a Voisin-built triplane in 1908, Baron de Caters flew this Voisin biplane at Châlons in April 1909.*
TOP LEFT: *One of the last of the classic Wright flyers was this Astra biplane of 1911, with 70hp Renault engine.*
CENTRE LEFT: *German aviator Alfred Friedrich, seen in the rear cockpit of his Blériot, flew from Germany to France in 1913.*
FAR LEFT: *This trim biplane was the Harvard I, of July 1910. It was based to some degree on the Blériot monoplane.*
LEFT: *The early aviator Colliex sitting at the controls of his Voisin aeroplane. The Voisin designs, though successful, were outdated by about 1910.*

ABOVE: *Alberto Santos-Dumont standing in front of one of his popular little Demoiselle monoplanes of 1907. The pilot sat under the warping wings and controlled flight by a combined rear rudder and elevator, side rudders and forward elevator.*
RIGHT: *After Phillips the next to test a flying machine in Britain was A. V. Roe, whose biplane made short hops at Brooklands in 1908. His 1909 triplane flew better.*
BELOW: *Trajan Vuia in his Vuia I tractor monoplane. Unfortunately he had no proper system of control, so his early aircraft – though they influenced others – were in themselves failures.*

Santos-Dumont, having, like other Europeans of the day, omitted to experiment with gliders or learn how to fly, found it difficult to test his machine without getting in and flying it. He did at least try. The 14-*bis* was finished in July 1906 and it was slung on a pulley from a long overhead cable, and pulled by a trotting donkey (why the wealthy pilot did not use one of his cars is a mystery). Then it was borne aloft slung beneath his *No 1* airship, but as it was not cast off to glide to the ground it is difficult to see the reason for this curious experiment. The 14-*bis* made a short hop of 7m (23ft) on 13 September, but was damaged in a bad landing. Santos-Dumont took out the 24hp Antoinette engine and put in one of the first 50hp versions of this superb purpose-designed French aero engine, and flew 60m (197ft) on 23 October. He then made the important modification of adding ailerons, intended to be worked by a body harness as he stood up in the machine's balloon basket. On 12 November, at Bagatelle, Paris, he made six flights, on the last of which he kept going strongly for over 21 seconds, covering 220m (722ft). This ranks as the first successful powered flight by anyone other than the Wrights, though how controllable or practical the 14-*bis* was is arguable.

Nobody copied it, and Santos-Dumont later turned to tricky little monoplanes called Demoiselles, with warping wings. Blériot experimented with the same pusher canard arrangement, but with monoplanes. He had been influenced by the first man

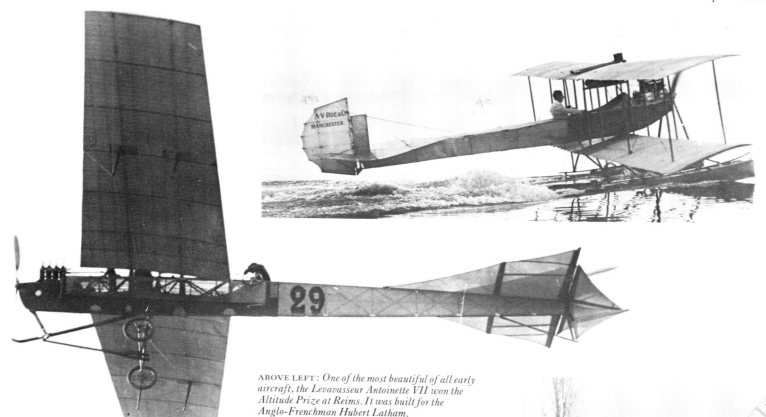

ABOVE LEFT: *One of the most beautiful of all early aircraft, the Levavasseur Antoinette VII won the Altitude Prize at Reims. It was built for the Anglo-Frenchman Hubert Latham.*
ABOVE RIGHT: *Commander Schwann's Avro of 1911, was one of the first seaplanes. The engine was a 35hp Green, one of the few British types.*
RIGHT: *Front view of the Santos-Dumont 19-bis, an unsuccessful form of Demoiselle with the engine between the wheels, with belt drive.*
BELOW: *Henri Farman was one of the greatest early aviator-designers. Like several of his rivals, he was a Briton living in France.*

ever to try to fly an aeroplane of the kind we have today, with tractor propeller, monoplane wing and rear tail. This man was the Paris-domiciled Romanian Trajan Vuia, and though his attempts to fly in March to October 1906 never achieved hops better than 10m (33ft) his sensible design made a great impression. One of those who was impressed was Blériot, and while in 1907 Voisin and Henri Farman at last achieved real success with classic box kite biplanes, Blériot built better and better tractor monoplanes. By this time aviation was at last beginning to 'take off'. Leon Levavasseur's Antoinette met the demand for a light and reliable engine, and after 1908 another engine, the Gnome rotary, came on the scene; soon there were many possible engines, and Blériot usually chose an REP or Anzani.

Farman, an Englishman living in France, was certainly the first European to fly a practical and fully controllable machine. Built for him by Voisin in the summer of 1907, the *Voisin-Farman I* first 'hopped' on 30 September. On 26 October it flew for almost a minute, on 8 November it made a definite turn, and on 9 November Farman flew the first complete circle in Europe. A little later Farman was flying with proper ailerons, because – apparently alone among Europeans – he had grasped the underlying importance of lateral control, as explained in the original patent of the Wrights. Once he had added ailerons Farman could fly as he wished, and on 30 October 1908, he made the world's first cross-country flight, 27km (16·78 miles)

from Bouy to Reims. On the next day Blériot flew his *VIII-ter* a slightly greater distance. And on 25 July 1909, Blériot flew his *XI* from France to England, to make the first international and first overseas flight.

Floatplanes

By this time aeroplanes could be designed to order, and industries had sprung up to build aeroplanes and teach the customers how to fly. Though heavier-than-air flight had begun to have useful applications, stalling and spinning were not yet understood, and experiments were still needed – as they still are. In March 1910 Henri Fabre successfully flew a float-seaplane, and this idea was enthusiastically taken up by American Glenn Curtiss. In January 1911 Curtiss flew a much better seaplane, and the following month he flew the first amphibian, which was equipped with both floats and wheels.

In 1912 the Royal Aircraft Factory at Farnborough flew the B.S.1 (B.S. for 'Blériot Scout', which meant it had a tractor propeller but did not indicate any other link with Blériot), which is generally regarded as the first purpose-designed military aircraft. In France the Deperdussin company built fast racers with radically different streamlined fuselages of monocoque construction. A. V. Roe in England built an enclosed cabin machine; the Short brothers built multi-engined aircraft, and experiments were made in taking off and landing on short platforms on warships. Perhaps the pinnacle of the pre-war achievements was the large, four-engined aircraft built by the young Igor Sikorsky in Russia in 1913, from which stemmed the world's first transports and strategic bombers.

Engines

All human flight depends on either buoyancy (achieved by lighter-than-air aerostats) or a source of power. With kites the source of power is the wind, reacted by a tethering line. With sailplanes it is a combination of gravity plus upcurrents of air, either in thermals or caused by wind over steep hillsides. Other aircraft need mechanical power, produced by muscles, by burning fuel or some other energy source such as stretched elastic, compressed air or electricity. Before 1895 nobody knew what was the best form of power for aviation. W. S. Henson in 1843 proposed a steam aeroplane and made a successful working model of its engine.

The first powered aircraft to fly, discounting impractical balloons supposed to be rowed by oars, was Henri Giffard's airship of 1852, which cruised at about 11km/h (7 mph) on a steam installation weighing 53kg (117lb) per horsepower. This relationship, which results in a figure called specific weight (total mass of an engine divided by its maximum power output), is an important one in comparing one aero engine with another. For long-range aircraft, an even more important variable is specific fuel consumption, which can be regarded as either the amount of fuel burned to generate unit quantity of energy (such as one horsepower-

hour; one horsepower developed for one hour) or the rate at which fuel must be burned to generate unit power (such as one horsepower).

Late 19th century

There seemed in the late 19th century to be little immediate prospect of making an aero engine light, powerful and reliable. Several experimenters tried gunpowder as a fuel for

an engine, though none tried to build a rocket aeroplane, which would have had a far greater chance of success. A major advance came in 1872 when Austrian Paul Hänlein built an airship filled with coal gas, of which part was fed to a Lenoir-type gas engine. This was the first internal-combustion engine to fly, though it weighed 93kg (205lb)/hp and of course reduced the available lift by gradually consuming the gas in the envelope.

In 1883 the Tissandier brothers made successful flights with an airship driven by a 1½-hp electric motor fed by 24 bichromate batteries. Though the specific weight was a poor 182kg (400lb)/hp, the results were satisfactory. A much better electric airship flew in 1884: *La France* by Renard and Krebs, with a 9-hp motor weighing 531kg (1,170lb), or 59kg (130lb)/hp. A further great advance came in 1888 when Gottlieb Daimler, pioneer of the petrol engine, saw one of his employees fly Wölfert's first powered airship, on the thrust of a propeller driven by a 2-hp single-cylinder Daimler engine.

In Australia, Lawrence Hargrave had been experimenting with rubber, clockwork and compressed-air motors since 1884, and in 1889 he built (but did not fly) a superb three-cylinder rotary engine fed from a long

BELOW: *This engine was the key to successful powered flight. It was designed by the Wright brothers with their gifted mechanic Charlie Taylor, and built mainly by the latter. It gave about 13 horsepower.*

ABOVE: *The five-cylinder radial engine used in S. P. Langley's Aerodrome. Built by Charles Manley, it was very much ahead of its time in having spark ignition and a carburetted petrol feed.*

FAR RIGHT: *When it burst on a delighted world of aviators in 1907–08 the Gnome appeared the answer to the engine problem. Designed by the French Seguin brothers, it was amazingly light and simple.*

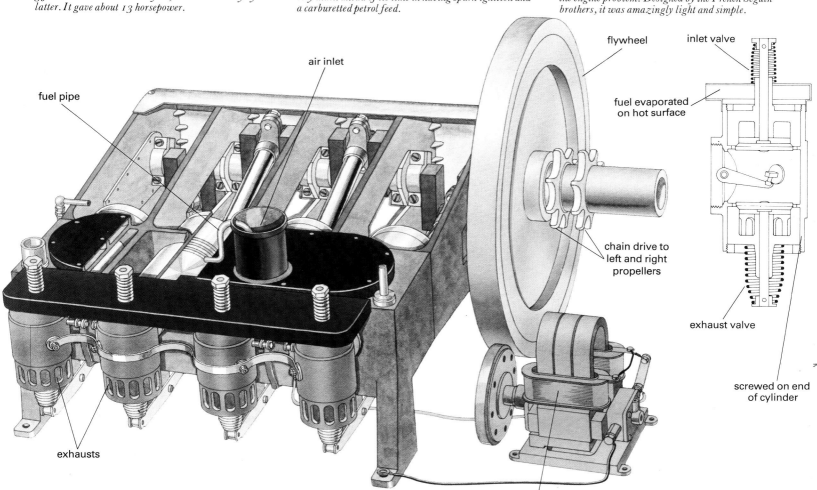

fuel pipe · air inlet · flywheel · inlet valve · fuel evaporated on hot surface · chain drive to left and right propellers · exhaust valve · screwed on end of cylinder · exhausts · friction drive magnets

compressed-air bottle intended to fit inside a fuselage. A rotary engine has a fixed crankshaft. Around it cylinders are arranged radially and, together with the crankcase and propeller, they rotate. One advantage was expected to be the ability of the cylinders to run at an acceptable temperature merely by having air-cooling fins, without heavy and troublesome water-cooling circuits (though this did not apply to a compressed-air engine, which hardly got warm). So far as is known no heavier-than-air machine ever flew with a compressed-air engine. Whether one flew on steam power is arguable. Sir Hiram Maxim's huge biplane of 1894 was powered by two compound steam engines of very advanced design, each calculated to develop 180hp at the remarkable steam pressure of 22·5kg/cm² (320lb/sq in), and with a specific weight of only 0·70kg (1·55lb)/hp exclusive of the heavy boilers and furnace. Clément Ader's *Eole* of 1890 had a much smaller steam engine, judged in 1960 to have developed 20hp for a net weight (without fuel or water) of only 23kg (51lb).

1900-1910

Undoubtedly the outstanding engine at the turn of the century was that of the American Charles Manley. Far ahead of its time, this was a five-cylinder radial, with carburetted petrol feed (the carburettor's function is to convert the bulk liquid into highly combustible petrol/air vapour) and spark ignition, and cooled by a water circuit. It developed more than 50hp and weighed

only 1·63kg (3·6lb)/hp, but achieved nothing because of the ridiculous launch system of the aircraft, S. P. Langley's *Aerodrome*, in which it was installed.

A contemporary engine was that first built by the Wright brothers. Though this was pedestrian and uninspired by comparison, it has a greater place in history because it powered the first successful aeroplane, the Wright 1903 Flyer. Features included four water-cooled cylinders in line, lying horizontally, fuel carburation by dripping on to a hot surface, and a large flywheel and two gearwheels driving the pusher propellers in opposite directions by chains. The weight was about the same as Manley's engine, 82kg (180lb); the power was said by the Wrights to be 12hp and estimated in 1961 at 'about 13hp'. Chief detail designer and constructor was the Wrights' mechanic, Charles Taylor.

A far more advanced engine was built in France. Léon Levavasseur named his engines (and subsequently his aeroplanes) after a friend's daughter, Antoinette. His first Antoinette engine dated from about 1901, most of the early ones being used in his speedboats. By 1905 a water-cooled vee-8 was in production rated at 50hp and weighing about 50kg (110lb), giving a specific weight of 1·0kg (2·2lb)/hp which was not generally to be surpassed for 25 years. Two features which were ahead of their time were evaporative (steam) cooling and direct fuel injection. These, however, harmed reliability, leaving the main commercial field to a yet more

revolutionary French engine, the Gnome rotary, designed by the Séguin brothers and first marketed in 1908. This differed in almost every respect from contemporary car engines and from other aero engines either before or since, but it offered a combination of high power/weight ratio and good reliability which no other engine of the day could match.

The first Gnomes had five cylinders, were

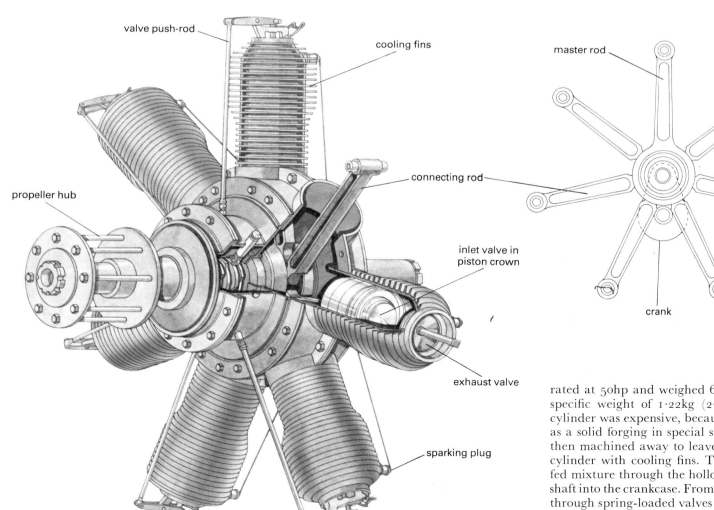

rated at 50hp and weighed 61kg (135lb), a specific weight of 1·22kg (2·7lb)/hp. Each cylinder was expensive, because it began life as a solid forging in special steel which was then machined away to leave a thin-walled cylinder with cooling fins. The carburettor fed mixture through the hollow fixed crankshaft into the crankcase. From here it escaped through spring-loaded valves in the heads of the pistons into the cylinders. After ignition,

ABOVE LEFT: *The dreaded Fokker E (Eindekker) was the first aircraft to go into production with an engine synchronized with a gun.*

ABOVE LEFT: *This Mercedes four-cylinder engine of 1909 started a classic breed.*
ABOVE: *One Mercedes offspring was the six-cylinder Benz Bz.IV, made in thousands.*

the hot gas escaped through valves in the cylinder heads (the closed ends of the cylinders) which were opened at the appropriate times by push rods driven by a fixed cam-ring in the crankcase. Though suffering from fundamental shortcomings, the Gnome rotary gave high power and, with the flywheel effect of the whirling engine, even torque. Furthermore, the rotating finned cylinders were self-cooling. Castor-oil lubricant was added to the fuel so that the combustible mixture passing through the crankcase and over the bearing surfaces obviated the need for a separate lubrication system. Burnt castor oil has a distinctive smell forever associated with these engines, which with designs derived from them made up 80 percent of the engines in World War I before 1917.

World War I

By 1910 more than 70 companies, virtually all of them in Western Europe, were making aero engines. Very few of them, however, saw war service in any numbers, for until 1916 the Gnome rotary dominated in the industry. This was made as the Bentley Rotary B.R.1 in Britain, the Thulin in Sweden and the Oberursel UR.I in Germany. The same basic engine was developed to have nine cylinders and then appeared in a two-row form, virtually two seven-cylinder engines joined together with the rear row 'staggered' by 25.7° so that its cylinders would be cooled by air passing between those of the front row. This two-row cylinder configuration was later to become the dominant form of high-power piston aero engines.

By 1911 the Séguin brothers had produced a different Gnome which attempted to overcome unreliability caused by the malfunctioning of the hot counterweight-operated valve in the head of the piston. In the Gnome Monosoupape (single-valve) the piston had no valve and the mixture escaped from the whirling crankcase into the cylinders via ports in the cylinder walls, uncovered by each piston near the bottom of its stroke. Many thousands of 'Mono' engines served the Allies throughout World War I.

Around 1911 other makers circumvented the Séguin patents by building other rotary engines. In the Le Rhône there were separate inlet and exhaust valves in each cylinder head, cunningly operated by a single rocker and rod which was pushed to open the exhaust valve and pulled to open the inlet valve, admitting mixture supplied through a pipe up the outside as in other engines. The Clerget was similar but had separate push-rods for the two valves in each cylinder head. Thousands of these engines were made between 1913 and 1918, with others by

BELOW: *This nine-cylinder Le Rhône rotary was a descendant of the Gnome.*

ABOVE: *A superb British 1917 engine, the 375hp Rolls-Royce Eagle VIII vee-12.*

Oberursel, BMW, Goebel and Siemens u. Halske in Germany, Bentley, Allen and other companies in Britain, and ten companies in France and the United States.

One outstanding engine of 1910 was the French Renault vee-8, of 80hp, with air-cooled cylinders and geared drive. It inspired the 90hp, Royal Aircraft Factory (RAF)

engine of 1913. From these were derived vee-8 air-cooled and vee-12 air- and water-cooled engines rated up to 260hp. The Salmson (Canton-Unné) was a unique static radial with water-cooled cylinders, usually made with 9 or 14 cylinders giving from 110 to 240hp. Nearly all German-designed engines were solid, robust and reliable, with six water-cooled cylinders in line; produced with large cylinders, they gave up to 300hp. Makers included Mercedes, Maybach, BMW, Benz and Austro-Daimler. Anzani, whose three-cylinder 20hp engine had powered Blériot's early machines, developed static radial and 'fan' engines with various numbers of cylinders. These could be anything up to 20 (2 radial rows of 10).

To provide Britain with powerful engines of native design, the Admiralty towed the 1914 TT-winning Mercedes car to Rolls-Royce where its engine inspired the Hawk (used in Royal Naval Air Service (RNAS) blimps) as well as the outstanding water-cooled vee-12 Falcon of 190-260hp and Eagle of up to 360hp. Most important of all

How a Piston Engine Works

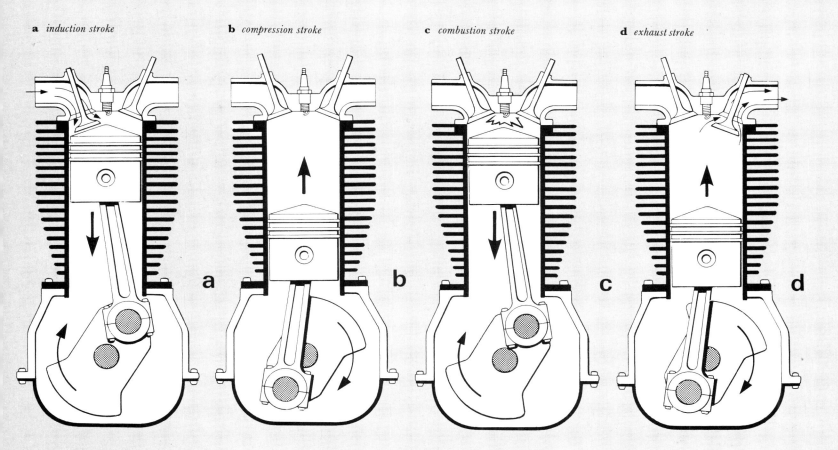

a *induction stroke*   **b** *compression stroke*   **c** *combustion stroke*   **d** *exhaust stroke*

In about 1850, Etienne Lenoir was developing an engine in which a piston was driven along a cylinder by the pressure of a burning mixture of air and coal-gas. At that time the only engines in common use were steam-driven, where the burning took place not in the cylinder but under a separate boiler. The new gas engine burned its fuel inside the cylinder, and so was called an internal-combustion engine. Compared with steam engines, both pressure and temperature inside the cylinder during the working stroke – the down-stroke of the piston, pushed by expanding gas – were much higher. This meant that the efficiency was greater, more power being developed for a given rate of consumption of fuel. In the petrol (gasoline) engine, developed in about 1880, the pressures and temperatures could rise even higher, particularly in the diesel engine where they were highest of all. Moreover, these newer engines were more useful because, when used in a vehicle, they could carry greater amounts of their liquid fuel.

In all these engines the basic principle is that a mixture of fuel and air is burned in a closed space, generating hot gas under high pressure. The expanding gas pushes down a piston, which via a connecting rod and crank turns a shaft. As the shaft continues to rotate, the crank and connecting rod then return the piston to the top of the cylinder. In the most common type of petrol engine, the form of more than 99 per cent of all aircraft piston engines, the piston makes only one working stroke in every four strokes, two up and two down. This is therefore called the four-stroke cycle, or

Otto cycle after the engineer who originally developed it.

On the first stroke (the induction stroke) the piston is pulled down by the crank, drawing in fresh fuel/air mixture through the open inlet valve. Near the limit of the piston's travel (called bottom dead centre), the inlet valve is closed by the mechanism which links the valves to the rotating crankshaft. The rotating crank then pushes the piston back up the cylinder on the second stroke (the compression stroke), squeezing the mixture into a small space (the clearance volume) at the top of the cylinder. Here it is ignited by a sparking plug, a device supplied with pulses of high-voltage electricity to make hot sparks at the appropriate moments. The mixture instantly begins to burn at the sparking plug, and the flame expands very quickly in a spherical shape (in the way a ripple grows on a pond, but in three dimensions) until it has consumed the whole mixture. Though the combustion is fast, it is not the same as an explosion. Sometimes, because of hot spots of carbon in the cylinder or the use of inferior fuel, the whole mixture ignites more or less at once. This is called detonation, or knocking and, because it results in excessive temperatures and violent pressures, it can soon cause severe damage. With proper combustion the whole mixture burns smoothly and cleanly. The hot gas does not damage the piston or cylinder, but its very high pressure imparts a large force on the top of the piston, pushing it down on the third stroke, the working stroke. Near the bottom of this stroke, the valve gear opens the exhaust valve so that the hot gas begins to escape. On the

fourth (upward) stroke, the exhaust stroke, nearly all the gas is expelled by the rising piston.

In most engines several cylinders are used to give smoother running with more frequent working strokes. There are many ways in which the cylinders can be arranged. Today the most common engines for light aircraft are of the horizontally opposed type – the flat-twin, flat-four or flat-six – with horizontal 'pots' (cylinders) on the left and right sides of the central crankcase. The opposing cylinders are often not quite aligned, so that their connecting rods can work side-by-side on the same crankpin. If they are in line, one con-rod has to have a Y-shaped end to fit on each side of its partner. In the higher power ranges, the radial is the preferred arrangement, with the cylinders disposed like spokes of a wheel around a central crankcase. There may be five, seven or nine cylinders, one of which will drive a master rod to which all the others are pivoted. For even more power, the two-row (in the pre-jet era, even the four-row) radial is necessary, with the second-row cylinders behind the gaps between the front-row cylinders. This is to improve cooling, because all modern radials have cylinders cooled by air flowing between deep, close-spaced fins which provide a large surface area over which to dissipate the excess heat. Nearly all light aircraft engines are also air-cooled, but today a few water-cooled engines, such as are fitted to most cars, are becoming popular for some small aircraft and helicopters. Water cooling can make an engine quieter, and some engineers claim that it can make it more

efficient. Conversely, such engines need a system of water pipes and a radiator which adds to bulk, weight and possibly drag, and which can leak. Air-cooled engines are generally considered to be superior when operating in conditions of extreme cold or extreme heat.

Not unnaturally, it does not suffice merely to pipe petrol or other fuel from the tank to the cylinder. For a petrol engine to run properly it must be supplied with a perfect mixture of about 15 parts of air by weight to 1 of fuel (about 9,000:1 by volume), and this needs either a carburettor or a fuel injection system. The carburettor measures a small, steady flow of fuel and allows it to mix with the flow of air entering the inlet manifold which ducts the mixture to the cylinders. A fuel-injection system measures the engine's exact needs and supplies extremely small doses of fuel to each cylinder in turn. In general, a fuel-rich mixture is used at take-off and a lean, or weak, mixture for economical running in cruising flight. The pilot can control the mixture in some engines, while in others it is adjusted automatically. As an aircraft climbs, air density decreases and so the amount of fuel supplied has to be decreased in proportion. Engines needing high power at high altitudes are supercharged by a fan which is driven by the engine at several times crankshaft speed in order to blow extra air into the cylinders. Other gears drive the valves, and in many engines the propeller is turned more slowly than the crankshaft by a reduction gear, because it may be more efficient and quieter to use a bigger propeller turning at reduced speed.

LEFT: *Pratt & Whitney built 35,000 Wasps in 35 years (1926–60). This one came off the line in 1935. Thousands are still in use.*
BELOW: *The DH Canada Otter is a utility transport flying on Wasp power.*
BOTTOM: *Another classic US engine was the Wright Cyclone which strove to take over from the British Jupiter as No 1 in the world.*

Allied engines was the American Liberty, designed for the US Army by a team led by the chief engineers of Packard and Hall-Scott in the first three months of America's participation in the war. The Liberty was designed to give 400hp and to be mass-produced with interchangeable parts. A water-cooled vee-12, it had separate cylinders and direct drive, and was of extremely simple design.

Swiss Marc Birkigt, chief engineer of a Spanish car company, designed a number of Hispano-Suiza products (including a famed aircraft cannon), among them the outstanding water-cooled vee-8 and vee-12 engines made in vast numbers in France, Britain and the United States. The later geared Hispanos were troublesome, and the RAF S.E.5a scout finally did well with a British development, the direct-drive Wolseley Viper. Curtiss in the United States supplied the OX–5 water-cooled vee-8 rated at 90hp for the JN–4 trainer, and in Italy the Fiat A–20 vee-12 and Isotta-Fraschini were important water-cooled types.

The 1920s

By the 1918 Armistice, the rotary had become outmoded, and refined engines of between 300 and 400hp were in production. Geared drives were becoming common, especially for bigger aircraft, to enable large-diameter propellers to turn more slowly, while the engine speed was increased to give greater power (power is force multiplied by speed). Much more advanced carburettors, able to

maintain the correct mixture strength for all powers and altitudes were also in use. Experiments began with variable-pitch propellers and with greatly improved cooling systems, using air or water, which instead of adding to drag actually increased propulsive thrust. Early engines were started by 'swinging the propeller', a dangerous and arduous task which was especially difficult with large aircraft and seaplanes. Electric starters began to appear, as did gas starting, in which the cylinders were pumped in sequence with fuel-rich mixture from a unit installed in the aircraft; this looked like a motor-cycle engine, one half actually being an engine and the other a pump. Much greater attention was paid to reducing weight and improving reliability, with strict control of material specifications, surface finish and fatigue properties.

Few wartime engines survived the need for long and reliable use. Liberty 12 engines were an exception, and so were German BMW and Junkers water-cooled six-in-lines. The British RAF 8a turned into the Siddeley (later Armstrong Siddeley) Jaguar, a small-diameter two-row radial with 14 cylinders rated at between 400 and 450hp and soon to appear with geared drive and supercharger. The Napier Lion had three banks of four water-cooled cylinders spaced at 60° in the W or broad-arrow configuration, and served reliably at around 450hp in large aircraft. Designed at Cosmos, the Bristol Jupiter was planned as a powerful and simple nine-cylinder radial, with four valves per cylinder,

and began life in 1920 at a little over 400hp. The Aircraft Disposal Co (ADC) produced engines using up some of the thousands of surplus Renault and RAF parts, and Frank Halford split this vee-8 in half to produce the Cirrus, weighing 122kg (268lb) and rated at 80hp, to give the first reliable lightplane engine. Its four upright cylinders were air-cooled. From it was evolved a completely fresh design, produced for de Havilland and called the Gipsy. Towards the end of the 1920s this was inverted, the cylinders hanging down to give the pilot a better view, under the name Gipsy Major. Rated at 130hp, it was the pre-eminent lightplane engine of the 1930s.

Rolls-Royce gradually came back in the 1920s with the Condor, designed in 1918 as a scaled-up Eagle with four-valve cylinders. In 1922 it was redesigned and a small number, producing 650 to 670hp, was used in heavy aircraft and airships. Lorraine, Hispano-Suiza and Farman in France continued with high-power water-cooled engines; the German Junkers L.5 and BMW.VI carried on the tradition of six-in-line engines of robust and reliable type; and several makers struggled in Italy. Most American engines were air-cooled, an outstanding series being the Wright Whirlwind radials which began with seven cylinders at 220hp and soon included five- and nine-cylinder models rated at 140 and 300hp respectively. The seven-cylinder model gained fame with Lindbergh's Atlantic crossing in 1927. Two years earlier, a group of engineers had left

BELOW: *The Schneider Trophy exerted a powerful influence on engine design by making designers concentrate on sheer speed. This made a streamlined-looking nose appear all-important, leading from the Curtiss V-1500 (D-12) in the Curtiss R3C (a) via the Rolls-Royce R in the S-6B (b) to the Fiat AS.6 in the Macchi 72 (c). Yet the air-cooled radial eventually proved superior for all uses.*

BOTTOM AND RIGHT: *These drawings of a Pratt & Whitney Wasp show basic features of a simple aircooled radial. Air is drawn in through a carburettor, where petrol vapour is added, and then compressed by the super-charger before being admitted to each of the nine cylinders in sequence. Only one cylinder is shown in the side view, driving the master connecting rod to which the rods from the other cylinders are pivoted.*

Wright to found Pratt & Whitney Aircraft; their completely new design, materializing as the Wasp, ran on Christmas Eve, 1925. Its fresh and skilled design soon put this 400hp engine into the forefront of US naval and military pursuit (fighter) markets, and began the world's biggest production of aero engines. By 1930 Wright had produced the outstanding new R–1820 Cyclone (1,820 cu in displacement: the piston area multiplied by the number of pistons, multiplied by the piston stroke from bottom to top dead centre). Pratt & Whitney was meanwhile producing the R–1340 Wasp and R–1690 Hornet, available with geared drive and super-charger. These swept away the big 800hp, water-cooled Packard and 650hp Curtiss Conqueror, the last early water-cooled engines in the United States.

This was remarkable because, in 1921, the Curtiss D–12 water-cooled vee-12 had led all other American engines and won the Schneider Trophy in 1923. Rated at 440–475hp, it could be installed, unlike almost all its rivals, in a streamlined, pointed nose. Fairey used it in the Fox bomber (faster than contemporary British fighters) and made the D–12 as the Felix. For the 1925 Schneider race the Curtiss gave 600hp, but was beaten by boosted Fiat engines, which in turn were beaten by a racing version of the Napier Lion giving 875hp for a specific weight of 0·48kg (1·06lb)/hp. To beat the D–12, Rolls-Royce produced the F, later named Kestrel, as a trim water-cooled upright vee-12 giving 480 to 525hp. An important feature was that

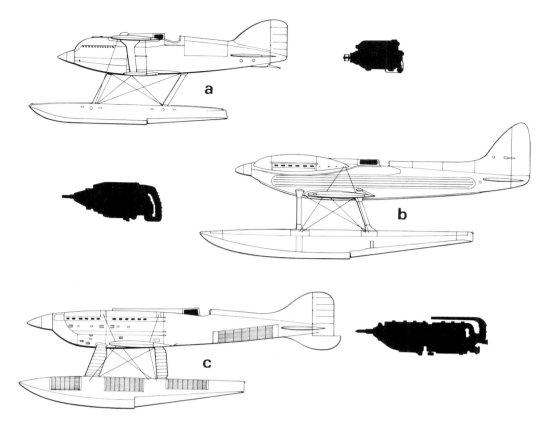

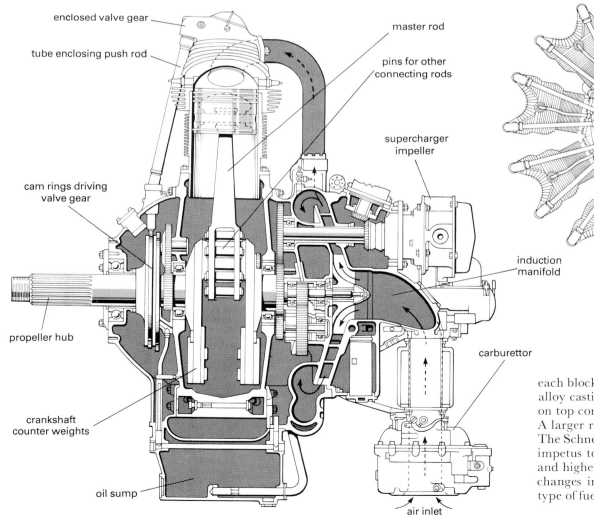

each block of six cylinders was a single light-alloy casting, with a separate casting bolted on top containing the heads and valve gear. A larger relative was the Buzzard of 825hp. The Schneider Trophy races gave increasing impetus to the push towards higher powers and higher stresses, and led to revolutionary changes in supercharging, cooling and the type of fuel used.

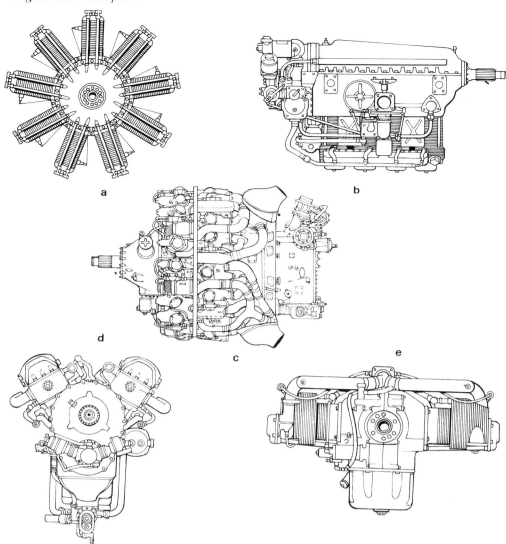

Throughout the heyday of the aircraft piston engine, from 1914 to 1950, there were heated arguments over which type was best. The broadest subdivision was into the water-cooled in-line (which included vee and other in-line arrangements) and the air-cooled radial. On balance, most of the advantages lay with the latter. It made more economical use of material, was cheaper to build, needed no heavy and complex cooling system, and for the same reason offered higher reliability. Installed in a fighter, its much shorter length improved the aircraft's power of manoeuvre. As the cooling needed no water, the air-cooled engine could work perfectly in the coldest climates; and it could also work much better in the hottest regions, because the cylinders could still dissipate their heat easily to the air, whereas the difference in temperature between the hot water and the air was much less. The installed weight was usually very much less, even though the weights of bare engines of equal power might not be very different.

The one advantage the water-cooled engine did have was that it often looked more streamlined, but even this was usually misleading because one had to take into account the drag of the cooling radiator. (Later, by about 1940, designers learned how to make both air-cooled and liquid-cooled engines have 'negative cooling drag' – in other words, to gain forward thrust from the cooling airflow.) It so happens that, during the crucial 1930s, aircraft designers achieved higher speeds with liquid-cooled

ABOVE: *Typical piston arrangements. (a) Clerget nine-cylinder rotary, mass-produced in World War I, but representing a dying species. (b) DH Gipsy Major inverted four in-line aircooled, designed in the 1920s for the first private-owner light aircraft. (c) Turbo-Compound 18-cylinder two-row radial with three exhaust turbines, an extremely complex and efficient engine representing the pinnacle of the piston-engine art around 1950. (d) Rolls-Royce Merlin liquid-cooled vee-12 (front view), typical of World War II engines. (e) Continental IO 360 'flat six' aircooled, typical of the final families of aircraft piston engines still being made.*

engines, but this was simply because they did not at that time know how to install the air-cooled radial engines properly. When they found out, such fighters as the Vought Corsair, the first in America to reach 645km/h (400mph), Republic Thunderbolt (the fastest piston-engined fighter, in one version), Focke-Wulf 190 and Hawker Tempest (faster with a radial) at last set the record straight. But today the world piston-engine speed record has been won back by an aircraft with a Merlin liquid-cooled engine.

1930–39

Since World War I it had been recognized that the aviation piston engine, because it was run so hard, would be a leader in the fight to find fuels less prone to detonation or 'knocking'. This is a condition in which, if an engine is run at too high a compression ratio or at too high a power, the mixture in the cylinders does not burn evenly over a period of a few thousandths of a second, but

explodes almost spontaneously throughout its whole bulk, causing excessive heat and a violent mechanical shock. Eventually, in 1930, the US Army Air Corps issued a specification for '87-octane' fuel (100-octane having anti-knock properties equal to those of a reference fuel, iso-octane), with an allowed small proportion of an anti-knock additive, Tetra-Ethyl Lead (TEL). At first TEL caused fouling of sparking plugs and corroded exhaust valves, but designers learned to make engines which could live with 'leaded' fuel. Over the remaining years of the high-power piston engine, the fuel was to be at least as important to power as was the engine itself.

Another major development concerned supercharging. Due to falling atmospheric density, engines draw progressively smaller masses of air into their cylinders as the aircraft climbs. To keep mixture strength correct, the fuel flow must therefore be reduced in proportion, thus reducing the power. A 500hp engine would give only 250hp at 6,688m (22,000ft) and less than 125hp at 12,160m (40,000ft). Various types of supercharger were experimented with in World War I to force in more air and maintain power. The favoured type was the centrifugal blower, geared up to run faster than the crankshaft. Rateau of France was the only company in production with a supercharger by 1918. Sanford Moss at General Electric (GE) in the USA tried the difficult task of driving the supercharger with a turbine spun by the white-hot exhaust gas. This increased not only power but also efficiency through the extraction of what otherwise would have been wasted energy.

Fuel and supercharger development enabled Britain in 1931 to win permanently the Schneider Trophy. From the Rolls-Royce Buzzard was developed the R racing engine, with higher compression ratio and much bigger gear-driven supercharger. Whereas the Buzzard gave about 825hp, the R gave 1,545hp on its first run in May 1929, using a fuel comprising 78 per cent benzol, 22 per cent light gasoline (petrol) plus a little TEL. By September 1929 the power was 1,900hp at 2,900rpm, at a boost pressure of $0 \cdot 86$kg/cm^2 (12$\cdot$25lb/sq in) above atmospheric pressure. In this case the supercharger was being used not to maintain power at altitude but to increase it at sea level, the most severe test of any engine. For the 1931 race the power reached 2,360hp at 3,200rpm, using 30 per cent benzol, 60 per cent methanol and 10 per cent acetone, plus a little TEL. For an attack on the world speed record the sprint version gave 2,783hp, which for a weight of 740kg (1,630lb) achieved the incredible ratio of $0 \cdot 266$kg ($0 \cdot 586$lb)/hp. This work facilitated the design of the Merlin, a slightly smaller engine of 1,650cu in which began life at 990hp in 1936 but gave 1,300hp with specially imported 100-octane fuel in the Battle of Britain..

During the 1930s Wright increased the power of the R–1820 Cyclone from 500 to 1,200hp by engineering refinement and by developing the ability to run on 100-octane fuel. In parallel came the R–2600 Cyclone 14-cylinder two-row engine of 1,300–1,900 hp, and the 18-cylinder R–3350 Duplex Cyclone rated at 1,750hp in 1939 and ultimately to give more than twice as much. Rival Pratt & Whitney concentrated on the R–1830 Twin Wasp, with two rows each having seven Wasp-size cylinders. Allison Division of General Motors boldly schemed a vee-12 of modern design with liquid cooling, the V–1710 (vee, 1,710cu in), which, like the Merlin, was cooled not by water but by a mixture of water and ethylene glycol, giving lower freezing point, higher boiling point and a smaller and lighter radiator and cooling system. Hispano-Suiza was preeminent among French liquid-cooled engine manufacturers, with the 12Y series rated at 860 to 1,000hp and made in the Soviet Union as well as other countries. German designers concentrated on inverted-vee 12-cylinder engines with direct fuel injection, a system at first regarded as complex and costly but which ultimately came to dominate over the traditional carburettor with its float chamber prohibiting inverted operation and a choke tube prone to icing. Daimler-Benz produced the DB600 of around 690hp in 1935 and the 1,000hp DB601 three years later. Junkers produced the similarly powered and timed Jumo 210 and 211 in addition to a series of economical but weighty two-stroke diesels having six double-ended cylinders fed in the centre and driving crankshafts along the top and bottom of a strangely thin and flat-sided engine. These were the only aircraft diesels made in large numbers, despite the widespread belief that the diesel's fuel economy and ability to burn cheap oil fuel would commend it for long-range use.

Among air-cooled radials the Jupiter had been dominant in the 1920s, being license-produced in every country possessing an aircraft industry and flying in 229 types of aircraft. From it were developed two even more important engines, the short-stroke Mercury of 550 to 950hp for fast aircraft and the Pegasus, which was the same size as the Jupiter, but greatly refined and ultimately giving powers in excess of 1,000hp. Roy Fedden, the Bristol chief engineer, could find no simple way to make a two-row radial with four valves per cylinder, but after many years of heartbreak perfected a new radial family with sleeve valves: the traditional poppet valve was eliminated and replaced by an oscillating thin-walled sleeve interposed between the piston and cylinder, with ports in the sleeve intermittently lining up with inlet or exhaust ports in the walls of the cylinder. The first sleeve-valve engines were the 900hp Mercury-sized Perseus, the 500hp Aquila, the 14-cylinder two-row Taurus of 1,000hp and the 14-cylinder Hercules of around 1,400hp.

By 1937 the desperate need to increase production led to the Mercury being selected for 'shadow production' by a consortium of car firms, and soon the Hercules was to follow suit. Napier made unusual air-cooled engines with four banks of cylinders in H configuration, the most powerful being the 24-cylinder Dagger of 1,000hp. In France Gnome-Rhône made Jupiters, sidestepped Bristol patents with engines having only two

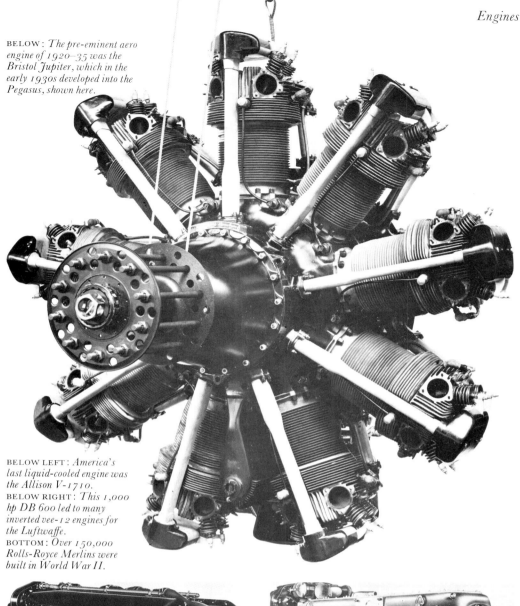

BELOW: *The pre-eminent aero engine of 1920–35 was the Bristol Jupiter, which in the early 1930s developed into the Pegasus, shown here.*

BELOW LEFT: *America's last liquid-cooled engine was the Allison V-1710.*
BELOW RIGHT: *This 1,000 hp DB 600 led to many inverted vee-12 engines for the Luftwaffe.*
BOTTOM: *Over 150,000 Rolls-Royce Merlins were built in World War II.*

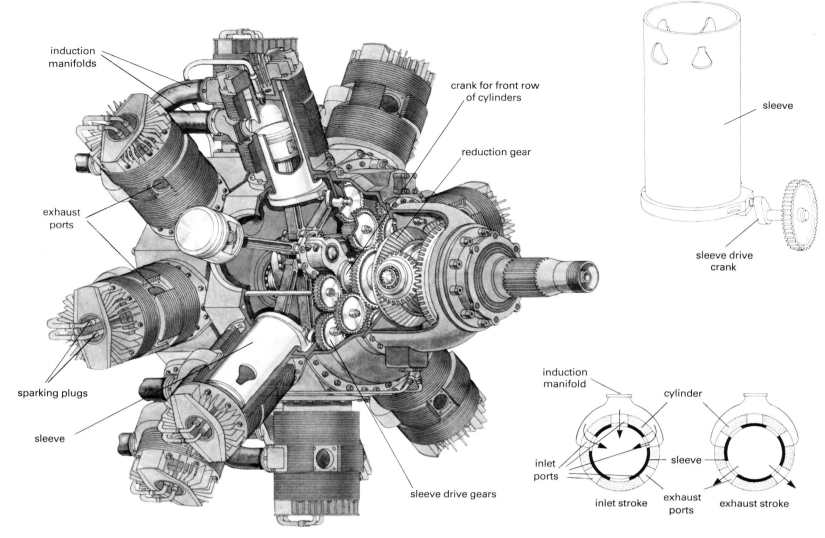

induction manifolds

crank for front row of cylinders

reduction gear

sleeve

exhaust ports

sleeve drive crank

sparking plugs

sleeve

induction manifold

cylinder

inlet ports

sleeve

sleeve drive gears

inlet stroke

exhaust ports

exhaust stroke

valves per cylinder and went into production with the 14-cylinder two-row Mistral Major of 900hp in 1931. In Germany the BMW Hornet (Pratt & Whitney) was developed into the 132 of 1,000hp, while Bramo (later merged into BMW) built the similar 323 Fafnir.

World War II

The demands of war were for ever-higher power and mass-production of established engines, two objectives in conflict. The Rolls-Royce Merlin was developed with two-stage mechanically driven supercharger and intercooler to double its power at high altitude, and both Rolls and American Packard V–1650 versions were rated at over 2,000hp on 115/145-grade fuel by 1945. The X-layout Vulture was a failure, but the Griffon of 2,239cu in went into production for fighters and remains in use to the present day in the Avro Shackleton. The Hercules likewise exceeded 2,000hp, and was joined by the 18-cylinder Centaurus of 3,270cu in Britain's most powerful piston engine. It powered many fighters and large aircraft, and also remains in use. The Hercules and Merlin were the leading British wartime engines, available in vast numbers as standard 'power eggs' needing no special installation design (for example, either could fit the Lancaster, Halifax or Beaufighter). Britain's other high-power engine was the 2,200hp Napier Sabre, which had 24 liquid-cooled sleeve-valve cylinders arranged in flat-H form.

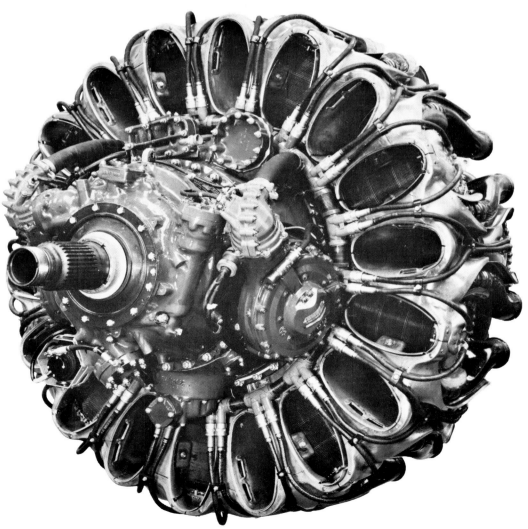

In the United States, sleeve valves did not reach production, the effort going into gigantic output of refined but established engines. In terms of horsepower manufactured per month, the leading engine was the Pratt & Whitney R–2800 Double Wasp, a superb 18-cylinder unit of 1,800 to 2,800hp and used in large numbers for military and civil aircraft right up to the present day (Canadair CL–215 fire-patrol amphibians and de Havilland Canada Caribou transports). Even larger numbers were made of the R–1830 Twin Wasp, used in the Consolidated B–24 Liberator with turbochargers, and the Wright R–1820 Cyclone, turbocharged for the Boeing B–17. The biggest US engine was the Wright R–3350, with two turbochargers per engine in the Boeing B–29. At the end of the war came the monster Pratt & Whitney R–4360 Wasp Major, with four rows each of seven cylinders, for which 115/145-grade fuel was created; it gave more than 3,000hp from the outset.

Little of note was done by Italy, Japan or the Soviet Union, though all conducted a vast amount of development work. Germany, however, produced the outstanding BMW 801 radial of about 2,000hp; the Daimler-Benz DB603 and 605; the Junkers Jumo 211 and 213 inverted-vee-12 engines, often as power eggs interchangeable with radials; and a wealth of unsuccessful double DB engines and complex prototypes intended to give much greater power at altitude. Piston engines since World War II are discussed later. (See pp. 69–71.)

engine is a turbojet, and one eventually ran in April 1937 after Whittle had spent years trying to interest the Air Ministry and private industry. But much earlier, in 1926, A. A. Griffith had begun to study a gas turbine geared down to a propeller, forming a direct successor to the piston engine and intended to give higher power and greater aircraft performance. After even greater delays this led to the test of a compressor in 1937 at Metropolitan-Vickers Ltd. This compressor was of the axial type, in which numerous rows of radial blades rotate like multi-blade propellers between interleaved rows of fixed stator blades; as in steam turbines the air under compression flows axially parallel to the shaft. After unnecessary delays amount-

FAR LEFT: *Rated at 1,500 to 2,000hp, the 14-cylinder Bristol Hercules was smooth and quiet because of its sleeve valves (see insets).*
LEFT: *One of the sleeves which fit closely inside the cylinder and are driven round by cranks as shown below. The ports in the sleeve alternately come opposite to the inlet and exhaust pipes in the cylinder.*
FAR LEFT BELOW: *The next size up beyond the Hercules produced the Bristol Centaurus, which was appreciated by Britain just too late.*
BELOW: *The Bristol Beaufighter was one aircraft powered by the sleeve-valve Hercules. By 1939 radial installations had made giant strides.*
BOTTOM LEFT: *The Pratt & Whitney R-2800 Double Wasp contributed more horsepower to World War II and postwar civil aviation than any other.*
BOTTOM: *Even today the Pratt & Whitney Double Wasp is still needed. The Canadair CL–215 is a 'water bomber' used for scooping up water from lakes and rivers and dumping it on forest fires.*

Early gas turbines

Gas turbines had been considered for aircraft propulsion since long before the Wright Brothers, but the first to propose a practical scheme was Sir Frank Whittle who, as a cadet in 1928, suggested using such an engine purely to discharge a propulsive jet. The engine he proposed comprised an air compressor delivering to a combustion chamber where fuel – usually of a kerosene type – was continuously burned at constant pressure to deliver a steady flow of hot gas to drive a turbine connected to the compressor. Except for that extracted by the turbine, all energy in the jet was used to propel the engine by direct reaction, the hot gas being accelerated through a suitably profiled nozzle. Such an

ing to more than 11 years, this work led to the Metrovick F.2 Beryl turbojet, first run in December 1941 and used to power a Gloster Meteor twin-jet in November 1943.

From the outset Whittle chose to use the centrifugal type of compressor. Such a compressor resembles a disc carrying curved radial vanes on its face (on both sides, in Whittle's engines, to handle twice the airflow), spun at very high speed. Air is ducted to the centre of the compressor, from where it is flung outwards between the vanes by centrifugal force. The air leaves the periphery of the compressor moving at extremely high speed, and this high speed is then converted into high pressure by slowing the air down in an expanding duct called a diffuser.

The centrifugal compressor, which is the usual kind of blower used in a vacuum cleaner, has the advantage of being robust, resistant to damage and relatively cheap to make.

In contrast, the axial compressor is usually made of hundreds of costly parts, and is also more likely to be damaged (for example, by the ingestion of stones or lumps of ice). At first, designers often tended to shrink from the axial compressor because it was technically more difficult, but for supersonic aircraft it was essential because the fat centrifugal type had too much frontal area for any given engine thrust.

A few early designers used radial-flow turbines, which work rather like a centri-

How a Gas Turbine Works

Section through a typical turbojet engine
(a) Front bearing (b) Inlet guide vanes
(c) Compressor blades (d) Stator blade

(e) Diffuser section (f) Centre bearing
(g) Fuel spray manifold
(h) Combustion chamber

(i) Rear bearing (j) Turbine guide
(k) Two-stage turbine blade
(l) Jet pipe

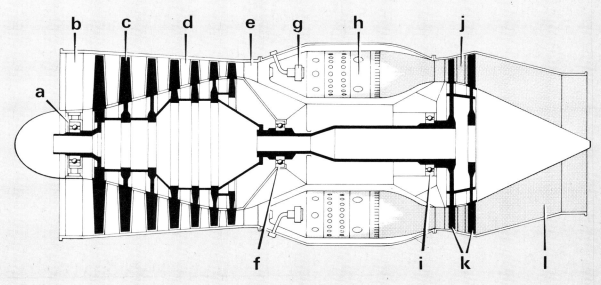

Most heat engines work by compressing a fluid – air, steam, combustible mixture or hot gas – and heating it; then it is allowed to expand, giving up energy to do useful work. In piston aero engines, the process is conducted in a series of oscillating cycles, giving an intermittent output. The gas turbine, on the other hand, operates continuously and gives a steady output. Apart from this, it operates in principle very like the piston engine, and it runs on air heated by having fuel burned in it to produce a flow of hot gas. It is also based on a rotating shaft, but in the simplest form of gas turbine this shaft merely transmits power inside the engine, and does not directly play a part in propelling the aircraft.

In the simplest gas turbine the shaft drives a compressor, which draws in fresh air and delivers it under pressure. The compressor may be of many forms, but nearly all aircraft gas turbines use either an axial or a centrifugal type. The axial has the form of a cylindrical drum, spinning at high speed. Around it project hundreds of blades, in size perhaps comparable to a playing-card, each of which is curved like a miniature wing. The spinning rotor is encased in a tight-fitting casing, from the inside of which stationary, or stator, blades project in rows interspersed with the moving blades, on the rotor. Altogether there may be as many as 1,000 blades, arranged in radial rows which are alternately fixed and moving. Air enters at one end of the compressor and is compressed by each stage of fixed and moving blades in succession, finally being delivered at the high-pressure end at a pressure up to 20 or even 30 times that of the surrounding atmosphere. The axial is tricky to perfect, and costly to build, but it has a small frontal area and extremely high performance. The rival centrifugal is simple and cheap, and less prone to damage, but seldom can it compress the airflow more than five to nine times; it also has a large frontal area. A centrifugal compressor comprises a spinning disc with radially arranged guide-vanes which accept the airflow near the centre and fling it off at high speed around the edge. The surrounding casing and ducting, called the diffuzer, slow the air down and turn the kinetic energy of its speed into pressure. Today, centrifugal compressors are favoured for the engines of light aircraft and small helicopters, while axials are preferred for transports and military aircraft.

From the compressor, the flow of hot, compressed air enters the combustion chamber where liquid fuel or fuel vapour is sprayed in and burned. The result is a steady flow of extremely hot gas, moving at high velocity and at a white heat which would soften or melt most materials. It passes first through a ring of fixed turbine/nozzle guide-vanes, or stator blades, made of special alloy which stays strong even when white hot. These direct the gas on to a row of

curved blades (often called 'buckets' in North America) fixed around the periphery of a disc in a similar way to the radial blades of an axial compressor. The disc and blades form the turbine rotor, and the centre of the disc is fixed to a shaft coupling it to the compressor, the whole being freely running in ball bearings, which are able to resist endloads, and roller bearings. The usual form of a turbine is of the impulse/reaction type, the blades being pushed round partly by the aerodynamic force (as if they were miniature wings) and partly by the change in momentum of the high-speed gas as its direction is changed by the curved blades.

This is the sum total of basic elements needed to make a gas turbine. Of course there is a bit more to it. The bearings have to be carefully designed to stand up to extremely high speeds, high temperatures and small errors of alignment or vibration. The compressor, combustion chamber and turbine must be linked by small air-gaps or pipes to provide high-pressure cooling air to the hottest parts, especially the turbine nozzles, rotor disc and, above all, the vital rotor blades. It was the lack of a material capable of withstanding the enormous loads imposed on the rotor blades at the high temperature of the gas flow which held back the gas turbine for many years. As in all heat engines, efficiency of a gas turbine is improved by running it hotter, but in the early jet engines the blades softened and stretched under the colossal centrifugal loads, or else broke off at the roots. Gradually, special high-nickel alloys were developed which remained strong even when almost white hot – between 1950 and 1977 the gas temperature of advanced engines rose from 810°C (1,490°F) to more than 1,300°C (2,372°F), equivalent in round figures to four times the power for twice the fuel consumption. Thus, specific fuel consumption – unit weight of fuel burned per unit force of thrust per hour, the best measure of efficiency – has been halved. In recent years great advances have been made in

the development of air-cooled rotor blades, containing complex air-cooling systems which keep the metal several hundred degrees cooler than the surrounding gas. They are made of 'directionally solidified' material which, in effect, makes the 'grain' in the metal's molecular structure run along the length of the blade instead of randomly or across it. Other features of a practical gas turbine include gear-driven accessories – at the very least a fuel pump, a control system to match the fuel supply to the engine's needs, and a starter to bring the engine up to the self-sustaining speed at which, when the fuel supply is turned on and ignited, the engine will continue to run.

This basic engine can be called a gas-producer; in some engines it is called the core, because it lies in the centre. But it can propel aircraft with the addition of nothing more than a suitable jetpipe and nozzle to convert as much as possible of the energy in the gas jet into propulsive thrust. Such an engine is a turbojet, and it is the simplest useful gas turbine. In the turbojet the turbine extracts from the gas flow only enough energy to drive the compressor. Turbojets are light and relatively cheap, but at low speeds have high fuel consumption. Because their propulsive jet has a high velocity they are noisy, and they are best suited to short-duration missiles and targets, sporting aircraft and trainers. They are also suitable for supersonic aircraft, but the need for high thrust and small frontal area can be better met by adding a reheat jetpipe or afterburner behind a turbojet in order to give increased thrust. This solution, however, carries the penalty of moderate increases in weight and cost and severe increases in fuel consumption and noise.

An afterburner is a swollen jetpipe in which extra fuel is burned. As there is no turbine to worry about, the gas can be raised to extremely high temperature, so that it issues as a supersonic jet in which glowing 'shock diamonds' are visible, marking the reflections of internal shock-

waves from the junction of the jet with the surrounding air. An afterburner needs a nozzle of increased area and convergent/divergent profile. Usually the afterburner is used only at take-off and for supersonic flight, so that at other times its fuel supply is switched off and the nozzle is closed down and changed in shape to suit unaugmented operation.

For subsonic aircraft the propulsive efficiency can be increased by making the gas generator (core) accelerate a much larger airflow, but more gently. This has the added benefits of reducing specific fuel consumption and greatly abating noise. One way of achieving this is to make the compressor bigger. Some of the air it delivers is then ducted straight to a propelling nozzle, which usually surrounds the core and blankets the hot jet, reducing noise still further. In modern transport aircraft the best arrangement is to use a separate Low-Pressure (LP) compressor, as described in the text, handling five to eight times the core airflow (i.e., with a by-pass ratio of 5 to 8). Intermediate forms of engine, with a by-pass ratio of about 1, are called by-pass jets and are used in modern combat aircraft.

All these engines need extra turbine power, obtained by adding extra stages on the turbine rotor(s). Another solution is to use this extra turbine power to drive a traditional gear and propeller. In this case the engine then becomes a turboprop.

Another kind of gas turbine is the turboshaft, in which the extra turbine power is taken to a gearbox through which an output shaft drives the rotor system of a helicopter (or a heavy road vehicle, train or ship). The fuel system is arranged in all these engines so as automatically to match the flow of fuel to the load on the output shaft. To prolong blade life, the fuel flow in most gas turbines is normally governed to keep the turbine temperature at less than the permitted maximum, but to allow it to be raised to the limit in case of sudden emergency (such as failure of another engine).

engines this gave terrible problems. Nobody had ever released heat at such a rate in such a small space, and engineers found it impossible to stop the thin sheet chambers from cracking and burning up. Some engines had tubular chambers, each with its own fuel burner(s) and inner flame tube, while others adopted the annular arrangement in which combustion occupies a continuous chamber wrapped around the engine. Today the annular type is dominant.

In Germany Pabst von Ohain did not begin turbojet study until 1935, but had the good fortune to interest Ernst Heinkel, who was able both to finance an engine and fly it in an aircraft, far sooner than the bitterly divided and frustrated British. The HeS 1 (not a practical engine, fed by hydrogen gas) ran in March 1937. The wholly practical HeS 3B ran at 500kg (1,100lb) thrust in 1939, and powered the world's first turbojet aircraft, the He 178, on 27 August of that year; it weighed some 363kg (800lb). Heinkel flew a twin-jet fighter, the He 280, with two HeS 8A engines, on 5 April 1941, but this work eventually came to nothing. Of potentially greater significance were the axial turbojets developed by BMW, Bramo and Junkers from 1938. By 1944 the BMW 003A was in mass-production at 800kg (1,760lb) thrust and the Junkers Jumo 004B in mass-production at 900kg (1,980lb). Though handicapped by poor materials, these were slim, simple and extremely useful axial engines which were made at rates more than 40 times greater than any of the Allied engines.

Apart from a wealth of more advanced turbojet, fan and turboprop developments, the Germans also produced aircraft rocket engines. The Walter company flew a rocket of 500kg (1,100lb) thrust in the He 176 in June 1939 and were in production with the HWK 509 engine, running on concentrated hydrogen peroxide and a mixture of methanol and hydrazine hydrate, in 1944. Fitted to the Me 163B interceptor, this engine was rated at 200 to 1,700kg (440 to 3,750lb) thrust at all heights. In 1941 an earlier version drove a prototype Me 163 at over 1,000 km/h (620mph), a speed never before reached by man.

After intense bickering, Whittle's outstanding work eventually bore fruit in the work of three companies, Power Jets, BTH and Rover, who tried not to collaborate but were eventually goaded into getting the W.1 engine, of 386kg (850lb) thrust, into the air in the Gloster E.28/39 in May 1941. Much happier work by Halford's team at de Havilland led to the H.1 (Goblin), and two

TOP LEFT: *The Walter 109–506 series of liquid-propellent rockets of World War II were the first to power piloted aircraft.*
TOP RIGHT: *Most famous of many aircraft powered by Walter 509-series rockets was the Me 163B Schwalbe tailless interceptor.*
CENTRE: *Whittle's pioneer original engine in its final much-rebuilt form, photographed in 1944. It is today in Washington.*
ABOVE: *First aircraft to fly with the Whittle engine, the Gloster E.28/39 proved a valuable aircraft. Two were built and flew intensively for four years (one crashed, one survives).*

fugal compressor in reverse, but today the axial-flow turbine is almost universal. Most gas turbines are of the combined impulse/reaction type, the turbine rotor being spun partly by the momentum of the hot gas hitting the blades (called 'buckets' in the United States) at high speed, and partly by the pressure variation of the flow round the blades like that round a wing which gives lift. But between the compressor and turbine comes the combustion system, and in early

of these, then rated at 680kg (1,500lb) each, powered the first Meteor in March 1943. But before this, in October 1942, the dynamic Americans had learned of Whittle's work, Americanized the W.2B as the GE (Schenectady) I–A of 590kg (1,300lb) thrust, and flown two in the Bell XP–59A. The Goblin powered de Havilland's own fighter, the Vampire, in September 1943, and the outstanding American Lockheed XP–80 Shooting Star in January 1944.

Gas-turbine programmes had been started in the United States before General H. H. Arnold imported the Whittle technology in May 1941. After much study in 1940, the NACA Army Air Corps and Navy set up the Durand Committee in February 1941 which recommended three engines: the Allis-Chalmers turbofan, which did not materialize, the GE (Lynn) XT31 turboprop and the Westinghouse 19A small axial turbojet, both of which were produced. Pratt & Whitney, like others in Europe, was trying to perfect a compound engine comprising a free-piston diesel feeding gas to a turboprop.

Other early gas turbines included the Japanese Ne-series axial turbojets, Swedish Lysholm turbojets and turboprops, Italian Caproni-Campini projects (the CC–1 jet aircraft was powered by a piston-engine driving a compressor) and French work which included a turbojet by Rateau, a turboprop by SOCEMA and a small turboshaft engine by Turboméca. The turboshaft engine is a turbojet fitted with one or more turbine stages whose function is to extract as much energy as possible from the gas flow, making it available at a high-speed drive shaft (coupled, for example, to a helicopter gearbox). The turboprop is similar but incorporates a gearbox driving a slow-speed output shaft matched to a suitable propeller which provides nearly all the propulsive thrust, especially at slow flight speeds. The turbofan is an intermediate between the turboprop and turbojet, and can have various forms. Today, virtually all have a front fan in the form of a large-diameter multi-bladed axial compressor functioning as a propeller (its inner portions also supercharging the core engine or gas generator which provides the power). Some early turbofans, often then called 'ducted fans', had the fan arranged at the rear, sometimes with double-deck blades around the original turbine and sometimes, as in the Metrovick F.3 and F.5, as a separate unit added behind a turbojet.

Turbojets of the 1950s

In October 1940, a production order was placed for 160 Power Jets W.2B engines for 80 Meteor fighters per month, but the British effort was so badly directed (despite great work by Mond Nickel on special turbine alloys and by Lucas on combustion problems) that progress was slow until the Power Jets/BTH/Rover responsibility was passed to Rolls-Royce in December 1942. Thereafter Power Jets did research only, and Rolls-Royce at last got the British programme moving. The Meteor F.1 with Rolls-Royce Welland turbojets of 772kg (1,700lb) thrust entered service in June 1944, the F.3 with

907kg (2,000lb) Derwent Is in December 1944 and the F.4 with 1,588kg (3,500lb) Derwent 5s in 1948, three years after a Mk 4 had gained the world speed record at 975 km/h (606mph). All these were pure Whittle-formula engines, with a double-sided centrifugal compressor, a large-diameter diffuser to convert air velocity to pressure and turn the air through 90°, a surrounding array of separate combustion chambers (cans) and a single-stage turbine. In September 1945 a Meteor flew with Derwents fitted with reduction gears driving small propellers (this first turboprop was called the Trent). Derwents rated at 1,634kg (3,600lb) thrust remained in production until 1955. An enlarged engine, the Nene of 2,270kg (5,000 lb) thrust, flew in a Lockheed XP–80 in July 1945. This powerful and reliable engine was the main base for jet technology in

France, the Soviet Union and many other countries, and was also extremely important in the United States where it was made as the Pratt & Whitney J42. The Rolls-Royce Tay, of 2,835kg (6,250lb) thrust, became the Hispano Verdon and Pratt & Whitney J48, representing the pinnacle of the Whittle centrifugal engine.

Both Verdon and J48 became the first production engines to be boosted by an afterburner (reheat jetpipe) in which, at the cost of greatly increased fuel consumption, thrust is augmented for short periods by

burning extra fuel between the turbine and nozzle. The gas temperature can be raised to more than 1,800°C (3,272°F) but the technique adds to engine bulk and mass and requires a nozzle whose area and profile can be varied. Actuation systems able to withstand the high temperature and resist the large forces were difficult to develop. A simple way of boosting thrust or shaft-power by a modest amount is water injection. Purified water, or a mixture of water and alcohol or methanol, is sprayed into the air entering the engine. The liquid quickly evaporates, cooling the air and increasing its density so that the mass flow (weight of air 'swallowed' each second by the engine) is increased. This not only raises thrust or power directly, but also allows more fuel to be burned without exceeding the permissible turbine temperature limits. In fact, in some

engines the water is injected not upstream of the compressor but into the combustion chamber, where its sole function is to cool the flow and allow more fuel to be burned. Of course, thrust or power is more or less directly proportional to the rate at which fuel is burned.

In 1943 Rolls-Royce had begun the development of axial engines, and these eventually led to the Avon which entered service at 2,950kg (6,500lb) thrust in 1951. By the time the Avon went out of production in Sweden in 1974, at an afterburning thrust rating of 7,761kg; (17,110lb), it had been produced in greater numbers than any other British gas turbine. In 1957 it had become the first engine in the world in service with turbine rotor blades cooled by multiple air holes. Early German engines had blades wrapped from sheet, but the modern forged or cast air-cooled turbine has dramatically improved engine performance by increasing the allowable difference between the metal temperature of the blade and the temperature of the gas; today, with far more refined cooling, the difference can exceed 340°C (644°F). In 1957 the Avon also became the first Vertical Take-Off and Landing (VTOL) jet engine to make a full transition, in the Ryan X-13 Vertijet. In 1958 it became the first jet in transatlantic airline service, the first in service with a noise-suppressing nozzle and the first to have a thrust-reverser to help slow the aircraft after landing. In 1959 it became the first jet in service on short airline sectors and the first to reach a time between overhauls of 1,000 hours, which was swiftly extended to 10,000.

Most Avons were of a later type, with the separate combustion chambers replaced by an annular chamber housing separate flame tubes (the so-called can-annular system) and with the superior compressor of a rival engine, the Sapphire. The Sapphire had begun life as the Metrovick F.9 and gone into production at Armstrong Siddeley (later Bristol Siddeley) and, in Americanized form, as the Wright J65. Rated at 3,266 to 4,980kg (7,200 to 11,000lb) thrust, it was an outstanding and important engine during the urgent re-armament at the time of the Korean War in 1950–53. Armstrong Siddeley had taken over Metrovick gas turbines in 1947, and via the Mamba turboprop produced the Viper as a short-life, simple turbojet for target Remotely Piloted Vehicles (RPVs) in 1951. Rated at 744kg (1,640lb) thrust, it was later turned into a very long-life engine for piloted aircraft and is still in production at ratings up to 1,814kg (4,000lb) thrust.

From 1941 the leader in the United States was GE. The Lynn (steam-turbine) plant gave up the T31 turboprop, but the Schenectady (turbosupercharger) team moved to Ohio and went from strength to strength. From the Whittle I–A came the J31 of 907kg (2,000lb) thrust and then the I–40, handed to Allison to build as the J33 at ratings up to 3,720kg (8,200lb) thrust with afterburner. In 1942 work began on a completely new engine with an axial compressor, owing only part of its combustion and turbine technology to Whittle. As the TG–180 this ran at 1,700kg (3,750lb) thrust and was built in batches as the J35 by the Chevrolet division of General Motors, powering the first North American XP–86 Sabre. Passed to Allison in 1946, the J35 was greatly developed and again made in very large numbers. Meanwhile GE refined the design into the TG–190, which as the J47 was made in larger numbers than any other gas turbine in history (36,500), more than half being installed in Boeing B–47s. Relatively small numbers were made of the Allison J71 and

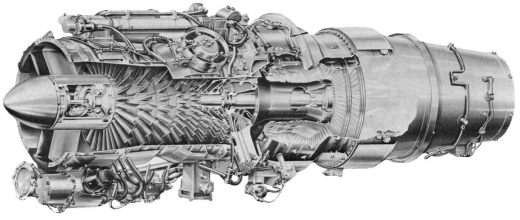

GE J73, both in the 4,536kg (10,000lb) thrust class. The big supersonic GE J53 reached 10,775kg (23,750lb) thrust on test, but was abandoned.

By 1948 the need for greater efficiency was leading to higher turbine gas temperature, made possible by improved alloys, and higher compressor pressure ratio. The centrifugal and early axial engines had been unable to compress the airflow more than about six times (a pressure ratio of six). Roughly double this figure seemed possible with new forms of compressor. Merely increasing the

ABOVE LEFT: *This sectioned Rolls-Royce Derwent of 3,500lb thrust shows a typical turbojet of 1945. A few are still flying.*
LEFT: *When the Derwent was new Rolls-Royce produced the axial-compressor Avon. This 17,000lb-thrust version is still serving.*
TOP: *Slightly larger than the Derwent, the Rolls-Royce Nene of 1945 gave 5,000lb thrust and was a world leader until about 1952.*
CENTRE: *Since 1970 the RPV has emerged as an important market for special families of small jet engines. This is a Teledyne Ryan 234.*
ABOVE: *This cutaway Rolls-Royce Viper 630 of 4,000lb thrust is today's descendant of a simple 1,640lb engine of 25 years ago.*

number of axial stages would not do; the engine became unwieldy and the compressor misbehaved and 'choked', the airflow stalling as on a wing operated at an excessive angle of attack and breaking down entirely. To make the air flow properly, existing axial engines needed spring-loaded safety valves to blow off air through the case, and pivoted inlet guide-vanes whose angle could be adjusted (especially when the engine was being started). Two new ideas were introduced: the two-spool engine and the variable stator.

GE adopted the latter, running the first J79 turbojet in 1953. Slightly larger than a J47, it had no fewer than 17 compressor stages, with the inlet vanes and next six stator stages all with variable incidence, driven by fuel-pressurized rams to maintain the correct airflow angles. Today, more than

the combustion chamber and ignited, and the gas flow then drives the LP spool. At all times the two spools run at their own best speed, called N_1 and N_2 and indicated to the pilot.

The first two-spool Pratt & Whitney engine ran in June 1949, and the prototype of a re-designed model, the wasp-waisted JT3, ran in January 1950. A robust and conservative engine, the JT3 received the military designation J57 and was made in large numbers at about 5,900kg (13,000lb) thrust, with water injection, for the Boeing B–52 and KC–135, the Douglas Skywarrior and many other aircraft, and at about 6,805 to 8,165kg (15,000 to 18,000lb) thrust with afterburner for many supersonic fighters. The civil JT3C was the original engine of the long-range 707 and DC–8 and, when JT3 production ceased in 1965, more than 21,200 had been delivered.

in the radical BE.53 engine with four swivelling nozzles to give lift or thrust and first run in 1959. Rolls-Royce pioneered simple, lightweight 'lift jets' for Vertical/Short Take-Off and Landing (V/STOL) aircraft.

Westinghouse developed its X19 (J30) into the 24C, which was built in numbers as the J34 of about 1,475kg (3,250lb) thrust; the bigger J40 was a dismal failure, causing Westinghouse to give up aircraft propulsion, though the J34 is still in service in booster pods on the Lockheed P–2 Neptune and Fairchild C–119. Wright likewise proved unable to compete, and this proud name faded from the scene. But in Europe a very pedestrian axial engine, based on wartime BMW designs, was taken to the production stage in 1954 by the French SNECMA company as the Atar 101C of 2,813kg (6,200lb) thrust. It was subsequently further developed

16,500 have been delivered, the engine being claimed to have logged more Mach 2 flight-time than all other engines combined, and with afterburning ratings of 6,805 to 8,120kg (15,000 to 17,900lb) thrust. Small numbers of non-afterburning commercial CJ–805 and CJ–805–23 aft-fan versions were also made.

Pratt & Whitney adopted the two-spool (also called two-shaft, dual-compressor or split-compressor) configuration. Instead of having one turbine driving one compressor, there are mechanically separate Low-Pressure (LP) and High-Pressure (HP) 'spools', the HP shaft being a tube surrounding the LP shaft, usually with a roller bearing between them. The starter drives only the HP spool, comprising the rear part of the compressor (smaller than the long-bladed LP spool), drive shaft and HP turbine. When this has reached the correct speed, fuel is supplied to

A closely similar British engine was the Bristol Olympus, run in 1949 and produced in 1954 at 4,990kg (11,000lb) thrust, developed in 1957 to 6,125kg (13,500lb) and by 1960 to 7,713kg (17,000lb) and then 9,074kg (20,000lb) thrust. Today, the same basic engine has been further developed as the power unit of the Concorde Mach 2 transport, with a rating of 17,240kg (38,000lb) thrust with afterburner. This is the largest growth in thrust of any aero engine.

Bristol later developed the very simple Orpheus single-shaft turbojet for tactical fighters. The de Havilland Engine Co continued building the Goblin, and the larger Ghost was chosen to power the Comet, the first jet civil airliner, which entered service in May 1952. There followed the Gyron Junior with an axial compressor and cast air-cooled blades, while Bristol cooled blades were used

as the Atar 9C of 6,400kg (14,110lb) thrust with afterburner and produced since 1959 for the Mirage III and 5.

In the Soviet Union, Vladimir Klimov refined the VK–1 centrifugal turbojet from the British Nene, production being later transferred to China. Arkhip Lyulka produced the impressive AL–7 and AL–21, big turbojets with afterburners. Alex Mikulin designed the AM–9 axial turbojet later mass-produced under Tumansky as the RD–9. Tumansky went on to produce the R–11, 13, 19, 25, 27, 29, 31, 32 and 33, the world's greatest and most numerous series of fighter engines. Back in 1950 Mikulin also managed the design of the big AM–3 turbojet, rated at around 9000kg (20,000lb) thrust, which made possible the development of the Tu–16 and M–4 bombers and the pioneer Tu–104 passenger jet.

At this time the US Air Force and Navy were interested in a turbojet system drawing energy from a nuclear reactor but, though enormous funds were applied, this difficult nuclear-propulsion programme was dropped before any system had flown. Likewise, a later project to use high-energy (so-called 'zip') fuel based on boron/hydrocarbon compounds was halted in August 1959, though the zip-fuel engine, the GE J93, was continued and, using special kerosene, flew in the Mach 3 North American XB-70 in 1964.

Turboprops of the 1950s

Bristol, a late starter in gas turbines in Britain, planned a complex turboprop in 1942 which featured axial and centrifugal compressors, reverse-flow combustion (air entered near the rear and flowed forwards) and a heat exchanger to improve economy by extracting heat otherwise wasted in the efflux. This ran in 1945 as the Theseus and was developed into the 4,500hp Proteus, of which a few are still flying in Bristol Britannias – many more serve in fast naval craft and other surface applications. Armstrong Siddeley's Python of 4,000hp, which powered the Westland Wyvern strike fighter from 1953 to 1957, was the sole outcome of the axial effort begun at Farnborough by Griffith nearly 30 years earlier. In contrast, Rolls-Royce in 1945 quickly produced a simple turboprop by scaling-up the supercharger of a Griffon piston engine to make the compressor and then adding a combustion chamber, turbine and gearbox. The result was the Dart, possibly Britain's most cost-effective engine. It is still in full production after more than 30 years, with nearly 7,000 engines delivered. Conversely, Napier turboprops were abandoned after the 2,800hp Eland was already in military and airline service, though the Gazelle of around 1,650hp survived as a helicopter turboshaft supported by Rolls-Royce.

In the United States, Allison took many years to get a reliable turboprop but by 1956 the T56 (civil, Model 501) was in production for the Lockheed C-130 Hercules military transport and for the civil Electra. A single-shaft engine, it carries the propeller gearbox on long struts ahead of the power section, and the C-130, Lockheed P-3 Orion, Grumman Hawkeye and other programmes have proved so successful that this 4,900hp unit is still in production with more than 10,000 delivered. At the other end of the scale, Allison produced the Model 250 (T63) in 1959 to meet a need for a simple engine of 250hp; this has today grown to the 350 to 650hp range. The engine is still in production for helicopters and light turboprop aeroplanes, with again more than 10,000 units delivered. GE produced the T58 in 1957 as a 1,000hp helicopter engine, and it remains in production with many thousands delivered (some made by Rolls-Royce as the Gnome) rated up to 1,800hp. Garrett AiResearch made about 25,000 small, non-flying gas turbines before launching the 600hp TPE331 (T76) single-shaft turboprop, with tandem centrifugal compressors. This is still in production at up to 1,000hp. In 1959 Pratt & Whitney Canada launched the 578hp PT6 free-

LEFT: *Flame blasts from the afterburner of a General Electric J79 at full power on a testbed.*
ABOVE: *This engine for a prototype B-52 was one of the first of the classic P&W J57 two-spool turbojets.*

BELOW: *Between 1949 and 1957 the Bristol Olympus increased in thrust from 9,140lb to 21,000lb (in this Mk 301 version). Today the Olympus 593 gives 38,000lb in Concorde.*

BELOW: *Engineers work on the General Electric J93 designed for the Mach 3 six-engined B-70. Giving scale is a Mach 2 J79.*

BOTTOM LEFT: *About to take off, this Spanish Mirage III DE is typical of many Mirage types.*
BOTTOM RIGHT: *One of the world's earliest supersonic combat aircraft, the MiG-19 has been produced under licence in China.*

RIGHT AND BELOW: *The world's first turboprop to go into use, the Rolls-Royce Dart is still in production. The smaller sketch shows a propeller with cropped and feathered blades.*
BOTTOM: *Still in use in warships and air-cushion vehicles, the 4,000hp Proteus had a reverse-flow layout. This Mk 765 powers a Britannia, several of which still fly.*
OPPOSITE PAGE
ABOVE LEFT: *Unlike the Proteus the Allison T56 is still in full production. Note the propeller drive gearbox at left.*
ABOVE RIGHT: *The 1,500hp General Electric T700 is assured of large military markets.*
CENTRE LEFT: *For 25 years the Lockheed C-130 Hercules has been produced with T56 turboprops.*

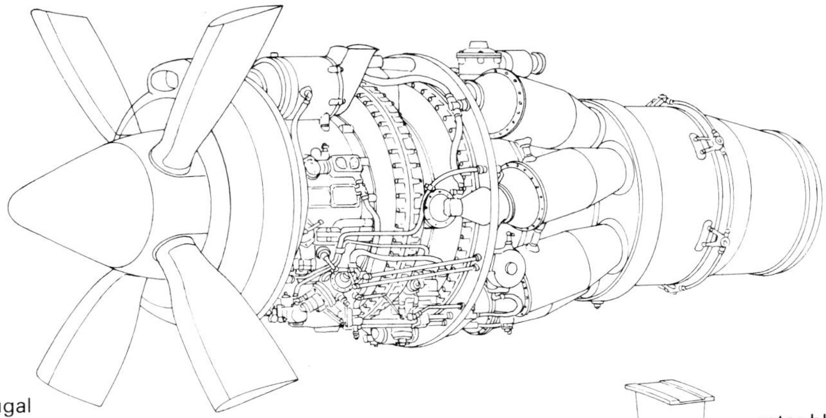

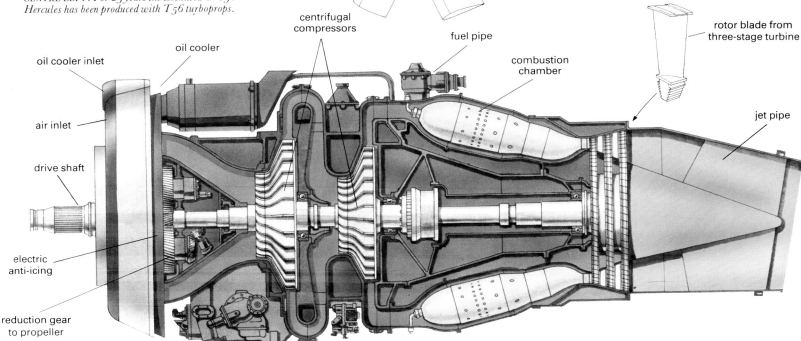

turbine engine (like the Bristol turboprops, it has one turbine driving the compressor and a free turbine driving just the output shaft); today there are 20 versions including turboprops rated up to 1,200hp, helicopter engines as single and twin units, models for airborne auxiliary services, trains, ships and other uses. Production for aircraft propulsion alone is running at 2,350 per year for more than 1,300 customers.

In 1946 Pratt & Whitney began development of the big single-shaft PT-2 engine

which flew in the nose of a Boeing B-17 in 1950 and finally went into production for the Boeing C-133 as the T34 of 7,500hp. Rolls-Royce's two-spool RB.109 of April 1955 became the 6,100hp Tyne, still in limited production by a European consortium for the Transall and used in many other aircraft. An even more powerful unit is the Soviet NK-12M, designed mainly by Germans working under Nikolai Kuznetsov in the period 1947 to 1952. It produced around 14,800hp for the Tupolev Tu-95, Tu-114 and Tu-126. Smaller Soviet turboprops include Alex Ivchenko's 4,000hp AI-20 and 2,500hp AI-24, both simple single-shaft engines, while

turboshaft units for helicopters include the 1,500hp Isotov TV2, Soloviev D-25V of between 5,500 and 6,500hp and Lotarev D-136 of 11,400hp. All three have a free-turbine drive at the rear.

Armstrong Siddeley developed the single-shaft Mamba into the Double Mamba, rated at up to 3,875hp and able to operate with either power section shut down and its propeller (front or rear of a double co-axial unit) feathered. But by far the most successful European producer of small gas turbines has been Turboméca, which since 1952 has sold 15,000 engines in 94 countries, with a further 12,000 built under licence, mostly by Teledyne CAE in the United States. Big-selling shaft engines include the single-shaft Artouste helicopter unit (400 to 600hp), Astazou turboshaft/turboprop (550 to 1,200hp), Bastan turboprop (1,000 to 1,400hp) and free-turbine Turmo (1,400 to 1,600hp). There are also the new 650hp Ariel and 1,800hp Makila.

Work by a German team between 1949 and 1954 led to the Avco Lycoming T53 free-turbine engine, since developed from 650hp to 1,800hp and used in numerous helicopters, fixed-wing V/STOL and turboprop aeroplanes; deliveries so far exceed 18,000 engines. The slightly larger T55 (LTC4) began life at 2,400hp and is now in production at ratings to 4,600hp. Pratt & Whitney added a free output turbine to a small turbojet, the J60 of 1,361kg (3,000lb) thrust, to produce the JFTD12 rated at 4,000 to 4,800hp. The latest shaft-drive engines are mainly for future

Diagram labels (centre cutaway):
oil cooler inlet
oil cooler
centrifugal compressors
fuel pipe
combustion chamber
rotor blade from three-stage turbine
air inlet
jet pipe
drive shaft
electric anti-icing
reduction gear to propeller

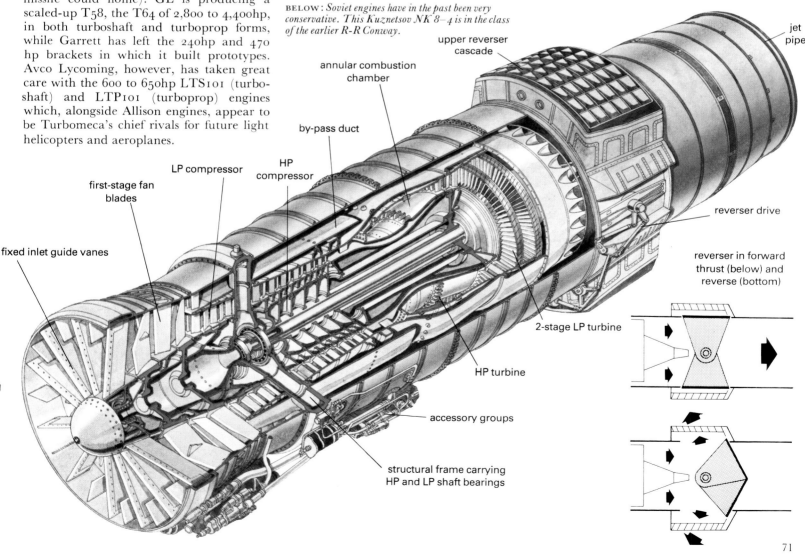

helicopters. The GE T700, of 1,536hp, is certain of large sales for many new civil and military helicopters. Like the Rolls-Royce Gem of 900hp, it has been designed to set new standards of low-cost and trouble-free operation while causing minimal environmental pollution (which means making less noise and pumping out less heat on to which a missile could home). GE is producing a scaled-up T58, the T64 of 2,800 to 4,400hp, in both turboshaft and turboprop forms, while Garrett has left the 240hp and 470 hp brackets in which it built prototypes. Avco Lycoming, however, has taken great care with the 600 to 650hp LTS101 (turboshaft) and LTP101 (turboprop) engines which, alongside Allison engines, appear to be Turbomeca's chief rivals for future light helicopters and aeroplanes.

Modern jet engines

The term jet engine loosely includes rockets, ramjets, turbojets, turbofans and an intermediate category called by-pass jets. In most modern turbofans, especially for subsonic aircraft, the by-pass ratio (ratio of cold air by-passed around the core to the hot core flow) is seldom less than 3 and may be 5·5 or up to

BELOW: *Soviet engines have in the past been very conservative. This Kuznetsov NK 8–4 is in the class of the earlier R-R Conway.*

8. The fan seldom has more than two stages, and usually only one; in a small engine it is driven by a reduction gear. The by-pass engine has a by-pass ratio of less than 1; it is essentially a two-spool turbojet with slightly oversized blades on the LP spool, or over the first three to five stages of it. Rolls-Royce failed to select the correct by-pass ratio in the early 1950s, and launched the Conway by-pass jet with a ratio of only 0·3. This was designed at 4,990kg (11,000lb) thrust, entered production at 6,805kg (15,000lb) and saw civil and military service at 7,940 to 10,205kg (17,500 to 22,500lb) thrust, later versions having the by-pass ratio raised to 0·6. This was the first turbofan of any kind in airline service, the first cooled-blade engine in airline service (1960) and the first aero engine to reach 10,000 hours between overhauls (1966). The Conway spurred Pratt & Whitney to produce a hasty conversion of the JT3C, the JT3D with three compressor stages replaced by two fan stages and with an extra stage on the LP turbine. By boldly choosing a by-pass ratio of 1·5, this conver-

jet pipe

upper reverser cascade

annular combustion chamber

by-pass duct

LP compressor

HP compressor

first-stage fan blades

fixed inlet guide vanes

reverser drive

reverser in forward thrust (below) and reverse (bottom)

2-stage LP turbine

HP turbine

accessory groups

structural frame carrying HP and LP shaft bearings

sion surpassed the Conway and the JT3D is still in production with more than 8,500 delivered at ratings up to 9,528kg (21,000lb). The Conway also triggered the Soviet Ivchenko AI–25, Kuznetsov NK–8 and Soloviev D–20 and D–30, all with by-pass ratios lower than would be chosen today.

The next generation after the Conway and JT3D were the smaller Rolls-Royce Spey and Pratt & Whitney JT8D, both planned as civil engines with by-pass ratios close to 1, pressure ratios around 18, and mass flows (maximum airflow at take-off) of about 93kg (205lb)/sec for the Spey and 145kg (320lb)/sec for the JT8D. Both have been built in large numbers, the US engine having gradually developed from 6,350 to 7,260kg (14,000 to 16,000lb) thrust and the Spey growing from 4,470 to 5,445kg (9,850 to 12,000lb) and with a Mach 2 military version giving over 9,300kg (20,500lb) thrust with afterburner. Refanned models of both engines are rated at around 7,713kg (17,000lb) thrust, with much larger single-stage fans (of the kind they

TOP: *Rolls-Royce Conway turbofans were among the first engines to be assembled upright. When this one has been 'dressed' with its accessories it will look quite different.*
ABOVE: *Virtually identical Pratt & Whitney F100 two-shaft turbofans power the F–15 and F–16 fighters. The nearer example of this 23,810lb-thrust engine shows the variable-profile afterburner nozzle.*
ABOVE RIGHT: *The extremely ordinary Pratt & Whitney JT8D has become the most widely used airliner engine. Rated at 14,000 to 17,000lb, it powers the Boeing 727 and 737, DC–9, Mercure and Caravelles.*
RIGHT: *Rated at 3,500 to 4,000lb thrust, the Garrett AiResearch TFE731 is a typical business jet and trainer engine. The large fan is driven by a reduction gear, so there is close affinity with turboprops.*
FAR RIGHT TOP: *Designed by Rolls-Royce and built by a partnership involving MTU of West Germany and Fiat of Italy, the Turbo-Union RB.199 is a modern combat engine. Its first application is to power the Tornado.*
FAR RIGHT BELOW: *General Electric CF6–50 turbofans being prepared for shipment from Cincinnati. These 49,000–54,000lb engines, power the Airbus A300, A310 and DC–10, and some Boeing 747s.*

should have had at the outset) raising by-pass ratio to almost 2 and dramatically cutting noise and specific fuel consumption. In an age of escalating development costs, refanning old engines is becoming more attractive than designing new ones.

The first modern High By-Pass Ratio (HBPR) turbofans stemmed from the US Air Force Lockheed C–5A freighter. GE won the C–5A propulsion contract with the TF39 of 18,780kg (41,400lb) thrust. It was a bold design with a by-pass ratio of 8, reflecting the modest 815km/h (507mph) speed of the C–5A, a 1·25 stage fan with a pressure ratio of 1·55 and a single-spool variable-stator compressor reaching 16·8 in 16 stages, giving an overall pressure ratio of 26·04. From this, GE developed the CF6–6 for the civil McDonnell Douglas DC–10, with ratings near 18,145kg (40,000lb) thrust, a by-pass ratio of 5·9 and a pressure ratio of 28·1. Adding booster stages behind the fan with variable by-pass doors resulted in the CF6–50 series, used in the DC–10–30, Airbus

Industrie A300B and some Boeing 747s. These engines have increased core airflow, giving a by-pass ratio of only 4·4 but a pressure ratio of 31·5 and thrusts of 22,680 to 24,500kg (50,000 to 54,000lb). Rival Pratt & Whitney produced the JT9D with a fan rotating on the LP shaft with three or four LP compressor stages, at a by-pass ratio close to 5, and with an 11-stage HP spool giving the more modest overall pressure ratio of 21 to 24. Thrusts of 19,504 to 24,040kg (43,000 to 53,000lb) are produced. In Britain Rolls-Royce chose three shafts in the RB.211–22 (single-stage fan, seven-stage Intermediate-Pressure, (IP), and six-stage LP) giving a pressure ratio of 25 for fewer stages and with better handling. The chosen by-pass ratio was 5 and thrust 19,050kg (42,000lb). Rolls then developed a new higher-flow core in the RB.211–524 giving better results than any rival at thrusts from 19,960 to 27,490kg (44,000 to 60,600lb), and with a large margin of extra thrust in climb and cruise. The giant imponderable,

the advent of the propfan, is discussed later in the section on propellers.

To meet an apparent need for smaller engines, American GE and French SNECMA are jointly producing the CFM International CFM56, which uses the core of the GE F101 supersonic bomber engine, an outstandingly advanced unit operating at high pressures and very high temperatures, matched with a new SNECMA fan giving a by-pass ratio of 6 and thrust around 9,980kg (22,000lb). Pratt & Whitney, with partners in other countries, is working on the extremely advanced PW2037 and IAE V.2500 to be rated at 11,115 to 18,145kg (24,500 to 40,000lb) thrust and also projected with cropped fan blades to meet the needs of aircraft requiring less thrust. In smaller categories are the Japanese FJR710 of 4,990kg (11,000lb), GE CF34 of about 3,855kg (8,500lb), Rolls-Royce M45 of about 3,402kg (7,500lb) and Avco Lycoming ALF502 of about 2,950kg (6,500lb) thrust. All have efficient cores and high by-pass ratios, but only the ALF502 is in airline

service, though the CF34 is a version of the TF34 military engine and is used in the Challenger business jet. In the Soviet Union the new bureau of Vladimir Lotarev has produced the D–36, in the 6,350kg (14,000lb) thrust class, and the D–18T rated at 23,430kg (51,655lb). Soloviev has the 16,000kg (35,275lb) D-90A.

Among military engines the outstanding leader in Europe is the Turbo-Union RB.199, two of which power the multi-role Panavia Tornado. The RB.199 has three shafts, a very efficient fan, an advanced afterburner which doubles thrust from 4,080 to 8,165kg (9,000 to 18,000lb), and an extremely efficient variable nozzle. Much more conservative is the Rolls-Royce Turbomeca Adour two-shaft engine, rated at 2,650kg (5,845lb) thrust in the British Aerospace Hawk and up to 3,810kg (8,400lb) thrust with afterburner for the SEPECAT Jaguar and Mitsubishi T–2 and F–1. Dominant engines in the United States are the powerful Pratt & Whitney F100 and GE F404. In the 1960s Pratt & Whitney produced the TF30 turbofan for the General Dynamics F–111, but this was not outstanding until the final version, the TF30–100 of 11,385kg (25,100lb) thrust with afterburner, became available for the last few aircraft. In contrast, the F100 is an outstanding engine in all respects, delivering nearly 10,890kg (24,000lb) thrust with afterburner and giving a combat thrust-to-weight ratio which substantially exceeds one-to-one in the twin-engined McDonnell Douglas F–15 and single-engined General Dynamics F–16. Bypass ratio is about 0·7, pressure ratio 23, and fuel consumption, size and weight are all less than in TF30 versions of lower thrust. The F404 is a low by-pass turbojet of some 7,258kg (16,000lb) thrust and, like the F100, has a thrust/weight ratio of about 8. Little is known of the many outstanding new Soviet engines, but technically they are catching up fast and it would be unwise to assume, that their design represents only a copy of the previous generation of Western engines.

Since 1959 GE has achieved considerable success with the small single-shaft J85 turbojet, of which more than 12,000 have been produced, almost all for Northrop T–38 trainers and F–5 fighters. This engine began life at 1,746kg (3,850lb) thrust with afterburner and is now usually rated at 2,268kg (5,000lb). Its success confirms the continuing appeal of old-fashioned but proven designs. The civil CJ610 has likewise proved a winner, rated at about 1,360kg (3,000lb) thrust, and with customers carefully protected by aftersales support and spares at guaranteed prices. About 2,000 of these simple engines are in use, plus 800 aft-fan CF700s rated at 1,928kg (4,250lb) in some Dassault Falcon and Rockwell Sabre business jets. But Garrett AiResearch is now taking a good slice of the business jet market with the excellent TFE731 of 1,588kg (3,500lb) thrust, in which the HP turbine drives a centrifugal compressor and the three-stage LP turbine drives a four-stage LP compressor as well as, via a 0·555 reduction gear, the single-stage fan which has a 2·66 by-pass ratio. AiResearch has also developed a radical three-shaft turbofan in the 2,268kg (5,000lb) thrust

class, the ATF3, in which the entire core airflow is twice reversed. Core air passes through the fan, which has a by-pass ratio of 3, then through the five-stage IP compressor, and then curls round and inwards at the rear of the engine where it encounters the centrifugal HP compressor – this brings overall pressure ratio to about 24. It then passes forwards through the combustion chamber, HP, IP and LP turbines, and finally turns another 180° to exit into the fan (by-pass) airflow through eight sets of cascades. Though seemingly complex, the ATF3 has proved a success. After some years of military flying as the XF104 it was in 1976 launched on the civil market in a new Dassault Falcon.

Modern piston engines

Almost all piston engines since 1945 have come from the United States and eastern Europe. It is especially sad that Britain's world-renowned Gipsy series of lightplane engines should have collapsed, leaving the field today almost wholly in the excellent hands of Avco Lycoming and Teledyne Continental. During the 1930s there were many US light-engine firms, including Warner, LeBlond, Kinner, Jacobs, Ranger, Aeromarine, Menasco and Franklin. They built engines of many shapes, but since 1945 the horizontally opposed flat-four and flat-six have accounted for more than 95 per cent of all light-engine sales, and almost all come from Avco Lycoming and Teledyne Continental. Engines range from 65 to about 400 hp, with a general growth among the best-sellers over the years from around 140hp to a current rating between 250 and 350hp. Most today have fuel injection and geared drives, and turbochargers are often available. Many engines can run upright in helicopter installations, and the larger types carry a freon compressor for cabin cooling in hot weather, or a cabin pressurization supercharger.

Large piston engines died slowly, the Pratt & Whitney R–2800 and Wright R–3350 being especially successful in both military and civil aircraft. The latter was finally certificated in 1954 as the Turbo-Compound, with its exhaust discharged through three high-speed blow-down turbines disposed at 120° around its rear. This added about 1,000 hp to the propeller shaft, turning a 2,500hp engine into an economical 3,500 to 3,700hp unit which, in the Douglas DC–7C and Lockheed L–1649, at last produced true transatlantic airliners. They had but two years of glory before the turboprops and turbojets swept them away almost overnight.

Today the main repository of the piston engine is the agricultural market, which needs a robust and economical 'farm truck' engine to work all day at sea level. Another worthwhile market is the helicopter, but for lower power engines below 500hp. A rapidly increasing proportion of the world output of piston aero engines comes from Poland and Czechoslovakia. Both produce engines of their own and Soviet design, carefully refined for helicopter and agricultural use. There are strong signs that water-cooled car-type engines producing about 300 to 500hp will break into this large market. The gigantic

ultralight and homebuilt market should also not be overlooked, although it has still to decide whether it prefers four-stroke or two-stroke engines and is being told that air-cooling will account for a rapidly decreasing proportion of the total. This market has also shown great interest in the well known Wankel-type Rotating Combustion (RC) engine, in which the piston rotates instead of oscillating up and down. Teledyne Continental in the USA and Norton in the UK are among the companies developing RC engines for aviation.

TOP LEFT: *Curtiss-Wright Wankel-type rotating combustion engine.*
TOP RIGHT: *The RFB/Grumman American Fanliner has Wankel-driven shrouded fan propulsion.*
CENTRE: *The Teledyne Continental 10520A is a 285hp flat-six with direct fuel injection.*

OPPOSITE TOP: *A true jet engine driving propellers, the UHB (ultra high bypass) developed by General Electric is seen here being demonstrated aboard a modified McDonnell Douglas MD-80 airliner.*
TOP CENTRE LEFT: *The Jaguar International, a version exported to Ecuador and Oman.*
ABOVE CENTRE LEFT: *Much bigger than the Jaguar, the GD F–111 flies similar missions.*
LEFT: *The Dassault Breguet Falcon 20 (Mystère 20) was the first business jet with turbofans.*
ABOVE: *The 600–650hp Avco Lycoming LTP 101 turboprop is used in many lightplanes.*

Airframes

Early aeroplanes were mainly built of traditional materials such as ash, birch, spruce and other hardwoods; bamboo, steel (almost always in thin sections such as tube, or as reinforcing plates at wood-joints) and small amounts of aluminium, which was then new and rather costly. The structure took the form of many parts joined together by glueing, pinning or bolting, to form a strong but light framework. This was then braced with numerous tensioned wires and the aerodynamic surfaces covered with good fabric such as varnished silk (the Wrights began with unbleached muslin) or rubberized waterproof linen fabric.

From the start it was realized that the main structural members were the spars which spanned the wing from tip to tip. These bear the main bending and shear loads in most wings, and for a spar to break would be catastrophic. Early spars were made of glued and pinned wood, usually ash, arranged either in the form of a box section or an 'I'. In either case the tension and compression loads are borne by the booms at top and bottom, while the vertical walls joining the booms are the webs. Along the spars were arranged ribs, lying in a fore-and-aft (chordwise) direction and usually built up from pieces of plywood, strip wood, and even aluminium strip or steel tube, to form the correct aerofoil section. Internal wires braced the wing to form a strong structure before covering. The fuselage (body) was a wire-braced girder framework, with a basis of four strong longerons of ash or spruce joined by vertical and transverse struts, though steel tubes were sometimes used as compression members.

The majority of aircraft produced up to 1935 were biplanes, although in World War I triplanes and quadruplanes were flown. The use of superimposed lifting planes enabled the classic boxkite structure to be used, with its combination of lightness and rigidity. By 1915 this structure had been refined in several ways. Interplane struts, linking the spars of upper and lower wings, were no longer of constant section but were tapered at the ends in a way forming an effective compromise between strength, weight, drag and ease of manufacture.

Round section piano wire for bracing had been replaced by the Rafwire (from the initials of the Royal Aircraft Factory) with lenticular (sharp-edged and flattened) section having much lower drag. Rafwires used to 'sing' in the air, the vibration soon leading to dangerous fatigue failure, until dampers were added, often in the form of small 'acorn' fairings where the wires crossed.

Apart from this, aerodynamic refinement was usually conspicuously lacking, and the structure was regarded as a basis on which to

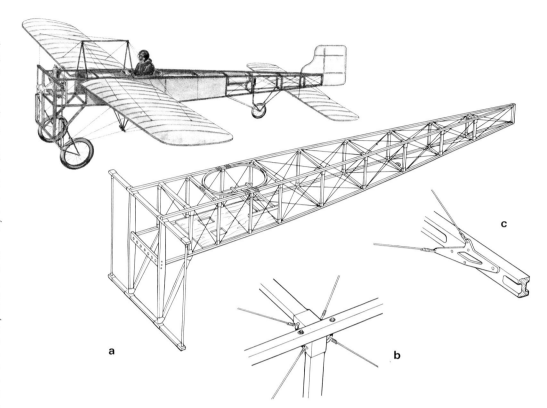

BELOW LEFT: *Blériot XI with details of (a) fuselage section (b) typical frame/longeron joint and (c) wing-spar wire bracing fitting.*
RIGHT: *A Sopwith Camel with detail of wooden wing structure with internal wire bracing typical of World War I aircraft.*
BOTTOM: *The Gastambide-Mengin experimental aircraft (a type of Antoinette) was built in 1908. Here the struts are being checked.*

hang an unstreamlined engine, fuel tank, water radiator and other items. There were a few notable exceptions, such as the racing French Deperdussins of 1912 to 1913 which introduced monocoque (single-shell) structure. The classic example of monocoque construction is a lobster claw, showing how a strong skin can obviate the need for any internal skeleton or supporting framework. The Deperdussin fuselage was eventually perfected as a finely streamlined body of circular section, made from multiple veneers of tulip wood glued together over a jig or former, a shaped mould to ensure that the finished assembly had the correct shape. During World War I many aircraft moved a small way towards monocoque construction by having plywood skinning, able to bear some of the load, and dispensing with internal bracing wires (but retaining an internal skeleton framework).

This traditional form of construction, often loosely termed 'mixed' because different materials are employed, gradually underwent complete change. The eventual choice of

most constructors was an all-metal 'stressed-skin' structure, and a few remarkable examples were built before 1920, but between these and the final development there were several intermediate stages. One of the major changes was the gradual adoption of an all-metal structure, often as a direct substitution for wood and with little change in the basic form and arrangement of the parts, and retaining a fabric covering. Another was the progressive reduction in the number of external bracing wires needed, and the eventual change from the braced biplane to the strut-and wire-braced monoplane and, in some cases, to the cantilever (unbraced) monoplane. A further fundamental change was the substitution of thin metal skin for fabric, though with stresses in the skin kept very low. The final change, which took 20 years to become established, was to make the metal skin bear the major part of the structural loads, and thus go most of the way towards realizing the structurally efficient monocoque form.

A major influence on aircraft structures was the accidental discovery by the German Durener Metallwerke in 1909 of the favourable properties of an aluminium alloy containing a little copper and magnesium. It was almost as light as aluminium, but much stronger, and it still had the good quality of stretching appreciably before breaking instead of snapping suddenly. The company called it Duralumin, and it was subsequently the subject of extensive research to obtain the best properties (surprisingly, it was not until 1934 that these properties were properly exploited).

One of the first to want to build an all-Dural aeroplane was Professor Hugo Junkers of Germany, and during World War I he designed several production aircraft of remarkably advanced cantilever monoplane design (plus an unusual cantilever biplane) with all-Dural structure. From then until 1933 Junkers favoured a skin stiffened by corrugations running fore-and-aft. Though the result was robust, and better able to stand up to rough usage than any rival machines, the drag was considerably higher than Junkers realized and performance was thereby held back. Several other builders used Duralumin construction, notably Short Brothers in Britain (with the Silver Streak biplane of 1920) and the German Rohrbach company (with a series of extremely clean monoplane flying boats built during 1923 to 1926, all of which featured smooth, stressed-skin construction.

Stressed-skin construction was only very slowly adopted generally, and a favoured structure for military machines in 1923 to 1925 was a framework of bolted and riveted steel strip, with wings spars of complicated built-up form achieved by riveting together a number of lengths of rolled strip to give the required cross-section with no flat areas which could buckle. Fabric covering was almost universal, though some large machines, such as the Soviet Tupolev TB–3 and French Amiot 143 bombers, had Duralumin skin to make them less prone to accidental damage.

During the late 1920s many great designers, especially the American John K. Northrop,

LEFT: *The Sopwith Schneider under construction at Kingston-upon-Thames. The Schneider was a floatplane version of the Sopwith Tabloid and was built specially for the Schneider Trophy Race of 1914. Despite the great speeds attained by the monocoque Deperdussin, the traditional biplane structure was decided upon for the Schneider. This was due to military discouragement of monoplanes in Britain. However, interplane struts and landing and flying wires were reduced to a minimum and special attention was paid to float construction.*

FAR LEFT BELOW: *The 1931 Boeing P–12C was one of the first fighters with a fuselage of bolted light alloy. It was a high-performance fighter built for both the Army and Navy. The Navy version was designated F4B–2 and was a forerunner of modern carrier fighters. Both were developed from earlier Boeing fighters with fuselages of welded steel tube. A total of 554 of the series were built, 366 as P–12s and 188 as F4Bs. The F4B had a span of 9m (30ft), a length of 6m (20ft) and was powered by one 500hp Pratt & Whitney R–1340. First deliveries were made in September 1929 and they continued to be built for several years.*

FAR LEFT BOTTOM: *The Junkers F13 was the first strut-free, all-metal civil aircraft. A low-wing cantilever monoplane, the F13 first appeared in 1919 and more than 40 were used by Lufthansa and other airlines. The F13 could operate on wheels, skis or floats.*
BELOW LEFT: *A DC–3 with details of (a) multi-spar wing section (b) outer wing bolted joint and (c) fuselage attachment. In addition to its stressed-skin monoplane structure, the DC–3 was the first aircraft to incorporate the failsafe multi-spar wing.*
BELOW: *Junkers Ju 52/3m and a detail of its structure of corrugated metal skin panels.*

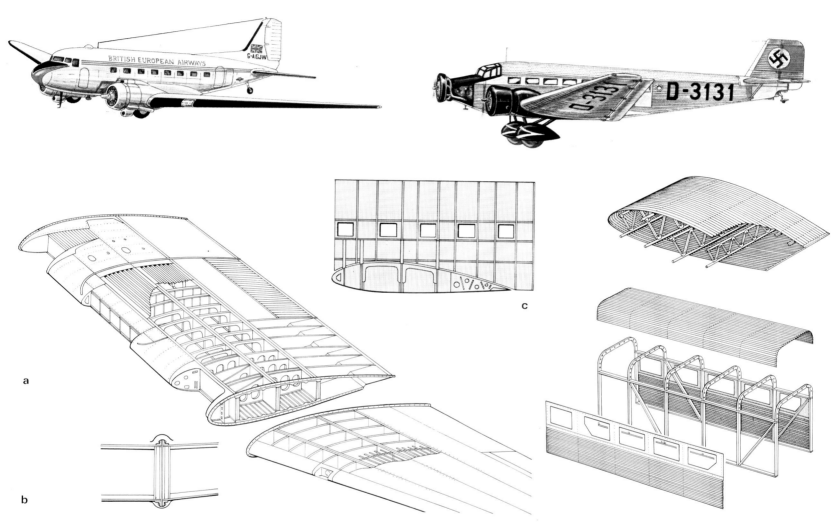

refined the all-metal airframe until, by 1932, it had reached almost its present form. The advances were numerous. The cantilever monoplane could have a thick high-lift wing which could transmit all the flight loads through its skin, supported by many spanwise 'stringers', yet still have less weight and drag than a braced structure. Instead of having just two spars, such a wing could have six or more. This eventually yielded quite unexpected dividends in the Douglas DC–3 by making the wing resistant to fatigue – the progressive weakening of a structure bent back and forth, as a wing is in turbulent air. Unlike the old biplane wings, the cantilever monoplane wing could be made of graduated strength and uniform stress from tip to tip by varying the thickness of the spar booms and skin. Higher structure loading resulted from the trend towards higher wing loadings (aircraft weight divided by wing area), so the stress in the material rose and the metal was used more efficiently.

Most aircraft in World War II were of all-metal stressed-skin construction, though it was still rare to find a truly efficient structure in which the metal in the skin carried the same stress as that in the most highly stressed members, such as the spar booms. Fast aircraft had smooth exterior surfaces obtained by the use of 'flush' (countersunk) rivets, which such advanced machines as the Boeing B–29 bomber pointed the way to the future by having such large loads in the wings that the skin was no longer mere commercial thin sheet, but formed from sheet more than 15·8 mm ($\frac{5}{8}$in) thick. There were also some developments in the 1930s which were quite outside the normal mainstream of aircraft structures.

One of these unusual forms of structure was the Geodetic type, devised by Barnes (later Sir Barnes) Wallis of Vickers as a result of his long experience with airships. (See page 339.) A geodetic (strictly, a 'geodesic') is the shortest line that can be drawn between two points on a curved surface. Wallis saw that an aeroplane can be made with fairly regular surface curvature all over, and he developed a metal basketwork in which the entire airframe is assembled from quite small geodetic members pinned together at the joints (structurally called nodes, because the loads are applied only at the nodes). Each member experiences pure tension, as a tie, or pure compression, as a strut, with no bending load at all. All of the flight loads could be carried wholly within this basketwork lying in the surface (thus forming a kind of monocoque), though spars and longerons were added to take the bending loads, because the unsupported basketwork could be bent an unacceptable amount.

The first two types of completely Geodetic aircraft were the Wellesley and Wellington bombers, both of which had long range and wings of high aspect ratio. Fabric covering was no problem because they were quite slow, and the Wellington, in particular, had so much structural redundancy (a term which means there were many load-paths, so that one or more could be removed without the

LEFT AND BELOW: *The Vickers Wellington was the only British bomber to remain in production throughout World War II, a tribute to its unique, durable construction. The cutaway diagram shows a wing section revealing its geodetic construction which consisted of a basket-like light alloy framework built up of short, curved sections (detail).*
BOTTOM: *Wellingtons under construction clearly show the geodetic lattice framework of the fuselage.*

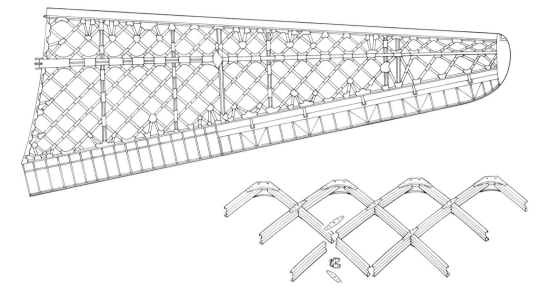

comb (a sandwich with an interior like a honeycomb) for structural purposes, had by World War II developed a completely new family of adhesives, which his company, then called Aero Research, marketed under the name of Redux. This was supplied either as a powder or in sheet form, cut to fit the joints to be bonded. The bonding was then made with both heat and pressure. With big airframe parts a large 'autoclave', a form of high-pressure oven was used. Redux made it possible to join metal parts by adhesive for the first time, and it was used in production of such successors to the Mosquito as the Vampire, Hornet and Dove. Later the successor to Aero Research, Ciba, developed a new adhesive, Araldite, which is familiar to everybody, as well as being important in aircraft construction.

One of the advantages of using adhesives is that joints are made with a continuous bond over an area, whereas rivets or bolts fasten parts only at small locations where there is a hole. Holes inevitably cause discontinuities in the material, and are likely to increase the stress in it when under load. Until after World War II this was no problem, because, provided designers and builders followed established codes of practice which had been built up over many years of often painful experience, they knew nothing would be likely to break. But from 1945 onwards aircraft, and especially civil aircraft, began to come apart in the air. Examination of the wreckage revealed the uncomfortable fact that the cause was metal fatigue. The popular way of demonstrating fatigue is to break off an open tin-can lid by bending it back and forth. Gradually the strip of metal along the bend seems to weaken, until the lid simply falls away without needing any perceptible pull. During the 1920s a few engineers admitted there might be fatigue problems in engines, but never in aircraft; as late as 1947 many loudly ridiculed Professor A. G. Pugsley when he pointed out that, even on the scanty evidence available at that time, the fatigue life of civil airliners might be similar to the planned lifetimes of such machines. In other words, just by normal flying through the turbulent sky, such aircraft might fall apart before they were retired from service.

In fact, trouble hit hard and soon. Those who lived through it will never forget the mysterious disappearance of two de Havilland Comet I jetliners, later shown in one case to have been caused by the explosive rupturing of the whole fuselage because the repeated pressurization had caused a crack to appear at a corner of a cut-out – foolishly, a square aperture instead of a round one – for an ADF loop aerial. The crack could have occurred at one of the Comet I's square windows, because when a complete airframe was tested under simulated flight conditions for the equivalent of a modest total lifetime it blew apart at one of the windows. This test was one of the first in which the whole fuselage was filled with, and immersed in, water. If air had been used the explosive failure would have badly distorted the specimen and caused havoc over a wide area, but water is almost incompressible.

Recognition that aircraft do have a fatigue

structure collapsing) that these aircraft gained a reputation for continuing to fly with airframe damage that would have caused catastrophic failure in a conventional structure.

The other unusual wartime structure was to be found in the Mosquito, an aircraft of extremely high performance built almost entirely of wood. The company responsible, Britain's de Havilland, had used wood for its Comet racer of 1934, and had subsequently experimented with a new form of construction called a 'sandwich', in which two load-bearing skins are firmly bonded to a low-density core. Each skin by itself would bend and buckle, but when the three layers are bonded together the result is thick enough to be completely stable under load. The Mos-

quito had wings with inner and outer skins of plywood bonded by strong adhesive to spanwise spruce stringers; the fuselage was made of plywood sandwich with a core of balsa wood, the very light wood often used by aeromodellers. The resulting structure was light and strong, and had the advantages of smooth inside and outside surfaces allowing the easy incorporation of cut-outs and attachment points. The main problem was that tropical environments reduced the strength of the adhesive.

Most of the adhesive bonding in the Mosquito was formaldehyde cement, a resin superior to the old casein glues used in earlier wooden aircraft. But N. A. de Bruyne, one of the pioneers of sandwich structures, and one of the first in the world to market a honey-

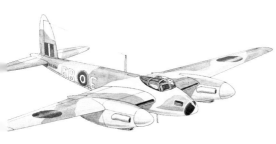

LEFT AND BELOW: *Besides the Wellington, the de Havilland Mosquito possessed an unusual structure. Although built almost entirely of wood, it had tremendous performance and survival capability. The airframe consisted of a unique sandwich form in which two load-bearing skins were firmly bonded to a low-density core. The details show (a) fuselage half-shell (b) wing section (c) fuselage top scarf joint and (d) end-grain balsa sandwich fuselage panel.*

problem came just in time to avoid massive trouble, because modern transports fly more hours each year than pre-1945 aircraft flew in their entire lifetime. In addition to the good practice of avoiding all 'stress-raisers' – sudden changes in section, sharp corners, scratches, roughness, etc. – modern structures are designed to fly a safe life greater than the expected aircraft lifetime, which in

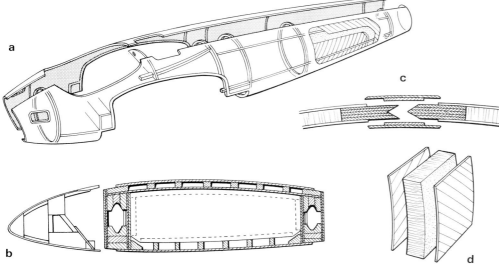

a

b

c

d

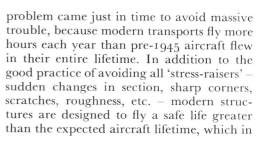

the case of airliners may be 40,000 to 60,000 hours. Such a life is achieved partly by attention to every detail of the structure, and limiting the maximum stresses in the most adverse conditions, partly by reducing the number of stress reversals by deliberately building in a static stress (for example, by tightening up joints so that they are under high tension even when the aircraft is not flying), and partly by using fatigue-resistant forms of structure with fewer joints and fewer discontinuities. On top of this, a form of structure called 'fail safe' has been devised in which, by use of multiple redundant load paths, the aircraft will not crash even if a

ABOVE RIGHT: *The scene at Hatfield on 27 July 1949 a few minutes before the first flight of the Comet prototype. The Comet was the first jet airliner and demonstrated tragically the disastrous effects of metal fatigue.*

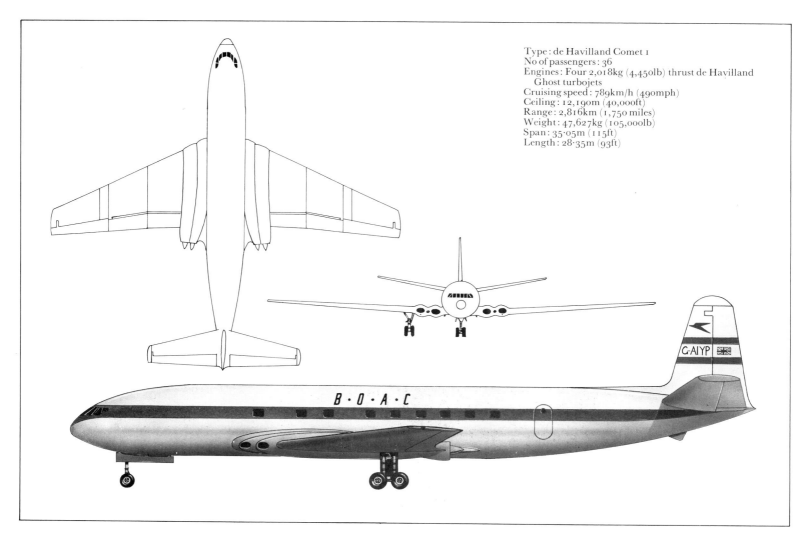

Type: de Havilland Comet 1
No of passengers: 36
Engines: Four 2,018kg (4,450lb) thrust de Havilland Ghost turbojets
Cruising speed: 789km/h (490mph)
Ceiling: 12,190m (40,000ft)
Range: 2,816km (1,750 miles)
Weight: 47,627kg (105,000lb)
Span: 35·05m (115ft)
Length: 28·35m (93ft)

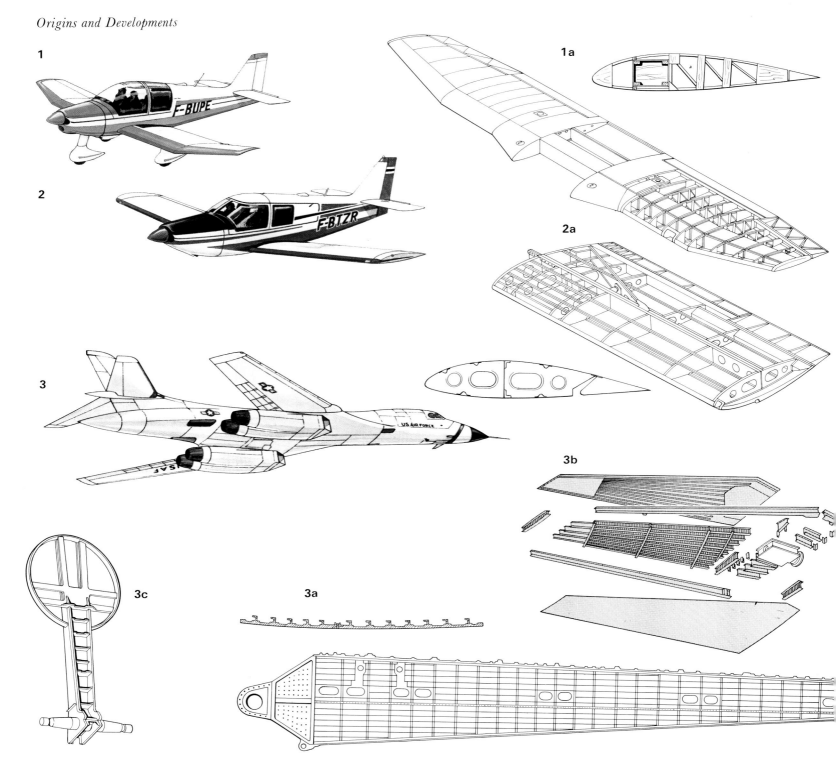

major fatigue crack should break one of the members.

The number of forms of structure introduced since 1950 is greater than in the previous 50 years. Sandwiches are today common, especially for secondary structure such as the area of wing behind the main wing box, fairings, large doors and access panels, and the whole exterior skin of some supersonic aircraft. Fillings in the sandwiches are often low-density foams, but more often are honeycombs of papers, plastics or various metals bonded to the face sheets by brazing or by adhesives. This form of structure is especially resistant to the high-frequency fatigue caused by intense noise, and is commonly used for making engine pods. Honeycombs with inner cells open to the atmosphere are used as noise-absorbing materials. Other structures, including large exterior skin panels in aircraft from bombers to lightplanes, are stabilized by bonding or spot-welding an inner skin of corrugated sheet, while most all-metal

lightplanes incorporate at least some exterior skin stabilized by chordal grooves or raised corrugations reminiscent of the old Junkers skins. As they are short, these stiffeners have low drag, and they allow the interior (of the rudder or other part) to be greatly simplified at a saving in cost.

During the 1930s it was realized that modern skin panels with 3-D curvature must be stretched or stamped by large presses, and the best answer, found in the United States, was the stretch-press. The sheet is gripped at each end by jaws held on hydraulic jacks and pulled bodily around a male die having the shape of the part needed. The sheet deforms plastically and springs back only very slightly. Almost all modern skin panels are stretch-pressed, except for the thinner and smaller panels for lightplanes which are pressed on more traditional tools. Subsequently the structure is assembled by automatic or manual riveting, adhesive bonding or spot-welding. But many of the largest skin panels, and

the most vital other parts, are made without any joints.

One way of forming a complicated part in a single piece is by casting, by pouring molten metal or alloy into a mould; today this is used to make the blades for advanced turbine engines, which have to be of heat-resistant alloys very hard to shape by other means, but castings have been important in only a few types of aircraft. Another method is forging, when a rough workpiece is squeezed or thumped into shape, either at room temperature or while very hot. Forging may be done by a tremendous, slow squeeze by a gigantic press, either between a pair of shaped dies (moulds) or open; alternatively it may be done by repeated mechanical hammer blows, which in High-Energy-Rate Forming (HERF) involve catching the workpiece between two equal masses driven together by gas pressure from rapid combustion or similar high-speed impulse. Forging can even be done by explosives or suddenly applied

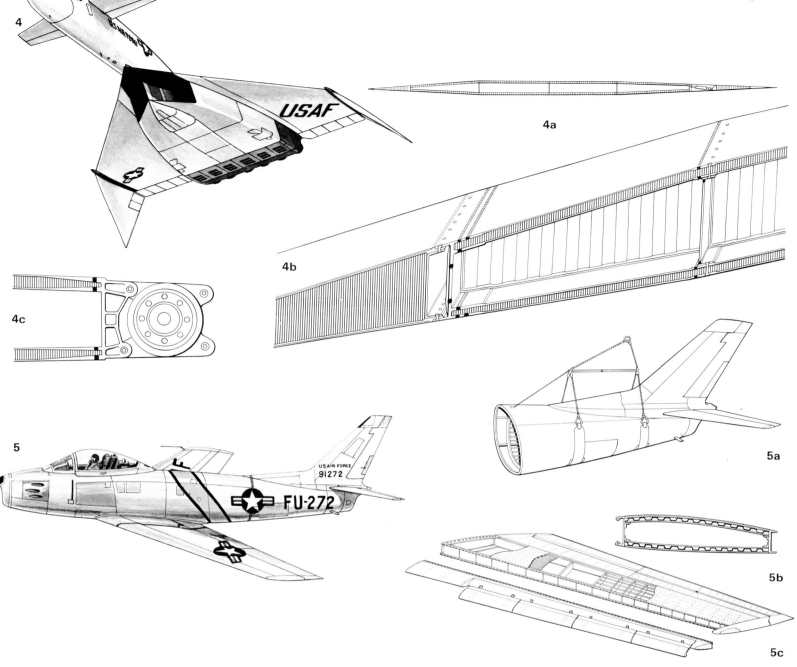

intense magnetic fields. Virtually all the parts of aircraft bearing the highest concentrated loads are forgings; examples are landing-gear shock struts, root ribs, spar booms, tailplane attachments and fin anchorages.

Among the first large aircraft forgings were engine mountings squeezed out in magnesium alloy by huge presses in Germany before World War II. After the war the US Air Force launched a 'heavy press program' as a national effort at a cost of over $570,000,000, as a result of which the US Air Force supersonic 'Century series' fighters were designed to have complete fuselage frames and wing spars squeezed out as single pieces. With closed-die forging such parts can be forged as an almost finished item, but a little machining (cutting to shape) is still needed. With such massive workpieces, the machine-tool industry, again led by the USA, developed monster machine tools of the family known as routers or millers, able to shape workpieces of a size never before seen outside shipyards. Parts

15·24m (50ft) by 9·14m (30ft) are today not exceptional, and a few are much longer. During the 1950s the aircraft industry led the world in Numerically Controlled (NC) machining. The workpiece is today designed with the aid of a digital computer, and the same computer then guides the machine tool(s) with perfect precision to shape the actual part. Aircraft still have to be designed by humans, and drawn in detail on paper, but the drudgery of doing the stress calculations and making the drawings is now all done by digital computers. The drawings are mainly for reference, and the cutting of the metal happens without human hands being involved–except, of course, to press the 'start' button.

Though advanced aircraft incorporate complex machined forgings, most of the machining is done on skins and detail parts. Until after World War II skin was merely uniform sheet, cut to shape. With such aircraft as the North American F–86 Sabre and

TOP LEFT: *The Robin is a typical example of a modern mass-produced light aircraft. The Robin DR 400/180 (1) is built of wood and the detail (1a) shows wing profile and section. The Robin Hr 100/Tiara (2) is built of metal. The detail (2a) shows the wing profile and section.*

ABOVE LEFT: *The Rockwell B–1 (3). The diagrams illustrate (3a) a wing skin panel integrally machined from thick aluminium billet (3b) all-composite tailplane structure (built by Grumman) pivoting in a box made of titanium and (3c) a forged and machined tailplane pivot frame.*

TOP: *The North American XB–70 (4) developed new construction techniques. Details show (4a) welded stainless-steel honeycomb wing section (4b) leading-edge spar and wing skin honeycomb panel joint and (4c) wing-tip fold rotary actuator fitting.*

ABOVE: *The F–86 Sabre (5) benefited from new construction technology. The details illustrate (5a) rear fuselage, which could be removed for engine change (5b) an inboard wing section with thick skin panel and (5c) wing construction showing leading edge attached by piano hinge for quick removal and access to controls and systems.*

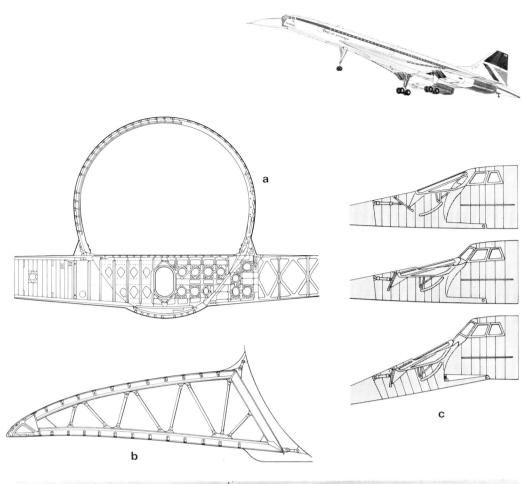

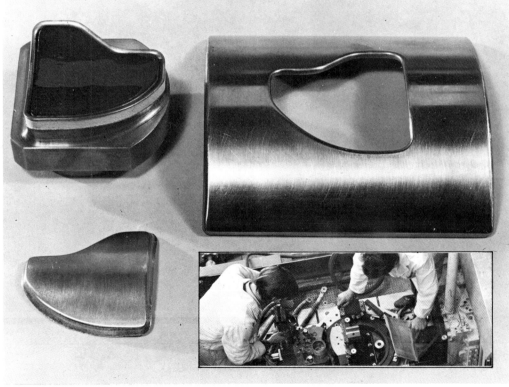

rear of the fixed part of the wing, was added as lighter 'secondary structure' made by traditional methods in thinner sheet, as were slats, flaps and control surfaces.

Wing boxes were the first parts to be made by some of the new methods in the 1950s. One new method was chem-milling, again pioneered in the United States. The entire skin is marked out into thinner and thicker portions, and the latter are protected by special coatings. The whole part, perhaps up to an inch thick and bigger than the floor of a living room, is put into a bath of acid or other corrosive chemical. After an exact time under controlled conditions the skin or slab is eaten away to the desired shape, usually with most of its area much reduced but the full thickness left where it was protected. Chem-milling is fast, does not distort the slab or harm its properties, and leaves no tool marks that could cause fatigue cracks.

It is very difficult to chem-mill deep grooves or complex forms, though since 1955 Britain has been in the forefront of a more versatile process called Electro-Chemical Machining (ECM), in which the acid bath is replaced by a bath of salty electrolyte, as in a battery, and the cutting is done by a soft copper tool carrying a powerful electric current. ECM can produce wonderfully detailed and exact results in the toughest and most challenging materials, and two of the ways in which it differs from ordinary machining are that the tool never touches the work and the liquid flow carries away the 'cut' material in the form of a sludge looking like fine mud. ECM is used for a high proportion of the parts of the latest engines, and is coming into use in airframe construction.

However, the bulk of modern airframe parts needing to be shaped in more than one dimension are not etched or ECM-shaped, but machined on special tools. Cutting from the solid, colloquially called 'hogging', became common in 1952, partly because avoiding joints improves fatigue life and partly because a one-piece component can be cheaper and usually lighter than one assembled from many parts. The usual name for this type of structure is integral construction. Today integral stiffening is found not only in wing boxes, where it began, but throughout advanced airframes.

During the 1950s aircraft speeds increased at a rate never equalled at any other time, and methods of construction were hard-pressed to keep pace. Aluminium alloys are adequate up to temperatures around 150°C (302°F), which can be reached after sustained flight at Mach 2 even in the −60°C (−76°F) air of the stratosphere. Acrylic plastics such as perspex become dangerously softened above this temperature. In the early 1950s there was no alternative but to make most of the exposed parts of the airframe of Mach 2-plus aircraft out of stainless steel. In America's Convair B−58 Hustler bomber, though it was not capable of exceeding Mach 2, most of the skin was made of stainless honeycomb sandwich, which called for years of intense research to perfect methods of brazing it and then inspecting to make sure the inaccessible interior was correctly bonded.

Far more useful was the North American

TOP: *The Concorde illustrates an aircraft of integral construction structure. Cutting from the solid-'hogging'-first became common in the early 1950s and was subsequently widely adopted due to its improvement of aircraft fatigue life. The cutaways show (a) a wing spar and fuselage bulkhead carved by large machines from fixed slabs of light alloy, (b) the forward wing section with flexible pin-jointed mountings and (c) droop nose and visor.*
ABOVE: *Electro-chemically machined parts with (inset) technicians inspecting the process.*

Boeing B−47 Stratojet it became worth while to use non-uniform sheet, thickest at the wing roots and tapered off towards the tips. By this time it had become universal in advanced aircraft to build the wing as an extremely strong 'box' consisting of upper and lower skins joined by front and rear spars, with many stringers and a few strong ribs, the whole assembly then being sealed by a rubbery lining to form an integral fuel tank. Sometimes this box extended to the leading edge, but usually the leading edge, like the

X–15, the fastest aircraft yet built, in which the whole airframe primary structure was welded from refractory high-nickel alloys, similar to those used for gas-turbine rotor blades, which had to be annealed at up to 1,095°C (2,003°F) and then 'aged' at 705°C (1,301°F) in furnaces big enough to swallow the whole aircraft. Almost all the 'plumbing' in this Mach 6·8 aircraft was welded or brazed, and virtually all the components and equipment had to be of unusual heat-resistant materials.

Progress with missiles and spaceflight in 1955 to 1965 led to a revolution in advanced structures, and also in fascinating methods of cooling the exterior to avoid their burning up like a meteor in the atmosphere. Manned aircraft, with 3,542km/h (2,200mph) perfordant but challenging metal titanium, which was so difficult to produce from its natural state it was more expensive than gold, and is still more costly than most other aircraft metals. In the early 1950s it was used only for such parts as exhaust pipes and engine nacelles, but gradually its valuable property of coming between aluminium and steel led to its widespread use, especially in supersonic machines.

Titanium's strength at temperatures of 150–250°C (302–482°F) is much greater than aluminium, and it is much lighter than steel, but it posed large problems. Among the first aircraft made chiefly of titanium were America's Lockheed Blackbird (YF–12 and SR–71) high-altitude reconnaissance aircraft, with 3,542km/h (2,200mph) performance, skin temperatures up to 320°C (608°F) over the whole airframe and engine nozzles white hot. Lockheed vice-president 'Kelly' Johnson recorded: 'Of the first 6,000 pieces we fabricated of Beta B–120 titanium we lost 95 per cent. With the help of Titanium Metals we attacked the problem vigorously, investigating such factors as hydrogen embrittlement, heat-treatment procedures, forming methods and design for production. We solved these problems, but at considerable cost.' At first B–120 seemed impossible to cut, and drills had to be re-ground after every ten holes; but by the end of development of these aircraft the alloy was being cut ten times faster, and drills were lasting 119 holes. The aircraft got their name from their exterior coat of high-emissivity black paint, a colour previously used on the X–15. Previously, slow aircraft had been kept cool in hot sunshine by painting the upper surface white.

Since 1966 there has been one more complete revolution in aircraft structures: the increasing use of composite materials. These are materials made up of two parts, which can either be thought of as a 'matrix' reinforced by very strong and stiff splinters or threads, or else as very strong and stiff splinters or threads stuck together with the adhesive that forms the matrix.

Composites were experimented with as early as 1920, when designers tried such combinations as brown paper layers stuck on top of each other with glue whilst being bent to the shape of the aircraft, and a sort of plastic made from seaweed which again was moulded with layers of wrapping paper. By 1950 glass

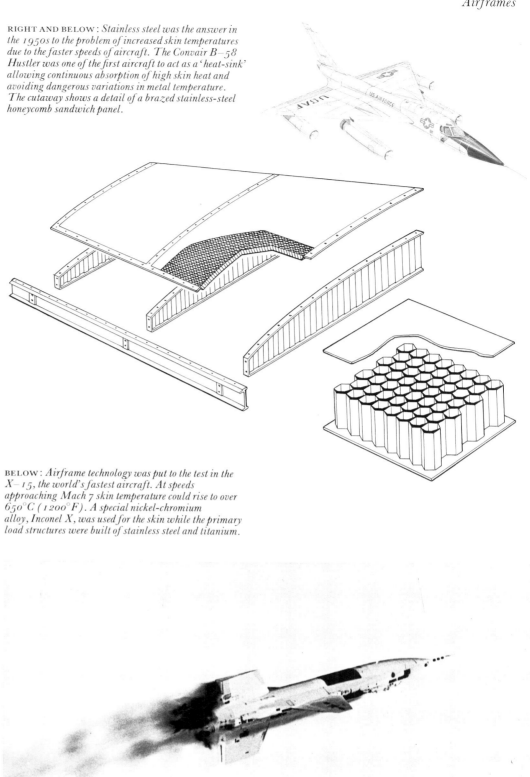

RIGHT AND BELOW: *Stainless steel was the answer in the 1950s to the problem of increased skin temperatures due to the faster speeds of aircraft. The Convair B–58 Hustler was one of the first aircraft to act as a 'heat-sink' allowing continuous absorption of high skin heat and avoiding dangerous variations in metal temperature. The cutaway shows a detail of a brazed stainless-steel honeycomb sandwich panel.*

BELOW: *Airframe technology was put to the test in the X–15, the world's fastest aircraft. At speeds approaching Mach 7 skin temperature could rise to over 650°C (1200°F). A special nickel-chromium alloy, Inconel X, was used for the skin while the primary load structures were built of stainless steel and titanium.*

fibre was already in production for airframe parts, but only for items such as radomes, fairings and wing-tips that did not bear severe stress. The spun glass is moulded with resin adhesive to form a neat and light one-piece shell which stands up to the action of the weather and can be made to have considerable strength and resistance to impacts. Several designers tried to build complete 'plastic aircraft', but they were hampered by lack of prior experience (so that airworthiness authorities imposed severe factors of safety to avoid failures, making the airframe heavy) and by the fact that glass is strong but not stiff and so the structure distorted more than a metal one.

At this point it is essential briefly to mention aeroelasticity. Thousands of people spend their whole lives solving aeroelastic problems, but it is a subject too complex to go into deeply here. It is impossible to avoid in modern aircraft, because no aeroplane or helicopter is rigid, but can distort or vibrate in an almost limitless number of ways. In

B-1 MATERIALS
AV-1

% OF DCPR WT	
ALUMINUM	41.3
STEEL	6.6
TITANIUM	21.0
COMPOSITES	.3
FIBERGLASS	
POLYIMIDE QUARTZ	30.8
OTHERS	

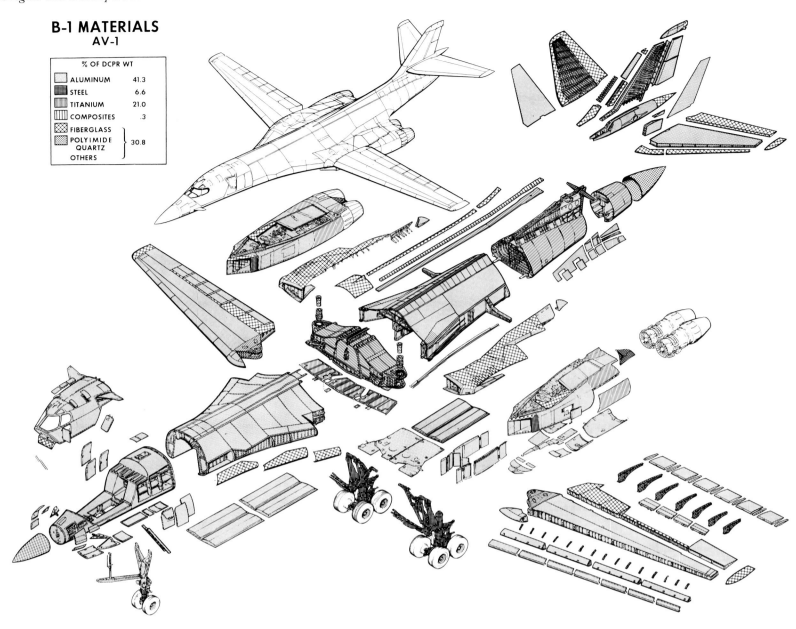

TOP: *Diagram of the different materials and sub-assemblies of the Rockwell B–1.*
ABOVE CENTRE AND ABOVE: *Glass-fibre and glass-reinforced plastics are increasingly attractive materials for certain aircraft structures, especially gliders and sailplanes. The Gläsflugel Hornet is typical of modern sailplane construction. The cutaway shows a wing section of glass-fibre and polystyrene foam–light, flexible and very smooth.*

early aircraft speeds were low enough for this to be generally ignored.

Aeroelasticity is caused by aerodynamic forces, elastic forces in the structure, inertia forces due to the masses of material, and certain other inputs, and can result in oscillation, buffet, control reversal (the surface has an effect opposite to that intended because it distorts the part of the structure to which it is hinged), vibration and, the most serious of all, divergence, in which the distortion has the effect of increasing the load causing it, so that a complete wing or tail may be broken. Today complete aircraft are resonance-tested before or during the start of the flight-test programme, by carefully suspending them and then 'exciting' them by vibrators which can shake the whole airframe at any desired frequency.

Many of us have discovered that if we shake a seemingly strong handrail barrier (or a high balcony, for example) at its natural resonant frequency we could quickly break it, though it would be many times too strong for us to break by a steady force. Likewise many aircraft, before aeroelasticity was fully recognized, oscillated and broke before the pilot could alter the condition that was causing it. Flutter can be induced in flight-testing either by deliberately exciting the airframe,

at different airspeeds, with controllable vibrators or by firing a quick-burning rocket fixed at the wing-tip or tail (called a 'bonker'), which imparts a vicious blow as if from a giant hammer. The resulting oscillations are measured both with and without such systems as autopilots and yaw-dampers in use.

In general, wooden aircraft have been less prone to aeroelastic problems than metal ones, but this is partly because they usually fly at lower speeds. Even today a significant proportion of lightplanes are built of wood, though this is influenced by the fact that some of the builders simply lack the facilities to handle modern metal construction. Where a good strength:weight ratio and a smooth streamline form are combined with low airspeeds, as in the sailplane, glass-fibre or Glass-Reinforced Plastic (GRP) is extremely attractive. A large and increasing proportion of sailplanes are 'glass', and their flexibility gives the pilot a smoother ride and results in none of the aeroelastic problems that would occur at higher speeds. GRP is also used for some parts of powered aircraft and even for modern propeller blades, fans and fan ducts.

In the early post-war era other composites were investigated for aircraft construction, notably Durestos, a resin-bonded material

Construction of the Airbus A.300B

The Airbus A300B entered airline service in 1974. Six countries have collaborated to design and build it. The main members of Airbus Industrie are Deutsche Airbus (MBB) in Germany, Aérospatiale in France, Fokker in the Netherlands, Spain's CASA and Britain's British Aerospace. Original design and testing was done by Hawker, Aérospatiale and Deutsche Airbus, and these companies employ 100,000 people in 39 factories. Of these, some 26,900 are engaged directly on the A300B programme. The largest single parts are the left and right wing boxes, and three million man-hours went into the design of the wing alone. Hawker later added a further 900,000 man-hours in refinement and improvement, though the basic design never changed. Indeed, much of the refinement was deliberately making the wing 'worse' by simplifying it, because in its original form it performed beyond all expectation, a fact first indicated during the 3,700 hours of wind-tunnel testing completed by the end of 1968 using 41 models, in 6 tunnels in England, 1 in France, 2 in Germany and 1 in the Netherlands.

Aérospatiale builds the centre section, which is the same width as the fuselage and forms an integral unit with the lower centre fuselage. Hawker builds the left and right wings at Chester, and it is there that the wing

aluminium-copper alloy. Each wing has three upper and three lower skins, the joints being above and below Rib 9, where the engine pylon is hung, and inboard from there along the centre spar. This spar, which runs across the aircraft between the engine ribs (Rib 9), is fabricated, which means it is built up from many parts. The rest of the primary structure is integrally stiffened and machined from large slabs or forgings.

Upper and lower skins are held while being machined by large numbers of suction cups. When one is finished, the air supply is reversed and the table of the machine tool, 18·29m (60ft) by 3·66m (12ft), becomes an air-cushion pallet on which the skin 'floats'. This is of great value in exactly positioning the fresh billet and in taking off the finished skin. After careful inspection by eye and by NC-controlled machines, to verify exact dimensions and ensure that there are no scratches or toolmarks, the skins are squeezed in a (1,221·82-tonne) 1,200-ton press to beyond the aerofoil profile of the wing. On release from the press they spring back to the exact curvature needed.

During World War II the popular figure of the aircraft industry was Rosie the Riveter, a hastily-trained girl with a rivet gun. In building modern aircraft such as the A300B, manual riveting of this kind is still possible, though it is hard to find. In making the wing the philosophy never changes: giant machines, computer control, and precise, repeatable results. Drilling is done automatically, with a surface finish of 152·4 micro-centimetres (60 micro-

having the same coefficient of expansion as the spars and skins of the wing, otherwise a wing that fitted in winter would not fit in summer.

The wing assembly jig consists basically of seven pairs of cylindrical columns, on which are mounted the many locating points that ensure not only the wing itself will be exactly the right shape, but also that it will precisely fit the centre section made in France. Most of the joints between the heavy spars and skins are made with Hi-Lok or Taper-Lok bolts, each of which fits like part of a watch. After assembly, the wing box has its root-end profile machined, and it also has to be sealed to form an integral fuel tank. Eventually these vast tanks become quite complex, with multiple covers, baffles, pipes, booster pumps, drains (the position of which was decided after exhaustive studies with one-fifth scale model wings) and manhole covers fitted into large holes, around which the skin has its full thickness of 25·4mm (1in).

When the wing box is finished, the pair for one Airbus are loaded into an Aero Spacelines Super Guppy aircraft and flown to Bremen, Germany, where the movable surfaces are added. Most of these are made in the Netherlands by Fokker-VFW and comprise, on each left or right wing: one inboard tabbed Fowler flap; two inboard airbrake/lift dumpers, an all-speed aileron (in line with the engine), two outboard tabbed Fowler flaps; five outboard airbrake/lift dumpers; a wing tip low-speed aileron; three large sections of powered slat, and, on the B2K and B4 models, an inboard Kruger flap and fuselage-mounted retractable 'notch' to give a perfect leading edge profile without the gap seen in other aircraft. These surfaces are fabricated by riveting and incorporate bonded and honeycomb structures. It is typical of modern-aircraft design that extensive studies with computers were

the parts to the assembly hall at Toulouse, where powerful jacks and fixtures lift each into exact position for mating with its neighbour. The towering Messier-Hispano landing gears are attached, and in a few days the aircraft is standing on its own ten tyres. The General Electric engines, 61·3 per cent shipped from the United States and the rest (including some of the most advanced technology portions) made in Germany and France, are packaged inside their enormous cowling pods which are mainly made by an American company's new French factory at Toulouse. The process of turning the bare airframe into a finished Airbus then occupies several weeks and represents much more than half the total cost. Thousands of items arrive from factories in France, the United States, Britain, Germany and other European countries, and the great size of the aircraft facilitates installation by allowing many personnel to be on board at the same time.

Tons of furnishing materials, including thermal and sound insulation, are among the last parts to be added before the customer's own choice of interior begins to take shape. Before furnishing is completed the 'green aircraft' (i.e., unfurnished

ABOVE: *Final assembly line of the Airbus at Toulouse, France.*
LEFT: *One of Air France's Airbuses.*

skins, the biggest single pieces of metal in the aircraft, are NC-machined by enormous millers. The skins begin life as slabs in Davenport, Iowa, USA. Computer tape controls the cutting heads as they carve out the original thick sheet to an integrally stiffened skin weighing about 19 per cent as much, the mountain of swarf and chips, representing 18 per cent of the original slab, being returned to the light-alloy supplier.

The skin materials are not all the same: upper skins are unclad aluminium-zinc alloy for compression, while the undersurfaces are strong

inches) or better than 76·2 micro-centimetres (30 micro-inches) if the holes are then reamed. The stringers that stiffen each skin are positioned and clamped, and a Drivmatic machine then runs along each stringer cold-forming the rivets from automatically supplied headless slugs. Each is automatically shaved flush by a small milling cutter as soon as it is formed. Then the stiffened skins are held in a gigantic vertical assembly fixture where the wing box is assembled. So big and so accurate are the operations, that the entire fixture has to be linked by light-alloy beams

undertaken just to decide on the materials for the tracks along which run the slats and flaps. The answer was machined forgings in titanium for the slats, and light alloy forgings for the flaps, with the track faces made from wear-resistant steel inserts.

The same kind of attention to detail, and extreme precision even in parts as big as a house, characterize the construction of the fuselage sections and vertical tail in Germany, the nose, flight deck and engine pylons in France, and the horizontal tail and doors in Spain. Finally the Super Guppy brings all

and without avionics) may have begun acceptance flying. It will spend a day or two in the world's most modern paint shop, where weather-resistant epoxy-based paints dress the exterior in the colourful scheme of a national airline. The customer's livery is exactly recorded and marked out partly manually and partly automatically, applied with spray-guns and brushes from gondolas mounted on telescopic rams travelling on overhead gantries in the roof. A painter can zoom from the nose to the tip of the fin almost 17m (55ft) above the exactly level concrete floor.

which a 'tow' of a few thousand filaments looking like black string is wound over a former into the shape of the workpiece whilst bonded with adhesive.

Unlike many fibres, carbon or graphite can be made in lengths measured in miles. Each filament is much finer than human hair, and it will not break until the stress has reached levels 5, 10 or even 20 times the limit for metals. Compared with glass, the black fibre is much stiffer, so a CFRP structure can be made not only amazingly strong and light but also rigid. Finished CFRP items can be made most readily by buying prepreg, the standard raw-material form, and then cutting it (thin sheet can be cut with scissors) and bonding it under heat and pressure to form the desired part.

From the start it was recognized that CFRP would not be cheap, though over the years costs have fallen in a predictable way as small pilot plants have given way to bulk industrial production. Carbon fibre is inherently much cheaper than its main rival, boron, developed mainly by the US Air Force, which has to be produced by depositing the boron metal as condensed vapour, in an atmosphere of boron trichloride, on a white-hot wire of tungsten! Boron filament thus grows on the white-hot wire like the ring-growths of a tree. It can thus be made thicker than carbon fibre and is very strong and stiff, and has been vigorously promoted in the United States in matrices of epoxy resin, polyimide high-temperature resin, phenolic resin, titanium, aluminium and magnesium. With the backing of numerous military programmes boron filament, mainly used in boron-epoxy composite, has been in production for US aircraft since 1973. The first batches of boron-epoxy prepreg cost $209 per pound, but the second year's production cost $107 per pound, at which level those selling it claim that airframe component cost becomes 'comparable to titanium'. Among the earliest uses for boron-epoxy composite were reinforcement of the General Dynamics F–111F wing hinge, stiffening of the Sikorsky CH–54B tailboom, complete manufacture of 50 rudders for McDonnell Douglas F–4 Phantoms as a test programme (the aircraft all entered service in the usual way), manufacture of the tailplanes of the Grumman F–14 Tomcat, and manufacture of both vertical and horizontal tails of the McDonnell Douglas F–15 Eagle. Since 1973 boron prepreg has not fallen significantly in price, and with the benefit of more American companies interested in CFRP boron now finds the going tough.

CFRP has not had the backing of big military programmes, because there have been none in Britain, but in a quiet way it has gradually taken over a vast number of markets. To show its versatility the biggest outlet for carbon fibre is in the shafts of golf clubs, with tennis rackets and fishing rods running close behind.

Carbon-fibre prepreg has also been used to stiffen countless structures in glass fibre or aluminium, including many aircraft structures, but its real worth is best seen in components designed to be made in CFRP. The almost ideal initial application was the manu-

reinforced by short filaments of asbestos. In bulk asbestos has little strength (we think of it as flabby fire insulation), but many substances can develop fantastic strength when in the form of fine filaments. This is because a single crystal of material has perfect properties, sometimes millions of times better than bulk material in which astronomic numbers of crystals are lumped together with weak joints. Even glass, which is not a crystalline material, is far stronger weight-for-weight as a fine filament than as a solid mass. Single filaments of asbestos are even better, but are difficult to produce in long lengths. Much research was done in 1945 to 1955 on various composites to try to get good adhesion between the filament and the matrix, and to get the highest proportion of strong filament into the final composite. Obviously, 100 per cent is impossible, because there has to be a matrix to stick the filaments or short fibres together. Among the other composites on which work began at that time were filaments of stainless steel in aluminium, silica

in aluminium and boron in magnesium.

Then, in 1963 to 1965, workers at Britain's Royal Aircraft Establishment at Farnborough produced fibres of carbon or graphite which appeared to exceed all others in their combination of strength, stiffness and low density. Such fibres had been studied in Japan, though not as structural material, and it was the British team that first pushed through 'carbon fibre' as an aerospace material that could be reliably produced in quantity and used efficiently. The first production method was to heat a commercial man-made fibre, Polyacrilonitrile (PAN), in a furnace until it had turned to almost pure crystalline carbon. It then had to be heat-treated and stretched, at white heat, to develop strength which, in comparison with metals, seemed like science fiction. Selected resins were then used to stick the fibres together to make either prepreg (prefabricated strip or sheet), of solid Carbon-Fibre Reinforced Plastic (CFRP), or a finished item by a laborious but ideal process called filament winding, in

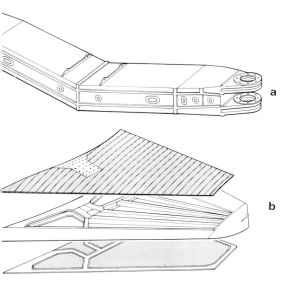

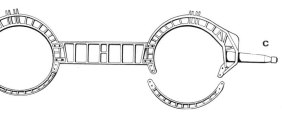

facture of the fan blades for the Rolls-Royce RB.211 engine, where CFRP saved a remarkable amount of total engine weight and had other advantages, but this was killed by the inability of Rolls-Royce to make the blade leading edge sufficiently resistant to wear and severe impact.

In a low key several European firms eventually got into production with CFRP helicopter rotor blades, transport aircraft floors, fighter control surfaces (such as Sweden's Saab Viggen canard), rocket-motor pressure vessels and many other parts. Most customers for the Boeing 747 specify British CFRP floors because they are much lighter than the alternative aluminium/balsa sandwich and last at least seven times as long. CFRP is routinely fitted to aircraft of quite modest performance; for example, the landing-gear doors of VFW-Fokker 614 are CFRP panels bonded over Nomex honeycomb, which gives a 30 per cent saving in weight over the alternative GRP construction.

Technology

None of man's creations has spurred the development of as much new technology as the modern aircraft. Gas dynamics (the behaviour of air or gas in motion), new materials, new forms of engine, new kinds of structure, new weapons, the processing and display of information, methods of navigation, systems for moving or controlling things remotely, and perhaps above all the vast field of electronics – all have been driven on at a cracking pace by the needs of aviation, and have subsequently percolated through industry and society into the whole spectrum of modern life. Some of the advances are discussed in special chapters (notably, engines and airframes). Others are explained in chapters dealing with particular classes of aircraft, or with military and civil aviation. In this chapter a broad overview is given of the main advances made in the field of aviation, with the emphasis entirely on the aircraft itself, although it is worth mentioning that aviation has also produced advances of other kinds.

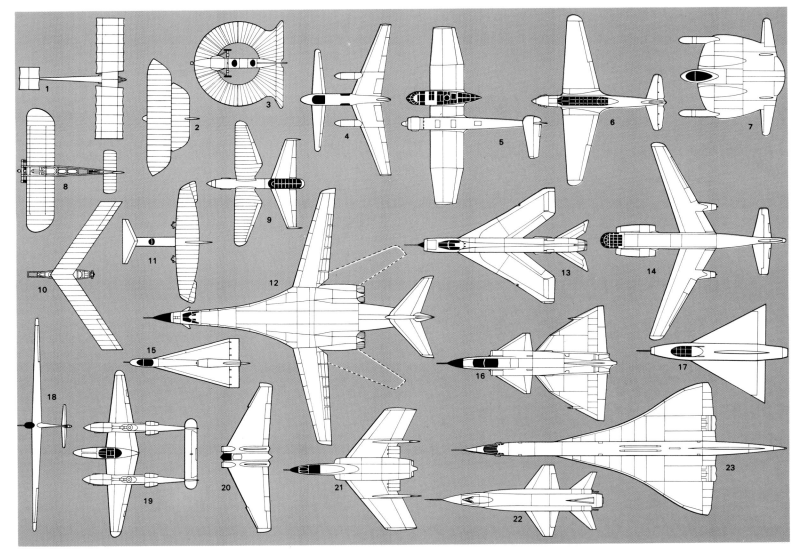

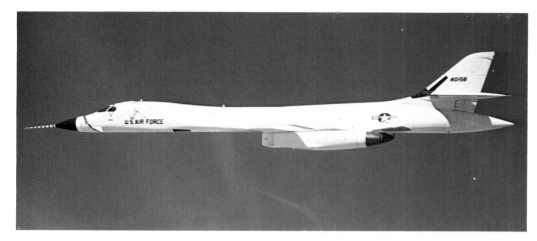

sian one, was that the main wing should be slightly higher than the rear one, to act like a slat. Miles produced Libellula (a species of dragonfly) aircraft between 1941 and 1943 which were again of the tandem-wing type. The main advantage of the tandem-wing aeroplane is that it has a much larger CG range than normal; in other words it can still fly safely with a lot of load near either the front or rear.

During World War II several canard fighters appeared, with vertical surfaces either ahead of the pusher propeller or on the wing trailing edges or tips. Conversely, the Messerschmitt Me 163 was of the so-called tailless family, with no separate horizontal surface at all. After 1945 jet fighters continued these unusual forms and the tailless delta, with more or less triangular-shaped wing, became quite common. The delta shape wing neatly solved several problems; the elevators could be far behind the CG for good control;

Shape

One of the arguments of those who attempt to scorn the work of the Wright brothers is that they settled on the wrong shape of aeroplane. They put the horizontal tail in front, the vertical tail at the back, used biplane surfaces everywhere, made the pilot lie on the wing on one side of the centreline to balance the engine on the other side, and used two sets of chains to drive two pusher propellers. This arrangement had many drawbacks, and several early aviators appreciated the simplicity of just fixing the propeller to the engine, but then they were not sure whether to make it pull or push.

One has only to study the aeroplanes of Louis Blériot to see how undecided designers were. Some Blériots were tractors (propeller in front) and some pushers, while the arrangement of tail and control surfaces varied from day to day. Many designers used tandem, or even circular, wings.

By 1910 most, but by no means all, designers had agreed that the classic aeroplane had one or more sets of wings near the Centre of Gravity (CG), and both vertical and horizontal tail surfaces. Until 1950 it was almost universal to have a fixed fin (vertical

stabilizer) carrying a hinged rudder, and a fixed tailplane (stabilizer) carrying hinged elevators. In general, the further away these surfaces are from the CG, the smaller they need to be, because the moment (turning or stabilizing effect) is roughly proportional to the surface area multiplied by the distance from the axis of rotation.

The Canard

Probably the main alternative aeroplane configuration is the 'canard' (French for duck), or tail-first, layout. It is rare to have a design with *all* the 'tail' surfaces in front; a canard usually has the horizontal tail in front and vertical tail at the rear. One reason for this configuration is that if the horizontal tail were to stall it would cause the aircraft automatically to pitch nose-down, thus unstalling and returning to controllable flight, whereas with a rear tailplane the stall of that surface would be disastrous. Professor Focke built a canard in 1909, and then exhaustively tried the *Ente* (German for duck) in 1930, finally concluding that there was no great advantage.

Many designers, including Delanne and Miles, experimented with a different shape in which there were essentially two wings. The Delanne idea, based on an earlier Rus-

the wing could have a very low thickness/chord (t/c) ratio (thickness divided by chord, i.e. the distance from leading to trailing edge); and the result came out light and with low transonic drag.

Eventually, fighter designers realized that the best wing is one of conventional form but with low aspect ratio. A fighter needs a wing of large area, for maximum power of manoeuvre, and low t/c ratio for supersonic performance. To carry a large load the wing needs high-lift devices while multi-role aircraft need a variable-sweep 'swing wing' mounted on pivots. In this arrangement, the wings can be spread out to give a long, slender surface for take-off, slow-speed loitering and landing, or folded back to give a very small span for supersonic flight.

It is especially important to have the smallest possible span and area for supersonic flight at low level, to avoid the crew having an impossibly rough ride. One large bomber, the Rockwell International B–1, even has a Low-Altitude Ride Control (LARC) system which senses atmospheric turbulence and works control surfaces on the aircraft to counter it and keep the crew compartment travelling in a smooth, straight line.

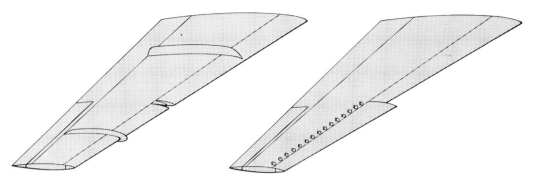

Aspect ratio

Aspect ratio is crucial to aircraft design. The slowest and most efficient aircraft need the highest possible aspect ratio; sailplanes and man-powered aircraft have extremely slender wings with a ratio of 25 to 35 in order to act on the greatest possible airflow. Most aircraft have an aspect ratio of between 5 and 11, while fighters and supersonic transports typically lie in the range 1·5 to 3·5. In the years immediately following World War II the arrow-shaped swept wing was almost universally adopted, along with the delta, to allow aircraft to reach higher transonic speeds. The first jet airliner, the de Havilland Comet, had a virtually unswept wing, but the Tupolev Tu–104, Boeing 707 and Douglas DC–8 all had wings swept back at about 35°. Later, the very fast Boeing 747 raised the angle to over 37°, but today highly swept wings are rare. The very fastest aircraft, such as the North American X–15 and Lockheed A–11/YF–12/SR–71 Blackbird family, have no sweepback at all. In general, long-range aircraft today have more sweep and short-range aircraft have less, but the trend with all subsonic transports is towards more slender wings with less sweep.

During the early 1950s a rash of problems with early transonic or supersonic wings led to a succession of 'fixes' which have in the main taken their place as permanent features of aircraft design. One is the turbulator or vortex generator, which is a small fin no bigger than a playing card, set at an angle to the airflow. Behind it streams a writhing

TOP: *In the 1950s a number of 'fixes' were devised to solve problems of high-speed wings. The two illustrated are the sawcut (left) and the dogtooth (right). The sawcut straightens the airflow and stops it slipping in a spanwise direction. The dogtooth acts as a vortex generator and fence combined.*
ABOVE RIGHT: *Another solution to the spanwise flow of air was the fence. Fences can be clearly seen on the MiG–15 shown.*

ABOVE CENTRE: *Swept wings were almost universally adopted after World War II. The DC–8 has wings swept back at 35°.*
ABOVE: *On 12 June 1979 the Gossamer Albatross was pedalled by Bryan Allen from Folkestone, England, to Cap Gris Nez, France, against the prevailing wind. Designer Dr Paul MacCready later built solar-powered aircraft, one of which flew to Kent from near Paris in 5h 23min.*
RIGHT: *The variable-incidence wing of the F–8 Crusader aids in take-off and landing.*

vortex which, though it increases drag, stirs up the air downstream and mixes the high-speed free-stream flow with the sluggish boundary layer covering the immediate proximity to the aircraft skin.

Airflow can also be improved by reducing the cross-section taper of control surfaces and terminating them with a blunt instead of a sharp trailing edge. The air is fooled into thinking the surface has a wider chord and sharp trailing edge, so drag is not significantly affected, but blunt rudders and ailerons are often the sign of a problem which might have been avoided. Another 'fix' is the flow-breaker or stall inducer, a small projection along the leading edge of a wing, which promotes breakaway of flow at high angle of attack (angle to the airflow). This ensures that the stall will start at the leading edge and not at some other portion of the wing causing possible loss of controllability.

A common fault of early swept or delta wings was sharp deflection of the airflow towards the wing tips, especially over the upper surface. This was often caused by a region of high pressure above the wing resulting from local shockwaves, the strong pressure waves formed when supersonic flow is present. The ideal wing has the airflow passing straight across from front to rear, as seen from above, and the simplest and crudest answer to the outward or spanwise flow was to add a fence, a thin wall aligned with the fore-and-aft direction round the leading edge and extending back across the upper surface. Some Soviet MiG fighters had as many as six large fences, and even today four is a common number.

Another of the 'fixes' of the early 1950s is the dogtooth. Early swept or delta wings were frequently improved by extending the chord of the outer portion, and drooping the extended leading edge slightly down to reduce the leading-edge angle of attack at high altitude.

In most aircraft the inboard end of the extra area forms a major discontinuity, called a dogtooth (because it looks like a curved canine tooth from end-on). This creates a powerful vortex which swirls back over the wing, especially at high angles of attack, and thus acts as a vortex generator and fence combined.

Yet another feature which became important in the early 1950s on fast fighters is the variable-incidence wing, which instead of being rigidly fixed to the fuselage is mounted on pivots. In fact, the attitude of the wing (its angle of attack or incidence) cannot alter for any given flight condition; the effect of the pivots is to cause the attitude of the fuselage to alter. In the Vought F–8 Crusader fighter, a variable-incidence wing enables the aircraft to land with the fuselage in a level, or even slightly nose-down, attitude, giving the pilot a good view ahead and allowing the main landing gears to be short.

Very fast aircraft

Extremely fast aircraft present their own problems. The North American XB–70 Valkyrie bomber, which happened to have a canard foreplane, was designed to make use of 'compression lift' caused by compressing the flow behind the inclined shockwaves under the wing. At supersonic speed, the air was forced to pass under the wing between the down-turned wing tips and a huge box containing the six engines. At subsonic speed, the wing tips (which were hinged) were raised to the normal position.

The XB–70 was one of the first modern aircraft to have two 'verticals' (tail fins). In the 1930s designers often gave aircraft two or

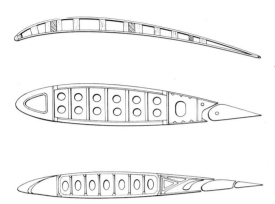

ABOVE: *The XB–70 Valkyrie was designed to make use of 'compression lift'.*
LEFT: *Wing cross-section profiles of (from top to bottom) the Blériot XI, the Avro Lancaster and the Boeing 727. Further information on wing construction can be found in the Airframes chapter.*
TOP RIGHT: *The McDonnell Douglas F–15 Eagle is fitted with two vertical tails. Either on its own could control the aircraft.*
CENTRE RIGHT: *Flaps and slats: (a) airflow at high angle of attack (b) airflow improved by the addition of a leading-edge slat (c) airflow in normal horizontal flight (d) addition of plain flap giving increases in lift and drag (e) plain flap (f) split flap (g) Fowler flap (h) double-slotted flap.*

more tail fins, sometimes in order to place fins and rudders in the slipstream from wing-mounted engines. Today, except for aircraft which for unrelated reasons have two tail booms (such as the Rockwell OV–10 Bronco and Israel Aircraft Industries Arava, which need two booms to carry the tail because freight must be loaded through a full-height rear door), virtually every aeroplane has a single vertical tail. Often, for structural or aerodynamic reasons, the horizontal tail is mounted on top, giving the so-called T-tail.

Most aeroplanes with two vertical tails are fighters, the first being the Mikoyan MiG–25. Two vertical tails can give better power of manoeuvre, especially at high angles of attack in a dogfight. They also give a measure of re-dundancy (in the McDonnell Douglas F–15 Eagle, either vertical tail could control the aircraft alone) and reduce overall height – they were needed on the Grumman F–14 Tomcat to hold height to 5·18m (17ft) to fit carrier hangars. Yet another feature of the latest dogfighters is a long forebody strake, a greatly extended wing root which has a sharp edge to create a strong vortex at high angles of attack and increase lift and improve con-trol. A similar feature is seen on the Anglo-French Concorde supersonic transport.

More lift

The lift of a wing is proportional to its shape (the cross-section profile or aerofoil section), to its area and to the square of the airspeed. Thus the lift at 200km/h (124·2mph) will be not twice the lift at 100km/h (62·1mph) but four times as great. It is also dependent upon its setting to the airflow, or angle of attack. Most wings have to be turned to a slight negative (nose-down) angle of attack before they will not give any lift at all. At a more pronounced negative angle they will give negative lift, in other words a down-force; this is the setting when an aeroplane flies inverted.

Slats

In the early days of aviation a stall was often fatal. It could lead to a spin, a stabilized con-dition in which the aircraft fell to the ground in a stalled spiral, with its controls more or less ineffective. In about 1920, Frederick Handley Page perfected a device to postpone the stall; he added a small auxiliary winglet called a slat ahead of the wing leading edge. At a high angle of attack, the lift on the slat pulled it away from the main wing on pivoted links, opening a gap called the slot (other wings had fixed slots built into their structure). The air was guided by the slot to keep flowing across the top of the wing, and thus the stall was delayed to an even more extreme angle. In modern aircraft the slat is often very large and strong, and instead of being allowed to pull open automatically it is pushed open by rows of hydraulic rams.

It so happens that supersonic aircraft with wings which are very long from front to rear (large root chord) but of small span, in other words with extremely low aspect ratio, do not follow the classic behaviour as angle of attack is increased. When the angle gets to about 20°, by which time ordinary wings would have stalled, the lift reaches a maxi-mum, but no stall is encountered. If the air-craft has sufficient power or momentum it can go on flying at angles as high as 60°, with the nose pointing skywards. Drag is then extremely high, but the lift is still equal to the weight and flight can be continued.

Early flaps

Even before the slat was invented, wings had been fitted with flaps. At first these were simple hinged surfaces. The plain flap is just a hinged part of the trailing edge which can be depressed before landing in order to in-crease lift and drag. The split flap is hinged to the underside of the rear part of the wing. This also increases drag but does not greatly increase lift. Both normally need consider-able force to operate because they have to be pushed down against the airflow, but with small and relatively slow aircraft a pilot can normally push them down by a hand-lever driving the flap by a linkage.

By 1935 aircraft were becoming larger and faster, and since then flaps have been usually driven by power (see Systems subsection). At this time several more efficient kinds of flap came into use. The principle of the slat, in which the airflow is accelerated through a narrow slot, is used in the slotted flap, in which the leading edge of the lowered flap leaves a slot under the streamlined lower rear part of the wing. The air flowing through this slot follows the upper surface of the flap, whereas with the plain flap it just breaks away (in other words the plain flap stalls when it is lowered). The slotted flap can therefore generate much more lift.

The next stage was obvious: by 1944 the first aircraft were in use with a double-slotted flap, and in .the best-selling Boeing 727 of 1962 a triple-slotted flap was used. This can give extremely high lift, and thus allows the aircraft to land relatively slowly despite hav-ing a small, highly loaded wing of thin aero-foil section. Other flaps include the Junkers double-aerofoil (virtually a narrow auxiliary wing carried behind the main wing on pivots); the Youngman (a much larger auxi-liary wing which could be rotated positively to increase lift and drag for landing or nega-tively to act as an air brake for dive bombing); and the very important Fowler which, though usually only a single surface, marked a major breakthrough in that it was not just hinged to the wing but mounted on guide tracks projecting behind the trailing edge. The initial movement of the Fowler was rearwards, increasing wing area and greatly augmenting lift. As the flap continued to extend its angle progressively increased, so that the final part of its travel did not add much lift but added a lot of drag.

Modern flaps, which are usually double-slotted or tabbed Fowlers (Fowlers with an auxiliary split or plain trailing-edge flap on the main flap), are carried by wheeled carriages running along strong forged steel or titanium tracks. Sometimes several hundred horsepower are needed to drive large flaps, and the common actuation system is a free-running cross-shaft inside the wing driving all flap sections by bevel gearboxes and ball-bearing screwjacks.

Such powerful flaps are almost always partnered by powered leading edge systems, usually driven by similar hydraulic cross-shafts energized automatically so that leading-edge and trailing edge surfaces operate together. Sometimes the leading edge merely 'droops', by hingeing downwards when the flaps are lowered. Often large slats are used, while some aircraft have leading-edge flaps such as the Krüger type, hinged to linkages which extend them out and forwards to provide a powerful, bluff, high-lift leading edge in place of the sharp surface needed for high-speed flight. Boeing and Airbus Industrie use a further refinement: the flexible Krüger, whose profile changes as it is extended.

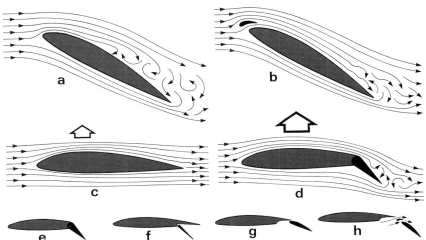

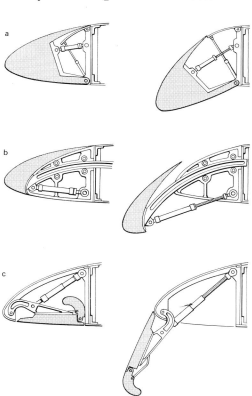

ABOVE: *Leading-edge devices: (a) drooped leading edge (b) slat (c) Krüger slat.*
RIGHT: *Flaps and slats on an Airbus.*

BELOW CENTRE : *The layer of air closest to the wing surface, the boundary layer, is subject to strong retarding frictional forces causing turbulence. Aerodynamicists have invented a number of methods to control the boundary layer. The MiG–21 (left) and the F–4 Phantom (right) are both fitted with the Attinello flap, a simple blown flap used to control the boundary layer and to increase the efficacy of the flap. Without blowing the flap would stall when lowered.*

BELOW : *Boundary Layer Control on the British Aerospace Buccaneer multi-role aircraft, which first entered service in July 1962. The tailplane possesses a slit along the leading edge while the wing has very powerful blowing slits across the upper leading edge and over the upper surfaces of the flaps and ailerons. The blowing system has a built-in failsafe device to allow all the slits to be fully blown in the event of the failure of either of the engines.*

BELOW : *Three examples of engine-augmented lift. From top to bottom, the Buffalo jet flap, Upper Surface Blowing on the YC–14, and the Externally Blown Flap (EBF) on the YC–15. The complex flap sections of the DHC–5 Buffalo receive four sets of engine-supplied air. The YC–15 has two-segment flaps in the wake of the four engines. The YC–14 relies on the Coanda effect to make use of Upper Surface Blowing.*
BOTTOM LEFT : *The Boeing YC–14.*

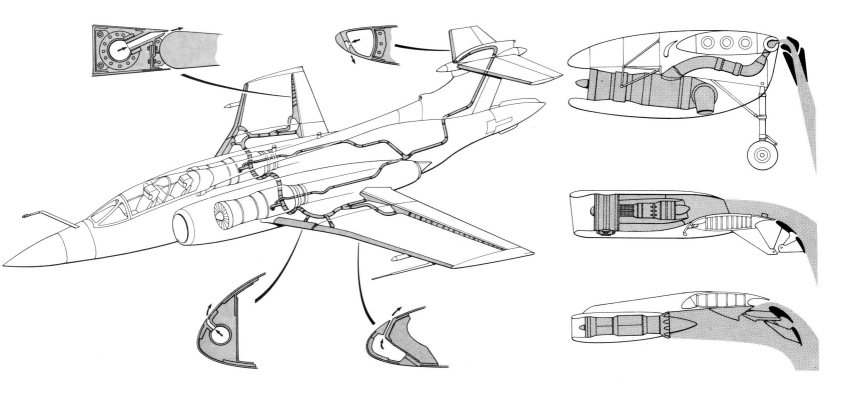

Boundary layer control

During the early 1950s, fighter designers began to use high-pressure air bled from the jet engines, by fastening a valve and stainless-steel pipe to the engine compressors, to blow at high speed (often faster than sound) through narrow slits in various parts of the airframe. This was often called 'Boundary Layer Control' (BLC), but in its simplest form it was used solely to increase the effectiveness of the flaps. So-called 'blown' flaps are usually of the plain type, but of wide chord and thus very powerful. Without blowing they would stall when lowered, and be less effective, but blowing a sheet of very hot, high-pressure air from a slit or row of holes just above the leading edge of the flap forces the air flowing over the wing sharply down across the depressed flap to give a dramatic increase in lift. Named the Attinello flap after its inventor, this simple blown flap is used on such important aircraft as the Lockheed F–104 Starfighter, McDonnell Douglas F–4 Phantom and the Mikoyan MiG–21.

The British Aerospace Buccaneer bomber has a much more advanced system in which the high-pressure air is blown over the flaps, along the leading-edges and along the leading-edge of the tailplane. This reconciles the need for a low-level attack aircraft to have a small wing (and, in this case, small tail) with the need to lift and control a heavy aircraft at modest speeds for carrier landing.

One shortcoming of simple flap blowing is that the need to reduce engine fuel-flow inevitably reduces the thrust available, so the

BELOW: *The Hunting H.126 was the only aircraft to use specially designed jet flaps, developed at the National Gas Turbine Establishment in the United Kingdom, to increase lift. The engine exhaust gases are conducted through pipes and ejected across the flaps from slots running the length of the wing's trailing edge. The detail (bottom) shows the ejector duct inside the wing. Reaction control nozzles are located at the wing-tips and rudder to improve control at low airspeeds.*

BELOW RIGHT: *The Br Michelin V was a WWI bomber of French design but constructed in Britain. The picture shows how everything was left out in the airstream; even the fuel tanks can be seen hung under the upper wing. It is hardly surprising that at full throttle this machine could only reach 138km/h (86mph). In those days reliability (which in the case of these aircraft was not very good) was far more important than streamlining to reduce aerodynamic drag.*

BOTTOM: *The Boeing 747 affords a staggering contrast to the Breguet Michelin. Though it weighs approximately 200 times as much, its aerodynamic drag is actually less when in cruising flight. The photograph, however, shows it on take-off, with leading-edge and trailing-edge flaps extended, landing gears down, and the auxiliary inlets around the engines opened to admit extra air. This particular aircraft is carrying a fifth engine as cargo to be taken to a distant engineering base.*

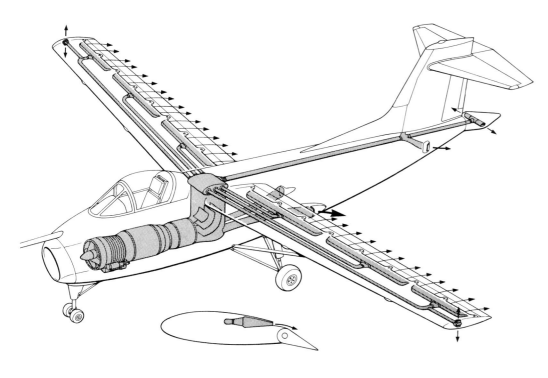

result is that the engines give less power at crucial times (landing, and possibly take-off). There are several other methods of using engine power to increase lift. One is to discharge the whole, or most, of the engine efflux through specially designed 'jet' flaps, made in the form of ejector ducts in heat-resistant material. First flown in the Hunting H.126 Jet Flap research aircraft in Britain in 1963, it is now flying in more advanced form in the National Aeronautics and Space Administration (NASA) 'augmentor wing' de Havilland Canada, DHC–5 Buffalo, with complex flap sections through which pass four sets of engine-supplied or entrained airflow. The entire fan airflow from this aircraft's Spey engines is ducted through large pipes to feed the flaps, blown ailerons and BLC on the fuselage, to reduce drag as well as increase lift. This aircraft can fly controllably at just over 75km/h (47mph).

Upper Surface Blowing

In the US Air Force Advanced Medium Short Take-Off and Landing (STOL) Transport (AMST) programme, two further ways of increasing the lift of a large transport have been investigated. The less advanced of these methods is the externally blown flap, explored on the McDonnell Douglas YC–15. In most aircraft with engine pods on the wings the flaps are deliberately omitted directly behind the engines to avoid the hot blast and buffet of the jet. In contrast, the YC–15 has extremely powerful two-segment flaps extended out and down directly in the wake of the four

engines, which are mounted close beneath the wing to blow right into the flaps.

The Boeing YC–14, on the other hand, has technically more difficult Upper-Surface Blowing (USB). This relies on the Coanda effect, i.e. the property of a fluid flow to cling to a surface even if that surface arches away. The Coanda effect can be demonstrated visibly by holding a bottle horizontally under a gently running tap; the water does not flow off the sides of the bottle but clings to the lower part and runs off at the lowest point. In the same way the YC–14 has two powerful CF6–50 turbofans mounted close above the wing roots. Their gigantic efflux, squashed against the wing by flattened nozzles and guided outwards by vortex generators, passes over inboard flap sections that can be depressed to 75°. The Coanda effect is so powerful that the engine jets likewise turn through the same angle, point almost straight down and thus give lift almost equal to the combined thrust of the engines. Special control systems automatically rearrange the flaps if either engine fails. A four-engined rebuilt de Havilland Canada Buffalo is now exploring USB for civil use.

Less drag

From early in the era of aeroplane flight, aviators have sought to reduce drag (which can loosely be translated as 'air resistance'). The fundamental method of achieving this objective is by streamlining, or making all parts of aircraft the shape of a streamline, the

path followed by smoothly flowing particles of air. In a large measure, drag reduction depended on structure, and real advances were made only with the adoption of the stressed-skin all-metal airframe. Smooth metal skin, flush rivets, carefully streamlined cockpits (preferably with canopies) and retractable landing gear all made major contributions. So did improved engine installations. Many arguments had revolved around the supposed lower drag of the water-cooled in-line engine compared with the bluff-fronted air-cooled radial, but these arguments were usually based on faulty reasoning which in turn rested on poor installations.

Retractable landing gears were sketched before 1914, but flown systems date from 1919 and production arrangements from 1931. In most cases the United States took the lead, and in early schemes the lower part of the main tyres projected to offer something on which to land in the event of failure (by the pilot or the system) to extend the gear before alighting. Early schemes were often complicated and heavy, and usually required arduous hand-cranking to operate. Not until 1935 did elegant geometries appear, the neatest fighter main gears folding inwards or outwards to be housed in the wing in front of or behind the wing front spar.

Reducing thickness/chord ratio

The coming of jets demanded a speeding-up of the process of reducing wing t/c ratio which had begun in the early 1930s. During the 1930s fighter aircraft were designed on the

LEFT: *The increase in aircraft speeds after World War II demanded a parallel reduction in wing thickness/chord (t/c) ratio. In fighters, such as the Hawker Hunter illustrated, t/c ratio fell to about 8 per cent compared with the 20 per cent common in the 1930s.*
RIGHT: *The t/c ratio fell even further in supersonic fighters. In the Dassault Mirage III it was reduced to about 3·5 per cent. Although the thin wing helped to reduce drag it correspondingly gave poor lift at low speeds, and consequently the aircraft had to take off and land at very high speeds.*
BELOW: *One of the most bizarre aeroplanes ever built, the AD-1 oblique-wing research aircraft tested the oblique-wing concept at the US National Aeronautics and Space Administration (NASA) Dryden Flight Research Center in California. A small twin-jet, its wing is a single unit pivoted at its mid-point, positioned at 90° to the fuselage at low speeds but rotated through 60° for high-speed flight.*
BELOW RIGHT: *In the North American P–51 Mustang the radiator was placed under the real fuselage and the duct profiled to give positive thrust.*
BOTTOM RIGHT: *This Curtiss Cox racer of 1921 was a glaring example of clumsy engine installation with the radiators projecting on each side. Two years later Curtiss streamlined his engines but could still do little about the radiators.*

assumption that it had to have a t/c ratio of about 20 per cent (i.e., the chord was five times the maximum thickness) and a bomber one of 25 per cent. As speeds and engine powers increased these figures fell. After World War II the laminar aerofoils became universal for fast aircraft, and t/c ratio fell to about 8 per cent in aircraft such as the Lockheed Shooting Star and Hawker Hunter, to 5 per cent in supersonic fighters and to a remarkable region between 3·3 and 3·5 per cent in the Convair B–58, Lockheed F–104 and Dassault Mirage III. These extremely thin wings helped early Mach 2 aircraft to keep drag lower than available engine thrust but had the drawback of giving poor lift at low speeds, so all three aircraft took off and landed at exceptionally high speeds.

The area rule

Richard T. Whitcomb postulated in 1953 a general rule for shapes with minimum transonic drag. Called the area rule, this states that the total cross-sectional area of an aeroplane intended to fly faster than sound should increase uniformly from nose to tail. In other words, a projection such as a canopy, wing, fin, drop tank or other excrescence should be compensated for by reduction elsewhere. Some early area-ruled aircraft, such as the F–102 (which thereafter went supersonic) and Buccaneer, showed obviously waisted mid-fuselages and bulged extremities, but modern supersonic aircraft obey the rule in subtler ways. Supersonic aircraft also tend to have shallow cockpit canopies, vee windscreens and modest taper angles on noses and other portions. Their overall fineness ratio, a measure of slenderness broadly equivalent to length divided by body diameter, has to be high. It is for this reason that passengers cannot sit ten-abreast in a Concorde as they can in subsonic aircraft. Yet some aircraft intended to lead to hypersonic (many times faster than sound) aerospace aircraft, able to fly in the atmosphere or in space, are very bluff and fat. Examples are the M2–F2 and HL–10 'lifting bodies', and the Space Shuttle and Hermes orbiter vehicles.

Laminar flow

All aircraft, supersonic or not, are today made as accurately as possible, with the smoothest attainable skin, to reduce the drag in the boundary layer due to skin friction. The laminar-flow wing helps because the airflow accelerates over the front part of the wing to the thickest part, and in an accelerating flow it is easier to keep the boundary layer laminar, or smooth. But over the rear part of the wing the flow slows down again, and the boundary layer then tends to break away from the surface and become turbulent, giving higher drag. This effect occurs wherever the flow slows down relative to the aircraft, and it was to avoid this that designers of early low-wing monoplanes often filled in the junction between the rear part of the wing and the fuselage with large fillets.

Propulsion

Early engine installations were crude. Petrol and water-cooling pipes had to be free from leaks, and the propeller was carved to the correct shape from multiple laminations of hardwood, but streamlining was conspicuously absent. Curtiss caused a stir with his streamlined D–12 engine installation of 1923, but even then the water radiator was merely a relative of a car radiator stuck in the airstream. By 1928 radiators were beginning to be enclosed in shaped ducts, which eventually had the form of a diffuzer – a duct fatter in the middle, to slow down the airflow – so that the air lingered as it passed the hot radiator surfaces and then accelerated back through the exit to the atmosphere. In many fighters there was no duct, but the radiator was mounted on vertical slides, with rubber water hoses, and the pilot could crank it out into the airstream just far enough to keep the temperatures within limits. But the ducted

radiator was eventually recognized to be essentially a propulsion system; by adding heat to an airflow its energy is increased, and by 1941 the radiator of the North American P–51 Mustang, under the rear fuselage, and the radiators of the de Havilland Mosquito, inside the inner wing leading edges, were giving not drag but positive thrust. By this time the coolant was not water but ethylene glycol, and the cooling circuit operated under pressure, with thermostatic control to adjust the position of the air exit flap(s).

Radial engines

Air-cooled radial engines never suffered the penalty of water circuits, but it took many years to discover how to instal one correctly. The first major step forward was the addition of the Townend ring in 1928. This American development was a circular aerofoil closely surrounding the engine and so shaped that it helped to pull the aircraft along. Throughout the 1930s, as engine designers added more

and deeper fins to the engine cylinders, the installation designers fitted longer and closer-fitting cowlings, usually with hinged flaps or sliding shutters at the rear called gills (cowl flaps in North America) to control the cooling airflow. These gills had to be partly or wholly open for taxiing and take-off but closed or under thermostatic control in cruising flight, to reduce drag and avoid tail buffet.

Getting rid of the exhaust was accomplished in the early days by having plain stacks or a long pipe. During the 1920s, radial engines began to gather the exhaust into a manifold, either around the front crankcase or behind the cylinders. But the installation that was a major advance was that of the BMW 801 in the Focke-Wulf Fw 190, and in this the individual stacks were taken to ejector exhausts projecting through the fuselage behind the engine. This big engine was closely cowled, and to keep it cool at low speeds a fan was provided with blades filling the gap between spinner and cowl,

ceeded that of engines in the pre-jet era. By 1930 many propellers had blades separate from the hub and individually replaceable. This made repairs easier and allowed the three-blade propeller (difficult with wood) to become common. But though these detachable blades could often be set in the hub at any of a range of angles, the far bigger advance was a propeller able to change its pitch in flight. For take-off and landing a fine pitch is needed. Like a nut on a fine-pitch screw-thread, this needs many turns and makes slow progress, so the engine can run at full throttle (or, for landing, can still run at high speed when throttled back) despite the low airspeed. In cruising flight the pitch needs to be much coarser, to allow a high airspeed to be matched with low engine speed and low fuel consumption. In a dive the pitch needs to be coarser still, while following engine failure the propeller blades should ideally be turned edge-on so that the engine is 'feathered' and does not rotate or cause much drag.

These advances came in stages. The first

(V/STOL) aircraft were using light-weight blades of glass-fibre, and today such blades are stiffened by carbon fibre. By 1944 propellers began to have an extra pitch setting: reverse pitch, giving backwards thrust when the throttle was opened to help slow aircraft after landing. Another feature of the modern propeller is beta control, by which the pilot can adjust pitch of each propeller directly to help in ground manoeuvring. Yet another device is a multi-engine synchronizer to cut out unpleasant aural 'beats' when several engines are running at slightly different speeds. Today the emphasis is strongly on increasing the size of propellers and reducing their rotational speed, to give a dramatic reduction in noise.

Modern commuter airliners enjoy quiet, and efficient propulsion with such propellers, typically with six lightweight blades of carbon and glass fibre. But their efficiency is limited to speeds of about 560km/h (350mph) and below. In 1973 the price of fuel began to rise to unheard-of levels, and in the USA NASA, working with Hamilton Standard, investigated how far advanced propellers could operate at jet speeds to cut fuel bills. HamStan registered the name Prop-Fan, but today the word propfan is used loosely for all kinds of advanced propeller.

By the end of the 1980s it was clear that fuel bills, and in theory the size of aircraft needed for a given transport task, could be reduced by either using a propfan or by greatly increasing the BPR (bypass ratio) of turbofan engines. The two merge together, so that the high-BPR turbofan is really a propfan engine with a surrounding duct. All propfans are characterised by having eight or more blades of thin, sharp-edged, scimitar-like form. Of course, they have variable pitch, and can feather or reverse. For best fuel efficiency two contra-rotating propfans are used. To reduce noise, interaction with the first-stage tip vortices and other reasons the rear fan is usually smaller and has a different number of blades.

A pioneer in this field is General Electric, which downstream of a fuel-efficient core engine has added two sets of free intermeshing turbines (eliminating the need for stators) carrying the propfan blades on their outer peripheries. In contrast, Allison and Pratt & Whitney have chosen to add a downstream gearbox to contra-rotating pusher propfans. Rolls-Royce has refined this concept, putting the gearbox in the propfan hubs and ducting the hot core gas through the pusher spinner, instead of letting it play on the blade roots. Others have gone for the tractor configuration, with or without a surrounding duct.

The first (GE36) UDF (UnDucted Fan) was flown on a 727 on 20 August 1986, and a refined version was flown on an MD-80 from 18 May 1987. A tractor propfan mounted on the wing of a Gulfstream II first flew on 29 April 1987. It is still too early to predict the timing for propfans. Boeing launched its 7J7 airliner with the promise of service in 1992, but then decided to delay by at least a year. Meanwhile, the A320, the most advanced aircraft in this class with conventional engines, continues to sell strongly.

Propellers can often be made to give reverse thrust, but for the turbojet a new device

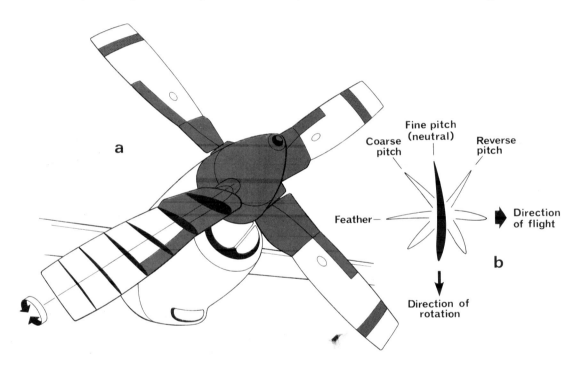

ABOVE: *The propeller of the Lockheed C–130 Hercules (a) showing blade sections and twist. Schematic view of pitch change and blade rotation (b).*
TOP RIGHT: *A view of the wooden paddle propellers of a Santos Dumont aircraft of 1907.*
ABOVE RIGHT: *This Piper Navajo of the Spanish Air Force is typical of modern light planes, with constant-speed anti-iced propellers.*
RIGHT: *The inlet system of the Concorde: (a) take-off (b) noise-abatement climb (c) supersonic cruise (d) reverse thrust.*

driven at 3·14 times the crankshaft speed.

The Germans also led the field in installing all their aircraft engines in a uniform manner, with common mountings and piping suitable for either liquid-cooled inverted-vee or air-cooled radial engines.

Propellers

If anything, development of propellers ex-

mass-produced variable-pitch (v-p) propeller was the Hamilton of 1926, first sold in numbers in 1933. Later, by 1936, the Hydromatic propeller came into use, with blades positioned by bevel gears turned by a drum-cam. The cam was rotated via curved slots and rollers by a hydraulic piston in the hub. Unlike the bracket-type propeller this piston was double-acting, and the blades could be set to any chosen angle. When a Constant-Speed Unit (CSU) was added to the engine, the propeller could be automatically controlled to maintain any desired engine speed, being called a constant-speed propeller.

There were many other types of propeller, some having hubs containing electric motors and gearboxes. Blades were made of aluminium 'densified wood' (multiple veneers bonded under very high pressure, with a tough plastic casing) or welded sheets of stainless steel with a hollow interior. By the 1960s, Vertical/Short Take-Off and Landing

had to be added: the thrust reverser. This matured in 1959 in two forms. One had upper and lower internal clamshells to seal off the nozzle and simultaneously open upper and lower cascade apertures (holes bridged by multiple deflector aerofoils), which ejected the jet in a diagonally forwards direction. The other had external upper and lower clamshells which either formed the skin of the engine nacelle or blocked the jet downstream of the nozzle and diverted it diagonally forwards.

Noise abatement

In recent years the increasing wish to avoid the nuisance of aircraft noise has been reflected in palliatives for existing propulsion systems and changed design features for new ones. The fundamental advance has been the high by-pass ratio fan engine, which makes today's 'Jumbo' of 362,875kg (800,000lb) gross weight many times quieter than earlier jets weighing less than one-tenth as much. Those older aircraft have been quietened by various noise-suppression methods. One is to line the walls of the inlet duct and nozzle with hollow sandwich material, one form being light-alloy honeycomb with a hole on the inner skin in the centre of each honeycomb cell; sound energy is constantly attenuated (used up) in each of the thousands of cells. Another palliative is the noise-suppressing nozzle, which typically is a multi-lobe shape rather like a flower or star with rounded rays. The long wavy boundary of the nozzle speeds mixing of the hot gas with the cool atmosphere and cuts down the violent shearing action which causes most of the noise in a high-velocity jet. Modern engines are quieter because designers select the best numbers of fan, turbine and compressor blades, adjust the axial spacings and use quiet-running gearwheels.

The variable nozzles needed for afterburners are discussed in the Engines chapter. Supersonic aircraft also need variable air inlets, though sometimes, as in the General Dynamics F–16, such complication is omitted to save money. A fixed inlet begins to lose efficiency at Mach numbers beyond unity, and is seldom seen in an aircraft capable of more than Mach 1·3.

The ideal solution is to have an inlet which at take-off is opened wide, to swallow all the air it can get, but which progressively closes down as altitude and Mach number are gained, until in supersonic flight there is a large wedge protruding into, and almost closing off, the inlet duct. In addition there must be various auxiliary inlet doors, dump and spill doors and even doors through which air is sometimes sucked in and sometimes blown out. At all times there is a large airflow round the engine inside its nacelle, and in supersonic flight this flow probably becomes a kind of hurricane, passing out round the nozzle which acts as a powerful ejector, or air pump.

Propulsion control system

There are many other facets of propulsion systems. One which deserves discussion is the control system. In the earliest aero engines this comprised a single electrical switch which

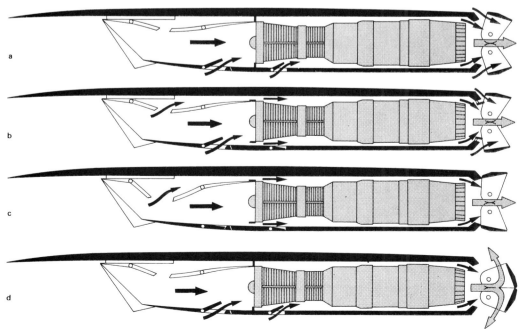

opened or closed the primary ignition circuit; a pilot had to adjust his landing approach by switching the engine off and then on again (it was known as 'blipping'). By 1920 virtually every engine had a throttle, like the accelerator of a car, and during the 1930s such extra features were added as automatic mix-ture control (to adjust the fuel : air ratio sup-plied to the cylinders), automatic compensa-tion for altitude and automatic boost control in supercharged engines to secure maximum power without the risk of overstressing the engine in denser air. Generally similar de-vices controlled early gas turbines, but gradu-ally the engine became more demanding, and supersonic powerplants, with variable inlets, variable nozzles, extensive air-bleed systems and other ancillary services, required a com-puter. Today, the most advanced aircraft have digital electronic control systems which constantly assess from hundreds to thousands

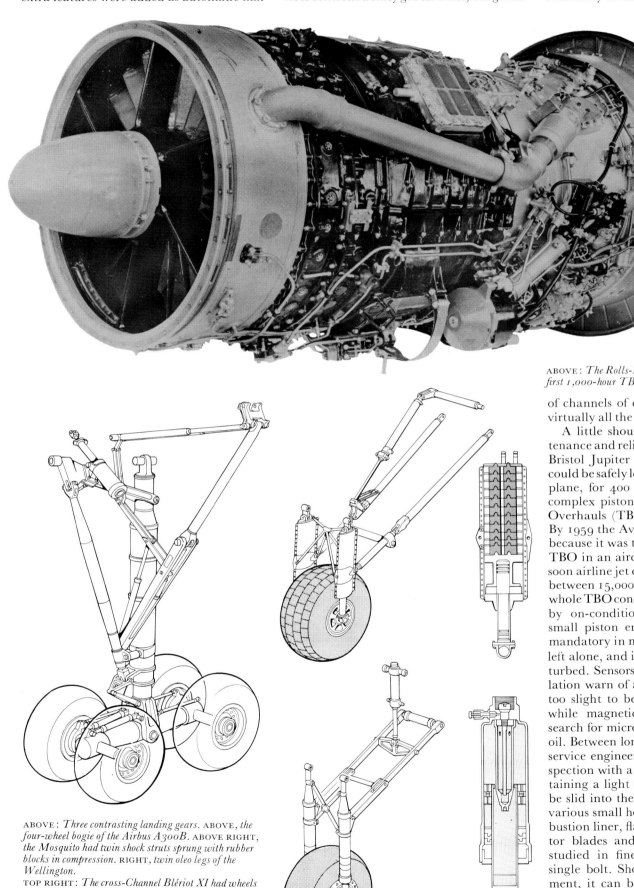

ABOVE: *The Rolls-Royce Avon RA.29/6 typifies the first 1,000-hour TBO civil jet engine.*

ABOVE: *Three contrasting landing gears.* ABOVE, *the four-wheel bogie of the Airbus A300B.* ABOVE RIGHT, *the Mosquito had twin shock struts sprung with rubber blocks in compression.* RIGHT, *twin oleo legs of the Wellington.*
TOP RIGHT: *The cross-Channel Blériot XI had wheels sprung by rubber belts in tension.*
CENTRE RIGHT: *The Hawker Fury fighters for Yugoslavia had Dowty internally sprung wheels.*
RIGHT: *The 28 wheels of the Lockheed C-5A are used to spread the load on unpaved airstrips.*

of channels of data and automatically take virtually all the work-load off the pilot.

A little should also be said about main-tenance and reliability. In the early 1930s the Bristol Jupiter gained headlines because it could be safely left to run, installed in an aero-plane, for 400 hours. By 1950 much more complex piston engines had Time Between Overhauls (TBOs) of 1,800 or 2,000 hours. By 1959 the Avon turbojet gained headlines because it was the first gas turbine to have a TBO in an aircraft of 1,000 hours, but very soon airline jet engines were achieving TBOs between 15,000 and 18,000 hours. Today the whole TBO concept has largely been replaced by on-condition maintenance, except for small piston engines, for which TBOs are mandatory in many countries. The engine is left alone, and if it works properly is not dis-turbed. Sensors built into the engine instal-lation warn of any vibration – even though too slight to be felt directly by the crew – while magnetic chip detectors constantly search for microscopic metal particles in the oil. Between long flights, or perhaps daily, a service engineer gives the engine a keen in-spection with a borescope – a thin tube con-taining a light source and optics which can be slid into the heart of the engine through various small holes – to allow the vital com-bustion liner, flame tubes, turbine rotor, sta-tor blades and other parts to be visually studied in fine detail without undoing a single bolt. Should anything need replace-ment, it can be done in the latest engines without taking the engine out of the aircraft; thanks to 'modular' construction, the rele-vant portion can be exchanged while leaving the other modules in place.

Landing and take-off

From the start of aeroplane flight it was clear that some form of sprung landing gear was needed. In early aircraft the springing was usually by bungee (rubber cord), but this had the drawback of bouncing badly. No proper answer was forthcoming until, around 1930, designers began to accept the cost penalty of landing-gear legs designed by engineers and made with precision. By far the most common was the oleo leg, and this is still the case for all except lightplanes. The oleo uses sliding pistons or cylinders, various fluids such as air, nitrogen, oil or special liquids with very large molecules (to make them more compressible), and valves to allow the leg to collapse readily without bouncing back.

When main gears were made retractable, most European countries chose single legs for fighters and lightplanes and twin-leg gears for bombers, the latter having one leg resting on each end of the wheel axle and folded either to the front or rear to lie in the nacelles of the engines. American aircraft, and the German Junkers Ju 88 normally had single legs, those of the Ju 88 being unusual in being sprung by stacks of chamfered rings of springy steel. Another unusual shock strut (leg) was that of the de Havilland Mosquito, which was sprung by a stack of rubber blocks. As aircraft weight grew, so did the number of tyres needed to spread the load on airfield surfaces. The Focke-Wulf Fw 200 Condor had twin

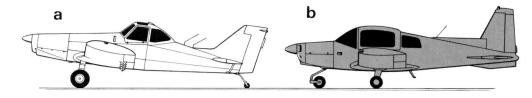

ABOVE: *From 1916 until about 1945 most aircraft had tailwheel-type landing gears, such as that of the Piper Pawnee agricultural machine (a). Today almost all have nosewheels, such as on the Grumman American Traveler (b).*
BELOW: *Lockheed F–104 Starfighter landing gears support 13,600kg (30,000lb) at over 320km/h (200mph) and occupy little space when retracted. Springing is by small 'Liquid Spring' shock struts pivoted near the tops of the hinged legs.*

RIGHT: *In the 1920s large, heavy aircraft had extremely large wheels, with vast low-pressure tyres. This Armstrong Whitworth Ensign typifies such gears. Today there would be numerous, much smaller wheels.*
BOTTOM: *In 1950 Boeing designed the B–52 heavy bomber to have unique landing gears with four two-wheel trucks, all on the fuselage, and small outrigger wheels under the outer wings. The main trucks can steer so that in cross-wind landings the B–52 can 'crab' along diagonally.*

wheels on complicated legs, the Messerschmitt Me 323 and Arado Ar 232 freighters had numerous small wheels along the fuselage, but the heaviest aircraft of World War II, the Boeing B–29, had single legs each supported by two tyres. Since 1945 there have been numerous landing-gear configurations, with an especially diverse number of schemes on V/STOL aircraft.

Wheels, tyres and brakes
In the earliest aircraft, wheels and tyres resembled those of cycles, but today the typical wheel is a squat drum of magnesium alloy, with the whole interior occupied by the brake. It is fitted with a special tyre with multiple plies of rubberized fabric inflated to a pressure which may be as high as 21·1 kg/cm² (300lb/sq in) – or ten times the pressure of most car tyres. Aircraft designed to operate from unpaved surfaces are usually light; if they are heavy they must have 'high-flotation' landing gear with multiple low-pressure tyres. Many aircraft take off and land at speeds as high as 322km/h (200mph) or more, and unless the runway surface is

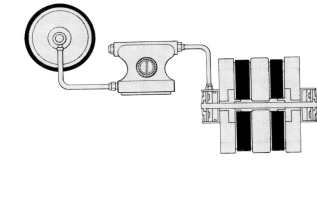

RIGHT: *One of the most widely used anti-skid brake systems is the Dunlop Maxaret. A small wheel spins at high speed, driven by its rubber rim in contact with the main wheel. If the latter locks, the small wheel stops rotating, and via special valves instantly eases pressure on the brakes, thus restoring rotation.*

BELOW: *Heaviest combat aircraft of World War II, the Boeing B–29 bomber set the fashion in having two wheels on each main and nose leg. Exactly the same arrangement is still seen on numerous aircraft. Another pioneer B–29 feature, now almost universal, was steerable nose landing gear.*

BELOW RIGHT: *The Soviet Sukhoi Su–7BM is one of many modern tactical aircraft designed to operate from unpaved airstrips under battle conditions. It has large low-pressure tyres, a braking-parachute installation and optional ATO (assisted take-off) rockets, which can be clipped on when needed and then jettisoned.*

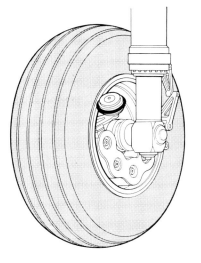

clean and dry this causes problems. Even a 13mm (0·5in) layer of water can make the aircraft 'aquaplane', skating across the water out of contact with the runway, and tyres have to be designed as high-capacity water pumps to eject water to each side. Many tyres have side ridges forming spray deflectors to prevent water in bulk from hitting the flaps or other areas at high speed. The tread may be plain, ribbed or covered with 'dimple' depressions to give an instant indication of wear.

Brakes in aircraft have to absorb hundreds to thousands of times more kinetic energy than those in cars or heavy trucks. The Rejected Take-Off (RTO), perhaps following sudden engine failure, of a Concorde involves stopping almost 203,000kg (200 tons) moving at over 322km/h (200mph), and to certificate such aircraft this must be demonstrated using wheel brakes alone.

Early aircraft brakes were simple friction bands, often pulled tight by a cockpit lever and cables rather like early car brakes. By 1940 hydraulic actuation was usual, with differential operation (more on one side of the aircraft and less on the other) to assist

turning, or to resist 'swing', the unwanted and unstable tendency to turn often experienced with tailwheel-type landing gear. If not smartly nipped in the bud, this caused a 'group loop' which could break the landing gear. Today virtually all aircraft have brakes controlled from foot pedals pressed down by the toes. Some landing gears still have pneumatic or hydraulic pressure applied to brake shoes pressed against the inside of the wheel drum, but nearly all fast or large aircraft use disc or multi-disc brakes in which the pressure pads squeeze one or more discs keyed to revolve with the wheel.

The problem of dissipating millions of foot-pounds of kinetic energy in the small space of an aircraft wheel is difficult. Brake discs can soon become bright cherry-red with heat, and fan cooling is needed. In extreme cases the high temperature could make the tyre explode, so the wheel incorporates fuzible plugs which, after a severe emergency stop, can melt and let the air out safely. All heavy or fast aircraft have anti-skid systems which allow the pilot to 'stamp on the brakes', even on an icy runway, without fear of skidding.

These systems exactly balance brake pressure against the speed of rotation of the wheel, relaxing pressure the moment the wheel tends to lock. The latest brake discs or rings are of special material able to absorb large amounts of heat and stay strong at high temperature. On the Lockheed C–5A they are of beryllium, and on the Concorde of reinforced carbon.

RATOG and JATO

Sometimes special equipment is needed to boost take-off or halt the aircraft after landing. Rocket-Assisted Take-Off Gear (RATOG) or Jet-Assisted Take-Off (JATO) was pioneered in the Soviet Union in 1934 and then taken up afresh in Germany, Britain and the USA ten years later. Solid-fuel motors (rarely, pressurized water or liquid-propellants) are either built into the aircraft or clipped on externally, with canted thrust axes passing through the aircraft centre of gravity to avoid any tendency to pitch the nose up or down. Sometimes the empty bottles are retained on board, but usually they jettison themselves as their short-time thrust decays. Extreme examples of boosted take-off were the experiments in zero-length (i.e., no run) take-off with supersonic fighters between 1956 and 1959, in which single enormous rockets thrust the aircraft diagonally off a missile-like launcher which could be inside a bombproof shelter.

Arrester systems began with various arrangements of wires on the decks of aircraft carriers; this is recorded on page 213, where the catapults used to launch aircraft from carriers and other vessels are also discussed. By 1955 arrester gear was being installed at military airfields. Alternatively, a safety net can be installed across the upwind end of the runway to arrest aircraft which for any reason overshoot. These nets usually incorporate numerous vertical strips of nylon webbing strong enough to pull the aircraft back without causing major airframe damage. Civil airports seldom use nets, but some have large overrun areas filled with particulate material such as Pulverized Fuel Ash (PFA), which is preferable to gravel because it does not dam-

increase in drag than the downgoing aileron, causing yaw opposite to that needed for a correct turn. One answer was to connect up the ailerons so that the upgoing one went up a lot while the other went down only a little, balancing out drag on each side. Another was the Frise aileron of 1926 which was hinged well back from the leading edge so that, when deflected up, the leading edge projected into the slipstream under the wing to cause more drag.

As aircraft became larger and faster, the forces on control surfaces naturally increased, until simple hinged surfaces became difficult to move. Well before World War I, designers had often eased the load by putting some of the area at the end of a surface ahead of the hinge axis. This was called a horn balance, and is still sometimes used, although a different form of aerodynamic balance is now much more common. The hinges are set back from the leading edge, as in the Frise aileron, so that the force on the leading part of the surface partly offsets that aft of the hinge.

The first tabs had actually been for trimming, the task of balancing out the masses and aerodynamic forces all over the aircraft so that the pilot could take his hands off the controls. Sometimes lengths of cord would be fixed along one side of the trailing edge of an aileron or other surface to get the trim right, while the elevators might be trimmed by an adjustable spring fixed to the bottom of the control column. Then, during World War I, it became common to fix tabs of aluminium to the trailing edge; these trim tabs were simply bent until the force of the tab on the control surface achieved the required perfectly trimmed condition.

By 1933, control surfaces were beginning to be fitted with hinged trim tabs, a much better idea because they could be driven through irreversible gears from handwheels in the cockpit. This allowed the pilot to retrim the aircraft in the air, greatly easing strain on long flights because previously he might have had to apply a large control force all the time to maintain straight and level flight.

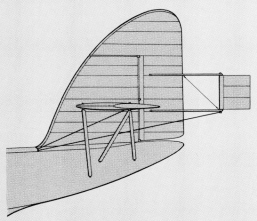

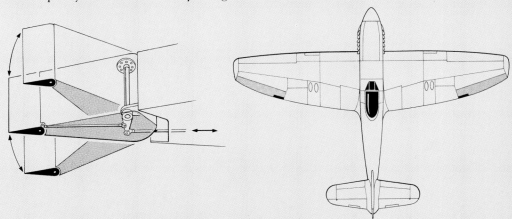

ABOVE: *The Short Calcutta flying boat (p.240) had a primitive servo tab to operate the rudder.*

ABOVE AND RIGHT: *Spring tabs were fitted to the Hawker Tempest fighter of 1944. These are seen as solid black on the ailerons in the plan view.*

age the aircraft. Many aircraft also have braking parachutes, or drag chutes, which can be released from an internal box on pilot command. The parachute increases drag and thus slows down the aircraft. When the speed has been reduced sufficiently the chute(s) can be jettisoned for repacking and re-use.

Flight control

The difficulty of controlling a flying machine posed more of a barrier to early aviation than lack of an engine. After the Wrights had shown how aeroplanes could be controlled laterally, by warping the wings, the aileron was resurrected and by 1912 almost all aeroplanes used the arrangement of aerodynamic surfaces most often seen today. For stability there are fixed surfaces at the back, like the tail on a dart. For control there are movable surfaces: rudder for yaw, ailerons for roll and elevators for pitch. In modern aircraft this arrangement is often modified, as described below.

From early days until after World War II, flight-control systems were worked manually, operated by the muscles of the pilot (sometimes two strong pilots working together). Problems were simple. One was that, when the aircraft was rolling, the upgoing aileron (on the downgoing wing) suffered a larger

It was also found necessary to use mass balance. As each control surface had weight, it could be deflected by sharply moving the part of the airframe to which it was hinged. Air turbulence and other disturbances could thus cause some surfaces to flutter, oscillating so violently that they could tear off or break the aircraft. One cure was mass balance. At first, streamlined weights were fixed on diagonal arms projecting ahead of the hinge axis but today mass balance is invariably internal, either at the leading edge (well ahead of the hinge axis) or on projecting arms which are normally flush with the fixed surface. To get more mass into a small space, patented heavy alloys or depleted uranium are used to make the balance weights.

Tabs

By the 1920s the biggest aircraft were appearing with 'servo surfaces' carried on arms behind control surfaces, notably the rudder. These auxiliary surfaces were small and easily moved by the pilot. The air load acting on them then moved the main surface. By 1933 this clumsy arrangement was being replaced by neat tabs carried directly on the control-surface trailing edge. As before, the pilot merely drove the servo tab(s) which in turn did the hard work in moving the main surface(s).

Thus many aircraft in World War II had control surfaces with both servo tabs to ease manoeuvres and trim tabs to enable the aircraft to fly 'hands off'. Alternatives to the servo tab also appeared. One was the balance tab, or geared tab. Here the tab was linked by pivoted arms to the fixed surface, and the pilot drove the main control surface. When he moved the main surface, the tab was automatically deflected to help him, and the gearing of the linkage was arranged to leave a pleasant residual control load. Too powerful a tab could lead to instability and overcontrol, and in some aircraft the controls were so light that anti-balance tabs had to be added, like geared tabs but moving in the opposite direction to add to the pilot's load.

By 1942, fighters were being fitted with a further refinement, the spring tab. Here the pilot drove both a servo tab and, via a springy link (such as a torsion rod), the main surface. This had the advantage of giving progressive assistance at higher speeds, because then the main surface was more difficult to move and the greater deflection of the spring-link resulted in greater movement of the tab. Attention was also paid to the leading edge, which was precisely matched with the fixed structure and was sometimes extended in a shallow beak or hinged flap moving inside close-fitting balance chambers.

ABOVE: *Zero-length launch by rocket of an F-104G Starfighter.*
RIGHT AND BELOW: *Cessna lightplane control surfaces, typical of those in most fixed-wing aircraft: (a) aileron (b) elevator (c) rudder (d) roll to the left, prior to turning left (e) roll to the right prior to turning right (f) nose-up pitch, to climb or if sustained to loop (g) nose-down pitch, to dive (h) yaw to the left, with 'left rudder' (i) yaw to the right.*

107

Autopilots

Back in the 1920s several pioneers had improved the automatic pilots (autopilots) which had been the subject of experiments since 1912. The objective was merely to keep aircraft on a straight and level course, irrespective of its trimming or possible upsets by air turbulence. What was needed was a gyroscope to sense rotation of the aircraft about each axis in which the autopilot exercised authority, plus a method of linking the sensed rotation to the flight-control system. Thus, should the aircraft begin to dive, the nose-down rotation would be detected and measured (almost always electrically, but sometimes pneumatically) and the resulting signal fed to a servo power unit capable of bodily moving the flying controls (in this case the elevators) to restore level flight. By 1939 autopilots were refined products available from specialist suppliers. They could command pitch, yaw and roll movements, not only to correct unwanted disturbances but also to make the aircraft do desired manoeuvres such as a correctly banked turn.

During and after World War II, autopilots acquired additional capabilities. They could be commanded to fly the aircraft to a particular compass heading, or to hold a particular airspeed or Mach number or fly to a chosen altitude and then hold it indefinitely, air turbulence notwithstanding.

Any installed autopilot system clearly had to have some carefully determined authority over the flight-control system. Its output signals were linked to a mechanical system capable of bodily moving the control surfaces, with or without causing corresponding movement of the cockpit flying controls. This mechanical system was sometimes electric, sometimes hydraulic and sometimes pneumatic, and could take the form of direct servo jacks connected either to whichever surfaces the autopilot had authority over or to the interconnecting linkage between the surfaces and the cockpit.

Establishing the authority was difficult. Suppose, for instance, that the autopilot were to encounter very turbulent air. To maintain straight and level flight at a constant altitude it clearly had to exert an immediate and powerful output, beyond the capability of a human pilot. Or suppose the autopilot were to go wrong, and issue a command to the flight controls to go 'hard over' in a way which could cause an unwanted manoeuvre violent enough to break the aircraft. The answer was to incorporate automatic protection circuits against incorrect outputs, or to enable the pilot to overcome such outputs, and to provide for almost immediate autopilot disengagement. Eventually autopilots became much more complex to enable bad-weather landing.

As the autopilot incorporated means for moving the flight-control surfaces independently of the pilot, it was a short step to adding mechanical boosters to help a pilot move the controls of extremely large aircraft. Early powered controls during World War II were cumbersome, fault-prone and needed extensive protection to limit their authority and prevent the aircraft being broken in flight or flown into the ground. Eventually, reliable

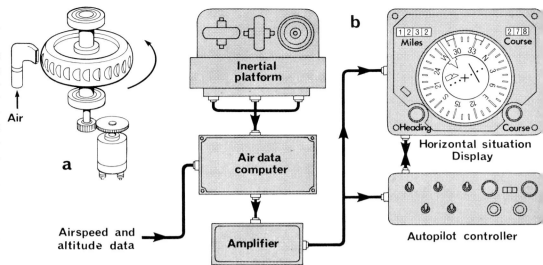

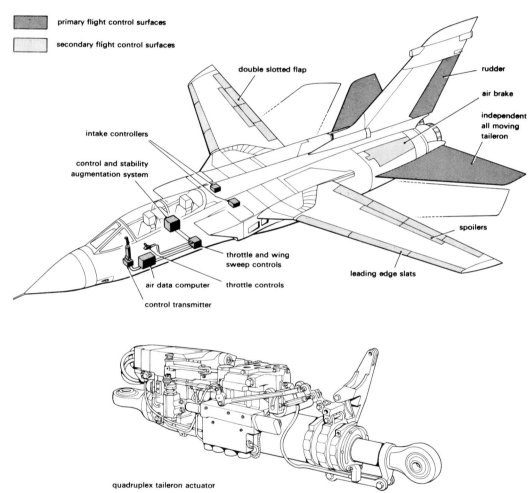

TOP: *A simplified inertial system showing a schematic air-driven gyroscope (a) with a geared synchro pick-off (electrically driven gyros are used today). In the centre (b) is the inertial platform with three gyros arranged with axes mutually perpendicular, feeding acceleration data to an air-data computer and amplifier which in turn drives the Horizontal Situation Indicator (HSI) and, via a controller, the automatic pilot. The HSI gives a readout of course, track, deviation from track and distance to destination or next waypoint. Automatic pilots have been the subject of experiments since the early days of aviation. However, the first successful system was developed by Elmer Sperry and used in a Curtiss flying boat in 1912.*

ABOVE CENTRE: *Panavia Tornado flight-control system, with dotted lines showing the position of the fully swept wing in which the trailing-edge flaps cannot be used and the spoilers are used to act as airbrakes. At all speeds the left and right tailerons (tailplanes or horizontal stabilizers) are the primary controls for pitch and roll, the spoilers being used differentially to increase roll control at low speeds. At low speeds, with wings unswept, the flaps and leading-edge slats increase wing lift greatly at high angles of attack for short-field (STOL) operation.*

ABOVE: *Tornado taileron power unit.*

RIGHT: *De Havilland Canada 'Dash 7' STOL transport, showing T-tail adopted to keep the tailplane in clear air at high attack angles.*

servo-boosters were devised with a mechanical pilot-input and a hydraulic output to the surface(s). The boosters were not irreversible, and could still be overcome by either the surface or the pilot; the latter still supplied part of the force needed, and the surface could flutter if not properly mass-balanced. But by 1955, despite such achievements as Boeing's all-manual 707 and Bristol's all-servo-tabbed Britannia, most large or fast aircraft had a completely new kind of system, the fully powered Flight-Control System (FCS).

Powered flight control system

In this the pilot's cockpit flight controls are merely input devices like the tuning knob on a radio. When the pilot moves any of them he commands a complex system to spring instantly to life and, via many limiters, checks and balances, causes Powered Flying-Control Units (PFCUs) to move the correct

need autostabilizers to damp out oscillations, caused by transonic or supersonic flight through rough air, at a frequency too rapid for an autopilot (or a human pilot) to handle. Autostabilizers usually act on the rudder and often on the horizontal tail surface, but seldom on the ailerons.

Some powered systems make provision for manual reversion in the event of failure, though the pilot has a hard task and the aircraft may be limited in permissible airspeed. Many modern systems make no such provision, so the powered system must be essentially foolproof. In a combat aircraft, the autopilot will usually be intimately linked with a radar and other sensors, weapons and their control/aiming systems, an air-data system (to measure with great accuracy the properties of the surrounding atmosphere), navigation systems, an angle-of-attack sensor and other devices, all tied into a powerful digital computer working in real time to manage the

fixed tailplane and no separate elevators. In lightplanes they are manually driven, but in supersonic aircraft they are fully powered, and in most fighters they can be driven together, to behave as elevators, or differentially (one up, the other down) to roll the aircraft. Roll is normally controlled by traditional ailerons, though some aircraft have them well inboard to avoid twisting the wing and most big airliners have 'high-speed ailerons' inboard, between sections of flap, and 'low-speed ailerons' outboard which are often locked centrally above a particular indicated airspeed or Mach number.

Combat aircraft and most airliners also have spoilers, which are flat surfaces driven by powerful rams to hinge up from the upper surface of the wing against the airstream. Used symmetrically they serve as powerful destroyers of lift and increasers of drag, for a fast let-down to a runway without increasing speed. After landing they can be flicked open

surface(s) exactly the required amount. The PFCU usually has mechanical input signalling and an output which may be a hydraulic jack, a hydraulic motor driving a ball-screwjack, or some other device.

This system has the advantage of being irreversible, so the surface cannot flutter and cannot move at all unless commanded. But the pilot, not being connected to the surface, can no longer 'feel' the things he is doing to the aircraft. The answer is an artificial-feel system, which may be a simple arrangement of springs but in any high-speed aircraft has to involve springs, bob-weights to increase the pilot load in proportion to aircraft acceleration and a hydraulic, pneumatic or electric servo-system to make the pilot load proportional to indicated airspeed and/or Mach number. In this way the aircraft is made to respond correctly in both dense air at sea level and thin air in the stratosphere.

Even this is still only the beginning of a modern FCS. Almost all high-speed aircraft

whole mission. The autopilot is often triplicated or quadruplicated in carefully planned ways, so that if any one 'channel' were to suffer a fault, the rest would outvote it. Even with faults in two channels automatic flight would be possible, but at some point it would be correct for the pilot to break into the closed loop of the control system (this is intended to occur infrequently, once in, say, 100,000 flights). The linkage between the pilot/autopilot/computer input and the PFCUs is increasingly not mechanical at all but electrical. In these Fly By Wire (FBW) systems the signals are transmitted along multi-core ribbons which are lighter than mechanical control runs, can go anywhere in the airframe and are much more accurate.

Before concluding this important section, the point must be made that the traditional arrangement of movable surfaces so carefully outlined earlier no longer applies to an increasing proportion of aircraft. Today's horizontal tails are often single surfaces, with no

to act as 'lift dumpers' to put the full weight on to the wheels to increase the power of the brakes. During the landing approach some aircraft have Direct Lift Control (DLC) in which the spoilers in effect become primary flight control surfaces to allow the pilot to adjust rate of descent instantly without altering speed or attitude. In many aircraft the spoilers can be used differentially to serve as the primary roll control, with ailerons locked inoperative (a swing-wing aircraft in the fully swept position invariably uses spoilers and/or a rolling tailplane, which is sometimes called a 'taileron').

Yet a further modern feature is subdivision of control surfaces into two to four sections, each driven by a PFCU in a different circuit to guard against failure. A visible feature of many aircraft today is the T-tail, with horizontal surfaces mounted on the vertical surface, while some supersonic machines have twin vertical surfaces which may slope outwards and often are one-piece (serving as

both fin and rudder). The list of new features is endless. For example, the Grumman F–14 Tomcat has a Mach/sweep programmer so that the wings automatically pivot to the optimum sweep angle without the pilot's attention.

Safety

Before 1920 some aircraft were equipped with navigation lights (red on the left wing with 110° arc of visibility, green on the right and white at the tail), and today this unchanged system has been supplemented by powerful landing lamps (usually retractable) and multi-million-candlepower rotating or flashing beacons visible for many miles in clear night air. By 1925 the first fire extinguishers were in wide use, and today aircraft incorporate sensors to detect flame or smoke, crash systems triggered by impact or mechanical deformation, and various other triggers which can set off extinguishing systems either generally or in specific locations such as a chosen baggage hold or engine pod. Fuel tanks are made self-sealing after a puncture, crashproof by design, and filled with a very light sponge-like substance called reticulated foam (this can be inside and/or surrounding the tanks) to inhibit fire. Inert nitrogen is pumped into the space above the fuel, and not only the fuel supply receptacle but the whole aircraft is 'bonded' with electrical conductors from stem to stern and from tip to tip (and, on the ground, to an electrical earth) to prevent any spark having to jump

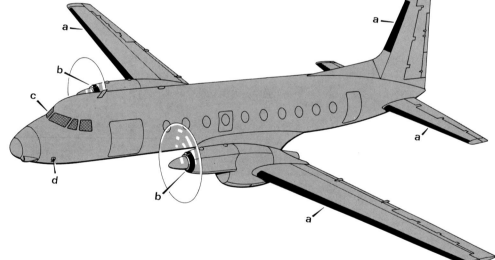

into an undulating surface which cracked ice as fast as it formed, so that it blew away in the slipstream. By 1942 many aircraft had electro-thermal ice protection, and a little later hot-air systems became common. From much earlier the extremely vulnerable choke tube of the carburettor, where the local drop in air pressure reduces temperature and promotes icing which stops the engine, was heated by circulating engine oil through its walls.

Gas-turbine engines made available ample supplies of hot air for anti-icing ('anti' prevents ice from forming; 'de' disperses it after it has formed) and today the preferred method on large aircraft is to duct extremely hot air from the engine compressors along the wing and tail leading edges and also to the engine inlets and pilot windscreens. Many areas, and a few complete aircraft, are protected by electric methods using elements embedded in rubber or sprayed on (for example, Spraymat can be sprayed on the airframe and Sierracin or Triplex methods give a transparent film on the windscreen).

Safety equipment carried by aircraft included an axe as soon as all-metal fuselages made emergency egress potentially difficult, and by the late 1930s all large aircraft included an inflatable dinghy (life raft) usually housed in the upper part of the wing or fuselage and released on a tether-link as soon as the aircraft settled in the water. Dinghies grew elaborate, with handholds, stabilizing pouches, anti-exposure screen, handcranked radio, sea-dye (fluorescein) tablet, concave-bladed knife on a tether, solar still for drinking water, rations and medical supplies. Crew flying clothing incorporated parachute harness and integral or separate parachute pack (plus dinghy for fighter pilots), whistle, marker, Mae West life jacket, torch and, by 1953, miniature radio beacon.

By 1943 certain fighters and early jet aircraft had led to the development of ejection seats fired by the occupier to make a clean emergency escape in high-speed flight. For further description see page 142. A few aircraft have complete jettisonable cockpits which, after soft landing, can be used as a radio-equipped survival shelter. It is hoped to devise crew capsules provided with rudimentary wings and propulsion so that, after separation from a crippled combat aircraft, the crew can fly perhaps 80km (50 miles) towards their base. A quite different emergency escape system is the quickly inflatable slide which extends from below the doors of airliners to allow passengers to slide perhaps 9m (30ft) to the ground.

Modern aircraft also incorporate literally hundreds of devices and subsystems to improve safety. Most are associated with particular functioning items, and serve duties similar to the fuzes that protect a domestic electrical appliance in preventing overspeed, overtemperature and similar faulty conditions. Throughout the whole aircraft a Built-In Test Equipment (BITE) installation may continuously monitor the performance and behaviour of very large numbers of devices and either give immediate warning or record any fault on a wire or tape recorder which is played back after each mission. There are

TOP LEFT: *A carrier landing by a Grumman F–14A Tomcat multi-role fighter. This versatile but costly aircraft can automatically adjust the sweep of its outer wings to suit the demands of air combat or other situations.*
CENTRE LEFT: *Anti-icing positions on the HS.748 transport: (a) pulsating pneumatic rubber de-icing boots (b) electro-thermal engine ice protection (c) electro-thermal windscreens (d) ice-detection probe.*
LEFT: *Inflatable rubber dinghies are now standard equipment on large aircraft. These are usually housed in the upper part of the wing or fuselage and released on a tether-link as soon as the aircraft is settled on the water.*
ABOVE: *Almost all large civil transports have to be equipped with inflatable escape slides so that all occupants can quickly evacuate a grounded aircraft. That shown is the one utilized on an Eastern Airlines TriStar.*

an air-gap. Static wicks in the form of billions of fine conductive filaments hang behind trailing edges to disperse into the atmosphere any electrical charge that could otherwise build up and eventually cause a potentially dangerous electrical flash.

De-icing systems

By 1932 de-icing systems were coming into use. One company supplied paste which when brushed on to external surfaces such as leading edges and control-surface hinges prevented ice from forming. Goodrich supplied tailor-made pulsating rubber de-icers in the form of flat tubes attached along the leading edges. Pneumatic pressure supplied to the centre and then the outer strips in rhythmic sequence turned the vulnerable leading edge

several families of recorders, all concerned with safety. The type just mentioned is usually called a maintenance recorder. The Cockpit Voice Recorder (CVR) makes a hard-copy record of all radio or intercom traffic during each flight – this may have significance in a court of law. There is a wealth of types of Flight-Data Recorders (FDRs) and Airborne Integrated Data Systems (AIDS) which monitor everything from crew procedures to engine health, while most commercial transports have to carry a special FDR protected against extremely severe impacts, crushing loads and temperatures. This crash recorder, which remains playable after an accident, has to record at least altitude, airspeed, vertical acceleration, pitch angle (nose up or down), heading and elapsed time, at intervals of one second throughout the flight. Most record 50 parameters, including control-surface positions.

Stall warning system

One of the most vital systems in advanced aeroplanes is the stall warner. Most wings stall whenever they exceed a critical angle of attack, and this can occur at any speed, the minimum stalling speed being in unaccelerated (i.e., straight and level) flight. A stall-warning system thus has to measure angle of attack, which it may do by sensing dynamic pressure at two small holes, one pointing $45°$ above the horizon dead ahead and the other $45°$ down, or at other lesser angles. In normal flight the pressures are approximately equal, but as angle of attack increases they diverge until at a preset value the alarm is triggered. The system must actually be more complicated, because the stall danger depends on the rate of increase of angle of attack; thus, it is less dangerous to fly steadily at just below stalling angle than to pull a tight turn so violently that this same angle is exceeded at a high rate of increase.

The warning may be visual in the cockpit, or a loud horn or other aural alarm. With continued increase in angle of attack, a 'stick shaker' can come into play, banging the pilot's control column or yoke to and fro, making a loud clacking noise and taking it out of his hands. If he is determined to stall the aircraft, and keeps pulling back, a stick pusher so powerful no pilot can fight it suddenly thrusts the stick forward, changing the nose-up command to a firm nose-down one. The only way the pilot can beat this is to 'dump' the entire system by a positive action on a clearly marked control.

Most aircraft probably crash flying into the ground while fully serviceable, usually in cloud or fog. The Ground-Proximity Warning System (GPWS) is an attempt to prevent this.

Yet another and even more difficult safety system is the Collision-Avoidance System (CAS). One company has had a CAS in constant use since 1964; it is called Eros, from 'Eliminate Range-Zero System', range-zero meaning a collision.

The cockpit

Every instrument in the modern cockpit is the result of many years of invention and re-

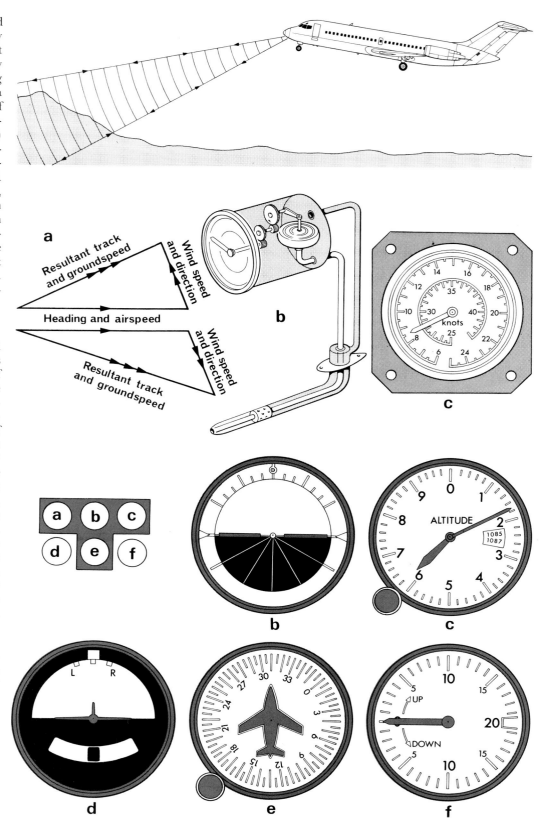

TOP: *This sketch of a DC-9-10 illustrates the basic principle of the Ground-Proximity Warning System (GPWS), now mandatory on most large civil transport aircraft. The system senses potential collisions with the ground or other obstructions and, while issuing loud audible warnings, pulls up the nose of the aircraft.*
CENTRE: *The basic navigational vector diagram (a) which old-time navigators were continually having to plot on charts, showing the effect of particular winds. The Airspeed Indicator (ASI) pitot/static system (b), showing how dynamic pressure inflates the capsule while static pressure from the side-facing small holes is fed to the instrument case. A typical traditional ASI presentation (c) (it would be in position (a) in the panel below it).*

ABOVE: *During World War II the basic blind-flying panel in British aircraft was standardized in this form, to ease the task of pilots who flew many types of aircraft. ASI (a) as shown at (c) above, but here used for the key diagram (b) artificial horizon (c) altimeter (d) turn/slip indicator, also called turn and bank indicator or turn co-ordinator (e) gyro compass or directional gyro and (f) rate of climb indicator, also called Vertical Speed Indicator or VSI.*
ABOVE RIGHT: *The cockpit of one of the first Spitfires, which was protected by a thick, bullet-proof screen.*
RIGHT: *In 1938 this Ensign transport was the last word in airline flight decks. In the centre are the autopilot panel and the engine throttle levers.*

finement. Apart from the tachometer, which measures engine speed (today not by a mechanical spinning drive but by electrical signals transmitted by a small generator on each engine), the first instrument was the the AirSpeed Indicator (ASI). Almost all operate by differential pressure between an air pipe sensing static pressure from sideways-facing perforations in the ASI pitot/static head (an electrically de-iced tube carried somewhere on the aircraft where the air is undisturbed), and dynamic or pitot pressure, sensed by a tube with an open end facing dead ahead. Since World War II the ASI has been backed up by, or integrated with, the Machmeter which indicates Mach number. This is basically an ASI capsule working in conjunction with an altitude capsule.

Altitude capsules are connected to static pressure only, and a stack of them can drive the altimeter which tells the pilot his 'pressure height' above the Earth. It actually indicates outside air pressure, and the altimeter must be corrected or re-set to take account of the fact that atmospheric pressure varies both with time and geographical position. The radio altimeter gives true height above ground by a radar method, measuring the time taken by a signal to reach the Earth and return.

Another height instrument is the Vertical Speed Indicator (VSI) which measures rate of climb or descent by sensing expansion or contraction of a capsule fed with static pressure and also open to the instrument via a small orifice (it is like an altimeter with a capsule having a slow leak). In steady flight the pressures inside and outside the capsule balance, but climbing or descending causes imbalance which drives the indicator needle.

Gyro instruments
There are many other pressure-type instruments, some of them in complicated air-data systems, but the other large family are gyro instruments. Most modern gyros are driven electrically. Some are 'floated' on liquid or gaseous bearings, without mechanical contact anywhere, to give perfect centring and almost zero friction. Some gyros measure angular displacement of the aircraft about any or all of its three axes; rate gyros are free in only two axes and thus measure not turn but rate of turn about the third axis, at 90° to each of the others.

One of the simplest gyro instruments is the Direction Indicator (DI). It contains no magnet and suffers from 'wander' over a period of time, and from error caused by the Earth's rotation, so it must be reset occasionally against one of the compass systems. It contains a single gyro spinning about an axis maintained horizontal relative to the carrier aircraft, which will 'topple' the gyro if it does aerobatics, and provides an accurate heading reference for short periods and, especially, for precise turns. Another instrument, the artificial horizon, contains a gyro with axis vertical and this drives a horizon bar which moves parallel with the Earth's horizon, either above or below, with or without tilt, in relation to a fixed gull-wing shape representing the aircraft seen from behind. Yet another basic gyro instrument is the turn-

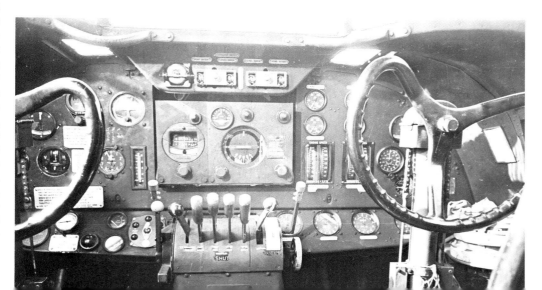

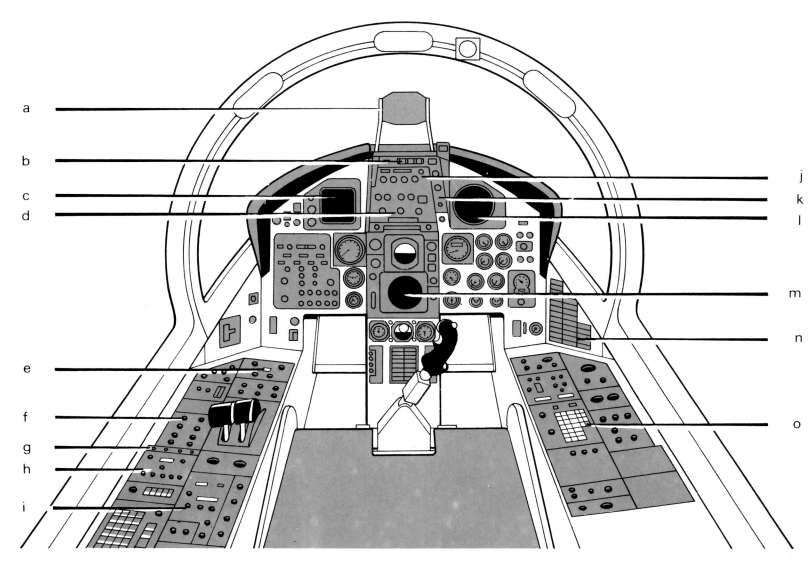

a
b
c
d
e
f
g
h
i

j
k
l
m
n
o

and-slip indicator; in its simplest form this contains a rate gyro with axis horizontal to indicate rate of turn, and a simple pendulous display (such as a heavy ball in a liquid-filled curved tube) to indicate any inward slip or outward skid.

In principle, aircraft magnetic compasses resemble those used as toys or on ships, but the remote compass is a rather different device with a 'fluxvalve' in one wing tip which is not free to rotate and senses the aircraft heading in relation to the local Earth's magnetic field. All aircraft compasses incorporate complicated devices for correcting errors caused by aircraft magnetism, aircraft manoeuvres and several other possible sources of error. Special compasses were devized for flying near the Earth's magnetic poles.

Display systems

On the basis of new technology the flight deck has been completely altered in the past 15 years. One major change is that the most obvious instruments today are big combined displays such as a Director Horizon (DH) or Horizontal Situation Indicator (HSI). A much bigger change is the gradual move towards even more integrated displays in which moving-map projected pictures, radar pictures, television pictures, alphanumeric data and even a real or imaged view of the Earth ahead are all presented on a single screen. Many advanced aircraft have at least one Head-Down Display (HDD) of this kind on the pilot's instrument panel, plus a Head-Up Display (HUD).

TOP: *A modern fighter cockpit, the F–15 Eagle: (a) Head-Up Display (HUD) which the pilot reads while also looking through it at what lies ahead (b) Identification Friend/Foe (IFF) controller (c) vertical situation display (d) HUD controller (e) navaids selector (f) radar controller (g) Tac Electronic Warfare System (TEWS) (h) IFF controller (i) communications panel (j) Ultra Hi-Frequency (UHF) selector (k) mode selector (l) TEWS display (m) HSI (n) warnings (o) navigation panel.*
ABOVE: *Flight deck of the modern B–1 bomber.*
OPPOSITE PAGE: *Three sketches of the S–3A Viking showing location of basic systems. Fuel system, Electrical System and Basic Flight Controls (top):*
  1 Wing integral fuel tanks
  2 Auxiliary tanks
  3 In flight refuelling probe (retractable)
  4 Fuel dump and vent pipes
  5 Engine driven electrical generator
  6 Emergency generator
  7 Electrical load distribution panels
  8 General Electric T–34 engine
  9 Auxiliary power plant
  10 Ailerons

11 Slotted flaps
12 Spoilers
13 Rudder
14 All moving tailplane with flaps
15 Drooping leading edges
16 Leading edge electric actuators
17 Electrically heated windscreen panels
Hydraulic System, Pneumatic System and Landing Gear (centre):
18 Engine driven hydraulic pump
19 Hydraulic accumulators
20 Aileron flap and spoiler servo controls
21 Wing fold rotary actuators
22 Rudder actuator
23 Tail fold jack
24 Tailplane actuator
25 Arrester hook jack
26 Main landing gear jack
27 Weapons bay door actuators
28 Nosewheel jack
29 Environmental control system
30 Avionics cooling ducts
31 Cabin air conditioning ducts
32 Wing leading edge de-icing ducts
33 Tailplane de-icing ducts
34 Rearward retracting main landing gear
35 Rearward retracting nosewheel leg
36 Catapult strop attachment
Operational Equipment (bottom):
37   Search radar
38 Pilot's seat
39 Co-pilot
40 Tactical control officer
41 Sensor control officer
42 Forward avionics bay (port and starboard)
43 Forward Looking Infra-Red scanner (FLIR)
44 Mission control avionics (port and starboard)
45 Univac control computer
46 Weapons bay (port and starboard) with two torpedoes each side
47 Sonobuoy launch tubes
48 Aft avionics bay (port and starboard)
49 Magnetic anomaly detector boom (MAD)
50 Wing tip sensors

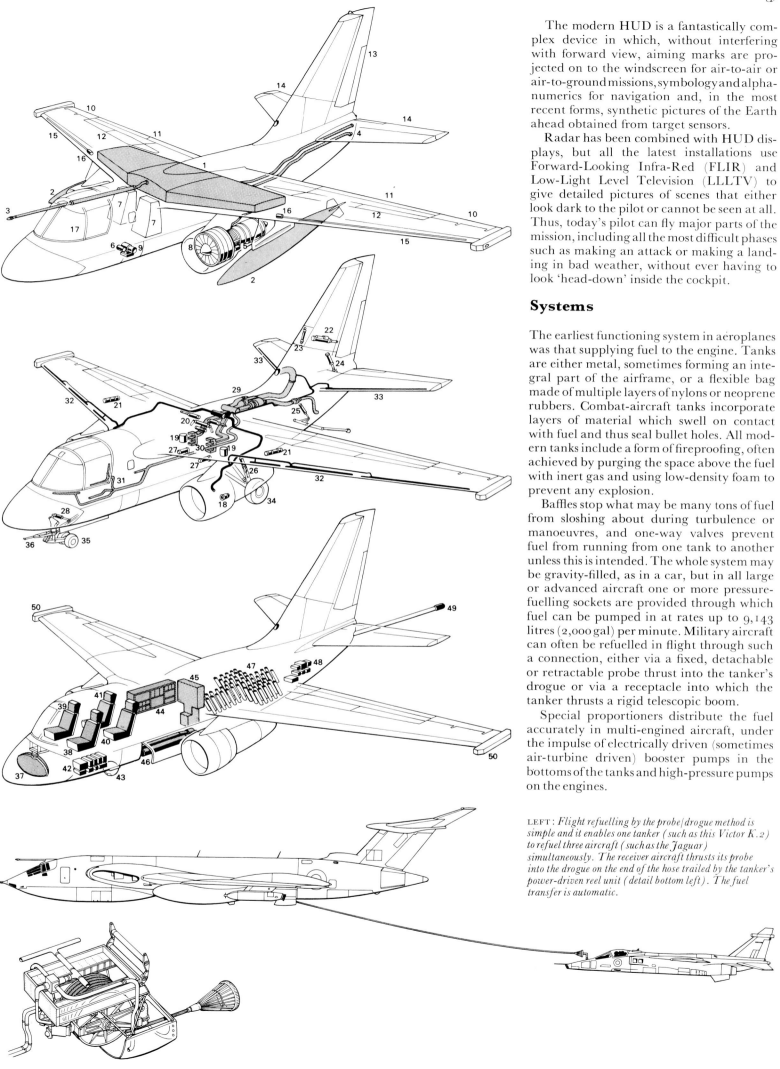

The modern HUD is a fantastically complex device in which, without interfering with forward view, aiming marks are projected on to the windscreen for air-to-air or air-to-ground missions, symbology and alphanumerics for navigation and, in the most recent forms, synthetic pictures of the Earth ahead obtained from target sensors.

Radar has been combined with HUD displays, but all the latest installations use Forward-Looking Infra-Red (FLIR) and Low-Light Level Television (LLLTV) to give detailed pictures of scenes that either look dark to the pilot or cannot be seen at all. Thus, today's pilot can fly major parts of the mission, including all the most difficult phases such as making an attack or making a landing in bad weather, without ever having to look 'head-down' inside the cockpit.

Systems

The earliest functioning system in aeroplanes was that supplying fuel to the engine. Tanks are either metal, sometimes forming an integral part of the airframe, or a flexible bag made of multiple layers of nylons or neoprene rubbers. Combat-aircraft tanks incorporate layers of material which swell on contact with fuel and thus seal bullet holes. All modern tanks include a form of fireproofing, often achieved by purging the space above the fuel with inert gas and using low-density foam to prevent any explosion.

Baffles stop what may be many tons of fuel from sloshing about during turbulence or manoeuvres, and one-way valves prevent fuel from running from one tank to another unless this is intended. The whole system may be gravity-filled, as in a car, but in all large or advanced aircraft one or more pressure-fuelling sockets are provided through which fuel can be pumped in at rates up to 9,143 litres (2,000 gal) per minute. Military aircraft can often be refuelled in flight through such a connection, either via a fixed, detachable or retractable probe thrust into the tanker's drogue or via a receptacle into which the tanker thrusts a rigid telescopic boom.

Special proportioners distribute the fuel accurately in multi-engined aircraft, under the impulse of electrically driven (sometimes air-turbine driven) booster pumps in the bottoms of the tanks and high-pressure pumps on the engines.

LEFT: *Flight refuelling by the probe/drogue method is simple and it enables one tanker (such as this Victor K.2) to refuel three aircraft (such as the Jaguar) simultaneously. The receiver aircraft thrusts its probe into the drogue on the end of the hose trailed by the tanker's power-driven reel unit (detail bottom left). The fuel transfer is automatic.*

ABOVE: *Flight refuelling by the Boeing Flying Boom method of a Boeing E-4B airborne command post (based on the Boeing 747). This aircraft along with the Lockheed C-5A Galaxy is the largest aircraft refuelled in the air.*

BELOW: *Sketch of a Boeing 707-3J9C of the Imperial Iranian Air Force (IIAF). Unlike American tanker versions it has two hosereel/drogue units at the wing tips as well as a Flying Boom under the rear fuselage. The two are needed to preserve compatibility with the IIAF's varied assortment of warplanes, which have several refuelling methods.*

RIGHT: *Typical multi-plunger hydraulic pump:*
1 Inlet from reservoir
2 1st stage pump
3/4 1st stage relief valve and compensator
5 Off-loading valve
6 Off-loading actuator
7 High pressure pistons
8 Cam drive
9 1st stage feed annulus
10 High pressure delivery annulus
11 High pressure outlet
12 Return to reservoir

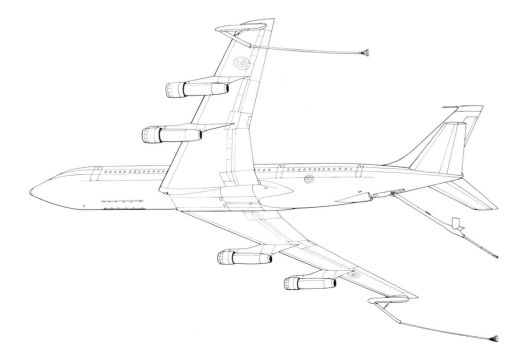

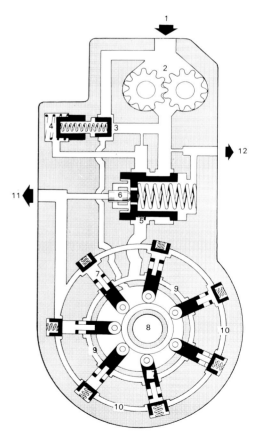

TOP RIGHT: *A unique form of drop tank was the Two-Component Pod (TCP) carried by the 2,250km/h (1,400mph) B-58 Hustler bomber until its retirement in 1969. The crane is lowering the nuclear-weapon portion into the fuel-carrying portion.*

Extreme care is taken to eliminate water from the fuel, and prevent any that may reach the tanks from getting into the system where, at high altitude, it would freeze and cause blockages. The aircraft filters are heated to melt any ice crystals which do enter. Special valves control the flow throughout the system, to allow fuel to flow no matter what the attitude of the aircraft and also to allow air or nitrogen into or out of the system to balance out the pressure differences between sea level and high altitudes. Even at extreme altitude the fuel must not be allowed to boil and cause bubbles in the system, despite the fact that the fuel may serve as the 'heat sink' to which nearly all waste heat in the aircraft is rejected.

The whole system is designed so that, as fuel is consumed, the aircraft's CG shifts by the smallest amount; conversely, in some aircraft (notably Concorde) fuel is pumped from one part of the aircraft to another deliberately to change the CG position and

counteract trim changes between subsonic and supersonic flight. Special features in supersonic aircraft are fuel systems to supply the afterburners and either special fuel or special provisions to stop the accumulation of tarry or 'coked' residues caused by long 'soaking' at high temperature because of kinetic heating. Many aircraft carry external drop tanks, which need self-sealing connec-

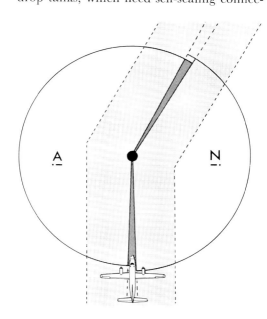

tions, valves for filling and supply through the same couplings, swivelling mounts for swing wings and, often, cartridge jettison plungers to throw an expended tank well clear of the aircraft to avoid damage.

Electrical systems

An electrical system is today carried on almost every aeroplane, though it may be no more complex than that of a car. But large and advanced aircraft often have electrical systems capable of generating current at a rate equivalent to many hundreds of horsepower, and the number of separate parts in the system may exceed 15,000. There are usually three kinds of current on board. Raw Alternating Current (AC) is like electricity in the home, and it is almost invariably three-phase at 115/200 volts generated by alternators (AC generators) on each engine. For emergency use another alternator may be driven by a Ram-Air Turbine (RAT): a windmill that can be spun by the slipstream when necessary; a Monofuel Emergency Power Unit (MEPU): a stand-by device burning special fuel; or an Auxiliary Power Unit (APU): a versatile package used either in emergency or when the main engines are not running, such as when the aircraft is in-between flights). Raw AC is used for many purposes, but by far the biggest loads are heating, for anti-icing the airframe and systems, for heating ovens in the galley and similar base loads.

The avionics (aviation electronics) need precisely controlled frequency, and this exactly controlled AC has to be generated by alternators kept running at precisely constant speed. This involves using either a Constant-Speed Drive (CSD) to turn the alternator(s) or a self-contained device called an Integrated-Drive Generator (IDG). The third kind of electricity is Direct Current (DC), used for many services and also stored in batteries, which in most advanced aircraft are of the nickel-cadmium type.

In most aircraft, all (or nearly all) the demands for mechanical power are supplied hydraulically. Typical examples are landing-gear retraction and locking, nosewheel steering, flaps, leading-edge devices, tailplanes, powered flying controls, swing-wings, cargo doors, airstairs (retractable stairways) and, in most large or fast aircraft, wheel brakes. Early systems used mineral oils at pressures up to $35 \cdot 15$kg/cm^2 (500lb/sq in), but today most use less inflammable fluids specially tailored for the job at pressures of 211 or 281 kg/cm^2 (3,000 or 4,000lb/sq in).

Compressed air is used in two main ways. In one there is high pressure in the order of hundreds or thousands of pounds per square inch and very little flow; in the other the pressure is in the order of tens of pounds per square inch and the flow may be enormous. The first type of pneumatic system is used either as a main power service in light aircraft, or (at lower pressure) to drive certain instruments, or to pressurize tanks, or where the environment is too hot for hydraulics (for example, to drive variable afterburning-engine nozzles), or as a local self-contained system and especially for one-shot emergency use. Often one-shot systems use gas from solid-fuel generators like completely enclosed rockets.

In large aircraft the biggest on-board system is that providing cabin pressurization and air-conditioning, and this either uses engine-driven compressors or, in jet aircraft, air 'bled' or 'tapped' from the engine compressors. Very large flows may be used, and the air passes through large duct systems where it is cooled, dried or moistened, and finally supplied to the inhabited areas and cargo holds at carefully controlled pressure and temperature.

Navigation Aids

The first airborne navaid was the Direction-Finding (D/F) loop aerial, a conductive coil which could be rotated until the received signal from a ground radio station was reduced to zero in the so-called null position, and the loop was then at 90° to the direction of the ground station. Navigators could obtain D/F bearings on two ground stations in quick succession and thus, by drawing the two lines on a map, obtain a 'fix' (a known aircraft position). Today the successor of the D/F loop is the Automatic Direction Finder (ADF), in which an immediate bearing to any selected ground station is indicated on an instrument.

Airways

In the late 1920s the United States began a nationwide system of 'airways' which survives today and now encompasses the globe. Each airway was at first a single invisible radio beam sent out from a ground station and pointing at the next. At the ground station were four aerials, two of which continuously transmitted the Morse letter A ($\cdot$ –) while the other two transmitted the letter N (– $\cdot$). As the dots and dashes were keyed precisely together the two signals merged along the centreline of the airway into a continuous note. If the aircraft was just off the centreline an A or N would be heard, while passing directly over the station caused the signal to disappear entirely for a short time.

This system was called Radio Range, and it was the basis for airways worldwide until after 1945. Its obvious drawback was that it provided only two, or four, routes to fly; the system was not much help to pilots who wanted to fly a different route. So the next step in this American scheme was to build a different kind of lighthouse called a Very-High-frequency Omni-directional Radio-range (VOR). This combines two sets of

BELOW: *Diagrams explaining aspects of VOR : (a) transmission pattern of a typical VOR beacon showing variation in signal in different directions (b) aircraft flying at 270° fixes position by VOR 1 at 204° and VOR 2 at 314° (c) a typical VOR compass display with needles driven by transmission from selected VOR ground stations.*

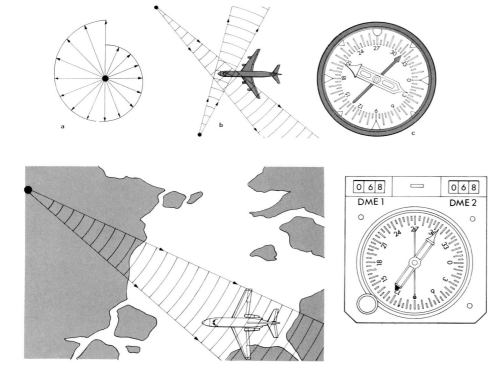

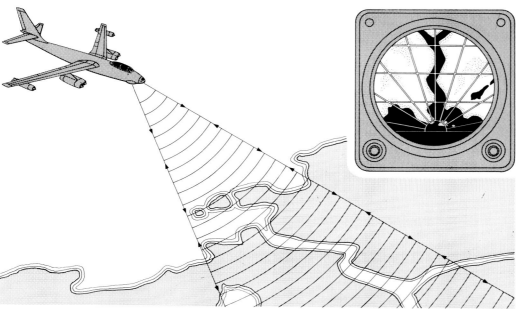

ABOVE CENTRE: *Distance-Measuring Equipment (DME) interrogates a ground station to give a distance readout, in this case 68 nautical miles.*
ABOVE: *First used by the RAF in late 1942, H₂S was the first airborne radar used for ground mapping. Modern radars can usually do the same job, though in wartime such duty is hazardous because the radar signals can attract hostile fighters. In this sketch a bomber is flying over a river estuary and picking up the water areas in very dark hues, while the land appears brighter. Cities would be brighter still.*
RIGHT: *Panavia Tornado seen from above showing the four beams from its doppler navigational radar. Measuring the frequency changes of the reflections from the Earth's surface of the four beams gives a precise measure of the speed of the aircraft along the four beam-directions, from which an airborne computer can continuously work out the true ground speed and track (and, if necessary, the current wind at the height at which the aircraft is flying).*

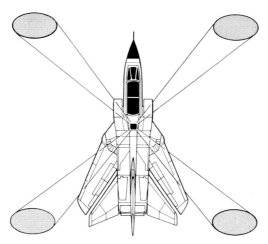

precisely timed radio signals in such a way that an indicator in the aircraft can immediately show the bearing from each VOR station. But again there are severe limitations: VOR gives no help in using the whole of the airspace and again restricts aircraft to congested airways (which may not be the routes each aircraft wishes to follow); and a separate device called Distance-Measuring Equipment (DME) has to be used to give range information.

Gee, Loran and Decca

In 1940 British radio engineers urgently produced a completely new navaid to help RAF bombers find their targets. Called ('Gee', it was uncanny in its accuracy; furthermore, it was able to fill the whole sky with guiding signals so that aircraft did not have to follow any prearranged route. It worked with one master ground station and two slave stations. In the original Gee system, the three stations were in different parts of England, but their signals could be used by bombers far over Germany. The master station sent out alternate single and double pulses controlled precisely by an electronic clock. These pulses were received by the aircraft and by the two slave stations, one of which automatically re-radiated the signals after each single pulse and the other re-radiated after each double pulse. In the aircraft a cathode-ray tube showed spikes of light which, by presenting a picture of the time differences of the three sets of signals, told the navigator his position.

Gee was the first R-nav (area-navigation) system, and the first hyperbolic system, so called because lines of equal time-difference from the ground stations appeared as hyperbolae on a map. Another was Loran (long-range navigation), which was developed a little later than Gee to use longer wavelengths, and thus give guidance over greater distances. Even today Loran is very important to civil and military craft flying over the oceans and more remote areas of the globe.

In the years after 1945 the British Decca company saw that the future need was for a refined R-nav system which would be usable over the whole airspace, right down to sea level, and which would have extremely high accuracy to assist in the same control of dense traffic right down to the Instrument Landing System (ILS) on the approach to the runway. The Decca navigation system was developed to meet these needs. A hyperbolic system like Gee, it differed in using continuous radio waves instead of pulses, emitted by a master and three slaves. At first, navigators had to plot their position on special hyperbolic maps, using three sets of dial instruments called Decometers. By 1951, however, the aircraft track could be automatically plotted on a moving map called a Flight Log, giving for the first time an exact picture of aircraft position at all times. Later, Decca grew into a wide family of navaids, all intrinsically more accurate than VOR/DME and far more flexible and useful.

Tragically, Decca did not happen to be American and, at the crucial meeting of the International Civil Aviation Organization (ICAO) in 1958 to decide on the standard world navaid, decisions were taken not on

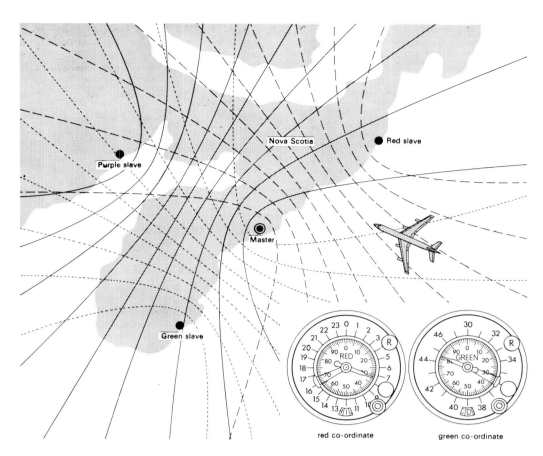

LEFT: *The principle of the original Decca Navigator system is illustrated by a map of the Nova Scotia chain. Continuous radio signals emitted in synchronization by the master and slave stations are picked up and analyzed by the aircraft, which compares the time-delays from the various stations. Matching the times from any two stations fix the aircraft on a hyperbolic position-line, some of which are plotted. In the 1945 system the navigator had to read the position-lines off dial instruments called Decometers, two of which are shown. Soon the system was made automatic, one method being to make the signals drive a pen across a roller-blind moving chart to give a continuous readout of aircraft position. Today Decca is combined with other systems to give foolproof navigation over any desired track at any desired height.*

the basis of merit but of political pressure. The consequence was a predictable win for VOR/DME, as a result of which the world has been covered with several thousand of these restrictive and outmoded stations which even today are the universal standard navaid. But commonsense has combined with increasing operating difficulty in the busiest airspace to force even the United States to take another look at the situation.

To take an ocean of air and carve it into narrow congested canals was not very clever, and today a wealth of R-nav systems, pioneered by Decca, are in use. Some are techniques which partly overcome the more serious shortcomings of the enforced system now in use, but most are completely new.

Two of the navaids on all large transport aircraft do not require any ground stations but are self-contained in the aircraft. Doppler radar uses the Doppler shift in the frequency of a signal caused by motion of its source relative to an observer; the classic case to help understanding of Doppler is the way the pitch of a fast-moving noise, such as a train whistle or a low-flying aircraft, sounds higher as it approaches and lower as it departs. The aircraft transmits radio waves in a fan of four narrow beams, which strike the ground and are reflected back to the aircraft. By exactly comparing the shifts in frequency of the reflected signals, the airborne navigator can determine track and speed over the ground. Thus, no matter what the wind might be, the position of the aircraft is continuously updated.

Inertial navigation systems

The other self-contained navaid is the most valuable of all. Inertial Navigation Systems (INS) could not be made until technology existed to make gyros and accelerometers with an almost incredibly high accuracy, far

more precise than the parts of the most expensive watch. Though a modern INS is too complex to attempt to describe, its principle could hardly be simpler. Imagine that inside the aircraft was a weight hung on a fine thread. With the aircraft parked on the apron the weight would hang straight down, but when the pilot released the brakes the weight would move to the rear. As the aircraft went round bends on the airport, the weight would swing out to the side. Such a system would be very rough and ready, but if we could measure continuously the position of the weight, and relate it to the passage of time, it would tell us how fast the aircraft was moving and in which direction.

The INS uses three sets of accelerometers which are in effect much more accurate counterparts of our simple pendulum. They continuously measure and record the acceleration of the aircraft in the fore/aft, left/right and up/down directions. Super-accurate gyros are used to hold the accelerometer platform exactly level relative to the surface of the Earth below, because even a very small tilt would be interpreted by the accelerometers as a small false acceleration. A modern INS platform is held level with the same accuracy as needed to point a light-beam at a small coin 0·5km (0·31 miles) away, while the accelerometers could measure the acceleration of a car which took three hours to go from rest to a speed of 40km/h (25mph).

Doppler and INS have made the most advanced aircraft able to navigate with great precision anywhere in the world without outside help. But there are many other navaids available for all kinds of customer. Some are simple, and aimed at the private pilot. Some use satellites orbiting the Earth. Omega is the latest radio navaid, a hyperbolic system using Very-Low-Frequency (VLF) signals and able to cover the whole globe with only

eight ground stations. Omega signals from the other side of the world are now guiding not only aircraft but submerged submarines.

Navstar GPS

In 1974 a concept was validated that is now revolutionizing the navigation problem – indeed, it eliminates it. The Navstar GPS (Global Positioning System) enables any receiver station, of any kind, anywhere on Earth, to know its position in three dimensions with near-perfect accuracy, as well as its exact velocity (if any) and the exact time.

It is a passive system; the receiver (user) sends out nothing. All he has to do is tune in to specially coded signals sent out on 1,575.42MHz and 1,227.60MHz by three orbiting satellites. The receiver instantly determines distance to each satellite by multiplying the signal transit time by the speed of light. The receiver has a clock synchronized to that in each satellite. To avoid error caused by lack of synchronization the receiver also computes range from a fourth satellite to calibrate its own clock error. By measuring the Doppler shift in each signal's carrier frequency the receiver's velocity (speed and direction) is obtained.

Accuracy transcends anything known previously. An accompanying table shows the contrast between the GPS and alternative systems. The gaps in the table mean that the system lacks a particular capability. Though this system began to meet the needs of the US military, it will ultimately eliminate problems in civil air-traffic control, blind landing at ill-equipped airfields, collisions between ships in fog and, for example, in exploring jungles or deserts.

The company chiefly involved has been Collins. During development, receivers were integrated on eight host platforms: an F–16, a B–52, an A–6E and a UH–60A, and a submarine, aircraft carrier, utility truck and infantry manpack. By year 2000 over 21,000 sets should be on US military platforms, and civil applications are open-ended.

Airports
their design and development

The first six decades of commercial aviation have seen the growth of the passenger airliner from a converted 120km/h (80mph) military biplane to the supersonic Concorde, travelling at twice the speed of sound.

Airport development has had to keep pace with new demands by the airlines, for there is little point in introducing high performance jets if time is wasted on the ground through badly designed or outdated handling facilities and procedures for both aircraft and passengers.

The first airports, in the strict sense of the

word, were commissioned in Germany in 1910 for the Zeppelin airships operated by the Delag company for passenger flights, and, by 1913, airship sheds, incorporating passenger handling facilities, existed at a number of German cities. All were served by rail links, and Delag carried almost 34,000 passengers in nearly 1,600 flights before war broke out in 1914, although it should be emphasized that these were not regular scheduled services.

A flying-boat service in 1914 heralded the dawn of passenger flying in the USA, but the first major step was the setting up, in the early 1920s, of a transcontinental air mail service which required the establishment of aerodromes across the country. Most regular aerodromes had two right-angle runways up to 762m (2,500ft) long, plus a 500,000 candle power beacon revolving on a tower, while emergency landing grounds had beacons a tenth of this power. By 1923 a lighted airway was set up, with acetylene-gas beacons at three-mile intervals along the routes.

Converted ex-military de Havilland D.H.9A and D.H.16 single-engined biplanes began the first scheduled passenger and cargo flights on 25 August 1919 from Hounslow Aerodrome, near London, to Le Bourget Airport, Paris. A reciprocal French service began the following month, using converted single-engined Breguet 14s, but these were soon followed by larger aircraft.

London's first real civil airport was opened at Croydon on 1 April 1920. It rapidly became one of the world's most famous airports until, overtaken by urban housing development and lacking space for expansion, it closed in September 1959.

As well as a rotating 'searchlight' beacon which could be seen from the air, Croydon also had wireless communication, and developed the rudimentary essentials of an air traffic control system which, in its initial form, involved a man with a red flag giving pilots the 'all clear' for take off! In spite of the spartan nature of these early operations, some 17,800 flights carrying nearly 47,000

passengers were made on cross-Channel services between August 1919 and March 1924.

As Croydon developed it was enlarged, and a new terminal came into use in January 1928. Airport boundary and obstruction lighting was installed, and a new red flashing beacon that could be seen up to 120km (80 miles) away in good weather conditions. Mobile and fixed landing floodlights and an illuminated wind direction indicator brought Croydon up to a high standard for its time, and airport development continued all over Europe in parallel with Croydon's upgrading. In many cases, hard runways were laid – Stockholm's Bromma Airport, for instance, opened in 1936 with Europe's first paved runways – but Croydon remained a grass airfield to the end of its days.

Le Bourget, the first international airport for Paris, was particularly well equipped for night-flying and in its time was one of Europe's most important airports.

Germany was well in the forefront of European airport development. Deutsche Lufthansa, the national airline, operated a large domestic network of air routes from Berlin, initially using Johannisthal Aerodrome. From the mid-1920s, the new Zentralflughafen Tempelhof, less than 3km (2 miles) from the city centre was also available.

Tempelhof, which had reached its fully developed form by 1938, was one of Europe's most advanced airports. It had a number of interesting features, not least of which was the massive terminal and hangar block, a curved structure which to this day has the distinction of being the world's third largest building. Designed to handle up to 300,000 passengers a year, Tempelhof had a unique advantage in that large aircraft could be taxied up beneath the terminal canopy for loading and unloading passengers. The inevitable necessity for longer runways led in October 1974 to the opening of the rebuilt Tegel Airport to replace urban, hemmed-in Tempelhof, although the latter will always be remembered with affection by West Berliners as the terminus for the Berlin Air Lift of 1948.

In the mid-1930s, there was a demand in certain European countries for seaplane and flying-boat operations. Some airports on the coast or near rivers were able to cater for these; Copenhagen's Kastrup, for instance, had a jetty, and Italy had a number of terminals for its considerable domestic and international flying-boat services to such cities as Rome, Venice, Genoa, Trieste, Barcelona, Lisbon, Brindisi, Palermo and Tunis. Operations of aircraft from water had certain advantages – no long runways were needed, although a buoy-marked area was necessary for take-off and landing. Flying-boat terminals could be incorporated in port areas.

New York's first airport, at Newark, was

FAR LEFT: *A Consolidated PBY at Edmonton Airport, Alberta. This is the oldest municipally-owned airport in Canada, and was opened in January 1927.*
FAR LEFT BELOW: *This photograph, which dates from the early 1930s, illustrates the rudimentary form air traffic control took at that time. A Lufthansa Junkers Ju 52/3m is being given instructions for take-off by a flag operator.*

BELOW LEFT: *An aerial view of the new ultra-modern airport Kansas City International. It was built at a cost of 250 million dollars, and has two runways each 3,200m (10,500ft) in length – the original city airport has one of 1,500m (4,900ft) and one of 2,100m (6,900ft). Its infrastructure was derived from the concept of a circular satellite space station, with a large central terminal and smaller subsidiary ones.*

BELOW LEFT: *View of Christchurch International Airport, New Zealand. This is a single runway airport handling a fairly large amount of traffic, and is now having its runway extended.*
BOTTOM: *One of the Deutsche Lufthansa buses which operated a shuttle service from Berlin city centre to Templehof Airport during the late 1920s, here seen picking up passengers from a Junkers G 24.*

opened on 1 October 1928 and boasted a 488m (1,600ft) hard-surfaced runway, claimed to be the first in the world. Combined hangars and passenger terminals were built, and the airport handled 20,000 passengers in 1930, a considerable figure for the time.

In spite of continued development at Newark, New York needed further airports for its rapidly growing airline traffic, and the New York Municipal Airport – La Guardia Field – was opened on 2 December 1939 to be followed in July 1948 by Idlewild Airport, now better known as John F. Kennedy International.

Without question, the world's busiest airport is Chicago/O'Hare, which in 1975 recorded 666,600 aircraft movements, giving an average daily total of take-offs and landings of almost 2,000! More than 37 million passengers were processed through the airport in the same period.

In Canada, airport development was slow, and by 1922 there were only 37 licensed 'air harbours' – 23 for landplanes, 12 for flyingboats and 2 which served both. Most landing grounds were very small, 6 of them being under 275m (900ft) square. By 1930, the number of licensed aerodromes had risen to 77, and a Prairie Air Mail Service had begun, linking Winnipeg with Calgary and Edmonton; the latter is Canada's oldest municipally-owned airport, licensed in June 1926 and opened the following January.

Development continued through the 1930s, the aim being to provide a lighted airport every 160km (100 miles), plus a number of emergency fields, and World War II provided a stimulus for considerable airport expansion.

Australia's two principal cities, Sydney and Melbourne, have had airports since 1921; by April 1936, 200 landing grounds had been prepared for civil aviation by the Commonwealth Government, and local authorities had established and licensed another 181 public aerodromes. An airline network was gradually built up and by 1976 there were 468 civil aerodromes, including service-owned facilities used by civil aircraft. Passenger traffic through the capital city air-

ports accounted for 78 per cent of all domestic airline passengers in 1975.

A brand-new airport, Melbourne/Tullamarine, was opened in the early 1970s, while Sydney's Kingsford Smith Airport was given new terminal facilities and runway extensions around the same time. A new international terminal was opened at Brisbane in December 1975, and other airport expansion work is being carried out.

Across the Tasman Sea, New Zealand was a slow starter in aviation; although there was some air taxi and charter work in the early 1920s, the first company was not registered until 1934. Subsequent growth in air services up to the outbreak of World War II provided a network of routes covering the North and South Islands; regular international services did not begin until October 1950, when Wellington and Sydney were linked.

Currently, New Zealand has some 80 public and 230 private aerodromes, including about 100 water landing areas. Auckland, Wellington and Christchurch are the three

main airports; runways at the first two have recently been extended and Auckland has a new terminal. A similar terminal is planned for Wellington, and Christchurch is having its runway extended.

Development and operating costs of the main airports are shared by central and local government. Management is provided by local authorities, while the Civil Aviation Division of the Ministry of Transport is responsible for aviation safety and the air navigation facilities.

South Africa's airport development was also slow, but by 1932 there were 42 licensed aerodromes, including two at Cape Town and one each at Durban, Johannesburg and Kimberley. By 1936, the total had risen to 53, and by 1939 Cape Town's Wingfield Aerodrome had the Union's first aerodrome blind approach beacon.

Johannesburg/Jan Smuts Airport, opened in 1963, is the country's main international airport, and is notable for its elevation of 1,690m (5,500ft) above sea-level. This naturally dictates a long runway (for 'hot and high' take-off conditions) with a length of 4,420m

(14,500ft). Other important airports, similarly named after South African statesmen, are Capetown/D. F. Malan, Durban/Louis Botha and Bloemfontein/J. B. M. Hertzog.

In the Soviet Union, the first regular air service began in the summer of 1922, linking Moscow with Nizhny Novgorod (now known as Gorky). Moscow was naturally the main hub for early services, but linked routes radiated from such cities as Leningrad, Kiev and Tashkent.

As in other countries, the early airports were of a makeshift character; most were grass fields, but flying-boats, operating from lakes and rivers, helped to establish air links.

In 1932, a new unit, Grazhdaviastroi, was established within the Civil Air Fleet for the planning of airports and installations; it was later renamed Aeroproyekt. After World War II considerable effort went into airport construction and Aeroproyekt planned for no less than 1,220 airports by 1990, to increase to 2,000 ten years after. Aeroflot, the Soviet state airline, carried 302 million passengers in the period 1965 to 1970, and forecast 500 million in the following five-year period. The airline now serves some 3,500 Soviet towns and cities.

Airport design

In commercial aviation's early days, the design and development of airports was less beset with problems than it is today.

Open areas on the outskirts of cities were

more numerous and land was relatively cheap. People rarely worried about aircraft noise – there were not that many aircraft anyway – and in fact, with air travel in its infancy, the opening of passenger services was welcomed as a prestige operation. A city on an air route had 'arrived'.

Because land could be acquired easily, airport developers gave little thought to planning for the future. They bought or leased the land but did not take options on adjoining areas for future extension. The predictable result was that as the airport flourished, so did the city, and buildings began to spring up around airports, gradually encroaching until, in some cases, the airport was totally surrounded. Too late, the airport developers realized that as aircraft got bigger they needed longer runways, and so the story began again.

The airlines, with new aircraft in prospect, were able to say in effect that they would be unable to serve a city because its runway was too short, so it became necessary to build a

BELOW LEFT: *Pan American Airways' 100 million dollar terminal at John F Kennedy Airport, New York, was opened in December 1973, at which time it was said to be the largest terminal in the world operated by a single airline. Its plan is basically circular, as are those of many other modern terminals.*

OPPOSITE PAGE TOP LEFT: *A Trans World Airlines Boeing 707 taxiing after landing at Hong Kong Airport. This is one major international airport where only one runway has been built, in contrast to most other large airports where there are usually several, either parallel or crossing.*

new airport outside the city boundary. If it was too far away there were complaints that the time gained in flying was being wasted in ground transportation, but if it was built too close the urban encroachment began again, and as aircraft became noisier so complaints increased.

Nowadays, it is necessary in some cases to build airports a considerable distance from the cities they serve. Montreal's new Mirabel Airport is a good example; it is 56km (35 miles) from the city centre, but it is important to remember that the main factor is not so much distance as time. An airport only 19km (12 miles) from a city but with poor surface transport may take longer to reach than one 48km (30 miles) away that is served by an expressway or direct rail link.

Runways and terminals

Grass runways were adequate for the earlier transport aircraft, but as aircraft became heavier hard runways became necessary.

Runways were orientated depending on local conditions, such as the prevailing wind direction and size, height and position of obstructions in the area. By 1940, the US Civil Aeronautics Board was specifying 910m (3,000ft) runways for scheduled operation of small transport aircraft, and the 1,830m (6,000ft) longest runway at New York/La Guardia was exceptional. The longest runways at commercial airports are now in the region of 4,570m (15,000ft)

Multiple runway layouts are common at many civil airports, but there are some which still have a single runway – not necessarily by choice, as there may not be room for expansion, or there may be financial reasons. The single runway airport is of course at a serious disadvantage. It may be closed for a number of hours by a small incident, such as an aircraft skidding off the edge and blocking the runway, or may have to close down completely for a matter of two or three weeks for its runway to be resurfaced.

A single runway airport may be upgraded by building a second runway parallel to the first, taking less additional land than a cross runway, but doubling the airport's aircraft handling potential – assuming, of course, that the terminal capacity is sufficient for the extra passengers.

Airport runway designs are numerous – runways may be parallel and in line, parallel but staggered, or at right or acute angles to one another; runway layout must be linked with the architect's terminal plan.

At London/Heathrow, the terminal area is in the centre of the airport between the various runways, and linked to outside roads by tunnels. Heathrow has been subject to so much extension since it opened in 1946 that the central area is now too small and con-

gested, but because of its 'island' location it is incapable of enlargement.

Use of a parallel runway system with runways about 1,524m (5,000ft) apart has much to commend it. One runway can be used for landings while take-offs are made simultaneously from the other, and the aircraft have less distance to travel between terminal and runways. A number of new airport designs now use this basic concept with slight local variations.

The enormous Dallas–Fort Worth Airport, opened in 1974, has a central 'spine' road giving access to the terminal areas situated on each side. By its final stage, due for completion in 2001, DFW will have 13 semi-circular terminal units, with four north–south and two north-west/south-east runways. The first phase, now in operation, covers five terminal units, two parallel north/south runways, and one north-west/south-east. The present terminals have 65 gate positions and handled nearly 11 million passengers in 1976.

The new Paris/Charles de Gaulle Airport opened in 1974 with a single runway, but in its final form (due to be completed about 1990) two pairs of staggered parallel runways and one cross runway are planned. There will eventually be two terminals, and the first of these, Aerogare 1, is built on the satellite concept. The central circular terminal, 200m (656ft) in diameter, is linked by tunnels with moving walkways to seven satellite terminals, in effect large waiting lounges, each of which can handle up to four aircraft simultaneously, depending on size.

The Aerogare has 11 floors, the top 4 of which are car parks, and the basement contains baggage handling facilities. Passengers coming by car may use drive-in check-in desks, leaving their baggage before proceeding to the car park, from where they descend by lift to the departure floor.

The idea of a circular terminal is certainly not new – one could cite London/Gatwick's 'Beehive', opened in 1936. With a diameter of about 55m (180ft) the building looked attractive but, in the opinion of contemporary American aviation writer John Walter Wood, the rigidity of the plan caused confusion among passengers and the circular design prohibited subsequent extension. There were, he said, problems with the aircraft circulation around the terminal because of the one-way system, with manoeuvring taking place in a confined area. Gatwick did, however, have one unique feature – covered walkways which could be run out on rails from the terminal to the aircraft in inclement weather, enabling passengers to reach the aircraft and remain dry. Gatwick also claims another first – the first airport piers to be constructed in Europe, which were a feature

of the new post-war terminal.

Toronto Airport also liked the idea of a circular terminal, but called it the Aeroquay. It proved highly successful in use, but when it came to building a second terminal the Canadian Department of Transport opted for a less expensive linear design instead of the three Aeroquays proposed.

Washington's Dulles Airport is interesting on several counts. With two parallel and one cross-wind runway, it was the first airport designed specifically for jet aircraft, and was also the first designed from the outset to use mobile lounges to transport passengers between the terminal and the aircraft. The terminal is 183m (600ft) long and capable of extension to three times this length. Designed by architect Eero Saarinen, the building is a superb example of what can be achieved in airport architecture, and is a fitting terminal for the US end of the western world's first transatlantic supersonic passenger services operated by Concorde from London and Paris to Washington.

Airport operation

Airports may be operated by a country's airport authority, its department of transport, by local authorities, by airlines, by private owners or by contractors.

In the last category, one of the largest companies specializing in staffing and managing airports under contract to governments is International Aeradio Ltd. (IAL), based in the United Kingdom, near London/Heathrow Airport.

IAL has 24 subsidiary and associate companies in 17 overseas territories from the West Indies to Singapore. With its staff of 3,500 it can, if required, provide a complete service from airport feasibility studies to design, construction and full operation.

Airport finance

In some cases, notably the USA, funding for airports comes from the issue of bonds, particularly where a large amount of money is required for development work. The Federal Aviation Administration operates an exten-

sive Airport Development Aid Program which has enabled considerable expansion of many US airports in the last two decades.

Ideally airports should be self-supporting, or better still make a profit. An airport derives its revenue from a number of sources, which generally come under two headings – traffic operations and commercial operations. Landing fees and concessions are the airport's most lucrative sectors. Concessions cover a wide range of items including fuel and oil, various shops, restaurants, car parking and car hire services, flight insurance and advertising.

Staff costs account for a great deal of expenditure, followed by depreciation. The balance is split between rent and rates, utility and general services, equipment and supplies, maintenance and repairs and other expenses.

An analysis of operating revenues of a group of US airports, each of which handled more than two million passengers a year, showed that average percentage distribution was 37 per cent airfield area (landing fees, fuel and oil sales, airline catering, etc); 10·2 per cent hangar and building areas; 2·7 per cent systems and services; 15·8 per cent terminal areas (airline rents, boarding guards, etc), and 34·3 per cent concessions.

Moving the passenger
Many of the early airports were close to city centres, and most were served by public transport. In some cases the airlines used their own buses to bring passengers from town terminals. With few private cars, there was no parking problem. Today, as airports are built further from the cities they serve, passenger movement becomes a major factor in the consideration of an airport's design.

airbridges the passenger will walk across the apron and up a mobile staircase into the aircraft. Some aircraft, such as the BAC One-Eleven, are equipped with a built-in retractable staircase in the rear fuselage.

In this comparatively short journey, the passenger will be using quite a lot of airport equipment, and it is worth describing this and explaining how it evolved.

Check-in and baggage handling
From the very first passenger-carrying air services, it was necessary to check-in passengers and baggage and to weigh both, as weights were very critical on early aircraft which did not have the reserves of power available on today's airliners.

trollies for towing to the aircraft.

An automatic baggage sorting system was built by the French Teleflex company at Paris/Orly Airport. The bag is placed on a deposit conveyor at the check-in desk while ticket details are verified by computer; it then moves to a second position for weighing and tagging. The coded destination is marked magnetically and the bag moves to a waiting position until the computer, sensing a space on the baggage conveyor, releases the bag. As it passes down the conveyor belt, detectors sense the magnetic coding and the bag is diverted at its terminal to await transportation to the aircraft.

Another French company, Saxby Otis, has developed an electric driverless tractor

There are still very few specially built rapid transit links between airport and city, and the best-served airports are those on a railway line or with an autoroute connection.

Even so, the airport's handling responsibilities begin when the passenger arrives, either incoming by air or outgoing by surface transport. Basically, the departing passenger will move into the terminal to check-in at the appropriate airline desk, be relieved of his baggage, pass through passport and customs controls (if he is departing on an international flight), and into the departure lounge. Here he will be able to make his duty-free purchases (again, only if he is leaving the country).

When his flight is called over the public address system and shown on the information boards, the passenger will leave the lounge, pass through a security check and walk, or travel by moving walkway, to the departure gate. At this point he will either be taken by bus to the aircraft if it is parked away from the terminal, or will board the aircraft through an airbridge or air jetty connecting it to the terminal building. In the case of a smaller airport not equipped with

Passengers were issued with tickets which described the airline's liability in case of accidents. The requirements for passenger handling remain the same today, except that passengers are not weighed, an average weight being allowed by the airline. Today's big difference, however, has been brought about by the necessity of handling loads of up to 400 passengers on a single flight; the difference can be summarized in one word – computerization.

The computer age has meant a revolution in the airlines' seat reservations and check-in systems. Many of the larger international airlines now operate a push-button reservation system, by which they can feed in information at Sydney, for example, and find out within seconds if a seat is available on a specific flight from any airport in the world linked to their reservation system. The system can also make a reservation and can be further adapted to provide automatic ticketing.

Once the passenger and his baggage are checked, the baggage goes on to a moving belt behind the counter and passes onto a baggage sorting area before being loaded on

for the new Paris/Charles de Gaulle Airport. The tractor tows luggage trailers from the central luggage sorting point in the terminal building through underground tunnels to the seven satellite terminals. It is steered automatically by means of a front-mounted probe which follows cables beneath the roadway surface, the load destination being pre-programmed. About 70 such tractors are in use at the airport.

Most baggage is carried in holds beneath the aircraft's floor, and is loaded by means of an inclined conveyor belt mounted on a light truck chassis, but in recent years there has been a growing trend towards pre-packing baggage in containers which may be carried on an aircraft's main deck. These are more often used on wide-body high-capacity aircraft, when passenger loads are lower than those for which the aircraft were designed. Some Boeing 747s, for instance, have been converted to passenger/cargo combinations (combis), where the fitting of removable bulkheads in the fuselage, combined with quickly removable seats, can very easily provide the required combination of passenger and cargo load.

BELOW LEFT: *One of the electric driverless tractors developed for Paris/Charles de Gaulle Airport being tested. They are used to tow baggage trailers from the main terminal via underground tunnels to the satellite terminals. Steering is by means of cables buried in the roadway surface.*
BELOW: *The US Department of Transportation has developed a uniform system of symbols to help passengers. The symbols illustrated are a selection of the basic types which may be seen at American airports.*
LEFT FROM TOP: *Ticket purchase and baggage check-in, baggage claim, customs, immigration.* CENTRE: *Car rental, restaurant, coffee shop, bar.* RIGHT: *Lost and found, baggage lockers, elevator, men's toilets.*

RIGHT: *Most airports have now installed security systems to cope with the enormous amount of passengers' baggage to be checked. One of these is the American Scanray X-ray system; here the image it produces clearly reveals two pistols concealed among the rest of the luggage.*
BELOW: *Security systems in use at Helsinki Airport. Passengers themselves are checked by walking through a Finnish Metor metal detector, and hand baggage is inspected by an American Bendix X-ray system between the coils of the Metor. Events of the past few years have made these elaborate precautions necessary to ensure the safety of the millions of travellers using the world's airlines every year.*

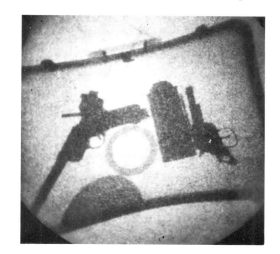

So much for departing baggage; arriving baggage begins the sequence of return to its owner by a reversal of the loading process. It is delivered from the aircraft to the terminal's baggage arrivals area, where it is off-loaded from the vehicle or container and placed onto a conveyor taking it to the baggage collection point in the arrivals hall. Here it is expelled onto a conveyor belt or 'carousel' bearing the flight number from which it has just come, and is collected by its owner.

There are several types of baggage conveyor, including a French system, the Teleflex Diplodocus, which, by use of a soft pressure cushion system, can carry baggage vertically upwards for short distances. This is valuable where constraints apply in a terminal design, making it difficult to provide space for a long, inclined conveyor.

Several variations on the baggage collection conveyor may be seen, ranging from the 'carousel' which, as its name implies, is a circular roundabout, to flat endless conveyors in a racetrack pattern. Here again, the type used will depend on the amount of space available, the degree of use the system may expect, and the amount of money available.

Information systems
To find his way to a seat in the aircraft, the passenger must be guided. It must always be assumed by the terminal planner that the passenger is making his first flight, and has no knowledge of airports and their systems. Because a large number of people of many nationalities use international airports, it is desirable to express certain basic information in the form of diagrams, or pictographs as they are called, and there has been considerable work on designs over the past decade. Details of flight times, destinations, etc, are now generally displayed on large information boards in departure and arrival halls.

These boards operate on several different systems – coloured lights, flap turn-over letters and so on – but the information they give is equally comprehensive and readily understandable. They are operated from a central point, and the information fed into the main boards is repeated on smaller boards in departure lounges, and sometimes at departure gates. In some cases, repeater displays are given via a closed-circuit television system at points around the terminal.

Security
While airport owners have always been conscious of the need for security from the very earliest days of civil aviation, events of the last decade involving sabotage, hijacking and other criminal activities have necessitated considerable expenditure on security systems for checking the passenger and his baggage. Initially, simple metal detectors were used, but later it was necessary to produce screening equipment which could detect explosives in addition to firearms carried in passengers' baggage.

Many screening systems are now in use throughout the world; most consist of a metal-detection framework through which the passenger walks and which gives the operator an audible or visual signal of the presence of metal objects on the passenger, and a unit for baggage which provides an X-ray picture of the bag's contents to the operator.

The full radiation dosage of all these systems is very low, and is carefully monitored by government departments to ensure that it does not even approach the danger level.

Reaching the aircraft

The larger airport terminals are usually connected to the aircraft by a telescopic covered walkway, known as an airbridge. These come in a variety of shapes and sizes – in single or double forms, the latter for use with wide-body aircraft demanding a large passenger flow. With a double bridge, it is possible to feed passengers into two entrances in the aircraft simultaneously, the forward entrance for passengers with seats at the front of the aircraft, and the centre or rear entrance for other passengers.

One of the early pioneers of the airbridge concept was Aviobridge of the Netherlands – now part of the huge Dutch/German VFW-Fokker group. Naturally enough, Amster-

dam/Schiphol, the Netherlands' premier airport, has a considerable number of Aviobridges.

Some airbridges or jetways have fixed anchorage points on the apron, the aircraft taxiing to a marked position and the end of the bridge extending out to meet its doorways, while other bridges have wheel-mounted ends and can be driven around an arc, enabling more flexibility in aircraft parking positions.

A simpler form of airbridge is the air jetty. This consists of a short, fixed tunnel connected to the terminal or passenger loading pier, and a very short telescopic front section which extends out to the aircraft's passenger door. The aircraft taxies up to the jetty, being

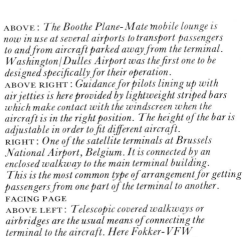

ABOVE: *The Boothe Plane-Mate mobile lounge is now in use at several airports to transport passengers to and from aircraft parked away from the terminal. Washington/Dulles Airport was the first one to be designed specifically for their operation.*
ABOVE RIGHT: *Guidance for pilots lining up with air jetties is here provided by lightweight striped bars which make contact with the windscreen when the aircraft is in the right position. The height of the bar is adjustable in order to fit different aircraft.*
RIGHT: *One of the satellite terminals at Brussels National Airport, Belgium. It is connected by an enclosed walkway to the main terminal building. This is the most common type of arrangement for getting passengers from one part of the terminal to another.*
FACING PAGE
ABOVE LEFT: *Telescopic covered walkways or airbridges are the usual means of connecting the terminal to the aircraft. Here Fokker-VFW Aviobridges are simultaneously serving a DC-9, three DC-10s and a Boeing 747 of KLM at Amsterdam/Schiphol Airport.*
ABOVE CENTRE: *In order to avoid an overlong chassis and steeply inclined staircase, these airliner steps made by the French company Douaisis have a 180° turn, and the height of the upper flight is adjustable to fit even the very large aircraft.*
ABOVE RIGHT: *The LTV Airtrains passenger-moving system in use at Dallas/Fort Worth Airport consists of remotely controlled vehicles which move along a specially built concrete trackway. It links not only the terminals but also outlying points such as car parks and cargo areas.*
BELOW: *New types of Aviobridges have been developed to serve special needs. Here an overwing Aviobridge is used to serve the rear door of a Boeing 747, while the front door is served by a standard type.*

aligned by the pilot by means of markings on the apron, on the terminal or, in some cases, by a coloured light system adjacent to the jetty, the correct colour sequence showing if the aircraft is properly lined up. A simple indicator is a lightweight striped bar extending from the jetty; the aircraft moves forward until the bar contacts the windscreen.

Where airbridges are not available, or where aircraft have to be parked some dis-

tance from the terminal, specially designed buses are used. In some cases, such as the German-built Neoplan vehicle, there are doors at each end to allow rapid loading and unloading of passengers.

However, the passenger still has to get into the aircraft, and while in the early days a small pair of steps sufficed, the high sill heights of the larger modern passenger aircraft necessitate wheel-mounted staircases.

These range from a simple, open staircase, with side rails which may be manually pushed into position, to sophisticated covered-in self-propelled stairways which may be adjusted hydraulically to the sill height. A French company, Douaisis, concerned by the increasing chassis length of mobile staircases as aircraft became bigger and the steps had to be made higher and therefore longer, has designed a unit which has a 180° turn in the staircase, enabling a greater height to be reached without increasing the length of the chassis.

Another way of getting passengers to their aircraft, although one which has taken much longer to establish itself, is the mobile lounge. Developed in the USA, this consists of a lounge with a capacity of about 150 passengers, mounted on a four-wheel chassis. The lounge section can be elevated from ground level to the height needed to load passengers into an aircraft and is self-propelled, operating from the terminal loading gate in much the same manner as an air jetty, with an extending connecting doorway.

Washington/Dulles was the first airport to be designed specifically for mobile lounge operation, and uses these exclusively. The lounges, built by Boothe Airside Systems and christened Plane-Mates, are considered to have several advantages over more conventional schemes for getting passengers and aircraft together. The lower cost of a smaller terminal (no in-built lounges required, as the passengers go straight from check-in to their mobile lounge) is one obvious saving, and mobile lounge makers claim that eliminating the use of airside buses makes a considerable saving.

Within the terminal, the passenger may have a fairly long walk between the terminal entrance or check-in desks and his departure gate. To alleviate this, some airports are installing moving walkways or travelators, continuous rubber belting moving at walking speed on which the passenger may stand or walk. These may be installed on a level or inclined plane, with the obvious proviso that the incline must not be too steep or the passenger would be unable to balance!

In some of the very large American air-

ports where there are a number of terminals, perhaps operated by different airlines, the transit vehicle has found favour. An example of this system is to be found at Tampa Airport, Florida, where the transit vehicles serve four satellite terminals, each accommodating up to seven aircraft. An elevated concrete trackway connects each satellite to the main terminal and carries electrically powered remotely controlled 125-passenger coaches locked on to a central guide beam. The trackways are duplicated, each operating two coaches moving simultaneously in opposite directions and taking 40 seconds to complete the satellite to terminal journey.

At peak periods, the eight-coach Tampa system is said to be capable of moving 3,000 passengers in 10 minutes. Operated from a sophisticated control point, the system is computer-controlled and monitored; the centre is continually manned. Similar systems with variations are in use at several US airports.

At Dallas–Fort Worth Airport, a transit system built by LTV and known as Airtrans provides a link not only within the terminals but also to outlying points around the airport such as car parks and cargo areas. With 21km (13 miles) of track in its initial form, Airtrans consists of rubber-tyred 40-passenger cars running under computer control in a reinforced concrete guideway. A maximum operating speed of 27km/h (17mph) maintains an 18-second separation between successive cars; there are almost 70 vehicles in use at the present time.

BELOW: *Mobile runway lighting systems were fairly primitive in the early days of passenger travel. This photograph shows a cart equipped with lamps which was used by Lufthansa when the airline first began its operations.*

RIGHT: *This giant snow-clearing unit used by the Russian airline Aeroflot in the severe winters of the Soviet Union has a bulldozer blade to push snow aside, a brush in the centre and a blower at the rear to blow snow clear of the swept runway.*

CENTRE RIGHT: *The French Bertin TS-2 runway ice-removal vehicle used hot air from a gas turbine.*
FAR RIGHT: *Test running the engine of a DC–9 with silencers (de-tuners) in position. In the background, de-tuners for the Boeing 747.*

Fire and rescue services

The fire and rescue service is a most important part of an airport's operation. Without a service of this type, which must reach certain defined standards, an airport will not be given a license to operate.

There has been a steady development in the field of airport fire and crash rescue vehicles. Early requirements could be met by virtually standard fire engines, but now there is a whole range of purpose-built fire-fighting equipment, from small rapid intervention vehicles (RIVs) to very large multi-wheel appliances.

The purpose of the high-speed RIV is to get to the incident in the fastest possible time, and to contain the fire until the arrival, some seconds later, of the large-capacity vehicles which use foam to smother the flames. While most fire-fighting vehicles are wheel-mounted, there are some interesting exceptions. The Canadian company Canadair Flextrac has supplied six caterpillar-tracked

fire tenders to the Indian Airports Authority, while Auckland Airport, New Zealand, uses an SRN.6 Hovercraft for traversing the mud flats and shallow waters around the airport at low tide.

Some airports have foam-laying equipment for use in emergencies. For instance, if an aircraft is unable to lower its undercarriage, a carpet of foam is laid on the runway so that the aircraft may touch down smoothly on its belly on the foam, which will smother sparks caused by friction of the aircraft's surfaces on contact with the runway.

A hazard less easy to deal with causes problems at some airports. Flocks of birds rising into the air at the approach of an aircraft create considerable danger. Their ingestion by an aircraft engine can cause it to stop, while 'bird strikes' on other parts of the aircraft have been known to cause crashes.

Various methods of scaring birds have been tried by the airport authorities, ranging from a vehicle with loudspeakers driving

down the runway and relaying tape recordings of bird distress calls, to the use of birds of prey, such as falcons and goshawks. But the only answer seems to be that the area must be made less attractive for birds to nest and roost. Such ornithological attractions as rubbish dumps must not be allowed in the vicinity of airports.

Snow and ice clearance are major problems at some airports; a closed runway can mean diversions to another airport or prevent aircraft from leaving, with subsequent loss of revenue. In the more temperate climates the problem is usually resolved fairly easily, except in cases of an exceptionally heavy or untypical fall, but in such countries as Canada, where high snowfalls can be expected, most airports are well equipped to deal with such a situation.

A paper by the Airport Facilities Branch of the Canadian Ministry of Transport describes the snow removal operation. A high-speed runway sweeper with a 4·27m (14ft)

wide brush clears a 3·66m (12ft) path at speeds up to 48km/h (30mph) on snow up to 7·6cm (3in) deep. Once several widths have been cleared, the ploughs are used in conjunction to pick up and cast aside the loose snow. Blowers, operating at speeds up to 40km/h (25mph), follow the plough and sweeper, blowing the snow at high speed over the runway edge, and this is finally removed at least 7·62m (25ft) beyond the edge of the runway lighting.

Too much slush or water on runways can cause an aircraft to aquaplane or skid, so a device is used to measure and record runway friction. Towed up and down a runway behind a vehicle, this device can assess whether or not the runway is safe for operations. One way to lessen the danger of water build-up on runways is to cut transverse grooves in the surface, enabling surplus water to drain away.

Runway lighting must be kept clean, particularly after snow, and specialized equipment is available for this purpose. The French Bertin company has produced a semi-automatic cleaning device named Fluxojet which, it is claimed, can clean 100 lights in an hour. The method is to place a 'head' connected to sources of air pressure over the light windows, blow a powder over the glass at high pressure to remove the dirt, and use suction to remove the residue into a waste tank.

Bertin also produces the Thermosoufflant, a large wheeled vehicle used for melting snow and ice on runways by means of heat generated by a gas turbine in the vehicle.

Fog has always been a source of concern at airports. During World War II experiments were carried out at certain RAF stations in England in an effort to clear fog sufficiently for aircraft returning from operations to land safely. The FIDO system (Fog Investigation and Dispersal Operation) consisted of petrol burners installed along the edge of runways and ignited when aircraft were approaching to land. Several thousand safe landings were made with the aid of FIDO, but it would have been prohibitively expensive in peacetime commercial operations, especially in view of the progress made in automatic landing. Another method is cloud seeding, used successfully by airlines on the US West Coast.

However, the versatile Bertin company has developed a fog dispersal system, Turboclair, which has been installed at Paris/Charles de Gaulle Airport following successful tests at Paris/Orly. This consists of gas turbines in bunkers alongside the runway, fog being dispersed by the heat from the turbines' efflux.

Moving damaged aircraft

The moving of aircraft which have suffered damage in emergency landings must be carried out carefully to avoid further damage, and must also be done as rapidly as possible, particularly if the aircraft is blocking a runway. One system, designed by a UK company, RFD Inflatables Ltd, uses pneumatic bags (elevators) which are inserted under the aircraft at certain points and then inflated to raise it. When a sufficient height has been reached, the landing gear may be lowered if it is undamaged, and the aircraft is then towed away. If the landing gear is damaged, inflatable tracks are laid and a jacking system lifts the aircraft clear of the ground, allowing it to be towed along the tracks.

Aircraft support

A wide range of equipment can be listed under this broad heading. Some of the vehicles are adapted by airports and airlines from standard vehicles, for instance the small 'follow me' cars and pick-ups used at some airports to guide aircraft to their allotted parking spaces. Others are purpose-designed.

The largest vehicles to be seen at an airport are the fuellers, of which there are two types: the standard tanker and the hydrant dispenser.

Tankers and the aircraft they serve have grown together. When commercial aviation began it was necessary to hand-pump fuel into an aircraft's tanks, often standing on the top wing of a biplane to do so, while in some more primitive areas it was necessary to strain all fuel before it went into the aircraft.

Today's large-capacity aircraft may demand more than 90,000 litres (23,775 US gallons) of fuel. This would require an exceedingly large and heavy tanker – for instance, a Gloster Saro articulated tanker fuelling Boeing 747s at Brussels National Airport in 1970 had an 82,000-litre (21,663 US gallons) capacity and was more than 21m (70ft) long. Such a vehicle needs considerable manoeuvring areas on the apron, and in these days when rapid aircraft turnrounds are demanded, very high fuel flow rates become necessary. The big Gloster Saro vehicle is capable of pumping 3,782 litres (999 US gallons) a minute.

With a hydrant system, fuel is piped from the airport's fuel farm to points on the apron area, and the hydrant dispenser, a much smaller vehicle than the tanker, is connected to the hydrant, forming the link between this and the aircraft's fuel tanks. The dispenser measures the fuel flow and quantity, and modern dispensers can achieve a flow rate of 5,550 litres/min (1,466 US gallons/min)

simultaneously through four hoses. Fuel filters ensure that the fuel is not contaminated.

Because of the necessity to provide underground piping, a hydrant system can be costly to instal, and it offers less flexibility than tankers but the disadvantages are offset by a higher flow rate. It is also cheaper to operate in terms of manpower. When an airport apron is being extended, enlarging of a hydrant system can be expensive, and for these reasons it seems likely that both tankers and hydrants will be used during the foreseeable future.

The need to move aircraft on the ground has led to a revolution in the design of very powerful towing vehicles, or tugs. When aircraft were smaller and lighter they could be towed by conventional tractors, but the advent of the large post-war airliner has demanded a radical re-think by the tractor designer. The T-800S jumbo tug built by the US International Harvester Company is typical of the large-capacity aircraft tugs; it is able to pull a Boeing 747 weighing about 362,850kg (800,000lb) and itself weighs 49,000kg (108,000lb). Powered by a pair of eight-cylinder turbocharged 300hp diesel engines, it has a driving cab at each end and can be driven from either. The cabs may be elevated or lowered flush with the body of the tractor, which is known as a 'low profile' type because it is only 1·57m (5ft 2in) high and can be driven beneath the fuselages of large aircraft.

All tugs need a towbar to tow aircraft; these are normally connected to the aircraft's nose-wheel undercarriage leg and come in a variety of sizes with varying endplate fittings which will adapt to various types of aircraft.

Air cargo is a fast-growing part of air transport operations, and special terminals with a considerable amount of computerized equipment are now in use at many airports. Much of the cargo-handling gear is computer-controlled and monitored, including cargo sorting in some cases.

When cargo is delivered to the terminal by standard road vehicles, specially designed cargo handling equipment is necessary to reach the loading doors of the aircraft. In the case of all-cargo aircraft, much of the load consists of goods ready packed in large oblong containers, the standard size being 6·1m long by 2·4m wide by 2·4m deep (20ft × 8ft × 8ft). The benefit of this type of container traffic is that it can travel equally easily by road, rail, sea or air, and there is considerable interchange traffic between these modes.

The largest cargo loaders currently in service are made in the USA; Cochran Airport Systems' Model 320 loader has a lifting capacity of 27,215kg (60,000lb), with a hydraulically powered cable-operated lift mechanism. It is a self-propelled vehicle with a six-cylinder engine. Some of the smaller loaders in use, lifting around 13,605kg (30,000lb), are raised and lowered by the scissor-lift principle – steel crossover beams which lay flat when the loading platform is at chassis level and become X-shaped as the platform is raised.

The platform decks of the loaders are fitted with a series of rollers or bearings on which containers are rolled into or from the aircraft, which have similar rollers on the floor of their cargo holds. Some of the smaller loading platforms are towed by tractors, while the smallest container trailers – 'dollies' – are tractor-towed and normally built to a fixed height. Some variants incorporate a turntable for easier movement of the containers onto the larger container loaders.

A wide variety of service vehicles may be seen clustered round an aircraft being prepared for flight; a Lufthansa description of

TOP: *Moving Sabena's Boeing 747s requires a large tug to avoid having to use the 747's own power. Here a French-built Secmafer SF700 is being used for the job. This is typical of the large capacity aircraft tugs, being very heavy itself and quite powerful enough to tow even a jumbo jet loaded with fuel and cargo.*

ABOVE: *Fire precaution is an important part of an airport's services. Here a turn-out by the Frankfurt Airport fire department shows the wide variety of vehicles available.*
TOP RIGHT: *A lorry-mounted 'cherry-picker' maintenance platform servicing the tail of a DC–10 belonging to the French airline UTA.*

CENTRE RIGHT: *Loading cargo containers on a Seaboard World Airlines Boeing 747 Freighter with a purpose-designed cargo transporter and cargo-loading elevator.*
RIGHT: *Refuelling an aircraft from a hydrant on the apron using a hydrant-dispenser vehicle as a go-between.*

'ramp lice' may sound unkind, but somehow aptly describes the scene. They include vehicles for cleaning out aircraft toilets, for replenishing water supplies, galleys and so on, and ground power units (GPU). These GPUs provide power for the aircraft's electrical systems when the engines are not running, and also for engine starting. Some aircraft are fitted with airborne power units (APUs), small engines which maintain the aircraft's electrical supply when it is on the ground, making it independent from the airport's electrical services. These are particularly useful at very busy airports, where GPUs may all be in use, or at more remote airports where they may be in short supply.

Some airports are now using electrical power systems which can supply power to the aircraft on the apron from a ring main fed into the airport's central generating plant. The Dallas–Fort Worth Braniff Airways terminal has a built-in power supply for this purpose, and there are others at London/ Heathrow, Brussels National and Stockholm/ Arlanda airports. While the idea has not yet gained overall acceptance, the general consensus of the airports using it is that it results in cleaner apron areas and cuts down noise and pollution caused by GPUs and APUs.

Aircraft maintenance

All the previously mentioned equipment is used when the aircraft is at the terminal apron. Another range of equipment comes into play in the airline maintenance areas.

The biggest piece of hangar equipment is undoubtedly the maintenance docking system. British Airways at London/Heathrow has a new dock for its Lockheed TriStar fleet. When a TriStar has to undergo routine maintenance, it is towed into the hangar along marked guidelines and the dock, consisting of working platforms at five levels, is moved into position around the aircraft, enabling engineers to work on all parts of the aircraft and its engines.

Some cleaning is carried out while the aircraft is docked, but when external cleaning is needed at other times it can be done on the parking apron by specially equipped vehicles which incorporate a mobile platform on an extending arm with a hose attachment. These 'cherry pickers', as they are known, can also be used for high-level maintenance, windscreen cleaning, fire-fighting and clearance of ice and snow from the aircraft's surfaces.

Airlines also need a considerable amount of test equipment, ranging from engine test cells to rigs for testing engine fuel flows. Engine testing is a necessary part of aircraft maintenance, and many airlines have built noise mufflers, or detuners, for this purpose. Swissair, at its Zurich maintenance base, has a massive installation capable of testing all four engines of a Boeing 747 simultaneously. The four large silencer tubes, each nearly 24m (78ft) long are enclosed in concrete foundations faired into a 7m (23ft) high wall which is 56m (184ft) long.

When mufflers are not available, blast deflectors are installed to enable aircraft to run up their engines to full power in the maintenance area without the jet blast affecting nearby vehicles and installations.

Monitoring and measuring

Perhaps the major complaint levelled at airports concerns noise, and unfairly so, since the airport itself does not generate the noise about which the complaints are made. This comes from its customers, the airlines, but airport authorities find themselves acting as buffers between airlines and the incensed local population, who may have legitimate cause for complaint.

The various means of quietening aircraft by the fitting of so-called 'hush kits' to their engines, changing the aircraft's angle of approach to landing, and so on, are described fully on page 101. Many airports now have noise monitoring equipment. This is set up at a number of points on and around the airport, and the noise measurements recorded are passed automatically to a central recording and control point for analysis.

As an incentive to airlines to reduce noise, Frankfurt Airport operated a scheme under which aircraft were awarded a number of points per landing, and at the end of the year the airlines with the best 'quietness' record were given a cash bonus.

Effective air traffic control at an airport demands continuous knowledge of weather conditions and visibility, and to obtain this two pieces of equipment are used. A ceilometer is an optical radar which gives the height of the cloud base above an airport, by measuring time taken for a pulse of light to travel from the transmitter to the cloud base and return to the receiver. The light source in the case of a newly developed ceilometer by the UK Plessey Company is a semiconductor laser.

To measure horizontal visibility, a system known as runway visual range (RVR) has been developed. Before this, markers were placed at fixed intervals along a runway within view of a runway control caravan. By counting the markers visible the observer could ascertain the range of visibility.

The most modern means of measurement in this field is instrumented visual range (IVR), the first fully automated system for the purpose, which has been developed by a UK company, Marconi Radar Systems. Operating from installations alongside the runway, IVR feeds range measurements automatically to a central monitoring point.

Movement of aircraft on the ground in bad visibility is helped by a radar equipment in the control tower known as ASMI – Aerodrome Surface Movement Indicator. Using a very short wavelength, this enables aircraft to be directed along taxiways.

Air Traffic Control

A description of Air Traffic Control (ATC) services on an international basis would take several books. This section therefore deals with ATC as it affects air traffic arriving at, departing from or overflying Australia.

In Australia, the National Air Traffic Control Service is a joint civil and military control service under the Departments of Transport and Defence. ATC has direct responsibility for the movement of civil and military aircraft over Australia. The Australian airspace is divided into five Flight Information Regions (FIRs) – Queensland, New South Wales, Victoria, Tasmania, South Australia, Northern Territory and Western Australia. Each region has its own ATC centre. The Sydney Area Approach Control Centre (AACC) at Sydney (Kingsford-Smith) airport, for example, covers airspace from Albury, in the south, to Coffs Harbour and Moree, in the north. Going east, it covers a large slice of the Pacific Ocean, and additionally as far north as the Equator and extending south of Tasmania. The western boundary reaches almost as far as Broken Hill. Airspace in the FIRs is divided into sectors – some for aircraft overflying en route, some scheduled into terminals, covering the principal airports of Australia.

Each sector has a Sector Controller (SECT) who manages operations and who is also responsible for the transfer of air traffic to adjoining sectors as it comes into and goes out of his sector. He has the assistance of a sector radar controller who is responsible for separation of traffic, chiefly by radar. Radar display presents a bright image on a screen, giving the aircraft's position and direction. The radar scanner linked to the centre, provides a sweep of a section of sky. A bright blip (or paint) on the controller's screen gives the aircraft's position. Radar controllers at the AACC can handle up to 12 aircraft simultaneously. At major airports in Australia radar controllers no longer work in the dark – they have been working in 'daylight' conditions for some years. Sector controllers maintain continuous contact with adjoining sectors to ensure correct separation between aircraft. A military controller at the ATC coordinates the movement of military aircraft crossing civil air routes.

The AACC Operations Room is under the supervision of a Senior Area Approach Controller (SAAC) who controls a watch of about 25 controllers and 5 flight data officers. There are 4 watch teams which provide a 24-hour service. Controllers monitor aircraft with flight progress strips which, for each aircraft, show the call sign, aircraft type, point of departure, destination, and estimated times over selected points en route. The strips are arranged in geographical order and time sequence and are prepared by the flight data officers. They are up-dated as fresh information comes in. The controller talks to aircraft by radio telephone in his particular area of responsibility. Continuous landline connections allow him to liaise with controllers in other centres. Placed above the controller's radar screen is a close-circuit TV screen. Displayed here are the locations of thunderstorms within a radius of 50 nautical miles.

Approach and landing control

Aircraft pass through various controls out of the airways system into an airport and to a selected terminal and terminal gate. Bright radar display units are provided for the approach/radar controllers in the AACC room. In the operations positions are the approach controller and three radar directors.

The approach controller accepts the aircraft which has been released and transferred by the sector controllers and takes his first radio call from the aircraft. He instructs the captain of the aircraft on landing procedures and runway use.

The radar directors, after the aircraft has been released by sector control, will sequence traffic, using surveillance radar, for bringing aircraft to the point of final approach about

Any listing of the world's airports based on the number of aircraft movements or passengers handled and placed in order of traffic would inevitably include a large proportion of US airports in the first 20. A conservative estimate of 180,000 civil aircraft in the US indicates why this is so.

The listing below therefore gives a more balanced picture, in alphabetical order, of airport traffic throughout the world, both in terms of passengers and cargo. It is based on statistics from the Aeroport de Paris and the British Airports Authority for 1975.

Traffic at Selected World Airports

Toronto Airport, Canada

Melbourne International Airport

Pretoria Airport, South Africa

JFK Airport, New York

Frankfurt Airport, West Germany

IATA code	Airport	Total aircraft movements	Total passengers	Cargo (tonnes)
ADD	Addis Ababa (Ethiopia)	29,400	178,100	6,053
ALG	Algiers/Dar el Beida (Algeria)	56,200	1,808,000	14,650
AMS	Amsterdam/Schiphol (Netherlands)	173,300	7,534,300	226,322
ATL	Atlanta/Hartsfield (USA)	469,000	25,268,900	201,942
ATH	Athens (Greece)	98,500	5,085,200	33,600
BKK	Bangkok/Don Muang (Thailand)	50,200	2,580,800	52,809
TXL/THF	Berlin/Tegel & Tempelhof (West Germany)	55,500	3,992,600	13,529
BLZ	Blantyre/Chileka (Malawi)	16,600	281,200	3,977
BOS	Boston/Logan (USA)	258,500	10,515,400	142,295
BRU	Brussels National (Belgium)	97,100	3,942,900	100,412
CCU	Calcutta (India)	30,400	1,008,600	10,682
CCS	Caracas/Maiquetia (Venezuela)	73,400	2,715,100	78,597
CHI	Chicago/O'Hare (USA)	666,600	37,123,000	544,745
CPH	Copenhagen/Kastrup (Denmark)	163,000	7,575,800	129,615
DFW	Dallas-Fort Worth (USA)	341,400	10,864,900	63,920
DEN	Denver/Stapleton (USA)	386,500	12,026,400	94,089
DLA	Douala (Cameroon)	23,200	295,800	18,010
DUB	Dublin (Irish/Republic)	79,400	2,193,600	42,300
DUS	Dusseldorf/Lohausen (West Germany)	113,100	5,124,100	27,198
FRA	Frankfurt (West Germany)	209,200	11,967,900	404,157
GUA	Guatemala/La Aurora (Guatemala)	83,400	596,700	13,765
HEL	Helsinki (Finland)	86,000	2,812,700	18,606
HKG	Hong Kong/Kai Tak (Hong Kong)	83,200	3,872,100	102,257
IST	Istanbul/Yesilkoy (Turkey)	49,800	2,807,100	16,866
JNB	Johannesburg (South Africa)	64,200	3,131,900	71,204
KUL	Kuala Lumpur (Malaysia)	42,400	1,362,600	11,239
LIS	Lisbon/Portrela (Portugal)	45,000	2,372,900	35,962
LGW	London/Gatwick (UK)	105,100	5,343,500	75,266
LHR	London/Heathrow (UK)	276,100	21,294,800	405,005
LAX	Los Angeles International (USA)	453,600	23,719,000	558,205
MAD	Madrid/Barajas (Spain)	118,700	7,966,400	108,830
MEL	Melbourne/Tullamarine (Australia)	96,000	4,536,000	67,900
MIA	Miami International (USA)	287,600	12,068,100	337,690
LIN	Milan/Linate (Italy)	89,400	4,032,700	52,043
YUL	Montreal/Dorval (Canada)	191,200	6,630,700	178,600
SVO	Moscow/Sheremetievo (USSR)	43,400	2,387,300	44,506
MUC	Munich/Riem (West Germany)	116,700	4,259,800	25,205
LGA	New York/La Guardia (USA)	299,200	13,185,500	45,862
JFK	New York/John F Kennedy (USA)	300,800	19,475,800	876,366
OSA	Osaka (Japan)	133,600	11,354,500	115,330
OSL	Oslo/Fornebu & Gardemoen (Norway)	132,400	3,037,400	23,099
PMI	Palma (Majorca)	74,500	6,802,600	22,919
PTY	Panama/Tocumen (Panama)	22,500	980,000	32,484
CDG	Paris/Charles de Gaulle (France)	85,800	6,009,800	220,244
LBG	Paris/Le Bourget (France)	84,200	1,463,600	16,315
ORY	Paris/Orly (France)	150,100	10,611,400	119,871
PHL	Philadelphia (USA)	298,100	7,515,100	109,725
PRG	Prague/Ruzyne (Czechoslovakia)	84,900	3,662,000	32,176
GIG	Rio de Janeiro/Galaeo (Brazil)	77,800	3,997,600	72,213
FCO	Rome/Fiumicino (Italy)	142,800	8,278,500	172,600
SFO	San Francisco International (USA)	326,700	17,503,800	294,432
CGH	Sao Paulo/Congonhas (Brazil)	109,300	3,609,900	33,989
SIN	Singapore/Paya Lebar (Singapore)	65,000	3,981,600	64,188
ARN	Stockholm/Arlanda (Sweden)	84,900	3,662,000	32,176
SYD	Sydney/Kingsford Smith (Australia)	156,700	6,692,500	96,900
YYZ	Toronto/Malton (Canada)	237,700	10,404,300	163,100
TUN	Tunis/Carthage (Tunisia)	27,100	1,500,600	7,642
VIE	Vienna/Schwechat (Austria)	62,800	2,022,400	23,625
DCA	Washington National (USA)	306,500	10,810,100	34,714
ZRH	Zurich/Klöten (Switzerland)	138,000	5,963,400	113,787

16km (10 miles) from the runway touch-down point. Radar directors have the responsibility of ensuring that correct separation between aircraft is maintained. Separation varies – it could be as much as 16km (10 miles) for smaller aircraft which may be behind wide-bodied jets like 747s, DC–10s, Tristars and others. These types produce considerable wake turbulence which could endanger smaller aircraft, unless adequate separation is maintained.

At busy periods over major airports in Australia – in much the same way as airports in other parts of the world – aircraft waiting to land may have to go into a 'stack', in a selected airspace near the airport. This could mean that an aircraft may have to circle a radio beacon some 50 nautical miles from

to Sydney, Australia's busiest domestic air route–sector control to approach control to tower control and finally to surface movement control which brings the passengers to their final debarcation.

Radar

So far as the airport is concerned, two radars are used: surveillance and precision approach.

Surveillance radar is used by the airport controller mainly to separate aircraft tracks and position the aircraft for their landing approach. Generally speaking, the controller brings in the aircraft with this radar to a point just before it is positioned on an imaginary line extended from the centre-line of the runway, and then hands over control to

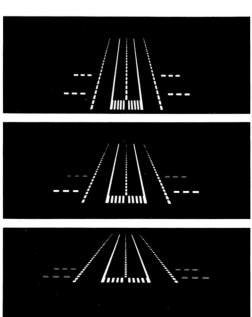

the airport at a prescribed altitude until told to descend by approach control. The top of the stack could be at 6,100m (20,000ft) and aircraft are called on to descend in steps of 300m (1000ft) until they can finally leave the bottom of the stack at 900m (3,000ft) for a final approach to land. An aircraft on final approach and lined up with the runway it has been instructed to use, is then handed over, or transferred, to the airport control tower which will actually bring it in to land.

The tower controller uses a radar display calibrated in miles. The blip on the screen indicates the aircraft's movements on the approach to land. Once the aircraft has landed and turned off the runway on to one of the taxiways, tower control hands it over to the surface movement controllers. They then marshal the aircraft on the ground to the appropriate terminal and terminal gate. So that is the sequence of controls for an aircraft flying, for example, from Melbourne

the pilot for either a visual or an instrument landing.

Precision approach radar (PAR) is, as its name implies, a more precise method of control, but, in its original form, is now used less frequently as 80 to 90 per cent of major international airports are equipped with instrument landing systems (ILS), which have a high degree of accuracy. When combined with surveillance radar, ILS renders PAR unnecessary.

Instrument Landing Systems (ILS)
Introduced in its basic form in 1929, an ILS now consists of a ground transmitter which projects a narrow radio beam into the air to the left and right of the runway centreline at an angle to the glide path – the path an air-

craft takes on its normal landing approach in a controlled descent.

The radio beam is received by equipment in the aircraft, and displayed on cockpit instruments as vertical and horizontal lines showing the pilot his position in relation to the beam. The pilot positions the aircraft merely by keeping the lines crossed at the centre of the instrument, thereby ensuring that his approach is at the right height and angle for landing.

Microwave Landing Systems (MLS)
ILS was adopted by the International Civil Aviation Organization (ICAO) in 1949, but with constantly increasing air traffic the scientists have been looking at ways to improve the basic system. Certain drawbacks

are evident with ILS – aircraft approaches are confined to a single narrow path, siting of the equipment at some airports is difficult because surrounding topography and buildings can cause reflections, and only a limited number of frequencies are available for transmission.

The planners are now looking at microwave landing systems (MLS) which would generate narrow beams more easily than ILS and would have far more channels available. A number of advantages are claimed – simple installation, none of the routine flight calibration necessary with ILS, no critical adjustments and high basic immunity from interference. Prototype equipment has been tested since the early 1970s and production is expected to follow.

A Typical Flight

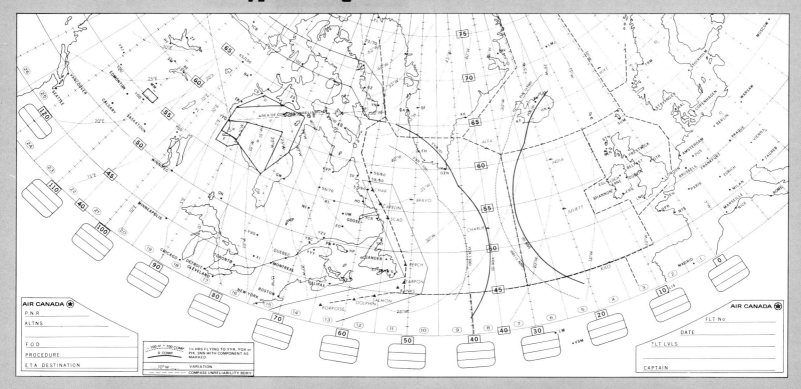

Air Canada Flight AC857, London to Toronto

Air Canada's Flight AC857 from London to Toronto on a day in late October 1976 is typical of many international flights.

It had featured in the airline's schedules months ahead, but this description takes up the story on the day of the flight.

The aircraft, Boeing 747 C-FTOE, the fifth of Air Canada's six 747s, arrives at London/Heathrow Airport on an inbound flight the day previous to departure, and is thoroughly checked over by Air Canada's London maintenance crew.

The first crew member to arrive at the airport is the Flight Director, 30 minutes ahead of the rest of the crew. His initial task is to check with the flight dispatch office on the forecast weather so that he can plan meal-times to avoid turbulent conditions as far as possible, and to get a list from terminal control of any VIPs, passengers who require special meals or special equipment, such as wheelchairs. The flight meals are supplied by Air Canada's own commissary unit at the airport.

The Flight Director is in charge of a cabin crew consisting of a purser, assistant purser and 11 flight attendants; during the flight he is responsible for looking after three accounts – for the on board duty-free goods, for the bar service, and for the hire of headsets. These, used for in-flight entertainment (music or the sound track of movies) are provided free in first class but because of an IATA regulation are charged in economy seating.

The flight crew, pilot (captain), co-pilot (first officer) and flight engineer (second officer) arrive at the airport about an hour before scheduled departure time. The captain and first officer go to flight dispatch for their briefing, while the second officer goes to the aircraft to oversee the fuelling.

Briefing is routine and therefore only takes about 10 minutes instead of the usual 20 associated with Atlantic weather; flight dispatch has

already received weather actuals and other information from the destination and specified alternative airports, and has prepared a weather chart showing recommended and available routes.

On acceptance of his route, the captain is given a printout with all details of the flight, including the course, times and distances to the various waypoints along the route etc. Armed with this information, the first officer plots the course on a flight progress chart. (Unlike most other airlines, Air Canada produces all its own airways, progress and landing charts.) The flight plan is then confirmed with air traffic control and filed with the National Air Traffic Service, with copies to the ATC regions through which it will pass, the destination airport and the alternatives.

Boarding of AC857 begins at 11.20 for 12.20 departure. Passengers pass through security checks with their hand baggage, their suitcases being X-rayed after baggage and ticket check-in.

The aircraft takes on 95,000kg (209,000lb/154·5 US tons) of fuel for its flight to Toronto and is expected to use about 75,900kg (167,000lb).

Push back by tug from the air jetty at Heathrow's No. 3 terminal begins at 12.25, and following the completion of the cockpit check list, C-FTOE moves out to the runway threshold to await take-off clearance from air traffic control.

At a take-off weight of 285,300kg (642,000lb/321 US tons) with 293 passengers aboard, the jumbo is airborne at 12.42 from runway 10R after a 28-second run, and turns left to join the airway at Brookmans Park, near Hatfield, at a height of 1,500m (5,000ft). Rate of climb is 400m/min (1,300ft/min) passing through slight turbulence between 6,400–7,000m (21,000–23,000ft) on the way to cruise altitude.

As the aircraft passes through the successive height bands assigned to it by air traffic control, the co-pilot reports the height and obtains clearance to the next required altitude.

Because of other traffic, it is not possible to obtain the desired cruising height of 11,900m (39,000ft) until somewhat later in the flight.

Cruising at Mach 0·84 (84 per cent of the speed of sound at the appropriate altitude), the aircraft becomes lighter as it burns off its fuel, consequently the flight engineer needs to keep the captain informed of the maximum power to be achieved by the engines to maintain the normal cruising speed. This is done by using a small manual calculator.

A cockpit display shows the distance to the next VOR along the airway and the aircraft is cleared to the Stornoway VOR, reaching 9,450m (31,000ft) over Stornoway.

Over the Midlands, the aircraft's inertial navigation system (INS) is brought into operation. Built by Litton Industries in the USA as the model LTN–72, this highly sophisticated computerized equipment guides the aircraft along its route. The waypoints plotted on the flight progress chart are fed into the computer, and once the INS has been activated it controls the aircraft's autopilot, making allowances for wind, drift and so on. The aircraft is flown along the selected track by use of 2-axis gyros. The only external input required for INS is the air temperature.

As each waypoint is reached, INS, through the autopilot, automatically turns the aircraft on to the appropriate course for the next waypoint, and a cockpit display shows the distance and time to the next point, as with the VOR beacons.

A maximum of 9 waypoints may be accommodated in the INS, and since 12 are required for this flight it is necessary to re-programme the system in flight to insert the remaining waypoints. This is achieved by switching the system to 'remote' and re-programming. The INS is duplicated to provide a cross-check and a fall-back system in case of failure, although this is very infrequent.

Another 747 contrailing 610m (2,000ft) above and to the left gradually diverges from our course and

disappears to the west as the airways are left and the aircraft climbs to 10,500m (35,000ft). ATC at Gander, Newfoundland, can be heard at 14.05, forty minutes before flying over Reykjavik, Iceland.

With long flights, the necessity to maintain a continuous listening watch on radio frequencies is eliminated by Selective Calling (Selcal). With this, the aircraft has a permanent call-sign – in this case BJEF – and when the ground controller wishes to communicate he calls this on the aircraft's frequency and a visible and audible warning is given to the crew to listen for a message.

In addition to standard instruments, Air Canada's 747s are fitted with weather radar, with a small presentation screen in front and to one side of the pilot and co-pilot. The radar can be adjusted to show the weather situation at ranges of 48, 160 or 480km (30, 100 or 300 miles).

Flying from east to west following the sun, southern Greenland and northern Canada are crossed. An eventual altitude of 11,300m (37,000ft) has been granted by ATC, but at 19.36 Toronto Control Centre authorizes the 747 to begin its descent, initially down to 7,300m (24,000ft). By 19.50 the aircraft is at 3,000m (10,000ft).

The 747's huge flaps are lowered at 2,600m (8,600ft) and the aeroplane edges out over Lake Ontario to begin its descent to runway 32 at Toronto International Airport for an ILS approach in good weather. The landing gear is lowered at 580m (1,900ft) and the fence is crossed. The 747 touches down at 257km/h (160mph) and taxies to the gate at Terminal 2, reaching the stand at 20.06 London time (15.06 Toronto time), 7 hours and 40 minutes from Heathrow, about 5,200km (3,230 miles).

Around 15,000kg (33,000lb/16·5 US tons) of fuel remains in the tanks and the landing weight is 212,000kg (446,000lb/233 US tons). More fuel has been used than estimated because it had not been possible to obtain the higher cruising altitude.

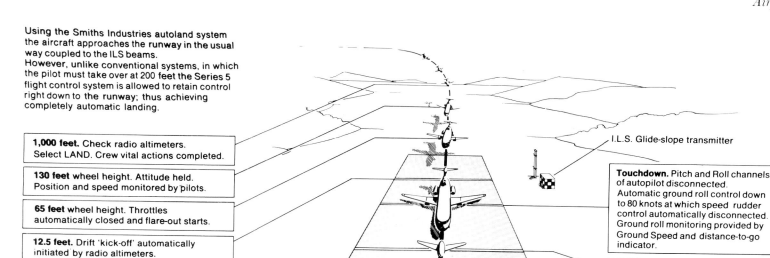

Using the Smiths Industries autoland system the aircraft approaches the runway in the usual way coupled to the ILS beams.
However, unlike conventional systems, in which the pilot must take over at 200 feet the Series 5 flight control system is allowed to retain control right down to the runway; thus achieving completely automatic landing.

1,000 feet. Check radio altimeters. Select LAND. Crew vital actions completed.

130 feet wheel height. Attitude held. Position and speed monitored by pilots.

65 feet wheel height. Throttles automatically closed and flare-out starts.

12.5 feet. Drift 'kick-off' automatically initiated by radio altimeters.

I.L.S. Glide-slope transmitter

Touchdown. Pitch and Roll channels of autopilot disconnected. Automatic ground roll control down to 80 knots at which speed rudder control automatically disconnected. Ground roll monitoring provided by Ground Speed and distance-to-go indicator.

80 knots. Automatic disengagement of rudder control. Below 80 knots PVD head-up guidance to give steering information down to taxying speed.

TAXYING SPEED

I.L.S. localiser transmitter

Autoland

An automatic landing system is primarily designed to enable an aircraft to land in conditions of poor visibility or even nil visibility from inside the aircraft.

Smiths Industries in the UK pioneered their Autoland system in collaboration with the Royal Aircraft Establishment's Blind Landing Experimental Unit in 1947, and it was the first fully automatic landing system to enter scheduled passenger service, doing so in May 1967 aboard BEA Tridents.

The Autoland system can be used at any ILS-equipped airport where the terrain and runway lighting are acceptable to the airworthiness authority and airlines using the system. Basically, the pilot selects the required radio aids for the runway on which he will land, programmes the system and sets the required approach speed. Autoland then controls the aircraft right on to the runway, with the pilot monitoring the cockpit instruments. The descent and runway alignment are directed by the airport's ILS beam, while the aircraft's speed is controlled by an automatic throttle. At 40m (130ft) altitude, the autopilot controls the aircraft at a constant altitude down to 20m (65ft), when three highly accurate radio altimeters level it out for touchdown. Should the pilot wish to abandon the approach, he merely selects full power on the throttles and the climb away sequence is carried out automatically. After touchdown, rudder control is disengaged

automatically at 148km/h (92mph) ground-speed and the pilot controls the steering, using the runway markings and centreline lighting as guides. This has proved adequate even in very low visibility.

Airport lighting

Airport runway lighting has been developed to an extremely high degree of efficiency. Approach lighting to the runways, centreline lighting on the runway, and runway edge lights all demand careful design. They must be intense, but not dazzling; recessed centreline lighting in particular must be able to withstand heavy shocks caused by aircraft landing gear, while runway edge lighting has to stand up to jet blast from aircraft with wing-mounted engines.

Part of the approach lighting is the VASI (Visual Approach Slope Indicator), designed by scientists at the UK Royal Aircraft Establishment, Farnborough, and initially developed by Thorn Lighting Ltd. It was adopted as a world standard by ICAO, the International Civil Aviation Organization, in June 1961.

Basically, the unit consists of two parallel rows of VASI lights on each side of the runway, just beyond the threshold. As the aircraft approaches to land, the pilot sees the two sets of lights. If he is approaching correctly, he will see the near lights as white, the farther as red. If both are white he is too high, if both red, too low.

A recent Thorn development, the mini-VASI, provides a similar guidance but uses two lights instead of three and is much lower to combat jet blast. This new system was commissioned at its first airport, London/Gatwick, in October 1976.

In order to differentiate between runways, taxiways and aprons at night, ICAO has declared a standard colour coding for airport lighting. White is used for approach lights, runway centreline and runway edges; blue is used only for taxiway edge lighting; green denotes threshold lights and taxiway centreline, while red appears as obstruction lighting and as colour coding for the last 914m (3,000ft) of the runway centreline.

TOP: *Smiths Industries Ltd Autoland system as used by British Airways' Trident airliners.*
ABOVE: *The new mini-VASI lighting system compared with a standard VASI installation.*

Aerospace Medicine

Aviation medicine embraces all the factors which affect the safety, efficiency and comfort of people who fly, whether aircrew or passengers. In its widest sense it includes the medical certification of private pilots, the fitness of patients to fly, standards of hygiene in the preparation of in-flight meals, and the spread of communicable diseases such as smallpox and yellow fever.

For the present purpose aviation medicine can better be defined as the study of the stresses imposed by a strange environment upon those who choose to enter it, with the aim of establishing the bounds of safe exposure. It is not enough, however, simply to specify limits which must not be exceeded, because the ideal may be unattainable, or

RIGHT: *Diagram to illustrate respiration. As the rib muscles contract, the ribs move outwards, increasing the capacity of the chest, and at the same time the diaphragm moves downwards. The lungs expand and air is sucked into them in an effort to equalize the pressure. Air is then forced out again as the rib muscles relax, moving the ribs inwards and the diaphragm upwards and thus decreasing the space in the lungs.*

FAR RIGHT: *The structure of the ear. The outer ear is connected by the auditory meatus to the ear drum, which separates it from the middle ear. The middle ear is filled with air and communicates with the outside air by way of the eustachian tube, which opens to let in air when we sneeze or swallow, keeping the air pressure equal on both sides of the eardrum. It is only if the pressure is unequal that we become conscious of discomfort.*

uneconomic, or just unpopular. Compromises are as inevitable in aviation medicine as in technology, and secondary safeguards must be available if primary systems should fail. The specification of these is also the responsibility of the practitioner of aviation medicine.

The environmental stresses faced by astronauts and cosmonauts are, with the single exception of weightlessness, very similar to those that confront the military pilot or the Concorde passenger. The fact that 'space medicine' is an extension of the older science is acknowledged in the title of this chapter, and in its final paragraphs.

The Effects of Altitude

Pressure changes

The pressure of the atmosphere falls progressively during ascent. At 5,485m (18,000 ft) it is reduced to one-half, at 10,060m (33,000ft) to one-quarter, and at 18,290m (60,000ft) – within the operating band of supersonic transport aircraft – to only one-fourteenth of the normal atmospheric pressure at sea level.

The composition of the air, however, does not change; at all altitudes it consists of about 21 per cent oxygen, the rest being nitrogen with a smattering of other inert gases. Un-

fortunately, the workings of the human body pay little heed to the mathematical niceties of percentages and fractions. The fundamental law of altitude is that the space occupied by a quantity of gas expands when the pressure is reduced. Conversely, a given volume of gas at a low pressure contains fewer molecules than the same volume at a higher pressure. These simple rules are responsible for all the aeromedical problems of altitude.

Large amounts of gas are contained in the lungs and in various cavities such as the nasal sinuses, the ears and the gut. These compartments must be held in pressure equilibrium with the atmosphere, and all are therefore affected by changes in altitude. Breath-

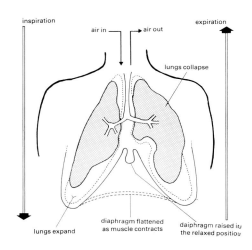

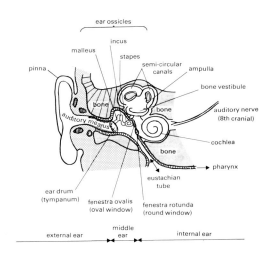

ing ensures that the balance is maintained in the lungs, but the expansion or contraction of gas in other sites can cause symptoms, embarrassment or pain. The escape of surplus gas from the cavity of the middle ear during ascent is often perceived as a 'popping', but it is rarely uncomfortable. Trouble may arise during descent, however, because the cavity of the middle ear is guarded by a flap valve, which closes as the external pressure rises and prevents easy equilibration with the atmosphere. Experienced air travellers learn to avoid the resulting deafness and pain by regularly clearing their ears – swallowing or yawning is usually enough. Babies can achieve the same end more noisily by crying during descent.

The nasal sinuses usually give no trouble during ascent, though a congested sinus may give rise to pain which is explosively relieved when the block is cleared by escaping gas. Re-inflation during descent is not easy, and sufferers from acute sinusitis who fly often have to pay for their misfortune with pain that lasts for some hours after their return to earth, as the sinuses only gradually refill with the normal volume of gas.

A feeling of fullness or distension results from the increase in the volume of intestinal gases at altitude, but speedy relief is denied only to the socially inhibited.

Decompression sickness

'Decompression sickness' is the omnibus term for a variety of symptoms caused by a reduction in the pressure surrounding the body. It was recognized long before aviation began; the older name of 'caisson disease' betokens its occurrence in tunnel workers, and the descriptive phrase 'the bends' derives from deep-sea diving. The first authentic case of decompression sickness at high altitude was described in 1930, but it was not until World War II that the need to fly prolonged missions at high altitudes resulted in an alarming incidence of 'bends' and of more serious symptoms in aircrew.

The most likely cause of decompression sickness is the release of small bubbles of gas from tissues and organs. At sea level, the gases dissolved in the body fluids are in equilibrium with those in the lungs. When the external pressure is reduced by ascent, oxygen and carbon dioxide are freely exchanged between the blood and the tissues but nitrogen, which is less soluble, cannot easily move to restore the pressure balance. As a result, some parts of the body become super-saturated with nitrogen and can be compared with a soda-water syphon which also contains dissolved gas under pressure and which, when disturbed or connected to the outside world, discharges a stream of small bubbles.

Decompression sickness rarely occurs at altitudes less than 7,620m (25,000ft) – equivalent to about one-third of an atmosphere – but it becomes increasingly common at greater heights. For this reason it is seen only in military aviators and in pioneers, such as balloonists, who ascend to high altitudes with inadequate precautions. Its commonest form (the 'bends') is characterized by pain in the joints, which is made worse by exercise. In serious cases the central nervous system may be affected, leading to 'Divers' paralysis' and sometimes to death.

The treatment of decompression sickness is to return the sufferer to a higher environmental pressure, at which the bubbles can once again enter solution. In mild cases of the bends it is often sufficient to reduce the altitude to 5,485m (18,000ft) or less, but occasionally it is necessary to expose an afflicted pilot to a pressure greater than one atmosphere before his symptoms disappear. For him, and for the stricken diver, the final return to sea level must be conducted very slowly and with great caution.

If an exposure to high altitude is known to be inevitable, the risk of decompression sickness can be reduced by breathing pure oxygen before and during ascent. This allows much of the nitrogen in the tissues to be flushed out by the more readily diffuzible gas, so that super-saturation is less severe. To remove *all* the nitrogen in this way would require at least 24 hours, but valuable insurance is given by even an hour or so of pre-oxygenation.

Oxygen lack

The body depends absolutely upon the delivery of oxygen to the tissues and organs in proportion to their need. This vital transport system can be disrupted by many factors but, to the aviator, the most critical problem is that of supply. Oxygen accounts for about one-fifth of the total barometric pressure, but within the lung, where exchange with the blood takes place, the 'pressure space' for oxygen is much less because the gas is saturated with water vapour and diluted with carbon dioxide. These two components remain almost constant irrespective of altitude, so that the oxygen pressure in the lungs falls more dramatically than does the barometer. By 3,050m (10,000ft) it has already fallen to a barely acceptable level, while at 15,140m (50,000ft) the pressures of water vapour and carbon dioxide combined are equal to the barometric pressure, leaving no 'room' for oxygen. At 19,200m (63,000ft) the pressure of the atmosphere is the same as that of water vapour, and the lungs are then filled with 'steam'.

Although the effects of such extremes will obviously be catastrophic, oxygen lack (hypoxia) presents a serious hazard at much lower altitudes, the signs and symptoms of mild hypoxia merging almost imperceptibly with normal behaviour. There is in fact a close similarity between the early stages of intoxication and of oxygen lack. Both can produce light-headedness, irresponsibility and a lack of insight, irritability and belligerence, slurring of speech and difficulty in focusing, and both can eventually progress to stupor, unconsciousness and even death. It is a generally accepted rule that hypoxia is not significant below 3,050m (10,000ft), but night vision suffers at heights as low as 1,525m (5,000ft). There is great individual variability both in tolerance and in the nature of the symptoms.

The fact that hypoxia is a hazard of altitude and that it can be prevented or cured by breathing oxygen has been known since the earliest days of aviation. In 1875 Tissandier and two companions made an epic flight in the balloon *Zenith*, and on the advice of the first true practitioner of aviation medi-

Altitude		Relative pressure	Limits	
FT	M			
0	0	100	Sea level – normal conditions	**Air required**
4,000	1,200	86	Impairment of night vision	
8,000	2,400	74	Maximum cabin altitude for airliners	
10,000	3,000	69	Safe air-breathing limit	
18,000	5,500	50	Half sea-level pressure	**Oxygen required**
24,000	7,300	39	Lower limit for decompression sickness	
33,000	10,000	26	100 per cent oxygen equivalent to breathing air at sea level	
40,000	12,000	19	100 per cent oxygen equivalent to breathing air at 10,000 ft	**Pressure suit required**
63,000	19,000	6	Water boils at body temperature	
65,000	20,000	5	Cruising altitude, supersonic transport aircraft	

RIGHT: *The balloon* Zenith *which on 15 April 1875 carried Gaston Tissandier and two companions to a height of 7,900m (26,000ft). The three striped bags contained supplementary oxygen, but not enough of it, because Tissandier himself was the only one of the three to survive the ascent – the other two succumbed to the effects of the extremely high altitude, lost consciousness, and died.*

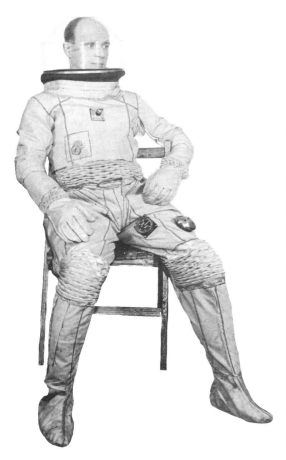

cine they took with them rubber bags containing oxygen. They intended to take a breath whenever they felt weak or affected by oxygen lack, but the impairment of judgment which is characteristic of hypoxia prevented them from using their supply (which was in any case woefully inadequate) when it was really needed. *Zenith* reached a height of more than 7,925m (26,000ft), and remained above 6,095m (20,000ft) for a considerable time. All its occupants lost consciousness, and when it returned to earth only Tissandier was still alive. His passengers were the first recorded victims of altitude hypoxia.

Oxygen lack can be prevented most simply by replacing *all* the inert nitrogen in the air with oxygen, but there are good physiological and economic arguments for keeping the oxygen pressure in the lungs close to its normal level. This principle of 'airmix' is adopted in most military aircraft. At 7,620m (25,000ft), for example, the required concentration of oxygen is only 50 per cent, and it is not until a height of 10,060m (33,000ft) is reached that pure oxygen is required to hold the gas mixture in the lungs in its sea-level state. However, just as an altitude of 3,050m (10,000ft) can safely be tolerated if air alone is breathed, 100 per cent oxygen can be used up to 12,190m (40,000ft).

Even this human limit for flight at great altitude is not an insurmountable one. The force of the atmosphere alone can no longer sustain the supply of oxygen to the blood, but the gas can be delivered to the lungs at a higher pressure – in engineering terms, the body can be 'super-charged'. With a well fitting mask and a simple oxygen system, the safe altitude can be extended in this way to about 13,715m (45,000ft), but higher breathing pressures lead to over-distension of the chest. This can be prevented by giving external support, and the combination of positive pressure-breathing and counter-pressure was used to great effect in World War II to achieve altitudes well in excess of 15,240m (50,000ft) for short periods of time.

Still greater altitudes can be tolerated if the entire man is encased in a pressurized bag, so that he is surrounded by, and breathes from, an artificial atmosphere maintained at the required pressure. Such 'pressure suits' have a long and honourable history, from the altitude records set in the 1930s to the lunar explorations of the 1970s. However, they are hot, bulky, uncomfortable and rigid, and they cannot be regarded, either in aviation or in astronautics, as anything more than a protection against an emergency situation.

The pressure cabin is merely a sophisticated extension of the pressure suit. It consists of a sealed compartment into which air is pumped to maintain an effective internal altitude considerably lower than that at which the aircraft flies. In civil airliners the cabin environment is usually not allowed to exceed 1,830m to 2,440m (6,000 to 8,000ft), giving an ample margin of safety from the 3,050m (10,000ft) limit already mentioned. Passengers and crew alike can enjoy 'shirt-sleeve' flight in an atmosphere which is adjusted for pressure, temperature and humidity. If the pressure cabin fails, however, the occupants are suddenly exposed to the low atmospheric pressure of the outside world.

The cruising altitude of jet airliners ranges from about 6,095 to 10,670m (20,000 to 35,000ft), and their aerodynamic characteristics are such that several minutes may be occupied in a descent to below 3,050m (10,000ft). During this time oxygen must be supplied to everyone on board. Considerable effort and ingenuity has been employed in the development of a supply system which is automatic in operation and relatively foolproof; the result is the 'drop-down' mask that most air travellers have seen demonstrated but which very few have had occasion to use.

Supersonic transport aircraft pose special problems, because they cruise at altitudes much beyond the capability of simpler oxygen systems. A loss of cabin pressure at 18,290m (60,000ft) and a slow descent to a safe height would result in unconsciousness in all cases and death in some. The only solution is to ensure that the cabin altitude can never exceed that at which the standard passenger oxygen system gives full protection, and a pressure equivalent to 7,620m (25,000 ft) has been chosen as this maximum. The design of Concorde makes provision for this requirement; the windows (which are the most vulnerable parts of a cabin) are small to limit both the rate and the degree of pressure loss, and 'flood-flow' equipment is installed so that large quantities of air can be supplied to overcome the breach. With these safeguards, the risk following a decompression is no greater for supersonic passengers than for less elevated travellers.

The price of a pressure cabin offering a low-altitude environment is paid in weight and complexity, and it is not feasible to apply civil standards in high-performance military aircraft. Moreover, the very small volume of the average fighter cockpit will, in the event of a failure, equilibrate very rapidly with the atmosphere and an 'explosive decompression' from, say, 2,440m (8,000ft) to 15,240m (50,000ft) represents an extreme hazard. Military pilots can, however, wear an oxygen mask throughout flight, and the cabin pressure need therefore only be high enough to prevent the occurrence of decompression sickness; that is, an equivalent altitude of about 7,620m (25,000ft) is sufficient. From that height a sudden loss of pressure is not catastrophic, and the fact that the pilot is already connected to an oxygen system allows the concentration and pressure of the gas in the lungs to be adjusted automatically and without delay to meet the new conditions. In some cases, two levels of cabin pressurization are available; a 'cruise' setting which provides a relatively low altitude, and a 'combat' setting to be selected when the risk of pressure loss is greater.

The Force Environment

The human body does not respond to unvarying speed but is very sensitive to changes in either the rate or the direction of travel. These produce forces which can conveniently be measured in terms of the Earth's gravity, g, which gives the body its normal weight, and they are sensed as alterations in the direction or degree of 'heaviness'. The forces induced by alterations of speed are usually small; less than 1g. They may last for a relatively long time (for example, during take-off), but they have little physiological effect. Only in emergencies such as crashes, when the change of speed is abrupt, are the forces large and the results disastrous. They are depressingly familiar, and can be seen on any highway.

The accelerations produced by changes in the direction of flight act outwards from the centre of rotation, and their magnitude is determined by the square of the speed and the tightness of the turn. The agile manoeuvres of military aircraft can involve large accelerations. At 483km/h (300mph) a turn with a radius of 0·8km (0·5 miles) results in a centrifugal force of 2·5g; at 724km/h (450 mph) the stress is more than twice as great. The times for which the forces act may be long; in the two examples given, turns

BOTTOM: *The 'drop-down' oxygen mask usually fitted in the passenger cabins of airliners. These are in case of an emergency where there is a loss of pressure in the cabin. Several minutes would probably elapse before the aircraft descended to a height safe for unassisted breathing.*

through 360° would take 38 seconds and 25 seconds respectively. By contrast, an airliner adjusts its course in a wide sweep, so that its passengers may hardly be aware of the acceleration involved. At a speed of 724 km/h (450mph) the turning radius might be 4·8km (3 miles), and the time needed to complete a full circle would be 130 seconds.

The earliest sensations of sustained accel-

BELOW: *The human centrifuge at the Royal Air Force Institute of Aviation Medicine. It has a radius of 9·15m (30ft) with the cabin at one end. The whole apparatus spins round at varying speeds and it is capable of producing accelerations of anything up to 30g for sustained periods.*

eration are those of increased weight and of being thrust into the seat. At 2g it is just possible to rise from the sitting position; at 3g the legs become almost too heavy to lift; at 6g the arms cannot be raised above the head. Tolerance for such forces is set by their effect upon the heart and circulation. The blood pressure at head level falls progressively as the g is increased, and the supply of

blood to the eye and the brain is correspondingly reduced. At first, vision becomes veiled and sight for peripheral objects is lost. As the acceleration rises, this 'grey-out' becomes more severe, passing through a stage of tunnel vision to the complete loss known as 'blackout'. (Although its existence had been known previously, blackout came into practical prominence for the first time in the Schneider Trophy race of 1931, when pilots were unable to see the pylons around which they were making their fast turns.) Still higher forces jeopardize the circulation to the brain, and cause a loss of consciousness.

Tolerance of acceleration varies between individuals and can be considerably modified by voluntary action. Tensing of the muscles, straining, shouting and crouching all raise the blood pressure and so delay the arrival of blackout. Most aircrew have learned to use one or more of these tricks; indeed, it is difficult to pull an aircraft into a tight turn without tensing the muscles. Helped only by his own efforts, an average pilot can withstand about 6g for a few seconds without loss of vision, but he must give most of his attention to 'fighting the g'.

The need to provide protection automatically is met by an anti-g suit, which essentially comprises a pair of inflatable trousers supplied with gas under the control of a g-sensitive valve, so that the pressure exerted on the abdomen and the legs is always pro-

LEFT: *The pilot of a North American F-100 Super Sabre wearing a special suit to protect him against the pressures and very large g-forces involved in flying this aircraft. The inflatable sections exert pressure on the body which helps to maintain the blood pressure. Without this aid, the pilot's vision may become distorted through lack of blood in the brain and eyes, and may possibly result in tunnel vision or complete blackout.*

BELOW: *The world is divided into several different time zones, most of which are one hour ahead or behind the next one. An imaginary line, known as the International Date Line, is drawn through the Pacific Ocean to indicate where one passes into the next day. This occurs because east of Greenwich the time is ahead, west of it the time is behind and as a result the zones would eventually coincide without the presence of this line.*

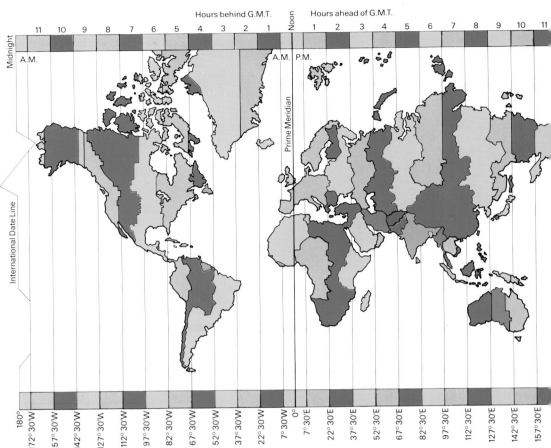

portional to the applied acceleration. An anti-g suit affords approximately 1·5g of protection and, if used in conjunction with crouching and straining, it raises the limit of tolerance to about 8g for 10 seconds or longer. Further gains can be achieved only by devices such as reclining seats, which reduce the vertical distance through which the heavy blood must be pumped from the heart to the brain. This principle is carried to its logical extreme in the almost supine posture adopted by astronauts.

Escape systems

The controlled application of high but brief accelerations in a precisely ordered sequence lies at the heart of the escape systems used in military aircraft. In the days of low-speed flight and of open cockpits it was possible for an aviator simply to climb from his seat, to jump clear of the structure and to pull the rip-cord of his parachute. But at speeds above about 322km/h (200mph) 'bailing-out' becomes impossible, because the would-be escaper does not have the strength to clamber out in the teeth of the wind. Moreover, a manually operated parachute cannot be used safely below 91 to 122m (300 to 400ft) because there is no time for it to open fully and begin to retard the descent before the wearer hits the ground.

Present-day automatic escape systems

have been successfully used at supersonic speed at high altitudes and at zero speed on the runway. The only action required of the pilot is to pull a handle mounted on his seat. The cockpit canopy is jettisoned to clear the path for the subsequent ejection and an explosive charge then propels the seat up guide rails attached to the aircraft structure. A rocket motor next ignites, to sustain guided thrust and so to increase the upward velocity of the seat. (It is this added velocity which makes escape possible from a stationary aircraft.) After a brief delay, the seat is stabilized by the deployment of a small drogue parachute; this in turn draws the main canopy from its container. The harness is released and the seat falls away separately, leaving the aviator to descend on his fully opened parachute.

Each phase of the escape (including the landing) involves acceleration or deceleration, but the greatest force is applied in the first tenth of a second, during which the seat reaches a vertical speed of about 24·4m (80ft)/sec. A properly restrained man can withstand an acceleration of 25g in the spinal axis, but the ejection force comes close to the limit of human tolerance. In fact, transient backache is an almost inevitable consequence of ejection and a crushed vertebra is a not uncommon complication, although it is rarely a serious injury.

The shock forces of parachute opening are normally less than the 20g or so which can be tolerated by an unsupported body, but in ejections at very high speeds deployment of the main canopy must be delayed so that the abrupt deceleration of the system does not break both the parachute and the man. A g-sensitive device therefore inhibits the sequence until most of the forward momentum has been lost. Parachute opening forces are also excessive at great altitudes, and not until the seat, stabilized by its small drogue, has descended to below 3,050m (10,000ft) is the main canopy allowed to blossom. This stratagem also ensures a rapid return to a height at which air can safely be breathed without needing extra oxygen.

Jet Lag

A business tycoon may leave London at 3PM and, after a tiring flight lasting 6 hours, arrive in Washington at 4PM local time. Two hours later, when his body clock denotes bedtime, his colleagues bid him welcome with cocktails and dinner. When he is allowed to sleep (which he probably does fitfully) he wakes at his 'normal' 7 o'clock to hear the American chimes of 2AM. After a week in which his biological pendulum, his alertness and his temper gradually adjust to match the imposed working and social schedules he

BELOW: *The sequence of events during automatic ejection. (a) the miniature detonating cord (MDC) blows open the canopy of the cockpit (b) the rocket-assisted seat leaves the aircraft (c) the seat is slowed down and steadied by the small drogue parachute (d) the pilot separates from the seat and the parachute begins to deploy (e) the aviator descends on his now fully opened parachute to make a safe, gentle landing.*

returns home, only to face the whole problem again in reverse. He now finds it hard to sleep at what he has learned to regard as 6PM, or to rise before mid-day local time.

The effects of this discrepancy between internal and external clocks are known as 'jet-lag'. (They could as logically be termed 'jet-lead', but we always seem to catch up with time and never to precede it.) The condition is real enough, and it is specific to the

as any, but few air travellers can follow it strictly. The majority must adapt as best they can, and it is fortunate that most of them are not required to take momentous decisions on the day after arrival. The best advice is to abide by 'home' time, but this is only possible/practicable given a brief visit, an iron will and a civilization that serves breakfast at any hour. At least one airline finds it worthwhile to impose this regimen on its crews.

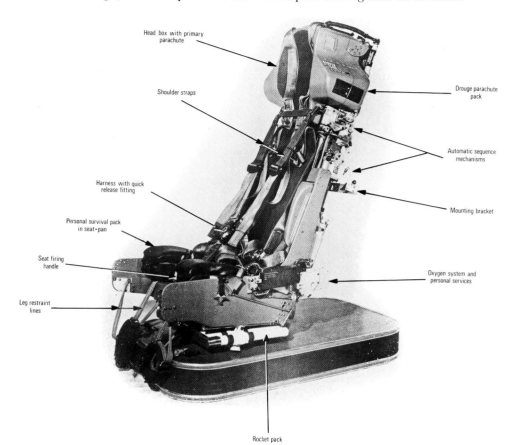

Head box with primary parachute

Shoulder straps

Harness with quick release fitting

Personal survival pack in seat-pan

Seat firing handle

Leg restraint lines

Drogue parachute pack

Automatic sequence mechanisms

Mounting bracket

Oxygen system and personal services

Rocket pack

rapid passage across time-zones that characterizes modern air transport. Complicating factors such as the tedium of a long air journey, exposure to noise and a 'canned' atmosphere, or the exchange of one climate for another can be ruled out because jet-lag is not induced by flights of similar length in a north–south direction, which do not involve time shifts.

The sleep clock is merely the most obvious of a number of biological rhythms having a period of around 24 hours. It is also the only one that can be manipulated to suit circumstance – the executive can force himself to stay awake when politeness or politics dictate, but he cannot control the cycles of the subtle biochemistries of mood or reasoning or creativity. The speed with which these sundry clocks can be re-set varies from person to person and within individuals; young people usually suffer less from jet-lag than do their elders, and their 'bio-rhythms' more rapidly reach a new equilibrium.

In the leisurely world of ocean travel clocks are set forward or back each night, so that passengers may gently adapt to the new time-frame that awaits them at their destination. The voyage to New York lasts about five days and traverses five time-zones. From this relationship has come the rule of thumb that one full day of recovery is needed for each hour of time-shift. It is as good a guide

Claims that supersonic airliners can reduce jet-lag are false, because the problem is one of geography. A regular supersonic shuttle service between London and New York would, however, allow the city magnate to fly to America, conduct his business for 4 hours, and return home within the day. He would undoubtedly be fatigued, but he would not be de-synchronized.

Space Medicine

The twin aims of space medicine are the same as those of aviation medicine: to ensure that the environment is habitable, and to provide protection against failures in the life-support systems. Flight in a vacuum, however, sets special problems. For example, the cabin atmosphere cannot be provided by compressing the outside air, because there is none. The simplest solution is to fill the spacecraft with pure oxygen at a low pressure, to give a cabin altitude of about 7,620m (25,000ft), but one disadvantage of an oxygen atmosphere was demonstrated tragically when three astronauts were incinerated in an Apollo capsule during a pre-flight rehearsal. In contrast to American practice, the Russians have used an air-filled cabin maintained at sea-level pressure, despite the attendant penalties of greater capsule weight and the higher risk of decompression. The best com-

ABOVE RIGHT: *A modern ejection seat for military aircraft provides for safe escape at all speeds and all altitudes, and even from a stationary aircraft on the ground. The only action required of the pilot is to pull the firing handle—everything else is operated automatically in this remarkable 'zero-zero' system.*

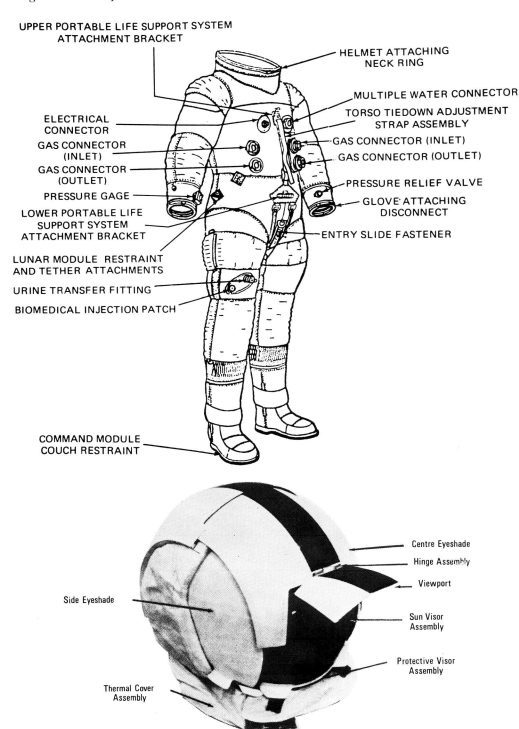

UPPER PORTABLE LIFE SUPPORT SYSTEM ATTACHMENT BRACKET

HELMET ATTACHING NECK RING

MULTIPLE WATER CONNECTOR

TORSO TIEDOWN ADJUSTMENT STRAP ASSEMBLY

ELECTRICAL CONNECTOR

GAS CONNECTOR (INLET)

GAS CONNECTOR (OUTLET)

GAS CONNECTOR (INLET)

GAS CONNECTOR (OUTLET)

PRESSURE GAGE

PRESSURE RELIEF VALVE

LOWER PORTABLE LIFE SUPPORT SYSTEM ATTACHMENT BRACKET

GLOVE ATTACHING DISCONNECT

ENTRY SLIDE FASTENER

LUNAR MODULE RESTRAINT AND TETHER ATTACHMENTS

URINE TRANSFER FITTING

BIOMEDICAL INJECTION PATCH

COMMAND MODULE COUCH RESTRAINT

Side Eyeshade

Centre Eyeshade

Hinge Assembly

Viewport

Sun Visor Assembly

Protective Visor Assembly

Thermal Cover Assembly

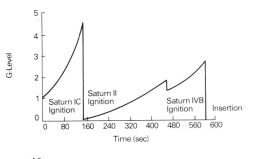
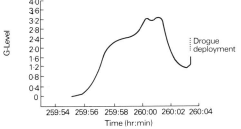
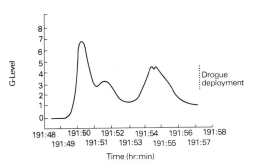

TOP LEFT: *Diagram of the type of space suit used by the crew of the Apollo spacecraft for landings on the Moon. It is really a personal pressure cabin containing its own built-in life support system, to meet every need of the astronaut from the moment of take-off to the landing.*
ABOVE: *The type of helmet used on the Apollo Moon landings. It incorporated several different pieces of equipment to act as a protection against heat, glare, radiation and the effects of a vacuum on the workings of the body.*

TOP RIGHT: *The accelerations of space flight. The upper graph shows a typical Apollo launch profile, the middle one is the re-entry pattern of Apollo 7 from Earth orbit, and the lower one illustrates the successful re-entry from lunar orbit of Apollo 10.*
OPPOSITE PAGE TOP LEFT: *Eugene A Cernan, the commander of Apollo 17, the sixth lunar landing mission, practising with a lunar drill during extravehicular activity (EVA) training at the Kennedy Space Center.*

promise probably lies between the two extremes, and consists of oxygen and helium in approximately equal parts with a cabin altitude of some 5,485m (18,000ft). Both nations have given much attention to variations of this theme.

The gas lost by leakage and the oxygen consumed in respiration must be replaced from supplies carried on-board, and exhaled carbon dioxide must be absorbed. Each man uses about 1kg (2·2lb) of oxygen a day, and produces an equivalent amount of carbon dioxide; thus, very long flights would entail the carriage of prohibitively massive stores. Electrical and chemical methods can be used to generate oxygen by the hydrolysis of water or by 'cracking' carbon dioxide, but these processes are prodigal of energy. They offer only partial relief to the problems of supply, and for inter-planetary flights the development of systems involving photosynthesis by green plants will be essential.

Space suit

The safeguard against loss of cabin pressure is a space suit, which is also required for extra-vehicular activities such as lunar exploration. It is similar to the pressure suit of the military aviator but is more complex because, like the space cabin, it must be an entirely self-contained source of all the astronaut's environmental needs.

To achieve orbital velocity (about 8km or 5 miles/sec) astronauts must be exposed to high accelerations, and the return to Earth involves forces of the same order as those applied during launch. Early space flights involved peaks of about 7g during launch and up to 12g during re-entry, both spread over several minutes. With the larger vehicles of the Apollo and Skylab missions, the forces

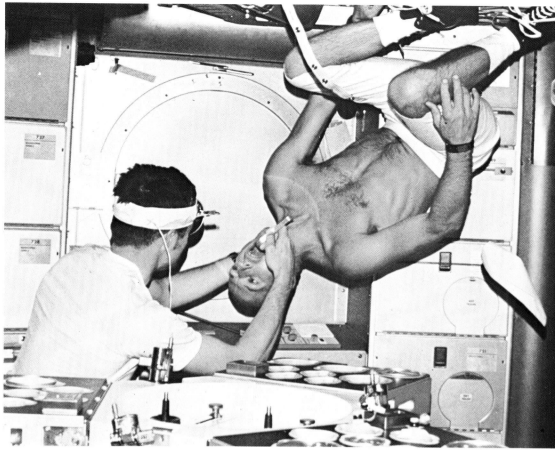

were lower and more prolonged, but even so the accelerations were well beyond the tolerance of pilots in conventional seats. For this reason, astronauts travel in a supine posture, so that the forces are applied at right angles to the long axis of the body. The Space Shuttle will carry mixed crews whose stamina may be less than that of professional astronauts, and the specification for that vehicle accordingly requires that the acceleration shall not exceed 3g. If this value were sustained throughout the launch period, orbital velocity would be reached in about $4\frac{1}{2}$ minutes.

Weightlessness

When the thrust from the launching rocket decays to zero, the spacecraft and its occupants become weightless, and remain so until external forces are again applied either for manoeuvring or for re-entry. The short-term effects of weightlessness upon the body are not serious, because all the vital functions (such as respiration, the circulation of the blood, speech, swallowing, digestion and excretion) are independent of gravity. Locomotion has its hazards, because a weightless body floats unless it is anchored, and attempts to walk may result in painful collisions with the surroundings. More seriously, the absence of a normal gravitational input produces an exaggerated response to rotational forces, and spacemen who are strongly resistant to other forms of motion sickness have suffered from nausea or even vomiting until they have found their 'space legs'.

Two other consequences of weightlessness become apparent only after the return to Earth. One is the tendency of astronauts to faint when standing, because their circulatory systems no longer compensate ade-

quately for the stress of gravity. The cause is complex, but a major factor is that weightlessness provokes a substantial reduction in the fluid content of the body regardless of the amount of liquid consumed. As a result, there is an insufficient volume of blood available to sustain the circulation at 1g. Normal tolerance for gravity returns within a few days as the fluid balance is restored.

The second hidden effect of prolonged weightlessness affects the skeleton. Weight-bearing appears to be essential to the stability of the mineral content of bones, and in the absence of gravity calcium salts are absorbed into the blood and excreted by the kidneys. The process can be partly but not entirely prevented by diet and by exercises designed to apply stress periodically to bones, and

TOP RIGHT: *The interior of Skylab. A medical inspection is being undertaken under the weightless conditions. A body will here float freely, thus helping the doctor to get at inaccessible parts of the body, such as the inside of the mouth.*

ABOVE: *Before man could travel in space, he needed to gain experience of weightlessness. In this photograph Soviet cosmonauts are seen training in a transport aircraft which is flying in a free-fall trajectory, providing a short period of zero gravity to simulate the conditions of Space.*

joints, and its implications for very long space flights remain a matter of some concern. Despite dire predictions of irreparable damage, however, there is no reason to believe that plans for interplanetary flight will be thwarted by insoluble medical or physiological problems.

Military Aviation

Air Power

Far-seeing men predicted the use of aircraft in war even before practical flying machines had been built. With the benefit of hindsight it appears obvious that the Wright Brothers' achievement radically altered the basic principles of warfare. At the time, however, this was not readily appreciated. For example, when Santos-Dumont made the first flight in Europe (in 1906) the British newspaper magnate Lord Northcliffe criticized his editors for not giving the event more prominence. 'The news is not that men can fly, but that England is no longer an island,' he said. How prophetically he spoke became evident soon enough in the conflict of World War I.

'See, not fight'

As World War I opened the major powers had done little to further the cause of military aviation.

'Aviation is good sport, but for the Army it is useless', declared the French General Foch. While in September 1914 the German General Staff reported, 'Experience has shown that a real combat in the air, such as journalists and romancers have described, should be considered a myth. The duty of the aviator is to see, not to fight.'

In short term policy the authorities were correct. The really important contributions of aircraft to the war effort were in reconnaissance and artillery spotting missions. Thus by spotting Russian troop movements before the Battle of Tannenberg the Germans were able to prepare a victorious counter-offensive. Hindenburg wrote in his memoirs that 'without airmen there would have been no Tannenberg'.

The British Army used four squadrons of aircraft to scout ahead of its advance from landings on the French coast in August 1914, and airborne reconnaissance played an extremely important part. Indeed, it was more effective in aiding strategic decisions such as where and when to take a stand rather than in day-to-day operations which required a rapid and continuous flow of information.

The German Army introduced special high-trajectory anti-aircraft guns to deter Allied reconnaissance machines, and the Allies followed suit. The resultant demand for fast, high-flying observation aircraft gave rise to some outstanding two-seaters. The obvious next step was to fight like with like.

Origins of air combat

France originated organized air combat. From January 1915 an *Escadrille de Chasse* was allocated to each field army, using types such as two-seat Morane-Saulnier Ls, in which the observer was armed with a machine-gun, and single-seat Morane-Saulnier Ns having forward-firing weapons and deflector plates on the propeller blades.

In the early summer of 1915 Major-General Sir David Henderson, commander

PREVIOUS PAGES:
The British Aerospace Hawk jet trainer.

of Britain's Royal Flying Corps on the Western Front, proposed the use of fighter packs in an offensive role.

Henderson's successor, Brigadier-General Hugh Trenchard, put the plan into practice at the Somme offensive in the summer of 1916. He allocated only 6 of his 27 fighter squadrons to support the British Fourth Army, and used four squadrons as a Headquarters Wing for air superiority. The plan might have succeeded if more aircraft had been used, but Trenchard also insisted on the available forces being employed on continuous dawn-to-dusk patrols rather than intermittent missions.

Some two weeks after the opening of the Somme offensive the German Army Air Service replied by grouping fighters into specialist squadrons, first the *Jagdstaffeln* (*Jastas* for short) and then the larger *Jagdgeschwader*. The latter were sent in at critical points on the front and moved up and down the lines of their own transport units. Their garishly painted aeroplanes (leading the Allies to dub them 'flying circuses') presented a formidable aspect in the air. And for a time they gained local tactical superiority.

The practice of using aircraft largely independently of the army was employed effectively by the Americans. In 1918 General John Pershing, Commander-in-Chief of the American Expeditionary Force, appointed Major-General Mason M. Patrick as commander of his air component. In August of that year Patrick concentrated all his frontline squadrons, plus some borrowed from the French, in the sector held by the newly formed US First Army. They were placed under the tactical command of Colonel William Mitchell, and when the Americans went on the offensive in September he successfully accomplished what Trenchard had first attempted. Only a small number of aircraft were assigned to the close support of assault troops and the rest were used to bomb and strafe airfields, communications centres, targets of opportunity and other aircraft.

Strategic bombing in World War I

Even while fighters and fighter tactics showed an increasing sophistication, important developments were taking place in other areas of aerial warfare. It must be remembered that fighters were produced in the first place to provide protection for reconnaissance and bomber aircraft. We have already noted the effects and early successes of reconnaissance missions. The future of air power lay, however, in the concept of strategic bombing.

Germany planned to destroy the French Army in six weeks, and accordingly a 36-aircraft strategic-bombing force was set up near Bruges in September 1914 in preparation for attacks on Britain. Dunkirk was bombed in January 1915, but the forward air bases necessary for worthwhile operations against Britain with the primitive and un-

reliable aeroplanes then available failed to materialize and the force was later withdrawn.

From May 1917 the Zeppelins which had been attacking Britain were replaced by fleets of up to 40 or so heavy bombers such as the Gotha G.IV and G.V and the Staaken R.VI. Their raids caused extensive damage but they were not employed in sufficient concentrations to have a serious effect on the course of the war.

France also set up a strategic-bombing force early in the war, under the direction of Commandant Barès. The 1ᵉʳ *Groupe de Bombardement*, equipped with 18 Voisins, was led by Commandant de Göys and laid the basis of long-range bombing. It was later joined by two other *Groupes* and was one reason for the German decision to take a more

ABOVE: *Pictured in July 1918, this RAF (Fighter Command) Squadron was equipped with one of the best aircraft of the time, the Royal Aircraft Factory's S.E.5a.*
LEFT: *This German Zeppelin, L.53, was shot down in August, 1918 by an RAF Sopwith Camel. Against well-armed fighters the dirigible's military days were numbered.*

obviously aggressive part in the air war.

In the spring of 1916, Britain established a strategic-bombing force in the shape of Sopwith 1½-Strutters, for day operations and Short Bombers and Caudron G.4s for night missions. From October 1917 Trenchard, commander of the Royal Flying Corps, was instructed to renew the bombing offensive, using D.H.4s, F.E.2bs and, later, Handley Page o/100s. When Trenchard resigned from the position of Chief of Staff he was appointed to command what became, in June 1918, the Independent Force. This incorporated the earlier strategic-bombing units and was planned to be expanded into the Inter-Allied Independent Air Force, including bombers from the US, France and Italy plus at least 40 squadrons of the RAF. France – and especially Marshal Foch, who was sure that the war would be won by surface forces alone – objected to the idea of an independent group on her soil, however, and the total strength by the Armistice was fewer than a dozen squadrons. The Independent Force dropped 550 tonnes of bombs during its career, half on airfields and the other half on industrial centres in the Rhineland.

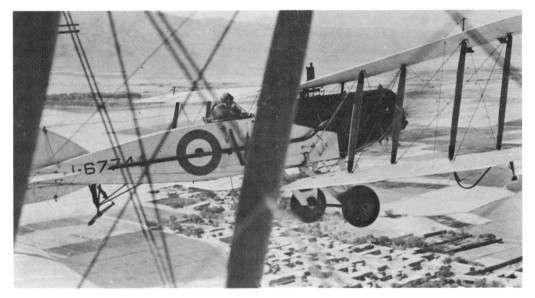

LEFT: *Arming a German Gotha heavy bomber. By early in 1917 Zeppelins had become deathtraps in the face of the new generation of Allied fighters. This signalled the dawn of the strategic bomber and among the first was this German type which could deliver up to 500kg (1,100lb) of bombs from a height of 3,650m (12,000ft).*
BELOW LEFT: *Photographed over Quetta along the North-West Frontier of British India, this is an example of the highly successful Bristol Fighter.*

An independent air ministry

It is difficult to overestimate the effect of establishing overall control of aircraft production and operation. General von Hoeppner, who took over command of the German Army Air Service at its lowest ebb, has been credited with having 'transformed a beaten, dispirited force into one which could claim a comparative superiority by the spring of 1917 and was still a formidable one later when outnumbered and outfought'. Yet both he and Inspector of Flying Troops Major Siegert, who had commanded the strategic-bombing force in Belgium during the autumn of 1914, despaired of matching Britain's organization. The British had established an independent Air Ministry, at the end of 1917, but attempts to set up a German Air Department failed. Britain also had an Aircraft Supply Department, whereas the German Army and Navy competed for aircraft and engines. In the last ten months of the war alone, the British industry produced 27,000 aircraft.

Between the wars

In 1919 Trenchard was reappointed Chief of the Air Staff, a position which he held for a further ten years, and Mitchell became Assistant Chief of the US Army Air Force. Mitchell admitted to having been profoundly impressed by Trenchard, whom he met in 1917, and with these two men at or near the top of their professions the doctrine of strategic air power held sway. They were supported by the Italian General Giulio Douhet, who in 1921 published *The Command of the Air*, a treatise reinforcing the need for independent air action and asserting that in future wars the most effective method of winning would be to attack the enemy's homeland with a rain of bombs. Two years later Mussolini did set up the *Regia Aeronautica* as an independent force, although its actions in Abyssinia and Spain neither denied nor confirmed Douhet's predictions.

Mitchell also campaigned for an independent US Air Force. To demonstrate the power of an air arm Mitchell masterminded the bombing of three surrendered German warships in 1921 and later similar attacks on three US battleships which were due to be scrapped. These actions were carried out in optimum conditions and against no defences, but many hailed the achievements as sounding the death knell of the capital ship. In 1925 Mitchell was court-martialled for insubordination as a result of persistent criticism of his superiors; he was sentenced to be suspended from duty for five years. Subsequently, a board set up to examine the future of US military aviation recommended against the establishment of an independent air service.

The Army Air Force was reorganized as the Army Air Corps (USAAC), however, and the Navy was left free to pursue its own airborne activities. The USAAC was concerned almost exclusively with army support until the construction of long-range bombers and air-defence fighters in the 1930s. Mitchell's ideas were later embraced by the United States – after his death – and the North American B–25 bomber was named after him.

In the early 1920s Britain developed a method, known as air control of underdeveloped areas, for policing Iraq and Transjordan. The move was surprisingly successful, but the targets of reprisal raids were undefended villages attacked in good weather under the best conditions – exactly the opposite of what would be found in Europe.

After France occupied the Ruhr in 1923, Britain's Salisbury Committee recommended that the RAF should always be strong enough to defend the homeland against the strongest Continental air force within striking range. The nebulous principle of 'air parity' was therefore adopted. This worked reasonably well as long as France was seen as the potential – if unlikely – aggressor. It was later to break down, however, when parity was required to be maintained with what Hitler deliberately made an unknown quantity: the *Luftwaffe*.

Germany deliberately fostered a false impression of the size of the newly formed *Luftwaffe*, encouraging potential enemies to believe that it was larger than it really was. This played an important role at the Munich conference in 1938, when Chamberlain and other negotiators believed that refusal to acquiesce to Hitler's demands would result in massive destruction from the air.

Under the terms of the Treaty of Versailles, Germany was prohibited from having an air force, but a Defence Ministry was permitted and this contained a small air staff. German manufacturers designed a range of airliners, trainers and fast communication aircraft which laid the basis for the later bombers, fighters and attack types. The national airline, Deutsche Lufthansa, ran training courses and airmindedness was encouraged by instruction in gliding and powered flying. When Hitler came to power in 1933 he initiated massive rearmament, with the *Luftwaffe* to the fore. Its first Chief of Staff, Lieutenant-General Max Wever, was influenced by Douhet's ideas and in 1935 began the 'Ural Bomber' project for a long-range strategic weapon. Wever died in an air crash in the following year, however, and his successors saw the *Luftwaffe*'s primary role as army support.

Operations during the Spanish Civil War showed that aircraft could be used very effectively as a tactical weapon, and this experience helped mould the *Blitzkrieg* combination of armoured forces, rapidly moving mechanized infantry and close-support aircraft.

France was not prepared for defence against the German onslaught. Haunted by memories of World War I her planners put their faith in the Maginot Line of fixed fortifications. The fighter force was certainly more effective than the bomber arm in 1940, but not sufficiently so.

Both Russia and Japan were building up their air forces, and in particular, the Japanese Navy's air arm expanded rapidly. But in neither country was there an outstanding aerial strategist who held sway over his fellows. The Soviet Air Force was being trained by tactically-orientated German officers. Japan did carry out strategic raids on the Chinese mainland from 1937, but her bombers were outfought by packs of Russian fighters in the Nomonhan war two years later.

Blitzkrieg

Air power came of age in World War II. Strategic bombing was the key to Allied victory, although the extent of its contribution is still hotly debated. The invasion of Poland appeared to vindicate the *Luftwaffe's* concentration on army support; a combination of Stuka dive-bombers and the Panzer tank forces rapidly brought about a victory.

In the ensuing lull on the ground, RAF Bomber Command mounted daylight attacks on German peripheral targets with its force of Vickers Wellington, Armstrong Whitworth Whitley and Handley Page Hampden medium bombers. The bombers were underarmed, had no self-sealing fuel tanks and were not escorted. They were, not surprisingly, badly mauled by German fighters and had to be transferred from day to night operations. The myth that bombers offering mutual support could make effective daylight attacks was effectively destroyed.

In the spring of 1940 German forces invaded Denmark and Norway, and Allied support for the latter country had to be withdrawn as a result of failure to gain local air superiority and because of disruption of communication and supply links by enemy air attacks. Then it was the turn of Belgium, Holland and France, where *Blitzkrieg* operations quickly overcame the defences.

It was during the Battle of Britain in the summer of 1940 that the *Luftwaffe* failed seriously for the first time. Its bombers were underarmed and carried too small a bomb-load. After sustaining heavy losses they were given fighter escorts, but these fighters might have been more effectively employed in taking on Fighter Command directly. The overall direction of what was intended to be a strategic offensive was blurred, and German resources were diversified to attack other targets, such as shipping, rather than concentrating on the job in hand. Then Hitler, in retaliation for a small raid on Berlin, ordered his bombers to be used against cities rather than RAF airfields and similar targets. Meanwhile, bombing of the German invasion fleet being assembled in Channel ports caused considerable damage, and Hitler was forced to postpone the invasion.

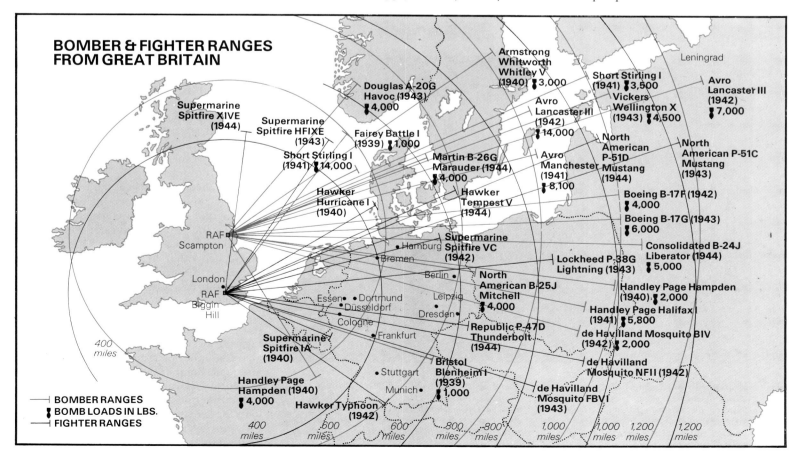

ABOVE: *A German V–2 rocket (officially known as A–4) photographed on its mobile launch-ramp. This forerunner of the modern ballistic missile could carry a 970kg (2,000lb) high-explosive warhead for up to 320km (200 miles) and was developed by the famous Dr Wernher von Braun.*

TOP: *A flight of Curtiss-Wright SB2C Helldivers returns to its base aboard the USS* Hornet *after an attack on Japanese shipping in the China Sea. These two-seat dive-bombers were one of the main carrier-based aircraft types used by the Americans during the Pacific war.*

Bomber Command

In November 1940 Air Chief Marshal Dowding was succeeded as head of RAF Fighter Command by Air Marshal Sholto Douglas, a supporter of the 'big wing' theory developed by Air Vice-Marshal Leigh-Mallory. This involved assembling wings, consisting of several squadrons, to intercept attacks. But gradually the British emphasis was switched from defence to offence, with bombers replacing fighters as the priority. Taking the war back to Germany was seen as being the only way to start winning, although the tide was still running in Germany's favour. In other areas tactical operations continued. Coastal Command was helping to win the Battle of the Atlantic and some 60 per cent of U-boat losses were caused by aircraft action (including those sunk by air-sown mines). Also the Fleet Air Arm, operating from aircraft carriers, was helping to safeguard supplies of arms, raw materials, fuel and food. Bomber Command resisted moves to allocate a sizable part of its force to anti-submarine operations, preferring to gird its loins for the strategic attack on Germany. It was, however, responsible for sinking or badly damaging a number of capital ships,

including the anchored battleship *Tirpitz*.

In February 1942 Air Marshal Arthur 'Bomber' Harris took charge of Bomber Command. He was convinced that strategic bombing would win the war, and at the end of May he mounted his first 1,000-bomber raid. The offensive continued unabated until March 1944, with the British night actions later being supplemented by US daylight raids. From the spring of 1943 the attackers were on top, but there were still unpleasant shocks. One of these was the mauling of an unescorted Boeing B-17 force at Schweinfurt, where the Flying Fortresses' close box formation and defensive armament of up to 13 machine-guns each failed to deter the *Luftwaffe* fighters.

It is widely accepted that D-Day would not have been possible without strategic bombing to disrupt arms production, fuel supplies and logistics. After the landings, heavy bombers were used in the close-support role as well as their major tasks, but this was only possible because Allied fighters had virtual command of the air. Retaliation by the *Luftwaffe* with V1 flying bombs and by the German Army with V2 ballistic missiles (see Chapter 19) was too late to be decisive.

Carrier power

If the strategic bomber was one of World War II's decisive weapons, the aircraft carrier was another. Swordfish torpedo bombers helped sink the *Bismarck* early in the war and virtually eliminated the Italian fleet in Taranto harbour in late 1940, but the most audacious attack by naval aircraft was the Japanese assault on the US base at Pearl Harbor, Hawaii, in December 1941. This action marked the transference of power from the battleship to the carrier, and yet the Japanese let the supreme prize – the US Navy's carrier fleet, which was not in Pearl Harbor at the time of the attack – slip from their grasp. After six months of runaway successes the Imperial Japanese Navy began to be driven back by the US Navy's carrier force. Strategic bombing of Japanese targets was the responsibility of Boeing B-29 Superfortresses, and the incendiary raids on Tokyo, largely constructed from wood, probably represented the nearest that such a policy ever came to destroying a population's will to resist. It is perhaps significant that the Pacific war was finally ended by the weapon which was to remove the last remaining doubts about the future possibilities of

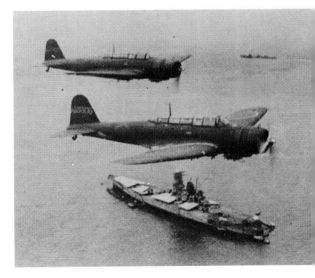

TOP: *Col Paul Tibbets (right) in front of the B-29 Superfortress he piloted to drop the world's first atomic bomb on the Japanese city of Hiroshima in August 1945. B-29s of the US Strategic Air Command were replaced by eight-jet B-52 Stratofortress bombers in the mid-1950s.*
ABOVE: *A pair of Japanese Nakajima B5Ns. At the time of Pearl Harbor these were Japan's main torpedo-bombers.*

effective strategic bombing: the atomic bomb.

Air power is not measured just in terms of heavy bombers and fighters, however. It also includes elements such as the ability to deliver invading forces rapidly into action against surprised defenders. This was graphically illustrated by the dropping of German paratroops on Crete and the Allies' use of gliders to transport men and equipment. Massive transport fleets were required in addition to the combat air forces.

Korea and Vietnam

In the two major, sustained conflicts since 1945 – Korea and Indo-China/Vietnam – air power has played an important but not decisive role. Carpet-bombing of Korean targets by Boeing B–29s was not notably successful, and while close air support was vitally necessary it too failed to destroy the enemy's ability to keep himself supplied and armed. Many of the types thrown into action were unsuitable for the tasks which they were called upon to perform, and the basic thinking behind their employment was also awry. The troops wanted close air support against pinpoint targets between a few metres and a few kilometres in front of them. Many planners, however, were still obsessed with the idea that interdiction against rear targets was the surest way of crippling the enemy.

The United States no longer had a monopoly on atomic weapons and could not threaten their use indiscriminately. In the spring of 1951 General Douglas MacArthur, Commander-in-Chief of United Nations forces in Korea, was dismissed for persistently advocating attacks on Chinese industrial centres. Later, with General Eisenhower as US President, more or less overt threats of atomic attacks on China may have contributed to the ceasefire negotiated in 1953.

Strategic Air Command

After the end of World War II the US Air Force gained a specialist heavy-bomber wing in the form of Strategic Air Command. The Convair B–36, originally designed to bomb Germany from the US, was equally capable of deployment against the Soviet Union. This was followed by the all-jet Boeing B–47 and B–52. Russia concentrated on long-range missiles (see Chapter 19) as its strategic striking force. Britain relegated her V-bombers to other, non-strategic, roles. Both France and the US, though, retain manned bombers alongside their missile arsenals. The latest aircraft to enter the strategic arena are the Rockwell B–1 and Russia's Tupolev Backfire, although the Soviet Union maintains that Backfire is not a strategic weapon and disputes that it should be included in strategic-arms limitation agreements.

Air power in the Middle East

In the Middle East, air power proved decisive – as far as it was allowed to go – during the 1956 Suez crisis, was instrumental in bringing about Israeli victory in 1967 but played a lesser role in 1973. On the first two occasions pre-emptive strikes destroyed the Arabs' air forces but, in the October war, ground forces were able to counter air action fairly effectively. The Israelis gained temporary

air superiority, but the Arabs declined to commit their fighters when surface-to-air missiles were proving an effective substitute. The one strategic force used in 1973, the US Air Force's fleet of C–5 and C–141 transports which took part in Operation Nickel Grass to resupply Israel, performed extremely creditably and reminded the world yet again that there is more to warfare than fighting. Both superpowers also used strategic reconnaissance during the conflict, with Russian Mikoyan MiG–25s and probably US Air Force Lockheed SR–71 Blackbirds supplementing satellite coverage.

NATO and the Warsaw Pact

One role of air power in the modern world is as a straightforward extension of the foot soldier, cavalry or artillery. Supersonic intercepters and ground-attack aircraft can destroy targets quickly and accurately, help ground and sea forces in both defence and attack, and protect the homeland. The air forces of the North Atlantic Treaty Organization (NATO) and the Warsaw Pact face each other in central Europe for just this purpose. The 'tripwire' doctrine formerly adopted by NATO has been replaced by one of flexible response. An incursion into the Western alliance's territory by Soviet Bloc forces would now be countered by a spoiling action rather than the immediate use of nuclear weapons. The Warsaw Pact greatly outnumbers NATO in tanks, so air power is the major balancing factor. The Soviet Union is also superior in numbers of aircraft, but until recently the gap has been substantially narrowed by the West's more advanced airborne electronics. Now that Russian navigation and attack equipment is improving, while the numerical preponderance is being maintained, the balance of air power in Europe is subtly changing.

NATO members such as Britain and Germany maintain a mixed force of offensive and defensive aircraft, as does France, which has withdrawn from the alliance. The United States and Canada, also members, divide their forces between their own territory and bases in Europe.

The Berlin Airlift showed that the aircraft

ABOVE: *As a Dakota comes in to land, this Douglas C–74 Globemaster unloads its valuable cargo during the Berlin Airlift of 1948.*
BELOW: *US Navy A–7 Corsair IIs lined up on the deck of USS* Kitty Hawk *during operations off the coast of North Vietnam in 1972.*

had come of age as a transport vehicle, and countries with overseas responsibilities today maintain large fleets of cargo aircraft. This is especially so in the United States, which would have to reinforce its troops in Europe if war broke out.

Off-shore patrols

Maritime aviation is also assuming ever greater importance. Specialized aircraft operating from land bases or aircraft carriers are used to track and destroy submarines and surface ships. Countries surrounded by water or with important shipping lanes passing close to their coasts, such as Australia, New Zealand and South Africa, place great emphasis on maritime patrol. As offshore re-

sources, such as minerals and fish, are exploited they have to be defended against saboteurs and illegal fishing.

So it can be seen that aircraft are of great importance on many levels. Perhaps the most important aspects of air power, however, are the ones which are often neglected – an industry able to provide the equipment, a training organization to supply operators of all types, a political organization which defines the overall objectives and ensures that they are met, and a military command and control structure that turns an air force into an organism which can change and respond to demands.

ABOVE CENTRE: *The Panavia Tornado is the subject of the biggest multinational development and production programme of all time, with almost 1,000 of different versions being made by three nations for eight customers. This is a Tornado F.3 interceptor fighter of the Royal Air Force.*

TOP RIGHT: *During a training exercise in Korea over 3,000 paratroops descend from Fairchild C-119 Flying Boxcars.*

ABOVE: *Two Dassault Mirage IIIs of the RAAF. These French fighters have been exported to many foreign air forces.*

Air Forces of the World

Abu Dhabi
Small but well equipped force in the Arabian Gulf. Jet aircraft include Mirage and Hunter fighter-bombers.

Afghanistan
Operates under military agreement with Soviet Union and all aircraft are supplied by Russia. Equipment includes MiG–21 fighters.

Albania
Uses Chinese-built aircraft, including F–7 fighters.

Algeria
Combat aircraft mostly supplied by Russia, although other types bought from the West. Fighters include MiG–21s, with MiG–17s and Su–7s for ground-attack.

Angola
Equipped by Soviet Union following independence in 1975. Modern types such as MiG–21s took part in civil war during 1976.

Argentina
Operates wide variety of Western-supplied combat aircraft, including Mirage and ex-Israeli Dagger interceptors and Skyhawks for ground-attack. Also uses Argentine-developed Pucara counter-insurgency type. Navy operates Skyhawks and Trackers from carrier *25 de Mayo*.

Australia
Powerful force with influence over large areas of ocean. Sole non-US operator of F–111 swing-wing bomber and has more than 70 Mirage interceptors and F/A–18 Hornets; P–3C Orion maritime-patrollers and C–130s. Navy operates Skyhawks, Trackers and helicopters from aircraft carrier HMAS *Melbourne*.

Austria
Small force with roles limited by treaty with Soviet Union. Operates Saab Draken and 105 fighter-bombers.

Bangladesh
Uses Russian combat aircraft, including MiG–21s, since breaking away from Pakistan.

Belgium
Member of NATO and is one of four European members of the alliance to use F–16 air-combat fighter. Collaborating with Holland, Denmark and Norway on F–16 production. Earlier types include Mirage 5s for the attack role and Sea Kings for search and rescue plus Mirage 5s for the latter role.

Bolivia
Uses mainly piston-engined types and early jets in counter-insurgency roles. Sabres operated as fighters and Embraer Xavantes used for attack.

Brazil
Largest South American air force. Mirage IIIs form part of integrated air-defence system, with AMX and F–5Es for ground attack. Brazil's thriving aircraft industry supplies the air force with several types, including the AMX, Xavante light strike aircraft and Bandeirante multi-purpose transports. Trackers and Navy helicopters are flown from the aircraft carrier *Minas Gerais*, and a large transport fleet is operated.

Brunei
Small helicopter force used to support the army.

Bulgaria
Member of Warsaw Pact, with all equipment supplied by Soviet Bloc. MiG–21 and MiG–19 fighters backed up by earlier MiG–17s.

Burma
Mainly counter-insurgency force operating early jet types such as Sabres, Vampires and Shooting Stars.

Cambodia
Being re-equipped with Russian types following Communist take-over in 1975.

Cameroun
Internal security force with no large combat aircraft. Receives aid and equipment from France.

Canada
NATO member, with forces based in Europe as well as in Canada. Equipment includes CF–18s in the fighter and attack roles, plus CF–5s, CP–140 Aurora maritime patrol aircraft, Challenger and 707 transports and many other types.

Central African Republic
Operates transport types and helicopters, but no heavy combat aircraft.

Chad
Another ex-French colony, with a small and mixed fleet of obsolescent types.

Chile
Recent acquisitions include F–5Es and additional A–37s from the US, following Britain's refusal to supply the military government with further combat aircraft to add to the substantial force of Hunters. A number of earlier types make up the second-line strength.

China, People's Republic of
One of the world's largest air forces. Operates mainly Russian-designed types, built under licence at first and then copied after the deterioration of relations with the Soviet Union. These include MiG–21s, MiG–19s (known as the Shenyang F–6), MiG–17s (F–4s) and MiG–15s (F–2s). The F–9 has been developed in China from the MiG–19, and Tu–16 Badgers are operated in the bombing role. Beriev Be–6s fly maritime-patrol missions and there is a large transport fleet. The Rolls-Royce Spey military turbofan is to be built under licence and may power Chinese-designed aircraft.

China (Taiwan)
Powerful air force operating US types, some of which are built in Taiwan itself. New F–5Es are joining earlier F–5A fighters and will replace nearly 300 Sabre fighter-bombers. Super Sabres and Starfighters also act as fighters, and large fleets of transports, helicopters and trainers are operated.

Colombia
A force of Mirage 5 fighter-bombers forms the front line, with older jets and piston-engined types backing them up. The air force runs its own airline, Satena, for internal services.

Congo Republic
Operates transports and helicopters of French, US and Russian origin. Provided support for Marxist forces in Angola during the civil war.

Cuba
Russian-supplied since the 1958 revolution. Types include MiG–21s, some of which are the modern MiG–21MF variant, together with MiG–19s and MiG–17s. Helicopters, transports and trainers have also been received from the Eastern Bloc.

Czechoslovakia
Member of the Warsaw Pact, operating a modern force of various marks of MiG, including about 170 MiG–21s, plus Su–7s for ground-attack. Sizable transport, training and helicopter fleets are also operated.

Dahomey
Uses the normal collection of transports and helicopters supplied as standard by France to its former colonies.

Denmark
Nato member, and one of the four European parties to the alliance to order F–16s in the 'arms deal of the century'. Denmark is one of only two export customers for the Swedish Saab Draken, which is used as a fighter-bomber and for reconnaissance. Other combat types include Super Sabres and Starfighters.

Dominica
Operates a variety of early jet and piston-engined types, such as Mustangs, Vampires, Invaders and Catalinas.

Dubai
Small but expanding force in the Arabian Gulf. The only combat aircraft are Macchi MB.326s, and the association with Italy has been continued by purchase of the G.222 transport.

Ecuador
One of the first export customers, along with Oman, for the Anglo–French Jaguar fighter-bomber. These are replacing Shooting Stars, and other equipment includes Canberras and Strikemasters.

Egypt
Since the October 1973 war with Israel, Egypt has turned towards the West rather than Russia to equip its air force. Types already supplied include helicopters and Mirage fighter-bombers, with negotiations continuing towards further purchases and agreements for licence production in Egypt. Other Arab countries – Saudi Arabia being the major contributor – will finance the operation and will receive aircraft from Egyptian production lines.

Ethiopia
Jet equipment includes F–5Es, A–37 ground-attack aircraft and Canberra bombers, plus F–5As supplied by Iran and some remaining Sabres. The combat aircraft have recently been used against forces trying to separate Eritrea from the rest of the country.

Finland
Finland receives military equipment from both Russia and the West, the former supplying MiG–21 fighters and Europe contributing Drakens from Sweden and Hawks from Britain. The Finnish aircraft company, Valmet, builds Drakens under licence and has developed its own basic trainer.

France
One of the most powerful air forces in Europe, despite the withdrawal of France from the military activities of Nato. The Mirage F.1 is replacing early jet types and the Delta Mirage 2000 is being developed to take over from Mirage IIIs. Some 200 Jaguar ground-attack aircraft are being supplied to the Armée de l'Air, as are the same number of Alpha Jet trainers. The land-based strategic force comprises ballistic missiles together with Mirage IVs and a supporting tanker force of KC–135s. The extremely successful French aerospace industry is able to supply almost all the air force's needs, including transports, trainers, helicopters and maritime-patrol aircraft, and it is a major exporter. The Navy has two aircraft carriers, *Foch* and *Clemenceau*, and the Super Etendard has been developed to extend their effectiveness into the 1980s.

Gabon
A small force of Mirage IIIs comprises the combat element, with a varied transport fleet consisting of helicopters and fixed-wing types from France, the US and Japan.

Germany (East)
A member of the Warsaw Pact, with substantial Russian forces stationed in the country as well as its own defences. Modern front-line equipment includes late-model MiG–21s and Su–7s as intercepters and fighter-bombers, backed up by MiG–17s and supported by fleets of transports, helicopters and trainers.

Germany (West)
The strongest European member of Nato, operating more than 250 Phantoms, several hundred Starfighters and Aeritalia G.91s for light attack. The last-named are being replaced by Franco–German Alpha Jets, and the Starfighters are to be superseded by the Panavia Tornado multi-role combat aircraft being developed jointly with Britain and Italy. The powerful naval air arm will receive 112 of Germany's 322 Tornados. The German Government is an enthusiastic advocate of collaborative development programmes, additional examples being the Transall transport and Atlantic maritime-patrol aircraft. Large fleets of trainers and helicopters are also operated.

Ghana
The air force's only combat aircraft are a handful of MB.326s, but these are supported by a variety of general-purpose transport, patrol and training types.

Great Britain
A major European contributor to Nato, with forces based in Germany and the Mediterranean as well as in the United Kingdom. Phantoms have largely replaced Lightnings in the interception role, and ground-attack duties have in turn been taken over by Jaguars. Buccaneers, Canberras and Vulcans are to be superseded by the tri-national Panavia Tornado, and some Victor former V-bombers have been converted into tankers. The Hawk jet trainer is joining the earlier Jet Provost, and maritime-patrol is the responsibility of Nimrods. Basic training is carried out on Bulldogs, and a wide range of helicopters, transports and subsidiary types support the main combat element. The Navy's last aircraft carrier, HMS *Ark Royal*, is being withdrawn but fixed-wing naval aviation will continue with the Sea Harrier, developed from the land-based variants operated since 1969.

Greece
The modernization of the Greek and Turkish air forces has been spurred by differences over Cyprus, with Greece receiving Phantoms, Corsairs and Mirage F.1s, together with new transports and trainers. Older types include Starfighters and Delta Daggers, plus F–5As used for reconnaissance.

Guatemala
A small number of A–37 attack aircraft make up the combat element. Jet trainers are also operated, along with a small fleet of transports and helicopters.

Guinea
Operates Russian-supplied aircraft, including a handful of MiG–17s plus small numbers of trainers and transports.

Guyana
Has no armed types, using only light transports and helicopters.

Haiti
Operates a number of obsolescent Mustang fighter-bombers and some armed Cessna 337 light aircraft. Also trainers, transports and helicopters.

Haute-Volta
Has no combat aircraft, acting as a transport force only. The standard range of types supplied to French ex-colonies is on strength.

Honduras
Operates a rarity in the 1970s, the F4U Corsair, together with Invader bombers. Shooting Star reconnaissance/trainers are the only jet types.

Hong Kong
A policing force only, using helicopters and light aircraft.

Hungary
Member of the Warsaw Pact, but possesses a relatively small air force. Front-line strength comprises MiG–21 fighters and Su–7s for ground-attack.

India
Buys aircraft from both East and West as well as building its own types. MiG–21s are constructed under licence, the first examples having been bought off the shelf, and the Indian-developed Ajeet is taking over from the Gnat on which it is based. Ground-attack squadrons operate Hunters, Su–7s and the indigenous Hindustan Aeronautics Marut, with Canberras used as bombers. The domestically designed Kiran jet trainer is being supplemented by the Polish Iskra, and the Il–38 May maritime-patrol aircraft is the latest acquisition from the Soviet Union. India's aircraft industry also provides basic trainers, transports and helicopters. The Navy has one aircraft carrier, *Vikrant*, from which Sea Hawks and Alizés are operated.

Indonesia
Now operates aircraft supplied by the US and Australia, with all Russian equipment grounded. The only jet combat types are Commonwealth Sabre fighters.

Iran
One of the world's most powerful air arms, and the only non-US operator of the F–14 swing-wing fighter. The present Phantoms and F–5Es are due to be supplemented by new lightweight fighter-bombers, and other modern equipment includes P–3F maritime-patrol aircraft, Boeing 707 and 747 tanker/transports, Hercules and Friendship transports, and one of the world's largest helicopter forces.

Iraq
Most aircraft have traditionally been supplied by the Soviet Bloc, but Western arms are now being bought. The most modern Russian type is the MiG–23 fighter, backed up by MiG–21s and –19s. Ground-attack machines include MiG–17s, Su–7s and Hunters, with Tu–16s and Il–28s used for bombing.

Ireland
This small defence force operates armed Magister light strike/trainers and a number of light aircraft fitted to carry stores.

Israel
A modern, powerful and well equipped air force with extensive recent combat experience. Ground-attack is carried out by some 500 Phantoms and Skyhawks, with Mirages, the Kfir J79-powered Mirage development and F–15 Eagles

flying as fighters. Israel Aircraft Industries builds the Arava light transport as well as Kfir, but virtually all other aircraft are supplied by the US. These include E–2C Hawkeyes for early-warning, Hercules transports, Boeing 707 tanker/transports and helicopters.

Italy
Member of Nato, with a small but healthy aircraft industry supplying many of its needs. The F–104S version of the Starfighter, developed by Aeritalia, is to be supplemented by the tri-national Panavia Tornado. The G.91 is used in a number of versions for training and ground attack, and the G.222 medium transport is being introduced to complement the larger Hercules. Macchi's MB.339 jet trainer has been ordered to follow the MB.326, and light aircraft are supplied by a number of companies. Atlantics and Trackers carry out maritime reconnaissance.

Ivory Coast
No combat aircraft are operated, the air force's roles being restricted to transport and liaison.

Jamaica
Light transports and helicopters are used for patrol, communication and liaison, but no armed types are operated.

Japan
Japan's aircraft industry is following licence production of US types such as the Phantom, Starfighter and Neptune with its own developments. These include the T–2 trainer and the FST–2 attack aircraft developed from it, together with the C–1 transport and the series of Shin Meiwa flying boats and amphibians. The Starfighters are due to be replaced by Eagles, Hawkeyes have been ordered for early-warning duties and a new maritime-patrol aircraft is required. Japan also supplies most of its own primary trainers, light aircraft and helicopters.

Jordan
A comparatively small but expanding force, with front-line equipment comprising Starfighters, F–5As and F–5Es. These are supplemented by a small transport force and a number of helicopters and trainers, both jet-powered and piston-engined.

Jugoslavia
More than 100 MiG–21s form the major part of the intercepter element, and the Orao – being developed jointly with Romania – is likely to be the main ground-attack type in the future. The helicopter fleet includes licence-built Gazelles, and a variety of Russian and Western transports are operated.

Kenya
Jet equipment comprises a small number of Hunter fighter-bombers and some Strikemaster light strike/trainers. The training fleet is being expanded.

Korea (North)
This Russian-equipped force operates the usual array of MiG and Sukhoi combat types, plus Il–28s and Soviet transports and helicopters.

Korea (South)
Almost totally US-equipped, with Sabre fighters complemented by Phantoms and F–5s for ground-attack. The Sabres are likely to be replaced by F–16s.

Kuwait
Modern jet equipment consists of Skyhawks and Mirage F.1s, the former supplementing the force of Hunters and Strikemasters with the latter taking over from Lightning intercepters.

Laos
US-supplied equipment is largely non-operational, following the Communist take-over. Russian aircraft may now be in service.

Lebanon
Hunters and Mirages are the only jet combat aircraft, although Syria may pass on some Russian equipment.

Libya
Operates modern Soviet aircraft in the form of MiG–23s and Tu–22s in addition to a variety of Mirages. Training aircraft, transports and helicopters have been supplied by Russia, the US, France and Jugoslavia.

Malagasy
No combat types, operating only a selection of light transports and helicopters.

Malawi
Operates only a handful of transports.

Malaysia
Commonwealth Sabres have been replaced by F–5Es, with Canadair CL–41 Tebuans doubling as light strike aircraft in addition to their role as trainers.

Mali
Supplied with Russian equipment, including MiG–17s and –15s.

Malta
This helicopter force has recently introduced its first fixed-wing light transports.

Mauritania
Has added Skyvans and Defenders to its force of French-supplied transports.

Mexico
No modern combat aircraft, relying on obsolescent types such as Vampires and Shooting Stars.

Mongolia
Uses Russian-supplied MiG–15s, transports and helicopters.

Morocco
Mirage F.1s and F–5Es are due to supplement the force of early-mark F–5s, armed Magisters and Russian fighter-bombers (now in storage). Modern transports and helicopters are operated.

Nepal
Operates transports and helicopters only.

Netherlands
An important member of Nato, operating more than 100 Starfighters as intercepters and for ground attack. These will be supplemented and replaced by F–16s built under licence in co-operation with Belgium, Denmark and Norway. Canadair NF–5s are also operated.

New Zealand
Small but up-to-date force, operating Skyhawks and Strikemasters. The Hercules fleet has been augmented by the purchase of ex-RAF Andovers, and P–3B Orions patrol the surrounding seas.

Nicaragua
Uses piston-engined and early jet types for ground-support and bombing, with more modern transports and helicopters.

Niger
A former French colony, supplied with a selection of transports.

Nigeria
MiG–21s have been added to the Russian aircraft supplied earlier, with the West contributing transports, trainers and helicopters.

Norway
Nato's northernmost member, and one of the alliance's four European participants to have ordered the F–16. This type will replace Starfighters and F–5s in the 1980s. Other aircraft include maritime-patrol Orions and Hercules transports.

Oman
Jaguar International fighter-bombers are augmenting the force of Hunters and Strikemasters used for ground-attack.

Pakistan
Operates aircraft from many sources, including Iran (surplus F–5s), France (Mirages and second-hand Atlantics), China (MiG–19s) and the US (B–57 bombers). A large number of trainers and helicopters are also used, and licence production of overseas types is building up.

Panama
Operates no combat aircraft but has a comparatively modern transport fleet.

Papua New Guinea
Uses transport aircraft donated by Australia.

Paraguay
Primarily engaged in transport and liaison duties, armed trainers being the only combat aircraft.

Peru
The first South American customer for Russian combat aircraft, having ordered swing-wing Su–20s. Other jet types include Mirage 5s, Canberras, Hunters and A–37s.

Philippines
Operates Sabre fighters and F–5 fighter-bombers, with transports and a variety of smaller aircraft supplied by several countries.

Poland
The largest Soviet Bloc air force after that of Russia itself. Some 200 ground-attack Su–7s are being supplemented by the Su–20 development, taking over from MiG–17s. About 300 MiG–21 fighters are also on strength.

Portugal
A Nato member, but lacking modern combat aircraft. The only aircraft committed to the alliance are Neptune maritime-patrollers. Sabres are used for interception, with G.91s for ground-attack.

Qatar
Four Hunters are the sole combat aircraft.

Romania
Member of the Warsaw Pact, operating the normal force of MiG and Sukhoi types for interception and ground-attack. Romania also has Alouette IIIs, used in the anti-tank role.

Rwanda
A small number of Magisters comprise the only jet equipment.

Salvador
Ex-Israeli Ouragans and Magisters have been supplied for fighter and strike duties, to add to a varied transport and training fleet.

Saudi Arabia
This rapidly expanding force is introducing US aircraft to augment its British-supplied types, including Lightnings and Strikemasters. Latest acquisitions comprise F–5E fighter-bombers, gunship helicopters and additional Hercules for transport and aerial refuelling.

Senegal
A small transport and liaison force with no combat aircraft.

Sierra Leone
Uses primary trainers and helicopters only.

Singapore
The attack force of Hunters and Strikemasters is being supplemented by Skyhawks modified to Singaporean requirements, and a new intercepter type is being sought.

Somalia
Largely equipped with Russian aircraft, including MiG–21s. Somalia also provides base facilities for Soviet aircraft.

South Africa
Various marks of Mirage III are used for interception and ground-attack, and these are being supplemented by Mirage F.1s in both roles. Canberras and Buccaneers are also on strength, with MB.326s being employed both for training and attack. Atlas Aircraft Corporation builds several types under licence, including the Mirages, MB.326s (known as Impalas) and a number of smaller types.

South Yemen
Operates Communist-supplied aircraft, including MiG–21s.

Soviet Union
Operates some of the world's most advanced aircraft. Swing-wing Backfire bombers are supplementing the force of Tu–16s, Tu–22s, Tu–95s and Mya–4s used for bombing and aerial refuelling,

the latter role also being carried out by Il–76 tankers. Variable-geometry Su–19 fighter-bombers and MiG–21 attack aircraft are the latest additions to the fighter-bomber force, which also has Su–17s and MiG–23s as well as older types such as Su–7s and MiG–17s. Perhaps most impressive of all is the MiG–25 Foxbat, operated both as a fighter and for reconnaissance. Other recce types include the Yak–28. The intercepter force has Su–9s, –11s and –15s in addition to some of the types already mentioned, plus Tu–28Ps. The same design bureau's Tu–126 is the Russian equivalent of AWACS, although not as advanced, and the Navy has Il–38s and a variety of flying-boats/amphibians for maritime-patrol in addition to many of the Air Force's types. The latest aircraft revealed to Western eyes is the Yak–36 vertical take-off fighter-bomber operated from *Kiev*-class aircraft carriers. Combat aircraft are backed up by massive fleets of transports, trainers and helicopters.

Spain
A well equipped force operating Phantoms and Mirages – IIIs and F.1s – with F–16s on order. F–5s are used for attack and reconnaissance, and Casa is developing the new C.101 jet trainer. The Navy operates Matadors (AV–8As) from the carrier *Dedalo*.

Sri Lanka
The only combat unit is a MiG–17 squadron, supported by jet trainers supplied by Russia and Britain.

Sudan
Supplied with combat aircraft by both Russia and China, but much equipment is grounded.

Sweden
Moving towards a force comprising only three combat types – Draken, Viggen and Saab 105 – all supplied by Swedish industry. Various marks of Viggen are used as fighters, for attack and on reconnaissance missions, over both land and sea.

Switzerland
The F–5E is being procured to replace the large force of obsolescent Venom fighter-bombers and to supplement the Hunters. Mirage IIIs are used as intercepters and for reconnaissance.

Syria
A major recipient of Russian aircraft, including MiG–23s and some 200 MiG–21s. MiG–17s and Su–7s are also used, and helicopters have been supplied by France.

Tanzania
Operates combat aircraft donated by China, including F–8s (MiG–21s). The transport and training fleets are composed of Western types.

Thailand
Almost all aircraft are of US origin, including F–5s, A–37s, Shooting Stars and a multitude of propeller-driven types.

Togo
Operates a varied fleet on transport and communication duties, plus a small number of Magisters.

Tunisia
F–5Es are replacing elderly Sabres, with MB.326s doubling as trainers and light strike aircraft.

Turkey
Rapidly building up strength, with Phantoms and Italian-built F–104S Starfighters bolstering the force of earlier Starfighters, Delta Daggers, Super Sabres and F–5s. A substantial transport fleet is operated, and more combat types – perhaps built in Turkey – are planned.

Uganda
Russia has supplied most of the combat aircraft, including MiG–21s.

United Arab Emirates
A small force operating helicopters. Further aircraft are under the control of member emirates.

United States
The Air Force's latest types include the F–15 Eagle air-superiority fighter, F–16 air-combat fighter and A–10 attack aircraft. Other newcomers are the E–3A AWACS and the winner of the AMST competition between the YC–14 and YC–15. Further in the future is the B–1 strategic swing-wing bomber. These will augment and partially replace B–52s operated by Strategic Air Command, which also has the smaller FB–111. Reconnaissance types include the SR–71 Blackbird and U–2. Tactical Air Command operates Phantoms, Corsairs, Thunderchiefs, F–111s and many other types, with the Air National Guard acting as a reserve. Military Airlift Command includes the C–5A Galaxy among its big-lifters and is stretching its StarLifters to increase payload. The US Navy has introduced the F–14 Tomcat shipboard fighter and plans the F–18 to take over from some Phantoms, Corsairs and Skyhawks. Other specialist USN aircraft include the Intruder heavy attack type, Orion and Viking maritime-patrollers and Hawkeye early-warning aircraft. The Marine Corps has AV–8A Harriers in addition to Navy types and plans to buy AV–8Bs to replace them, while the Army is by far the world's largest helicopter operator.

Uruguay
The single combat squadron operates Shooting Stars, and Trackers are used for shore-based anti-submarine missions. Several types of transports and helicopters are also on strength.

Venezuela
Mirages, Sabres and F–5s are used as fighters, with Canberras and Broncos for attack. A wide variety of transports, trainers and helicopters are also operated.

Vietnam
Equipped with aircraft of Russian and Chinese origin. Has expanded its operations to include the whole of Vietnam, following the withdrawal of US forces from the South.

Yemen
Western types are replacing Russian-supplied equipment.·

Zaire
Mirage fighter-bombers have bolstered the combat element, which also includes MB.326 light strike/trainers and propeller-driven types. Modern transport aircraft are operated.

Zambia
The major combat force comprises MB.326s, supported by piston-engined trainers which can also carry stores.

War Machines

BELOW: *The French Morane-Saulnier Type L parasol scouting monoplane was an early attempt to combine the monoplane's speed with the strength and reconnaissance capabilities of the biplane.*

BOTTOM: *The B.E.2c, designed by Britain's Royal Aircraft Factory, was one of the most widely used reconnaissance aircraft of all time. It was extremely stable, making it an excellent observation platform.*

Reconnaissance aircraft

Military commanders have always wanted to know what the enemy was doing. The invention of the balloon in the late 18th century provided the first opportunity for reconnaissance to be made from the air, and tethered balloons and kites were widely used for observation. They had the disadvantage, however, that they could not provide more than a local view of the enemy's movements, and they were cumbersome to set up and operate.

Mobile airborne platforms were much better suited to the task, with both aeroplanes and airships being quickly adopted for this role. In 1911 a small force of Italian aircraft was used for observation in the dispute with Turkey about control of Libya, and in the spring of 1912 these aircraft carried out the first photographic reconnaissance.

Initial developments

When World War I began in 1914 the main task of aircraft on both sides was to glean information for use by the army and navy. As the air war developed, reconnaissance types played the central role, with the early fighters and bombers supporting them. Fighters were intended to protect the reconnaissance aircraft and to shoot down the enemy's airborne observers, while bombers were used to attack the bases from which reconnaissance aircraft and airships were operating. Knowledge of the enemy's movements at sea was especially important, and the German Navy's Airship Division, for instance, was intended mainly for reconnaissance over the North Sea.

Observers could not remember everything they saw on a long mission, so cameras were installed to bring back a complete record. France, with a strength of 160 aircraft at the beginning of the war, set up specialized photo-reconnaissance units. Main types were Morane-Saulnier parasols and the pusher Farmans. Strategic aerial reconnaissance proved itself definitively, when on 3 September 1914, French airmen noted a growing gap between the German First and Second Armies near the Marne. Knowledge of the fatal opening led to the Allied victory in the subsequent battle.

The German Army encouraged aerial observation, and especially photography, from the very beginning. Germany produced the world's best lenses and had an extensive camera industry. Soon after the war began the German High Command had a complete photographic record of the Western Front which could be updated every two weeks. By 1916 almost every German Army headquarters had its own detachment of three specially equipped camera aircraft, and types such as the L.F.G. Roland C.II were even

designed specifically for photo-reconnaissance. General Ludendorff, the German Commander-in-Chief, said: 'Complete photographic reconnaissance with no gaps must be ensured. This is of decisive importance.' By the time of the last great offensive of the war, the German Army Air Service was taking about 4,000 photographs a day.

Britain relied on lumbering observation aircraft such as the Royal Aircraft Factory B.E.2c, with a maximum speed of little more than 113km/h (70mph). They were extremely stable, allowing details on the ground to be recorded without the pilot having to concentrate on flying a high-performance machine. But they were sitting ducks for German fighters. The German Fokker E I fighter entered service in 1915, armed with a machine-gun synchronized to fire through the propeller disc, and it was the B.E.2cs which took the brunt of their attack. Of the 80 kills scored by Manfred von Richthofen (the Red Baron), for example, 47 were reconnaissance aircraft.

By 1918 R.E.8s had replaced the B.E.2cs. More than 2,000 R.E.8s were used on the Western Front alone, mainly for artillery spotting (reporting by radio to correct the gunners' aim). The Airco D.H.9, with a range of nearly 500km (300 miles), could photograph strategic targets east of the Rhine, and the RAF ended the war with about 9,000 airborne cameras in service.

Germany prepares for World War II

Most of the hard-won expertise in airborne reconnaissance techniques was dispersed soon after the war. Germany was defeated, and the Allies had no need for the elaborate network of interpreters and collators which is part and parcel of any comprehensive intelligence network. As the aviation industry gradually built up again in Germany, however, emphasis was once more placed on the reconnaissance units. The first was formed in November 1936 to act as the aerial-observation wing of the Condor Legion in Spain.

About 80 units, comprising one-fifth of all German military aircraft, were operational when World War II began in September 1939. Reconnaissance squadrons were divided into short-range units, for tactical army support, and a long-range strategic element. The main strength of the army-support squadrons was the Henschel Hs 126 high-winged monoplane carrying fully automatic cameras in a fuselage bay. Almost 260 were in service at the beginning of the war. The *Luftwaffe* also operated a large number of long-range Dornier Do 17P reconnaissance aircraft and had more than 5,000 ground processing and interpretation staff dispersed in support of the Army's *Blitzkrieg* strategy.

Development of specialized aircraft

Britain's Royal Air Force, in contrast, was ill equipped. Credit for building up the RAF's reconnaissance forces belongs largely to Squadron Leader Sidney Cotton, who had been carrying out clandestine photographic missions before the outbreak of war. Within a few weeks of the start of hostilities he was appointed head of the newly formed Photographic Development Unit (PDU) at Heston, and the following month he managed to obtain two Supermarine Spitfires to augment his force of converted Bristol Blenheims. The requirement was for an aircraft which could exceed 483km/h (300mph) at 9,145m (30,000 ft) to avoid anti-aircraft fire and Messerschmitt Bf 109 fighters. A camera fitted with a 12·7cm (5in) lens was mounted in each wing of one of the Spitfires. This plane, the Spitfire PR.1A, could reach nearly 644km/h (400mph) when stripped of its guns and armour-plating. It was painted duck-egg blue to make it difficult to see against a sky background. Later experiments showed that the best colour for high-altitude aircraft was a greyish blue known universally as 'PR blue' (PR – Photographic Reconnaissance). Surprisingly, pale pink was the most effective colour for camouflage when an aircraft was operating against a background of sun-lit clouds.

March 1940 saw the appearance of the Spitfire PR.1C, the first with a fuselage-mounted camera. In the late summer of that year it was thought that the Germans might be assembling an invasion fleet in the Baltic, and Cotton enlisted the help of the Chief of the Air Staff in obtaining his first specialized reconnaissance Spitfire, the PR.1D. Filling the wing leading edges with 523 litres (115 Imp gal) of fuel gave the aircraft a range of 2,815km (1,750 miles) and the PR.1D, nicknamed 'the Bowser' because it was considered to be a flying petrol tanker, made its first sortie at the end of October.

Cotton's approximate opposite number in the *Luftwaffe*, Oberstleutnant Rowehl, experimented throughout 1939 to find the best photo-reconnaissance machine. In his position as head of the German High Command's secret reconnaissance wing he tried pressurized and unarmed variants of the Junkers Ju 86 medium bomber, and the Dornier Do 217. The Ju 86P–1 and P–2 were fitted with three automatic cameras, and in the summer of 1940 a P–2 was coaxed up to 12,500m (41,000ft). The Spitfire PR.1 could reach only 9,755m (32,000ft) and was far more easily observed by the enemy.

The long-range Do 217A–os had their fuselages extended to house two vertical cameras, and the *Luftwaffe* thus obtained excellent high-altitude photographs of every important RAF airfield and British city. By

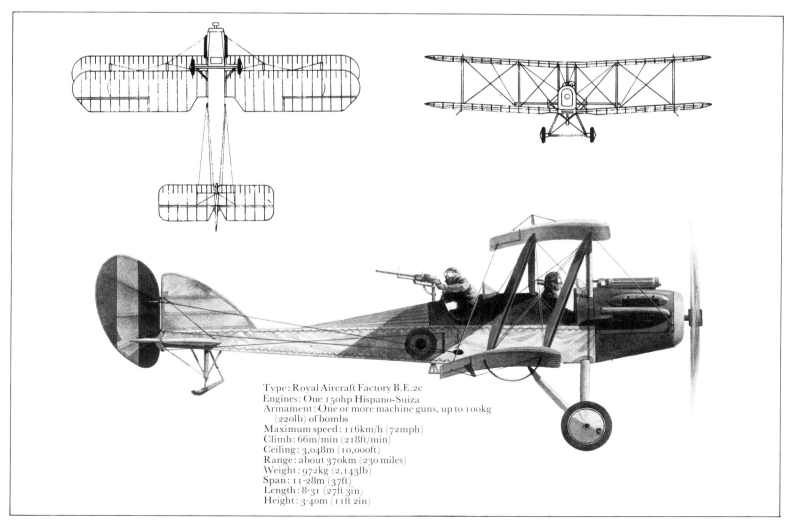

Type: Royal Aircraft Factory B.E.2c
Engines: One 150hp Hispano-Suiza
Armament: One or more machine guns, up to 100kg
  (220lb) of bombs
Maximum speed: 116km/h (72mph)
Climb: 66m/min (218ft/min)
Ceiling: 3,048m (10,000ft)
Range: about 370km (230 miles)
Weight: 972kg (2,143lb)
Span: 11·28m (37ft)
Length: 8·31 (27ft 3in)
Height: 3·40m (11ft 2in)

the end of 1940 these aircraft were operating over the Soviet Union in preparation for Operation Barbarossa, the invasion of Russia.

Designers did not strive for higher cruise altitudes merely because fighters would then have to climb further to make an interception. At altitudes between 8,230 and 12,190m (27,000 and 40,000ft), depending on humidity and temperature, the water vapour in the exhaust from an aero engine will condense to form a visible trail. There is little point in painting an aircraft to merge with its background if it is producing a vapour trail, so machines which could cruise undetected in the stratosphere were continually being sought. There was also the added advantage that any fighter climbing to intercept such an aircraft would itself make a vapour trail, giving warning of its approach.

Morale in the *Luftwaffe* declined as interpreters were transferred to other posts and venerable aircraft were not replaced. Planners concentrated on a range of exotic but unsuccessful designs, although one specialized reconnaissance type – the Arado Ar 234 could have had a decisive impact if it had been deployed earlier during the war. Its range and ceiling were modest, but two turbojets gave it a top speed of more than 740km/h (460mph).

The Soviet Union had used aerial reconnaissance to good effect during World War I, with the Sikorsky Ilya Mourometz carrying two observation officers on sorties up to 240km (150 miles) behind the German lines. Photo-reconnaissance missions were later carried out during the Russo-Finnish war, but at the time of the German invasion in 1941 fewer than one Russian aircraft in 20 was equipped for such work. The force was soon built up, however, with modified Lavochkin and Yakovlev fighters being employed on short-range army-support missions and converted bombers being used on strategic sorties.

The United States made extensive use of bombers for reconnaissance, including the Boeing B–17 (as the F–9), Consolidated B–24 (F–7 and US Navy PB4Y), Boeing B–29 (F–13) and Douglas A–20 (F–3). Maritime-patrol types also carried out surveillance missions, and more than 1,000 Lockheed P–38 Lightnings were modified specifically for reconnaissance. The most important Japanese aircraft in this category was the Mitsubishi Ki–46, a twin-engined type with a top speed of 600km/h (375mph), allowing the defensive armament to be dispensed with until Allied fighters could match its performance. Converted bombers such as the Mitsubishi G4M1 were also employed in this role after being replaced by more modern designs in their primary capacity.

The Royal Air Force's PDU, renamed Photographic Reconnaissance Unit (PRU) in July 1940, continued to operate Spitfires, but other types were used in different theatres. In November an American-built Martin Maryland operating from Malta plotted the position of 5 Italian Navy battleships, 14 cruisers and 27 destroyers in Taranto harbour. The Fleet Air Arm was able to disable half the Italian battle fleet with only 11 torpedoes, making this one of the most graphic

demonstrations of the value of reconnaissance in warfare. Hawker Hurricanes and Bristol Beaufighters also served in this role in the Middle East.

In the spring of 1941 the first Spitfire PR.IVs were delivered. These were production 'Bowsers', having the fuselage of the Mk V fighter, special wings and a Merlin 45 engine. The leading edges held 605 litres (133 Imp gal) of fuel, giving a range of 2,900km (1,800 miles), and cameras were fitted in the wings and fuselage. The reconnaissance units were always pressing for greater range, and the first production de Havilland Mosquito to leave the line in 1941 was a recce variant. With a range of more than 3,200km (2,000 miles) and a top speed exceeding 612km/h (380mph), the Mosquito was exactly what the Air Force needed.

The Allies had good cause to thank the reconnaissance crews for their work during the war. The discovery of German radar installations, the V1 and V2 revenge weapons and many other objectives of strategic importance depended on fast and accurate reconnaissance work. Specialist targets such as the dams which supplied water to the Ruhr could not have been attacked successfully without an excellent knowledge of their defences, and the necessary information was supplied to a large extent by the reconnaissance crews.

A new kind of war in Korea

With the end of hostilities, however, the intricate wartime reconnaissance network began to disintegrate, just as it had in 1919. The United States Air Force (USAF) toyed with specialized aircraft such as the twin-boom Hughes XF–11, but strategic reconnaissance remained the responsibility of converted – and still armed – bombers such as the Con-

vair RB–36 and Boeing RB–47. The RB–47, with up to seven cameras in its bomb bay, could photograph 125,000 square km (48,000 square miles) in a single three-hour flight. When the Korean War broke out in 1950 the USAF found itself with jet reconnaissance aircraft in the form of modified fighters, but the cameras left over from World War II were designed for slower operating speeds and could not perform adequately. Also, the North Koreans and Chinese moved by night, and the standard of night photography was not good. A rapid re-equipment programme was needed.

Soon after the end of World War II the US Army carried out experiments with captured A4 (V2) ballistic rockets, one of which was fitted with a 35mm film camera. These tentative beginnings led to what is now probably the most advanced field of 'aerial' observation – reconnaissance satellites.

The U-2

At the July 1955 summit conference President Eisenhower proposed an 'Open Skies' policy as a means of defuzing international tension. The United States and Russia would, under this plan, allow each other to carry out unrestricted reconnaissance flights over their territory. Marshal Bulganin rejected the proposal, but in the previous December the USAF had already given the go-ahead for an aircraft suitable for high-level surveillance missions. This was the Lockheed U–2, designed by Kelly Johnson and built virtually by hand at the company's secret 'Skunk Works' in Burbank. The U stood for 'Utility', one of the many steps taken to disguise the real purpose of the aircraft. The first U–2 made its maiden flight in August 1955, and the later U–2B, with its glider-like wings and lightweight construction, could cruise at well

FAR LEFT ABOVE: *The giant Convair B–36J was powered by four turbojets as well as six Wasp Major piston engines.*
FAR LEFT BELOW: *The twin-engined Mitsubishi Ki–46 was the most important land-based Japanese reconnaissance aircraft of World War II.*
LEFT: *The Russian Tupolev Tu–20 Bear is a typical example of a former strategic bomber finding a new lease of life in the recce role. A McDonnell Douglas Phantom acts as escort.*
BELOW: *The Boeing B–17 Flying Fortress – this example is a B–17F – is remembered mainly as a bomber but was also used, as the F–9, for long-range reconnaissance.*
BOTTOM: *The Dornier Do 217 medium bomber was one of the Luftwaffe's major reconnaissance types in the early part of World War II. A Do 217E is illustrated.*

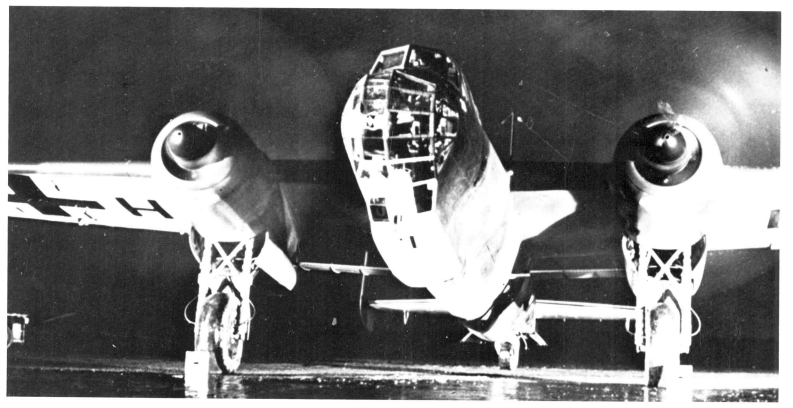

RIGHT: *The Grumman OV–1 Mohawk was used extensively by the United States Army in Vietnam. It can carry pods housing Infra-Red (IR) sensors or Sideways-Looking Airborne Radar (SLAR), or both, to map large areas of enemy-held countryside. Moving-Target Indication (MTI) allows mobile targets to be picked out from the background on the radar.*

FAR RIGHT: *The McDonnell RF–101 Voodoo was a reconnaissance version of the normal fighter, with the cameras housed in the distinctive chisel-shaped nose. RF–101s were heavily involved in surveillance of Cuba during the 1962 crisis, making low-level runs over the island to detect missile sites and other military installations.*

over 21,335m (70,000ft) for more than 6,435-km (4,000 miles).

In April 1960 a set of U–2 prints revealed what looked like the first Russian ICBM (Intercontinental Ballistic Missile) installation. A U–2 was accordingly ferried from Incirlik Air Force Base in Turkey to Peshawar in Pakistan, ready for a long-range overflight covering about 4,700km (2,900 miles) of the Soviet Union. At 1100hr the aircraft, flown by Gary Powers, was knocked into a spin by the explosion of a surface-to-air missile detonated by a proximity fuze. Powers bailed out and was captured by the Russians. At his trial, it was revealed that the U–2 carried equipment capable of taking 4,000 pairs of photographs covering a strip 160km to 200km (100 miles to 125 miles) wide and 3,540km (2,200 miles) long.

High- and low-level reconnaissance

The news that a U–2 had been knocked down by a missile while flying at 20,725m (68,000ft) came as a blow to Britain as well as to the United States. The RAF was only just introducing its English Electric Canberra PR.9, which had a broad-chord centre section to the lengthened wing and uprated Avon engines, allowing it to cruise at 18,285-m (60,000ft). The USAF later exploited the Canberra even further, in the form of the licence-built RB–57F with a wingspan of 37·2 m (122ft). Its range was reported to be 6,435-km (4,000 miles) and its cruising altitude 30,480m (100,000ft).

The RAF had developed the art of low-level high-speed photography in Palestine and Malaya from 1950. Canberra PR.3s and PR.7s replaced the venerable Mosquitoes, but the unarmed Supermarine Swift PR.6, which was to have taken over from the Gloster Meteor PR.10, was cancelled.

Russia, like Britain and the United States, continued to use converted bombers for high-level reconnaissance. Tupolev Tu–20 Bears, Myasischev Mya–4 Bisons and Tupolev Tu–16 Badgers became familiar sights to Western fighter pilots sent up to escort them away from their airspace.

Cuban Crisis

U–2s did not fade from the scene after the Powers incident. In August 1962 a U–2 from Laughlin Air Force Base at Del Rio in Texas discovered that Russian surface-to-air missiles were being installed in Cuba. Overflights were stepped up, and the presence of Soviet SS–4 Sandal intermediate-range ballistic missiles, which could be fitted with nuclear warheads, was detected. So began the Cu-

ban crisis. Aerial reconnaissance really came into its own, with Strategic Air Command Boeing B–47s and B–52s spotting the movements of ships in the western Atlantic, carrier-based Vought RF–8A Crusaders of the US Navy making low-level runs over the island, USAF McDonnell RF–101 Voodoos carrying out similar missions and the U–2s continuing surveillance from on high. On 27 October one of the U–2s was shot down by a surface-to-air missile, and the situation became very grave. The next day, however, the Russians agreed to dismantle the SS–4 sites and low-level operations by RF–101s confirmed that their word was being kept.

Capability extended by new sensors

By the mid-1960s the United States was heavily engaged in the war in Vietnam. Reconnaissance was no longer based on aircraft making high-speed dashes at low level or cruising at high altitude, out of reach of enemy intercepters or missiles. Missions of this type were still very important, but the aircraft had become just one part of a complicated surveillance network.

The normal complement of optical cameras had now been joined by infra-red sensors capable of detecting minute heat differences. Images recorded on film allowed interpreters to say whether a particular aircraft parked on the apron of an enemy air base had just returned from a mission, whether it had recently emerged from its hangar and was being prepared for a sortie, or if it had been standing outside all day and was therefore probably either a dummy or unserviceable. All this could be derived by noting different heat patterns – hot engines showed up a different colour, and an aircraft newly emerged from a cool hangar would also stand out against the warmer background of the aerodrome's tarmac.

Infra-red sensors could also be made to produce an image like a television picture. Such Forward-Looking Infra-Red (FLIR) equipment allowed lorries to be spotted at night by their warm engines, or ships could be detected because of the warm wake which remained visible for hours afterwards. Radar could produce maps at night when optical cameras could not be used without dropping flares and announcing the presence of a reconnaissance aircraft. At dawn and dusk Low-Light-Level Television (LLLTV) came into its own, and image-intensifiers allowed objectives to be found in almost total darkness. Also in use are air-delivered sensors such as Acoustic and Seismic Intrusion Detection and Commandable Microphone.

Vietnam War

Powered aircraft developed from sailplanes were used to gather intelligence at low altitude over the jungle without being heard, and special listening probes derived from naval sonobuoys could be dropped to listen for enemy movements. Men were increasingly removed from the danger areas completely, with pilotless drones carrying out missions over heavily defended areas. These drones, which came to be known as Remotely Piloted Vehicles (RPVs), could bring back a record of their flight, or even relay it directly to ground controllers while still in mid-air.

Modified fighters and bombers such as the RF–101 Voodoo and B–57G of the USAF, alongside the Navy's North American RA–5C Vigilante, were joined by types developed specifically for the war in Vietnam. The Rockwell OV-10 Bronco, powered by a pair of turboprops, met the Light Armed Reconnaissance Aircraft (LARA) specification, while the larger Grumman OV–1 Mohawk could be fitted with radar or infra-red pods, or both, to map large areas of enemy-held countryside. The little Cessna O–1 Bird Dog, carrying a Forward Air Controller (FAC) for spotting potential targets, was joined by the Cessna O–2 with more advanced sensors and greater capability. Light-armed helicopters also played their part, finding targets to be attacked by other helicopters or fixed-wing aircraft.

SR–71 Blackbird

After the United States ended its bombing of North Vietnam in January 1973, only two reconnaissance types were permitted to continue overflights. One of these was the Teledyne Ryan AQM–34L remotely piloted vehicle and the other was the Lockheed SR–71 Blackbird. The Blackbird is a remarkable vehicle – the ultimate in a search for ever higher and faster reconnaissance aircraft. Built mainly of titanium, and designed by the same 'Skunk Works' which was responsible for the U–2, the SR–71 can cruise at more than Mach 3 (three times the speed of sound) at a height of 24,400m (80,000ft) or more. In September 1974 one flew from New York to London in less than two hours, at an average speed of more than 2,900km/h (1,800mph). Maximum range at this sort of speed is some 4,825km (3,000 miles). The Blackbird's cameras, radars and infra-red sensors can map 155,000 square km (60,000 square miles) in an hour, and its missions have included keeping an eye on trouble spots in other parts of the world, such as the Middle East, as well as in Vietnam.

BELOW: *The Voisin was the most widely used French bomber in the first two years of World War I, and more than 2,000 were built in all. The type had a steel airframe to which increasingly powerful engines could be attached, but the pusher arrangement fell from favour when fighters evolved and could move in unopposed from the rear. Voisins were sturdy and dependable, however, and the gunner/bomb-aimer had an excellent view from his front cockpit.*

VIII engines of 375hp each gave a maximum speed of 145km/h (90mph) and up to six Lewis guns, including one in the tail, were carried for self-defence. The Vickers Vimy was a contemporary of the Handley Page bomber, being designed to attack Berlin from bases in France, but it likewise failed to see war service and became best known for its use on pioneering and record-breaking flights such as Alcock and Brown's crossing of the Atlantic and Ross and Keith Smith's race to Australia, both in 1919. Large numbers of D.H.9s and B.E.8s were also used during the war, but not so effectively.

German airships and giant aircraft

The Imperial German Military Aviation Service had in September 1914 formed a strategic bombing force to attack England from Belgium, but the 100hp two-seaters with which it was equipped proved to be unsuitable for sustained overwater operations. They carried out the first large-scale night attack, against Dunkirk in January 1915, but after they were transferred to the Eastern Front during that spring the responsibility for attacks on Britain fell even more heavily on the shoulders of the airship crews. In the early part of the war both Germany and Austro-Hungary relied heavily on Zeppelins and other airships as their main bombing vehicles, and the German Naval Air Service continued with lighter-than-air operations when the military arm turned to aircraft.

When Germany began daylight raids with aircraft in early 1917, the RFC and RNAS were forced to pull back aircraft from the front for use on home-defence duties. One of the mainstays of the long-range bomber fleet was the Gotha series. The G V, which took over from the G IV as a night bomber in August 1914, could carry 300kg (660lb) of weapons across the English Channel, representing about half its total load. The two Mercedes water-cooled in-line engines of 260hp each gave a maximum speed of 140-km/h (87mph). The G V's ceiling was 6,400m (21,000ft) and its range 840km (520 miles). About 230 of the earlier G IV model were built, but there were far fewer G Vs.

The main force of Gothas and AEG bombers was backed up by smaller numbers of Friedrichshafen and Zeppelin types. The Friedrichshafen G III, which was used mainly as a night bomber against French and Belgian targets as well as in the offensive against Britain, had the same engines as the Gothas and virtually identical take-off weight – 4,000kg (8,800lb) – and speed. Ceiling was 4,570m (15,000ft) and endurance 5hr. The G III appeared in early 1917, and more than 300 were eventually built. Its normal bomb load was 500kg (1,100lb), of which one-fifth could be carried internally, and the three-man crew was provided with one or two Parabellum machine guns in the front and rear cockpits for self-defence.

The giant Zeppelin Staaken R VI had double the engine power of the Gothas and Friedrichshafens, allowing a bomb load of 2,000kg (4,400lb) to be carried over a short range. Endurance could be as much as 10hr with smaller payloads. The R VI's wings spanned 42·2m (138ft 6in), the fuselage was 22·1m (72ft 6in) long and maximum take-off weight was 11,790kg (26,000lb). The type bombed France and Britain from September 1917, but it was the only one of the Zeppelin R series to enter production and a mere 18 were built.

Use of bombers by other air arms

The Imperial Russian Air Service had the distinction of operating the first four-engined bomber in the world, the Sikorsky Ilya Mourometz. The type equipped a specialist bomber group which began offensive operations from a Polish base in February 1915. Although fewer than 80 were built, they had a significant impact. Many different engines were tried, the Type V being equipped with Sunbeam water-cooled models producing 150hp each, and the low top speed of only 120km/h (75mph) was offset by the use of extremely effective Russian-designed bomb sights and formidable defensive armament. Up to seven machine guns could be fitted, and the only Ilya Mourometz to be shot down in air combat accounted for three enemy fighters before succumbing.

When Italy joined the war in May 1915 it already had a bomber force in the shape of

Caproni Ca 2s. They were sufficiently reliable for missions over rugged country, and operations against Austro-Hungary began on 15 August 1915. Austro-Hungary retaliated with standard reconnaissance two-seaters manufactured by the Lloyd and Lohner concerns. Their missions include a daylight raid on Milan in February 1916. Gothas were later introduced.

Between the wars

In the years from 1919 to 1939, bombers progressed from fabric-covered wooden biplanes to all-metal monoplanes. The Junkers D.1, which first flew in 1916, had many of the features that did not become standard until the 1930s – all-metal construction, cantilever-monoplane layout – but few were built because of production difficulties. Post-war cuts in defence spending and the lack of an official re-equipment policy meant that innovation was generally kept to a minimum and well established layouts were retained.

French bombers

France ended World War I with more military aircraft than any other state, and despite post-war cutbacks the *Aviation Militaire* retained nearly 40 bomber squadrons equipped with the Breguet 14, 16 and later the 19. The two-seat Br.19 light bomber, which replaced the Br.14, had a fabric-covered metal framework. One of the most successful metal biplanes, it entered service in 1925 and was still operation in small numbers 15 years later. A variety of powerplants with outputs of up to 550hp gave a top speed of about 230km/h (143mph) and a range of 800km (497 miles). Up to 440kg (970lb) of bombs could be carried, and four machine guns were fitted for offensive use as well as self-defence. Some 1,100 were built in France and the type was widely constructed under license.

The duralumin Lioré et Olivier LeO.20 three-seat twin-engined night bomber was widely deployed with French squadrons during 1927–37, rivalling the Farman F.160 to F.168 series. The three-seat single-engined Amiot 122 BP3 bomber/escort biplane, also constructed from light alloy, served during 1928–35 and could carry as great a load as the much larger twins. One of the earliest stressed-skin bombers was the Amiot 143, a twin-engined high-wing monoplane. The Bloch 210, developed from the Model 200 five-seat night bomber introduced in 1934, was used in support of Republican forces during the Spanish Civil War as was one Bloch 200. Other bombers were the Farman 60, 63 and 221, and the LeO 206.

Developments in Italy and Germany

Italy reduced its military strength after World War I but set up an independent air service in 1923. Single-engined Caproni Ca 73s replaced multi-engined bombers and the later Ca 101 and 111 high-wing monoplanes were among types used in Ethiopia. Smaller European countries such as Czechoslovakia and Holland were also active in bomber development. The Czech Aero A 11, which entered service in 1923, was produced in many variants. It carried a crew of two at 215km/h (133mph) over 750km (466 miles),

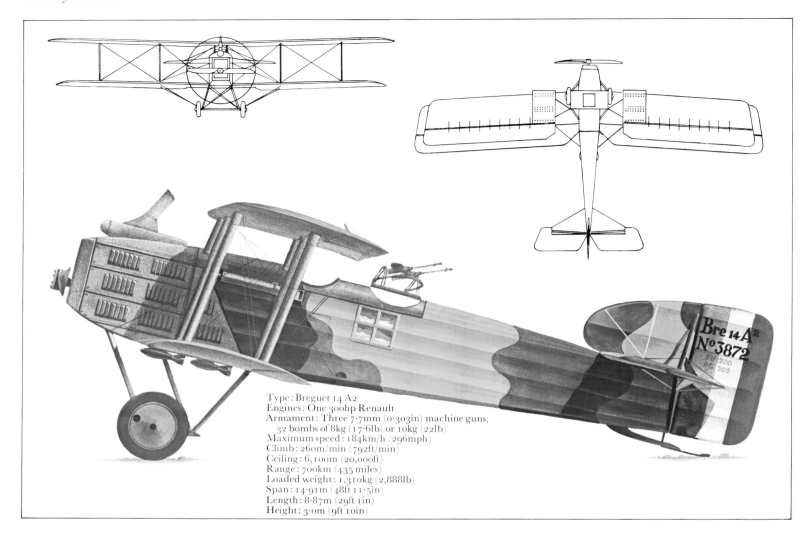

Type: Breguet 14 A2
Engines: One 300hp Renault
Armament: Three 7·7mm (0·303in) machine guns,
    32 bombs of 8kg (17·6lb) or 10kg (22lb)
Maximum speed: 184km/h (296mph)
Climb: 260m/min (792ft/min)
Ceiling: 6,100m (20,000ft)
Range: 700km (435 miles)
Loaded weight: 1,310kg (2,888lb)
Span: 14·91m (48ft 11·5in)
Length: 8·87m (29ft 1in)
Height: 3·0m (9ft 10in)

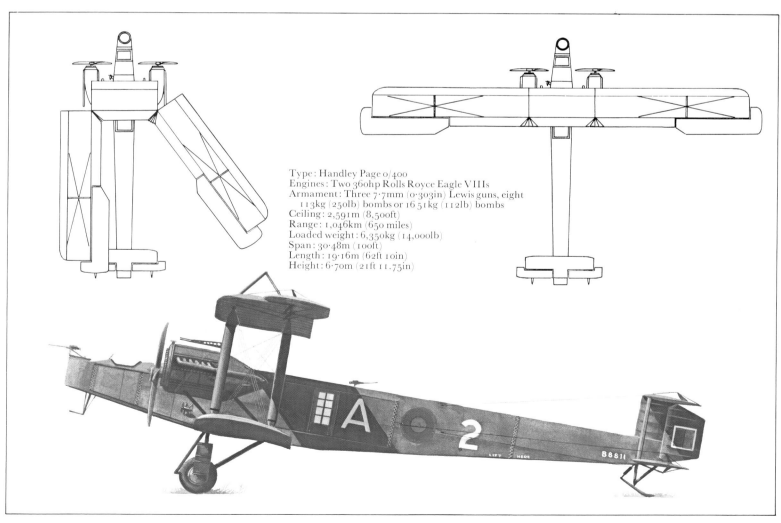

Type: Handley Page 0/400
Engines: Two 360hp Rolls Royce Eagle VIIIs
Armament: Three 7·7mm (0·303in) Lewis guns, eight
    113kg (250lb) bombs or 16 51kg (112lb) bombs
Ceiling: 2,591m (8,500ft)
Range: 1,046km (650 miles)
Loaded weight: 6,350kg (14,000lb)
Span: 30·48m (100ft)
Length: 19·16m (62ft 10in)
Height: 6·70m (21ft 11.75in)

although its bomb load was only 50kg (110-lb). More than 400 of the later Letov S–238 were built, and a number captured by the Germans were transferred to the Bulgarian and Slovak air forces, the latter using them during the invasion of Poland and Russia. In Holland, Fokker produced types such as the C.V series, which served from the mid-1920s until the late 1930s. A light bomber and reconnaissance aircraft, it was very reliable and easy to fly. The type was designed to accept a variety of wings and powerplants for different roles, and it was bought by a dozen countries as well as being built under licence.

In Germany, the aircraft industry re-established itself after World War I and the Dornier Do 23, developed from the Do F mail and freight transport of the late 1920s, was the first medium bomber to equip the newly formed *Luftwaffe*. The very advanced Heinkel He 70, which made its maiden flight in December 1932, was designed and built in just over six months but the twin-engined Do 17 was preferred for long-range duties. The *Luftwaffe* used the Spanish Civil War to test its equipment and techniques, discovering for example that the defensive armament of the Ju 86 was inadequate when faced by modern fighters. Also used as a bomber in that conflict was the Ju 52/3m 'Iron Annie', later to become famous as a transport.

Advances in Britain

In Britain, the Royal Air Force had been formed as the first independent air arm in Europe by merging the RFC and RNAS. It was assumed by the British Government that no major war would break out for at least ten years, so the RAF's strength was reduced from 188 squadrons to 33, and types such as the D.H.9A, D.H.10 Amiens and Vimy continued to see action in India and the Middle East. The first post-war two-seat light day bomber was the Fairey Fawn, which served during 1923–26. It was handicapped, however, by having to conform to restrictive official specifications and its performance was inferior to that of its predecessor, the D.H.9A.

In 1917 it had been officially decreed that the standard new British aero engine would be the ABC Dragonfly, a nine-cylinder radial intended to produce 400hp. It proved to be a technical disaster, so when Richard Fairey decided to develop a light bomber as a private venture he went to the United States for

the powerplant. He chose the Vee Curtiss D12, which had helped give the Curtiss seaplane its victory in the 1925 Schneider Trophy. The engine, built under licence as the Fairey Felix, gave the Fox a top speed of 250km/h (156mph) when it appeared in 1925. This was faster than most contemporary fighters, and the Fox also had the edge in manoeuvrability, but only one RAF squadron was equipped with the aircraft – fitted with Rolls-Royce Kestrels in the production version – and it was in Belgium that the type achieved its well deserved acclaim. Later variants, powered by Hispano-Suiza engines as well as Kestrels, were capable of 360km/h (225mph), and the Belgian production line turned out nearly 200 to add to the 28 built in Britain.

The RAF's elderly D.H.9A was replaced by the Westland Wapiti, using the same wing and undercarriage as its predecessor. The prototype and initial production machines had fabric covering on the wings and rear fuselage, but later variants were all-metal. The main production version was the Mk IIA, powered by a 500hp Bristol Jupiter, and 415 were built. The air-cooled Jupiter, originally a back-up for the Rolls-Royce Eagle, dominated the aero-engine world in the 1920s and was built under license in no fewer than 16 countries. In 1918 it turned out between 375 and 400hp. Later it produced more than 500hp and the Pegasus, developed from it, was capable of more than 1,000hp by 1938. The Wapiti Mk V had a lengthened fuselage and was exported to several countries as well as being built under licence in South Africa. It was developed into the Wallace, of which only about 100 were built and which served solely with Auxiliary squadrons. The Mk II version was powered by a 680hp Pegasus IV engine, and in 1933 a Wallace was one of two aircraft to make the first flight over the summit of Everest, the world's highest mountain.

The mainstay of the RAF's fleet of heavy night bombers from 1924 to 1937 was the Vickers Virginia, a conventional wooden biplane. It was powered by two 450hp Napier Lions, this engine being the other powerplant originally developed as a back-up for the Rolls-Royce Eagle. A water-cooled 'Vee', as opposed to the air-cooled Jupiter, it produced up to 1,320hp in racing form and was also used in the Handley Page Hyderabad. The Virginia Mk X, powered by 580hp Napier Lions, had a gross take-off weight of 7,983kg (17,600lb) and could carry 1,360kg (3,000lb) of bombs. It cruised at 160km/h (100mph) and carried a crew of two, using three 0·303in machine guns for self-defence. The Avro Aldershot also entered RAF service in 1924, and in the same role, but it saw only limited service because the Air Staff were opposed to single-engined heavy bombers. The Aldershot nevertheless had a respectable 900kg (2,000lb) bomb load.

One aircraft which enjoyed a reasonable production run was the Fairey III series used by the RAF, Fleet Air Arm and a number of overseas air forces. The most numerous version was the IIIF, which entered RAF service in 1927 on light bombing and general duties. The first British metal heavy bomber

was the Handley Page Hinaidi, developed from the Hyderabad. It entered service in 1929 and served until 1933, when it was replaced by the Heyford. This latter aircraft was unusual in having the fuselage attached to the upper rather than lower wing, and it also featured a rotating ventral dustbin turret which was housed inside the fuselage when not in use.

In 1928 the RAF had introduced its first medium bomber, the Boulton-Paul Sidestrand, which was sufficiently fast and manoeuvrable to be used in daylight. In 1934 it was replaced by the same manufacturer's Overstrand, the first British bomber with a power-operated nose turret. Apart from these two examples, all British day bombers were single-engined two-seaters until the monoplanes arrived in the late 1930s. The standard RAF light bomber during 1930–37 was the Hawker Hart, which has been described as 'one of the most technically and strategically significant' bombers designed in the ten years following the end of World War I. Nearly 1,000 were built, about half being used as trainers. Its top speed of 280km/h (175mph) was better than that of many fighters, and the Hart was also highly manoeuvrable. It was replaced on an interim basis from 1935–38 by the Hind, the RAF's last biplane bomber. Nearly 600 were built, and the type was widely exported. It in turn was superseded from 1937 by the Fairey Battle and Bristol Blenheim all-metal cantilever low-wing monoplanes. The Fairey Hendon of 1936 had been the first RAF twin-engined cantilever monoplane, but its metal basic structure had fabric covering.

The RAF's bomber force in the 1930s also included the Vickers Vildebeest, delivered from 1932 to replace the Hawker Horsley torpedo bomber of mixed wood and metal construction. The Horsley had itself taken over from the Fairey Fawn. The Vildebeest still equipped two squadrons of Coastal Command at the outbreak of World War II. The Handley Page Harrow heavy bomber was unusual in having a mixture of fabric and metal covering, but the private-venture Vickers Wellesley was even more innovative. It was constructed on the geodetic principle developed by Dr Barnes Wallis, using a basketwork of members carrying loads in compression or tension but not bending. This was covered with an unstressed fabric skin, allowing battle damage to be repaired by replacing deformed members. The Wellesley entered service in 1937, but only 136 were built because the Wellington, using the same method of construction, was already becoming available.

Bomber aircraft of the Soviet Union

The Soviet Air Force was re-established in 1924, and Russian industry began to turn out a vast number of aircraft to equip it. About 6,000 Polikarpov R–5s were built between 1928 and the late 1930s, the type being used in the Spanish Civil War, against the Japanese in Mongolia during 1938 and 1939, and in the 1939–40 Winter War with Finland. The assault version carried 500kg (1,100lb) of bombs plus seven 7·62mm machine guns. The Polikarpov U–2 was even more remark-

able, more than 20,000 being built between the late 1920s and 1952. Until the outbreak of World War II, it served on general-purpose duties but then, as the U–2VS, it was used for light bombing and close support. The U–2VS could carry 250kg (550lb) of bombs and was defended by a 7·62mm machine gun. Its operations included the defence of Stalingrad, and rocket projectiles were later carried in place of bombs.

The first Russian metal bomber was the Tupolev R–3 of 1926. A two-seat single-engined biplane, it was covered in corrugated aluminium sheeting. In the following year the TB–1 twin-engined monoplane appeared, setting the pattern for Russian bomber design until the end of the 1930s. It had a crew of six and carried 3,000kg (6,614 lb) of bombs. The Tupolev TB–3 entered service in 1932 and remained operational throughout World War II, about 800 being built. It was powered by four Vee-12 engines of 730hp each. Various M–17 or M–34 V12s of outputs ranging from 600 to 1,200hp were the standard powerful engines used in Russian combat aircraft until 1941. The TB–3 followed the tradition of the Ilya Mourometz as a four-engined bomber, several years before America's Boeing B–17 came on the scene. It could carry 2,200kg (4,850lb) of bombs, some internally and some externally, was equipped with six 7·62mm machine guns for self-defence and took part in experiments with parasite fighters. The later Tupolev SB–2 entered service in 1936 and operated throughout the war as a night bomber. Production exceeded 6,000.

United States bombers

The United States Army Air Service was set up in April 1918, but with the coming of peace its money was cut by Congress and only 27 squadrons could be established instead of the planned 87. Of these 27, a mere four were assigned the bombing role and but a single squadron was equipped with the Martin MB–2. The MB–2 was a development of the MB–1, the design of which had been influenced by features of Handley Page and Caproni types built under licence during the war. The MB–1's radius of action, at 280km (175 miles), was only half that of its European contemporaries and the bombload was a paltry 455kg (1,000lb), so it was developed into the MB–2 – regarded as the United States' first indigenously designed bomber – by fitting strengthened and slightly larger wings which raised the payload to 818kg (1,800lb). The MB–2 replaced 0/400s from 1919 and remained in service until 1927.

The MB–2 was superseded by a series of

aircraft designed by Huff-Daland (later Keystone). The layout of the B–4A – followed by the B–5A and B–6A – was changed from single-engined to twin for safety reasons and to give the crew a better view and freedom of action by leaving the nose clear. This was at approximately the same time as the Avro Aldershot was failing to find favour because of its single engine. In fact the Keystone bombers offered little improvement on their predecessors.

British D.H.4s, built under licence as Boeing DH–4As, continued to operate as light bombers. More than 5,000 were built in the United States, many being converted to the improved DH–4B and 4M standards. They remained in service as bombers until 1928 and were finally retired in 1932.

From then on the pace of development began to accelerate, and the heavy bomber was transformed within the space of a decade. The Douglas Y1B–7 of 1930 (later B–7) was the first US monoplane bomber. In the following year the Boeing YB–9 made its appearance. This was an all-metal cantilever low-wing monoplane with a semi-retractable landing gear and variable-pitch propellers for its two engines. It could carry a 1,025kg (2,260lb) bombload entirely within its semi-monocoque fuselage. Only seven YB–9s were built. Although chronologically a little earlier than the B–9, the Martin B–10 has a significant place in US Army history. The use of powerful 775hp Wright R–1820–3 engines, combined with streamlining, gave the B–10B a 343km/h (213mph) top speed at 1,980m (6,500ft) – faster than most contemporary fighters. A fully retractable landing gear and enclosed cockpit were fitted, and the B–10, of which more than 150 were built (including the B–12 and B–14), became the first mass-produced US monoplane all-metal bomber. It was also the first to have a gun turret.

This series was replaced at the end of the 1930s by the Douglas B–18, developed from the DC–2 and DC–3 airliners. The switch to monoplane design in the civil field had begun with the Douglas DC–2, Boeing 247 and Lockheed Electra airliners. The B–18A could carry a maximum bombload of 2,950kg (6,500lb) and had a speed of 345km/h (215 mph). It was powered by two 1,000hp Wright R–1820–53 Cyclones, one of the new breed of air-cooled radials – alongside the Pratt & Whitney R–1830 Twin Wasp – which had been becoming increasingly dominant in US combat aircraft. The B–18 was later overshadowed by the longer-ranged and much more expensive Boeing B–17 Flying Fortress four-engined bomber, which had flown in 1934 as the Model 299 but which was not ordered until 1938.

Japan prepares for war

Across the Pacific, the country that was to bring the new generation of US bombers into action had been building up its own forces. Japan had flown against German and Austro-Hungarian forces on the coast of China during World War I and against Russian forces in Siberia from 1918 to 1921, using imported and license-built types. Japanese companies employed European designers to help build up an indigenous industry, and also asked

FAR LEFT: *The Tupolev SB–2 was one of two Russian medium bombers to be operated for the duration of World War II.*
LEFT: *The advanced Martin B–10 was the first monoplane all-metal bomber to be mass-produced, and the first to have a gun turret. A fully retractable landing gear was used.*
BELOW: *Some 6,000 Polikarpov R–5s were built in the 1920s and 1930s; the type saw action in Mongolia, Spain and Finland.*
BOTTOM: *The Japanese Kawasaki Type 88 was used for reconnaissance and, as the Type 88–II, as a light bomber. More than 400 were built for the latter role, seeing action in Manchuria.*

overseas concerns to design types for production in Japan. The Kawaski Type 87 heavy bomber, for example, was designed by Dornier and powered by a pair of 500hp BMW engines. Only 28 were built, and the Mitsubishi 2MB1 (Type 87 light bomber) did not fare much better, with production totalling 48 units. The latter aircraft, a three-bay biplane, was powered by a single Hispano-Suiza water-cooled engine of 450hp and was replaced by an adaptation of the Kawasaki Type 88 reconnaissance biplane in 1929, only two years after entering service.

Designed by Dr Richard Vogt, later of Blohm und Voss, the Type 88–II – as it was designated in the light bombing role – could carry a payload of only 200kg (440lb) and was rather slow, but it could withstand heavy punishment. More than 400 were built as light bombers, many serving in the Manchurian conflict. This campaign showed up

the deficiencies of the Japanese bombers, so the Army ordered three new aircraft in the Type 93 series: the Mitsubishi Ki–1 twin-engined heavy bomber, the Mitsubishi Ki–2 twin-engined light bomber and Kawasaki's Ki–3 single-engined light bomber. The two Mitsubishi aircraft were all-metal designs influenced by Junkers. The four-seat Ki–1 had a gross take-off weight of 8,165kg (18,000 lb) and was powered by two 940hp water-cooled Ha–2 engines. The three-seat Ki–2 grossed at 4,535kg (10,000lb) and was fitted with a pair of 450hp air-cooled Bristol Jupiters, while the two-seat fabric-covered Ki–3 – the last Japanese Army Air Force biplane bomber – was powered by a single water-cooled BMW IX engine of 800hp and had a maximum take-off weight of 3,100kg (6,834lb).

The first and third of the series were slow and unreliable, mainly because of problems with their water-cooled engines, and they were soon withdrawn from front-line service. The Ki–2 was popular, however, especially in the later Mk II version with retractable landing gear and two 550hp Ha–8 engines, and was used widely up to the outbreak of World War II. The JAAF's first really modern bomber was the Mitsubishi Ki–21 (Type 97 heavy bomber), which replaced the Ki–7. An all-metal monoplane powered by two 950hp Ha–5 engines, it had a gross weight of 9,700kg (21,400lb) and was the principal Army bomber in the Chinese and Pacific wars. More than 2,000 were built. The Ki–3 was replaced by the Kawasaki Ki–32 (Type 98 light bomber) and Mitsubishi Ki–30 (Type 97 light bomber). The former, a mid-wing monoplane powered by an 850 hp in-line Ha–9, proved unpopular with its crews because of the unreliability of its engine. The Ki–30, also a mid-wing monoplane, introduced wing trailing-edge flaps and an internal bomb-bay and was powered by a

conventional 950hp Ha–5 air-cooled radial engine. Both types, with combat ranges greater than 1,600km (1,000 miles), were widely used on strategic missions against China.

The Ki–2 was eventually replaced by the Kawasaki Ki–48 twin-engined light bomber, of which 2,000 were built. It was short on range, however, and had a small bombload. In its operations against China the Ki–21 was shown to have inadequate range and poor defensive armament and it was superseded by the Nakajima Ki–49 (Type 100). Eight hundred were built, but these aircraft proved difficult to handle and there was still no Japanese bomber capable of carrying an adequate payload.

Meanwhile the Japanese Navy had been building up its land-based forces as well as its

carrier fleet. In the early 1930s it had asked Mitsubishi to design a twin-engined biplane torpedo-bomber. The resulting G1M (Type 93) was withdrawn from service, after only a handful had been built, because of stability and engine problems. The Navy turned to the Hiro arsenal for a replacement, but the resulting G2H (Type 95) also had serious defects. Mitsubishi had developed a mid-wing twin-engined reconnaissance aircraft with retractable landing gear which first flew in May 1934, and this led to the G3M (Type 96) medium bomber. It entered service in 1937 and was used in the war against China, bombing cities from bases in Japan and Formosa. It also took part in World War II, more than 1,000 having been built by the end of 1943. The final version had two 1,200hp Kinsei engines. The G3M was replaced by the Mitsubishi G4M (Type 1), of which some 2,500 were constructed. Armour-plating was sacrificed for the sake of extra range, and self-sealing fuel tanks were not fitted – a common

failing of Japanese military aircraft at that time – so the bomber was apt to catch fire once hit by gunfire.

German bombers of World War II

Germany was undoubtedly the country which made the greatest strides in bomber design during World War II, even if many of the more ambitious proposals failed to see the light of day or appeared too late to affect the course of the war. For example, a *Luftwaffe* medium bomber such as the Dornier Do 17Z of 1939 could carry 1,000kg (2,205lb) of bombs at 360km/h (225mph) and 4,000m (13,125ft) altitude. Five years later the jet-powered Arado Ar 234B light bomber could carry 1,500kg (3,300lb) at twice the speed and height of the earlier aircraft.

The first chief of staff of the *Luftwaffe*, Lt-

Gen Wever, supported long-range strategic bombing and began the 'Ural Bomber' project in 1935. After his death in June 1936, however, he was succeeded by Gen Kesselring, who believed that the primary role of the *Luftwaffe* was tactical support of the Army. He therefore concentrated on short-range, lightly armed day bombers, and although the *Luftwaffe* had a front-line strength of 1,750 bombers in September 1939 it lost about 2,000 aircraft and 5,000 crew on daylight attacks on Britain before the force was transferred to night operations.

Germany's mainstay bombers during the Battle of Britain in the summer of 1940 were the Dornier Do 17 and Heinkel He 111. The former, originally designed as a commercial transport, was first delivered in early 1937; by the outbreak of war some 370 were operational. They were first used in the Polish campaign, later bombing convoys in the English Channel before being switched to attack RAF airfields and British cities. The Do 17Z–2, powered by a pair of 1,000hp Bramo radial engines and carrying a crew of four or five, cruised at a mere 300km/h (186mph) at an altitude of 4,000m (13,125ft) when fully laden and had a tactical radius of only 320km (200 miles) in this condition. Maximum internal bombload was 1,000kg (2,205lb), and six machine guns were fitted for self-defence. During the spring of 1941 Do 17s were used for anti-shipping operations in the eastern Mediterranean, and by the time of Operation Barbarossa, the invasion of Russia in July that year, only one unit was equipped with the type. Do 17s were later used as glider tugs and were replaced in the bombing role by Do 217s and Junkers Ju 88s (see below).

The Heinkel He 111 remained in service so long mainly because no suitable replacement could be found; production exceeded 7,300, with peak output being reached in 1939 and 1943. It was used widely in the invasions of Poland, Denmark and Norway, and in attacking Britain's Royal Navy North Sea fleet. It suffered heavy losses in the Battle of Britain, however, and by mid-1940 was being replaced by Ju 88s. Defensive armament, despite including cannon as well as machine guns, was insufficient for the unescorted missions on which the He 111 was employed. Bombload, at 2,500kg (5,500lb), was also inadequate, and the 2,050km (1,275 miles) range at 385km/h (240mph) and 5,000m (16,400ft) altitude was too short for effective strategic operations.

The He 111 continued occasional night bombing, anti-ship strikes and mine-laying, was used during Barbarossa and in the spring and summer of 1942 was employed on torpedo strikes against Arctic convoys, flying from Norwegian bases. The type was increasingly used as a transport and glider tug on the Eastern Front, and in the summer of 1944 was given the new role of air-launching V1 flying bombs. More than 1,200 of these missiles were fired from He 111s.

The Do 217, scaled up from the Do 17, had both longer range and a better bombload than its predecessor. It entered service in 1941, carrying out missions over Britain and against ships in the North Sea in the spring of 1942. From August 1943 the type attacked shipping in the Bay of Biscay and Mediterranean, using Hs 293 and Fritz–X guided bombs. More than 1,300 were operated as bombers and missile carriers, with a further 300-plus being used as night fighters and intruders. The Do 217M night bomber was powered by two Daimler-Benz engines of 1,750hp each at take-off, giving a cruise speed of 400km/h (250mph) and a range of 2,400 km (1,500 miles) on internal fuel. Bombload was 4,000kg (8,800lb), of which 2,500kg (5,500lb) was internal, and six machine guns were fitted for self-defence.

The Junkers Ju 88, of which more than

entered service, and the need for a bomber able to attack any British target from bases in France and Norway receded as the tide turned against Germany. The Heinkel He 177 Greif (Griffon) heavy bomber evolved too late and also failed to live up to its promise for a variety of reasons. The original design was very advanced, but shortage of a suitable 2,000hp powerplant meant that its designer, Siegfried Günter, was compelled to rely on the expedient of coupling two 1,000 hp Daimler-Benz engines together, using two of the units to power the He 177. They proved unreliable and were prone to overheating.

15,000 were built, was one of the outstanding aircraft of the war and operated in a wide variety of roles. In many ways it was the German equivalent of the Mosquito. The Ju 88A–4 was powered by a pair of Junkers Jumo in-line engines rated at 1,340hp each, had a normal range of 1,790km (1,110 miles) and carried 1,800kg (4,000lb) of bombs plus a combination of machine guns and cannon. A slightly larger development, the Ju 188, entered service in 1942. The two 1,775hp Jumo engines of the Ju 188A–2 version provided a maximum speed of 520km/h (325 mph).

The search for a high-altitude bomber to replace the Ju 88 and He 111 – the 'Bomber B' – lasted four years and failed to turn up an effective answer. The various proposals, such as the Ju 288, Do 317 and Fw 191, never

Mention must also be made of the Arado Ar 234B Blitz, the second jet-powered aircraft to enter *Luftwaffe* service and used as a bomber as well as for reconnaissance. In March 1945 the type took part in an all-jet attack on the Ludendorff bridge over the Rhine at Remagen, with Messerschmitt Me 262 fighter-bombers strafing the flak emplacements. Only 210 were completed, but they ushered in the era of the jet bomber.

British bombers of World War II
The RAF substantially modernized its bomber force in the late 1930s. Deliveries of Bristol Blenheim light bombers began in January 1937, and over the next few years the aircraft was widely exported and built under license. Despite being some 64km/h (40mph) faster than contemporary fighters

when it appeared, the Blenheim was not as successful as might have been expected. The Mk I, of which more than 1,500 were built, was followed by nearly 2,000 Mk IVs and some 200 of the original aircraft were converted into night fighters. The Mk IV was built in Canada as the Bolingbroke, where 629 were used for coastal patrol and target-towing. Douglas Bostons and de Havilland Mosquitoes replaced the Blenheims in Europe from 1942, but the type continued to serve in the Middle East and Far East until the end of the following year. The Mk IV was powered by a pair of 920hp Bristol Mercury XV radials, giving a speed of 428km/h (266mph) over a range of 2,350km (1,460 miles). The Blenheim could carry 599kg (1,320lb) of bombs and was fitted with a defensive armament of five 0·303in Browning machine guns.

Bristol used its experience with the Blenheim to develop the Beaufort torpedo bomber, which entered service with RAF Coastal Command in December 1939. More than 1,100 were built, the type being replaced in 1943 by the Beaufighter, based on the Beaufort and regarded – in its radar-equipped Mk X form – as one of the best maritime strike fighters of its day.

The distinction of carrying out the first RAF bombing raid of the war went to Bristol Blenheims and Vickers Wellingtons, which attacked the German Fleet in the Schillig Roads, off Wilhelmshaven, on 4 September 1939. The Wellington was transferred to night operations that December as the main

had a good range and bombload, and was easy to fly. Eight squadrons were equipped with Hampdens at the outbreak of war, operating them as heavy day bombers, but inadequate defensive armament caused the aircraft to be transferred to night operations and in 1942 they were passed on to Coastal Command as torpedo bombers.

The first Allied four-engined bomber of the war was the Short Stirling, which entered service at the end of 1940 and began operations over enemy territory the following February. The engines of the Mk I version were 1,590hp Bristol Hercules XI radials, giving a speed of nearly 418km/h (260mph) and a range of 3,100km (1,930 miles). A payload of nearly 6,400kg (14,000lb) could be carried, maximum gross weight being 26,950-kg (59,400lb). The Stirling was well defended – it carried eight machine guns – and could absorb punishment, but most had been transferred to night operations by the beginning of 1942 and a year later the type was outdated as a bomber.

The RAF's main 'heavies' from 1942 were the Handley Page Halifax and Avro Lancaster. The former was designed to meet the same specification as the Avro Manchester, the twin-engined bomber on which the Lancaster was based. The Halifax entered service at the end of 1940, and as well as operating over Germany it was also the only RAF four-engined bomber to serve in the Middle East. More than 6,000 were produced, the most numerous version being the Mk III. This was powered by four 1,615hp Bristol Her-

RAF bomber until the arrival of the four-engined heavy types. It was also the first aircraft to drop the 1,814kg (4,000lb) blockbuster bomb. The most numerous version was the Mk X, produced from 1943. Powered by two 1,675hp Bristol Hercules VIs or XVIs and carrying a crew of six, it had a maximum take-off weight of 16,555kg (36,500lb) and carried 2,040kg (4,500lb) of bombs as well as eight 0·303in Brownings for self-defence.

One of the mainstays of Bomber Command in 1939 was the Armstrong Whitworth Whitley. In its initial role as a night bomber and leaflet-dropper it became the first British aircraft of World War II to bomb Berlin and to attack Italy. Whitleys later operated with Coastal Command on reconnaissance and anti-submarine duties, as well as being used to drop paratroops. The major production version was the Mk V, of which some 1,500 were delivered between 1939 and June 1943. This was powered by a pair of Rolls-Royce Merlin Xs of 1,145hp each, giving a speed of 357km/h (222mph) and a range of 2,655km (1,650 miles) with 1,360kg (3,000lb) of bombs. Maximum load was 3,175kg (7,000lb).

Handley Page's first monoplane bomber, the Hampden, was faster than the Whitley,

cules XVIs and had a range of 3,860km (2,400 miles) at a maximum of 454km/h (282mph). A bombload of about 5,900kg (13,000lb) could be carried, and normally eight Browning machine guns were fitted for self-defence.

Nearly 7,400 Lancasters were built, the first entering service at the beginning of 1942. The type took part in many famous raids, including the dam-busting mission by 617 Squadron in May 1943, the aircraft carrying Barnes Wallis's 'bouncing bombs'. In November 1944 thirty-one Lancasters, each carrying a 5,450kg (12,000lb) Tallboy bomb, sank the German battleship *Tirpitz* in a Norwegian fjord. The Lancaster Mk I (Special) was fitted with the even larger – 10,000kg (22,000lb) – Grand Slam. Maximum take-off weight with a Grand Slam aboard was 31,750kg (70,000lb), slightly more than when the normal load of 6,350kg (14,000lb) was carried. The four Rolls-Royce Merlins of up to 1,640hp each gave a top speed of 462km/h (287mph) and a maximum range of 2,670km (1,660 miles) with normal

FAR LEFT ABOVE: *The Handley Page Halifax entered service with the RAF in 1940.*
FAR LEFT BELOW: *The first Allied four-engined bomber of World War II was the Short Stirling.*
BELOW LEFT: *The Savoia-Marchetti S.M.79 Sparviero (Hawk) medium bomber comprised well over half the Italian Air Force's bomber fleet at the outbreak of World War II.*

BELOW: *The Vickers Wellington medium bomber used the Barnes Wallis geodetic method of construction and was the first aircraft to drop the 1,814kg (4,000lb) blockbuster bomb.*
BOTTOM: *The Avro Lancaster, one of the classic aircraft, was developed from a failure – the underpowered twin-engined Manchester. The 'Lanc' dropped Barnes Wallis's bouncing bombs.*

bombload. A formidable defensive armament of ten machine guns was installed.

The de Havilland Mosquito was designed as a light bomber, despite its use in a multitude of other roles. More than 13,000 were built, with the aircraft entering service in 1941 and still performing useful hack duties well into the 1960s. The Mk XVI high-altitude bomber with its pressurized cabin was powered by a pair of 1,680hp or 1,710hp Rolls-Royce Merlins, a top speed of 668km/h (415mph) and ceiling of 12,190m (40,000ft) allowing defensive armament to be dispensed with. The two-crew Mosquito had a range of 2,205km (1,370 miles) in this variant and could carry a 1,811kg (4,000lb) bombload.

Developments in France, Italy and Russia during World War II

French bombing activities in World War II were obviously limited, as the country was invaded in 1940. On the night of June 7–8 in that year, however, a Centre NC223.4 four-engined bomber – originally designed as a transatlantic mailplane – became the first Allied plane to Bomb Berlin in World War II.

When Italy joined the war in 1940 its air force had nearly 1,000 bombers, of which well over half were Savoia-Marchetti S.M.79 Sparviero (Hawk) medium bombers. These trimotors, were thought by many to be among the best land-based torpedo bombers of the war. They could carry 1,250kg (2,750 lb) of bombs internally or two torpedoes. Also active as a medium bomber around the Mediterranean and on anti-ship duties was the

Cant Z.1007*bis* Alcione (Kingfisher), production of which began in 1939. It also was a trimotor, powered by 1,000hp Piaggio radials, and it carried four machine guns for self-defence as well as up to 2,000kg (4,410lb) of bombs or two torpedoes.

Fiat's B.R.20 Cicogna (Stork) medium bomber operated briefly against British targets from bases in Belgium in 1940 but was soon withdrawn, and only a handful of the production run of some 600 remained in service at the time of the Italian surrender in 1943. The four-engined Piaggio P.108B heavy bomber, of which only about 160 were

built, entered service in 1942 and was employed against Mediterranean targets including Gibraltar. It could carry 3,500kg (7,700lb) of bombs.

On the Eastern Front two Russian medium bombers which served throughout the war were the Ilyushin Il–4 and Tupolev 2B–2*bis*. The former was the first Russian aircraft to bomb Berlin and was also widely used as a torpedo bomber in the Baltic, its payload being either 2,000kg (4,410lb) of bombs or a single torpedo. Two 1,100hp radial engines gave the Il–4 a range of about 4,000km (2,485 miles). The SB–2*bis*, also twin-engined and

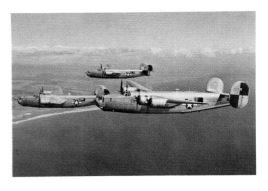

with a crew of three, had a smaller bombload – 600kg (1,320lb). The Soviet Air Force did not invest in strategic bombers as heavily as did the RAF and US forces, but the four-engined Petlyakov Pe–8 was used to attack German and Balkan targets with up to 4,000 kg (8,820lb) of bombs. From the same manufacturer came the much smaller Pe–2 light bomber, which served on all Russian fronts from 1941 to 1945. Power was provided by a pair of Klimov in-line engines of 1,000hp each, giving an impressive top speed of 540 km/h (335mph). The two-crew aircraft could carry 1,000kg (2,205lb) of bombs. Towards the end of the war the Pe–2 was replaced by the Tupolev Tu–2, regarded as one of the best Russian designs of the period. The four-man aircraft, powered by a pair of 1,850hp Shvetsov radials, could carry 2,270kg (5,000 lb) of bombs and was defended by two 23mm cannon and five machine guns.

America's long-range bombers

The United States produced bombers in massive quantities to meet the needs of its own forces and those of its allies in the Pacific, North Africa, the Middle East and Far East as well as in Europe. More than 12,700

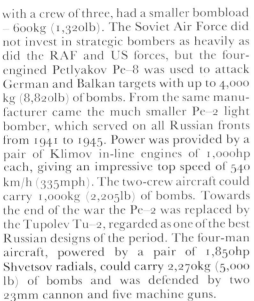

Boeing B–17s were built, of which nearly 8,700 were the B–17G model. Four 1,200hp Wright Cyclones gave a speed of 462km/h (287mph) and a range of about 3,220km (2,000 miles) with a 2,720kg (6,000lb) bombload. The B–17G, with its crew of ten, could carry a maximum bombload of nearly 8,000 kg (17,600lb) over short distances and was armed with 13 0.50in machine guns for self-defence.

The Consolidated B–24 Liberator was built in greater numbers than any other US aircraft – more than 18,000 were completed – and is said to have dropped 635,000 tons of bombs and destroyed more than 4,000 aircraft in its career. Operated by the RAF and Commonwealth air forces as well as by the US, the Liberator saw service as a transport and reconnaissance aircraft in addition to its bombing role. The B–24J, of which 6,678 were built, was powered by four 1,200hp Pratt & Whitney Twin Wasps and could carry its 12 crew for 3,380km (2,100 miles). Normal maximum bombload was 3,990kg (8,800lb), although 5,800kg (12,800lb) could be carried over short ranges, and 10 machine guns were installed for self-defence.

Medium bombers included the North American B–25 Mitchell and Martin B–26 Marauder, both of which had a normal bombload of 1,360kg (3,000lb). The Mitchell was also very heavily defended, the B–25H carrying 14 machine guns. Production ran to 4,300-plus and 5,000-plus respectively. The principal Mitchell variant was the B–25J powered by two 1,700hp Wright Cyclones. The B–26 used 2,000hp Pratt & Whitney Twin Wasps. A 1,180kg (2,600lb) bombload was carried by the Douglas A–20 Havoc – half internally and the rest externally – of which more than 7,000 had been built when production ended in September 1944. Also known as the Boston, it was used by the RAF as well as the US and was widely employed as a torpedo bomber by the Soviet Union.

The heaviest combat aircraft of the war was the Boeing B–29 Superfortress, which entered service in the early summer of 1944. Intended as a strategic bomber for use against Germany, it in fact operated only against Japan. Maximum take-off weight of the B–29A was 64,000kg (141,100lb) and the four immensely powerful – 2,200hp each – Wright Cyclones gave a speed of 575km/h (357mph) and a normal range of 6,600km (4,100 miles). Bombload was normally 5,440kg (12,000lb)

and the defensive armament, largely remotely controlled by the standard crew of 10, comprised a 20mm cannon as well as 10 machine guns. It was a B–29, the *Enola Gay*, which dropped the atomic bomb on the Japanese city of Hiroshima.

Japanese bombers

The Japanese Navy's land-based bombers gave a good account of themselves early in the war, G3Ms sinking the battleships HMS *Repulse* and *Prince of Wales* in December 1941. The later G4M was used both for normal bombing and for torpedo operations, and from October the Navy's torpedo units began operating what was in fact an Army aircraft, the Mitsubishi Ki–67 Hiryu (Flying Dragon). Some 700 were built, and it was regarded as the Army's best bomber of the war. Two Mitsubishi Ha–104 radials of 2,000hp each gave a top speed of 540km/h (335mph), and the range was an impressive 3,780km (2,350 miles). A 20mm cannon and four machine guns were carried in addition to the 795kg (1,750lb) of bombs.

1945 to the present

The B–29, with its operational ceiling of more than 9,000m (30,000ft), radar-controlled defences and pressurized crew positions, bridged the gap between World War II heavy bomber and strategic jet types. Nine B–29 squadrons operated in the Korean War, and by 1951 most Strategic Air Command front-line units had re-equipped with the B–50 development powered by Pratt & Whitney Wasp Majors. Top speed approached 650km/h (400mph). The B–29 was also built in Russia between 1946 and 1949 as the Tupolev Tu–4 Bull, and the defensive armament of the next generation of Russian

bombers was based on that fitted to the Tu–4.

An American legacy from World War II was the Convair B–36, the requirement for which was originally drawn up when it seemed likely that the whole of Europe would be overrun and strategic missions would have to be flown direct from the United States. The B–36, which first flew in August 1946 and entered service two years later, was therefore enormous. Its wings spanned 70·1m (230ft), it was 49·4m (162ft) long and a crew of 15 was embarked. Design range was 16,100km (10,000 miles) with a bombload of 4,500kg (10,000lb). In fact, the B–36B carried a normal bombload of 32,660kg (72,000lb) over a range of 13,150km (8,175 miles). The six 3,500hp Pratt & Whitney Wasp Majors buried in the wings gave a speed of 600km/h (372mph) at 12,200m (40,000ft), but this was insufficient so four turbojets were added in the B–36D version to raise the speed to 690km/h (430mph). A total of 350 B–36s was built, with the type equipping 33 Strategic Air Command squadrons at its peak.

The United States had a monopoly on intercontinental strategic bombers for a decade after the end of the war, until the Tupolev Tu–20 Bear entered service in 1956. The Bear is unusual in being a turboprop, the four engines delivering nearly 15,000hp each to give a top speed of more than 800km/h (500mph). Maximum take-off weight of the Bear is about 170,000kg (375,000lb); most of the type have now been converted for reconnaissance and missile mid-course guidance. The smaller Tu-16 Badger medium bomber, using defensive armament based on that in the Tu–4, entered service two years earlier. Roughly equivalent to the American B–47 and British Valiant, it is powered by a pair of turbojets. In Russian terms, small numbers

of Myasischev Mya–4 Bisons, the Soviet Union's first four-jet strategic bomber, were also deployed in the late 1950s, but are now used mainly as flying tankers.

In the US the mixed-powerplant B–36 led on to the all-jet Boeing B–47, of which more than 2,000 had been delivered when production ended in 1957. The six-engined B–47 had much lower drag than had been predicted, giving it a sufficient turn of speed to outrun many contemporary fighters. Boeing followed the B–47 with the B–52 Stratofortress, which entered service in 1955. The original Pratt & Whitney J57 turbojets, each of 5,080kg (11,200lb) thrust, were more economical than earlier engines, and in later models the fuel capacity was greatly increased, which reduced the need for aerial refuelling. The turbofan-powered B–52H has a range of 16,093km (10,000 miles) with two Hound Dog stand-off missiles, and some 200 aircraft out of the total production run of 744 have been converted to operate the SRAM air-to-surface missile which is used to attack defences in the bomber's path as well as the main target. The B–52 can also carry a conventional weapon load such as 54 340kg (750lb) bombs internally plus a further 12

LEFT: *The Handley Page Victor, with its distinctive crescent wing, was one of the Royal Air Force's three types of V-bomber. The others were the Vickers Valiant and Avro Vulcan.*
BELOW LEFT: *The Rockwell B–1B strategic bombers are painted in dark low-level camouflage. The USAF has 100, and will later receive 132 Northrop ATBs, the first bombers designed to be undetectable in flight.*
BELOW: *France's Dassault Mirage IV-P, a Mach-2 strategic bomber.*

crew of four were housed in an ejectable capsule, and the engine inlets were fully variable for flight at speeds up to Mach 2 at high altitude. This bomber was cancelled in June 1977, amidst much controversy, only to be re-ordered a few years later.

By this time the Soviet air and naval air forces were receiving large numbers of a rather smaller aircraft developed from the Tu–22 (NATO name 'Blinder'). Like its predecessor it had two powerful afterburning engines at the tail, but in the new Tu–26 'Backfire' they were fed from long inlet ducts along the sides of the fuselage. Moreover, the outer wings were pivoted as in the B–1. The Soviet government insisted this was a theatre rather than intercontinental aircraft, though with its nose refuelling probe in use it could launch cruise missiles against targets in the USA. Most 'Backfires' operate in the maritime surveillance and attack role, with bombs, cruise missiles and comprehensive electronics.

In 1981 it was announced that the USAF would receive 100 B–1Bs, to be delivered between May 1985 and June 1988. Compared with the B–1A, the B–1B is redesigned to carry much greater loads of fuel and weapons at subsonic speeds in the low-level role. The engine inlets are fixed, but like the rest of the airframe specially designed and treated to minimise radar signature. The crew have ordinary ejection seats, and the avionics systems have been totally redesigned to assist penetration of hostile airspace. The internal weapon bay can accommodate a movable bulkhead and an eight-round rotary launcher, and typical weapon loads can include eight ALCMs, 38 SRAMs or nuclear bombs, or 128 Mk 82 conventional bombs.

Before this book appears the USAF will have seen its first Northrop ATB (Advanced Technology Bomber), probably later to be designated B–2, make its maiden flight. This all-wing machine is a total 'stealth' or low-observables design, intended to be almost undetectable by hostile defence systems. Meanwhile, the Soviet Union is in production with 'Blackjack', looking like an enlarged B–1.

under each wing. In missions over Vietnam B–52s carried loads of 31,750kg (70,000lb) or more. Today the 263 surviving B–52s, all of them Gs or Hs, have been repeatedly rebuilt and re-equipped with new avionic systems for all-weather navigation and weapon delivery at low level. An eight-round rotary dispenser in the fuselage can launch SRAMs or the AGM–86B ALCM (air-launched cruise missile), and 12 more of either missile can be carried under the wings. B–52Gs in the maritime support role carry AGM–84 Harpoon anti-ship cruise missiles.

The Soviet Union introduced its Ilyushin Il–28 medium bomber in 1950 and Britain's English Electric Canberra followed a year later. The Canberra was also built under licence in the US as the Martin B–57. The RAF's three V-bombers – the Vickers Valiant, Avro Vulcan and Handley Page Victor – were introduced during the mid-1950s, replacing the Lincoln derivative of the Lancaster. Britain's primary nuclear strike role has since been taken over by missile sub-

marines. France retains a strategic nuclear bomber in the form of the Dassault Mirage IV carrying free-fall weapons or cruise missiles.

The USAF introduced the first supersonic strategic bomber, the Convair B–58 Hustler, in 1960. This three-man aircraft, built mainly of stainless steel, had a range of more than 8,000km (5,000 miles) on internal fuel. A belly-mounted pod containing fuel for the outward journey, together with the nuclear weapons, was jettisoned over the target to allow the B–58 to return home 'clean'. The Hustler was retired in 1970.

Following ten years of study the USAF awarded Rockwell a contract for a new strategic bomber, the B–1A, in 1970. The first of four prototypes flew in 1974. Rather smaller than the B–52, the B–1 had a low-mounted wing whose outer panels were pivoted to adjust sweep between 15° and 67.5°. Under the fixed centre section were four GE F101 turbofans in the 13,600kg (30,000lb) class, with full afterburner. The

Bombs

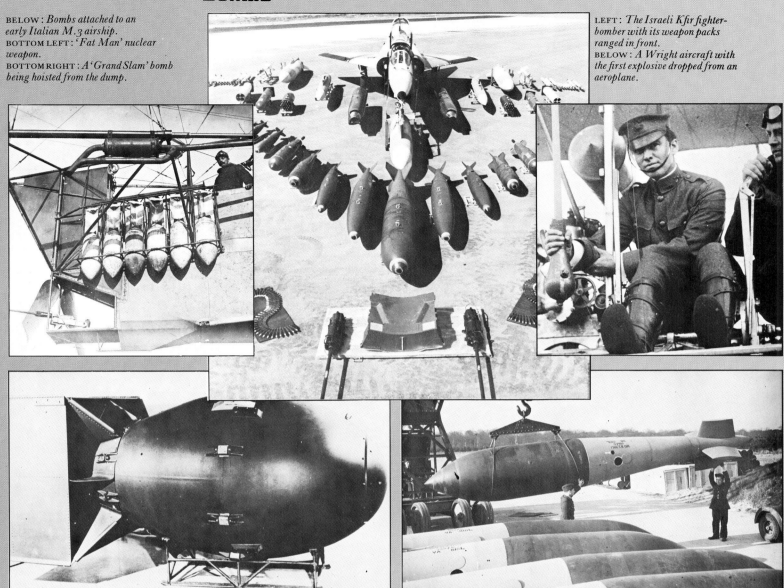

The first bomb designed specifically to be dropped from an aeroplane was the German APK of 1912, a simple cast-iron sphere filled with explosive. At 5kg (11lb) or 10kg (22lb), it was too light to have any great effect and was not used in World War I. It led on, however, to the Carbonit series of weapons employed throughout 1914 and 1915. These weighed up to 50kg (110lb) and had begun to take on the streamlined shape which later became characteristic. The pear-shaped body was fitted with a steel nose to assist target penetration and carried a stabilizing tail unit.

Britain's equivalent, the 9-kg (20-lb) Hale bomb, was developed for the Royal Naval Air Service and carried by its Avro 504s during early wartime raids on Zeppelin sheds. Soon afterwards, versions weighing up to 45kg (100lb) were in use by the Royal Flying Corps as well as the RNAS. France adopted modified artillery shells of 75mm, 90mm and 155mm calibre and fitted with tail fins and fuses by Canton-Unné. Although the French later developed weapons specifically for aircraft use, the 75-mm shell was the most widely employed bomb throughout the war.

As the war progressed and bombers became larger it was possible to carry heavier weapons. By 1918 the four standard British bombs were the 9-kg (20-lb) Cooper, which replaced the Hale weapon of the same weight;

the Woolwich Royal Laboratory's bombs of 23kg (50lb) and 51kg (112 lb); and the Royal Flying Corps weapon of 104kg (230lb). Heavier bombs, weighing up to 816kg (1,800 lb), were dropped on occasion from the Handley Page 0/100 and 0/400, and a development weighing 1,500kg (3,300lb) was built for the same company's V/1500 but never used in action.

Germany's 300-kg (660-lb) PuW bomb of 1915 was extremely advanced for its day, having a steel rather than cast-iron casing. Its canted fins produced a high rotational speed, allowing the weapon to be armed centrifugally in the manner of an artillery shell instead of by a propeller. The PuW's streamlined shape reduced aerodynamic drag, and the weapon was carried horizontally in racks or in a bomb bay. The Allies, in contrast, transported their weapons on the fuselage sides or underneath the aircraft. The Germans also produced the heaviest bomb used operationally during the war – a monster of 1,000kg (2,200lb) dropped by Zeppelin (Staaken) Giants on London.

Other World War I weapons included Le Prieur rockets, used mainly against balloons; various anti-Zeppelin weapons; steel darts some 13cm (5in) long which were dropped from drums containing 500; and incendiaries which developed from

petrol-filled canisters to specialized weapons burning at very high temperatures.

By the end of World War II, bomb development had reached an advanced stage. Weapons were available for specific purposes, such as piercing heavy naval armour or concrete bunkers. General-purpose bombs between 45kg (100lb) and 900kg (2,000lb) were commonplace, and in the spring of 1941 the first 1,800-kg (4,000-lb) weapon was dropped on Emden by a Wellington. Three years later the 5,450-kg (12,000-lb) Tallboy, an 'earthquake bomb' developed by Dr Barnes Wallis, entered service and in 1945 it was joined by the 9,980-kg (22,000-lb) Grand Slam. The first Grand Slam destroyed two spans of the Bielefeld railway viaduct when dropped from a Lancaster of the RAF's 617 Squadron, the same unit which two years previously had attacked the Mohne, Eder and Sorpe dams with Barnes Wallis's 'bouncing bombs'.

A weapon which made its debut in 1944 was napalm, a jellied mixture of naphtha and palm oil (hence its name) in a droppable canister. It is ignited on contact with a target, such as fuel dumps or soft-skinned vehicles, and adheres to everything that it touches.

Torpedoes were first used in action by the RNAS in 1915, against

Turkish shipping. By the later stages of World War II the US Mk 13 torpedo had been improved as a result of experiences in the Pacific and could travel in excess of 73km/h (40 kt) for several miles. It could seek out its target and carried a proximity fuse to detonate the 272-kg (600-lb) warhead. Weighing 900kg (2,000lb) and measuring 3·96m (13ft) long by 56cm (22in) in diameter, it could be dropped at altitudes of 300m (1,000 ft) and speeds of 555km/h (300kt) compared with the precise launching at 15m (50ft) required by the Mk 7.

World War II also saw the service introduction of the atomic bomb. Since the first such weapons were dropped on Hiroshima and Nagasaki, the maximum yield of air-dropped nuclear weapons has risen from about 20 kilotons to 25 megatons or more, and thermonuclear bombs have replaced the first fission weapons. Conventional bombs are available with low-drag profiles in weights up to 1,360kg (3,000lb) and others are fitted with parachutes or flip-out brakes to allow the aircraft to drop them from extremely low levels without risk of being damaged by the explosion. Cluster bombs dispense hundreds of bomblets in a cloud to cover a target, and a host of guided missiles (see chapter 10) have been developed for airborne use. Other air-launched weapons include grenades, rockets and mines.

Fighters

At the beginning of World War I in 1914 there were no specialized fighters, only a handful of scouts capable of 145km/h (90-mph) at best. Aircraft were regarded as being useful only for observation – if indeed they were thought to have any worthwhile military application at all. The first Royal Flying Corps aircraft assigned an offensive role were unarmed and intended to attack Zeppelins by ramming them; this was not greeted with much enthusiasm by the crews, who were not supplied with parachutes. Aircraft of both sides flew over enemy territory on reconnaissance missions, reporting back information such as troop movements and trench positions for use by the army. They acted also as artillery spotters, correcting the aim of ground gunners. But very soon it was realized that there could be another role for aircraft; machines on surveillance missions soon found themselves coming under attack by enemy two-seaters in which the observer carried a hand-aimed rifle or pistol. So began the development of air-to-air fighting.

Aircraft acquire machine guns

At the beginning of 1915 air warfare took a big step forward with the introduction of aircraft machine guns. Before the war Colonel Isaac N. Lewis had fitted a machine gun on a Wright biplane of the US Army Signal Corps, but he met with so little official enthusiasm that in 1913 he left the country in disgust and set up a factory in Liège, Belgium. Lewis's machine gun was paralleled by developments in Britain, France and Germany. The Vickers 0·33in and German weapons were variants of the Maxim infantry machine gun, while

TOP LEFT: *The two-seat Vickers F.B.5 Gunbus equipped the first British all-fighter squadron in France – in July 1915 – and helped counter the 'Fokker scourge'.*

ABOVE: *France's SPAD S.XIII carried twin Vickers machine guns, twice the armament of its predecessor, and the powerful Hispano-Suiza engine gave a reasonable top speed.*

TOP CENTRE: *The Fokker Eindekker series carried the first synchronized machine guns.*
TOP RIGHT: *The Bristol Scout was Britain's first type with synchronized guns.*

France produced the Hotchkiss. The German 7·92mm LMG.08, operated like the Vickers by the pilot, was commonly known as the Parabellum, and the LMG.08/15 was an observer's gun in the same mould as the Lewis.

It was left to the French to pioneer aerial fighting in their *Escadrilles de Chasse*, formed to protect reconnaissance aircraft and to escort bombers. These squadrons were equipped with Morane Type Ns, with a machine gun firing forward through the propeller, and two-seat parasol-wing Type Ls, in which the observer was similarly armed.

Before the war Franz Schneider of LVG and Raymond Saulnier of Morane-Saulnier had worked on interrupter gear which would allow guns to be fired through the propeller disc without damaging the blades. The equipment was installed in an LVG E.VI in 1915, but the aircraft was destroyed on its way to the front. In his experiments with interrupter gear, Saulnier had been troubled by faulty ammunition and, as a result, had fitted steel wedges to the propeller of his aircraft in order to deflect any bullets striking the blades. Roland Garros, who adopted the steel-wedge

idea on a Morane scout, accounted for five victims in under three weeks, largely because the enemy did not expect to be attacked with forward-firing weapons. On 18 April 1915, however, Garros was shot down and the secret was revealed.

German designers tried to reproduce the arrangement but were unsuccessful, possibly because steel-jacketed bullets were being used, but three of Anthony Fokker's engineers went one better and developed an interrupter mechanism along the lines of those pioneered before the war. This was installed

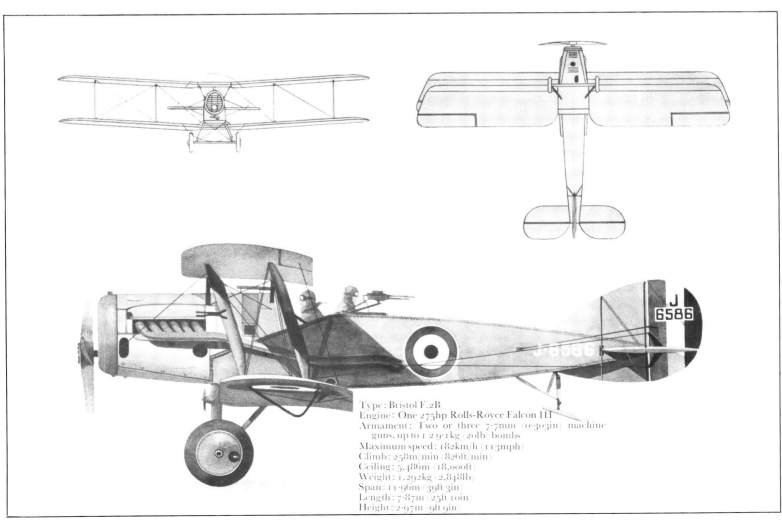

Type: Bristol F.2B
Engine: One 275hp Rolls-Royce Falcon III
Armament: Two or three 7·7mm (0·303in) machine guns, up to 129·1kg (20lb) bombs
Maximum speed: 182km/h (113mph)
Climb: 258m/min (826ft/min)
Ceiling: 5,486m (18,000ft)
Weight: 1,292kg (2,848lb)
Span: 11·96m (39ft 3in)
Length: 7·87m (25ft 10in)
Height: 2·97m (9ft 9in)

in a Fokker E.I – E = Eindekker (monoplane) – and accounted for its first victim on 1 August. The Eindekkers were used at first to escort two-seat observation aircraft and for local defence, but the Germans soon realized that they had a superior weapon and, stung by criticism following sustained unhindered bombing raids by Voisins and Farmans, began to fight back for the first time.

The 'Fokker scourge'

The most famous of the Eindekker series was the E.III, and the six months or so of domination by this type was known by the Allies as the 'Fokker scourge'. In fact the E.I with its 89hp engine giving a top speed of only 128 km/h (80mph) was not a particularly impressive machine, but – as has so often been the case since – superior armament made up for basic deficiencies. Numerous methods of

mounting a forward-firing gun had been tried before the Eindekkers appeared. As early as May 1915 a Martinsyde biplane had been fitted with an upward-firing gun on the centre section of the top wing, so that the stream of bullets missed the propeller disc, and this method was later used on the Nieuport XI (*Bébé*). Another solution was to mount a Lewis gun so that it fired abeam at 45°, its bullets thus clearing the propeller.

The Eindekkers' success forced the French to abandon their daylight bombing missions by December 1915, but their main victim was the British B.E.2c two-seat observation aircraft. In the summer of 1915 the Germans had introduced the C series of observation types, such as the Rumpler C.I, Albatros C.I and Aviatik C.I, in which the observer occupied the rear cockpit instead of the front. This allowed them to carry a 'sting in the tail' in the form of an LMG.08/15 machine gun (commonly known as a Spandau, after its place of manufacture). The C series were a great improvement on the Bs, and their 160 hp Mercedes D.III engines, coupled with the fact that the Spandau was superior to the Hotchkiss, gave them quite an edge. Placing the observer at the rear gave him a much better field of fire and, although the aircraft were vulnerable to frontal attacks, the generally slow and unmanoeuvrable pusher biplanes encountered at first did not provide formidable opposition.

The British, however, refused to learn and continued to turn out great numbers of extremely stable and hence unmanoeuvrable

B.E.2cs with the observer hemmed in at the front. He fired the Lewis gun from his shoulder or from one of the mountings disposed about his cockpit.

The British temporarily overcame the problem of firing through the propeller by producing a series of pusher aircraft, thus removing the problem altogether. The first British all-fighter squadron in France had already received the new F.B.5 Gunbus – a two-seat pusher – in July 1915, and single-seat fighters joined them in February 1916. The D.H.2 and F.E.8 – both single-seat pushers – helped put an end to the Fokker scourge, and the two-seat pusher F.E.2b with its 120hp or 160hp Beardmore engine further augmented the force.

By the spring of 1916 the ageing Eindekkers were being replaced. Immelmann, for one, had extracted the last ounce of aggression by fitting three Spandaus to his aircraft. The interrupter gear was not completely reliable, however, and it was a quite common occurrence for a pilot to shoot off his own propeller. The Germans re-equipped with Halberstadt D.1 biplanes and Pfalz monoplanes, together with two-seat LFG Roland C.IIIs which, at 165km/h (103mph), were extremely fast for their day. In March 1916 the Bristol Scout, the first British aircraft with synchronized guns, appeared in France. And by the time of the Battle of the Somme in July the Sopwith 1½-Strutter, the first British tractor two-seater with the observer at the rear, had at last entered service. The 1½-Strutter was slow, but its forward-firing

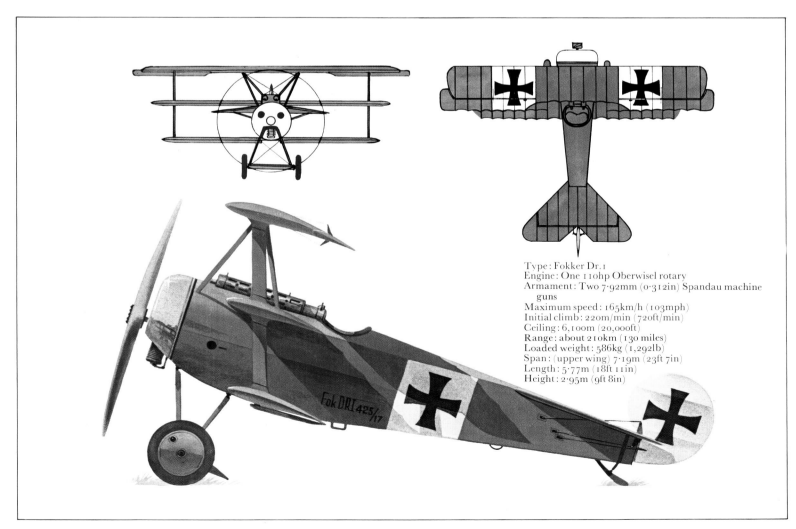

Type: Fokker Dr.1
Engine: One 110hp Oberwisel rotary
Armament: Two 7·92mm (0·312in) Spandau machine
 guns
Maximum speed: 165km/h (103mph)
Initial climb: 220m/min (720ft/min)
Ceiling: 6,100m (20,000ft)
Range: about 210km (130 miles)
Loaded weight: 586kg (1,292lb)
Span: (upper wing) 7·19m (23ft 7in)
Length: 5·77m (18ft 11in)
Height: 2·95m (9ft 8in)

Vickers and rear-mounted Lewis gun largely overcome this deficiency.

More disappointing were the B.E.12, based on the B.E.2, and the R.E.8. The latter was insufficiently manoeuvrable to escape enemy attacks, and when it first entered service was prone to spin and had an unreliable engine – an unfortunate combination. The most effective British two-seater remained the F.E.2 series, especially the F.E.2ds fitted with 250 hp Rolls-Royce Eagles. The D.H.2s and F.E.2s reigned supreme until towards the end of 1916, even though the former were capable of no more than 123km/h (77mph) at 3,050m (10,000ft). The F.E.8 proved to be obsolete by the time it joined the fighting, but generally the pushers, with an unobstructed forward view and a high rate of fire from their unsynchronized guns, proved to be good enough.

Not to be outdone, France contributed the single-seat Spad S.VII and various Nieuport developments. The former, which entered service in September 1916, was strong and fast – 190km/h (118mph) – but carried only a single gun. It was later developed into the S.XIII, however, with twin Vickers and an Hispano-Suiza 8B engine of 220–235hp, giving a top speed of 220km/h (138mph). The Nieuport 16 and 17 retained the single Lewis gun on top of the upper wing, and the design was later copied by the Germans in the Albatros D.III, D.V and D.Va.

The Fokker biplanes developed to replace the Eindekkers were disappointing, but the Halberstadt D.II and D.III, together with the Albatros D.I and D.II with their twin Spandaus, wrested back the initiative for the Germans. From then on, superiority in the air swung back and forth between the contestants for the duration of the war. The S.E.5, with its Hispano-Suiza engine, could achieve 182km/h (114mph) at 3,000m (10,000ft) compared with the D.H.2's 123km/h (77mph), and Sopwith produced the remarkable series of Pup, Triplane and Camel. The Pup, developed from the Tabloid, had a low wing loading and was consequently very manoeuvrable, being able to out-turn an Albatros D.II. The Triplane went one better, with improved rate of climb and better visibility, while the Camel eventually destroyed over 3,000 enemy aircraft to become the most successful fighter of World War I.

In the F.1 Camel – the first British fighter with twin Vickers – weight was concentrated as near as possible to the centre of gravity. This, combined with powerful controls, conferred excellent manoeuvrability. With a 110 hp Le Rhône, the Camel had a maximum speed of 195km/h (122mph) at sea level, and engines of up to 150hp could be fitted. The torque of the powerplant gave the Camel its famous lightning turn to the right, but it also led to an undeserved reputation for unsafe handling during take-off and landing. The F.2B Bristol Fighter was one of the types fitted with a powerful Rolls-Royce engine, in this case a 275hp Falcon, which gave a top speed of 198km/h (123mph) at 1,210m (4,000ft). Indeed, by the end of World War I the two-gun fighters were as nimble as the lighter single-gun aircraft of 1916–17 and could often reach 192km/h (120mph).

FAR LEFT: *Germany's Albatros scouts – a D.I is illustrated here – were an improvement on the first generation of biplane fighters but were less manoeuvrable than types such as the Sopwith Pup. The D.II, with its twin Spandaus, was nevertheless formidable.*
LEFT: *The French Nieuport 17 carried on the sesquiplane layout, with the lower wing reduced to a bare minimum. A single Lewis gun was mounted on top of the wing and needed no interrupter gear. This arrangement was copied in the Albatros D.III, V and Va.*
BELOW: *The Royal Aircraft Factory's S.E.5a (Scouting Experimental No 5) was one of several types designed round the Hispano-Suiza engine. Armament was a synchronized Vickers, plus a drum-fed Lewis on the upper wing.*

In the final year of the war Fokker produced a new fighter biplane, the D.VII, which was outstanding in all respects. Strong and agile, it had a 160hp Mercedes or 185hp BMW engine, the latter giving a speed of 193km/h (120mph). Wily Fokker saw to it that he sold hundreds of the D.VIIs after 1918. Aircraft had begun to be adapted for more specialized roles such as ground attack, one example being the Hannover CL.IIIa, which carried out close-support missions as well as acting as an escort fighter. It was compact and often mistaken for a single-seater – sometimes with disastrous results, because a rearward-firing Parabellum was carried in addition to the fixed Spandau. This kind of specialization was carried further after the end of the war.

French inter-war fighters

The European countries followed much the same development path – ever more powerful and faster biplanes of conventional configuration, then the major step to modern monoplanes. In many cases the transition came too late, since the performance of most other fighters had to be judged against that of the Messerschmitt Bf 109. The first French monoplane fighter with an enclosed cockpit and retractable undercarriage was the Morane-Saulnier M.S. 405, which made its maiden flight in August 1935. The M.S. 406 development entered service in 1939, and more than 1,000 had been completed when France fell in the summer of 1940. An 860hp Hispano-Suiza 12Y engine gave a top speed of 490km/h (304mph), and the armament of two 7·5mm machine guns and a 20mm cannon allowed M.S. 406 pilots to account for more than 250 enemy aircraft, including some Bf 109s.

Another fighter used extensively towards the end of the Battle of France was the Bloch MB. 152, late examples of which were fitted with a 1,100hp Gnome-Rhône 14N–25 radial engine and had a maximum speed of 512km/h (318mph). Armament comprised a pair of 7·5mm machine guns and two 20mm cannon, but the weapons were not always reliable. Nearly 700 MB.152s were built, in addition to 85 of the earlier MB.151, and the type served in Romania after the fall of France. On a larger scale was the Potez 631 three-seat heavy fighter, one of the 63 series which also included developments for bombing, attack and reconnaissance. Powered by a pair of 700hp Gnome-Rhône 14 radials, it lacked the speed which could have made it a potent fighting machine – armament comprised two 20mm cannon and up to eight 7·5mm machine guns. France's best fighter was undoubtedly the Dewoitine D.520, which entered service too late to be decisive but accounted for more than 100 victims in the Battle of France. It was slower than the Bf 109, but more manoeuvrable.

Italian inter-war fighters

Italy's best known scout of World War I had been the S.V.A.5, which carried out long-range missions against Austria. Its 220hp engine gave it an impressive top speed of 228 km/h (143mph). The biplane tradition continued via the single-seat Fiat C.R.32, which

entered service in 1933 and was also widely exported. Its 600hp Fiat A.30 gave a top speed of 350km/h (217mph) and armament comprised the virtually mandatory pair of machine guns. Later Fiat biplanes such as the C.R.42 saw extensive war service.

Other European inter-war fighters
The standard Czech pre-war fighter was the single-seat Avia B–534 biplane which entered service in 1935. A 650hp Hispano-Suiza 12Ybrs engine propelled the aircraft at up to 363km/h (225mph). Of similar vintage was Poland's P.Z.L. P–11c, a single-seat monoplane advanced for its time but hopelessly outclassed against the *Luftwaffe* in 1939. In Holland, Fokker's single-seat monoplane D.XXI of 1937, with its fixed undercarriage and maximum speed of 435km/h (270mph), was also no real match for the invading forces. The same manufacturer's G–1A heavy fighter, which appeared two years later, was the first modern type with twin booms, allowing it to carry the formidable armament of nine machine guns in a central fuselage nacelle. Two 830hp Bristol Mercurys gave a speed of 475km/h (295mph), but the aircraft was too late to save Holland.

Russia's Polikarpov I–16 Ishak (Little Donkey) of 1933 was the first monoplane fighter with a fully enclosed cockpit and retractable undercarriage, but by 1941 it was outclassed by the Bf 109. The *Luftwaffe* had also espoused biplanes, operating the Heinkel He 51 and Arado Ar 68 in the Spanish Civil War, but the monoplane Bf 109 had made its maiden flight in 1935 and by the outbreak of World War II the Bf 109 had become the standard single-seat fighter of the Luftwaffe.

British inter-war fighters
A rapid run-down of fighter strength in Britain followed the Armistice in 1918, with the Sopwith Snipe and some Bristol Fighters forming the backbone of the remaining force. The Snipe, with its two forward-firing machine guns and a top speed of 191km/h (119 mph), had reached the Western Front in September 1918. Its equipment included electrical heating, oxygen and pilot armour, indicating just how far the fighter had come in four years of war. The Snipe was the RAF's

only home-defence fighter between April 1920 and October 1922, and it remained the force's major type until the mid-1920s. It was also used in Iraq from 1922, when the RAF took control there, and was still operational in the Middle East as late as 1926.

A modest expansion programme was authorized in 1923, allowing 52 squadrons to be allocated to air defence alone. The Gloster Grebe and Hawker Woodcock replaced the Snipe, but they were no more than extensions of the same formula, with fabric-covered wooden fuselages, bulky uncowled radial engines, wooden propellers and twin Vickers machine guns mounted directly in front of the pilot's face. The Gloster Gamecock and Bristol Bulldog followed, along with contemporaries such as the Armstrong-Whitworth Siskin, but the principles stayed the same – despite the fact that the Bulldog remained in service until 1936.

Bulldog deliveries to No 3 Squadron began in June 1929, the IIA being powered by a 440hp Bristol Jupiter VIIF and having a top speed of 286km/h (178mph). It was the RAF's main intercepter in the early and mid-1930s. Engine powers and thus maximum speeds increased steadily throughout this period. The Hawker Fury was the first British fighter capable of more than 320km/h (200 mph), its 525hp Rolls-Royce Kestrel producing up to 333km/h (207mph). The Fury was widely exported but did not enter full production for the RAF until 1936, five years after becoming operational, and by then it was obsolete. The V12 Kestrel, which was

also the powerplant for the Hart bomber and its derivatives, had each bank of cylinders cast in a single block rather than separately, but it retained water cooling, needing a separate radiator.

The two-seat Hawker Demon, whose ancestry can be traced right back to the Bristol Fighter, was a contemporary of the Fury. The logical progression of ever-faster single-seat fighters continued with the Gloster Gauntlet, a popular and manoeuvrable type which served from 1935 to 1939. Its 640hp Bristol Mercury VI gave a top speed of 370 km/h (230mph). Then came the type which in many ways represented a transition. The Gloster Gladiator, which entered service in January 1937, was the RAF's last biplane fighter and only now was the standard World War I armament of two machine guns doubled to four. An 840hp Bristol Mercury IX gave the Gladiator a top speed of 407km/h (253mph), but this was soon to be put in the shade by the new generation of monoplane fighters, which had armament doubled again to eight Brownings and top speeds in excess of 480km/h (300mph).

The Hawker Hurricane prototype flew in November 1935 and the first Spitfire followed it four months later, but by the time of the 1938 Munich crisis only two of the RAF's 30 fighter squadrons – including Auxiliaries – were equipped with Hurricanes, and Spitfires had yet to enter service. By the summer of 1940, however, these types had replaced the biplanes in time – but only just in time – for the Battle of Britain.

American inter-war fighters

In the United States, the Army Air Corps had to content itself almost exclusively with European types during World War I, but in the early 1920s 200 Boeing-built MB–3As were ordered to add to the original 50 Thomas Morse MB–3s. This constituted the largest single order for US military aircraft for the next 17 years. The MB–3A was powered by a Wright-Hispano engine of 300hp, giving a top speed of 225km/h (140mph) and a remarkable rate of climb – from take-off to 3,000m (10,000ft) in under five minutes. Despite suffering from engine overheating, a cramped cockpit and maintenance difficulties, it was an impressive beginning for the Army Air Corps series of pursuit ships (derived from the French practice of calling their fighters *avions de chasse*).

Boeing followed the MB–3A with the PW–9, of which more than 100 were built. That company's main rival, Curtiss, produced the Hawk series, which remained in service until 1939. The best known model was the P–6E of 1932, with an engine of up to 600hp giving a top speed of 318km/h (197mph).

Meanwhile Boeing had moved on to the P–12, similarly armed with a pair of 0·303in machine guns. The P–12E variant, entering service in 1932, was fitted with a 525hp Pratt & Whitney Wasp and could achieve 305km/h (190mph). The Air Corps retained the P–12 until 1934, and the type served the US Navy for a further two years. US pursuit ships then took a major step forward at the same time as the first really modern bombers also made their appearance. The Boeing P–26, of which 111 were ordered in 1933, was the first US monoplane fighter to feature all-metal stressed-skin construction. Dubbed the Peashooter by its pilots, it was one of the fastest fighters of its time. The 600hp Pratt & Whitney Wasp of the P–26A drove it through the air at 376km/h (234mph). The P–26 was used in China against Japanese forces and later took part in the early stages of World War II.

In 1935 there was a brief resurgence of the two-seat fighter. Two USAAC squadrons were equipped with Consolidated PB–2s, but in the following year the single-seat Seversky P–35 appeared. It was an improvement on the P–26, its 950hp Pratt & Whitney R–1830 giving a top speed of 450km/h (280mph) and a range of nearly 1,600km (1,000 miles), but it still carried two guns only and had no protective armour. It did, however, form the basis for the P–47 Thunderbolt.

Japanese inter-war fighters

In the years after World War I the Japanese Army relied on imported fighters and overseas designs built under licence. Then in 1931 the Nakajima NC single-seat high-wing monoplane was ordered as the Army Type 91. Powered by a 450hp Bristol Jupiter, it saw widespread use in Manchuria as the main fighter of the early 1930s: 450 were built. Kawasaki, defeated in the competition in which the Type 91 was selected, designed the single-seat KDA–5 biplane, powered by a 750hp BMW. It was later adopted by the Army as the Type 92, and 380 were constructed. Kawasaki went on to develop the Ki–10 which, as the Type 95, was the Army's last combat biplane. Nearly 600 were produced and saw widespread combat, but the Army realized in 1935 that fabric-covered biplanes were outdated and therefore launched a competition to find a successor. The winner was the Nakajima Ki–27 (Type 97 fighter), of which more than 3,300 were built and which remained in front-line service until the beginning of the Pacific war. A low-wing all-metal monoplane, it was light and manoeuvrable. Its excellent record in the Sino-Japanese war hid the fact, however, that modern fighters needed armour plating and self-sealing fuel tanks, with the result that later types proved vulnerable in combat.

German fighters of World War II

In September 1939 the *Luftwaffe* had a front-line strength of 1,200 fighters, including the formidable Messerschmitt Bf 109. As the war progressed, and the advantage swung away from Germany, greater emphasis was placed on intercepters and night fighters than on bomber escorts, and in the last year output was concentrated almost entirely on last-ditch defenders. The Bf 109E was the *Luftwaffe's* standard single-seat fighter for the first three years of the war and was able to outfight or outrun virtually all opposition. From the summer of 1942 the Bf 109G, powered by a Daimler-Benz DB 605D producing 1,800hp with water-methanol injection and giving a speed of 685km/h (428mph), entered service in Russia and North Africa before being de-

FAR LEFT: *The Boeing P–26 'Peashooter' was the first US monoplane fighter and one of the fastest in its class anywhere in the world. It saw action in China in the 1930s and fought early in World War II.*
LEFT: *The Hawker Fury was obsolescent by the time it entered full-scale service with the Royal Air Force, but it was the first British fighter capable of more than 320km/h (200mph) and was widely exported. The Fury was powered by the outstanding Rolls-Royce Kestrel V12 engine.*
BELOW: *The Bristol Bulldog was the RAF's main intercepter in the early and mid-1930s. Engine power and thus speed were steadily increased – this example is a Mk IIIA – but armament was restricted to twin Vickers guns.*
BOTTOM: *The Bf 110 was Messerschmitt's first twin-engined fighter and served throughout World War II, being employed in less demanding theatres after ignominious failures in the Battle of Britain. Major roles included night fighting and bomber interception.*

ployed in every other theatre. With its standard armament of a cannon and two machine guns the Bf 109, like the Spitfire, saw action throughout the war.

From 1941 it was joined by the Focke-Wulf Fw 190, widely regarded as Germany's best fighter. Its introduction has been compared with the 'Fokker scourge' of 1915, and it was more than a match for the contemporary Spitfires. The Fw 190 was armed to the teeth – four 20mm cannon plus two machine guns – and later versions could carry pairs of cannon slung under the wings. Early Fw 190s, powered by an air-cooled BMW radial, were Germany's first radial-engined monoplane fighters, but in the Fw 190D–9 of 1944 a Jumo 213F 2,060hp liquid-cooled vee-12 engine replaced the earlier powerplant and conferred a top speed of 730km/h (453mph). The long-nosed Fw 190D was also developed into the Ta 152 – after its designer, Kurt Tank – in which the installation of a 2,300hp (with boost) DB 603 engine pushed the speed up to 745km/h (463mph). The Ta 152, armed with a 30mm cannon and four of 20mm, equipped only a handful of squadrons at the end of hostilities but might have been decisive if widely deployed earlier in the war.

The Bf 110, Messerschmitt's first twin-engined design, served throughout the war but was increasingly used for intercepting bombers and as a night fighter after its mediocre display in the Battle of Britain. The first fighter version of the much more successful multi-role Junkers Ju 88 was the Ju 88C,

armed with three 20mm cannon and the same number of machine guns. A fourth crew member was added in the specialist Ju 88G night fighter, and the Ju 88H–2 was employed as a long-range escort. Many Dornier Do 217 bombers were built for the night-fighting role from 1942, becoming Do 217J–1s and J–2s, the latter being radar-equipped. Do 217Ms were also the basis of 217N night fighters. All were armed with four 20mm cannon and up to five machine guns.

Messerschmitt's proposed Bf 110 successor, the Me 210, was a failure and the Me 410 Hornisse (Hornet) developed from it was not much better, even when operated as a heavy

FAR LEFT: *The Boulton-Paul Defiant II was the world's first fighter to dispense with forward-firing armament and had to rely instead on a power-operated gun-turret.*
FAR LEFT BELOW: *One of the greatest fighters of World War 2, the Fw 190 also became the Luftwaffe's top ground-attack aircraft. These Fw 190G-3s were seen over Romania.*
LEFT: *The Supermarine Spitfire was the only British fighter in production throughout World War II.*
BELOW: *The Messerschmitt Bf 109 was the Luftwaffe's standard single-seat fighter in the first half of World War II.*
BOTTOM: *The Gloster Gladiator was the first RAF fighter to carry four machine guns, but it was outclassed by German types.*

LEFT: *The revolutionary Messerschmitt Me 163B Komet, powered by a rocket engine, was a last-ditch bomber interceptor.*
CENTRE LEFT: *The twin-jet Messerschmitt Me 262 could have had a major impact if it had been brought into service sooner.*
BOTTOM LEFT: *The Heinkel He 162 Salamander was built from 'non-strategic' materials and was to be flown by hastily trained young pilots.*
TOP RIGHT: *The Gloster Meteor was the only Allied jet aircraft to see action in World War II. This is a 1954 night fighter MK 14.*
CENTRE RIGHT: *The Hawker Hurricane was far more important than the Spitfire in the Battle of Britain. It was a robust gun platform.*

fighter with a 50mm cannon in the bomb bay as well as four 20mm weapons. A specialist night fighter which could have been devastating if official bickering had not disrupted its development was the Heinkel He 219 Uhu (Owl). Powered by a pair of 1,900hp DB603s, giving a speed of 670km/h (416mph), it carried the ultimate in firepower – six 30mm and two 20mm cannon. Yet another victim of bad planning and muddled thinking was the Me 262, the first jet aircraft to enter service. Adolf Hitler ordered that it should be used as a bomber, despite the fact that this negated its speed advantage over Allied piston-engined types. Only when Germany's position was hopeless did he give the Me 262 priority over all other projects.

Last-ditch attempts to hack down the overwhelming bomber fleets were typified by the Me 163B Komet and Heinkel He 162 Salamander. The former, the only rocket-powered aircraft ever to see service, could reach its service ceiling of 12,100m (39,700ft) in 3·35 minutes under 1,700kg (3,750lb) of rocket thrust. It carried a pair of 30mm cannon but its high speed – 960km/h (596mph) – and endurance of only eight minutes combined to make interceptions difficult. The undercarriage was jettisoned to save weight, and more Komets were lost as a result of explosions following landings on skids than were shot down by the Allies. More than 350 Me 163s were built, but it was the same story all over again – too little and too late. Equally remarkable – and hopeless – was the He 162, otherwise known as the Volksjäger (People's Fighter). Making its maiden flight only ten weeks after the beginning of design work, the Salamander was planned to be produced by the thousand from non-strategic materials such as wood. It was to be flown by young pilots with the minimum of training, but it proved difficult to handle and only a hundred-odd were built.

British fighters of World War II

At the outbreak of war in September 1939 the re-equipment of RAF Fighter Command with the new generation of monoplanes was well under way. Front-line fighter strength was more than 1,000, of which well over half comprised Hawker Hurricanes (18 squadrons) and Supermarine Spitfires (nine squadrons). The former type shot down nearly half the enemy aircraft accounted for by the RAF in the first year of war, and it made up 60 per cent of Fighter Command in the Battle of Britain. Early models were armed with eight machine guns, with later variants carrying 12 of these weapons or four 20mm cannon.

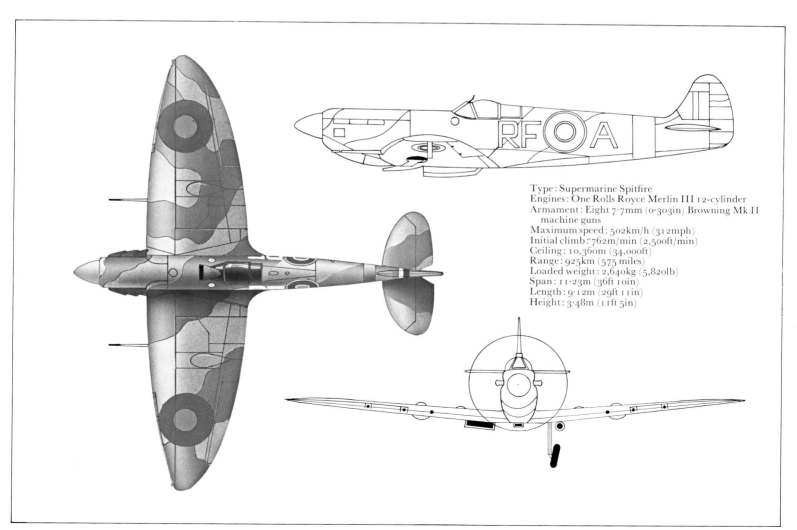

Type: Supermarine Spitfire
Engines: One Rolls Royce Merlin III 12-cylinder
Armament: Eight 7·7mm (0·303in) Browning Mk II
    machine guns
Maximum speed: 502km/h (312mph)
Initial climb: 762m/min (2,500ft/min)
Ceiling: 10,360m (34,000ft)
Range: 925km (575 miles)
Loaded weight: 2,640kg (5,820lb)
Span: 11·23m (36ft 10in)
Length: 9·12m (29ft 11in)
Height: 3·48m (11ft 5in)

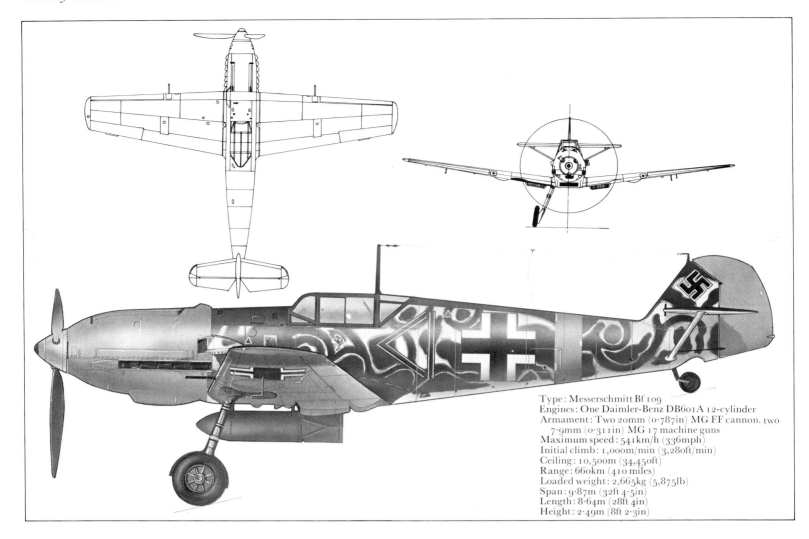

Type: Messerschmitt Bf 109
Engines: One Daimler-Benz DB601A 12-cylinder
Armament: Two 20mm (0·787in) MG FF cannon, two
7·9mm (0·311in) MG 17 machine guns
Maximum speed: 541km/h (336mph)
Initial climb: 1,000m/min (3,280ft/min)
Ceiling: 10,500m (34,450ft)
Range: 660km (410 miles)
Loaded weight: 2,665kg (5,875lb)
Span: 9·87m (32ft 4·5in)
Length: 8·64m (28ft 4in)
Height: 2·49m (8ft 2·3in)

The Hurricane was slightly slower than the Messerschmitt Bf 109, so its major role in the summer of 1940 was bomber interception. It was later used for a wide variety of other missions, including ground attack and operation from catapult-equipped merchant ships, with more than 14,500 of all variants being built.

The Spitfire, one of the classic aircraft of all time, ran to more than 40 versions and was the only British fighter to remain in production throughout the war. More than 2,000 were on order in September 1939, and production exceeded 20,000, making the Spitfire the most widely produced British aircraft of all time. The first variants were powered by the ubiquitous Rolls-Royce Merlin – the same engine was used in the Hurricane as well as many other types – and later models had the follow-on Griffon installed, giving a top speed of more than 724km/h (450mph) in the fastest fighter version. The original armament of eight machine guns was replaced by four cannon in some of the later models.

A contemporary of the Hurricane and Spitfire was the Boulton-Paul Defiant, the world's first fighter to dispense with forward-firing armament and rely on a power-operated gun turret behind the cockpit. The Defiant entered service in early 1940 but was not particularly successful, being relegated to night-fighting in August 1941 and then to target-towing.

Two other major types widely used as night fighters among a multiplicity of roles were the Bristol Beaufighter and de Havilland Mosquito. The former, equipped with an early version of AI (Airborne Interception – later known as radar), strengthened the night-fighter force from November 1940. It combined a high speed, 515km/h (320mph), with a range of 2,415km (1,500 miles) and carried the devastating armament of four 20mm Hispano cannon and seven machine guns. The Beaufighter was later replaced as a night fighter by the de Havilland Mosquito.

Also armed with four Hispanos was the Hawker Typhoon, which entered service in September 1941 to counter Focke-Wulf Fw 190s. The 2,200hp Napier Sabre engine gave a top speed of 663km/h (412mph) to the Typhoon IB, and the aircraft later served extensively in the ground-attack role.

The Tempest followed the Typhoon, with the Sabre uprated to 2,420hp and speed increased to 700km/h (435mph) in the Tempest II. The Mk V was the only version to see war service, and between June and September 1944 Tempests shot down 638 V1 flying bombs before moving into Europe and accounting for 20 Me 262s among a large number of other victims.

Also used against V1s from July 1944 was the Gloster Meteor, the only Allied jet aircraft to become operational during the war. In the F.4 version, which entered service in 1948, the two Rolls-Royce Derwent turbojets of 1,590kg (3,500lb) thrust conferred a maximum speed of 885km/h (550mph) at an operational ceiling of approximately 9,150m (30,000ft).

Italian fighters of World War II

The other European Axis power, Italy, was not so well prepared at the beginning of the war. Its weak link was the force of obsolescent biplane fighters and underpowered monoplanes, and it was not until German-designed engines began to appear in 1942 that the quality improved. The single-seat biplane Fiat C.R.42 Falco (Falcon) served as an escort and night fighter from 1940 until replaced by monoplanes at the end of 1941. More than 1,750 had been built when production ended, but the 840hp Fiat A.74 gave a top speed of only 440km/h (274mph). The same engine powered the Fiat G.50 Freccia (Arrow) and Macchi C.200 Saetta (Lightning) monoplanes, which could achieve 472 km/h (293mph) and 502km/h (312mph) respectively. Like most Italian fighters, they were light and easy to handle but underarmed (they carried no cannon) and underpowered. Production ran to 730 G.50s, including 450 of the *bis* version, and more than 1,090 C.200s.

The C.200 was developed into the C.202 Folgore (Thunderbolt), powered by a 1,200 hp Daimler-Benz in-line engine in place of the heavy and drag-inducing Fiat A.74. This type entered service in 1941, and then both it and the G.50 were re-engined with license-built 1,475hp DB605As to become the C.205V Veltro (Greyhound) and G.55 Centauro respectively. These fighters at last had adequate top speeds–645km/h (400mph) and 620km/h (385mph) – but did not enter service until 1943 and were consequently too late to affect the course of the war.

Aces

The ace system originated in France during World War I, any pilot with five confirmed victories in air combat being mentioned in an official communiqué. The United States later adopted the practice of conferring 'ace' status on a pilot with five aerial victories, but the Germans regarded ten as the cut-off point and the British shied away from singling out individuals. Methods of verifying scores also varied, so the totals given here cannot strictly be compared.

* = *approximate total.*
† = *radar operator.*
Note: *other jet aces include Pakistani and Israeli pilots.*

World War I

Name	Nationality	Score
Rittmeister Manfred von Richthofen	German	80
Capitaine René P. Fonck	French	75
Major E. C. Mannock	British	73*
Major W. A. Bishop	Canadian	72
Oberleutnant Ernst Udet	German	62
Major R. Collishaw	Canadian	60
Major J. T. B. McCudden	British	57
Captain A. W. Beauchamp-Proctor	South African	54
Capitaine Georges M. L. J. Guynemer	French	54
Captain D. P. MacLaren	Canadian	54
Oberleutnant Erich Loewenhardt	German	53
Major W. G. Barker	Canadian	52
Captain P. F. Fullard	British	52
Major R. S. Dallas	Australian	51
Captain G. E. H. McElroy	Irish	49
Leutnant Werner Voss	German	48
Captain A. Ball	British	47
Captain R. A. Little	Australian	47
Hauptmann Bruno Loerzer	German	45
Lieutenant Charles E. J. M. Nungesser	French	45
Leutnant Fritz Rumey	German	45
Hauptmann Rudolph Berthold	German	44
Leutnant Paul Bäumer	German	43
Major T. F. Hazell	Irish	43
Leutnant Josef Jacobs	German	41
Capitaine Georges F. Madon	French	41
Hauptmann Oswald Boelcke	German	40
Hauptmann Godwin Brumowski	Austro-Hungarian	40
Leutnant Franz Buchner	German	40
Major J. Gilmour	British	40
Captain J. I. T. Jones	British	40
Oberleutnant Lothar von Richthofen	German	40
Leutnant Karl Menckhoff	German	39
Leutnant Heinrich Gontermann	German	39
Lieutenant Willy Coppens de Houthulst	Belgian	37
Captain F. R. McCall	Canadian	37
Captain W. G. Claxton	Canadian	36
Captain J. S. T. Fall	Canadian	36
Leutnant Max Müller	German	36
Captain H. W. Woollett	British	36
Captain A. C. Atkey	Canadian	35
Lieutenant Maurice Boyau	French	35
Leutnant Julius Buckler	German	35
Leutnant Gustav Dörr	German	35
Captain S. M. Kinkead	South African	35–40*
Hauptmann Eduard Ritter von Schleich	German	35

Also

Maggiore Francesco Baracca	Italian	34
Captain Edward V. Rickenbacker	US	26
Major A. A. Kazakov	Russian	17*

(*The above three pilots were the top scorers of their nationalities*)

World War II, Spanish Civil War, Russo-Finnish Winter War and Japanese action on the Asian mainland

Germany (day fighters)

Major Erich Hartmann	352
Major Gerhard Barkhorn	301
Major Gunther Rall	275
Oberleutnant Otto Kittel	267
Major Walter Nowotny	258
Major Wilhelm Batz	237
Major Erich Rudorffer	222
Oberst Heinz Bär	220
Oberst Hermann Graf	212
Major Heinrich Ehrler	209*

Germany (night fighters)

Major Heinz-Wolfgang Schnauffer	121
Oberst Helmut Lent	102
	(plus 8 by day)
Major Heinrich Prinz zu Sayn-Wittgenstein	84
Oberst Werner Streib	65
	(plus 1 by day)
Hauptmann Manfred Meurer	65
Oberst Gunter Radusch	64
Hauptmann Heinz Rökker	64
Major Rudolph Schönert	64
Major Paul Zorner	59

Also

Major Wilhelm Herget	57
	(plus 14 by day)

Japan

Warrant Officer Hiroyoshi Nishizawa	JAAF	87
Lieutenant Tetsuzo Iwamoto	JNAF	80*
Petty Officer 1st Class Shoichi Sugita	JNAF	70*
Lieutenant Saburo Sakai	JNAF	64
Warrant Officer Hiromichi Shinohara	JAAF	58
Petty Officer 1st Class Takeo Okumura	JNAF	54
Master Sergeant Satoshi Anabuki	JAAF	51
Lieutenant Mitsuyoshi Tarui	JAAF	38
Warrant Officer Isamu Sasaki	JAAF	38
Petty Officer 1st Class Toshio Ohta	JNAF	34

Russia

Ivan N. Kozhedub	62
Aleksandr I. Pokryshkin	59
Grigori A. Rechkalov	58
Nikolai D. Gulayev	57
Arsenii V. Vorozheikin	52
Kirill A. Yevstigneyev	52
Dmitri B. Glinka	50
Aleksandr F. Klubov	50
Ivan M. Pilipenko	48
Vasilii N. Kubarev	46

British Commonwealth

Squadron Leader M. T. St J. Pattle	South African	51*
Group Captain J. E. Johnson	British	38
Wing Commander B. Finucane	Irish	32
Group Captain A. G. Malan	South African	32*
Flight Lieutenant G. F. Beurling	Canadian	31⅓
Wing Commander J. R. D. Braham	British	29
Wing Commander R. R. S. Tuck	British	29
Wing Commander C. R. Caldwell	Australian	28½
Group Captain F. R. Carey	British	28*
Squadron Leader N. F. Duke	British	28

United States

Major Richard I. Bong	USAAF	40
Major Thomas B. McGuire	USAAF	38
Commander David S. McCampbell	USN	34
Lieutenant Colonel Gregory Boyington	USMC	28
Colonel Francis S. Gabreski	USAAF	28
Major Robert S. Johnson	USAAF	28
Colonel Charles H. MacDonald	USAAF	27
Major Joseph J. Foss	USMC	26
Lieutenant Robert M. Hanson	USMC	25
Major George E. Preddy	USAAF	25

Finland

Lentomestari Eino Ilmari Juutilainen	94
Kapteeni Hans Wind	78
Majuri Eino Luukkanen	54
Lentomestari Urho Lehtovaara	44
Kapteeni Risto Olli Puhakka	43
Lentomestari Oiva Tuominen	43
Lentomestari Nils Katajainen	36
Luutnantti Kauko Puro	35
Luutnantti Lauri Nissinen	32½
Majuri Jorma Karhunen	31

Also

Captain Prince Constantine Cantacuzene	Romanian	60
Commondante Joaquin Garcia Morato y Castano	Nationalist Spanish	40
Leutnant Cvitan Galic	Croatian	36
Second Lieutenant Rotnik Rezny	Slovakian	32
Sergeant Joseph Frantisek	Czech	28
Capitano Franco Lucchini	Italian	26
Capitaine Marcel Albert	French	23
Wing Commander Stanislav F. Skalski	Polish	21*

(*The above pilots were the top scorers of their nationalities*)

Korean War

Captain Joseph McConnell Jr	USAF	16
Major James Jabara	USAF	15
Captain Manuel J. Fernandez	USAF	14½
Major George A. Davis Jr	USAF	14
Colonel Royal N. Baker	USAF	13

Vietnam War

Colonel Tomb	Vietnamese People's Air Force	13
Captain Nguyen Van Bay	Vietnamese People's Air Force	7*
Captain Charles D. DeBellevue	USAF	6†

TOP LEFT: *Charles Nungesser.*
TOP RIGHT: *Edward Rickenbacker.*
CENTRE LEFT: *Manfred von Richthofen.*
CENTRE RIGHT: *Stamford Tuck.*
BOTTOM LEFT: *W A Bishop.*
BOTTOM RIGHT: *Oswald Boelcke.*

Russian fighters of World War II

In the early months following the invasion by German forces, beginning in July 1941, the USSR relied on fighters such as the Lavochkin LaGG–3 and the Mikoyan MiG–1/MiG–3 series. The latter were very fast for their time – 650km/h (405mph) – but suffered from a number of shortcomings and were relegated to reconnaissance by the end of 1943. The all-wood LaGG–3 was strong but not particularly manoeuvrable, being used for roles such as escorting ground-attack types. It was replaced by the La–5, one of two new fighters making their combat debut at the Battle of Stalingrad in October 1942 (the other being the Yakovlev Yak–9). The La–5 was in effect an LaGG–3 with an aircooled radial engine of 1,600hp. The later La–7, which carried three 20mm cannon in place of the La–5's two, could achieve 660km/h (413mph) with a 1,775hp powerplant. The Yak–9, of which many thousands were built, was light and easy to handle. The smaller Yak–3, which followed it in 1944, could reach 645km/h (403mph) and was widely used as an escort for ground-attack types.

American fighters of World War II

More than 50 per cent of the US Army Air Force's fighter strength in the first half of World War II was made up of two types, the Bell P–39 Airacobra and Curtiss P–40 Warhawk/Kittyhawk series. Both were powered by variants of the Allison V–1710 engine ranging from 1,040hp to 1,360hp. The P–39

was unusual in having the powerplant mounted centrally in the fuselage and driving the propeller through an extension shaft. It was also one of the first production fighters with a tricycle undercarriage. The type entered service with the RAF in October 1941, and about 9,500 – of which some 5,000 were supplied to the Soviet Union – were eventually built. The most numerous version was the P–39Q, of which nearly 5,000 were produced. This carried four 0·50in machine guns, in addition to the 37mm cannon, whereas previous variants had been armed with a mixture of 12·7mm and 0·30in weapons. The Curtiss P–40 series, which followed the P–36 Mohawk, was the first mass-produced US single-seat fighter. More than 14,000 were built, being supplied to the RAF (which called it the Tomahawk), Dominion air forces and the Soviet Union as well as to the USAAF. More than 5,200 P–40Ns alone, armed with six 0·50in machine guns, were constructed.

Nearly 10,000 Lockheed P–38 Lightnings were built, this type being responsible for destroying more enemy aircraft in the Pacific war than any other fighter. Nearly 4,000 of these were P–38Ls, capable of 665km/h (415 mph), and the radar-equipped P–38M night fighter saw service in the closing stages of the war. The Douglas A–20 Havoc was also used as a night fighter, but the first US aircraft designed specifically for this role was the Northrop P–61 Black Widow, which became operational in the summer of 1944.

Two remarkable fighters which did ster-

ling duties as long-range bomber escorts as well as their other duties were the Republic P–47 Thunderbolt and North American P–51 Mustang. Originally designed as a light-weight fighter, the P–47, affectionately known as the Jug, turned out to be the biggest and heaviest single-seat fighter then built. The 9,615kg (21,200lb) P–47N, with strengthened and extended wings for Pacific operations, was almost as heavy as the twin-engined Beaufighter. It was powered by the 2,800hp engine which had given the earlier P–47M a top speed of more than 750km/h (470mph). Most of the lighter P–47Ds, of which 12,000 were built, had the Pratt & Whitney Double Wasp of 2,300hp, giving a speed of 687km/h (429mph). The Mustang was designed originally to a British specification and was powered by a Rolls-Royce/Packard Merlin engine. It was later adopted by the USAAF, and nearly 8,000 of the P–51D model alone were built. Its 1,590hp powerplant conferred a top speed of 703km/h (437mph); armament comprised up to six 0·50in machine guns. The P–82 (later F–82) Twin Mustang, a pair of P–51s joined in Siamese twin-fashion, entered service just before the end of the war and later operated as a long-range escort.

Japanese fighters of World War II

Across the Pacific, Japan had introduced its first low-wing monoplane fighter, and the first with an enclosed cockpit, in mid-1937. The Nakajima Ki–27 (Type 97 Fighter), of which more than 3,300 were built, was a

main Japanese Army Air Force fighter until replaced by the Ki–43 at the end of 1942. The same manufacturer's Ki–43 Hayabusa (Peregrine Falcon) became the most widely produced JAAF fighter, nearly 5,800 being built. Late models had nearly twice the power of the Ki–27, raising top speed from 458km/h (285mph) to 512km/h (320mph). The Ki–43, armed with two machine guns, saw service in every theatre of the Pacific war. Nakajima's Ki–44 Shoki (Demon), of which more than 1,200 were constructed, was heavy and unpopular with its pilots but had a good rate of

climb and, at 600km/h (375mph), was fast.

The Kawasaki Ki–45 Toryu (Dragon Slayer) entered service in October 1941 as a JAAF night fighter and was widely used in the south-west Pacific and the defence of Japan, about 1,700 being built. It was heavily armed, carrying one 37mm and two 20mm cannon. The same manufacturer's Ki–61 Hien (Flying Swallow) followed in the late summer of 1942, and its close resemblance to the Messerschmitt Bf109, coupled with the fact that an in-line engine was most unusual in a Japanese aircraft, gave rise to speculation that the Ki–61 might be a German type built under license. In fact, its engine was a licence-built version of the German Daimler-Benz DB601A. It was armed with two 20mm cannon and a pair of machine guns, and had a top speed of 556km/h (385mph). About 2,750 were built, and a shortage of in-line engines led to some being fitted with radials from the beginning of 1945 for use in the defence of Japan. These were designated Ki–100.

The formidable Nakajima Ki–84 Hayate (Gale) entered service only a year before the end of the war, but more than 3,500 were built and they made their mark. Later versions carried four cannon, two 20mm and the others 30mm, and the 1,900hp engine was capable of propelling the aircraft at 612km/h (380mph). The same company's J1N1 had a chequered career, being designed as a fighter, seeing service on reconnaissance duties and finally, as the Gekko (Moonlight), operating as a night fighter with primitive radar. Not to

ABOVE FAR LEFT : *North American P–51B Mustang.*
TOP LEFT : *The Yakovlev Yak-3 entered service with the Soviet Air Force in 1944.*
TOP RIGHT : *Lavochkin's LaGG–3, built of wood, was strong but relatively unmanoeuvrable.*

CENTRE : *The Curtiss P-40 was known as the Kittyhawk in the Royal Air Force.*
ABOVE : *The Curtiss P–36 Mohawk laid the foundations for the P–40, the first US mass-produced fighter, with more than 14,000 built.*

Guns and Cannon

designed brilliant guns of up to 37 or even 45-mm calibre. Japan was slow in introducing weapons of heavier calibre – a 12·7-mm machine-gun was requested by pilots in 1939 but not received for two years, and the 40-mm cannon did not appear until 1943.

The first major innovation for some time in aircraft gun armament was the General Electric M61 Vulcan, introduced in the Lockheed F–104 Starfighter and since adopted as the standard fighter gun for the US forces. A six-barrelled 20-mm weapon, it uses the Gatling-gun principle to spit out up to 6,000 rounds per

it fires shells weighing 360 grams (three-quarters of a pound). The combination of high velocity and weight gives the shell six-and-a-half times the kinetic energy at impact of a round from the previous-generation Aden or DEFA 30-mm cannon fitted to such types as the Hawker Hunter, Harrier and Hawk, and the Dassault Mirage series. The time of flight to 1,000m (3,300ft) is 1·23 seconds, and the drop in trajectory at that distance is only as much as that experienced in earlier 30-mm weapons at 400m (1,300ft).

For air-to-ground operations the latest state of the art is represented by the General Electric GAU–8/A Avenger fitted in the Fairchild A–10. The muzzle velocity, 1,050m/sec (3,450ft/sec), is not much greater than that of the Vulcan but the shell's flight time is reduced by 40 per cent because of improved ballistics. The Avenger's 30-mm shell weighs 0·7kg (1·6lb) and a single hit will destroy a tank. Sufficient ammunition is carried for ten two-second bursts at the maximum firing rate of 4,000 rounds per minute.

The aircraft machine-gun was introduced at the beginning of 1915, armament previously having been restricted to hand-held pistols and rifles. Typical observer's guns were the US 0·303-in Lewis and the German 7·92-mm LMG.08/15. Ammunition was fed to the air-cooled Lewis gun from a revolving drum containing 47 rounds at first and 97 in later versions. Cartridge cases were collected in a receptacle to prevent damage to the aircraft which would have been possible if they had been ejected overboard. By 1918 the Lewis's rate of fire had reached 850 rounds per minute, while the Parabellum could fire 700 rounds a minute. The most commonly used pilot's guns, the British 0·303-in Vickers and the German 7·92-mm LMG.08, were both developments of the Maxim infantry machine-gun. The LMG.08, almost universally referred to as the Spandau (after its place of manufacture), was slightly heavier than the 10kg (22lb) of the similar Parabellum.

Other weapons included France's 8-mm Hotchkiss, which produced a lot of drag and used an ammunition drum containing only 25 rounds; it was replaced by the Lewis gun. The Austrian 8-mm Schwarzlose had a short range and low rate of fire, and Italy's 9-mm Villa Perosa was not particularly successful, despite having two barrels and a high muzzle velocity. Some Voisins were fitted with 37-mm or 47-mm Hotchkiss cannon for ground-attack work, and these weapons were also employed occasionally in the air-to-air role, but not to any great effect.

Gun development between the wars was generally slow, and a pair of

forward-firing Vickers machine-guns remained the standard armament of RAF fighters until the mid-1930s. The US forces espoused the machine-gun even longer. Hitting power was improved from the mid-1930s, however, by the introduction of 0·50-in heavy machine-guns to supplement or replace the 0·30-in weapons. It was not until after the beginning of the Korean war that the switch to cannon armament became general. The cannon fires a heavier shell than the machine-gun, and range and muzzle velocity are also generally greater. The destructive effect is enhanced by the use of explosive rather than solid shells.

French, British and German fighters usually used 20-mm cannon at first, but calibres of up to 75-mm could be found in ground-attack types and the *Luftwaffe* tried a 50-mm weapon for air-to-air work. Russia

minute at a muzzle velocity of 1,000 m/sec (3,300ft/sec). The Soviet Union has recently fitted a 23-mm weapon operating on the same principle to some of its latest fighter-bombers.

A modern example of a single-barrel cannon is the Oerlikon KCA developed to arm the Swedish Air Force's new JA37 Viggen fighter. A 30-mm weapon with a muzzle velocity of 1,030m/sec (3,400ft/sec) and a rate of fire of 1,350 rounds a minute,

TOP LEFT: *The observer of an FE.2b and his Lewis gun.*
TOP RIGHT: *Twin 50-calibre machine guns arming a Boeing B-29.*
CENTRE LEFT: *The six-barrelled 20-mm Vulcan gun (in a pod).*
CENTRE: *Tail of a Boeing B-29.*
CENTRE RIGHT: *The GAU-8 cannon mounted in the Fairchild A-10.*
ABOVE LEFT: *The pioneer turret of an RAF Overstrand bomber.*
ABOVE RIGHT: *7·62-mm Minigun in the Fairchild AU-23A.*

199

be outdone, the Japanese Navy introduced the Kawanishi N1K2–J Shiden (Violet Lightning), a land-based fighter with a rapid rate of climb and good manoeuvrability. It was too late to affect the course of the war, however.

Post-World War II fighters

Although piston-engined fighters remained in service after World War II and for a few years new ones were developed – the last Russian propeller fighter, the Lavochkin La–11, did not enter service until 1948 – the post-war era is definitely the age of the jet. Britain and Germany had established an early lead, but this was soon lost to the United States and Russia, both of whom absorbed experience and hardware previously developed by British and German engineers. Europe re-emerged later as a centre of jet-fighter production, while other countries such as Japan, India and Israel have since joined the exclusive band.

The first US jet was the Bell P–59 Airacomet, which made its maiden flight in October 1942, but its performance was unimpressive and it was relegated to training duties. Lockheed's F–80 Shooting Star just failed to see war service, but it became assured of a place in aviation history when, on November 8, 1950, Lt Russell Brown, USAF, shot down a MiG–15 over Korea in what is thought to have been the first-ever all-jet combat. The Republic P–84 Thunderjet first flew as early as February 1946, but it was destined to be the USAF's last straight-

winged subsonic fighter. Wartime work in Germany had proved that swept wings were more efficient for high-speed flight, and the North American P–86 Sabre was redesigned to take a swept wing before its maiden flight in October 1947. Republic went on to build more than 4,400 Thunderjets, armed with six 0·50in machine guns or unguided rockets, and a further 2,700 of the swept-wing F–84F Thunderstreak development were constructed.

Meanwhile the Soviet Union built a number of jet-fighter designs, beginning with the Yakovlev Yak–15 and MiG–9, which both made their first flights on the same day in April 1946. The former's single RD–10 axial-flow turbojet of 900kg (1,980lb) thrust gave it a top speed of just over 800km/h (500mph); the MiG–9, however, was powered by a pair of axial-flow turbojets. The Lavochkin La–15, another jet design, entered service in 1949 but it was the MiG–15 which really made a permanent mark in history.

MiG–15s and F–86 Sabres tangled over Korea in a faster, more sophisticated repeat of the Spitfire-versus-Bf109 dogfights of the Battle of Britain. The Russian aircraft, armed with one 37mm and two 20mm cannon, outgunned the Sabre with its six 0·50in machine guns (later variants carried four 20mm weapons and, ultimately, 24 2·75in folding-fin rockets). The MiG–15 was also more manoeuvrable above about 10,000m (33,000ft), but at low levels the Sabre had the edge. The decisive difference between the two

TOP: The Republic F–84 Thunderjet was later developed into the swept-wing RF-84F photo-reconnaissance aircraft seen here.
ABOVE LEFT: The MiG–15 had just about everything its pilots could ask – speed, manoeuvrability and heavy armament.
ABOVE: The North American F–86 Sabre was outgunned by MiG–15s in Korea, but the battle-hardened US pilots won through.
TOP RIGHT: The de Havilland Vampire was under development from early in World War II but entered service only in 1946.
TOP FAR RIGHT: North American's F–100 Super Sabre was supersonic in level flight.
ABOVE CENTRE RIGHT: The Hawker Hunter is still used throughout the world.
BELOW CENTRE RIGHT: The variable-geometry Dassault G.8 with its wings swept back.
BOTTOM RIGHT: The experimental G.8 with its wings swept forward for low-speed flight and during take-off and landing.

turned out to be that the USAF pilots were battle-hardened veterans, whereas those flying the MiGs were comparatively raw recruits. The USAF and USN claimed to have shot down 792 MiG–15s over Korea, though the figures were later revised. The improved MiG–17 joined its stablemate in 1952, and the combined production run of these two types is estimated at some 20,000.

The Lockheed F–94 Starfire, developed from the F–80, joined the USAF inventory in 1950 but by then supersonic fighters were becoming the name of the game, North American took the Sabre, swept the wings back by a further 10°, made many other changes and improvements, and turned out the F–100 Super Sabre. It entered service in 1954 as the first non-Soviet fighter able to exceed the

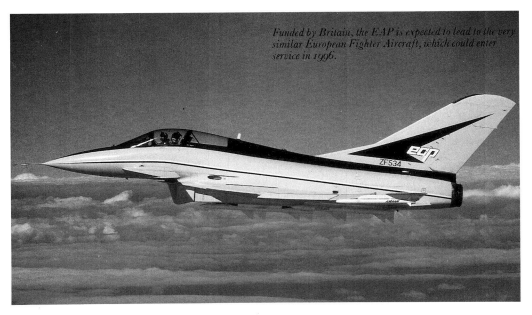

Funded by Britain, the EAP is expected to lead to the very similar European Fighter Aircraft, which could enter service in 1996.

Planned as a light fighter of limited capability, the General Dynamics F-16 Fighting Falcon has developed into an all-weather, multi-mission aircraft.

speed of sound in level flight. The Pratt & Whitney J57 turbojet ultimately produced nearly 7,700kg (17,000lb) of thrust, enough to propel the Super Sabre at a little over 1,350km/h (850mph), or Mach 1·3. The USAF continued its practice of switching to cannon rather than machine guns, four 20 mm weapons being installed. More than 2,200 F-100s were built. In Russia, the MiG-19 first flew in 1953, providing the other superpower with its first supersonic fighter.

Britain lagged behind, relying on developments of the wartime Gloster Meteor and de Havilland Vampire, although the latter type did not enter service until 1946. The more advanced Supermarine Swift and Hawker Hunter became operational in 1954, but both were essentially subsonic and the former was soon confined to photo-reconnaissance duties. The Hunter, with its four 30mm Aden cannon, was nevertheless a fine fighter and remains in service today. Britain's first supersonic interceptor – and only Mach 2 warplane – was the outstanding English Electric Lightning, which remained in front-line RAF service until it was finally retired in late 1987.

Europe's first swept-wing jet fighter was the Swedish SAAB J29, nicknamed the Flying Barrel, which entered service in 1951. Its de Havilland Ghost engine gave it a top speed of 1,024km/h (636mph) and it carried an armament of four 20mm cannon. The first production European fighter which was supersonic in level flight, France's Dassault Super Mystère B.2, did not enter service until 1958. Its SNECMA Atar turbojet of 4,500kg (9,920lb) thrust gave it a speed of Mach 1·25. The Super Mystère's ancestry can be traced back through the Mystère IVA, which saw action during the Suez crisis of 1956; the earlier Mystère IIC, which was basically a swept-wing version of the Ouragan; and the Ouragan itself. France followed the Super Mystère with the Mirage III; and Sweden introduced the SAAB Draken in 1960. All these types are capable of Mach 2, but this milestone had already been reached by the Lockheed F-104 Starfighter and the MiG-21. The Starfighter, designed originally as a day fighter with stunning performance in response to the USAF's demands after its experience in Korea, was later converted into a multi-role type with ground-mapping radar. It was the first aircraft to be armed with the General Electric M61 Vulcan six-barrelled Gatling gun, giving an extremely high rate of fire.

Several countries developed specialized fighters designed to operate at night and in bad weather, necessitating installation of radar and provision for two crew members. The USAF's Northrop F-89 Scorpion was a typical early example; France's Sud-Ouest Vautour carried four 30mm cannon and rockets. Canada contributed the Avro CF-100, and the Soviet Union relied on the Yak-25. The RAF's specialist in this field, the Gloster Javelin, was the first twin-jet with a delta wing when it entered service in 1956. In the United States, Convair had carved out a niche in the delta-wing market, developing the supersonic F-102 Delta Dagger and the F-106 Delta Dart. The latter, powered by a J75 engine of 11,130kg (24,500lb) thrust, entered service in mid-1959 with a top speed of Mach 2·3.

Many aircraft often thought of as fighters, such as the F-111, Harrier and Tornado, are really attack aircraft and discussed later. There is, however, a long-range all-weather interceptor version of the Tornado, originally ordered by the RAF as the Tornado F.3; it has a longer fuselage, more fuel and is able to carry tandem pairs of Sky Flash or AMRAAM missiles in ventral recesses. It has advanced interception radar and a new avionics suite.

France's Dassault-Breguet company followed the best-selling Mirage III with the Mirage F1, with a high wing and rear tailplane. In turn this was followed by the Mirage 2000, a greatly improved tailless delta which goes a little way towards the modern philosophy of making the aircraft longitudinally unstable (potentially lethally uncontrollable) and then controlling it with fast-acting computers through a digital FBW (fly by wire) electrically signalled flight-control system. The 2000 entered service in 1983. In 1986 testing began of the first of the next generation, the Rafale, a totally new design with a rear delta wing, canard foreplanes and twin engines.

In 1970 the US Navy's F-14 made its first flight; this is described in the section on carrier aircraft. In 1972 the USAF's McDonnell Douglas F-15 Eagle began its flight test programme. Powered by two 10,800kg (23,830lb) Pratt & Whitney F100 afterburning turbofans, the F-15 is a large aircraft with a wing big enough (56.5m², 608sq ft) to give outstanding manoeuvrability without slats or other devices. Engine thrust exceeds clean gross weight, and today well over 1,000 Eagles are regarded as among the world's best fighters. The F-15E is a new multirole version.

In 1974 two much smaller LWFs (light weight fighters) were built, and in 1975 a developed version of one was selected by the USAF as the General Dynamics F-16 Fighting Falcon. Powered by a single F100, the F-16 quickly established a great reputation for agility, versatility, air/ground attack capability and many other qualities. Adopted reluctantly, as a kind of 'cheap' alternative to the F-15, the F-16 found customers all over the world. The USAF and Air National Guard alone are buying well over 3,000, and about 4,200 are so far on order, later versions often having the more powerful GE F110 engine.

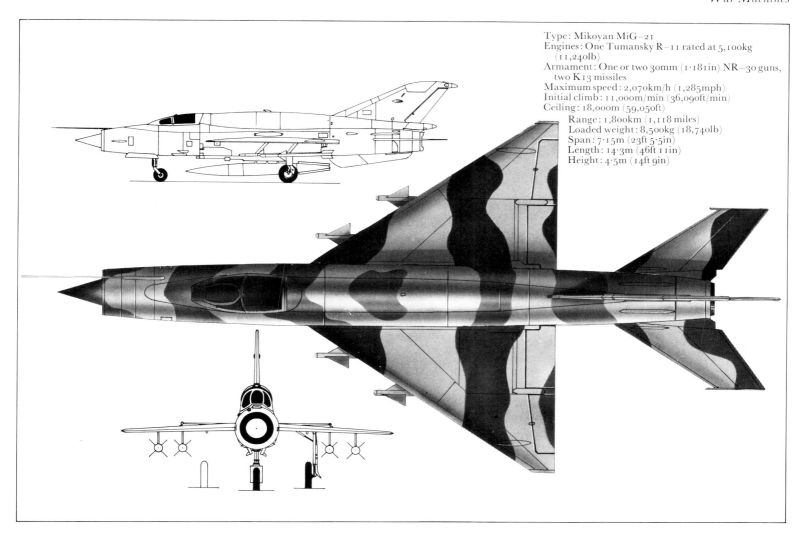

Type: Mikoyan MiG–21
Engines: One Tumansky R–11 rated at 5,100kg
  (11,240lb)
Armament: One or two 30mm (1·181in) NR–30 guns,
  two K13 missiles
Maximum speed: 2,070km/h (1,285mph)
Initial climb: 11,000m/min (36,090ft/min)
Ceiling: 18,000m (59,050ft)
  Range: 1,800km (1,118 miles)
  Loaded weight: 8,500kg (18,740lb)
  Span: 7·15m (23ft 5·5in)
  Length: 14·3m (46ft 11in)
  Height: 4·5m (14ft 9in)

Northrop spent its own money building the little N-156F in 1959, but this led to the F–5 series, powered by twin GE J85 turbojets, of which 2,610 were built for customers worldwide. Northrop again spent its own money on the outstanding F–20 Tigershark, but just failed to find enough customers. It is, however, one of the winners of the potentially gigantic USAF ATF (Advanced Tactical Fighter) programme, with the YF–23A being built in partnership with McDonnell Douglas. The rival ATF is the Lockheed/GD/Boeing YF–22A. The ATF is to set wholly new standards in supersonic cruise without afterburner, low-observability 'stealth' qualities, reliability and many other attributes, on planned entry to service in 1994.

Sweden continued its unbroken run of successful programmes with System 37, the Viggen, built as the Saab AJ37 (attack), SF and SH37 (recon), SK37 (dual trainer) and JA37 (fighter), all powered by the RM8 turbofan of some 12,700kg (28,000lb) thrust and notable in being designed for sustained operations away from vulnerable known airfields (using dirt roads and farm tracks). In 1987 Saab rolled out the first of the next generation, the small JAS39 Gripen powered by a Volvo RM12 afterburning turbofan developed from the GE F404. With a normal gross weight of only 8,000kg (17,635lb) the Gripen is the lightest and potentially least costly of all new combat aircraft despite having complete all-weather multirole capability.

In the Soviet Union the same tailed-delta configuration was used in 1955 for what became the small MiG–21 and the large Su–9. These were in turn greatly developed over the next 20 years to 1975. The MiG–21 grew with a series of more powerful Tumansky engines, ending with the 7,500kg (16,535lb) R–25. The Su–9 and 11 were single-engined all-weather interceptors, replaced in production by the twin-engined Su–21 with a large radar and mix of missiles and gun pods (NATO name 'Flagon').

In 1967 the prototype was displayed of a new MiG fighter with a high-mounted variable-sweep wing. This led to the mass-produced MiG–23 (NATO 'Flogger') used by many countries as an all-weather interceptor, and to the MiG–27 series of attack aircraft. A common MiG–23 engine is the Tumansky R–29B rated at 12,475kg (27,500lb), giving supersonic speed down to sea level.

A totally different and even faster interceptor is the MiG–25 ('Foxbat'), designed in 1969 to shoot down the cancelled B–70. The MiG–25 was not cancelled, and resulted in a fighter with poor manoeuvrability but a speed up to Mach 3·2! Various versions were built, leading to the MiG–31 ('Foxhound'), a formidable interceptor with a crew of two, steerable dual mainwheels and many advanced features including the ability to carry eight missiles.

In the 1970s a totally new configuration with twin engines slung under a tapered high wing, with a high forward fuselage and twin fins, was adopted for the MiG–29 ('Ful-crum') and the much bigger Su–27 ('Flanker'). Both are totally capable all-weather aircraft able to shoot down hostile aircraft at any altitude or attack ground targets. The Su–27 is believed capable of carrying ten missiles, plus a large gun. Some idea of the Su-27's performance is given by a 1986 series of climb records including reaching 12km (39,370ft) in 55 seconds!

Virtually all new fighters are deliberately made longitudinally unstable, like a dart thrown tail-first, and are then controlled with extreme precision by fast-acting computers. The safety of the aircraft is totally dependent upon the computers and the digital flight-control system. The result is an aircraft with extreme powers of manoeuvre, the limits always being set by the pilot, and increasingly with the ability to aim a fixed gun exactly at the point in space to hit a manoeuvring target.

A key factor in the design of today's fighters is advanced fibre-reinforced composite structures, using such fibres as graphite (carbon) and Kevlar (which somewhat resembles spider web). One research aircraft of fighter type, the Grumman X–29, could never have been built without such structure. It is the first fast jet to have a forward-swept wing (FSW). The FSW promises to make fighters more agile, smaller and safer, but using traditional materials the wings would quickly be torn off. Thanks to 'tailoring' the directions of the fibres in the wing skins the X–29 has wings that stand up to combat manoeuvres at the highest jet speeds.

RIGHT: *The Lockheed F–104 Starfighter,*
nicknamed the 'missile with a man in it', carries a
20-mm Vulcan cannon and can fly at more than
twice the speed of sound.

BELOW: *The variable-geometry MiG–23 fighter is one of*
the Soviet Air Force's most numerous combat types. It can
carry missiles as well as its built-in gun.

BELOW RIGHT: *The Swedish SAAB Draken, with*
distinctive double-delta wing, has equipped the
Swedish Air Force since 1960. It is capable of
Mach 2.

BOTTOM: *A visit to Finland by Mig-29s in 1986 gave the*
world the first public view of one of the latest and most
formidable of the Soviet Union's fighters.

FAR RIGHT ABOVE: *The British Harrier entered*
service in 1969, becoming the world's first
operational V/STOL combat aircraft.

FAR RIGHT CENTRE: *The Panavia Tornado IDS*
(interdiction strike) version is the world's most
formidable all-weather attack aircraft. These are serving
with the RAF.

FAR RIGHT BELOW: *The Junkers CL.1 is regarded*
as the best German ground-attack aircraft of
World War I.

Attack aircraft

The definition of an attack aircraft has never been clearly laid down. Some authorities include under this heading all bombers which do not deliver their weapons in level flight – in other words dive-bombers are counted as attack aircraft. This is debatable – for example, essentially low-level types such as the BAC/Dassault-Breguet Jaguar drop their bombs in level flight but would be counted as attack aircraft even if they carried no cannon armament as well. Perhaps the best definition of an attack aircraft is one which is designed specifically or primarily to attack ground targets with guns, rockets or guided weapons, or which is intended to deliver free-fall bombs from low level.

Almost every fighter aircraft has at one time or another been used to drop bombs. In fact, the McDonnell Douglas Phantom II, officially designated F–4 (F = Fighter), has been very often used to carry a heavier bomb load than could be lifted by erstwhile 'heavy bombers'.

World War I

Ground attack, otherwise known as close support or army co-operation, began to emerge as a specialist task in the last year or

so of World War I. From the very beginning it was obvious that the forward-firing guns in the early fighters could be used effectively against surface targets as well as against other aircraft, especially as the main role of aircraft at that time was to support the army and not to fight an air war *per se*. In those days of small bombs it was also very difficult to distinguish between specialist bombers and observation types which dropped bombs as a secondary role.

Germany's CL types were intended originally as escorts for C-type observation aircraft, but it was soon found that their forward-firing machine guns made them extremely

useful for ground attack. From the middle of 1917, the Halberstadt CL.II was widely used for harassing ground forces, the observer being supplied with four or five small grenades or mortar bombs, weighing about 10kg (22 lb), which he dropped by hand. Similar missions were later carried out by the Halberstadt CL.IV and the Hannover CL.IIIa. In early 1918 Albatros and AEG J types, the latter developed from the C.IV, entered service as specialized ground-attack aircraft. The AEGs, of which some 600 were built, carried twin Spandaus mounted in the floor of the rear cockpit, pointing downwards at about 45°. The engine and crew were protected by 390kg (860lb) of armour plating, and an experimental single-seat version with six downward-firing guns was also constructed. It was found, however, that such armament was very difficult to aim accurately at low level.

Junkers developed the J.I to replace the interim J types, and the same manufacturer's CL.I was widely regarded as the best German ground-attack aircraft of the war. Grenades were stored in racks alongside the rear cockpit, but the type's introduction was too late to have any great effect.

On the Allied side, some 200 Voisin 4s were built with a forward-firing Hotchkiss 47mm cannon for strafing. The later Voisin 8 LBP could be fitted with either this weapon or one of 37mm. Some F.E.2bs were converted for ground attack, with a Vickers one-pounder installed in place of the Lewis machine gun, and the Martinsyde G.100/102 Elephant was employed in a light bombing role after being found unsuitable as a fighter. The Bristol Fighter could carry 12 bombs of 9kg (20lb) each. The D.H.5 was used from the autumn of 1917 as an attacker with 11kg (25lb) of bombs under its fuselage, and the S.E.5 had the same payload. The R.E.8 could carry four weapons of 30kg (65lb) each.

The first specialized British ground-attack aircraft was the Sopwith Camel, which incurred heavy losses when used as a light bomber with four weapons of 9kg (20lb) under its fuselage. An experimental version with downward-firing machine guns was tried, and experience gained with the Camel led to the Sopwith Salamander. Designated the T.F.2 (T.F. = Trench Fighter), the Salamander was similar to the Snipe but carried 295kg (650lb) of armour around the pilot and the fuel, plus a pair of forward-firing Vickers machine guns supplied with 1,000 rounds each. This development was too late to see widespread service.

World War II

In the inter-war years, with money generally at a premium, air forces tended to concentrate on fighters and bombers rather than ground-attack types. Those specialized aircraft that were developed for this role were not bought in large numbers. In the United States, Curtiss produced the A–3 version of the O–1B, which had an additional machine gun in each lower wing to supplement the pair in the nose, and later turned out the A–8/A–10/A–12/ Shrike series of all-metal low-wing monoplanes armed with four 0.30in machine guns and able to carry four bombs of 45kg (100lb)

or two weighing 14kg (30lb). Northrop A–17/A–33s entered service with the USAAC in 1936 but were soon withdrawn and eventually passed on to South Africa via Britain.

Italy used the Breda Ba65, armed with four machine guns, as a specialist ground-attack aircraft in the invasion of Abyssinia and later in North Africa, while the *Luftwaffe* tested its Henschel Hs 123 biplane in Spain during the civil war. This type later saw service during the invasion of Poland and on the Russian front, the Hs 123C version having two additional machine guns in the lower wings for strafing. Another type used successfully in Spain and Poland was the Junkers Ju 87 Stuka dive-bomber. The Stuka, with its distinctive cranked wings and high-pitched whine during the attack, came to symbolize the inhumanity of modern warfare. It fared badly against eight-gun fighters such as the Hurricane and Spitfire, however, and lost much of its aura of omnipotence. As a dive-bomber the Stuka could carry a weapon load of 1,800kg (4,000lb) in addition to four 7·9mm machine guns; later specialist ground-attack developments included the heavily armour-plated Ju 87D–3 and D–4, with underwing gun packs, and the anti-tank Ju 87G fitted with a 37mm cannon under each wing.

The Stuka's 'tank-busting' role was taken over in North Africa and on the Russian front by the Henschel Hs 129, designed specifically for this role. It carried a single 30mm and two 20mm cannon, machine guns and two bombs of 50kg (110lb) each. In the B–2/R4 variant the 30mm weapon was replaced by one of 75mm. Eight hundred Hs 129s were built. The Ju 88 was also employed in the anti-tank role, the Ju 88P having one or two heavy-calibre cannon in a housing under the fuselage, and the later Ju 188S specialized in low-level operations with a variety of weapons including ventrally mounted 50mm cannon. Smaller fighters were also pressed into service in the ground-attack role, the Fw 190F – developed from the Fw 190A fighter – having extra armour-plating and fewer guns. On fighter-bomber missions the Fw 190G could carry a 500kg (1,100lb) or 1,000kg (2,200lb) bomb, and the Bf 109F was transferred from air-to-air operations when Bf 109Gs were introduced. The 109Fs were in this case armed with rocket projectiles; later on, the 109G was also employed on ground-attack work. Hitler's apparent insistence that the jet Me 262 be used for attack roles resulted in misuse of an aircraft which was undoubtedly far more successful as a pure fighter.

From October 1941 the RAF used 'Hurri-bomber' versions of the Hurricane IIB and IIC, the aircraft carrying a 114kg (250lb) bomb under each wing. These were followed in mid-1942 by Mk IIDs armed with a pair of 40mm cannon under a strengthened wing. Each shot pulled the Hurricane's nose down by 5°, so the aircraft then had to be re-aimed, but the type was used with great success in the Western Desert. Rocket-carrying vari-

ants of the Mks IIB, IIC and IV became available from September of the following year, and the Hurricanes supplied to Russia were modified to operate from skis in winter. These aircraft were employed by the French *Normandie* squadron operating in Russia, as well as by the Soviet Union itself. Another type which doubled as a fighter and attacker was the Bristol Beaufighter. Some Mk VIs were operated as 'Flakbeaus' armed with eight rockets or up to 454kg (1,000lb) of bombs. Similarly, the Mosquito F.B.XVIII could carry eight rockets and two 225kg (500lb) bombs as well as a 57mm (six-pounder) gun under its nose. The earlier F.B.VI, of which 2,584 were built, could also be fitted with eight 27kg (60lb) rocket projectiles.

Two very effective types used against tanks, trains and shipping were Hawker's Typhoon and Tempest. Their four built-in Hispano 20mm cannon were augmented by eight rockets and up to 900kg (2,000lb) of bombs.

The French Air Force's specialized two-seat ground-attack Potez 633s saw action in the Battle of France, but were too sparsely deployed to have any great effect. Liore-Nieuport 401s and 411s were also used at this time. The United States' fighters in the early part of the war, especially the Bell P–39 and Curtiss P–40, were also operated as fighter-bombers. The former could carry a 225kg (500lb) bomb, the latter as many as three of the same weight or one under the fuselage and six of 9kg (20lb) each under the wings. The P–38L version of the Lockheed Lightning, of which nearly 4,000 were built, carried ten 127mm (5in) rockets, as could the North American P–51D Mustang and late models of the Republic P–47 Thunderbolt. Five hundred A–36As, specialized ground-attack Mustangs with dive-brakes above and below the wings, had been supplied earlier.

On the larger side, a formidable attacker was the Douglas A–20 Havoc. Most of the 2,850 A–20Gs were supplied to Russia, where they were an invaluable addition to the air-to-surface strike force. They were fitted with up to eight nose-mounted 0·50in machine guns, or two of these weapons and four 20mm cannon, and could carry a bombload of up to 1,814kg (4,000lb). The A–20 was developed into the A–26B (later B–26) Invader, a heavily armoured attacker with six machine guns in the nose; the fire from these could be augmented by locking the upper turret in the forward position and firing its guns under control of the pilot. Even heavier was the disappointing North American B–25G/H series, armed with a massive, nose-mounted 75mm cannon as well as up to 14 0·50in guns. Some B–25Js were fitted with no fewer than 18 0·50in machine guns, plus eight rocket projectiles, for ground attack.

The Soviet Union's best known attacker was the Ilyushin Il-2 *Shturmovik*, which was designed specifically for that role. It entered service in 1941 as a single-seater armed with two 20mm cannon and a pair of forward-firing 0·30in machine guns. The pilot was protected by armour-plating, but the aircraft was found to be vulnerable from the rear. In 1942, therefore, a two-seater with a pair of rearward-firing 0·30in weapons for self-defence was introduced. The Il–2 was one of the first types to use rocket projectiles—up to eight weighing 25kg (56lb) each—and could carry 400kg (880lb) of bombs. It had the lowest wartime attrition rate of any Russian combat aircraft. The machine-gun-armed Petlyakov Pe–2 was also used for ground attack, as was the Tupolev Tu–2 that replaced it. Fighters such as the Polikarpov I–16, Lavochkin LaGG–3 and Yakovlev Yak–3 were employed on strafing duties and could be fitted with rockets, while Yak–9Ts were armed with 75mm cannon for anti-ship operations.

In Japan the Mitsubishi Ki–46–IIIB was a ground-attack adaptation of the type normally used for reconnaissance, and about 2,000 of the same manufacturer's specialized Ki–51 – sometimes called the Japanese Stuka – were built. The Kawasaki Ki–102, only a small number of which were completed by the end of the war, was formidably armed with a 57mm cannon, two of 20mm, a 0·50in machine gun and 500kg (1,100lb) of bombs. Kawasaki's earlier Ki–45 had also been heavily armed, carrying one 37mm and two 20mm cannon or a 75mm weapon for anti-ship strike. This form of attack was the forté of Japanese *Kamikaze* pilots in the Pacific. With some success they tried to stem the tide of Allied shipping by the simple expedient of crashing into it. This was a desperate tactic and another case of that war-time cliché: 'too little, too late'.

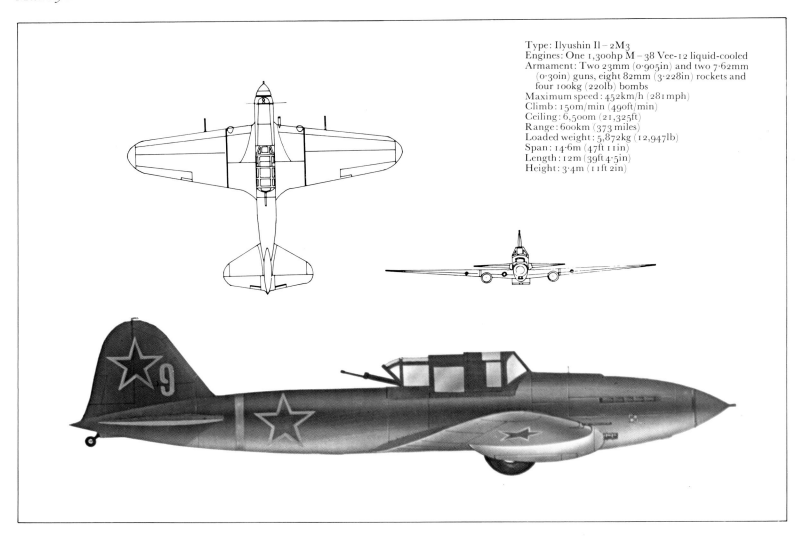

Type: Ilyushin Il–2M3
Engines: One 1,300hp M–38 Vee-12 liquid-cooled
Armament: Two 23mm (0·905in) and two 7·62mm (0·30in) guns, eight 82mm (3·228in) rockets and four 100kg (220lb) bombs
Maximum speed: 452km/h (281mph)
Climb: 150m/min (490ft/min)
Ceiling: 6,500m (21,325ft)
Range: 600km (373 miles)
Loaded weight: 5,872kg (12,947lb)
Span: 14·6m (47ft 11in)
Length: 12m (39ft 4·5in)
Height: 3·4m (11ft 2in)

Post-war developments

In the immediate post-war years most effort was concentrated on applying the new jet engines to ever faster fighters and bigger bombers. Specialized ground-attack types were generally ignored, with the result that the outbreak of the Korean War in 1950 caught all combatants with ageing attack forces. In fact, jet power was not necessary for this role because a stable weapon platform was usually more important than a high top speed. United Nations forces continued to rely on such types as the North American Mustang and Douglas B–26 (formerly A–26) Invader. Probably the most effective close-support aircraft was the Douglas A–1 Sky-raider, which remained in production from the end of World War II until 1957, so difficult was it to find something better to replace it. The Skyraider had an endurance of up to ten hours and could be fitted with a wide variety of weapons on 15 attachments. A typical load was a dozen high-velocity 127 mm (5in) rockets, two 450kg (1,000lb) bombs and a napalm canister. A–1s, of which more than 3,000 were built, could be flown with 6,350kg (14,000lb) of underwing stores in addition to their four 20mm cannon in the wing. For really hard targets such as bridges they could mount 12in Tiny Tim rockets. Both the A–1 and B–26 were later used in Vietnam.

As had been the case in the war, virtually all fighters were also used for air-to-surface missions. Types such as the Republic F–84 and North American F–86 carried bombs in addition to their built-in cannon and machine guns, while the Soviet Union employed Mikoyan MiG–17s and –19s against ground targets. The Russian practice of fitting large cannon–up to 37mm–for air-to-air work gave the fighters a good punch against tanks, but design of the aircraft made them suitable for high-speed flight and bombload was severely limited. Britain had produced the de Havilland Venom, armed with four 20mm cannon and up to 900kg (2,000lb) of bombs, and in the Hawker Hunter provided a true multi-role aircraft tailored to the needs of many emerging air forces. The Hunter combined 30mm cannon with heavy rockets–16 of 27kg (60lb) each, for example –and bombs. In France, the Sud-Ouest Vautour could carry 3,200kg (7,000lb) of bombs as well as four 30mm cannon; the Dassault Super Mystère, mounted bombs and rockets plus its two 30mm guns.

The United States continued to develop fighters which performed equally well, if not better, as bombers. The North American F–100 Super Sabre is still in use, toting up to 2,720kg (6,000lb) of bombs and delivering a hail of shells from its four 20mm cannon. The McDonnell Voodoo, in its F–101C variant, was modified for ground attack. The M61 Vulcan six-barrelled Gatling gun was first brought to bear on surface targets in the Lockheed F–104G Starfighter. By then a wide range of missiles were available, allowing attacks to be made with pin-point accuracy from outside the range of the target's defences. The Republic F–105D Thunderchief also packs the punch of a Vulcan gun, and despite its fighter designation can carry, internally and externally, a load of 6,350kg (14,000lb). With a typical arrangement of 16 bombs of 340kg (750lb) each under its wings, the F–105 bore much of the responsibility for USAF tactical bombing in Vietnam. The F–105G version operated on Wild Weasel missions, attacking enemy radars with missiles which homed on to their radiation.

In the late 1950s the aircraft gun came to be regarded as old-fashioned, and many new fighters were designed around missile armament only. Early versions of the McDonnell F–4 Phantom, for example, relied on Side-winder and Sparrow air-to-air missiles. It was soon realized, however, that the gun added versatility and in many close-in dog-fighting cases was superior to the missile. The F–4E version of the Phantom therefore incorporates a Vulcan, enhancing its effectiveness in the air-to-surface role as well as in air combat. In fact, the Phantom is no mean bomber in its own right, being able to carry a load of 7,257kg (16,000lb).

In the 1950s, Fiat (now Aeritalia) developed the G.91 to meet a NATO requirement for a light strike fighter, and in its twin-engined G.91Y variant it can carry 1,800kg (4,000lb) of bombs, missiles or machine gun pods in addition to the pair of built-in cannon. Russia also produced a ground-attack specialist in the form of the Sukhoi Su–7, carrying rockets and cannon. This type has seen extensive action in the Middle East, where

RIGHT: *Up to 42 General Dynamics F-111A attack bombers of the USAF are being completely rebuilt by Grumman as EF-111A electronic-warfare aircraft. Their* BELOW: *The single-seat version of the French SO.4050 Vautour (Vulture) attack aircraft.* BOTTOM CENTRE: *The Douglas A-1 Skyraider was one of the most successful attack types.* BOTTOM: *The Vought A-7 Corsair was a standard US Navy and Air Force bomber.*

and also in India and elsewhere in Asia.

France's Dassault Mirage III can be employed on ground-attack duties, carrying a pair of 450kg (1,000lb) bombs or missiles in addition to two DEFA 30mm cannon in a gun pack. Dassault has also developed the same basic aircraft into the Mirage 5, with simplified electronics, more fuel, easier maintenance and a far bigger – 4,000kg (8,800lb) – weapon load. In Sweden, SAAB followed the Lansen with the Draken, a multi-role type able to act as a fighter-bomber as well as a Mach 2 intercepter.

Vietnam

The war in Vietnam changed the face of the American armoury. Attack aircraft of all shapes and sizes proliferated; Republic F-105s delivered guided and unguided ordnance by the ton, light aircraft weighed in with rockets and machine guns, and in between came the specialized attackers. The USAF took the Navy's Vought A-7A Corsair II carrier-based bomb truck and turned it into the A-7D with a new engine and improved weapon-delivery systems. The Cessna T-37 trainer was beefed up, given twice the thrust and became the A-37B, with a multi-barrelled 0.30in Minigun and 2,270kg (5,000lb) of bombs. A requirement for a light armed reconnaissance aircraft resulted in the Rockwell OV-10A Bronco, carrying four 0.30in machine guns and 1,100kg (2,400lb) of other weapons.

More unusual were the converted transport aircraft used to pour a hail of fire on to

BELOW: *The McDonnell Douglas F/A-18A Hornet is seen here in a two-seat version serving with the Spanish AF.*
BELOW CENTRE: *Powered by a British Spey turbofan, the AMX attack aircraft is a joint product of Italy and Brazil.*

BOTTOM CENTRE: *Argentina's IA 58A Pucará is an unusual close-support aircraft, being powered by twin turboprops.*
BOTTOM: *The AJ37 was the first of five versions of the Swedish Saab 37 Viggen. All can operate from straight highways.*

ground targets. The Douglas AC–47D, known unofficially as 'Puff the Magic Dragon', could carry three 0·30in Miniguns each able to fire 6,000 rounds a minute out through the side of the fuselage. At a later stage, Fairchild AC–119s and Lockheed AC–130s were fitted with cannon and Gatling guns for the same role.

The European battlefield

The Soviet Union has 20,000 tanks available for a thrust against Western Europe, and NATO's ground-attack aircraft would have the task of helping to stop such an assault. Land-based types would also be called on to assist carrier aircraft in missions against shipping. The RAF deploys the BAe Buccaneer, Harrier and Jaguar in these roles. The Buccaneer S.2, designed originally for the Royal Navy, can carry four 450kg (1,000lb) bombs internally and an additional 5,450kg (12,000lb) of various weapons under its wings – these could include rockets or Martel guided missiles. Being designed specifically for low-level attack, it is a stable weapon platform ideally suited to such a task. The V/STOL (Vertical/Short Take-Off and Landing) Harrier GR.3 can have two pods containing 30mm Aden cannon plus underwing stores such as a pair of 450kg (1,000lb) bombs and two containers each holding 19 68mm rockets. The RAF's Jaguar, carrying the computer-based NAVWASS (Navigation and Weapon-Aiming Sub-System), is designed to fly a ground-hugging mission to make surprise, one-pass attacks using its two built-in Aden guns and up to 4,500kg (10,000lb) of weapons such as the BL755 cluster bomb, which dispenses 147 bomblets in a dense, destructive cloud.

France also uses Jaguars, with different weapons, as well as ground-attack Mirage IIIs and some Mirage 5Fs. Germany's Lockheed F–104Gs were replaced by the Panavia Tornado multi-role combat aircraft, being built in collaboration with Britain and Italy. A British variant of the swing-wing Tornado is used as an interceptor but the basic common version is designed for ground attack, anti-shipping duties and reconnaissance. Germany also co-operated with France, in this case on the Dassault-Dornier Alpha Jet light attack/trainer. In the *Luftwaffe* the Alpha Jet replaced G.91s, capable of being fitted with a pod containing a Mauser 27mm cannon, the same gun as used in Tornado, or up to 2,200kg (4,850lb) of bombs or other weapons on a total of five pylons. The Alpha Jet has been operated by Belgium as well as its two sponsoring countries. Britain's BAe Hawk – which the RAF uses as an advanced flying and weapons trainer – is competing in the same export market in single- and two-seat versions which can carry up to 3,085 (6,800lb) of mixed weapons.

Belgium is also one of several countries, including Holland, Denmark and Norway, to have ordered the General Dynamics F–16 for ground-attack duties as well as air combat. Sweden continues to produce modern weapons of extremely high quality, including the SAAB AJ37 Viggen for anti-ship strike in the Baltic as well as overland operations.

The USAF's main tank-buster in Europe is the Fairchild A–10. Its top speed is modest – a significant departure from the 'fast is good' thinking which has largely prevailed since World War II – but it is a stable and yet highly manoeuvrable weapon-delivery platform. The A–10 is built round 'a General Electric GAU–8/A Avenger seven-barrelled cannon, which can fire 1,350 rounds of armour-piercing ammunition at a rate of 4,000 rounds a minute and with a muzzle velocity of 1,050m/sec (3,450ft/sec). The shells weigh 0·73kg (1·6lb) each and can disintegrate a tank. The A–10 also has 11 pylons for up to 7,250kg (16,000lb) of guided and unguided ordnance.

Rather oddly, the Soviet Union seems to have copied the Northrop A–9A (the design which lost to the A–10) in the Sukhoi Su–25 (NATO 'Frogfoot'). This is a basically lighter aircraft than the A–10, but with more powerful R–13–300 engines it is considerably faster. It too has a powerful 30mm gun and ten weapon pylons, and like the Grumman A–6 (described later) has split airbrakes at the wingtips. In Afghanistan Su–25s developed techniques in which they made precisely co-ordinated close-support attacks in partnership with Mi–24 gunship helicopters.

Few countries have built dedicated jet attack aircraft in this close-support category, but two recent jet attack types are the IAI Lavi and the AMX. The Lavi was to have been Israel's first wholly home-designed warplane (though with extensive US help), resembling a small F–16 with a PW1120 turbojet of 9,353kg (20,620lb) afterburning thrust. The Lavi promised to be a fast and agile attack fighter, but in 1987 it succumbed to budgetary pressures.

In contrast, the AMX, developed jointly by Aeritalia and Aermacchi of Italy and Embraer of Brazil, is a dedicated close-support attack and reconnaissance aircraft with subsonic speed but good rough-field capability, versatility and low cost. Powered by a Rolls-Royce Spey turbofan, it can carry a wide range of weapons, including cannon, or three alternative reconnaissance pallets in an internal bay.

A unique close-support aircraft is the Argentinian IA 58 Pucará. It almost resembles World War II aircraft in having a speed of 500km/h (311mph) on twin 978hp Turbomeca turboprops. Pilot and copilot sit in tandem ejection seats, and weapons include many kinds of bombs, rockets, cannon and machine guns. About 24 were lost in the Falklands in 1982.

In a slightly different category are the numerous kinds of trainer which can also carry weapons. At the top end of this group are the Hawk and Alpha Jet already mentioned. Both have been developed in dedicated attack form, the latter as the Alpha Jet 3 with inertial platform, forward-looking infra-red and other devices for day/night attack, and the Hawk as the MK 200 single-seater with tremendous all-round capability which includes a bombload of 3,175kg (7,00lb) and range up to 3,610km (2,244 miles).

Elsewhere in the world there are more than a dozen types of jet trainer which can be used for weapons training and thus, by implication, for actual close-support and attack missions. One type, the Spanish C–101CC Aviojet, has an internal bay in the fuselage for cannon, machine guns, cameras, electronic counter-measures or a laser designator, the main weapon load being carried on six wing pylons.

An even larger number of countries is producing propeller-driven armed trainers. Of these the fastest and most powerful is the RAF Shorts Tucano, but this carries weapons (up to 625kg, 1,378lb) only in export versions. By far the most commercially successful aircraft in this category is the piston-engined (260hp) SIAI-Marchetti SF.260 family from Italy, over 800 of which are in service with many air forces. This is despite the low power, which limits weapon load to 300kg (661lb), flown as a single-seater. A generally similar machine is the French Aérospatiale Epsilon, with 300hp and an equally wide range of light weapons. Even lighter machines are the French Socata Guerrier, with 235hp and various machine guns, rockets or light bombs, and the Swedish Saab Supporter (mass-produced in Pakistan as the Mushshak) with only 200hp but six weapon pylons.

Many people passionately believe that even the most powerful air forces and armies have a need for small, agile, low-cost aircraft to fly such battlefield missions as reconnaissance, close support and, especially, anti-helicopter attack. The American Rutan Long-Eze (similar to the VariEze, p.294) did trials in this role, while in December 1987 British Aerospace announced its belief in the SABA (Small Agile Battlefield Aircraft). SABA could weigh 4,536kg (10,000lb), carry a tremendous variety of weapons and operate from 304m (1,000ft) dirt strips on the thrust of a 12-blade contraprop driven by a 4,500hp T55 turboprop.

Right at the other end of the attack spectrum, the General Dynamics F–111 was ordered and designated as a fighter, but used as a long-range attack aircraft. The first combat aircraft to feature pivoting 'swing wings', the F–111 first flew in 1964, and altogether 562 were built, of which 24 were supplied to Australia and about 380 remain in USAF service. All the survivors are being subjected to a major avionics updating, though the TF30 engines of some 8,392–9,072kg (18,500–20,000lb) thrust are not being changed. Unusual features include side-by-side seating and giant main landing gears which, together with a belly airbrake, make it impossible to carry anything under the fuselage except a small counter-measures pod. Heavy bombloads can be carried under the wings, and the long-span FB–111A version (which can take off at weights up to 51,846kg, 114,300lb) serves with Strategic Air Command as a strategic bomber over shorter missions.

In the Sukhoi Su–24 (NATO 'Fencer') the Soviet Union has what looks almost like a copy of the F–111, though with the US aircraft's faults sidestepped (for example, by fitting long inlet ducts to the engines and landing gear which still allow heavy weapon loads to be hung under the fuselage). Maximum bombload is estimated at 11 tonnes (24,250lb), as well as a six-barrel gun of 30mm calibre. Flying from Warsaw Pact airfields the Su–24s can carry out night or bad-weather attacks throughout Western Europe.

Should war come to Europe the only aircraft likely to survive would be the RAF's few Harrier GR.3s and improved GR.5s (to be upgraded with night equipment to GR.7 standard). They need never go near vulnerable airfields because of their unique Pegasus jet-lift engines, yet they have flashing speed and agility. The GR.5 can carry a weapon load of 7,711kg (17,000lb) and still pull 6g and fly a long mission radius. The Sea Harrier is described in the next section.

BELOW: *Fairchild delivered 713 A-10A Thunderbolt II close-support aircraft to the USAF. Armed with a devastating gun and various missiles, the A-10 is big and rather slow, but designed to kill tanks.*

BOTTOM: *An exact Soviet equivalent of the USAF's big swing-wing F-111 bomber, the Su-24 can carry 11 tonnes of bombs and menaces the whole of Western Europe without in-flight refuelling.*

Carrier-based aircraft

From the very earliest days of fixed-wing flying it seemed natural to use water as a runway – it was generally flat, unrestricted and readily accessible. Glenn Curtiss made the first flight in a practical seaplane during February 1911 (a seaplane has floats, whereas a flying boat has a planing boat-like hull). It is significant that Curtiss's flight took place from alongside a warship, the cruiser USS *Pennsylvania* anchored in San Diego Bay, as military planners were to prove more enthusiastic about seaplanes than about aircraft with normal undercarriages but based aboard ship. It must be remembered, however, that Curtiss's pioneering flight immediately followed two other very significant events. In November 1910 a Curtiss biplane piloted by Eugene Ely, a demonstration flyer, had taken off from a platform on the bows of the light cruiser USS *Birmingham*, anchored off Virginia. This was the first time that an aircraft had been operated from a ship, and two months later Ely made the first shipboard landing, touching down on a platform built on the stern of the cruiser *Pennsylvania* anchored in San Francisco Bay.

In 1898 US Naval officers had been on an inter-Service board set up to investigate the possibilities of Langley's Aerodrome, which at one time looked likely to be the first successful aeroplane. It was a failure, however, and interest waned until two naval officers attended a demonstration of the Wright Model A to the US Army in 1908. In September 1910 a Captain Chambers was appointed officer in charge of aviation for the US Navy, and he ordered two aircraft from Glenn Curtiss. By the time the first of these, the A–1 Triad, flew in July 1911, Chambers had arranged Ely's demonstrations and two Lieutenants had been taught to fly.

the conflict, 19 months later, the number had grown to more than 2,000. The only shipborne wheeled aircraft, however, was a Sopwith Camel operated from a turret on the battleship *Texas*.

British naval aviation had begun in the spring of 1911, when three Royal Navy officers and one from the Royal Marines Light Infantry were taught to fly at the Isle of Sheppey. In October of that year the Admiralty bought the two Short pusher biplanes being used for instruction, and in the same month a naval flying school was set up at Eastchurch under Lt C. R. Samson, one of the original four students. The following January Samson flew a Short S.27 from a platform on the stationary battleship HMS *Africa*, and in May he took off from the battleship *Hibernia* while she was under way, this being the first sortie from a moving vessel. The Royal Flying Corps had been set up the previous month, and in July 1914 its naval wing became the Royal Naval Air Service.

Pups had first been embarked on the converted Isle of Man packet *Manxman*, which joined the fleet at the end of 1916. The Pups were modified for anti-Zeppelin patrols and were also used in deck-arresting trials at the Isle of Grain and for the development of flotation gear, as well as equipping a total of five aircraft carriers and seven cruisers.

Sopwith Camels later replaced the Pups, serving on five aircraft carriers, two battleships and 26 cruisers. The 2F.1 shipboard variant had an upward-firing Lewis gun in place of the starboard Vickers, and the wings were of shorter span. The type became operational in the spring of 1918, and 340 were built. Some acted as diver-bombers with two 23kg (50lb) bombs under the fuselage. On 17 July 1918 six Camels operating from *Furious* destroyed the Zeppelins L.54 and L.60 in their shed at Tondern.

carried 28 seaplanes or flying boats and four catapults. It remained in service until 1950. Both Germany and Italy were later to lay down carriers, but they were never completed. The first flush-deck carrier, HMS *Argus*, had joined the British fleet just before the end of World War I. She carried the Sopwith/Blackburn Cuckoo, the first purpose-designed torpedo bomber. The single-seat Cuckoo was powered by a 200hp Sunbeam Arab engine, giving a top speed of 140 km/h (87mph), and it could carry a 450kg (1,000lb) torpedo. The cruiser *Cavendish* was converted in the same way as the *Furious* to become the *Vindictive*, and the *Furious* herself was given a flush flight deck at the end of the war. The world's first purpose-built carrier was HMS *Hermes*, and the first with the superstructure offset to starboard as an island was *Eagle*, completed in 1920.

British inter-war carrier aircraft

The Fairey Flycatcher, the Fleet Air Arm's first specialist carrier-borne fighter, entered service in 1923. The RFC and RNAS had been amalgamated to form the Royal Air Force in April 1918, and to a certain extent naval aviation entered the wilderness. The Blackburn Skua, for example, entered active service just before World War II but had

In 1914 seven pilots and five aircraft participated in two detachments aboard the cruisers USS *Birmingham* and *Mississippi*, being used to observe Mexican positions. In July 1915 the Office of Naval Aeronautics was set up within the Navy Department, and the Naval Flying Corps (150 officers and 350 men) was authorized in August 1916. When the United States entered World War I in 1917, the NFC had 54 aircraft. By the end of

The Royal Navy was soon joined by those of the United States and Japan in operating aircraft carriers, and France had the *Béarn*. The French had begun experiments with carrier-based aircraft as early as 1918 when Lieutenant de Vaisseau Georges Guierre flew a Hanriot HD 1 from a platform on board the battleship *Paris*.

In 1932 the French Navy commissioned a seaplane carrier, the *Commandant Teste*, which

TOP: *The Fleet Air Arm's first specialist carrier-based fighter was the Fairey Flycatcher, which entered service in 1923. It carried a pair of synchronized machine guns and remained in service for a decade.*
ABOVE: *Nearly 2,400 examples of the popular Fairey Swordfish torpedo bomber were built, and the type outlasted its intended successors. The 'Stringbag' saw action at Taranto, where the Italian fleet was badly damaged, and took part in the sinking of the* Bismarck.

Various ways of accelerating aircraft on take-off have been tried from the earliest days of aviation. In 1904–05 the Wright brothers used a tower supporting a block-and-tackle arrangement which carried a weight to give the aircraft its initial impetus when released. The first true catapult launch was made by Lieutenant Ellyson at the Washington Navy Yard in 1912; a wheeled dolly supporting his aircraft was propelled along a track by compressed air. Deck-mounted catapults were widely used on seaplane carriers and true aircraft carriers, although the low flying speed of aircraft right up to the time of World War II meant that a normal take-off could generally be made.

In 1934 the US Naval Aircraft Factory began developing a flush deck catapult, using a bridle to accelerate the aircraft and do away with the need for launching dollies. France, Britain and Germany also worked on catapults, but Japan lagged behind despite her rapid building programme for carriers. The need for an effective catapult became critical when escort carriers were developed during World War II. These were converted merchant ships with confined flight decks, a low top speed and a tendency to wallow badly in heavy seas. The use of catapults confers many other advantages also, including the ability to carry more aircraft, launch at night without extensive deck lighting, and conduct operations with less regard for wind direction.

A modern US Navy carrier such as the USS *John F. Kennedy* has four catapults, two at the bow and a pair at the waist. The longest one is 99m (325ft) and the other three are 81m (265ft). They operate under constant steam pressure, valves opening at different rates to give variations in acceleration according to launch weight. An aircraft taxis up to the catapult and its nose-tow launch bar, which replaces the bridle (strop) used previously, engages the shuttle. A T-bar forms the weak link between the aircraft and a hold-back bar which ensures that there is no slack in the system; it snaps when the catapult fires. Some aircraft, such as the Grumman F–14 Tomcat, use a different arrangement – a reusable rod which stretches and disengages on firing. The aircraft is launched with its engines at full power and is accelerated to its single-engine safe flying speed plus a margin of about 30km/h (18·75mph).

Arrester gear

When Eugene Ely made the first landing aboard a ship he was brought to rest by 22 ropes laid across the deck and attached to 23-kg (50-lb) sandbags at each end. Hooks between the aircraft wheels engaged the ropes.

A development of this eminently practical method was later used on the US Navy's first aircraft carrier, the *Langley*, with the addition of longitudinal wires to engage hooks on the axle and keep the aircraft straight.

An improvement of the sandbag arrangement was introduced by running the arrester cables to weights at the bottom of towers. When the aircraft snagged the wires the weights were lifted, automatically applying tension. The Norden gear introduced by the US Navy carried this a step further, the cables being wound at each end round a drum fitted with a brake to ensure that paying-out was slow and smooth. An electric motor rewound the drum after use. The problem of wire being paid out unequally if an aircraft landed off the centre-line was solved by using a drum at one end of the cable only.

This is the essence of the modern arrangement. In the *John F. Kennedy*, the four steel-cable arrester wires are each 3·8cm (1·5in) thick and 33m (110ft) long. They are attached to 640-m (2,100-ft) purchase cables driven by engines below deck. The tension of the arrester wires is adjusted for each aircraft according to its weight, so that all are brought to rest in the same distance. Pilots always aim for the third wire.

During World War II just about the most important man on a carrier was the Landing Signals Officer (LSO). He hand-signalled with 'paddles' (light-coloured discs) to indicate to pilots how they were doing on the approach. Modern carriers still carry LSOs in case the automatic equipment breaks down, but his paddles have been replaced by the Manually Operated Visual Landing Aid System (MOVLAS) – a series of lights which he illuminates to aid pilots on the approach. Normally, landings are made with the aid of a combination of lenses and coloured lights. The pilot flies to keep the image of a yellow light positioned between two horizontal bars of green lights. This ensures that he is approaching at the correct vertical angle, and he aligns his aircraft with the deck centre-line markings to keep himself correctly positioned in azimuth.

ABOVE LEFT: *A biplane making a landing on HMS* Furious *in 1917.*
ABOVE RIGHT: *A Grumman Avenger landing on an aircraft carrier. A similar type of arrester cable is in general use today.*
BELOW LEFT: *A French Etendard IV on the aircraft carrier* Foch *before launching.*
BELOW RIGHT: *A Vought F–8 Crusader being prepared for launch by catapult.*

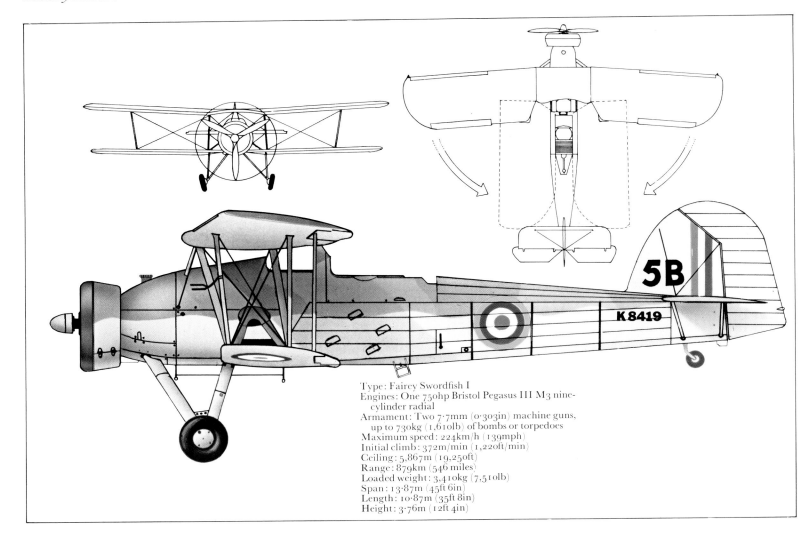

Type: Fairey Swordfish I
Engines: One 750hp Bristol Pegasus III M3 nine-
    cylinder radial
Armament: Two 7·7mm (0·303in) machine guns,
    up to 730kg (1,610lb) of bombs or torpedoes
Maximum speed: 224km/h (139mph)
Initial climb: 372m/min (1,220ft/min)
Ceiling: 5,867m (19,250ft)
Range: 879km (546 miles)
Loaded weight: 3,410kg (7,510lb)
Span: 13·87m (45ft 6in)
Length: 10·87m (35ft 8in)
Height: 3·76m (12ft 4in)

been requested as a Flycatcher replacement in 1924. The Flycatcher itself, armed with two synchronized 0·303in machine guns and powered by a 400hp Armstrong Siddeley Jaguar giving a top speed of 214km/h (133 mph), remained in service until the early 1930s. The Blackburn Dart followed the Cuckoo as a torpedo bomber and eventually led to the last of the Blackburn biplane torpedo bombers, the Shark. It was first delivered for service aboard HMS *Courageous* in May 1935, and replaced by the Swordfish in 1938.

The Fairey IIIF Mk III was a popular general-purpose type which entered service in 1928 and remained operational until 1940. Operated as a three-seater by the Royal Navy (the RAF's Mk IV had two seats), the IIIF was replaced by a developed version known as the Seal by the Navy and Gordon by the RAF. The Seal differed from its predecessor in being powered by an Armstrong Siddeley Panther rather than a Napier Lion. Ninety were built, the type serving on several carriers from 1933 until being replaced by the Blackburn Shark in 1935–38. Fairey continued its naval tradition with the Swordfish, affectionately known as the Stringbag. Deliveries began in February 1936, and nearly 2,400 were built; in fact the Swordfish outlived its intended replacements. The 750hp Bristol Pegasus 30 gave a top speed of 224km/h (139 mph), and if desired a number of small bombs could be carried in place of the normal 680 kg (1,500lb) mine or 730kg (1,610lb) torpedo. The Swordfish was used as a floatplane

in addition to its role as a carrier-based aircraft, and took part in many famous engagements during World War II.

American inter-war carrier aircraft

The US Navy's first aircraft carrier, a converted 11,000-ton collier renamed the USS *Langley*, was commissioned in September 1922. The first US aircraft designed specifically for carrier operations was the Naval Aircraft Factory TS–1, used aboard the *Langley* and as a seaplane. It entered service in 1922 and was later built as the Curtiss F4C. A 200hp Wright J4 engine gave a maximum speed of 198km/h (123mph), and armament comprised a single 0·30in machine gun. Curtiss's later F6C–4 Hawk fighter equipped a single squadron on *Langley* until the beginning of 1930, only 31 being built. A 410hp Pratt & Whitney R–1340 conferred a top speed of nearly 250km/h (155mph), and armament was doubled to a pair of machine guns. The same basic powerplant, but uprated to 425hp, was installed in the Boeing F2B–1 single-seat shipboard fighter, deliveries of which began in January 1928 to equip the *Saratoga*. The slightly modified F3B–1 followed seven months later, serving aboard the *Langley* and later the *Lexington* and *Saratoga*. Production of the F2B–1 and F3B–1 totalled 32 and 74 respectively, and the latter type remained in service until 1932.

The next in the series, the F4B–1, was likewise powered by an R–1340 (this time of 450 hp) and had a top speed of 283km/h (176 mph) compared with the F3B's 254km/h

(158mph). Deliveries of the 27 F4B–1s began in June 1929, the aircraft equipping *Lexington* and *Langley*. They were followed by 46 F4B–2s from early 1931 (*Saratoga* and *Lexington*), 21 F4B–3s from the end of that year (*Saratoga*) and 92 F4B–4s from July 1932 (*Langley* and *Saratoga*). The F4B–4's R–1340 engine of 550 hp gave it a speed of just over 300km/h (188 mph), and the type remained in service until 1937. All F4Bs were armed with a pair of 0·30in machine guns or with one weapon of that calibre and one of 0·50in.

Curtiss, meanwhile, followed the unsuccessful F8C–4 Helldiver two-seat fighter, of which 25 were built and which served aboard the *Saratoga*, with the F11C single-seat fighter. Twenty-seven F11C–2s were delivered from February 1933, equipping one squadron on *Saratoga*, and in March of the following year their designation was changed to BFC–2 to reflect their fighter-bomber role. They remained in service until February 1938. The next off the line, 27 BF2C–1s (formerly F11C–3s) which served aboard *Ranger* for a few months from October 1934 until being withdrawn because of undercarriage failures, were the US Navy's last Curtiss fighters. One of the companies which inherited the Curtiss mantle was Grumman. The two-seat FF–1, of which 27 were built, was the US Navy's first fighter with a retractable undercarriage. Its 700hp Wright R–1820 gave a speed of 332km/h (207mph), and armament comprised a forward-firing 0·30in machine gun plus two more on flexible mountings in the rear cockpit. The FF–1s were later modified

RIGHT: *The Chance Vought Corsair was the US Navy observation aircraft of the late 1920s and early 1930s.*
BELOW RIGHT: *The Boeing F4B–4 shipboard fighter served aboard the US Navy carriers* Langley *and* Saratoga *from 1932.*
BELOW: *A Chance Vought Ve–7 off the USS* Langley, *the first American aircraft carrier, converted from a collier, which entered service with the US Navy in September 1922.*

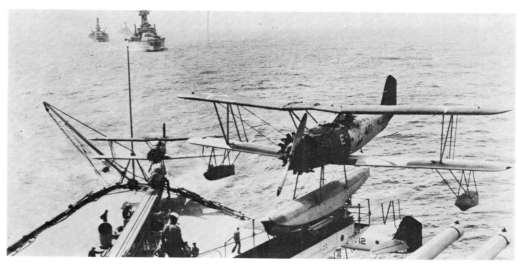

to FF–2 standard with dual controls. The type served aboard the *Lexington* for three years from 1933.

Grumman then moved on to a series of single-seat fighters. The F2F 'Flying Barrel', a scaled-down FF–1, was the first of its type to have a fully enclosed cockpit. It served from 1935 to 1940, at various times equipping the *Ranger*, *Lexington*, *Yorktown* and *Wasp*. The F3F–1 had a longer fuselage and wings to improve manoeuvrability, and was powered by a 650hp Pratt & Whitney R–1535 Twin Wasp. It was in service for four years from 1936, on the *Ranger* and *Saratoga*. The F3F–2, with the Wright R–1820 of 950hp, entered service aboard the *Enterprise* in 1938. At the end of that year it was joined by the F3F–3, capable of 425km/h (264mph). The F3F series were the last American biplane fighters, and were finally replaced by the Wildcat (see below) in October 1941.

While Curtiss, Boeing and Grumman were supplying fighters, Martin produced a series of bombers. The T3M–1 biplane torpedo/bomber/scout, based on the earlier CS–1/2 but having a steel-tube fuselage and uprated Wright T–3B engine, was delivered from September 1926. Wheels or floats were interchangeable. It was followed in mid-1927 by the T3M–2, in which the crew was carried in three individual cockpits and the power-plant was changed to the 770hp Packard 3A–2500, giving a speed of 175km/h (109-mph). One hundred were built, and embarked aboard the *Lexington* and *Langley*. In August 1928 the T4M–1 entered service on

the *Lexington* and *Saratoga*. It had the less powerful 525hp Pratt & Whitney R–1690 Hornet, but the top speed was 183km/h (114 mph). The Great Lakes company later took over production, the aircraft being redesignated TG–1 and TG–2 as they were improved. The latter served from 1930 to 1937 as the US Navy's last biplane torpedo bomber. Its 620hp Wright R–1820 gave it a maximum speed of 204km/h (127mph). Before the take-over Martin had also built 32 BM–1 and BM–2 two-seat dive/torpedo bombers, which were embarked on the *Lexington* and *Langley* for six years from 1931.

Great Lakes used the experience gained during production of the TG–2 to design the BG–1 two-seat dive-bomber. The 60 aircraft served from October 1934 until 1938 aboard the *Ranger* and *Lexington*. They were powered by a 750hp Pratt & Whitney R–1535 and could carry a 450kg (1,000lb) bombload. Results with the Curtiss F8C–4 had been unsatisfactory, so the Vought SBU two-seat scout bomber entered service in 1935. It could carry a 225kg (500lb) bomb and was powered by a 700hp Pratt & Whitney R–1535 which gave a maximum speed of 330km/h (205mph).

The distinction of being the last combat biplane built in the United States (production ended in April 1940) went to the Curtiss SBC Helldiver scout bomber. Deliveries of the SBC–3, of which 83 were built, began in July 1937 and the first of 174 SBC–4s followed in 1938. The latter, powered by a 950 hp Wright R–1820 giving a top speed of

380km/h (237mph), could carry a 450kg (1,000lb) bomb and two squadrons of Hell-divers were still operational aboard the *Hornet* at the time of Pearl Harbor.

Japanese inter-war carrier aircraft

The Japanese Navy's first aircraft carrier, the 9,500-ton *Hosho*, was commissioned in 1922 and could carry 19 aircraft. It was followed in 1927 and 1928 respectively by *Akagi* and *Kaga*, converted from unfinished 34,000-ton battle-cruisers. The Navy's first fighter squadrons operated the Mitsubishi 1MF (Type 10 Carrier-based Fighter), a single-seat biplane capable of 220km/h (138mph) on its 300hp Hispano-Suiza engine. The 2MR was developed from it to become the Type 10 Carrier-based Reconnaissance Aircraft, of which more than 150 were built, and Mitsubishi was also responsible for the 2MT (Type 13 Carrier-based Attacker), a two/three-seat biplane of which nearly 450 were constructed.

The 1MF was replaced by the Nakajima A1N1–2 (Type 3 Carrier-based Fighter), powered by a 420hp Bristol Jupiter. Designed for Nakajima by Gloster in Britain, who called it the Gambet, 150 were modified for naval use from 1929 by Eng Takao Yoshida. The 2MT was superseded by the Mitsubishi 3MR4, otherwise known as the BsM1–2 and designated Type 89 Carrier-based Attacker. A three-seat biplane designed by Blackburn, who also built a prototype, it was powered by a 600hp Hispano-Suiza. Just over 200 were constructed, from 1932.

TOP LEFT: *The Grumman TBF Avenger torpedo bomber, which replaced the Devastator, was one of the most successful torpedo bombers.*
TOP RIGHT: *The Aichi D3A2 dive-bomber took over from the D3A1, which had caused havoc at Pearl Harbor, and served until 1944.*

ABOVE CENTRE: *The Grumman F4F Wildcat was the US Navy's standard fighter at the time of the Japanese attack on Pearl Harbor.*
ABOVE: *The Vought F4U–2 carried a radar in one wingtip, and in October 1943 it carried out the US Navy's first radar interception.*

The Type 3 Carrier-based Fighter was replaced by a development of itself, the Nakajima A2N1–3, designated the Type 90 Carrier-based Fighter and powered by a 580hp Kotobuki II. In all, 106 were built.

Type 3 fighters and Type 13 attackers took part in the Shanghai conflict of January and February 1932, the first time that Japanese aircraft had been involved in air combat. The Navy aircraft operated from the carriers *Notoro*, *Hosho* and *Kaga*. Experience gained in these operations showed that the Type 89 torpedo bomber was in need of replacement and a design competition led to selection of the Yokosuka Arsenal B3Y, an improved version of the Type 13 with a 600hp Type 91 engine, which then entered service as the Type 92 Carrier-based Attacker. It saw widespread service at the beginning of the Sino-Japanese war, but the engine proved to be unreliable and the Navy therefore ordered about 200 Yokosuka B4Ys (Type 96 Carrier-based Attackers) powered by an 840hp Hikari. The Type 96 had a good range and was relatively trouble-free; it also proved to be the Japanese Navy's last fabric-covered biplane torpedo bomber.

News of American interest in dive-bombers led to Japanese Navy procurement of the Aichi AB–9, also known as the D1A1 (Type 94 Carrier-based Bomber), its first aircraft of this kind. It had a 580hp Kotobuki II and could achieve 500km/h (310mph) in the dive, but its payload was only 300kg (660lb). It was very reliable, however, and was developed into the Type 96 which was used extensively in the Sino-Japanese war. Total production of the Type 94/Type 96 was 590. In the Mitsubishi A5M (Type 96 Carrier-based Fighter) Japanese Navy pilots found the answer to their prayers. A low-wing, fixed-undercarriage monoplane, it was fast – 450 km/h (280mph) – and light, weighing less than 1,800kg (4,000lb) fully loaded. It could climb to 3,000m (10,000ft) in less than four minutes and was extremely manoeuvrable, with an endurance of more than three hours. The powerplant was a 560hp Kotobuki V, and armament comprised a pair of 0·303in machine guns; nearly 1,000 were built. Its successor, the Zero, was even better (see page 218).

World War II

In 1939 the US Navy's standard fighter was the Grumman F3F, a biplane but nonetheless able to reach 4,700m (15,000ft) in just over six minutes. A two-stage supercharger gave it a speed in excess of 416km/h (260mph), and the type introduced the heavy 0·50in machine gun as a permanent fixture. The US Navy's first monoplane fighter, and the first capable of exceeding 480km/h (300mph), was the Brewster F2A introduced in 1941. It was otherwise not very impressive, however, and by the time of Pearl Harbor the standard USN fighter was the Grumman F4F Wildcat. This had already been in service with the Fleet Air Arm for more than a year as the Martlet, the aircraft being 'inherited' after the fall of France. With a 1,200hp Pratt & Whitney R–1830 Twin Wasp, the F4F–3 had a top speed of 523km/h (325mph), and armament comprised four 0·50in machine guns. It was replaced from January 1943 by the F6F Hellcat, able to extract a speed of 605km/h (376mph) from its 2,000hp Pratt & Whitney R–2800 Double Wasp and with an extra two guns. More than 12,000 were built, the type also equipping the Fleet Air Arm.

The last US piston-engined fighter, and at the time the most powerful naval fighter ever built, was the Chance Vought F4U Corsair. Again, more than 12,000 were built, including those turned out by Brewster and Goodyear, and the type was responsible for destroying

three-crew Avenger had a power-operated gun turret and could carry a 22in torpedo internally. Its 1,700hp Wright R–2600 Double Cyclone gave it a top speed of more than 435km/h (270mph) and its range was an impressive 1,600km (1,000 miles) with a full weapon load.

The Douglas SBD Dauntless dive-bomber was by far the most important naval attack bomber of the war, sinking more ships than any other dive bomber. It could absorb heavy punishment and had the lowest attrition rate of any US carrier-based type operat-

1937 to 1942 and the D3A2 from that year until 1944. Nearly 1,300 were built in all. The later variant was powered by a 1,300hp Mitsubishi Kinsei 54, conferring a top speed of 426km/h (266mph). The payload was 370 kg (816lb) and maximum range was nearly 1,600km (1,000 miles). The three-crew Nakajima B5N served from 1937 to mid-1944, also in two versions. The B5N1 was powered by a 770hp Hikari 3, giving a top speed of 369km/h (229mph), and was armed with a 0·303in machine gun in addition to its bomb-load of 800kg (1,760lb). It was replaced by

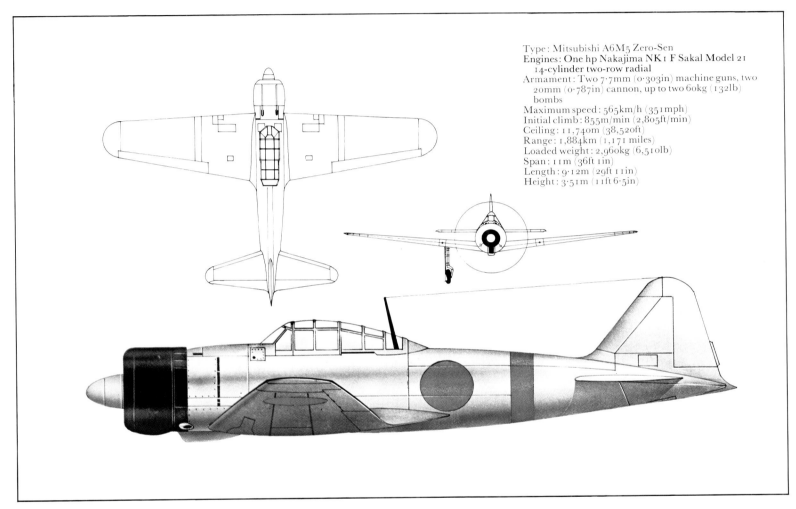

Type: Mitsubishi A6M5 Zero-Sen
Engines: One hp Nakajima NK1 F Sakai Model 21 14-cylinder two-row radial
Armament: Two 7·7mm (0·303in) machine guns, two 20mm (0·787in) cannon, up to two 60kg (132lb) bombs
Maximum speed: 565km/h (351mph)
Initial climb: 855m/min (2,805ft/min)
Ceiling: 11,740m (38,520ft)
Range: 1,884km (1,171 miles)
Loaded weight: 2,960kg (6,510lb)
Span: 11m (36ft 1in)
Length: 9·12m (29ft 11in)
Height: 3·51m (11ft 6·5in)

troying more than 2,000 enemy aircraft as well as surface ships and submarines. The Corsair entered service in 1943, and the F4U–4 fighter variant could reach 718km/h (446mph) from its 2,450hp Pratt & Whitney R–2800 Double Wasp. It also had a range of just over 2,400km (1,500 miles) and could carry two 450kg (1,000lb) bombs or eight rockets as well as six machine guns.

The Douglas TBD Devastator torpedo bomber, designed initially to equip the USS *Ranger*, was the US Navy's first carrier-based monoplane to be produced in quantity. Deliveries for embarkation on *Saratoga* began in November 1937, and by the time of Pearl Harbor the type equipped a further five carriers. Its 800hp Pratt & Whitney R–1830 Twin Wasp gave the Devastator a top speed of only 208km/h (129mph), and the type was withdrawn after being badly mauled during the Battle of Midway in June 1942. The TBD was replaced from the spring of that year by the Grumman TBF Avenger, one of the most successful torpedo bombers of the war. The

ing in the Pacific. The Dauntless could carry 450kg (1,000lb) of bombs or half that weight of depth charges, and had a range of 1,770km (1,100 miles). Nearly 6,000 were built from 1939 to mid-1944. The Curtiss SB2C Helldiver joined the fleet in November 1943 and could carry 450kg (1,000lb) of bombs or a torpedo internally. Nearly 1,000 SB2C–1s were followed by SB2C–3s with an uprated (1,900hp) Wright Cyclone R–2600, and by SB2C–4s with underwing bomb and rocket attachments. The two-crew Helldiver had a top speed of 473km/h (294mph) and a maximum range of nearly 3,200km (2,000 miles). It carried two 20mm cannon and a machine gun in addition to its complement of bombs.

In December 1942 the Imperial Japanese Navy attacked Pearl Harbor with 353 warplanes from six aircraft carriers. The air armada comprised the 'unholy trinity' of D3A dive-bombers, B5N torpedo bombers and A6M fighter-bombers. The Aichi D3A was produced in two versions, the D3A1 from

the Nakajima B6N2 Tenzan (Heavenly Mountain), of which more than 1,200 were built. It carried a crew of three, was capable of 480km/h (300mph) and had a range of 1,450km (900 miles). Its weapon load was a single torpedo or six 100kg (220lb) bombs externally in addition to a pair of 0·303in machine guns. The D3A dive-bomber was replaced by the Yokosuka D4Y Suisei (Comet), which in its D4Y2 form was powered by a 1,400hp Atsuta 32 and could achieve 575km/h (360mph). More than 2,300 were built, including land-based and reconnaissance versions.

The outstanding Japanese aircraft of World War II was probably the Mitsubishi A6M Reisen, commonly known as the Zero (it was the Navy's Type O carrier-based fighter). The Japanese Navy issued its specification for a new fighter in October 1937, and pre-production Zeros were used in China from August 1940. Very few problems were encountered, and the A6M2 had replaced all A5Ms by the time of Pearl Harbor. This was a devastating fighter based on the Japanese Navy's 20 years of experience. It was fast – 520km/h (325mph) – and highly manoeuvrable, could reach 4,700m (15,000ft) in five minutes and was comparatively heavily armed (two 20mm cannon and two 0·303in machine guns). More than 10,000 were built in all, the major version being the A6M5. In this variant a Nakajima Sakae 21 gave a top speed of 560km/h (350mph), and a 0·50in machine gun replaced one of the 0·303in weapons.

Seven new aircraft carriers were ordered for the Fleet Air Arm in 1938, but at the outbreak of war the force was deficient in both ships and aircraft. Its two fighter types were the Blackburn Skua, which had to double as a dive-bomber, and the biplane Gloster Sea Gladiator. The Skua had been very advanced for its day when it was designed in 1934, being an all-metal monoplane with a retractable undercarriage, but it was a heavy two-seater and was underpowered. The Blackburn Roc, supposedly a fighter derivative of the Skua, was on the point of entering service, but if the Skua was disappointing, the Roc was disastrous. A four-gun power-operated turret replaced the Skua's four wing-mounted 0·303in Brownings and a rearward-facing Lewis. The Roc had a maximum speed of 354km/h (220mph), making it slower than the Sea Gladiator, and it took no less than 25 minutes to reach 4,700m (15,000ft).

In June 1940 the Fleet Air Arm's first eight-gun fighter, the Fairey Fulmar, joined the fleet. It was also underpowered, its 1,080hp Merlin giving it a top speed of only 448km/h (280mph), but it carried a lot of ammunition for its eight Brownings and had a four-hour endurance. It was replaced by the Fairey Firefly and Supermarine Seafire from 1943. The former was powered by a Rolls-Royce Griffon of up to 1,990hp, giving it a maximum speed of 509km/h (316mph), and it had a 1,600km (1,000-mile) range. Armament comprised four 20mm cannon plus bombs or rockets. More than 1,200 Seafires were built, the most important version being the Mk III, with folding wings, which entered service in 1943. The Royal Navy also made extensive use of US fighters and bombers.

An unusual application of maritime air power was the deployment of Hawker Sea Hurricane Mk IAs – known as 'Hurricats' – on merchant ships equipped with catapults. The fighter had to be ditched after each mission, but this tactic was successful in combating maritime-patrol aircraft such as the Focke-Wulf Condor. The Mk IB also served as a normal carrier-based fighter equipped with arrester gear.

The Fairey Swordfish soldiered on in a variety of roles, and in 1943 the force was joined by the Mk II, with a strengthened wing for firing rockets, and the Mk III, carrying ASV (Air-to-Surface-Vessel) Mk X radar. It was planned to replace the Swordfish with the Fairey Albacore, which entered service in March 1940 but flew only from shore bases for the first year of its career. Its first carrier action was from HMS *Formidable* at the Battle of Cape Matapan in March 1941, and this was also the first time that the type dropped torpedoes in anger. Eight hun-

dred Albacores were built, and they also served on convoy-escort duties in the Baltic and on anti-submarine operations in the Mediterranean and Indian Ocean. The type was in fact outlasted by the Swordfish.

The Royal Navy's first monoplane torpedo bomber was the Fairey Barracuda. It entered service in September 1942, supported the Salerno landings a year later and attacked the *Tirpitz* in April 1944. More than 2,500 were built.

Post-war carrier aircraft

The US Navy emerged from World War II as one of the most powerful fighting arms anywhere. Grumman supplied the F7F Tigercat and F8F Bearcat to follow its long line of piston-engined carrier fighters, but the jet age had dawned and Grumman's next effort was the straight-winged F9F Panther powered in its initial version by a licence-built Rolls-Royce jet engine of 2,270kg (5,000lb) thrust. The USN's first jet fighter had been the

McDonnell FH-1 Phantom, also straight-winged and with a pair of Westinghouse turbojets in the roots. It was succeeded by the F2H Banshee, virtually a scaled-up Phantom with more power and greater range. Both the Banshee and Panther carried four 20mm cannon, and both acquitted themselves well in the Korean War as fighter-bombers.

The Panther was replaced by the F9F-6 Cougar, in effect a swept-wing development of its predecessor but also with a new tail. The company's later F11F Tiger, designed as a day interceptor, was late in entering service and was overshadowed by the F-8 Crusader (see next column). Another type in action over Korea – alongside piston-engined machines such as the F4U-4 and -5 Corsair, and AD Skyraider – was the Douglas F3D Skyknight, a radar-equipped night fighter powered by a pair of belly-mounted Westinghouse turbojets. Douglas was additionally responsible for the F4D Skyray tailless interceptor/fighter and later produced the A-3 Sky-warrior and A-5 Vigilante nuclear bombers, since adapted for reconnaissance duties. Sandwiched between these was the A-4 Sky-hawk light bomber, in production from 1954 until 1979!

North American developed the F-86E Sabre into the FJ-2 Fury for the US Navy, and followed it with the FJ-3 and FJ-4/4B. McDonnell built the F3H Demon, but it was dogged by troubles and was eventually used

as a night fighter and missile-armed interceptor. Vought's F7U Cutlass tailless twin-engined intercepter was also less than successful, but the company's F-8 Crusader did more than make amends. It was the first carrier-borne aircraft capable of exceeding 1,600km/h (1,000mph) in level flight, having a top speed of up to Mach 2·0. The F-8 served from the late 1950s to the mid-1970s, carrying an armament of four 20mm cannon and Sidewinder missiles.

The McDonnell Douglas Phantom II, described in the attack section, was originally designed as an all-missile-armed naval fighter, and it gave sterling service until replaced by the Grumman F-14 Tomcat. One of the world's most advanced fighters, the variable-geometry Tomcat can attack six airborne targets simultaneously with Phoenix missiles, and it also carries Sparrows, Sidewinders and a 20mm gun. After 18 years the F-14 remains in production and still offers unrivalled capabilities. Its disastrous TF30

engine has at last been replaced by the GE F110 in the F-14A–Plus and F-14D, the latter also having upgraded avionics.

An even longer production run is being enjoyed by another Grumman, the A-6 Intruder, which equips all the US Navy medium attack squadrons. Powered by two J52 turbojets, each of 4,218kg (9,300lb) thrust, the A-6 has long-span wings with no engine-bleed blowing system as in the British Buccaneer but with full-span slats and high-life flaps, plus split airbrakes at the wingtips. Pilot and bombardier/navigator sit side-by-side (the pilot slightly higher and further forward), and up to 8,164kg (18,000lb) of bombs or other weapons can be hung under the wings and fuselage. In 1990 production switches to the A-6F, with F404 turbofan engines and totally new avionics, 30 years after the A-6's first flight!

Newest combat aircraft in the Navy and Marine Corps, the MacDonnell Douglas F/A-18 Hornet was developed from the Northrop YF-17 light fighter of 1974. Powered by two afterburning GE F404 engines, each rated at 7,258kg (16,000lb), the F/A-18 was finally developed as a dual-role fighter/attack aircraft (hence the designation) to replace the F-4 and A-7. An advanced cockpit with three large electronic displays (instead of dial instruments) enables the missions to be flown by one man. Though the F/A-18 came out slightly short on range

LEFT: *The Grumman F–14 Tomcat is the US Navy's most lethal carrier-based fighter.*
BELOW: *Intended as a cheap alternative to the F–14, the F–18 Hornet has actually turned out to be extremely expensive.*

BELOW CENTRE: *Dassault has developed the Super Etendard naval strike fighter from the earlier Etendard by redesigning the wing, installing a new engine and radar.*

BOTTOM: *The Dassault-Breguet Alizé anti-submarine aircraft operates from the French Navy's carriers Foch and Clemenceau, and from Indian Navy carriers.*

For shipboard stowage the rotors are quickly stopped, the engines pivoted horizontal, the blades folded along the leading edge and the entire wing rotated to lie fore-and-aft. Substantial sales are expected to the US Army and many foreign customers.

In the Lockheed S–3A Viking the US Navy for the first time has a carrier aircraft able to carry out the complicated task of seeking out and destroying submarines at great ranges and without aid from the ship or other aircraft. Another compact type is the Grumman E–2C Hawkeye early-warning aircraft, a flying radar station which extends the eyes and ears of the ship and its fighter defenders.

After World War II the Fleet Air Arm operated navalized versions of many RAF aircraft, such as the de Havilland Sea Mosquito torpedo bomber, Sea Hornet all-weather/night fighter, Sea Venom two-seat all-weather fighter and Hawker Sea Fury single-seat fighter-bomber. The Royal Navy's first purpose-built carrier-borne jet fighter was the Supermarine Attacker, armed with four 20mm cannon and powered by a Rolls-Royce Nene of 2,300kg (5,100lb) thrust. The same engine was installed in the similarly armed Hawker Sea Hawk, which was in turn replaced by the innovative Supermarine Scimitar, with its four 30mm cannon, area-ruling, powered controls and blown flaps. The Scimitar's powerplant of two Rolls-Royce Avon turbojets was the same as that in the twin-boom de Havilland Sea Vixen fighter, armed with cannon and later missiles. The Fleet Air Arm's last purpose-designed fixed-wing type was the Blackburn (now British Aerospace) Buccaneer, an outstanding low-level attack aircraft with tandem seats, great range and the ability to carry heavy bomb loads internally at speeds faster than an F–4 or F–15 with the same load. In Red Flag exercises in Nevada, Buccaneers have proved difficult to detect and almost impossible to intercept.

In 1967 the British government decided to eliminate Royal Navy fixed-wing airpower, with the lunatic suggestion that the Fleet could be protected by the homebased RAF! Fortunately in 1975 a small order was sanct-

and much more expensive than had been hoped, its ability to fire radar-guided Sparrow missiles was a factor in beating the F–16 in sales to Canada, Australia and Spain (none of which have conventional aircraft carriers). The F/A–18(R) and RF–18D are reconnaissance versions.

This is as good a place as any to describe the Bell/Boeing V–22 Osprey, the first tilt-rotor aircraft to go into production. Such a machine takes off vertically (or with a short run) with the engine power devoted to lifting the aircraft, as in a helicopter. In the hovering mode the fuel consumption is slightly higher than that of a helicopter but much lower than for jet-lift aircraft. In flight the

rotors are tilted through 90° forward until in cruising flight the V–22 is in all respects an aeroplane, able to cruise at 630km/h (391mph) with fuel consumption about one-third that of a helicopter. Bell flew a tilt-rotor machine in the mid-1950s but it has taken until 1988 to get a production type off the ground. The V–22 has two 11·58m (38ft) rotors driven by Allison T406 engines in the 6,000hp class pivoted to the ends of a 14·19m (46ft 7in) wing. The latter is mounted above a fuselage which can be adapted for the Marines' MV–22 assault transport, Navy HV–22 combat search/rescue aircraft, Navy SV–22 anti-submarine aircraft and Air Force CV–22 long-range special forces transport.

ioned for the Sea Harrier, developed from the RAF Harrier but with simple radar and able to fly fighter as well as attack missions. In May 1982 this tiny force operating from small decks in atrocious weather enabled a British task force to retake the invaded Falklands. Today further Sea Harriers have been ordered, one batch going to India, and British aircraft are being upgraded with new avionics and weapons. Their ability to operate without a carrier or a land airfield gives these small aircraft enormous importance in any real war.

France has stuck to conventional carriers and is building two costly nuclear-powered examples. Current aircraft are the Alizé turboprop anti-submarine aircraft, Super Etendard attack aircraft (able to launch the Exocet anti-ship missile) and Crusader fighter, which in 1995 is to be replaced by the Rafale M.

The Soviet navy has commissioned a series of carrier-type ships and in 1988 was completing a giant attack carrier. The latter may operate a version of the Su–27, while earlier ships carry Kamov helicopters and the Yak–38 ('Forger') jet STOVL attack fighter.

Maritime-patrol aircraft

There are three basic types of maritime-patrol aircraft: land-based machines employed to seek out and destroy enemy shipping and additionally carry out reconnaissance; flying boats and seaplanes which carry out similar tasks from coastal bases or seaplane carriers; and fleet spotters operated from capital ships. The first category is now easily the biggest, but seaplanes and flying boats played an important part in both World Wars.

Germany operated no flying boats, relying on seaplanes. The Friedrichshafen series was much used from coastal bases and seaplane carriers, nearly 500 being built. The Austro-Hungarian Navy's principal type was the single-engined Loehner L flying boat used against Allied shipping in the Adriatic and to attack Italian land-based targets. It carried a crew of two and 200kg (440lb) of bombs or depth charges. Franco-British Aviation produced a range of flying boats for the French, British, Italian and Russian navies. The role of ship-based seaplanes is described in the carrier aviation section.

The most successful of the Allies' big flying boats was the Felixstowe F.2A, derived from the Curtiss America series but having an improved hull design. Deliveries began in November 1917 and the type accounted for a number of submarines and airships during

its North Sea patrols. The F.2A was very manoeuvrable, and its armament of up to seven machine guns was enough to deter even single-seat fighters. An endurance of up to nine hours was possible, and two bombs of 100kg (220lb) each could be carried. The pair of 345hp Rolls-Royce Eagle VIIIs gave a top speed of just over 150km/h (95mph).

World War II

Britain turned out a range of big biplane flying boats including the Supermarine Stranraer, Saro London and Short Singapore. The Short Sunderland was a military development of the C-class Empire boats and entered service in the summer of 1938. More than 700 were built, and they became known as 'Flying Porcupines' because of their heavy defensive armament – a pair of 0.50in machine guns and up to a dozen of 0.30in. In later versions the four Bristol Pegasus engines were replaced by 1,200hp Pratt & Whitney R–1830 Twin Wasps. The Sunderland carried a crew of up to 13, had a maximum range of about 4,800km (3,000 miles) and a bombload of 900kg (2,000lb).

Another aircraft well liked by its crew was the Consolidated PBY Catalina, which established an enviable reputation for ruggedness, reliability and adaptability. More than 2,000 were built for the US Navy and a further 650-plus went to the RAF, the type serving in virtually every theatre of World War II; 749 of an amphibious version, the Canso, were built in Canada for the RCAF. Late models were powered by a pair of 1,200hp Pratt & Whitney Twin Wasps and could transport their crew of eight over 5,000km (3,100 miles). Defensive armament usually comprised two 0.50in and two 0.30in machine guns. In Germany the Blohm und Voss Bv 138 was employed for long-range reconnaissance, convoy patrol and U-boat co-operation. Powered by three 880hp Junkers Jumos, it had a range of more than 4,000km (2,500 miles) and could deliver six 50kg (110lb) bombs and four depth charges.

Italy had some 200 Cant Z.501 Gabbiano (Gull) parasol-winged single-engined flying boats in service at the outbreak of war, while the Soviet Union employed the monoplane flying boat Beriev Be–2 (MBR–2) on duties such as coastal reconnaissance. One of the most successful flying boats of World War II was the Kawanishi H8K2, of which 167 were built. It replaced the obsolescent H6K, which had a crew of nine and could carry a 1,600kg (3,500lb) payload over a range of 5,000km (3,100 miles). The H8K2 could carry 2,000kg (4,400lb) of weapons, plus five 20mm cannon and four 7.7mm machine guns

for self-defence, and had a maximum range of 6,400km (nearly 4,000 miles).

World War II seaplanes included the four-crew Heinkel He 115, of which more than 300 were constructed. Powered by a pair of 960hp BMW radials, it was used in such roles as U-boat co-operation over ranges of up to 2,000km (1,250 miles). The He 115's offensive warload comprised 1,000 kg (2,200lb) of bombs, mines or torpedoes. It also carried a pair of 7.9mm machine guns for protection. Italy contributed the Cant Z.506B Airone (Heron), which served throughout the war. It had a crew of five, and its three 750hp Alfa Romeo engines could power it over a range of 2,300km (1,430 miles). Two machine guns – one of 7.7mm calibre and the other of 12.7mm – were carried, and the Airone's warload was up to 1,200kg (2,640lb) of bombs or torpedoes.

Many types were used partly or exclusively for fleet spotting and general reconnaissance from heavy warships. The ungainly Supermarine Walrus entered service in July 1936 and equipped Royal Navy battleships and cruisers until succeeded by the same manufacturer's Sea Otter. The standard German Navy catapult seaplane in the war years was the twin-float Arado Ar 196, of which more than 400 were built. Up to four were carried by a battleship, one of the famous vessels to use the type in battle being the *Graf Spee*. The two-crew Ar 196, which had a range of 1,070km (665 miles), was also used for coastal patrol. It was fitted with two 20mm cannon and a pair of machine guns, and could carry a 50kg (110lb) bomb under each wing.

Japan built more than 1,400 Aichi E13A1s for ship-based reconnaissance and general duties, the type carrying out the pre-Pearl Harbor surveillance and serving throughout the Pacific. It was succeeded by the E16A1 Zuiun (Auspicious Cloud), used as a dive-bomber as well as for reconnaissance. Its armament consisted of two 20mm cannon and a pair of machine guns, including one of 13.2mm calibre. Battleships and cruisers of the US Navy carried the two-seat scout/observation Curtiss SOC Seagull. The SOC-1 version, of which 135 were built, entered

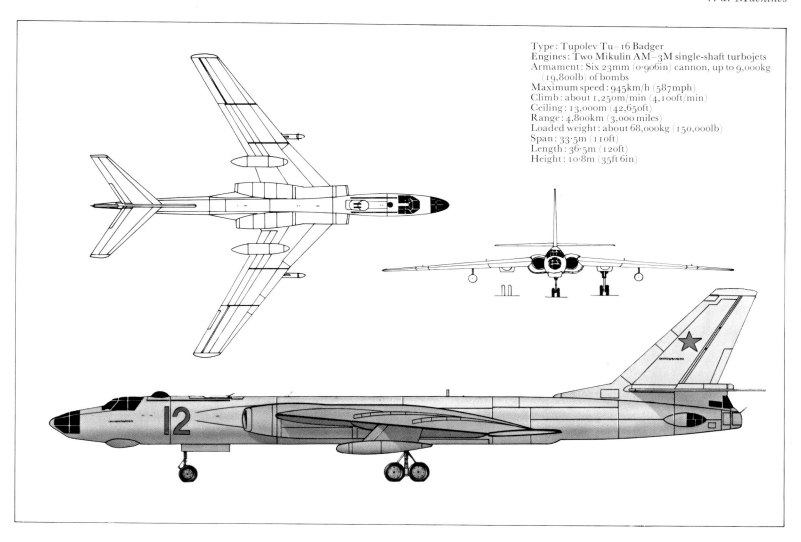

Type: Tupolev Tu–16 Badger
Engines: Two Mikulin AM–3M single-shaft turbojets
Armament: Six 23mm (0·906in) cannon, up to 9,000kg (19,800lb) of bombs
Maximum speed: 945km/h (587mph)
Climb: about 1,250m/min (4,100ft/min)
Ceiling: 13,000m (42,650ft)
Range: 4,800km (3,000 miles)
Loaded weight: about 68,000kg (150,000lb)
Span: 33·5m (110ft)
Length: 36·5m (120ft)
Height: 10·8m (35ft 6in)

service in November 1935 as the USN's last Curtiss biplane.

The Allies used many bombers, such as the Armstrong Whitworth Whitley, Boeing B–17, Bristol Beaufort, Consolidated B–24, Handley Page Halifax and Vickers Wellington, for maritime patrol in addition to their main role, but all the major combatants also developed specialist land-based maritime aircraft. The Avro Anson, although perhaps better known as a trainer, was originally designed for shore-based reconnaissance and served in this capacity from March 1936 until replaced by Lockheed Hudsons from 1939. The Anson's performance was fairly modest, range being 1,225km (760 miles) and its payload 163kg (360lb). Defensive armament consisted of two 0·303in machine guns. The Lockheed Hudson, also used by the USAAF as the A–28/A–29, was developed from the Model 14 airliner to meet a British requirement. First deliveries were made in the summer of 1939, and RAF Coastal Command alone operated more than 2,000. The Hudson, powered by two 1,200hp Pratt and Whitney Twin Wasps or Wright Cyclones, also served with many other air forces and saw action in every theatre of the war. It carried a 635kg (1,400lb) bombload and five crew over ranges up to 4,500km (2,800 miles), and was armed with as many as seven 0·30in machine guns.

Lockheed then developed the PV–1 Ventura, again from an airliner (the Model 18 Lodestar), to meet Britain's request for a Hudson replacement. It entered RAF service

in 1942, while 1,600 were operated by the US Navy as PV–1s. In Canada, Venturas supplemented Liberators on coastal patrols. When the type was adopted by the USAAF it was designated B–34. The PV–1, powered by two Pratt & Whitney R–2800 Double Wasps of 2,000hp each, carried a crew of four and had a range of 1,750km (1,100 miles). Defensive armament was six machine guns of 0·50in calibre, and the PV–1 could carry a 22in torpedo, bombs or depth charges up to a total of 1,600kg (3,500lb). The PV–2 Harpoon had an additional four machine guns, 225kg (500lb) of payload and 1,450km (900 miles) of range.

Perhaps the most famous maritime-patrol aircraft of the war was the Focke-Wulf Fw 200 Condor. Originally designed as a commercial transport, it was delivered from the spring of 1940 and began attacking shipping from French and Danish bases.

The Fw200C had seven or eight crew, five 7·9mm machine guns and a 20mm cannon for ship attack, and could carry up to 2,100kg (4,625lb) of bombs, although 1,500 kg (3,300lb) was more normal. The four BMW-Bramo engines produced 1,000hp each at sea level, and the Condor could cruise at 250km/h (160mph) for 3,500km (2,200 miles), giving an endurance of 14 hours. Maximum range was 6,400km (nearly 4,000 miles). The Junkers Ju 290 was proposed as a Condor successor, and it did operate on U-boat co-operation missions as well as launching anti-ship glide bombs against merchant ships.

Post-war developments

Only the Soviet Union, China and Japan now operate large maritime-patrol flying boats. The United States' involvement ended with the Martin P–5 Marlin developed from the wartime Mariner. More than 200 were built, a pair of Wright R–3350 Turbo-Compound engines of 3,400hp each giving a speed of 400km/h (250mph) and a range of 3,220km (2,000 miles). The ultimate basic variant, the P–5B, made its maiden flight in 1954 and ended its career as the SP–5B with advanced sensor equipment. Weapons and sonobuoys were carried in the engine nacelles, each of which could accommodate two torpedoes, or a pair of 907kg · (2,000lb) bombs; a combination of smaller weapons could be carried on underwing pylons, taking maximum load up to 3,630kg (8,000lb). Self-defence in the SP–5A was provided by a radar-directed twin 20mm cannon in the tail.

The Soviet Union produced the Beriev Be–6 flying boat in 1949. It was powered by a pair of 2,300hp Shvetsov radials and had a range of about 4,800km (3,000 miles). All armament – four depth charges, bombs, etc – was carried externally, and 23mm cannon were installed in the nose and in the dorsal position. This was followed in 1960 by the Be–12 amphibian powered by two 4,000hp Ivchenko AI–20D turboprops, giving a range of 4,000km (2,485 miles). Sensors include a nose-mounted radar and a MAD (magnetic anomaly detector) sting in the tail, and a range of weapons can be carried on six pylons as well as in the internal bay.

ABOVE: *The Breguet Atlantic is a European collaborative maritime patrol and anti-submarine type. It entered service in 1965.*

BELOW LEFT: *Britain's British Aerospace Nimrod is the world's only long-range, jet-powered anti-submarine and maritime-patrol aircraft.*

BELOW RIGHT: *The United States Navy's standard shore-based anti-submarine type is the Lockheed P–3 Orion, used by many countries.*

Japan's big flying boat is the Shin Meiwa PS–1, which is used for anti-submarine patrol. The design philosophy is unusual in that a dunking sonar is employed rather than the more common sonobuoys which are sown and then interrogated from the aircraft. The PS–1 has high-lift devices to give it STOL (Short Take-Off and Landing) performance – it can clear a 15m (50ft) obstacle after a take-off run of less than 300m (1,000ft), and can touch down as slowly as 76km/h (47mph) – and is designed to operate in rough seas. Power is provided by four General Electric T64 turboprops built under licence in Japan, and patrol endurance is up to 15 hours. A wide variety of weapons may be carried.

The newest and most powerful marine aircraft is the Chinese Harbin SH–5. This impressive amphibian was first flown in 1976, but did not enter service until September 1986. Powered by four 3,150–hp WJ–5A–1 turboprops, the SH–5 is used for every kind of maritime duty including ASW, search/rescue and transport. With a length stretched to 38.9m (127ft 8in) by the nose radar and tail MAD boom, it has a crew of eight and can carry a wealth of special equipment, underwing torpedoes or C–101 cruise missiles and many other stores. Maximum speed is 555km/h (345mph).

The vast majority of modern maritime-patrol missions, however, are carried out by aircraft operating from land bases or from aircraft carriers. Some carrier-based types, such as the Fairey Gannet and Breguet Alizé,

could also fly from shore bases, but long-range work is left to airliner-sized machines. Early post-war patrol aircraft included the Avro Shackleton and Lockheed P–2 Neptune. The Shackleton was based on the Lancaster/ Lincoln series of bombers, and in its MR.3 variant was powered by a pair of Viper turbojets in addition to four Rolls-Royce Griffon piston engines of 2,500hp each. The Shackleton, which entered service in 1951, had a maximum range of 5,800km (3,600 miles) and could carry 12 450kg (1,000lb) bombs or a variety of other ordnance such as nine depth charges, three torpedoes and 12 sonobuoys. The Neptune predated its British counterpart by some three or four years. Later versions, beginning with the P–2H, also used auxiliary turbojets – two Westinghouse J34s of 1,540kg (3,400lb) thrust each – in addition to a pair of 3,500hp Wright R-3350 Turbo-Compounds. Maximum range and speed were 3,540km (2,200 miles) and 565km/h (350mph) respectively, and more than 1,000 Neptunes were built. In Japan, Kawasaki produced the P–2J variant with licence-built 2,850hp General Electric T64 turboprops and IHIJ3 auxiliary turbojets of 1,400kg (3,085lb) thrust.

The Wright Turbo-Compound had further applications in the Lockheed C–121 series, based on the Constellation airliner, and in Canadair's CL–28 Argus. The C–121s were used for a wide variety of duties, but the Argus was developed specifically for maritime patrol. It incorporated parts of the Bristol Britannia airliner, with a new fuselage

and a change of powerplant. A search radar was installed in a chin-mounted bulge, and a MAD sting in the tail. Maximum range was just over 9,000km (5,600 miles), speed 505km/h (315mph) and weapon capacity 1,800kg (4,000lb) in two internal bays plus 3,450kg (7,600lb) on pylons. The Argus entered service in 1958 and was replaced by the Lockheed Aurora (P–3 development).

A smaller type which has been operated from shore bases as well as from carriers is the Grumman S–2 Tracker. Powered by a pair of 1,525hp Wright R–1820s, it had a normal patrol range of 1,450km (900 miles) and carried weapons such as four torpedoes or rocket projectiles. The same manufacturer's HU–16 Albatross, which also uses the same powerplant was a general-purpose type which included anti-ship operations among its duties. Maximum range was 5,280km (3,280 miles) when external fuel tanks were carried.

The current generation of specialist maritime-patrol types comprises the Lockheed P–3C Orion, British Aerospace Nimrod, 38. All except the Atlantic were adapted from airliners: the Electra, Comet and Il–18 respectively. The original P–3A entered service in 1962, and more than 400 of the Orion series have since been built. The type's maximum weapon load is a little over 9,000kg (20,000lb), of which more than a third can be carried internally. Power is provided by four Allison T56 turboprops of 4,900hp each, giving a patrol endurance of up to 18 hours. The Nimrod is the only jet-powered type in this class. Four Rolls-Royce Spey turbofans of 5,445kg (12,000lb) thrust each are installed to allow a high-speed dash to be made to the search area, but two of these can be shut down if necessary to conserve fuel on a long mission. At 925km/h (575mph), the jet-powered Nimrod's top speed is naturally appreciably higher than that of its competitors, and normal endurance is 12 hours. The Nimrod entered service in 1969.

The Atlantic, built by a European consortium, entered service in December 1965 and remained in production for ten years. A Mk 2 version has now been developed to take its place. The Mk 1, powered by a pair of 6,105hp Rolls-Royce Tyne turboprops, has an exceptionally long endurance of 18 hours at 320km/h (200mph); it normally cruises at 555km/h (345mph). The CSF radar is installed in a retractable 'dustbin' and the customary MAD boom is mounted in the tail. Weapons and sonobuoys are carried both internally and on underwing pylons. The Mk 2 became known as the Atlantique 2 and has completed updated avionics and an improved airframe. The first of 42 for the French navy was to fly in 1988.

A Soviet navy patrol type is the Il–38 (NATO 'May') in the class of the P–3 and likewise derived from a turboprop transport (the Il–18). A related machine is the Il–20 ('Coot–A') Elint and ocean reconnaissance machine. It is festooned with antennas, as are four special Elint/ECM versions of the An–12 'Cub', likewise powered by four 4,250hp AI–20 engines.

Among many other patrol types are versions of turboprop commuter liners, the Gulfstream business jets, the 737 jetliner and the piston- or turboprop-engined Defender family.

The field of maritime-patrol aircraft is a very stable one in that designers are not continually striving for more speed, payload or manoeuvrability—merely improved equipment in the same basic shell. The service life of each type therefore tends to be very long, with updating going on almost all the time. The present generation will thus remain operational for a good many years. Despite its lack of glamour, however, maritime patrol is a vital role and one which becomes ever more important as the technology of submarines and surface vessels advances.

Trainers

Many World War I front-line aircraft were allocated to training duties once they had been overtaken in performance by the next generation. In some cases one specific sub-type, such as the B.E.8a, was used exclusively for training. That war, however, also saw the introduction of aircraft designed for this role from the outset. One of the most famous of these was the Curtiss JN series, widely known as Jennies. More than 5,500 of the JN–4 variant alone were built, in Canada as well as the United States. Probably the best known trainer of this era was the Avro 504, although its early career was as a light bomber and ground-attack machine.

The first version intended specifically for instruction was the 504J introduced as late

ABOVE: *The ubiquitous Polikarpov Po–2.*
BELOW: *BAC's Jet Provost.*
BOTTOM LEFT: *A North American Harvard.*

ABOVE: *An Avro Anson trainer.*
BOTTOM RIGHT: *The USAF's aerobatic team, the Thunderbirds, fly the Northrop T–38 Talon.*

as the autumn of 1916; it became the Royal Flying Corps' first basic trainer. The later 504K was adapted to take a wide variety of engines and the type remained in production for more than ten years, in excess of 10,000 being built. Some were still in service at the outbreak of World War II.

The inter-war years saw the development of several basic trainers. In Russia the Polikarpov Po–2—which was to cause headaches for jet-fighter pilots in Korea more than 20 years later—made its maiden flight as the U–2 in 1928. Britain's similar de Havilland D.H.82 Tiger Moth followed soon after, entering service with the RAF in February 1932. In France there were several Morane-Saulnier types, chiefly the 230, from 1930 to 1939.

Monoplanes soon followed. North American developed the AT–6 Texan (known as the Harvard in service with air forces of the British Commonwealth) as a cheap trainer with the handling characteristics of a fighter. About 15,000 were built. The RAF's first

monoplane basic trainer was the Miles Magister. The Arado Ar96B was adopted as the *Luftwaffe's* standard primary and advanced trainer in 1940, and in the same year production began in America of the Fairchild PT–19/23/26 series. Built in Canada as the Cornell, the PT–26 replaced the Fleet Finch and de Havilland Tiger Moth as the standard elementary trainer of the British Commonwealth Air Training Plan.

The RAF's first twin-engined monoplane trainer was the Airspeed Oxford, which joined the Central Flying School in November 1937. Production exceeded 8,700, but this figure was easily beaten by the Avro Anson, with more than 11,000. The 'Faithful Annie' was used initially for maritime reconnaissance, but it was much more widely employed as a crew trainer. In Germany the twin-engined Siebel Si 204D replaced the Focke-Wulf Fw 58 in this role, carrying five students in addition to its normal crew. The type was built in Czechoslovakia and France to leave German production capacity avail-

BELOW: *Nearly 2,000 Lockheed C-130 Hercules have been delivered to 57 Nations. This RAF C-130K is dropping food packages from near-zero height.*
BELOW CENTRE: *A version of the McDonnell Douglas DC-10-30, the KC-10 Extender is a long-range air-refuelling tanker and transport of the US Air Force.*

BELOW: *Antonov's An-124 Ruslan is the biggest and heaviest aeroplane in the world, yet with 24 wheels it can operate from unpaved airstrips.*
BOTTOM: *British counterparts of the KC-10, Lockheed TriStars of the RAF were bought secondhand after airline service.*

able for front-line aircraft.

The United States forces' main transitional and advanced trainers were the Beechwood Beech AT–10 Wichita and the Cessna AT–7 and –11, the nimble Curtiss AT–9, the plywood Beech AT–10 Wichita and the Cessna AT–17 Bobcat. The RAF introduced the single-engined Miles Master as an advanced trainer in the spring of 1939, and more than 3,300 of them were turned out. Japan's main twin-engined advanced trainer was the Tachikawa Ki–54. Radio/navigation instruction was carried out with such types as the D.H. Dominie, a military version of the Dragon Rapide small biplane airliner; the Mitsubishi K3M high-wing monoplane which had been adapted from a Dutch design in 1931/32; and the single-engined Percival Proctor.

The Bristol Buckmaster, one of the fastest and most powerful trainers of its time, served the RAF for ten years from 1945. Its two 2,500hp Bristol Centaurus VIIs conferred a top speed of 566km/h (352mph). The first jet aircraft in this category also made its appearance – the Bell P–59 Airacomet. It was intended to be a fighter in its own right,

but mediocre performance led to it being employed for instruction of pilots destined to fly the Lockheed P–80 Shooting Star.

The Shooting Star itself, as the two-seat T–33, was to become one of the most successful post-war jet trainers. The piston-engined North American T–28, designed to seating rather than the tandem arrangement used previously. The world's first jet trainer to enter production, however, had been France's Fouga Magister.

In Britain the Percival Provost succeeded the Prentice, and in turn was replaced by the succeed the Texan, was itself replaced by the Cessna T–37. This was the first USAF jet *ab initio* trainer, and introduced side-by-side Jet Provost, loosely based on its piston-engined forebear. The Yakovlev Yak–11, with 700hp, had entered service in 1947 and was replaced by the Czech Aero L–29 Delfin, selected as the Warsaw Pact's standard basic jet trainer. This in turn has been followed by the L–39 Albatros.

The United States Navy adopted turbo-prop Beech T–34C Mentors to replace the earlier piston-engined variants in the T–34 series, while the Service's standard multi-

stage jet trainer is the Rockwell T–2 Buckeye. The same engines as in the Buckeye – General Electric J85 turbojets – also power the USAF's Northrop T–38 Talon advanced trainer, the first in the world capable of supersonic speeds. Sweden's SAAB 105G is a further application of the J85. Another much-used engine is the Rolls-Royce Viper, installed in the Italian Aermacchi MB.326/MB.339 series and other overseas trainers as well as in the Jet Provost.

Today's top trainers, the Hawk and Alpha Jet, have been described as attack aircraft. A special version of the former, the BAe McDonnell Douglas T–45 Goshawk, is replacing the Buckeye as the US Navy pilot trainer. The USAF failed to replace the T–37 with the new Fairchild T–46 and is updating the old machines, but almost all over the world companies are busy producing military trainers with jets, turboprops and piston engines. The RAF's replacement for the Jet Provost is the Shorts Tucano, based on a Brazilian turboprop aircraft, while a Swiss rival is the Pilatus PC–9, both having engines of around 1,150hp and tandem ejection seats with the rear seat higher to give a good view. Of course trainers are also needed for flight personnel other than pilots. The T–43, a version of the Boeing 737, is the USAF's navigator trainer, while the RAF and Royal Navy use radar-equipped versions of the BAe Jetstream.

Transport aircraft

Although many World War I bombers could double as transports, their accommodation was strictly limited and large-scale movement of troops and equipment by air was impossible. After the war specialized transport adaptations of heavy bombers appeared, but it was the airlines that paved the way by demanding more speed, accommodation and comfort. Modified bombers such as the Vickers Victoria and its successor, the Valentia, were easily outclassed by the new generation of all-metal monoplane airliners. These included the famous Douglas DC–2/DC–3 series, which later became the military C–47 Skytrain and C–53 Skytrooper. More than 10,000 were built as wartime transports, and the type was constructed under licence in Russia as the Lisunov Li–2, with 1,000hp Shvetsov radial engines replacing the pair of 1,200hp Pratt & Whitney R–1830 Twin Wasps. The C–47s and C–53s, which entered service in 1941, were used as ambulances and glider tugs in addition to carrying troops and equipment. They were known as R4Ds by the US Navy and by the RAF as Dakotas – a name that has stuck ever since.

During World War II, the United States was given prime responsibility for building transports, allowing Britain's aircraft industry to concentrate on combat types. Bombers and patrol aircraft as well as civilian machines were used as the basis for transports, examples being the Consolidated LB–30 and C–87 Liberator Express versions of the B–24 bomber and the Consolidated PB2Y–3R flying boat, able to carry 44 troops or 7,260kg (16,000lb) of cargo.

The RAF operated a number of home-designed types as well as those supplied by the US. These included the Handley Page Harrow; designed as a transport for 20 troops, then operated as a bomber, it finally reverted to its original role when replaced by Wellingtons. The latter type was itself one of a series of bombers used also as transports.

European developments

Germany produced one of the war's outstanding transports in the trimotor Junkers Ju 52/3m, but even this started life as a bomber in the Spanish Civil War. 'Iron Annie' or 'Auntie Ju' took part in many airborne invasions, carrying up to 18 paratroops, and about 4,800 were built. Its proposed successor, the Ju 352 Herkules, proved to be little or no improvement. The Blohm und Voss Bv 222 Wiking (Viking), designed as a transatlantic passenger flying-boat, could carry 110 fully equipped troops and had a range of 6,000km (3,730 miles). It was overshadowed, however, by the Me 323. Derived from the Me 321 Gigant (Giant) glider, the Me 323 was powered by six 990hp Gnome-Rhône radials and could carry 130 troops or 10,000kg (22,000lb) of equipment. Only 40 of the specialist Arado Ar 232 were built, but the type could carry 4,500kg (9,920lb).

Italy's Savoia-Marchetti S.M.81 Pipistrello (Bat) served as a bomber in Abyssinia and Spain but was later relegated to transport duties, and the S.M.83 Canguru (Kangaroo), able to carry 40 troops, was the country's largest wartime transport.

The Russian Air Force employed a variety of erstwhile bombers as transports. These included the four-engined Tupolev ANT–6, which as the G–2 could carry 30 paratroops; the Petlyakov Pe–8; and the PS–41 version of the Tupolev SB–2.

Two of the Japanese Army Air Force's major transports were the Kawasaki Ki–56 and Mitsubishi Ki–57. The former was copied from the US Lockheed 14, more than 700 replicas being followed by 100 plus of the Ki–56 built to Japanese standards, and some 500 of the 11-passenger Ki–57 were constructed. The Ki–57 resembled the company's Ki–21 bomber, some of which were also used as transports, and other dual-role types included the Ki–49, Ki–54 (up to nine passengers) and the Japanese Naval Air Force's G3M and G4M.

Post-war transports

The Avro York, using a new fuselage and tail attached to the Lancaster's wings and engines, gave sterling service during the Berlin Airlift, and the demands of the Korean War occupied a fleet of Boeing C–97s (134 troops) and Douglas C–124 Globemasters (200 troops, or bulky equipment which could not previously be carried by air). Smaller types with twin piston engines included the twin-boom Fairchild C–119 Flying Boxcar (Wright R–3350s, 62 troops) and Nord Noratlas (2,000hp Bristol Hercules), plus the

Fairchild C–123 Provider which was still going strong in the Vietnam War. Updating of the USAF's transport fleet resulted in the introduction of turboprop types – the 200-troop Douglas C–133 Cargomaster (four Pratt & Whitney T34s) and the Allison T56-powered Lockheed C–130 Hercules, able to carry 92 troops or 20,000kg (44,100lb) of equipment.

Strategic transports kept getting bigger. The Lockheed C–141B StarLifter, powered by four Pratt & Whitney TF33 turbofans of 9,525kg (21,000lb) thrust each, can swallow 154 troops or 31,750kg (70,000lb) of cargo and has a range of 6,440km (4,000 miles). The stretched C–141B was produced by modifying the present fleet. Another type powered by four TF33s is the Boeing C–135, the military version of the 707 airliner. The biggest military transport ever is the Lockheed C–5 Galaxy, which has two decks able to accommodate 120,200kg (265,000lb) of cargo made up of 270 fully equipped troops, two tanks, five armoured personnel carriers or a selection of awkward and bulky machinery. The Galaxy, powered by four General Electric TF39 turbofans of 19,500kg (43,000lb) thrust each, has a range of just over 6,000km (3,750 miles). The RAF's strategic freighter until 1976 was the Short Belfast, which could carry 270 troops or 35,380kg (78,000lb) of equipment. The Soviet Union relies on mammoth lifters such as the Antonov An–22, capable of transporting 45,000kg (99,200lb) of cargo on the power of four 15,000hp Kuznetsov turboprops and with a maximum range of 11,000 km (6,835 miles).

Tanker/cargo types

More recently the USAF has introduced the KC–10 Extender, a very capable tanker/transport version of the DC–10, while the RAF has bought secondhand TriStars and converted them as tanker/cargo aircraft. In the Soviet Union extensive use is made of both military and civil versions of the Il–76 ('Candid'), a 200-tonner with four turbofans and tailored to short rough airstrips. This aircraft also exists in tanker and AWACS versions. In 1982 the world's biggest and heaviest aircraft, an An–124 Ruslan, made its

ABOVE LEFT: The four-engined Tupolev ANT–6, designated TB–3 when used as a bomber and G–2 in the transport role, could carry 30 paratroops. It was obsolescent by the time Russia entered World War II.

first flight. Again tailored to unpaved runways, this 405-ton monster has four D–18T engines of 23,400kg (51,590lb) each, with reversers, and can carry a 150-ton load over long ranges.

On the smaller side are the Antonov An–12, powered by four 4,000hp Ivchenko turboprops and having a typical load of 100 paratroops; the collaborative European C.160 Transall with two Rolls-Royce Tyne turboprops of 6,100hp each, able to carry 93 troops and with a maximum range of 4,550km (2,830 miles); and the Japanese Kawasaki C–1, which uses a pair of Pratt & Whitney JT8D turbojets. Battlefield transports include the 44-troop Aeritalia G.222 and 41-passenger de Havilland of Canada DHC–5 Buffalo, both using a pair of General Electric T64 turboprops. The British equivalent was the Hawker Siddeley Andover, a military derivative of the HS.748 airliner powered by a pair of Rolls-Royce Dart turboprops.

While production of the evergreen C–130 continues unabated, with nearly 2,000 delivered to air forces all over the world, various designs have been prepared for powered-lift STOL successors. More conventional short-field machines include military versions of the BAe 146, with four turbofans, and the giant McDonnell Douglas C–17, in the size class of the DC–10 but able to take the biggest and heaviest items direct into front-line airstrips. Powered by four 16,783kg (37,000lb) F117 (PW2037) engines, the C–17 can extend giant flaps direct into the jets from the engines to get powered lift. First flight is due in 1990.

Civil and Maritime
Aviation

Airlines and Airliners

In the first decade of this century, to be precise on 16 November 1909, the world's first airline was founded. This was Deutsche Luftschiffahrts AG, generally known as Delag, with headquarters at Frankfurt-am-Main, Germany, and its purpose was to operate passenger flights with Zeppelin airships and to train Zeppelin crews. Count Ferdinand von Zeppelin had flown his first rigid airship, LZ 1, at Manzell on Bodensee (Lake Constance) on 2 July 1900, but it was not a complete success. But on 19 June 1910, his seventh airship, the 147·97m (485ft 6in) long LZ 7 *Deutschland* (*Germany*) made its first flight. It had been ordered by Delag and, although short-lived, was the first powered aircraft to carry passengers.

Delag had plans for a system of Zeppelin-operated air services, built a number of airship stations with sheds, and even published maps of the routes, but activities were limited to voyages from these stations and no scheduled services were operated. By the time

PREVIOUS PAGES : *A biplane class racer.*

World War I began Delag had operated seven passenger Zeppelins, made 1,588 flights covering 172,535km (107,211 miles) and carried 33,722 passengers and crew without injury.

It was on the other side of the Atlantic, in Florida, that the only pre-World War I scheduled air services were operated. On 4 December 1913, the St Petersburg–Tampa Airboat Line was established, and at 10·00hr on New Year's Day 1914 a single-engined 75hp Benoist biplane flying boat, piloted by Tony Jannus, left St Petersburg for the cross-

ing of Tampa Bay and alighted at Tampa 23 minutes later to complete the world's first scheduled airline service. The city of St Petersburg subsidized the service, and traffic demand made necessary the use of a second flying boat. Operations continued until the end of March, 1,024 passengers being carried.

The next important step in establishing air services was taken in Britain, when the Royal Air Force set up a Communication Wing to provide fast transport, mainly between London and Paris, for members of the Government and other officials attending the Peace Conference. Using mostly D.H.4s, 4As and, later, Handley Page O/400 twin-engined bombers, regular London–Paris services began on 10 January 1919. This was the start of cross-Channel air services, and it is claimed that the modified O/400 *Silver Star* carried the first non-military cross-Channel passengers and also made the first passenger night flight across the Channel. The RAF services continued until September 1919, 749 flights were made with 91 per cent regularity, and 934 passengers and 1,008 bags of mail were carried.

The start of commercial air transport

Aircraft Transport and Travel (AT and T), although founded in 1916, had to wait for the return of peace and for government permission before it could begin commercial operation. However, all was eventually ready, and a civil Customs airport was established at Hounslow, near the present London Airport at Heathrow, for the start of scheduled services on 25 August 1919. At 12.40 that day Major Cyril Patteson took off with four passengers in a de Havilland 16 on the first scheduled London–Paris service. On that same morning at 09.05 Lieut E. H. 'Bill' Lawford had left Hounslow for Paris in an AT and T D.H.4A with one passenger and some goods; Lawford's flight has often been reported as the inaugural service, but this is not true.

A second British airline, Handley Page Transport, had been formed on 14 June that year, and on 25 August one of the company's O/400s flew from London to Paris with journalists. Regular Paris services did not begin until 2 September, when Lt-Col W. Sholto Douglas (later Lord Douglas of Kirtleside) flew an O/7 from Cricklewood to Le Bourget.

Handley Page opened a London–Brussels service on 23 September 1919, and a London–Amsterdam service on 6 July 1920, although AT and T had already begun a London–Amsterdam service on 17 May in conjunction with the Dutch KLM.

A third British airline was that operated by the Aerial Department of the shipowners S. Instone and Co. This company had begun a private Cardiff–Hounslow–Paris service on 13 October 1919, for the carriage of staff and shipping documents, and on 18 February 1920, began a public London–Paris service when Capt F. L. Barnard flew a D.H.4A on the inaugural service. From 12 December 1921, operations were under the title The Instone Air Line.

By the end of 1920 the British airlines were in financial trouble. French companies had begun Paris–London services in September 1919, and the Belgian SNETA had begun flying between Brussels and London in June 1920. These British companies were now competing with each other and with sub-sidized continental companies for very limited traffic. At the end of October Handley Page ceased operating the Amsterdam route, in November gave up regular operation of the Paris route, and then on 17 December AT and T stopped all its operations.

On 28 February 1921, all British air services came to an end through lack of finance. As a result, small Government subsidies were granted to Handley Page and Instone and they resumed London–Paris services in the second half of March at the same £6 6s single fare as the French airlines. The situation was not helped when the Daimler Airway began London–Paris services on 2 April 1922, so that three British airlines and two French competitors were all fighting for the same traffic; in the year ending 31 March 1922, a total of 11,042 passengers had crossed the Channel by air, 5,692 flying in British aircraft and 4,258 in French.

From April 1922 government subsidies were granted to Daimler, Instone and Handley Page for the Paris route and to Instone for a London–Brussels route which was opened on 8 May. However, it soon became

obvious that the competitive operations could not continue, and in October a revised scheme was introduced and routes allocated. Daimler was allocated £55,000 for a Manchester–London–Amsterdam route with a connection to Berlin; Instone £25,000 for London–Brussels–Cologne; Handley Page £15,000 for London–Paris; and British Marine Air Navigation (not then formed) £10,000 for a Southampton–Channel Islands –Cherbourg flying boat service. The Southampton–Guernsey service opened on 25 September 1923, but no regular service was flown to Cherbourg.

Operations on these lines, with some changes, continued, and Handley Page opened a London–Paris–Basle–Zürich service in August 1923; but the financial situation was still unsatisfactory and the government decided to set up a national airline. This came into existence on 31 March 1924, as Imperial Airways with £1 million capital and guarantee of a subsidy of £1 million spread over ten years. Imperial Airways took over the fleets, staff and operations of the four earlier airlines and was due to begin operation on 1 April but unfortunately a pilots' strike prevented the working of any services until 26 April.

Most of the aircraft used by the pioneer British airlines were of two families: Airco and Handley Page. AT and T, having close associations with Airco, chose de Havilland designs. The RAF Communication Squadrons had used single-engined D.H.4 two-seat bombers, and some of these were converted to D.H.4As with two passenger seats over which there was a hinged cover with windows. AT and T had four of these aircraft and eight D.H.16s, which were four-passenger adaptations of the D.H.9 bomber design. Both types were of wooden construction, powered by 350/365hp Rolls-Royce Eagle water-cooled engines, and cruised at about 160km/h (100mph).

Following the D.H.4A and D.H.16, de Havilland designed the purely civil D.H.18, with 450hp Napier Lion engine and eight seats. This type appeared in 1920 and was used by AT and T and Instone. A much improved development was the Lion-powered eight-passenger D.H.34 which first flew on 26 March 1922, and went into service with Daimler Airway on 2 April! It also served with Instone and one was exported to Russia.

ABOVE LEFT: *One of the fleet of more than 100 Breguet 14s used by the Latécoère line to develop the air services between Toulouse and West Africa. Mail containers can be seen beneath the lower wings. The services of this line were gradually extended during the early 1920s, and carried both mail and passengers.*

TOP: *The famous Instone Vickers Vimy Commercial* City of London. *The Instone house flag appears on the nose. This aeroplane flew mainly on the London–Brussels–Cologne route.*

ABOVE: *One of the Daimler Airways' de Havilland D.H.34s, seen after passing to Imperial Airways in the spring of 1924.*

ABOVE RIGHT: *The twin-engined Farman Goliath. This type of aircraft was widely used by the early French airlines on many of their European routes and made a Paris–London return night flight in June 1922. They were also used on the first regular scheduled international passenger service, operated by Farman from Paris to Brussels. This was a weekly flight starting in March 1919.*

The RAF Communication Squadrons had also used the large Handley Page O/400 with two Eagle engines and a span of 30·48m (100ft). A few had passenger cabins in place of the bomb cells, while others only had the military equipment removed and seats installed. Handley Page Transport used O/400s with austere cabins, but developed a number of sub-types with improved accommodation. These included the O/7, O/10 and the O/11 freighter. In all of these aircraft two passengers could be carried in the open nose cockpit. The maximum weight was 5,470kg (12,050lb), cruising speed about 112km/h (70mph) and passenger accommodation was 14 in the O/7, 12 in the O/10 and 5 in the cargo O/11.

Based on the O/400 series, Handley Page

designed and built the W.8 with reduced span, better passenger accommodation and single fin and rudder in place of the O/400's box-like structure. The sole W.8, with 12 to 14 seats and Lion engines, first flew on 4 December 1919, saw limited service with Handley Page Transport and was followed in 1922 by three 12/14-seat Eagle-powered W.8bs.

On 30 April 1920, Instone took delivery of an aeroplane that was destined to become the best known of all in those pioneering days. It was the Vickers Vimy Commercial named *City of London*. This was basically the same as the modified Vimy bombers which in 1919 made the first nonstop transatlantic flight and the first England–Australia flight, but incorporated a new fuselage having seats

operations until the following year, starting a Paris–London service with Farman Goliaths on 29 March.

The network of French air routes rapidly expanded and services were begun between France and Corsica and France and North Africa. But the great French dream was an air service to South America, and Pierre Latécoère, the aircraft manufacturer, set about its achievement. The route through Spain to West Africa was opened in stages for the carriage of mail and operated as far as Casablanca by April 1920. Services were initially worked under the title Lignes Aériennes Latécoère, but in April 1921 the name was changed to Cie Générale d'Entreprises Aéronautiques, to be succeeded by Cie Générale Aéropostale in 1927. Operating

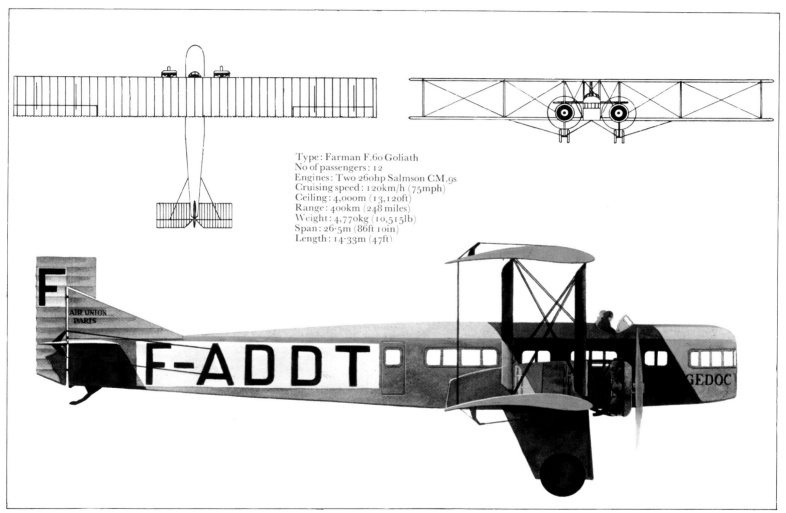

Type: Farman F.60 Goliath
No of passengers: 12
Engines: Two 260hp Salmson CM.9s
Cruising speed: 120km/h (75mph)
Ceiling: 4,000m (13,120ft)
Range: 400km (248 miles)
Weight: 4,770kg (10,515lb)
Span: 26·5m (86ft 10in)
Length: 14·33m (47ft)

for as many as ten fare-paying passengers.

British Marine Air Navigation, for its cross-Channel services, used three wooden Supermarine Sea Eagle amphibian flying boats. These had a Lion or Eagle engine with pusher propeller and a small six-seat cabin in the bows. Only the three Sea Eagles were built, and two passed to Imperial Airways along with the Vimy Commercial, the three H.P.W.8bs, seven D.H.34s and, un-airworthy, a D.H.4A and an O/10.

France quickly set about the development of air services after the war, formed a number of airlines and made plans for trunk routes to South America and the Far East. On 8 February 1919, a Farman Goliath flew from Paris to Kenley, near London, with 11 military passengers. It is often claimed that

this was the start of French cross-Channel services, but, although this is untrue, France did open the first regular scheduled international passenger service as early as 22 March 1919, when Farman began a weekly Paris–Brussels service.

Early in 1919 an association of French aircraft manufacturers established Compagnie des Messageries Aériennes (CMA), and this airline developed a system of air services beginning with a daily link between Paris and Lille from 1 May. Breguet 14 single-engined biplanes were used and it was this type with which CMA began Paris–London services on 16 September, later working in pool with Handley Page Transport.

Compagnie des Grands Express Aériens was also founded in 1919, but did not begin

under severe climatic conditions and crossing stretches of desert where a forced landing could mean death or torture for the crew at the hands of hostile tribesmen, the line was pushed forward through Agadir, Cap Juby, Villa Cisneros, Port Étienne and St Louis to Dakar – the full route to West Africa being opened in June 1925.

French airlines were also looking towards the Orient, and the first step was the founding, on 23 April 1920, of Cie Franco–Roumaine de Navigation Aérienne. This airline opened the first sector of its eastward route, from Paris to Strasbourg, on 20 September 1920. Prague was reached in that October and Warsaw in the following summer. In May 1922 the Paris–Strasbourg–Prague–Vienna–Budapest route was opened,

RIGHT: *A Luft-Verkehrs Gesellschaft C VI of the Deutsche Luft-Reederei airline at Gelsenkirchen. This was very probably on the occasion of the inauguration of an air service.*

BELOW: *German airlines also began to spring up very soon after the end of World War I. One of the smaller ones, Rumpler-Luftverkehr, first opened a service from Augsberg to Berlin via Munich, Nuremberg and Leipzig, in 1919. Ex-military Rumpler C1 biplane fighters were used on the route. In this photograph the passengers are seen just before climbing into the open rear cockpit.*

BOTTOM: *KLM (Royal Dutch Airlines) established its operations during the early 1920s with fleets of Fokker monoplanes. On the left is a Fokker F.II and in the background the F.III.*

BELOW RIGHT: *A Blériot Spad 33 of Cie Franco-Roumaine de Navigation Aérienne, seen at Le Bourget, Paris, in the early 1920s. There were several versions of this aircraft which could accommodate from four to six passengers and attain a speed of 170km/h (106mph).*

BOTTOM RIGHT: *Post Office DH–4s, US-built versions of the de Havilland two-seat bomber, at Omaha on the US transcontinental mail route.*

FAR RIGHT: *The first all-metal civil aircraft, the Junkers F 13, appeared in June 1919. It was one of the first specially designed transport aeroplanes. A low-wing cantilever monoplane, the aircraft was powered by a single 185hp B.M.W. engine and could carry four passengers. A number of airlines used the aircraft, including some float-plane versions, throughout the 1920s.*

Bucharest was reached in September and on 3 October the entire route was open to Constantinople (now Istanbul). This route across Europe involved flying in mountainous terrain and through some of the continent's worst weather, and therefore represented a great achievement. Up to 1923 all flights were made with single-engined aircraft and confined to daylight, but in that year three-engined Caudron C.61s were introduced and night flying was pioneered with the first such service between Strasbourg and Paris on 2 September 1923, and between Bucharest and Belgrade on 20 September. On the first day of 1925 the company's name was changed to CIDNA, and by the end of 1927 the line was operating no less than 76 aircraft.

Night flying was also pioneered on the Paris–London route in June 1922 when Grands Express made a return night flight with a Goliath, but it was not until April 1929 that regular night services were flown over the route – by Air Union, which in 1923 had been created by the merger of CMA and Grands Express.

The numerous French pioneer airlines employed a very wide range of aircraft. Wartime Salmson 2–A.2 and Breguet 14 single-engined reconnaissance and bomber biplanes were used in large numbers, Latécoère employing well over 100 of the latter on the route to West Africa.

Franco-Roumaine used Salmsons and some single-engined Potez biplanes before acquiring a main fleet of Blériot Spad cabin biplanes. The first of these was the Spad 33 of 1920, and it was developed into a whole family of similar aircraft with seats for four to six passengers and a variety of engines including, in later models, the air-cooled Jupiter. More than 100 of these attractive little biplanes were built, and some could cruise at 170km/h (106mph).

Widely used by the Farman Line, CMA, Grands Express and some non-French airlines, was the Farman Goliath. This was designed as a twin-engined bomber but converted to have two passenger cabins with seats for up to 12 passengers. The wing span was 26·5m (86ft 10in), most had two 260hp Salmson water-cooled radial engines, and cruising speed was about 120km/h (75mph).

Germany also made an early start in establishing regular services. Numerous com-

panies began operating, but the two most important were Deutsche Luft-Reederei (DLR) and Deutscher Aero Lloyd. DLR began Europe's first sustained regular civil daily passenger services when it opened the Berlin–Weimar route on 5 February 1919. The operation grew rapidly and employed a large fleet of ex-military single-engined biplanes, mostly the L.V.G. C VI with room for two passengers in the open rear cockpit, but there were also numbers of A.E.G. J IIs and some twin-engined A.E.G. G Vs and Friedrichshafen G IIIas.

Junkers began flying a Dessau–Weimar service in March 1919 with a modified J 10 two-seat attack aircraft. This was a low-wing cantilever monoplane of all-metal construction with corrugated metal skin, and is almost certainly the first all-metal aeroplane to have operated an air service. In 1921 Junkers set up Junkers-Luftverkehr to operate air services and promote its F 13 cabin monoplane. The F 13 was the first all-metal aeroplane designed and built as a transport, having an enclosed cabin for four passengers, and made its first flight on 25 June 1919. More than 300 were built, in many versions.

In January 1926 Deutsche Luft Hansa was created by the merging of Aero Lloyd and Junkers-Luftverkehr and operations began on 6 April. Thereafter Luft Hansa (written as one word from the beginning of 1934) was the German national airline, but numbers of small airlines continued to exist.

Although short-lived, there was one German airline operation which must be mentioned. This was the postwar revival of Delag, with ambitious plans for airship services within Germany and on international routes, and two new Zeppelins, the LZ 120 *Bodensee* (*Lake Constance*) and LZ 121 *Nordstern* (*North Star*) were built. The *Bodensee* flew on 20 August 1919, and with accommodation for 21 to 27 passengers began working a Friedrichshafen–Berlin service on 24 August, flying in opposite directions on alternate days. *Bodensee* flew until December, made 103 flights (including one from Berlin to Stockholm) and carried 2,253 revenue passengers, but the Inter-Allied Control Commission would not allow the service to restart in 1920.

Numerous other European countries began air transport operations. In Belgium SNETA did pioneering work leading to the formation

RIGHT: *The Fokker F.XII* Leeuwerik (Lark) *was the first of the type to fly over the Amsterdam–Batavia route operated by KLM.*
BOTTOM: *Imperial Airways' Handley Page H.P.42* Hannibal *at Baghdad. A biplane with wings mounted above the fuselage, the H.P.42 used four 475hp radial engines and could carry up to 38 passengers in a luxurious interior.*
FAR RIGHT: *Imperial Airways' Armstrong Whitworth Argosy* City of Glasgow *seen over London.*

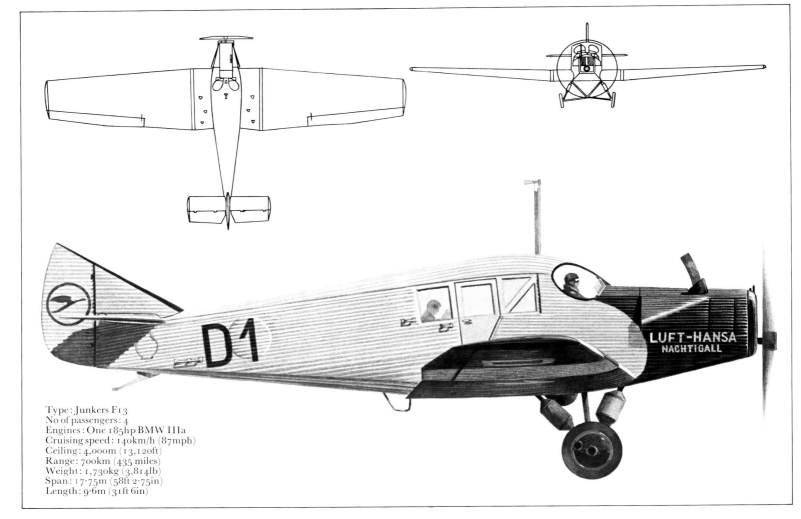

Type: Junkers F13
No of passengers: 4
Engines: One 185hp BMW IIIa
Cruising speed: 140km/h (87mph)
Ceiling: 4,000m (13,120ft)
Range: 700km (435 miles)
Weight: 1,730kg (3,814lb)
Span: 17·75m (58ft 2·75in)
Length: 9·6m (31ft 6in)

of Sabena in 1923. KLM Royal Dutch Airlines was founded in October 1919. Although it did not initially operate its own services, KLM was to play a major part in developing world air transport and still operates under its original title. DDL – Danish Air Lines – was another very early European airline, beginning a Copenhagen–Warnemünde seaplane service on 7 August 1920. Today DDL is a constituent of SAS – Scandinavian Airlines System.

United States concentrates on the mail

The first stage in opening up US nationwide air mail services was the start on 15 May 1918, of a mail service linking Washington, Philadelphia and New York. Curtiss JN–4 biplanes were used, flown by Army pilots.

The Post Office took over the US Aerial Mail Service on 12 August 1918. Specially built Standard JR–1B mailplanes were introduced and by the end of the year the service had achieved an average 91 per cent regularity. The Washington–New York mail service was closed down at the end of May 1921, by which time the entire transcontinental mail service was in being.

Late in 1918 the Post Office acquired a large number of war-surplus aircraft, including more than 100 US-built DH–4Bs with 400hp Liberty engines. This fleet made it possible to start establishing the coast-to-coast service and on 15 May 1919, the Chicago–Cleveland sector was opened, saving 16hr on the Chicago–New York journey. On 1 July the New York–Cleveland sector was opened, with through New York–Chicago flights from September, and the San Francisco–Sacramento section was opened on 31 July. On 15 May 1920, the mail route was opened from Chicago to Omaha via Iowa City and Des Moines, and the full route came into operation on 8 September with the opening of the Sacramento–Salt Lake City and Salt Lake City–Omaha stages. Branch lines were also opened between Chicago and St Louis and Chicago and the twin cities Minneapolis/St Paul.

In order to save time it was decided to make experimental flights in each direction, with night flying over some stages. On 22 February 1921, two aircraft took off from each end of the route and so began a saga that is now part of American history. One of the eastbound aircraft crashed in Nevada, killing the pilot, and only one of the westbound aircraft managed to reach Chicago because of extremely bad weather, and at that point the flight was abandoned. The surviving eastbound mail was taken over at Salt Lake City by Frank Yeager, who flew through the night to Cheyenne and North Platte, where he handed over to Jack Knight for the flight to Omaha. Because of the termination of the westbound flight there was no aircraft at Omaha, and Knight secured a place in history by flying on to Chicago. Ernest Allison flew the last stage to New York, and the total coast-to-coast time was 33hr 20min.

In 1922 a start was made on lighting the mail route. The aerodromes were equipped with beacons, boundary and obstruction lights and landing-area floodlights. At the

regular stops there were revolving beacons of 500,000 candle-power, and at emergency landing grounds 50,000 candle-power beacons. Flashing gas beacons were installed every three miles along the route, and by the end of 1925 the entire 3,860km (2,400 miles) of route had been lit at a cost of some $550,000.

During 1926/27 the Post Office Department turned over its air mail services to

private contractors. When the Post Office's own operations ceased it had flown more than 22 million km (13¾ million miles) with 93 per cent regularity and carried more than 300 million letters; but there had been 4,437 forced landings due to weather and 2,095 because of mechanical trouble. Worse still, there had been 200 crashes with 32 pilots killed, 11 other fatalities and 37 seriously injured.

Pioneering the trunk routes

In spite of sparse ground facilities, harsh climatic conditions, and low-performance aircraft, Britain, Belgium, France and the Netherlands were all eager to establish air communication with their overseas territories, and this was to lead to the opening of trunk routes which ultimately grew into our present globe-encircling system uniting what were remote corners of the world.

The first attempt to operate commercial air services in the tropics was that by the Belgian SNETA in the Congo (now Zaïre) when it opened the first stage of the Ligne Aérienne Roi Albert (King Albert Air Line) between Kinshasa and N'Gombé on 1 July 1920, using a three-seat Lévy-Lepen flying boat with a 300hp Renault engine. The N'Gombé–Lisala sector was inaugurated on 3 March 1921, and by that July the entire Congo River route was open between Kinshasa and Stanleyville (now Kisangani).

In Britain, at the end of October 1919, the Advisory Committee on Civil Aviation had recommended the establishment of trunk air

routes linking the Commonwealth with the United Kingdom. They stated that 'the proper place for initial action' was the route to India and ultimately thence to Australia, 'to be followed by a service to South Africa...' and in October 1926 there was an agreement between Air Ministry and Imperial Airways for operation of an Egypt–India service. But things did not quite work out like that.

In March 1921 there was a Cairo Conference to examine administration and control of mandated territory, and Sir Hugh Trenchard, Chief of the Air Staff, proposed policing Mesopotamia (now Iraq) with air forces instead of the orthodox ground forces. The proposal was adopted and this in turn led to the establishment of the Desert, or Baghdad, Air Mail to speed communication between the United Kingdom and Baghdad. In June a survey was made of the route, landing grounds were marked out in the desert at intervals of 25 to 50km (15 to 30 miles) and a furrow was ploughed over some sectors as a navigational aid to the pilots. The route between Cairo and Baghdad was opened by the RAF on 23 June 1921, and maintained until commercial operations began in January 1927.

Ground and air surveys were made of the route between Cairo and India, and a fleet of special aircraft was ordered for Imperial Airways so that commercial service could replace the Desert Air Mail and implement the plan for a service to India. The route from Cairo involved long desert stages, several mountain crossings and the use of comparatively high-elevation aerodromes in extreme temperatures. The limited range and low speed of the aircraft also meant that aerodromes had to be provided at frequent intervals. Perhaps the most serious obstacle was political, for it was the task of obtaining permission to overfly and land in various countries that was to delay the opening of services, and for many years prevent through flights from the United Kingdom.

It should be realized that the early aircraft found the Alps a formidable barrier; they had to fly round them over France or the Balkans, and this still applied to many transport aircraft into the 1950s. France was in dispute with Italy, and would not allow British aircraft to fly via France into Italy and this, apart from a few experiments, enforced a railway sector in British trunk air routes until late in the 1930s. Persia did not wish to allow air services through her territory, and so the first British trunk air route was confined to the Cairo–Baghdad–Basra sectors.

Five de Havilland 66 Hercules biplanes were specially built for the route. These had three 420hp Bristol Jupiter air-cooled engines, accommodation for eight passengers, cruised at 176km/h (110mph) and had a range of about 640km (400 miles).

On 7 January 1927, one of the Hercules left Basra to open the fortnightly service to Cairo via Baghdad and Gaza. It arrived in Cairo on 9 January and three days later the first eastbound service left Cairo.

Eventually Persian permission was granted for a coastal route, and this enabled Imperial Airways to open its long awaited England–India service. The Armstrong Whitworth Argosy City of Glasgow took off from Croydon on 30 March 1929, and carried its passengers and mail to Basle. From there they travelled by train to Genoa, then in a Short Calcutta flying boat to Alexandria and on to Karachi in a Hercules. The Egypt–India sector involved eight scheduled intermediate stops, the total journey took seven days and the single fare was £130.

The service was extended to Jodhpur and Delhi in December 1929, and in association with Indian Trans-Continental Airways to Calcutta in July 1923, to Rangoon that October and to Singapore by the end of the year. Because of political and operational difficulties the route was transferred from the Persian to the Arabian shore of the Gulf in October 1932.

The setting up of a route through Africa was much more difficult. The continent rises slowly from the coast and then steeply from southern Sudan to over 1,525m (5,000ft) at Nairobi and about 1,830m (6,000ft) at Johannesburg. Temperatures can be extremely high, and tropical storms of great intensity led to flooding of aerodromes as well as severe turbulence in flight. Aircraft performance falls off sharply in high temperatures and at high elevations. Ants were also apt to build quite solid obstructions on landing areas. In the 1920s and early 1930s Africa posed a formidable challenge to the operation of air services. However, all the problems were tackled, aerodromes constructed, rest houses built and fuel supplies laid down, and on 28 February 1931, the inaugural service to Central Africa left Croydon, although at that stage passengers were only carried to Khartoum.

The Central Africa service which left Croydon on 9 December 1931, was extended experimentally to Cape Town for the carriage of Christmas mail, a Hercules being used south of Nairobi. It was this type which operated the Nairobi–Cape Town sector when the entire route was brought into opera-

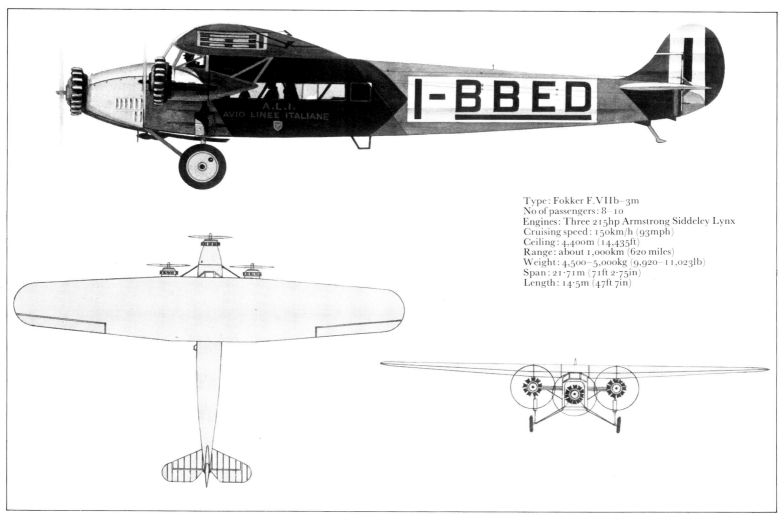

Type: Fokker F.VIIb–3m
No of passengers: 8–10
Engines: Three 215hp Armstrong Siddeley Lynx
Cruising speed: 150km/h (93mph)
Ceiling: 4,400m (14,435ft)
Range: about 1,000km (620 miles)
Weight: 4,500–5,000kg (9,920–11,023lb)
Span: 21·71m (71ft 2·75in)
Length: 14·5m (47ft 7in)

tion in the following January.

In 1925, while employed with the French company CIDNA, Maurice Noguès made a Paris–Teheran survey flight with a Blériot Spad 56, and in 1927 CIDNA began negotiations for the extension of its services to Beirut, Damascus, Aleppo and Baghdad. Noguès later transferred to Air Union's Ligne d'Orient which was founded in 1927, that organization joined with Air Asie to form Air Orient, a regular Marseilles–Beirut mail service was started in June 1929 and the route was extended to Baghdad at the end of the year. On 17 January 1931, the entire line to Saïgon was opened with a tenday schedule, although at first passengers were carried only as far as Baghdad. In 1938 the route was extended to Hong Kong, and in 1939 260km/h (161mph) three-engined Dewoitine D.338 monoplanes were working the entire route, by then operated by Air France.

On 1 October 1931, KLM inaugurated a regular Amsterdam–Batavia (now Jakarta) passenger service using Fokker F.XIIs. These aircraft had luxury-type seats for 4, although in Europe they could carry 16 passengers. The scheduled journey time was ten days and the flying time 81hr for what was almost certainly at that time the world's longest air route. In 1932 Fokker built five slightly larger F.XVIIIs to replace the F.XIIs, and the new aircraft had sleeper seats. In 1934 a KLM Douglas DC–2 took part in the England–Australia race and, carrying three passengers, covered the 19,795km (12,300 miles) to Melbourne in just over 90hr. In June 1935 DC–2s were introduced on the Far East service with a six-day schedule.

Consolidation in Europe

In Europe the pioneer years of air transport continued through the 1920s and, according to definition, into the 1930s. For most of this time many routes were operated only in the summer, and night services were only gradually introduced. There was steady development of aircraft, particularly in Germany and the Netherlands, and the aircooled engine largely replaced the watercooled engine, with its attendant plumbing problems and heavy radiators. Radio came into increasing use, airport lighting was developed, and Germany, in particular, gave considerable attention to developing ways of navigating and landing in bad weather.

The British national airline operated a very small European route system. Nevertheless, Imperial Airways did improve its fleet, at first with developments of the early Handley Pages but from 1926 with a small number of three-engined Armstrong Whitworth Argosy 18/20-seat biplanes. Unlike their British predecessors, these had metal structures although they still employed fabric covering.

A much greater improvement in standards came in 1931 when Imperial Airways began using a fleet of Handley Page H.P.42 Hannibal and H.P.45 Heracles biplanes. These were large four-engined aircraft with Jupiter engines, 38 seats on European services and 16 to 24 on the Egypt–India and Egypt–Central Africa routes. On Heracles two

stewards were carried, full meals were served, and during peacetime this fleet operated with perfect safety. The only problem was their low speed of around 160km/h (100mph).

The Handley Pages remained in operation until World War II although some smaller and faster aircraft had been added to the fleet by this time, including the beautiful de Havilland Albatross monoplanes, known as the Frobisher class by Imperial Airways, which cruised at 335km/h (210 mph).

In 1937 a second British operator, British Airways, began using fast Lockheed Electras on European services and later added Lockheed 14s. They flew to Germany, Poland and Scandinavia as well as Paris and, in terms of speed, Imperial Airways was shown in a bad light. It was finally agreed to amalgamate the two airlines and BOAC – British Overseas Airways Corporation – resulted.

France continued expansion of its air routes and used a wide variety of indigenous aircraft, mostly biplanes, including the

OPPOSITE PAGE
LEFT: *An Air France Lioré et Olivier 213. It was painted red and gold and bore the fleet name* Rayon d'Or *(Golden Ray).*
CENTRE: *Air France's Dewoitine D.338* Ville d'Orleans *(City of Orleans) at Le Bourget Airport, Paris, embarking passengers for Lyons, Marseilles and Cannes. In the background the Bloch 220* Provence *is positioned to operate a London service. The photograph was taken in 1938 or 1939, and the large letter F on the roof of the Dewoitine was for recognition while flying through the war zone in the Far East.*
RIGHT: *A Savoia Marchetti S.73 of Sabena, built under licence by SABCA in Belgium.*
THIS PAGE
ABOVE LEFT: *The Ad Astra Aero base at Zürich-horn in the early 1920s, with a Macchi M.3 flying boat on the slipway.*
ABOVE RIGHT: *Swissair's Fokker F.VIIa CH 157 on the ice in front of the Grand Hotel, St Moritz. This aircraft had originally been one of the Balair fleet. This design was the first notable civil aircraft from the Royal Netherlands Aircraft Factory after its very successful models of WWI.*

BELOW: *Nordmark, one of Lufthansa's Focke-Wulf FW.200 Condors, seen at Tempelhof Airport, Berlin, shortly before World War II. It was a four-engined monoplane capable of a speed of about 368km/h (230mph).*

single-engined Breguet 280T and twin-engined Lioré et Olivier 21. But CIDNA had introduced the three-engined low-wing Wibault 280T monoplane on its Paris–Istanbul route and Air Union had at least two on the London–Paris route before Air France was formed in 1933 as the successor to these airlines and Aéropostale, Air Orient and the Farman Line.

The new organization introduced a further wide range of types including the twin-engined Potez 62 and Bloch 220 and the three-engined Dewoitine series.

Sabena expanded its routes with Fokker F.VIIb–3ms, and later introduced three-engined Savoia Marchetti monoplanes. KLM used a whole series of Fokker mono-planes for its European passenger and cargo network, including the four-engined F.XXII and F.XXXVI which appeared in 1935 and 1934 respectively. But the Fokkers, with their wooden wings and welded steel-frame fuse-lages and, except in one case, non-retractable undercarriages, were outmoded by the Douglas DC–2 and DC–3, and KLM became the first European operator of these advanced aeroplanes.

It was Germany, in the form of Lufthansa, which dominated the air transport map of Europe. When the airline was founded it took over 162 aircraft of 19 different types. The largest batches of standard, or reasonably standard, aeroplanes were 46 Junkers-F 13s, 19 Fokker-Grulich F.IIs and 13 F.IIIs. There were also numbers of Dornier Komets, and it was one of these that operated the

airline's first service. Three-engined Junkers-G 24s were used for the first night service on 1 May 1926, and three-engined Rohrbach Rolands pioneered trans-Alpine services from 1928.

Lufthansa expanded rapidly and in its third year had a European network of more than 33,000km (20,500 miles) over which it flew 10 million km (6,214,000 miles) and carried 85,833 passengers and 1,300 tons of cargo, mail and baggage. The airline had the biggest domestic network of services in Europe, and also served most of the continent's main cities.

In 1932 Lufthansa began operation of that very famous aeroplane, the Junkers-Ju 52/3m, This was a three-engined low-wing mono-plane with accommodation for 17 passengers, was of all-metal construction with corrugated skin and, although not particularly fast, was extremely reliable. Lufthansa was to use about 230 with a maximum of 78 at any one time. The Ju 52/3m was widely used as a civil and military transport; nearly 5,000 were built, and most prewar European airlines used them at some time – two on floats working in Norway until 1956.

Lufthansa also introduced a number of limited-capacity high-speed aircraft including the four-passenger single-engined Heinkel He 70, commissioned on *Blitz* (*Lightning* or *Express*) services in Germany in June 1934, and the twin-engined ten-passenger He 111 which had a smoking cabin and began service in 1936. The four-engined 40-passenger Junkers-Ju 90 and 25/26-passenger Focke-

Wulf Fw 200 Condor were prevented by war from showing their full potential.

Almost every European country operated its own air services. Austria's Ölag was founded in 1923 and flew Junkers aircraft. Poland began very early, using mostly French and German aircraft; LOT was founded in 1929 and flew Junkers and Fokker types and a few Polish designs, and later adopted the Lockheed Electra and Model 14. Czechoslovakia's CSA and CLS used Czecho-slovak aircraft and Fokkers before CLS adopted the DC–2 and CSA the Savoia Marchetti S.73. Hungary used Fokkers and Junkers; Greece used Junkers types; Rumania used a mixture of British, Czechoslovak, French and German aircraft; and Yugo-slavia used mostly British and French aircraft. Most of these airlines operated domestic services and services to other European destinations, although LOT had a route to Palestine.

Spain began mail services in October 1921, when CETA began operating D.H.9s between Seville and Spanish Morocco. In 1925 UAE was formed and, using Junkers aircraft, opened services linking Madrid with Seville and Lisbon. A Madrid–Barcelona route was opened in 1927 by the original Iberia using Rohrbach Rolands, but in March 1929 the three pioneer Spanish companies were amalgamated to form CLASSA, in turn superseded in 1932 by LAPE, which operated domestic and international services with Fokker F.VIIb–3ms, Ford Tri-motors and, later, Douglas DC–2s.

A number of small Swiss companies were founded and at the end of 1919 one of these was renamed Ad Astra Aero. Initially it operated a number of small flying boats from the Swiss lakes, but did not begin regular scheduled operations until June 1922 when it began Geneva–Zürich–Nuremberg services with Junkers F 13s. This airline expanded its operations, mostly with Junkers aircraft, and, in 1931, amalgamated with Balair to form Swissair. Balair had begun services in 1926 and was equipped with Fokkers. Since its inception Swissair has been one of Europe's most technically progressive airlines, has introduced a range of advanced aircraft and been among the first to adopt such types as the DC–2 and DC–3. It was Swissair's introduction of the Lockheed Orion monoplane with retractable undercarriage in May 1932 that led Germany to produce the high-performance Heinkel He 70 and Junkers Ju 60 and Ju 160.

Scandinavian air transport differed from that in other parts of Europe because of terrain and the difficulty of providing aerodromes. DDL, formed in Denmark in 1918, began operation with a seaplane, but otherwise was able to develop its routes with landplanes but Sweden and Norway had to rely mainly on seaplanes.

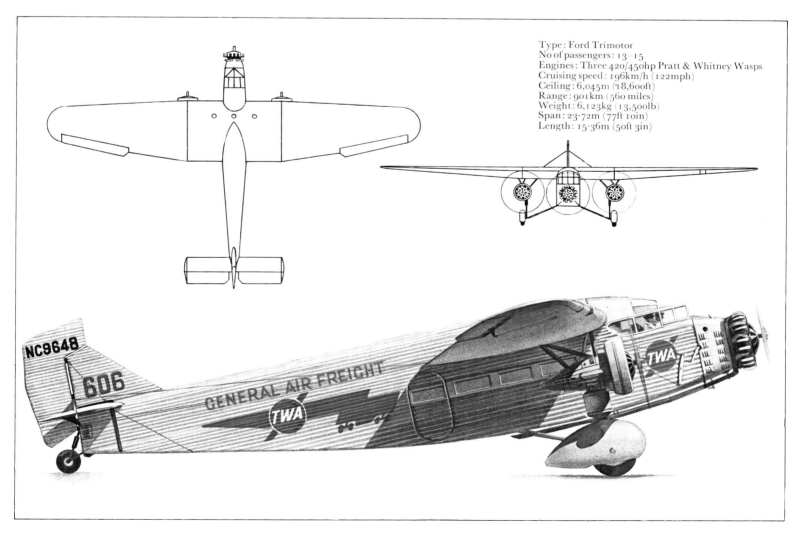

Type: Ford Trimotor
No of passengers: 13–15
Engines: Three 420/450hp Pratt & Whitney Wasps
Cruising speed: 196km/h (122mph)
Ceiling: 6,045m (18,600ft)
Range: 901km (560 miles)
Weight: 6,123kg (13,500lb)
Span: 23·72m (77ft 10in)
Length: 15·36m (50ft 3in)

BELOW LEFT: *One of the high-speed Lockheed Orions introduced by Swissair in May 1932 on the route from Zurich to Vienna via Munich. A single-engined low-wing monoplane, it could carry up to six passengers at a maximum speed of about 360km/h (225mph).*
BELOW CENTRE: *DDL's (Danish Air Lines) first Junkers Ju.52/3M, Selandia, at Kastrup Airport, Copenhagen. This aircraft first flew in 1932 and remained in production throughout the 1930s and World War II. An all-metal, low-wing monoplane, it used three B.M.W. radial engines to attain a top speed of 288km/h (180mph).*

BELOW RIGHT: *One of Aero O/Y's Junkers F.13s fitted with skis for winter operation. This was the first all-metal civil aircraft, and first appeared in 1919. A low-wing cantilever monoplane, the aircraft was powered by a single 185hp B.M.W. engine and could carry four passengers. A number of airlines used the aircraft throughout the 1920s.*
BOTTOM: *The first Short Calcutta flying boat, seen on the Medway at Rochester, England, before delivery. It first entered service with Imperial Airways in 1928, and it was used principally on the Mediterranean section of the air route to India.*

RIGHT: *One of Ala Littoria's Savoia-Marchetti S.66 twin-hulled flying boats alighting on the Tiber near Ostia, close to Rome's present Leonardo da Vinci Airport. This was one of a series of flying boats built by the company during the 1920s and 1930s, while it was primarily a manufacturer of sea planes. The S.66 was a development of the S.55. It was equipped with three Fiat engines of 550 or 750hp and could accommodate up to 18 passengers. Some 24 were built and used in service by, among others, Aero Espresso, SAM and SANA. A few survived to be used in World War II.*

AB Aerotransport – Swedish Air Lines – began services in June 1924 between Stockholm and Helsinki using Junkers F 13 floatplanes. As traffic grew a three-engined Junkers G 24 was added and in August 1932 a Ju 52/3m floatplane – almost certainly the first regular airline operation of the Ju 52/3m, even before Lufthansa. Lack of a land aerodrome at Stockholm forced seaplane operation to continue until May 1936, when Bromma Airport was opened. However, ABA had established services to Denmark, Germany, Amsterdam and London from the airport at Malmö on the west coast. ABA was an early operator of DC–3s, acquiring three in 1937. Like Swissair, ABA was a technically advanced airline, as its fleet composition shows, and it also opened the first experimental night mail service in Europe when, on the night of 18/19 June 1928, an F 13 with onboard sorting facilities flew from a military aerodrome near Stockholm to London in four stages via Malmö, Hamburg and Amsterdam.

The real beginning of air transport in Norway came with the founding in 1932 of DNL. A coastal route from Oslo through Kristiansand, Stavanger and Bergen was established into Arctic Norway. It was mainly operated by the Ju 52/3m on floats, although a few other types were used, including Short Sandringham flying boats after World War II.

Aero O/Y, in Finland, began its services in 1924 with a Helsinki–Reval route. F 13 seaplanes were used, followed by a G 24 and Ju 52/3ms, and pre-World War II operations were mainly confined to routes across the Baltic and the Gulf of Finland. In winter many of the aircraft had to be fitted with skis because ice prevented the use of seaplanes. Aero O/Y now operates as Finnair.

Most of the pioneer Italian services, too, were operated by marine aircraft, but Transadriatica, based in Venice, began operation in 1926 with a fleet of F 13 landplanes and, with three-engined G 24s, opened a Venice–Munich service in 1931.

Avio Linee Italiane was founded in 1926, used Fokkers on its early services, and had established a route to Munich by 1928 and Berlin by 1931. It also opened a Venice–Milan–Paris service and in June 1938 extended this to London, with Fiat G.18s.

Ala Littoria was formed in 1934 to take over most of the Italian airlines and it operated numbers of Savoias and other Italian aircraft but long remained a large operator of flying boats and seaplanes, although flying three-engined Caproni landplanes on its services to Italian East Africa.

Marine aircraft in Europe

Much use of marine aircraft was made in Europe in the pioneer years of air transport. There was an erroneous belief that seaplanes and flying boats offered greater security on overwater crossings.

Britain used flying boats on the trans-Mediterranean section of the Empire routes. France and Italy, with many Mediterranean and Adriatic routes, were Europe's biggest users of commercial seaplanes and flying boats, although Germany used such aircraft on coastal resort services and in the Baltic.

When the England–India route was opened in 1929, Imperial Airways used Short Calcutta flying boats on the trans-Mediterranean section. They were 12-passenger metal-hulled biplanes with three 540hp Bristol Jupiter engines, and the first example flew in February 1928. The Calcutta was followed by the Short Kent, of similar configuration but with four 555hp Jupiter engines and improved accommodation for 16 passengers. There were only three Kents, they comprised the Scipio class and entered service in May 1931.

In July 1936 Short Brothers launched the first of the S.23 C class flying boats, the *Canopus*. These had been designed for implementation of the Empire Air Mail Programme and were very advanced high-wing all-metal monoplanes powered by four 920hp Bristol Pegasus engines, had accommodation for 16 to 24 passengers and a top speed of just on 320km/h (200mph). *Canopus* made the

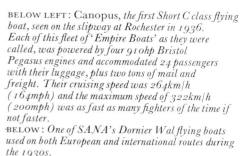

BELOW LEFT: *Canopus, the first Short C class flying boat, seen on the slipway at Rochester in 1936. Each of this fleet of 'Empire Boats' as they were called, was powered by four 910hp Bristol Pegasus engines and accommodated 24 passengers with their luggage, plus two tons of mail and freight. Their cruising speed was 264km/h (164mph) and the maximum speed of 322km/h (200mph) was as fast as many fighters of the time if not faster.*
BELOW: *One of SANA's Dornier Wal flying boats used on both European and international routes during the 1930s.*

first scheduled flight of the type, from Alexandria to Brindisi on 30 October 1936. Eventually they were to work the entire routes between Southampton and Sidney and Southampton and Durban, and some remained in service until after World War II.

Sandringham and Solent developments of the C class operated some of BOAC's post-war routes, and the last British flying boat services were those operated by Aquila Airways' Solents to Madeira until the end of September 1958.

The German resort services were mostly flown by small seaplanes and the Baltic services by Dornier Wals. The first Dornier Wal flew in November 1922 and, because of Allied restrictions on German aircraft production, most early Wals were built in Italy and Italian airlines employed them in considerable numbers. The Wal was an all-metal monoplane with the wing strut mounted above the hull and braced to the stabilizing sponsons, or sea wings. Two engines were mounted back to back on top of the wing,

and the passengers occupied a cabin in the forward part of the hull. There were many versions of the Wal, with different engines, weights, performance and accommodation, but the early examples had seats for eight to ten passengers and were usually powered by Rolls-Royce Eagle or Hispano Suiza engines. SANA was the biggest operator of passenger Wals, and put them into service in April 1926 on a Genoa–Rome–Naples–Palermo service.

There were three major flying boat constructors in Italy – Cant, Macchi and Savoia Marchetti. The Cants were biplanes, mostly of wooden construction, and SISA used the single-engined four-passenger Cant 10ter on its Trieste–Venice–Pavia–Turin service, which opened in April 1926, and on its Adriatic services. To supplement them SISA had a fleet of three-engined ten-passenger Cant 22s. Aero Espresso used a twin-engined eight-passenger Macchi M.24bis biplane flying boat on its Brindisi–Athens–Istanbul service.

Although the biplane served well into the

1930s, it was the Italian monoplanes that proved of most interest. In 1924 Savoia built a twin-hulled monoplane torpedo-bomber to the design of Alessandro Marchetti, and from this developed a passenger aircraft for Aero Espresso's then planned Brindisi–Istanbul service. This was the S.55C with thick-section 24m (78ft 9in) span wooden wing, accommodation for four or five passengers in each hull, triple rudders and two tandem-mounted 400/450hp Isotta-Fraschini engines. These aircraft went into service in 1926 and were followed in 1928 by the more powerful S.55P version, of which SAM had 14 or 15 on trans-Mediterranean services.

Developed from the S.55 was the 33m (108ft 3¼in) span S.66 of 1932. This had three 550 or 750hp Fiat engines mounted side by side, and the twin hulls could accommodate up to 18 passengers. It is believed that 24 were built. They were operated by Aero Espresso, SAM and SANA, and some passed to Ala Littoria, working services between Italy and Tunis, Tripoli and Haifa, and some survived to be taken over for war service.

In a completely different category was the Cant Z.506 which saw widescale service with Ala Littoria on Mediterranean and Adriatic routes. This was a three-engined twin-float low-wing monoplane with seating for 12 to 16 passengers.

As early as 1923 L'Aéronavale started using twin-engined four-passenger Lioré et Olivier 13 wooden flying boats between Antibes and Ajaccio, and in May that year a Latécoère affiliate opened a Marseilles–Algiers service with them. More than 30 were built.

Also in 1923, CAMS had begun production of a series of civil and military flying boats designed by Maurice Hurel. One of them was the wooden CAMS 53 biplane with two tandem 500hp Hispano Suiza engines and a small cabin for four passengers. It was introduced by Aéropostale on the Marseilles–Algiers route in October 1928.

The biggest French biplane flying boat in passenger service was the 35·06m (115ft) span Breguet 530 Saïgon, with three 785hp

Hispano Suiza engines, three-class accommodation for 19 to 20 passengers and a maximum weight of 15,000kg (33,069lb).

As a replacement for its CAMS 53s, Air Union had also ordered a fleet of four-engined 10 to 15 passenger Lioré et Olivier H 242 monoplane flying boats. These had their 350hp Gnome Rhône Titan Major engines in tandem pairs above the wing, and 14 went into service with Air France beginning in 1934.

The last trans-Mediterranean flying boat services of Air France were those operated to Algiers immediately after the war by two Lioré et Olivier H 246 four-engined 24 to 26 passenger monoplanes.

Mail and passenger services in the USA
On 1 March 1925, Ryan Airlines opened its Los Angeles–San Diego Air Line with modified Standard biplanes and a Douglas Cloudster, and this is claimed to have been the first regular passenger service wholly over the US mainland to be maintained throughout the year.

The first major legislative step in creating an airline industry was the passing of the Contract Air Mail Act (known as the Kelly Act) in February 1925. This provided for the transfer of mail carriage to private operators, and it was followed in May 1926 by the Air Commerce Act, which instructed the Secretary of Commerce to designate and establish airways for mail and passengers, organize air navigation and license aircraft and pilots. This Act came into force at midnight on 31 December 1926.

The history of United States air transport over the next few years was extremely complex, with numerous airlines competing for mail contracts. Some were small concerns, but others were backed by large financial organizations and closely linked with the aircraft manufacturing industry. Here only the briefest details of this intricate picture can be given.

The first five mail contracts were let on 7 October 1925, and 12 airlines began operating the services between February 1926 and April 1927 as feeders to the transcontinental mail route which was still being operated by the Post Office.

The Ford Motor Company had begun private daily services for express parcels between Detroit and Chicago on 3 April 1925, using single-engined all-metal Ford 2–AT monoplanes. Ford secured Contract Air Mail Routes (CAM) 6 and 7 covering Detroit–Chicago and Detroit–Cleveland and was the first to operate, beginning on 15 February 1926. Passengers were carried from August that year.

Next to start was Varney Air Lines, on CAM 5, between Pasco, Washington State, and Elko, Nevada, via Boise, Idaho. Leon Cuddeback flew the first service on 6 April 1926, using one of the airline's fleet of Curtiss-powered Swallow biplanes, but the operation was immediately suspended until the Swallows could be re-engined with air-cooled Wright Whirlwinds.

Also in April 1926 Robertson Aircraft Corporation began flying the mail on CAM 2 between St Louis and Chicago, and this operation is claimed as the first step in the eventual creation of the present American Airlines. Western Air Express began working CAM 4 between Los Angeles and Salt Lake City with Douglas M–2 biplanes, and opened its first passenger service over the route on 23 May that year.

CAM 1, New York–Boston, was awarded to Colonial Air Transport, but the services did not start until June 1926 and passenger service, with Fokkers, began almost a year later in April 1927.

PRT – Philadelphia Rapid Transit Service – obtained CAM 13 for Philadelphia–Washington and operated three flights a day from 6 July 1926. Whereas most of the mail carriers used passengers as fill-up load or ignored them altogether, PRT, with its Fokker F.VIIa–3ms, catered for passengers and used mail as fill-up.

The prize routes were those covered by CAM 17, from New York to Chicago, and CAM 18, San Francisco to Chicago. National Air Transport (NAT) was founded in May 1925, secured CAM 17, and began operating it with Curtiss Carrier Pigeons on 1 September 1927, having opened CAM 3, Chicago–Dallas, on 12 May the previous year. NAT was not very interested in passenger traffic and soon acquired a fleet of 18 ex-Post Office Douglas mailplanes; but, later, did buy Ford Trimotors.

Boeing, however, secured the San Francisco–Chicago route and was very interested in passengers. The company built a fleet of 24 Model 40A biplanes for the route. They were powered by 420hp Pratt & Whitney Wasp engines and could carry pilot, two passengers and 545kg (1,200lb) of mail, and were superior in payload and performance to the aircraft operated by NAT and most other mail carriers. By midnight on 30 June 1927, the Boeing 40s were deployed along the route and service began the following day.

Western Air Express (WAE) had hoped to get the San Francisco–Chicago mail contract, but Boeing's tender was much lower. Instead WAE developed a Los Angeles–San Francisco passenger service with Fokkers and

established a high reputation for reliability and safety.

In 1926 and 1927 two events took place which had a marked effect on the growth of air transport in the United States. The first was the appearance of the Ford Trimotor, which first flew in June 1926. This was an all-metal high-wing monoplane powered by three 200/300hp Wright Whirlwind engines, and had accommodation for 10 or 11 passengers in its original 4–AT version. Later came the 5–AT with 400/450hp Pratt & Whitney Wasps and seats for 13 to 15. The other event, in May 1927, was Lindbergh's flight from New York to Paris. This was the first solo flight across the North Atlantic, and it created enormous interest in aviation.

Brief mention must be made of three other airlines. Pacific Air Transport was organized by Vern Gorst in January 1926 and opened service over CAM 8, Seattle–San Francisco–Los Angeles, on 15 September; Maddux Air Lines opened a Los Angeles–San Diego passenger service with Ford Trimotors on 21 July 1927; and Standard Airlines under Jack Frye began a Los Angeles–Phoenix–Tucson passenger service with Fokker F.VIIs on 28 November 1927.

The Standard Airlines operation, by using a rail connection, provided a 70-hour air-rail transcontinental service from February 1929 and further development cut the time to 43hr 40min. Boeing Air Transport acquired Pacific Air Transport in January 1928, and this gave Boeing a connection between the Chicago–San Francisco route and Seattle. Boeing's passenger traffic was developing, but its aircraft were unsuitable, so the company built the three-engined Model 80 with Pratt & Whitney Wasps and accommodation for 12 passengers. Four Boeing 80s were built and began service in October 1928, to be followed by ten more powerful Model 80As with 18 seats.

In May 1928 Transcontinental Air Transport (TAT) was formed, and began planning a transcontinental route using Ford Trimotors on which meals could be carried, although the first air stewardesses appeared on Boeing's Model 80s in 1930. TAT came to an agreement with the railways for an air-rail coast-to-coast service. It began on 7 July 1929. Westbound passengers left New York's Pennsylvania Station in the *Airway Limited* and travelled overnight to Port Columbus in Ohio, where a special combined air-rail terminal had been built. From there they flew to Waynoka in Oklahoma where they transferred to the Atchison, Topeka and Santa Fe Railroad for the night journey to Clovis in New Mexico. The final stage by air took them to Los Angeles, and from there

they could continue free to San Francisco by train or by Maddux Air Lines. The New York–Los Angeles journey took exactly 48hr and the fare ranged from $337 to $403 one way. In November that year TAT acquired Maddux.

In the east, on 1 May 1928, Pitcairn Aviation began operating CAM 19 between New York and Atlanta with Pitcairn Mailwings and, using the lighted airway, did some night flying. That December the airline took over Florida Airways' Atlanta–Miami route, and the company became Eastern Air Transport in January 1930. The route was extended to Boston and passengers were carried over part of it from August. In December 1930 Curtiss Condors were introduced, and were also used by TAT. These were large twin-engined biplanes with 600hp Curtiss Conqueror engines and 18 seats. They led in 1933 to the T–32 model, which could carry 12 passengers in sleeping berths.

In 1929 Walter Folger Brown became Postmaster General, and he had strong views on how US airways should be developed. The Post Office contract for the central transcontinental route – CAM 34 New York–Los Angeles – was up for tender, but before TAT could secure the contract Brown forced the merger of TAT with Western Air Express, thus forming Transcontinental and Western Air (TWA). The new airline began the first coast-to-coast all-air through service on 25 October 1930, using Ford Trimotors on the New York–Los Angeles route with a nightstop at Kansas City and an overall time of 36hr.

American Airways, formed on 25 January 1930, obtained the southern route CAM 33 via Nashville and Dallas by a series of route extensions and company take-overs.

Then in July 1931 United Air Lines was organized to take over officially Boeing Air Transport, NAT, PAT and Varney, which for some time had been working under the United title.

Thus, by mid-1931, the Big Four had been created out of nearly 30 airlines of varying

OPPOSITE PAGE

LEFT: *A Ryan Los Angeles–San Diego Air Line Standard biplane of 1925.*
CENTRE: *Maiden Dearborn IV, one of the Ford 2–ATs operated by the Ford Motor Company on the routes from Detroit–Chicago and Detroit–Cleveland.*
RIGHT: *One of National Air Transport's fleet of Curtiss Falcon mail planes, used on the New York to Chicago route during the 1920s.*
THIS PAGE
ABOVE LEFT: *One of the Boeing Air Transport fleet of Boeing 40Bs, a development of the Model 40As introduced in 1927. The aircraft bears CAM 18 on its fuselage, the number of the San Francisco–Chicago contract mail route. It could carry both passengers and mail and was powered by a 420hp Pratt & Whitney Wasp radial engine.*
ABOVE CENTRE: *One of the four-engined Fokker F.32s used by Western Air Express, before it was amalgamated with Transcontinental Air Transport to form TWA. These aircraft were used on the Los Angeles to San Francisco passenger service route, which they operated with great reliability and safety.*
ABOVE RIGHT: *A Pitcairn PA–5 Super-Mailwing of Colonial Western Airways, used on the CAM 20 Cleveland to Albany route.*

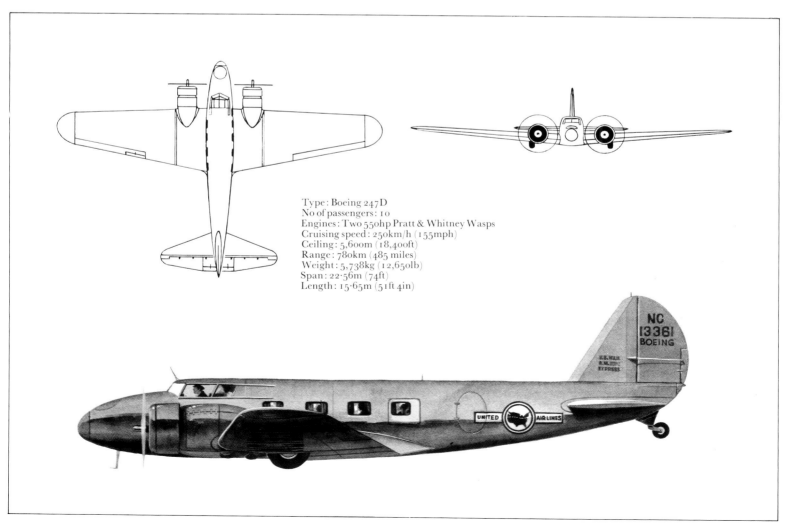

Type: Boeing 247D
No of passengers: 10
Engines: Two 550hp Pratt & Whitney Wasps
Cruising speed: 250km/h (155mph)
Ceiling: 5,600m (18,400ft)
Range: 780km (485 miles)
Weight: 5,738kg (12,650lb)
Span: 22·56m (74ft)
Length: 15·65m (51ft 4in)

size. They were American Airways, Eastern Air Transport, TWA and United Air Lines.

Although this is but a brief summary of the early development of US air transport, mention must be made of the Ludington Line and of two more transport aircraft. Ludington was an airline which really believed in passengers, and on 1 September 1930, began a service 'every hour on the hour' from 08.00 to 17.00hr over the Newark (New York)–Camden (Philadelphia)–Washington route, mainly with Stinson Trimotor monoplanes. In the two years before the company was taken over by Eastern it carried 124,000 passengers.

An event of great importance was Boeing's production of the Model 247 ten-passenger low-wing all-metal monoplane with two Pratt & Whitney Wasp engines and retractable undercarriage. This can be claimed as the prototype of the modern airliner. The Boeing 247 first flew on 8 February 1933, 60 were ordered for the United group, and they went into service with Boeing Air Transport, National Air Transport and Pacific Air Transport and became United Air Lines' standard equipment up to the end of 1936. The Boeing 247 was able to climb with one engine inoperative and cruise at 250km/h (155mph), and it made obsolete all other US airline equipment.

Boeing 247s were not available to non-United group airlines, and so TWA asked Douglas to produce a competitive aircraft. This was the DC–1, of similar layout to the Boeing. It first flew on 1 July 1933, and TWA ordered 20 of the improved production DC–2s.

But before the DC–2 could enter service, in July 1934, President Roosevelt, believing that some airlines had been unduly privileged in securing contracts, cancelled all the mail contracts on 9 February 1934, thus ending an important chapter in US airline history. Under White House instructions the last mail flight had to be completed on 19 February. That day, as a dramatic gesture, Jack Frye of TWA and Eddie Rickenbacker flew the DC–1 from Los Angeles to Newark, via Kansas City and Columbus, in a transcontinental record time of 13hr 4min.

Development in the Commonwealth

On 12 November 1919, Capt Ross Smith and Lieut Keith Smith with two crew took off from Hounslow in a Vickers Vimy and made the first flight to Australia. They arrived at Darwin on 10 December, after which they continued across Australia to Sydney, Melbourne and Adelaide. In preparation for the flight across Australia a route was surveyed, aerodromes prepared and fuel and oil provided. Responsible for this work on the Darwin–Longreach section were W. Hudson Fysh and P. J. McGinness, and it was these men, with others, who founded Queensland and Northern Territory Aerial Services (QANTAS) on 16 November 1920.

It was to Norman Brearley's West Australian Airways that the honour went of starting the first subsidized air service in Australia. On 4 December 1921, a Bristol Tourer took off from Geraldton, the railhead north of Perth, to inaugurate a weekly mail service to Carnarvon, Onslow, Roebourne, Port Hedland, Broome and Derby. The route was extended to Perth in January 1924, and from Derby to Wyndham in July 1930. In 1929 a weekly Perth–Adelaide service began using de Havilland Hercules biplanes, and Vickers Viastra monoplanes were introduced in 1931, covering the journey in less than 24hr.

On 2 November 1922, QANTAS opened its first scheduled service when P. J. McGinness flew an Armstrong Whitworth F.K.8 from Charleville to Longreach with mail, and on the next day Hudson Fysh flew on with mail and one passenger to Cloncurry.

QANTAS steadily expanded its operations in Queensland, and in 1931 took part in the first experimental England–Australia mail flights, carrying the mail from Darwin to Brisbane. QANTAS was chosen as the partner to Imperial Airways to carry passengers and mail between Singapore and Brisbane when the England–Australia service opened in December 1934. A new company, Qantas Empire Airways (referred to hereafter as Qantas), was registered in January 1934 with Imperial Airways and QANTAS each holding half the share capital.

Qantas ordered a fleet of D.H.86 four-engined biplanes for the Brisbane–Singapore operation, but, when the first service left Brisbane on 10 December 1934, they had to use a single-engined D.H.61 and D.H.50 because of the late delivery, and Imperial Airways worked the Darwin–Singapore sector until February 1935. Passengers were carried over the entire England–Australia route from April.

In September 1932 the Holyman brothers began a Launceston–Flinders Island service with a Fox Moth, and soon after merged with another concern to form Tasmanian Aerial Services, which a year later became Holyman's. In October 1935 Adelaide Airways began operations, and sometime that year Airlines of Australia began services based on Brisbane and Sydney. Then, in July 1936, Australian National Airways (ANA) was incorporated to include Holyman's, Adelaide Airways, Airlines of Australia and West Australian Airways.

Also in 1936 came an insignificant event which, years later, was to have a major influence on Australian air transport. Reginald Ansett had been refused a licence for a bus service between Hamilton and Melbourne, so, on 16 February 1936, he began an air service between these points, for which a licence was not required. After World War II Ansett was to get control of ANA and take over other companies to form Australia's largest private airline.

In Canada, during 1920, about 12,070km (7,500 miles) of landplane and seaplane routes were surveyed, the first flight to penetrate the Northwest Territories was made in the following year, and in 1923 the first air mail flight was made between Newfoundland and Labrador.

On 11 September 1924, Laurentide Air Service and Canadian Pacific Railway established an air service linking the railway at Angliers with the Quebec goldfields at Rouyn. This was the first regular air service introduced in Canada for the carriage of passengers, mail and freight.

Much of Canada's early air transport was on these lines, with air services linking remote areas with the nearest railhead. In summer the aircraft operated as landplanes and seaplanes, but the harsh winter climate enforced a change to skis.

Many concerns were involved in these early operations, mostly using single-engined aircraft. On 26 December 1926, Western Canada Airways began operations at Sioux Lookout and regular services were begun to Rolling Portage and Red Lake mining districts, and over the next few years a fairly extensive route network was established. On 3 March 1930, the airline began the nightly *Prairie Air Mail* service over the Winnipeg–Calgary and Regina–Edmonton routes. The initial service was flown by a Fokker Universal, and these were the first scheduled night flights in Canada.

On 25 November 1930, Canadian Airways Ltd began operations, having been formed

by Canadian Pacific Railway, Canadian National Railway, Western Canada Airways and a group of airlines controlled by Aviation Corporation of Canada. The new concern established services in many parts of the country and undertook a lot of mail and freight carriage. But although it linked many pairs of cities, it did not provide a transcontinental service.

However, work went ahead on preparing airports and navigational services for a transcontinental route and on 10 April 1937, Trans-Canada Air Lines (TCA) was created by a Government Act. Survey flights over the route began from Vancouver in July 1937 and on 1 September that year TCA began operation with a Vancouver–Seattle service flown by Lockheed Electras. Early in 1938 Vancouver–Winnipeg mail and cargo services began, and that October the route was extended, for cargo only, to Toronto, Ottawa and Montreal, mail being carried from the beginning of December.

On 1 March 1939, the full-scale official transcontinental mail service was inaugurated, and passengers were carried from 1 April on this route and on the Lethbridge–Calgary–Edmonton route. TCA now operates as Air Canada.

The other major Canadian airline is CP Air, which was formed on 30 January 1942, as Canadian Pacific Air Lines. This was founded as a subsidiary of Canadian Pacific Railway, and represented the merger of Arrow Airways, Canadian Airways, Dominion Skyways, Ginger Coote Airways, Mackenzie Air Service, Prairie Airways, Quebec Airways, Starratt Airways, Wings Ltd and Yukon Southern Air Transport. The initial CPA fleet consisted of 77 single- and twin-engined aircraft.

On 31 July 1929, Mrs F. K. Wilson founded Wilson Airways in Nairobi, and when Imperial Airways' Central Africa route was opened Wilson Airways operated connecting flights between Kisumu, on Lake Victoria,

and Nairobi, using a Puss Moth. In August 1932 Wilson Airways began Nairobi–Mombasa–Tanga–Zanzibar–Dar-es-Salaam mail services, and subsequently developed a network of services in the region.

With the introduction of the Empire Air Mail Programme in 1937, Imperial Airways introduced flying boats via a coastal route to Durban, and Wilson Airways opened a weekly Kisumu–Nairobi–Moshi–Dodoma–Mbeya–Mpika–Broken Hill–Lusaka mail service to connect with Imperial Airways at Kisumu.

In 1931 two airlines were formed in Central Africa, Rhodesian Aviation and Christowitz Air Services (in Nyasaland). The former opened a weekly Bulawayo–Salisbury service, subsidized by the Government of Southern Rhodesia and the Beit Trustees. The service was only operated as required and used South African-registered Puss Moths. Christowitz opened a Blantyre–Beira service, also using a Puss Moth. In 1933 Rhodesian Aviation began a weekly passenger and goods service over the Salisbury–Gatooma–Que Que–Gwelo–Bulawayo–Johannesburg route with a Fox Moth and Christowitz began a Salisbury–Blantyre service. Then in October 1933 Rhodesia and Nyasaland Airways (RANA) was formed. This acquired the assets of Rhodesian Aviation, taking over the Salisbury–Johannesburg service, which was terminated at Bulawayo, and in February 1934 the Christowitz Salisbury–Blantyre service. RANA also developed new routes and, like Wilson Airways,

operated services connecting with Imperial Airways flying boat operations.

The first Government air mail flight in the world took place in India when, on 18 February 1911, Henri Pequet, in a Humber biplane, flew mail from Allahabad to Naini Junction about 8km (5 miles) away. The first actual mail service in India began on 24 January 1920, when the RAF opened a weekly service between Bombay and Karachi. This was maintained for only a few weeks, and there was no Indian air service until 15 October 1932, when Tata Sons opened a Karachi – Ahmedabad – Bombay – Bellary – Madras mail service to connect with Imperial Airways' England–Karachi flights. The Tata aircraft was a Puss Moth, and it was flown from Karachi to Bombay by J. R. D. Tata.

In May and June 1933 two airlines were formed in India. The first, Indian National Airways (INA), was established to participate as a shareholder in the second, Indian Trans-Continental Airways, and to develop services in northern India. Indian Trans-Continental was set up to operate the trans-India route in association with Imperial Airways, beginning the Karachi–Calcutta service on 7 July 1933, with Armstrong Whitworth Atlanta monoplanes.

On 1 December 1933, INA started a weekly Calcutta–Rangoon service with Dragons, and on the same day opened the first daily air service in India – between Calcutta and Dacca.

INA and Tata Air Lines (as successor to Tata Sons) continued to expand their opera-

LEFT: *Union Airways' de Havilland D.H.86 Karoro flying on the Palmerston North-Dunedin route in New Zealand. This aircraft, a development of the D.H.84 Dragon, was known as the Dragon Express: it had four engines and could carry up to 16 passengers.*
RIGHT: *The Boeing 247, the forerunner of the modern airliner. A well streamlined low-wing monoplane, powered by two 550hp Pratt & Whitney radial engines, it had retractable landing gear to reduce drag in flight and was the first transport aircraft to have a de-icing system for the wings and tail unit. It was also able to climb on only one engine.*

tions, Tata opening a Bombay–Delhi service on 6 November 1937, and INA establishing a Karachi–Delhi service a year later.

A number of short-lived air services were operated in New Zealand during the 1920s, and in 1930–31 Dominion Airways operated a Desoutter monoplane on about 100 flights between Hastings and Gisborne before the loss of the Desoutter ended the undertaking. Air Travel (NZ) started a Hokitika–Okuru service with a Fox Moth on the last day of 1934, East Coast Airways began a twice daily Napier–Gisborne service with Dragons in mid-May 1935, and at the end of that December Cook Strait Airways opened a Wellington–Blenheim–Nelson service with two Dragon Rapides.

The year 1935 saw the founding of Union Airways of New Zealand as an offshoot of Union Steam Ship Co, and this airline began a daily Palmerston North–Blenheim–Christchurch–Dunedin service on 16 January 1936, with D.H.86s. The airline commissioned

three Lockheed Electras in June 1937 and put them on a daily Auckland–Wellington service. Union Airways purchased East Coast Airways in July 1938, and the following March opened a Palmerston North–Napier–Gisborne–Auckland service. After the war Air Travel (NZ), Cook Strait Airways and Union Airways were all absorbed by the newly founded New Zealand National Airways Corporation.

South Africa was slow in setting up any form of air transport, and it was only in March 1925 that the South African Air Force began a weekly experimental mail service over the Cape Town–East London–Port Elizabeth–Mossel Bay–Durban route, using D.H.9s. Only 32 flights were made, with 276 bags of mail, and it was not until more than four years later that a private air service began when Union Airways opened a subsidized Cape Town–Port Elizabeth service with extensions to Durban and Johannesburg.

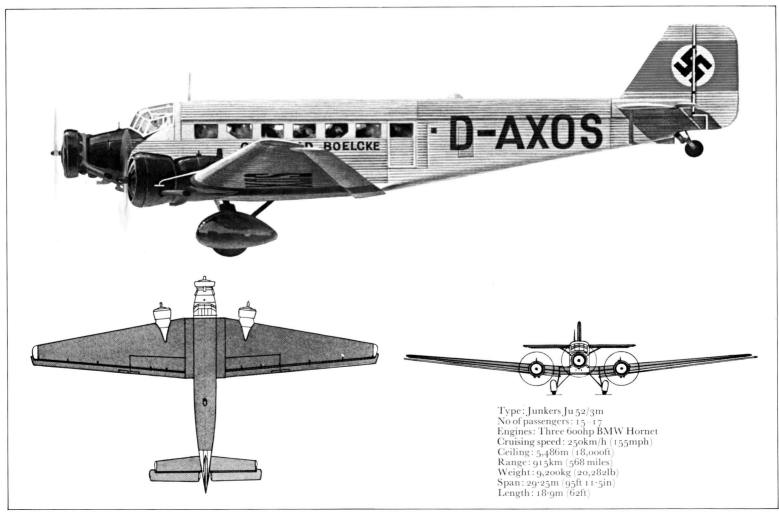

Type: Junkers Ju 52/3m
No of passengers: 15–17
Engines: Three 600hp BMW Hornet
Cruising speed: 250km/h (155mph)
Ceiling: 5,486m (18,000ft)
Range: 915km (568 miles)
Weight: 9,200kg (20,282lb)
Span: 29·25m (95ft 11·5in)
Length: 18·9m (62ft)

During 1930 the administration of South West Africa concluded an agreement with Junkers to form South-West Africa Airways and operate a weekly passenger, freight and mail service between Windhoek and Kimberley. This and a number of other services were begun in 1931 using F 13s and Juniors.

A change in policy came with the founding, on 1 February 1934, of South African Airways (SAA), which began operation with aircraft and staff taken over from Union Airways. SAA introduced Junkers–Ju 52/3ms, and these were the first multi-engined aircraft used by a South African airline.

At the beginning of February 1935 SAA took over South-West Africa Airways. The airline steadily expanded its network and on 1 April 1936, took over the Cape Town–Johannesburg sector of the England–South Africa route. Junkers–Ju 86s were added to the fleet in 1937, and by the time all civil flying was suspended in May 1940, SAA had opened routes to Lusaka, Broken Hill, Nairobi, Kisumu and Lourenço Marques.

BELOW: *The* Graf Zeppelin *at Recife, Brazil. This airship made a historic flight round the world in 1929, taking 21 days for the journey. During this trip it became the first aircraft of any type to make a flight across the Pacific Ocean.*

CENTRE: *One of Air France's South Atlantic Farman F.2200s,* Ville de Montevideo. *This was a large four-engined aeroplane.*
BOTTOM: *A Wilson Airways' de Havilland D.H.84 Dragon at Zanzibar.*

The South Atlantic

Having opened the air route between Toulouse and Dakar, France set about establishing services between Natal in Brazil and Buenos Aires. Airports were prepared and radio equipment installed, at an estimated cost of \$1·5 million. Aeropostal Brasileira and Aeroposta Argentina were established as subsidiaries of Aéropostale, and the Natal–Buenos Aires services was opened in November 1927. The dream of a service linking France and South America finally came true on 1 March 1928, when the entire route from Toulouse to Buenos Aires was opened for mail, with a transit time of eight days, but the ocean sector had to be operated by ships.

In 1928 Aeroposta Argentina opened services from Bahía Blanca, south of Buenos Aires, to the oil centre of Comodoro Rivadavia 950km (590 miles) further south, and from Buenos Aires to Asunción in Paraguay. In July 1929 the great barrier of the Andes was conquered with the opening of the Buenos Aires–Santiago service, and French

air services had reached the Pacific coast.

For several years the services in South America were operated by a series of single-engined Latécoère monoplanes, the Laté 17, 25 and 26, and open-cockpit Potez 25 biplanes. Much of the flying was done at night and weather was frequently appalling. On the route to Comodoro Rivadavia there were very strong winds to contend with, but it was the service through the Andes which called

for a very high standard of flying and a great amount of bravery. One incident is enough to illustrate the character of the route. In June 1930 Henri Guillaumet left Santiago in a Potez 25 with the mail for Europe. He encountered blinding snowstorms and violent turbulence and was forced to land close to Laguna Diamante at an elevation of 3,500m (11,480ft). The little biplane overturned, but Guillaumet survived and it took

BELOW: *One of the Latécoère 25 aircraft operated by Aeroposta Argentina. A single-engined monoplane, this type was used on the routes from north to south of the continent and also for the hazardous journeys across the Andes.*

CENTRE: *The four-engined Blériot 5190 flying boat Santos-Dumont. This aircraft made its first crossing of the South Atlantic from Dakar to Natal on 27 November 1934, and began regular service with Air France on this route early in 1935.*

BOTTOM: *Lufthansa's 10-ton Dornier Wal Boreas being lifted on to the catapult of the depot ship Schwabenland. This aircraft used two engines, which were placed in line, one driving a pusher propeller and the other a tractor propeller.*

him five days and four nights to struggle to safety.

Great as were the achievements of what was known as The Line, the South America route could not be regarded as satisfactory until the entire route could be covered by air. The first attempt was made in 1930, when on 12–13 May Jean Mermoz flew the ocean crossing from St Louis, Sénégal, to Natal in 21hr in the Latécoère 28 seaplane *Comte de la Vaulx*. But a single-engined floatplane was not a suitable aircraft for the ocean crossing, and the French Government ordered the three-engined Couzinet 70 *Arc-en-Ciel* landplane and the four-engined Blériot 5190 flying boat *Santos-Dumont*.

On 16 January 1933, Mermoz flew the *Arc-en-Ciel* from St Louis to Natal in 14hr

27min. After modifications, as the Couzinet 71, it began regular South Atlantic mail flights at low frequency from 28 May 1934, completing eight ocean crossings by the end of the year. Flown by Lucien Bossoutrot, the *Santos-Dumont* made its first crossing, from Dakar to Natal, on 27 November 1934, and began regular service with Air France early in 1935. It made at least 22 ocean crossings and cut the Toulouse–Buenos Aires time to 3 days 20hr.

The construction of the big Blériot flying boat had been delayed by financial problems, and as a result it had been beaten into service by the four-engined Latécoère 300 flying boat *Croix du Sud* (*Southern Cross*). This boat made its first crossing on 3 January 1934, and continued as far as Rio de Janeiro. Thereafter it shared the route with the *Arc-en-Ciel* and made six crossings by the end of the year; but on 7 December 1936, radio contact was lost with the *Croix de Sud* about 4hr after it left Dakar. The flying boat, its famous commander Jean Mermoz, and his crew

disappeared and were never found.

The overall success of the *Croix du Sud* led to construction of three similar Latécoère 301s, and these began service early in 1936.

Air France also employed a number of large four-engined Farman landplanes on the South Atlantic route, beginning with the F.220 *Le Centaure*, which made its first Dakar–Natal flight on 3 June 1935. In November 1937 the last of these, the F.2231 *Chef Pilote L. Guerrero*, owned by the French Government, flew from Paris to Santiago in an elapsed time of 52hr 42min and made the Dakar–Natal crossing in 11hr 5min.

Although passengers were carried on the route to West Africa and in South America, none were carried on these prewar ocean crossings.

and the entire route from Berlin to Buenos Aires was scheduled for four to five days. The 10-ton Wal was introduced on the route and, after trials over the North Atlantic in 1936, the much-improved Do 18s began working over the South Atlantic. Four-engined Do 26s made 18 mail crossings before war stopped the service, and during their operations the Wals made 328 crossings.

The third country to begin South Atlantic air services was Italy. Ala Littoria Linee Atlantiche was set up as the Atlantic division of Ala Littoria, and had begun taking delivery of a special fleet of Savoia-Marchetti S.M.83 three-engined monoplanes before Linee Aeree Trascontinentali Italiane (LATI) was established to operate the services. After proving flights, a regular

BELOW LEFT: *Two of SADTA's Junkers W 34s on a river in Colombia. The airline used these aircraft on several of the routes in the northern part of South America from 1919 onwards. It was founded with German capital immediately after World War I and gradually built up until eventually in 1940 it became part of the new airline Avianca.*
RIGHT: *One of the Boeing Model 314 flying boats, more usually known as 'Boeing Clippers'. This resulted from their individual names, such as Yankee Clipper, bestowed upon them by Pan American Airways. A total of 12 were built, 6 as 314s with 1,500hp Wright Double Cyclone engines, and a further 6 314As with increased passenger accommodation and 1,600hp engines.*

Germany used different methods in establishing its services to South America. Luftschiffbau Zeppelin decided to use the LZ 127 *Graf Zeppelin* on the route and on 18 May 1930, this airship left Friedrichshafen on a trial flight to Rio de Janeiro. This was followed by further trials in 1931. As a result the airship left Friedrichshafen on 20 March 1932, to open a regular service to Recife in Brazil, and this flight carried paying passengers – the first ever to fly on a transocean air service. Four flights were made that spring and five at fortnightly intervals in the autumn, with the last three continuing to Rio de Janeiro. In 1933 the *Graf Zeppelin* made 9 flights over the route, in 1934 twelve, in 1935 sixteen, and in 1936 nine, with an additional seven by the larger *Hindenburg*. From March 1935 this service was operated by Deutsche Zeppelin-Reederei, which was founded by Lufthansa and the Zeppelin company.

Germany's second method of flying to South America was to use landplanes between Germany and Africa and in South America and flying boats over the ocean. A system was devised for catapulting the flying boats from depot ships because they could not take off under their own power with sufficient fuel and while carrying a payload. The Dornier 8-ton Wal was used initially and the first experimental crossing made on 6 June 1933, when the *Monsun* was catapulted from the depot ship *Westfalen*. Regular mail services began in 1934, when *Taifun* made the first crossing on 7–8 February,

Rome–Rio de Janeiro service was inaugurated in December 1939. Avoiding British and French territory, the route was via Seville, Villa Cisneros, the Cape Verde Islands, Natal and Recife.

US air mail upheaval and a new start

After President Roosevelt's cancellation of the mail contracts and the last mail flight on 19 February 1934, the US Army Air Corps was given the task of flying the mail. The 43,450km (27,000 mile) network of mail routes was reduced to 25,750km (16,000 miles), and about 150 aircraft of various types were allocated to the operation. The weather was bad, the crews were inexperienced and ten pilots were killed before the last flight on 1 June.

The President admitted that he had been wrong, and in April 1934 the Postmaster General, James Farley, called in the airlines to find ways of salvaging what remained of the nation's airways. Temporary mail contracts were awarded, and among the conditions required to qualify for a contract was the stipulation that no contract carrier could be associated with an aircraft manufacturer.

The Big Four were reorganized as American Airlines, Eastern Air Lines, TWA Inc and United Air Lines, and of the 32 new mail contracts they were awarded 15, with TWA and United both getting mail contracts for transcontinental routes – Newark–Los Angeles and Newark–Oakland respectively. Eastern got three important routes, Newark–New Orleans, Newark–Miami and

Chicago–Jacksonville, while American's routes included Newark–Chicago and Newark–Boston. By combining Newark–Fort Worth and Forth Worth–Los Angeles they gained a transcontinental route, although it was not as direct as TWA's and United's.

Latin America

Most of the early airlines in South America were established by German nationals and with German capital, the first being SCADTA, which was founded in Colombia in 1919. This airline had a fleet of Junkers–F 13 seaplanes and, after a period of experimental operation, opened a regular service in 1921 over the Magdalena River route linking the port of Barranquilla with Girardot, the rail-head for Bogotá. The distance was 1,046km (650 miles) and the flight took 7hr, compared with a week or ten days by steamer. New routes were added and more modern equipment acquired until finally, in 1940, the airline was merged with another small company to form today's Avianca.

The next airline to be formed, again by Germans, was Lloyd Aéreo Boliviano (LAB). Equipped with F 13s, this company was founded in August 1925, and by the end of the year was running regular services between Cochabamba and Santa Cruz, taking three hours against the surface time of four days. This company was almost certainly the first to operate Junkers–Ju 52/3ms, although it did so on military operations during the Gran Chaco war in 1932. LAB is still the Bolivian national airline.

In 1927 two German influenced airlines were founded in Brazil, Varig and Syndicato Condor. Condor opened a Rio de Janeiro–Pôrto Alegre–Rio Grande do Sul service in October, and during the year handed over the Pôrto Alegre–Rio Grande sector to Varig. Both companies used German aircraft, mostly Junkers, with floatplanes predominating, and built up a route system, initially in the coastal regions. In 1942 Condor was reorganized as Serviços Aéreos Cruzeiro do Sul and, although retaining its identity, was taken over in 1975 by Varig, which remains the principal Brazilian airline and now operates intercontinental services.

In Peru Elmer Faucett founded Compañia de Aviación Faucett in 1928, and by the following year had established air services extending the length of the country, from

Ecuador in the north to Chile in the south. Faucett was unusual in that it built many of its own aircraft, based on Stinson single-engined monoplanes. Although now under different ownership, Faucett is still operating.

A military air service was introduced in Chile in 1929. This ran north from Santiago to Arica near the Peruvian border. It was taken over by Linea Aérea Nacional (LAN) in 1934, at first flying mail and then passengers, and since that time a considerable route system has been developed, including transatlantic services. LAN is the only airline to serve Easter Island, which it includes on its Chile–Tahiti route.

The Unites States airline which became deeply involved in Latin America was Pan American Airways. This company, led by Juan Trippe for nearly 40 years, began regular contract mail services between Key West and Havana on 28 October 1927, using Fokker F.VIIa–3m monoplanes, and carried passengers from January 1928. PAA developed a network of services in the Caribbean, mostly using Sikorsky amphibians, and its routes reached as far as Cristóbal in the Panama Canal Zone. Extension southward down the west coast of South America was blocked by the Grace shipping line; so in 1929 Panagra was formed, with Pan American and Grace each holding 50 per cent of the shares. Panagra secured a mail contract for the route from Cristóbal to Santiago, Chile, and across the Andes to Buenos Aires. The first mail left Miami on 14 May 1929, and from Cristóbal was flown by Sikorsky S–38 and then Fairchild FC–2 to Mollendo in Peru, where the route then terminated. The extension to Santiago was inaugurated on 21 July and, after the acquisition of Ford Trimotors, Buenos Aires was reached on 8 October and Montevideo on 30 November. A speeding up of the services in April 1930 enabled mail to travel from New York to Buenos Aires in 6½ days. Passengers were carried to Santiago from 15 August 1931, and through to Montevideo on 5 October.

Although Pan American had to settle for a half share in South American west coast operations, the company would not compromise on the east coast. In March 1929 NYRBA (New York, Rio and Buenos Aires Line) was founded, and that August it began a Buenos Aires–Montevideo service. The

BELOW: *Pan American Airways' Martin M–130 flying boat* China Clipper, *which flew on the trans-Pacific route. It carried mail from November 1935, and up to 14 passengers from October 1936.*
BOTTOM: *A Panagra Fairchild FC–2 flying over the Andes.*
BELOW RIGHT: *One of NYRBA Line's 14 Consolidated Commodore flying boats, which were used on the east coast. By February 1930 the entire route from Miami to Buenos Aires had been opened.*
CENTRE RIGHT: *Pan American Airways' flying boat, the Sikorsky S–40* Southern Clipper. *Although amphibians, the S–40s were normally operated as pure flying boats. This type was designed specially for NYRBA Pan American as an aircraft with a longer range and bigger payload than the Consolidated Commodore. These features were needed for long routes over the sea.*
BOTTOM RIGHT: *One of Mexicana's Douglas DC–3s from the post-war period.*

Hornets, tail units carried on twin booms, and accommodation for 32 passengers.

A much more advanced and more important flying boat was the Sikorsky S–42, powered by four 700hp Hornets and having accommodation for up to 32 passengers. The S–42 cruised at 274km/h (170mph) and had a normal range of 1,930km (1,200 miles). Ten examples of three models were built, and the type was introduced on the Miami–Rio de Janeiro route on 16 August 1934.

It was also in 1934 that VASP came into being and today, with Cruzeiro, Transbrasil and Varig, this company operates the Ponte Aérea (Air Bridge) between Rio de Janeiro and São Paulo, with 370 flights a week in each direction.

In Central America a New Zealander,

1939 Yerex founded British West Indian Airways (BWIA), and the airline opened its first service, Trinidad to Barbados via Tobago, in November 1940, using a Lockheed Lodestar.

Mexico has had a large number of airlines, its first being Cia Mexicana de Aviación (CMA), which began work in August 1924, carrying wages to the oil fields near Tampico with Lincoln Standard biplanes. This operation was devised to circumvent the activities of bandits. Under the title Mexicana, CMA is now a major international airline. Mexico's other major airline, Aeromexico, was created in 1934 as Aeronaves de Mexico. During the course of its long history this airline has absorbed a considerable number of other Mexican carriers.

next month a mail and passenger service was started between Buenos Aires and Santiago with Ford Trimotors – five weeks ahead of Panagra's trans-Andes service.

Before the east coast route could be opened to the United States it was necessary to obtain suitable aircraft. There were long overwater sectors and few landing grounds, so a fleet of 14 Consolidated Commodore flying boats was ordered. These were large monoplanes powered by two 575hp Pratt & Whitney Hornet engines, and could carry 20 to 32 passengers. They cruised at 174 km/h (108mph), and had a range of 1,600km (1,000 miles). Four had been delivered by the end of 1929 and on 18 February 1930, the entire route between Miami and Buenos Aires was opened, the 14,485km (9,000 miles) being flown in seven days. NYRBA also set up NYRBA do Brasil to undertake local operations in Brazil.

Pan American desperately wanted the east coast route, and finally managed to acquire NYRBA in September 1930, when it took over the Commodores – including those still on order – and changed NYRBA do Brasil to Panair do Brasil. Postmaster General Brown awarded Pan American the mail contract for the east coast route, FAM–10, on 24 September 1930.

Through NYRBA Pan American had acquired a fleet of Commodores, it urgently needed long-range aircraft with bigger payloads, and the need was met by three specially designed Sikorsky S–40s. These were large flying boats with four 575hp Pratt &

Lowell Yerex, was responsible for much of the early air transport development. In 1931 he founded Transportes Aéreos Centro-Americanos (TACA) in Honduras, and built up a main airway linking British Honduras with the capitals of Costa Rica, Guatemala, Honduras, Nicaragua, Panama and Salvador. In 1933 he opened the first service between Tegucigalpa and San Salvador. Networks of routes were set up in the Central American republics, and eventually TACA associated companies were established in Colombia, Costa Rica, Guatemala, Honduras, Mexico, Nicaragua, Salvador and Venezuela.

Today this vast TACA empire has shrunk to the present TACA International, which in the spring of 1976 was operating three BAC One-Elevens and three DC–6As. In

Conquering the Pacific

The first air crossing of the Pacific was not made until 1928, when Charles Kingsford Smith, C. T. P. Ulm, Harry Lyon and James Warner flew from Oakland, across the Bay from San Francisco, to Brisbane in Queensland in the three-engined Fokker F.VII *Southern Cross*. They left Oakland on 31 May and arrived in Brisbane on 9 June with a flight time of 83hr 38min.

In July 1931 Charles Lindbergh and his wife made a survey of a northern route in a Lockheed Sirius single-engined seaplane, flying to Japan via Alaska, Siberia and the Kuriles, but political problems prevented the establishment of such a route.

The only alternative was to use island stepping stones which were United States territory, and bases were therefore prepared on Wake Island and Guam to enable a service to operate from San Francisco to Manila via Honolulu, Wake and Guam. This gave stages of 3,853km (2,394 miles) from San Francisco to Hawaii, 3,693km (2,295 miles) to Wake, 2,414km (1,500 miles) to Guam and 2,565km (1,594 miles) to Manila.

To obtain aircraft for the operation of a regular service, Pan American issued a specification for a flying boat capable of flying 4,023km (2,500 miles) against a 48km/h (30mph) headwind while carrying a crew of four and at least 136kg (300lb) of mail. To this specification Martin built three M–130 flying boats, each powered by four 800/950hp Pratt & Whitney Twin Wasp engines. The boats had a span of 39·62m (130ft), weighed 23,700kg (52,252lb) fully loaded and could carry 41 passengers, although only 14 seats were installed for the Pacific route. Cruising speed was 253km/h (157mph) and the range 5,150km (3,200 miles), or 6,437km (4,000 miles) if only carrying mail.

The M–130s were named *China Clipper*, *Philippine Clipper* and *Hawaii Clipper*, and the first was delivered in October 1935. The *China Clipper*, under the command of Capt Edwin Musick, inaugurated the trans-Pacific mail service when it left Alameda on 22 November 1935, and it alighted at Manila 59hr 48min later. Paying passengers were carried from 21 October 1936.

Early in 1937 the Sikorsky S–42B *Hong Kong Clipper* was delivered, and that spring it made a survey of the southern Pacific route to Auckland, subsequently being used to extend the trans-Pacific operation from Manila to Hongkong, the first service being on 27–28 April. This gave Pan American a direct link to China through the China National Aviation Corporation's Hongkong–Canton–Shanghai service. From that time the Martin M–130s were completing the San Francisco–Manila out and back flights in 14 days.

On 23 December 1937, Pan American inaugurated a San Francisco–Auckland service via Hawaii, Kingman Reef and Samoa, but the S–42B, with its commander, Edwin Musick, and crew, was lost on the second flight and the service had to be suspended. But on 12 July 1940, a fortnightly service was opened via Hawaii, Canton Island and New Caledonia using one of the new Boeing 314s, with passengers being carried from 13 September.

The modern airliner

The principal passenger aircraft in use in the United States in the late 1920s and early 1930s were Fokker F.VIIs and Ford Trimotors. They had accommodation for 8 to 15 passengers, but cruised at only a little over 160km/h (100mph), and for much of the time very few of their seats were occupied. Some people believed there was a need for smaller and faster aeroplanes, and as a result there was a period when numerous airlines were operating fleets of small single-engined monoplanes capable of cruising at more than 241km/h (150mph).

John Northrop and Gerrard Vultee designed a very clean four-passenger high-wing monoplane known as the Lockheed Vega. It was a wooden aeroplane, powered by a Wasp engine, and it first flew in July 1927. The Vega entered service with International Airlines on 17 September 1928, and subsequently there were several versions, including one with a metal fuselage. Cruising speed was 217–241km/h (135–150mph), and the type was used by a number of airlines, including Braniff and TWA. Very similar was the parasol-winged Lockheed Air Express, which was produced for Western Air Express and is believed to have entered service in 1929.

The last of Lockheed's single-engined high-speed transports was the six-passenger Orion, which had a low-mounted wing and retractable undercarriage. It went into service with Bowen Air Lines in May 1931. With either a Wasp or Cyclone engine the Orion cruised at 289–313km/h (180–195 mph), and is claimed as the first transport aircraft capable of 320km/h (200mph). American Airways, Northwest Airways and Varney Speed Lines were among the Orion users, two were exported to Swissair, and Air Express Corporation operated them on a US transcontinental freight service, achieving coast-to-coast times of 16–17hr.

The most advanced of these single-engined monoplanes was the Vultee V–1A, which could carry eight passengers and cruise at 340km/h (211mph). The Vultee had an 850hp Wright Cyclone engine, was of all-metal construction and had a retractable undercarriage. American Airlines introduced the type in September 1934.

There were also two Boeing prototypes. These were the Model 200 and 221, both named Monomail. They were all-metal low-wing monoplanes, each powered by a 575hp Pratt & Whitney Hornet. The Model 200 was originally a single-seat mail and cargo carrier. It first flew on 6 May 1930, and its semi-retractable undercarriage made possible a cruising speed of 217km/h (135mph). This first Monomail was followed in August 1930 by the slightly longer Model 221 with non-retractable undercarriage and a cabin for six passengers. It went into service with Boeing Air Transport, and as the Model 221A was later lengthened to provide two extra seats. The original Monomail was modified to Model 221A standard and put into service on the Cheyenne–Chicago route. Experience gained with these aircraft was incorporated in the twin-engined Model 247, described earlier.

On 19 February 1934, the Douglas DC–1 had made its dramatic coast-to-coast flight, and TWA had already ordered 20 of the 14-passenger production DC–2s. These were each powered by two 720hp Wright Cyclones, giving a maximum cruising speed of 315km/h (196mph) and a range of just over 1,600km (1,000 miles). The first DC–2 was delivered to TWA on 14 May 1934, and four days later made a proving flight from Columbus to Pittsburgh and Newark. On 1 August DC–2s began transcontinental operation over the Newark–Chicago–Kansas City–Albuquerque–Los Angeles route to an 18hr schedule. Apart from providing much improved transcontinental services, the Newark–Chicago sector was the first nonstop operation over the route.

On 5 May 1934, American Airlines had begun transcontinental sleeper services with Curtiss Condors, and had also used them to build up frequency on its New York–Boston route. Condors were no match for the DC–2, and so American Airlines asked Douglas for a sleeper development of the DC–2.

Douglas enlarged the DC–2 by widening and lengthening its fuselage and added 3·04m (10ft) to its wing span. Powered by two 1,000hp Cyclone engines and having 14 sleeping berths, this new type was known as the DST – Douglas Sleeper Transport.

American Airlines had ordered 10 DSTs, but after its first flight on 17 December 1935, increased the order and changed it to cover eight DSTs and 12 dayplanes – DC–3s with 21 seats. The first DST was delivered at the beginning of June 1936 and, used as a dayplane, went into service on the New York–Chicago route on 25 June. American took delivery of its first DC–3 in August, and on

15 September was able to inaugurate its DST *American Mercury* skysleeper transcontinental service with an eastbound schedule of 16hr.

Thus was launched one of the world's great transport aeroplanes, the DC–3, of which by far the biggest percentage was powered by 1,200hp Pratt & Whitney Twin Wasps. It was to be built in numerous civil and military versions, with a total of 10,655 produced in the United States and others being built in Japan and, under licence, in the Soviet Union. After the war large numbers of surplus military DC–3s became available to civil operators, and in the postwar years almost every airline operated them at some period. There are still several hundred flying.

Lockheed also embarked on production of a series of fast twin-engined monoplanes. The first was the ten-passenger Model 10A Electra with 450hp Pratt & Whitney Wasp Junior engines and a cruising speed close to 322km/h (200mph). The Electra was introduced on 11 August 1934, by Northwest Airlines. The original British Airways had 7, and also bought 9 of the more powerful 12-passenger Model 14s, which were about 64km/h (40mph) faster. The Lockheed 14 was also known as the Super Electra and the Sky Zephyr and, like the Electra, was first operated by Northwest Airlines, from September 1937.

The major United States airlines required an aeroplane with greater capacity and range than the DC–3 and, in March 1936, the Big Four and Pan American each came to an agreement to share the cost of developing the four-engined Douglas DC–4E, with a span of 42·13m (138ft 3in) and a maximum weight of 30,164kg (66,500lb). It was powered by 1,450hp Pratt & Whitney Twin Hornets, and cruised at 322km/h (200mph). The DC–4E had a nosewheel undercarriage, the first on a big transport, triple fins, and production models were to be pressurized. United Air Lines ordered six 52-passenger sleeper DC–4Es in July 1939, having put the prototype into experimental operation in the previous month. But the aeroplane was found to be unsuitable, the order was cancelled, and the only example exported to Japan.

The last American transport landplane to go into service before the Japanese attack on Pearl Harbor was the Boeing 307 Stratoliner. Although only ten were built, it has an important place in history as the first pressurized aeroplane to go into airline service. The Stratoliner was a low-wing monoplane with a wing span of 32·64m (107ft 3in), powered by four 900hp Wright Cyclones and

LEFT: *One of TWA's Lockheed Orions, with, in the background, one of the airline's single-seat Northrop Alpha mail-planes. The Orion was the first Lockheed aircraft to be fitted with fully retractable landing gear. Very successful in the US, it was also exported to Europe.*
CENTRE: *A metal-fuselage Lockheed Vega used on Braniff Airways' Chicago–Dallas services.*
RIGHT: *One of American Airlines' Vultee V–1A fleet. This was a single-engined, eight-passenger monoplane airliner.*

LEFT: *The Boeing Model 200 Monomail, an all-metal low-wing monoplane fitted with semi-retractable landing gear.*
BELOW: *A Douglas DC–3 of CP Air. A 21-seater plane with a maximum speed of 320km/h (200mph) and a range of up to 2,400km/h (1,500 miles), it was an immediately successful plane from its introduction in 1936.*
BOTTOM: *Progenitor of the ubiquitous DC–3 Dakota, the DC–1 was the Douglas Aircraft Company's answer to the Boeing 247.*

having a maximum weight of 19,051 kg (42,000 lb). It cruised at 354 km/h (220 mph) and had a range of 3,846 km (2,390 miles). This design employed the wings, nacelles, powerplant and tail surfaces of the B–17 Flying Fortress bomber, but had a completely new circular-section fuselage with pressurized accommodation for 33 passengers and 5 crew. The first aircraft flew on the last day of 1938, but was lost before delivery to Pan American. PAA had three, TWA five, and Howard Hughes had a modified aircraft for record breaking.

TWA introduced the Stratoliner on its transcontinental route on 8 July 1940, and cut eastbound times to 13 hr 40 min; Pan American based its Stratoliners in Miami for Latin American operations; but at the end of 1941 these aircraft were ordered into war service, with TWA's fleet being used over the North Atlantic. After the war the TWA aircraft, much modified, were put back into civil operation with 38 seats and increased

take-off weight. Some Stratoliners were sold to Aigle Azur in 1951, and some saw service in Latin America and the Far East for several years, a few remaining in use until the mid-1960s.

United Kingdom domestic airlines

When sustained air services were established in the United Kingdom, most that were to prove successful were based on routes which involved a water crossing. In the spring of 1932 British Amphibious Air Lines began irregular operation of such a route, between Blackpool and the Isle of Man, using a Saunders-Roe Cutty Sark amphibian. From June until the end of September the service operated regularly, and there was a bus connection between Blackpool and a number of towns in Yorkshire.

In April 1932 Hillman's Airways opened a service between Romford and Clacton, and by June had achieved a frequency of every three hours from 09.00 until dusk. A de

Havilland Fox Moth was used for the first time on this route.

June 1932 saw the start of a ferry service between Portsmouth and Ryde. This was operated, at very low fares, by Portsmouth, Southsea and Isle of Wight Aviation, using a three-engined Westland Wessex monoplane. The company was to work up a network of services in the south of England, use a wide variety of aircraft and carry a thousand passengers a day at some periods before World War II closed down all operations.

Two other events of importance in 1932 were the opening of a twice-daily Bristol–Cardiff service, which began on 26 September and was operated by Norman Edgar with a Fox Moth; and the first flight of the de Havilland D.H.84 Dragon on 24 November. Norman Edgar's service was to develop into a sizeable operation under the title Western Airways, and the Dragon was to make possible economic short-haul airline operations in the United Kingdom and

RIGHT: *The Portsmouth, Southsea and Isle of Wight Aviation's Westland Wessex at Portsmouth Airport. This was a light airliner used on short range routes.*
OPPOSITE PAGE
ABOVE: *The fleet and staff of Midland and Scottish Air Ferries lined up at Glasgow's Renfrew Airport. The nearest aircraft is the Avro Ten.*
BELOW LEFT: *Six of Jersey Airways' de Havilland Dragons on the beach in Jersey in April 1934. The D.H.84 Dragon was an eight-passenger transport biplane using two de Havilland Gipsy Major engines.*
RIGHT: *Jersey Airways' de Havilland D.H.89 Dragon Rapide St Ouen's Bay at Jersey. This, the most famous of the Dragon class, was also a twin-engined eight-passenger aeroplane and was first introduced into service in 1934. It was used during World War II under the name of* Dominie.

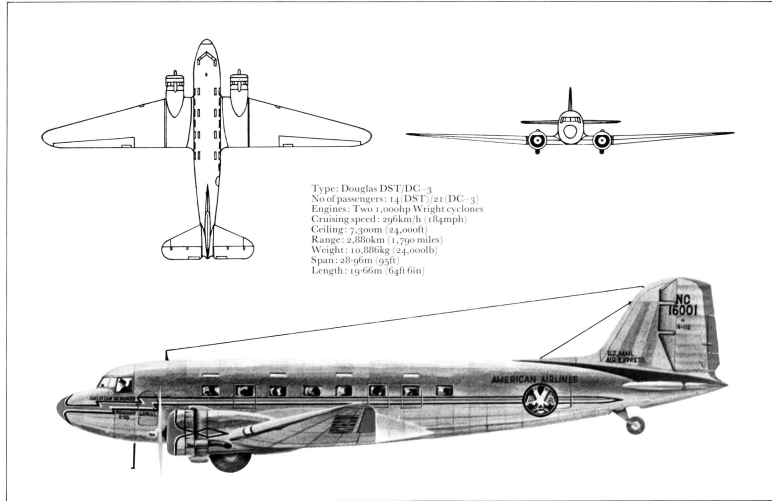

Type: Douglas DST/DC–3
No of passengers: 14(DST)/21(DC–3)
Engines: Two 1,000hp Wright cyclones
Cruising speed: 296km/h (184mph)
Ceiling: 7,300m (24,000ft)
Range: 2,880km (1,790 miles)
Weight: 10,886kg (24,000lb)
Span: 28·96m (95ft)
Length: 19·66m (64ft 6in)

many other parts of the world. The Dragon was a six-passenger biplane with two 130hp de Havilland Gipsy Major engines, and 115 were built in the United Kingdom and 87 in Australia. The first Dragon was delivered to Hillman's on 20 December 1932.

The entry of the railways into UK airline operation took place in 1933, and the same year witnessed the start of air transport in Scotland and the beginning of large-scale air services to the Channel Islands. On 12 April the Great Western Railway (GWR) began a public service, twice each way on weekdays, between Cardiff and Plymouth, with a stop at Haldon to serve Torquay and Teignmouth. The aircraft used was a Westland Wessex chartered from Imperial Airways. The service was extended to Birming-

ham in May, the frequency cut to once a day, and at the end of September closed for the winter.

On the day that the GWR began its services, Spartan Air Lines began a service between Cowes and Heston with a Spartan Cruiser three-engined monoplane, and when it was withdrawn at the end of the summer 1,459 passengers had been carried. This operation was continued in 1934, grew considerably, and became part of the railway group's operations.

Of much greater importance was the founding of Highland Airways by Capt E. E. Fresson on 3 April 1933. On 8 May this airline opened a regular service linking Inverness and Kirkwall, in Orkney, via Wick. A Monospar S.T.4 monoplane was

used and, in spite of some appalling weather, set such a high standard of regularity that a year later the airline was awarded the first domestic mail contract. The airline extended its routes to include Aberdeen and Shetland and operated the first Orkney inter-island services.

In Glasgow, John Sword had established Midland and Scottish Air Ferries (M&SAF), and on 8 May it made its first recorded ambulance flight, from Islay to Glasgow. The air ambulance is still an essential element of Scottish air transport. Regular passenger services were begun on 1 June over the Glasgow–Campbeltown–Islay route, with Dragons, and M&SAF were to develop a number of routes before closing down at the end of September 1934.

periods they catered for holiday traffic.

Hillman's Airways had opened a London–Liverpool–Isle of Man–Belfast service on 16 July 1934, and this was extended to Glasgow on 1 December, when the company got the mail contract instead of RAS.

Following the ending of Midland and Scottish Air Ferries' operations in September, George Nicholson gave up his short-lived Newcastle–Carlisle–Isle of Man service, moved to Glasgow and founded Northern and Scottish Airways. Operations began on 1 December with twice-weekly Glasgow–Campbeltown–Islay services, and the company steadily developed, with Skye being served from 5 December 1935. By July 1936 a circular route was linking Glasgow with Skye, North Uist and South Uist. Most

services were flown by Dragons or Spartan Cruisers, and in August 1937 the company amalgamated with Highland Airways to form Scottish Airways, the link between the two networks coming in May 1938, when a Glasgow–Perth–Inverness–Wick–Kirkwall–Shetland service began.

On 30 September 1935, three domestic airlines, Hillman's Airways, Spartan Air Lines and United Airways, were merged to form Allied British Airways. The name was changed to British Airways on 29 October, and in the following August the company absorbed British Continental Airways. British Airways operated some domestic routes and controlled a number of United Kingdom domestic airlines, but its main effort was concentrated on developing services to the continent.

E. Gandar Dower founded Aberdeen Airways, later renamed Allied Airways (Gandar Dower) Ltd, and operated services in Scotland, including routes to Orkney and Shetland, and on 12 July 1937, started a service between Newcastle and Stavanger. Between March 1935 and the summer of 1936 Crilly Airways operated a number of domestic services; North Eastern Airways developed routes down the eastern side of the country between London and Scotland; Blackpool and West Coast Air Services operated over the Irish Sea; and there were others which operated for varying periods.

Although Aer Lingus is the Irish national airline, its beginning was closely involved with the United Kingdom domestic operations. The airline was founded on 22 May 1936, and a week later began a daily Dublin–Bristol service. In the same month it started working between Dublin and the Isle of Man, and in September opened a Dublin–Liverpool service and extended its Bristol service

On 18 December 1933, Jersey Airways began a daily Portsmouth–Jersey service. There was no airport on Jersey until 1937, so the fleet of Dragons operated from the beach at St Aubin's Bay near St Helier. Sometimes the whole fleet was on the beach at the same time, because schedules were governed by the tides and the fleet flew in a loose formation.

Having suffered as a result of increasing road competition, the four mainline railways had obtained rights to operate domestic air services. On 21 March 1934, they registered Railway Air Services, with the London Midland and Scottish Railway, London and North Eastern Railway, Great Western Railway, Southern Railway and Imperial Airways as the shareholders and Imperial

Airways responsible for undertaking flying operations. The first RAS service was opened on 7 May. This was the previous year's GWR operation, and the route was extended to Liverpool and operated by a Dragon. At the end of July RAS began a Birmingham–Bristol–Southampton–Cowes summer service, but the main RAS operation was to be the *Royal Mail* trunk route linking London, Birmingham, Manchester, Belfast and Glasgow. This was to be operated with D.H.86s and the opening was set for 20 August, but the weather was atrocious and only part of the route could be flown. Full working began the next day. This route, with modifications, continued throughout the years up until the war, but RAS's other routes were mainly confined to summer months, during which

to London. All these operations were in association with Blackpool and West Coast Air Services, and under the title 'Irish Sea Airways'. Dragons were used on the first three routes, and D.H.86s flew to London.

A variety of aircraft served the United Kingdom routes, but the biggest contribution was made by the de Havilland biplanes – the D.H.84 Dragon, D.H.86 and D.H.89 Dragon Rapide. The Dragon Rapide had first flown in April 1934; it was a much-improved Dragon with two 200hp Gipsy Six engines, there were seats for six to eight passengers, and the cruising speed was about 210km/h (130mph). Several hundred Dragon Rapides were built; they served airlines in many parts of the world and a few are still flying on short routes today.

The Empire air mail programme

In December 1934 HM Government announced that from 1937 all letters despatched from the United Kingdom for delivery, along what were then the Empire routes, would, as far as practicable, be carried by air without surcharge. The carrier was to be Imperial Airways, with certain sectors to be covered by Commonwealth airlines.

A fleet of 28 Short S.23 C class flying boats was ordered, and 12 Armstrong Whitworth A.W.27 Ensign landplanes, although some of the latter were for European operation. The first C class flying boat was launched at Rochester on 2 July 1936, making its first flight on 4 July, and the first Ensign flew on 24 January 1938.

The Ensigns were large, high-wing monoplanes powered by four 850hp Armstrong Siddeley Tiger engines, and had accommodation for 27 passengers on Empire routes and 40 on European. Cruising speed was 274km/h (170mph) and range 1,287km (800 miles). The Ensigns suffered numerous troubles and, although introduced in Europe towards the end of 1938, played little part in Empire operations until they had been re-engined with Wright Cyclones during World War II.

The C class flying boats began operation in October 1936, and the first stage of the Empire Air Mail Programme was inaugurated on 29 June 1937, when the *Centurion* left Southampton with 1,588kg (3,500lb) of unsurcharged mail for the Sudan, East and South Africa.

On 23 February 1938, the Mail Programme was extended to cover Egypt, Palestine, India, Burma and Malaya when the *Centurion* and the Qantas 'boat *Coolan-*

gatta left Southampton. By early April the schedules had been improved to give a Southampton–Karachi time of 3 days, a Singapore time of $5\frac{1}{2}$ days and Sydney time of $9\frac{1}{2}$ days. At that time the all-up mail did not apply to Australia, and the flying boats did not work through to Australia until June.

The Mail Programme was extended to Australia, New Zealand, Tasmania, Fiji, Papua, Norfolk Island, Lord Howe Island, Nauru, Western Samoa and certain other Western Pacific territories with the departure of *Calypso* from Southampton on 28 July 1938. Not all these places were served by air, the mail sometimes being sent over its last stages by sea.

The North Atlantic

Today there are some 30 airlines operating scheduled passenger services across the North Atlantic, and in 1975 the airlines made more than 80,000 scheduled flights over the ocean, carrying 8,782,176 passengers and more than half a million tons of cargo. In 1947 the airlines carried 209,000 passengers across the North Atlantic, whereas 415,000 travelled by sea. In 1957 the numbers using sea and air services were just about equal, at slightly over one million by each type of transport, and in 1958 the airlines carried 1,292,000, while sea passengers had dropped to 964,000. In 1970 the airlines carried 10,038,000 passengers on scheduled and charter services while sea traffic had fallen even further, to 252,000. Total North Atlantic airline passengers in 1975 numbered nearly $12\frac{1}{2}$ million.

It is not surprising that from the earliest days of aviation there were dreams of North Atlantic services, but the problems of making these dreams reality were formidable. The Great Circle distance between the present Shannon Airport, in Ireland, and Gander, in Newfoundland, is 3,177km (1,974 miles), and that is the shortest direct ocean crossing. North Atlantic weather is notoriously bad, with frequent fog in the Newfoundland area, and very strong westerly prevailing winds add considerably to the still-air distance.

The first aerial crossing of the ocean was made in May 1919 by US Navy flying boats, NC–1, NC–3 and NC–4. They left Newfoundland on 16 May, and the NC–4, commanded by Cdr Albert Read, reached the Azores after a flight of 15hr 18min. Thereafter it continued via Portugal and Spain to alight at Plymouth on 31 May.

On 14 June John Alcock and Arthur Whitten Brown took off from Newfoundland in a modified Vickers Vimy bomber, and they landed at Clifden in Ireland the next day after spending 16hr 27½min in the air and making the coast-to-coast crossing in 15hr 57min. It was the first non-stop flight across the North Atlantic. Then on 2 July the British airship R34 left East Fortune, in Scotland, and flew to Long Island. The airship was commanded by Maj G. H. Scott and carried 28 officers and men and one stowaway, and the flight took 108hr 12min. The return flight took 75hr 3min. This was the first out and return flight over any ocean, the first North Atlantic crossing by an airship and the first westward flight by any aircraft.

It was not until April 1928 that an aeroplane crossed the Atlantic Ocean from east to west, when Baron von Hünefeld, Cmdt J. Fitzmaurice and Hermann Köhl flew from Ireland to Greenly Island off Labrador in the Junkers–W 33 *Bremen*.

As early as 1928 Pan American Airways began investigating possible North Atlantic routes. Numerous surveys were made, including one by Lindbergh, who studied a northern route via Greenland and Iceland, making his flight in a Lockheed Sirius seaplane. The very big problem was aircraft range, and Pan American and Imperial Airways did not begin trial flights until 1937.

France began designing a large transatlantic flying boat in 1930. During the design stage it had to be considerably modified, and emerged in January 1935 as the Latécoère 521, to be named *Lieutenant de Vaisseau Paris*. The Laté 521 was a 42-tonne monoplane powered by six 800/860hp Hispano Suiza engines. It was designed to carry 30 passengers on Atlantic routes or 70 over the Mediterranean, and its range was 4,000km (2,485 miles). Its first successful North Atlantic flight, from Biscarosse to New York via Lisbon and the Azores, was made in August 1938, and it subsequently made trial flights over various North Atlantic routes.

Germany undertook North Atlantic trial flights during 1936 using two Dornier Do 18 twin-engined flying boats – *Zephir* and *Aeolus*. These were catapulted from the depot ship *Schwabenland* near the Azores, and from it the *Zephir* flew to New York in 22hr 12min

on 11 September, arriving with 10 hours' reserve fuel. By 20 October, when the trials ceased, the two Do 18s had flown 37,637km (23,386 miles) on eight flights over various North Atlantic routes. In the periods August–November 1937 and July–October 1938 a further 20 flights were made by three Blohm und Voss Ha 139 four-engined seaplanes operating from the *Schwabenland* and the *Friesenland*.

But the honour of operating regular North Atlantic air services still went to Germany, when the Zeppelin LZ 129 *Hindenburg* began operation in 1936. Operated by Deutsche Zeppelin-Reederei, the *Hindenburg* left Friedrichshafen for New York with the first paying passengers on 6 May 1936, and on its return flight landed at Frankfurt, which became the regular European terminal. Ten return flights were made before the service was suspended for the winter, and so great was the demand that the airship's passenger accommodation had to be increased from the original 50. During that first season 1,309 passengers were carried, and the fastest flight from Lakehurst, New Jersey, to Frankfurt was made in 42hr 53min.

A programme of 18 return services was announced for 1937, and on 3 May the airship left Frankfurt with 97 passengers and crew. But on 6 May (local time) fire broke out as the ship was landing at Lakehurst, 35 people lost their lives, and German Zeppelin services came to an end, although three new Zeppelins had been ordered and the LZ 130 was test flown in September 1938.

FAR LEFT ABOVE: *In the summer of 1939 Jersey Airways used the prototype de Havilland D.H.95 Flamingo. This aircraft was a twin-engined high-wing airliner designed for short-range routes.*
FAR LEFT BELOW: *The Blohm und Voss Ha 139a Nordmeer (North Sea) with, in the background, the Nordwind (North Wind) on board the depot ship Friesenland. This flying boat was used on the North Atlantic route, where the depot ship stationed off the Azores could retrieve, refuel and catapult the aircraft off. In 1937 14 scheduled flights were operated successfully by this method.*

Germany did not re-establish North Atlantic air services until 1955, but in August 1938 made a spectacular non-stop flight from Berlin to New York in 24hr 56min and back in 19hr 55min. This flight was made by a special Focke-Wulf Condor, and was the first across the ocean by a four-engined landplane.

The first regular service by heavier-than-air craft to be operated over part of the North Atlantic began on 16 June 1937, when Imperial Airways opened a Bermuda–New York service with the C class flying boat *Cavalier*, and Pan American started working a reciprocal service with the Sikorsky S–42 *Bermuda Clipper*.

Then on 5–6 July that year the two airlines made their first North Atlantic survey flights. The special long-range C class 'boat *Caledonia*, commanded by Capt A. S. Wilcockson, flew from Foynes on the Shannon to Botwood in Newfoundland in 15hr 3min, and the Sikorsky S–42 *Clipper III* (Capt H. E.

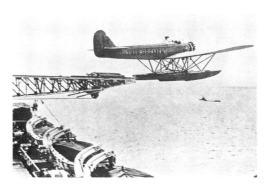

ABOVE: *The Dornier Do 18 flying boat* Aeolus *which made several trial flights across the North Atlantic during 1936. The route was from Berlin to New York, with a refuelling stop in mid-Atlantic alongside the depot ship* Schwabenland. *Mail was carried on these trips but there were no passengers on these experimental flights.*
LEFT: *A Heinkel He 58 seaplane, used to carry transatlantic mail from ship to shore. It was launched in mid-ocean from a catapult on board ship, and carried mail from both merchant ships and passenger liners, such as the* Europa.

Gray) flew in the opposite direction. *Caledonia* continued to Montreal and New York, and the Sikorsky flew on to Southampton.

Neither the C class 'boats nor the S–42 were suitable for commercial operation because they did not have the ability to carry sufficient fuel and a payload. Pan American ordered the large Boeing 314 for its services, and Imperial Airways undertook experiments designed to increase the range of its aircraft. One experiment involved refuelling in the air. An aircraft can carry a greater load than it can lift off the ground or water, so modified C class 'boats were built which could take off with a payload and then receive their main fuel supply from tanker aircraft via a flexible hose.

The other experiment involved launching a small aircraft from the back of a larger one. The Short-Mayo Composite Aircraft was therefore built. This consisted of a modified C class 'boat, on top of which was mounted a small floatplane. Take-off could be achieved with the power of all eight engines and the lift of both sets of wings, and then at a safe height the aircraft would separate, the smaller mailplane flying across the ocean and the flying boat returning to its base. The launch aircraft was the Short S.21 *Maia*; the mailplane the S.20 *Mercury*, powered by four 340hp Napier Rapier engines and having a crew of two. The first separation took place successfully near Rochester on 6 February 1938, and on 20–21 July the *Mercury* made the first commercial crossing of the North Atlantic by a heavier-than-air craft when,

commanded by Capt D. C. T. Bennett, it flew from Foynes non-stop to Montreal in 20hr 20min carrying mail and newspapers. From Montreal *Mercury* flew on to Port Washington, New York, but it played no further part in North Atlantic air transport.

On 4 April 1939, Pan American's Boeing 314 *Yankee Clipper* arrived at Southampton on its first proving flight from New York, and on 20 May the same aircraft left New York on the inaugural mail service. Flying via the Azores, Lisbon and Marseilles it arrived at Southampton on 23 May and left

BELOW: Coriolanus, *the last of the Q.E.A. Empire flying boats, was able to carry about 24 passengers plus mail at a cruising speed of 264km/h (165mph).*
BOTTOM: *The Short-Mayo composite aircraft on the Medway at Rochester. It consisted of a Short S.21 Maia flying boat and a Short S.20 Mercury seaplane. The S.21 flying boat carried the smaller seaplane on top of its fuselage, releasing it to complete the journey with mail and newspapers.*
BELOW RIGHT: *The Short S.30 C class flying boat* Cabot *being refuelled from a Handley Page Harrow tanker during tests over Southampton Water in July 1939.*

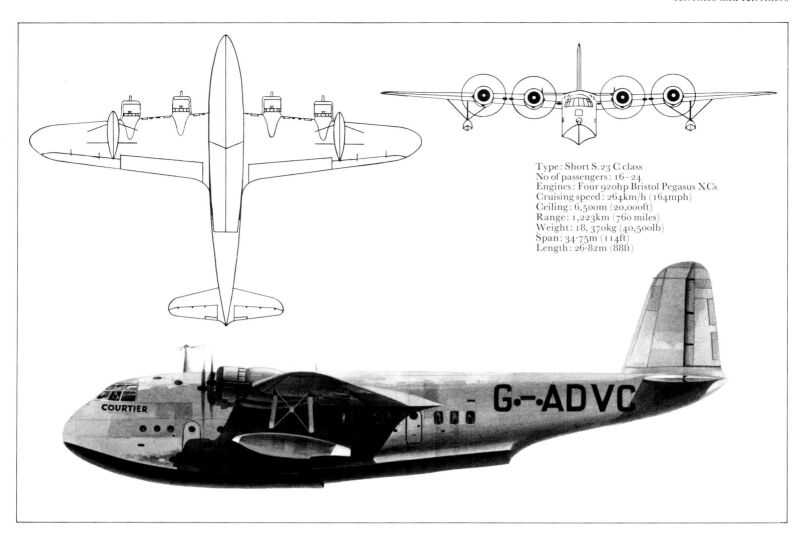

Type: Short S.23 C class
No of passengers: 16–24
Engines: Four 920hp Bristol Pegasus XCs
Cruising speed: 264km/h (164mph)
Ceiling: 6,500m (20,000ft)
Range: 1,223km (760 miles)
Weight: 18,370kg (40,500lb)
Span: 34·75m (114ft)
Length: 26·82m (88ft)

on the first westbound service the next day.

The northern mail route was opened on 24 June by the same aircraft, and on 28 June the *Dixie Clipper* left Port Washington on the inaugural southern-route passenger service – the first by a heavier-than-air craft. On 8 July, with 17 passengers, the *Yankee Clipper* left Port Washington on the inaugural service over the northern route, via Shediac, Botwood and Foynes. The single fare was $375.

On 4 August Britain began a weekly experimental mail service between Southampton and New York via Foynes, Botwood and Montreal, using S.30 C class flying boats which were refuelled in flight from Handley Page Harrow tankers based at Shannon and Botwood. The first service was flown by the *Caribou*, commanded by Capt J. C. Kelly Rogers. The full programme of 16 flights was completed on 30 September in spite of war having started. Because of the war Pan American terminated its services at Foynes, and then withdrew altogether on 3 October.

BOAC, as successor to Imperial Airways, operated the C class flying boat *Clare* on four round trips between Poole, in Dorset, and New York via Botwood and Montreal during the period 3 August–23 September 1940, and one round trip by the *Clyde* during October. These flights, most of which were made while the Battle of Britain was in progress, carried mail, despatches and official passengers, but the first truly commercial British North Atlantic services did not begin until 1 July 1946.

Airlines at war

BOAC had the wartime task of maintaining communication between the United Kingdom and the Commonwealth. This was largely achieved in spite of the German occupation of Europe and Axis control of much of the Mediterranean. The C class flying boats operated the *Horseshoe* route between Durban and Sydney via the Middle East, and connection to this route was provided by services between the United Kingdom and West Africa and thence via a trans-African route to Khartoum. The *Horseshoe* services were maintained until Japan cut the route at the beginning of 1942, after which they terminated at Calcutta. But by a brilliant operational feat Qantas reopened the route by introducing a service between Perth and Ceylon. This began on 10–11 July 1943, and was operated by Consolidated Catalina flying boats, which had to maintain radio silence while flying the 5,654km (3,513 miles) overwater route. On the inaugural flight the *Altair Star* took 28hr 9min from Koggala Lake to the Swan River, but the longest crossing made took 31hr 35min. Qantas also played a very important part in the fighting around New Guinea and suffered heavy casualties.

In May 1941 the Atlantic Ferry Organization of the Ministry of Aircraft Production began operating the North Atlantic Return Ferry service. This was flown with Consolidated Liberators, and had the task of returning crews to North America after they had delivered military aircraft to the United Kingdom. In September that year BOAC took over the operation.

BOAC also maintained North Atlantic services over various routes with three ex-Pan American Boeing 314s, operated a considerable number of vital routes in the Middle East, and flew regularly between Scotland and Sweden – flying unarmed aircraft across occupied Norway to maintain essential communication and import much-needed ball bearings to Britain. It was on this last operation that the airline used the all-wood de Havilland Mosquito. An outstanding BOAC operation was the evacuation of 469 British troops from Crete to Alexandria in April and May 1941 by the flying boats *Coorong* and *Cambria* in the course of 13 return flights.

The war had a major effect on many of the United States airlines. Within the US they had to maintain vital services with what fleets they were allowed to keep – mostly DC–3s – and they had to provide a wide range of services for the government. Until that time Pan American had been the only US airline to operate outside North America, and it undertook global operations. But many others undertook long-haul transoceanic flying for the armed services, mostly with military Douglas DC–4s (the C–54 Skymaster). TWA operated North Atlantic flights with Boeing Stratoliners, and American Export Airlines began a New York–Foynes service on 26 May 1942, with Sikorsky VS–44 flying boats.

One of the epic operations of the war was that over the 'Hump'. This was the supply route from India across the mountains into

China from 1942 until 1945. This was really a military operation, but was flown by the crews of several US airlines, and of US Air Transport Command and China National Aviation Corporation. DC–3s were used as well as some Consolidated C–87 Liberator transports, but the bulk of the work was done by Curtiss C–46 Commandos. The scale of the 'Hump' operation can be appreciated from the fact that 5,000 flights, carrying 44,000 tons, were made in one month. At some stages of the operation aircraft were taking off at two-minute intervals.

Finally, mention must be made of the New Zealand–Australia link. Before the war three C class flying boats had been ordered for Tasman Empire Airways (TEAL), although the airline was not founded until April 1940,

with New Zealand, Australia and the United Kingdom holding the shares. Only two boats, *Aotearoa* and *Awarua*, were delivered, but not until March 1940, and on 30 April *Aotearoa* flew the first Auckland–Sydney service. In June 1944 the 1,000th Tasman crossing was completed, and throughout the war these two flying boats provided the only passenger service of any kind between the two countries.

The return to peace

The war had brought about three major changes in air transport. These were the design and construction of higher-capacity longer-range aircraft capable of transoceanic operation, the large-scale construction of land airports, and worldwide operations by United States airlines. A further change was to follow

FAR LEFT ABOVE: *Pan American World Airways' Douglas DC–4* Clipper Union. *This was the first four-engined Douglas airliner and it went into service in 1939, but with the beginning of World War II many DC–4s were used by the USAAF under the designation C–54.*

FAR LEFT BELOW: *BEA's first Vickers-Armstrongs Viking service leaving Northolt for Copenhagen on 1 September 1946. This was the first post-war British airliner design, and was based on the Wellington wing and tail.*

LEFT: *A Scottish Airways' de Havilland Dragon Rapide, in wartime camouflage and with red, white and blue identification stripes, at Stornoway in Lewis. The eight-passenger Dragon Rapide stayed in production throughout World War II serving civil and military users all over the world.*

BELOW: *AB Aerotransport (Swedish Air Lines) Douglas DC–3* Falken *declares its Swedish nationality. Later ABA DC–3s were painted orange overall.*

shortly, as the overseas territories of European countries gained their independence and set up their own international airlines.

Most of the immediate postwar airline operations was undertaken with the DC–3, most of them war surplus, but the four-engined Douglas DC–4, mostly ex-military Skymasters, began to appear in some numbers and to be used to inaugurate transatlantic services, although with intermediate fuelling stops.

In Britain the decision had been taken to nationalize the air transport industry, with three corporations each having its own area of responsibility. BOAC already existed and, on 1 August 1946, British European Airways Corporation (BEA) and British South American Airways Corporation (BSAA) were

established under the Civil Aviation Act which was passed in 1946.

BOAC was to be responsible for all long-distance services except those to South America, BEA's sphere was mainly Europe and the United Kingdom, and BSAA was to operate all services to Latin America. BOAC had already opened a number of European routes and set up a BEA Division, and these were handed over on 1 August. But the domestic airlines retained their identities until early 1947, working the routes on behalf of the new BEA. Before the war there had been no British air services to South America, but in January 1944 British Latin-American Air Lines had been founded by a number of shipping companies. The name was changed to British South American Airways in

October 1945 and, following a series of proving flights, BSAA opened regular London–Buenos Aires services with Avro Lancastrians on 15 March 1946 – the first scheduled operations from the new Heathrow Airport. A route had been opened down the west coast of South America to Santiago in Chile, and the east coast route was also extended to that city by the time the company was changed to a State corporation.

BSAA was later absorbed into BOAC, and over the years numerous private companies were to establish regular air services. Finally, in 1972, BEA and BOAC were to be amalgamated to form British Airways, while British Caledonian Airways – resulting from the mergers of several private airlines – was to become the second major British airline,

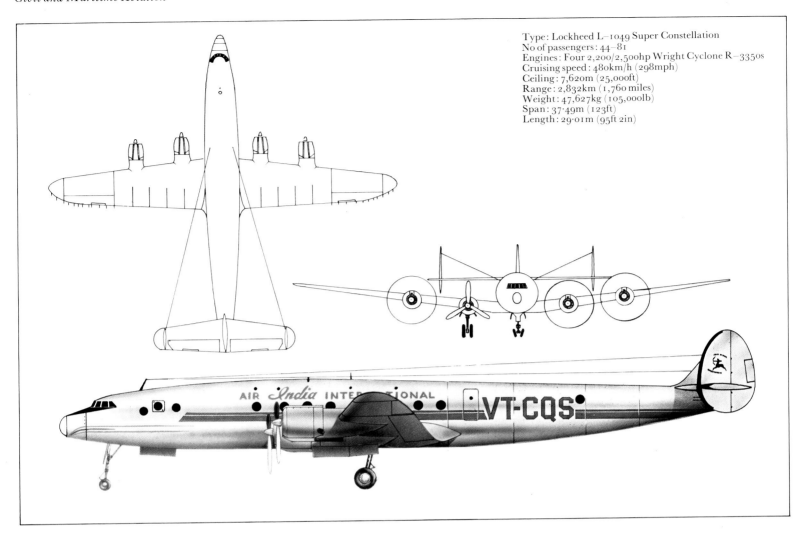

Type: Lockheed L–1049 Super Constellation
No of passengers: 44–81
Engines: Four 2,200/2,500hp Wright Cyclone R–3350s
Cruising speed: 480km/h (298mph)
Ceiling: 7,620m (25,000ft)
Range: 2,832km (1,760 miles)
Weight: 47,627kg (105,000lb)
Span: 37·49m (123ft)
Length: 29·01m (95ft 2in)

with responsibility for South American operations from October 1976.

Throughout Europe the airlines were rebuilding, mostly with DC–3s on short routes and DC–4s on intercontinental services. Air France, Sabena, KLM, Swissair and others established transatlantic services and other long-haul routes into Africa or the Far East. Sweden established SILA to operate Atlantic services and then eventually formed SAS – Scandinavian Airlines System – as a consortium of Danish, Norwegian and Swedish airlines. Some airlines were established with outside help – Alitalia was set up in Italy with BEA assistance, and TWA was to be involved in assisting airlines in Greece, Ethiopia and Saudi Arabia.

In Eastern Europe the prewar CSA in Czechoslovakia and LOT in Poland restarted services, and in Bulgaria, Hungary, Rumania and Yugoslavia airlines were set up with Soviet participation – BVS in Bulgaria, Maszovlet in Hungary, TARS in Rumania and JUSTA in Yugoslavia. JAT was also established in Yugoslavia. Later these East European airlines became completely nationally owned as TABSO (later Balkan Bulgarian Airlines), Malév (Hungary) and Tarom (Rumania). JUSTA ceased to exist in 1948 and Tarom was renamed LAR at the end of 1976. These airlines developed domestic and international services and in the cases of CSA, JAT, LAR and LOT have North Atlantic services.

The year 1955 saw the rebirth of Lufthansa, and it has now developed into one of the

line, developed its system, opened up networks of services in the New Guinea area, introduced round-the-world services and pioneered the Indian Ocean route to South Africa. ANA was the main Australian domestic operator and, on behalf of the newly formed British Commonwealth Pacific Air Lines, worked a Sydney–San Francisco–Vancouver service until BCPA could undertake its own operations. But a change in Australian civil aviation policy in 1945 was to have a profound effect on ANA. The Australian National Airlines Act of 1945 led to the founding of the State-owned Trans-Australia Airlines (TAA), and all main domestic air services had to be shared equally between TAA and ANA. Over the years Ansett had been acquiring various Australian airlines, and in October 1957 acquired ANA to form Ansett-ANA, now renamed Ansett Airlines of Australia. Meanwhile, TAA has been renamed Australian Airlines. These two major operators compete vigorously, but have virtually unblemished safety records and set high standards.

In New Zealand the newly created New Zealand National Airways ran the domestic services, and TEAL (now Air New Zealand) developed its services between New Zealand and Australia, retaining flying boats until June 1954. TEAL also opened a number of South Pacific routes – on which flying boats survived even longer – and eventually a trans-Pacific service to the United States.

India's airlines rapidly developed after the war, new companies were formed, and

world's major airlines. In East Germany a separate Deutsche Lufthansa had been established the previous year, but it was forced to abandon this name and since 1963 has worked as Interflug, which had been set up in 1958 to work services between East Germany and Western Europe.

The United States rebuilt its domestic networks and started developing worldwide services on a big scale. Pan American resumed commercial North Atlantic services, and was joined by TWA and American Overseas Airlines (AOA), successor to American Export Airlines. TWA opened up routes to the Middle East and Far East, and AOA was absorbed by Pan American which even-

tually inaugurated round-the-world services. United Air Lines and Pan American found lucrative traffic between the USA and Hawaii, and Northwest Airlines established a northern Pacific route to the Orient. Braniff opened services to South America and in more recent times National Airlines opened a Miami–London route.

In Canada Trans-Canada Air Lines and Canadian Pacific Air Lines made rapid progress, both building up services to Europe and the latter establishing routes to Australia and Japan. They are Canada's biggest airlines and now trade as Air Canada and CP Air respectively.

Qantas, as Australia's international air-

FAR LEFT CENTRE: *The Qantas de Havilland Canada Otter* Kerowagi *on the hillside aerodrome at Wau, New Guinea, in 1959.*
FAR LEFT BELOW: *The Douglas DC–6 prototype, a development of the DC–4, with a pressurized fuselage and more powerful engines. The C–118 was a military transport version of the DC–6.*
TOP: *Transatlantic services to the UK were started by Pan American on 4 February 1946 and by TWA on the following day, both using the Lockheed L–049. Illustrated are the much bigger L–1049C Super Constellation, the first civil version with Turbo-Compound engines, and the long-range L–1649 Starliner which entered service in 1957.*
ABOVE: *A TWA Lockheed L–1049G Super Constellation over New York. This aircraft is typical of the final stage of development of the long-range piston-engined airliners that preceded the jet age.*

in July 1946 Air-India was founded as the successor to Tata Air Lines. In March 1948 Air-India and the Indian Government formed Air-India International, and this airline began a Bombay–Cairo–Geneva–London service on 8–9 June 1948, using Lockheed Constellations. Now State-owned as Air-India, this airline enjoys a remarkable reputation for efficient operation and has a jet-operated network stretching from New York through India to Japan and Australia. By 1953 India had eight domestic airlines operating an extensive route system with a total fleet of nearly 100 aircraft, but on 1 August that year they were amalgamated to form the State-owned Indian Airlines Corporation which now carries in excess of three million passengers a year.

With the creation of Pakistan in August 1947, that country rapidly had to develop its air transport in order to provide communication between its widely separated west and east wings (the latter is now Bangladesh). The first airline was Orient Airways, which had been founded in India the previous year on the initiative of Jinnah, the founder of Pakistan. This airline established vital services including the supply routes through the Indus Valley and the world's highest mountains to Gilgit and Skardu. Pakistan International Airlines (PIA) was set up, and with three Lockheed Super Constellations began Karachi–Dacca nonstop services in June 1954 and Karachi–Cairo–London services on 1 February 1955. The airline was reorganized as a corporation with majority government holding in March 1955, and took over Orient Airways.

South African Airways greatly expanded operations after the war, worked the *Springbok* services to London in association with BOAC, opened a trans-Indian Ocean route to Australia, another to Hongkong and one to New York via Ilha do Sul.

In January 1947 the Governments of Kenya, Tanganyika, Uganda and Zanzibar established East African Airways, initially with a fleet of six de Havilland Dragon Rapides. Owned by Kenya, Tanzania and Uganda, East African Airways was a jet-equipped airline flying regional services, and routes to Europe, India and Pakistan. It was on the point of dissolution in early 1977.

Starting in 1951, Japan Air Lines began in a small way with the help of Northwest Airlines, which supplied aircraft and crews – Japanese nationals still being forbidden to fly as pilots. This airline rapidly established itself, and is now widely regarded as a major international carrier, having a very extensive route system and carrying about nine million passengers a year. There is now an enormous volume of air traffic in Japan, and this is shared between JAL, All Nippon Airways, TOA Domestic Airlines and a few small companies. Japan Asia Airways was formed to operate services to Taiwan after JAL began operating to China.

A new generation of transport aircraft
In the first full year of peace after World War II, the scheduled services of the world's airlines carried 18 million passengers, double

BELOW: *The flight deck of a Boeing 377 Stratocruiser, with the engineer's panel just visible at right. This was Boeing's first post-World War II airliner, entering service in 1947. Four 3,500hp Pratt & Whitney engines gave the aircraft a speed of about 560km/h (350mph) and a range of up to 6,400km (4,000 miles). Only 55 civil versions were built – but nearly 1,000 military!*

BOTTOM LEFT: *The Douglas DC–4 Sverker Viking of SAS (Scandinavian Airlines System).*
BOTTOM RIGHT: *A Pakistan International Airlines' Douglas DC–3 at Chitral in the Hindu Kush, one of the many examples of DC–3s still in service with airlines throughout the world 40 or more years after their first introduction in 1936. Some may go on to 2000.*

the 1945 total, and in 1949 the total for the year was 27 million. For some time to come the passenger total was to double every five years, and after a while cargo tonnage was to grow even faster.

These enormous traffic totals were largely due to the production of a remarkable series of transport aeroplanes – mostly designed and manufactured in the USA. Ever since the war the DC–3 has played a major role, although its numbers have declined; but even in the early 1940s it was obvious that larger, longer-range aircraft were required.

The four-engined Douglas DC–4E prototype had been abandoned as unsuitable in 1938, but Douglas designed a smaller, unpressurized DC–4. Before any could be delivered the United States was at war, and DC–4s were put to military use, mostly as C–54 Skymasters. US airlines gained experience with these aircraft, flying them on military duties, but had to wait for peace before they could acquire surplus ex-military aircraft and some of the 79 civil examples built after the war.

The DC–4 was a low-wing monoplane with nosewheel undercarriage and four 1,450hp Pratt & Whitney R-2000 engines. Initially it had 44 seats, but high-density seating for up to 86 was later installed. American Overseas Airlines introduced the DC–4 on New York–Hurn (for London) services at the end of October 1945. There were two intermediate stops and the scheduled time was 23hr 48min. On 7 March 1946, DC–4s went into US domestic service on the New York–Los Angeles route with American Airlines. Subsequently they were used by most US carriers and by many airlines throughout the world. A total of 1,242 civil and military DC–4s was built.

The DC–4 was a magnificent aeroplane, but it was unpressurized and its cruising speed of 352km/h (219mph) was no match for the other US wartime transport, the Lockheed Constellation. This had been designed for TWA as a long-range aircraft, but it did not fly until January 1943. TWA and Pan American had each ordered Constellations, but the small number built went to the United States Air Forces as C–69s, entering service in April 1944.

After the war civil production was resumed and the military aircraft were brought up to

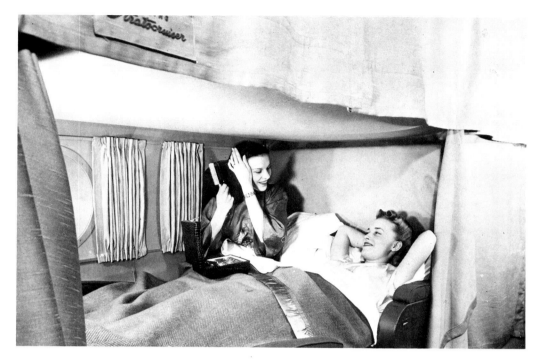

At about that time Curtiss-Wright's 3,250/3,400hp Turbo-Compound engine became available, and both Douglas and Lockheed developed their aircraft to make use of the new engine. Douglas slightly lengthened the DC–6B, but retained the same wing and produced the longer range DC–7, while Lockheed retained the basic Super Constellation airframe to produce the Turbo-Compound-powered L.1049C, D, E, G and H.

The DC–7 was the first aircraft capable of operating US transcontinental services in both directions without stops, and American Airlines introduced it on non-stop New York–Los Angeles services on 29 November 1953, scheduled to take 8hr 45min on the westbound flight, and 8hr eastbound – although it did not always make these times. Only 110 DC–7s were built, all of which were initially purchased by US airlines. Some later served as DC–7F freighters.

TWA had begun non-stop coast-to-coast services with L.1049C Super Constellations between Los Angeles and New York on 19 October 1953, but these had to stop at Chicago for fuel on westbound flights. However, the L.1049G, with wing tip fuel tanks, enabled TWA to match American Airlines' non-stop performance in both directions.

The DC–7 was developed into the longer-range DC–7B. This entered service in 1955, but only 112 were built – 108 for US airlines and four for South African Airways.

The DC–7 and L.1049G had made US transcontinental non-stop operation a practical undertaking, but at that time non-stop North Atlantic operation in both directions was rarely possible. In order to attain this desirable goal, Douglas added 3m (10ft) to the span of the DC–7 and increased its fuel tankage. It thus became the DC–7C Seven Seas with 7,412km (4,606 miles) range and North Atlantic non-stop capability. Pan American introduced them on the route on 1 June 1956, and numerous other airlines adopted the type, 121 being built.

Lockheed went further than Douglas and designed a completely new 45·72m (150ft) span one-piece wing, containing fuel for more than 9,655km (6,000 miles), and mated this with the Super Constellation fuselage to form the L.1649A Starliner, with up to 99 seats. It entered service with TWA on the New York–London service on 1 June 1957.

These two very advanced piston-engined aircraft also made possible one-stop services between Europe and Japan over the Polar route, with a call at Anchorage in Alaska. SAS was first with the DC–7C on the Copenhagen–Tokyo route on 24 February 1957, and Air France followed about a year later, with L.1649As flying between Paris and Tokyo.

In addition to the Douglas and Lockheed types, there was one other large American four-engined airliner – the Boeing Stratocruiser. This was bigger than the other types, and had a lounge and bar on a lower deck. It had seats for up to 100 passengers, was powered by four 3,500hp Pratt & Whitney Double Wasp engines, and entered service over the North Atlantic with Pan American in 1949. Only 55 civil Stratocruisers were built, and they earned a poor reputation

civil standard. Pan American began a Constellation service between New York and Bermuda on 3 February 1946, and in the same month TWA introduced Constellations on its New York–Paris and New York–Los Angeles services. The original Constellation had four 2,200/2,500hp Wright Cyclone R–3350 engines and 44 seats – later increased to as many as 81 in some configurations. It had a pressurized cabin, and at 480km/h (298mph) its cruising speed was at least 127km/h (79mph) higher than the DC–4's. Several improved models were built, and total production amounted to 233.

The superior performance of the Constellation led Douglas to build the comparable DC–6, with pressurized cabins and a maximum cruising speed of just over 482km/h (300mph). The DC–6 had the same 35·81m (117ft 6in) span as the DC–4 but was longer, had accommodation for 50 to 86 passengers, and was powered by four 2,400hp Pratt & Whitney R–2800 engines.

American Airlines and United Air Lines were the first to operate DC–6s, introducing them on 27 April 1947, on the New York–Chicago and transcontinental routes re-

spectively. Its eastbound transcontinental scheduled time was 10hr. It was used by many major airlines and 170 were built.

In Montreal Canadair produced versions of the DC–4 and DC–6 modified to take Rolls-Royce Merlin liquid-cooled engines.

Douglas followed the DC–6 with the longer fuselage DC–6A and DC–6B, which had increased range. The DC–6A was a cargo aircraft mainly used by the US Air Force and Navy, but the DC–6B proved to be one of the finest piston-engined passenger aircraft ever produced. It entered service on 29 April 1951, on American Airlines' transcontinental route, began service with 54 seats but, later, in high-density configuration could seat 102 passengers. The DC–6B cruised at 507km/h (315mph) and had a range of just over 4,830km (3,000 miles). It was used by a large number of airlines, a total of 288 being built.

Lockheed developed the Constellation into the L.1049 Super Constellation, with 5·48m (18ft) increase in fuselage length and 2,700hp Cyclone engines. It went into service as a 66-seat aircraft with Eastern Air Lines on 17 December 1951.

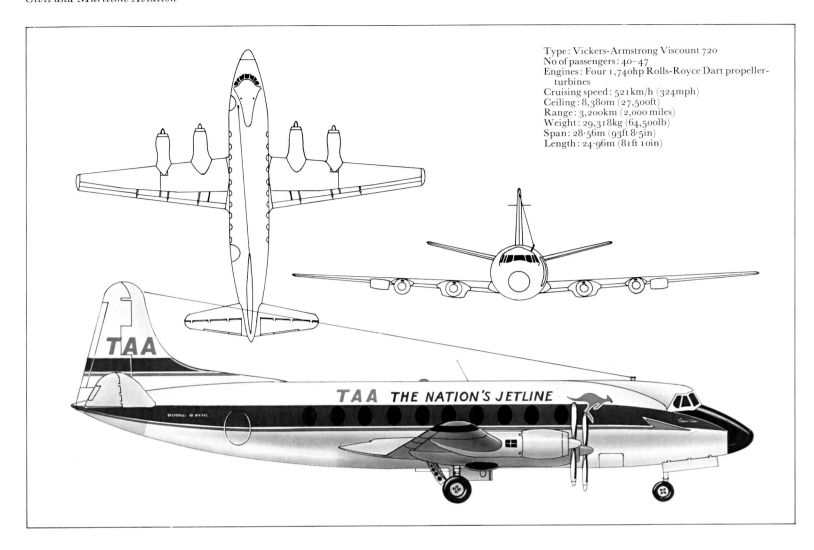

Type: Vickers-Armstrong Viscount 720
No of passengers: 40–47
Engines: Four 1,740hp Rolls-Royce Dart propeller-
    turbines
Cruising speed: 521km/h (324mph)
Ceiling: 8,380m (27,500ft)
Range: 3,200km (2,000 miles)
Weight: 29,318kg (64,500lb)
Span: 28·56m (93ft 8·5in)
Length: 24·96m (81ft 10in)

because of engine and propeller problems. In spite of this they were popular with crews and passengers, and maintained BOAC and Pan American first-class North Atlantic services until the introduction of jet aircraft in 1958.

In the immediate postwar years there was an urgent need for a twin-engined short-haul aeroplane to replace the DC–3, and several countries designed aircraft to meet this requirement. In the United States, Martin and Convair both produced aircraft in this category. They were low-wing monoplanes with nosewheel undercarriages and 2,500hp Pratt & Whitney R–2800 engines.

The 42-seat Martin 2–0–2 was unpressurized and entered service in autumn 1947 with Northwest Airlines in the United States and LAN in Chile. The type was involved in a number of serious accidents, and one suffered a wing failure which led to the type being withdrawn. After strengthening the 2–0–2 returned to service in 1950 as the Martin 2–0–2A. It was followed by the 48 to 52-seat pressurized Martin 4–0–4, which entered service with TWA in October 1951. A total of 149 aircraft of both models was built for several airlines.

Much greater success was enjoyed by the Convair-Liner. The CV–240, with 40 seats and pressurized cabin, entered service in June 1948 with American Airlines. It quickly found favour, and 176 civil examples were built. The improved 44-passenger CV–340 followed in 1952, and in 1956 the 52 to 56 seat CV–440 Metropolitan made its debut.

More than 1,000 civil and military Convair-Liners were built, of which more than 240 were later re-engined with propeller-turbines.

In Britain in the early postwar years BOAC was using DC–3s, Avro Lancastrians and Yorks developed from the Lancaster bomber, and Short flying boats. But plans had been made for new aircraft, and two of them flew before the end of 1945. They were the Vickers-Armstrongs Viking, developed from the Wellington bomber, and the Bristol 170 Freighter/Wayfarer.

The Viking was a mid-wing monoplane with two Bristol Hercules engines, unpressurized cabin, and a tailwheel undercarriage. It cruised at about 320km/h (200 mph), and on entering service with BEA on 1 September 1946, had 27 seats. For several years it was the airline's principal type. It also served with many other airlines, and 163 were built.

The Bristol 170 was very different. It was a high-wing monoplane with a slab-sided fuselage, non-retractable undercarriage and two Hercules engines. The Wayfarer was a passenger aircraft, but the Freighter undertook the duties which its name implied – the best known operation being Silver City Airways' cross-Channel vehicle ferries, which it and the longer-fuselage Superfreighter undertook for many years. The type also operated cargo services between North and South Island in New Zealand.

A very handsome British aeroplane was the high-wing Airspeed Ambassador, with two 2,625hp Bristol Centaurus engines and

pressurized cabin. It had accommodation for 47 to 49 passengers and was introduced into scheduled service as the Elizabethan by BEA in March 1952. Only 20 were built.

France's first postwar transport aircraft to enter airline service was the 33-seat Sud-Est SE.161 Languedoc which, as the Bloch 161, had made its first flight in September 1939, but could not be produced until after the war. It was originally powered by four Gnome Rhône engines, but most had Pratt & Whitney R–1830s. A total of 100 Languedocs was built, entering service with Air France on the Paris–Algiers route in May 1946 and serving a number of airlines in Europe and the Middle East.

Italy continued to produce three-engined Fiats and four-engined Savoia Marchetti S.M.95s. One of the latter inaugurated Alitalia's Rome–London service on 3 April 1948, but these types made no significant contribution to air transport outside Italy.

There is one other European transport

FAR LEFT BELOW: *One of Finnair's Convair CV–340s at Helsinki. Soon after the end of World War II Convair began to produce a series of twin-engined, short-range airliners which were exported to a number of European airlines in addition to being used by American operators.*
BELOW: *A Vickers-Armstrong Viscount of Trans-Australia Airlines. This aircraft was the first successful turboprop airliner.*

BOTTOM LEFT: *A Vickers-Armstrong Viking of Indian National Airways. The design of its wing and tail was based on that of the Vickers Wellington bomber, and the prototype made its first flight in June 1945. It was soon in service with BEA and many independent airlines. Two 1,675hp Bristol Hercules 634 radial engines produced a maximum cruising speed of 366km/h (210mph) and a range of up to 1,840km (1,150 miles).*

BOTTOM RIGHT: *A Bristol 170 Freighter flying with the port engine shut down and the propeller feathered. This high wing monoplane started its development during the closing years of World War II and made its first flight in December 1945. Most production models went to military users, and a few were used for car-carrying transport by commercial operators. All-freight versions had a clam-shell nose-door.*

aircraft of the postwar era worthy of mention. This was the Swedish Saab Scandia, a low-wing monoplane with 32 to 36 seats and two 1,800hp Pratt & Whitney R–2180 engines. It first flew in November 1946 and 18 were built, including six by Fokker.

Turbine power

In Britain the Brabazon Committee was set up in December 1942 to make recommendations for the development of postwar transport aircraft. Among its recommendations was the design and production of a North Atlantic turbojet aeroplane, Brabazon Type IV, and a short-haul propeller-turbine type, Brabazon Type IIB.

The Type IV evolved through a number of designs to become the de Havilland Comet – the world's first turbojet airliner. Two types were designed to meet the Type IIB recommendation: the Armstrong Whitworth Apollo, of which only two prototypes were built, and the Vickers-Armstrongs Viscount. Tur-

bine power was to change the whole standard of airline flight by reducing vibration and interior noise levels. In the case of the turbojet, flight times were to be approximately halved in one step.

The prototype Viscount, the V.630, made its first flight on 16 July 1948. It was a low-wing monoplane with a pressurized cabin for 32 passengers, and was powered by four 1,380hp Rolls-Royce Dart turbines driving four-blade propellers. Although the V.630 was considered too small to be economic, BEA did use it to operate the first ever services by a turbine-powered aeroplane. On 29 July 1950, it flew from Northolt, west of London, to Le Bourget, Paris, carrying 14 revenue passengers and 12 guests. The V.630 stayed on the Paris route for two weeks and then, from 13 August until 23 August, operated over the London–Edinburgh route on the world's first turbine-powered domestic services.

Rolls-Royce increased the power of the

Dart and enabled Vickers to stretch the Viscount to become a 47- to 60-seat aircraft as the V.700. BEA placed an order for 20 and regular Viscount services began on 18 April 1953, when RMA *Sir Ernest Shackleton* operated the London–Rome–Athens–Nicosia service. The Viscount immediately proved a success, being produced in several versions and sold in many parts of the world, including the USA and China. It frequently doubled the airline's traffic, and 445 were built.

The Comet 1 was a very clean low-wing monoplane with four 2,018kg (4,450lb) thrust de Havilland Ghost turbojets buried in the wing roots. There was accommodation for 36 passengers in two cabins and pressurization enabled it to fly at levels over 12,190m (up to 40,000ft). The first prototype flew on 27 July 1949, and soon made a number of spectacular overseas flights. BOAC took delivery of ten Comet 1s and on 2 May 1952, operated the world's first jet service – over the London–Johannesburg route. With their cruising speed of 788km/h (490mph) Comets covered the 10,821km (6,724 miles) in less than 24hr. On London–Singapore they cut the time from $2\frac{1}{2}$ days to 25hr, and reduced the London–Tokyo time from 86hr to $33\frac{1}{4}$hr.

Air France and UAT introduced Comets, and they were ordered by several other airlines. But exactly a year after their introduction a Comet broke up in flight near Calcutta, and in January 1954 another disintegrated and fell into the sea near Elba. After modifications the Comet was put back into service,

but less than three weeks later, on 8 April, a third Comet broke up, and the type was withdrawn from service.

Fatigue failure of the pressure cabin was said to have been the cause of the last two failures, and some fuselage redesign resulted. Comet 2s, already under construction, were modified and went to the RAF. Work went ahead on the Rolls-Royce Avon-powered Comet 4 with longer fuselage, seats for up to 81, and extra wing-mounted, fuel tanks.

BOAC ordered 19 Comet 4s and on 4 October 1958, operated the first ever North Atlantic jet services, the London–New York flight being made in 10hr 22min with a fuel stop at Gander. The eastbound flight was made nonstop in 6hr 11min. A shorter-span longer-fuselage Comet 4B, with seats for up to 101, was introduced by BEA on 1 April 1960, and in the same year the Comet 4C was commissioned – this combined the Comet 4 wings with the Comet 4B fuselage.

The second European turbojet airliner to enter service was the French Sud-Aviation SE.210 Caravelle. It was a twin-engined aircraft and the first to have its turbojets mounted on either side of the rear fuselage. The Caravelle was designed primarily for operation between France and North Africa. It first flew on 27 May 1955, was powered by two 4,762kg (10,500lb) thrust Rolls-Royce Avons and, as the Caravelle I, had seats for up to 80 passengers. The Caravelle I entered service with Air France in May 1959 on the Paris–Rome–Istanbul route, and with SAS between Copenhagen and Cairo.

piston engines to meet the requirement. Subsequently it was agreed that Bristol Proteus propeller-turbines would be used and in this form, as the Britannia, the prototype flew on 16 August 1952. The Britannia was a large low-wing monoplane with seats for up to 90 passengers and a take-off weight of 70,305kg (155,000lb).

BOAC introduced this Britannia 102 on the London–Johannesburg route on 1 February 1957, and the longer-range Britannia 312 on London–New York services on 19 December 1957. The latter were the first North Atlantic services by turbine-powered aircraft. The Britannia was an extremely good aircraft, but it was delayed by BOAC and only 85 were built.

Canadair produced a number of Britannia variants, one of which, the CL–44, went into airline service. These aircraft generally resembled the Britannia, but had 5,730hp Rolls-Royce Tyne propeller-turbines and lengthened fuselages. Known as the CL–44D, and provided with swing-tail fuselage for cargo loading, the type entered service with the US cargo carrier Seaboard World Airlines in July 1961. Then came the CL–44D, and provided with a swing-tail fuselage modation for up to 214 passengers, for the cheap-fare North Atlantic services of the Icelandic airline Loftleidir.

The large propeller-turbine aircraft built for BEA was the Vickers-Armstrongs Vanguard, also powered by Tynes. It could carry 139 passengers and had considerable underfloor cargo capacity. The Vanguard first

Airlines and Eastern Air Lines in January 1957, and entered service in 1959.

The most successful of all the western propeller-turbine transports has been the Fokker F27 Friendship. This is a high-wing monoplane powered by two Rolls-Royce Darts and normally accommodating about 48 passengers. The first prototype flew on 24 November 1955, and the aircraft went into production in the Netherlands and also in the United States as the F–27, built by Fairchild (later Fairchild Hiller).

The type first entered service in the United States, with West Coast Airlines, on 28 September 1958. In Europe Aer Lingus began operating the Dutch-built aircraft on Dublin–Glasgow services on 15 December 1958. Numerous versions have been built, to a final (1986) total of 786. Its place on the production line has been taken by the Fokker 50, with 2,250hp Pw125B engines driving quiet six-blade propellers and with many improvements to the structure and systems. Normally seating 50, the Fokker 50 entered service in 1987.

Britain's attempts to break into this market have been less successful. Handley Page produced the twin-Dart Herald, which was very similar to the F.27, but only 48 were completed. More successful was the low-wing Avro (later British Aerospace) 748, also powered by two 3,060hp Dart engines, of which 381 were built in Britain and India. This was powered by Darts similar to those of the F27, but for the successor ATP (Advanced Turbo Prop) BAe selected the 2,653hp

Numerous versions of the Caravelle were produced, up to the 128-passenger Caravelle 12B, and they were operated in many parts of the world, including the United States, where United Air Lines had 20. The Caravelle was an exceptional aeroplane, and of the 280 built nearly 200 are still in service.

Although BOAC introduced the pure-jet Comet as early as 1952, the airline was to commission a large propeller-turbine aircraft nearly five years later, and BEA began using a similar type several months after it had introduced Comets. BOAC issued a specification early in 1947 for a Medium Range Empire (MRE) transport, and Bristol designed the Type 175 with four Centaurus

flew on 20 January 1959, and began operation with BEA on 17 December 1960, with regular operation from March 1961. About a month earlier it had entered service with Trans-Canada Air Lines, the only other customer for new Vanguards. Only 43 were built, and both airlines converted some into freighters.

Only one type of large civil propeller-turbine aeroplane, and Lockheed L–188 Electra, was built in the United States. This was smaller than the Vanguard, having accommodation for up to 99 passengers, and was powered by four 3,750hp Allison 501 engines. The Electra first flew on 6 December 1957, and entered service with American

PW 126, driving six-blade quiet propellers. First flown in August 1986, the ATP is certainly the largest and quietest and probably the most efficient of all new turboprops, seating up to 72 passengers.

There has also been one Japanese turbine-powered transport, the NAMC YS–11. This is a 46 to 60 passenger low-wing monoplane powered by two 3,060hp Dart engines and 182 were built.

The first 'big' jets

Not until the Comet 1 had been withdrawn from service did the first United States turbojet transport make its maiden flight. At Seattle, in Washington State, on 15 July

ABOVE LEFT: *A BOAC Bristol Britannia 102, a turboprop airliner famed for its low level of noise and therefore very popular with passengers.*
ABOVE CENTRE: *The Aer Lingus Fokker F.27 Friendship* St Fergal, *used on short-range European routes. The first prototype flew in 1955.*

ABOVE RIGHT: *One of Piedmont Airlines' Japanese NAMC YS–11s, powered by two Rolls-Royce Dart 3,060hp turboprop engines, which gave a maximum cruising speed of 475km/h (297mph) and a range of up to 2,360km (1,475 miles).*
OPPOSITE PAGE
LEFT: *A BAe 748 of Royal Nepal Airlines.*

ABOVE RIGHT: *A Lockheed L–188 Electra II. This was a medium-range turboprop airliner which came into service in January 1959, with several US airlines and KLM.*
CENTRE RIGHT: *A de Havilland Comet 4B of BEA, a version of the Comet 4 specially tailored to short European sectors.*

1954, a large brown and yellow swept-wing monoplane with four pod-mounted engines took to the air and began a new era in air transport. It was the Boeing 367–80, known affectionately as the Dash 80, and was in fact the prototype Boeing 707. This was a much more advanced aeroplane than the Comet: its wings were swept back 35°, and it embodied the experience gained with Boeing B–47 and B–52 jet bombers.

Pan American placed the first order, and the type went into production as the Model 707–100 with the customer designation 707–121. The production aeroplane had a span of 40·18m (131ft 10in), a length of 44·04m (144ft 6in), accommodation for up to 179 passengers, and a maximum weight of 116,818 kg (257,000lb). It was powered by four 5,670kg (12,500lb) thrust Pratt & Whitney JT3C turbojets, cruised at 917km/h (570 mph) and had a maximum range of 4,949km (3,075 miles) with full payload. The 707–121 entered service with Pan American on the

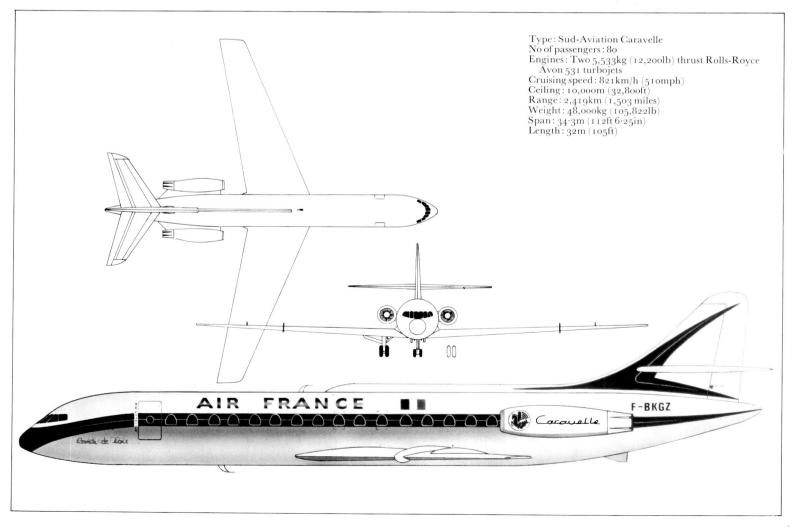

Type: Sud-Aviation Caravelle
No of passengers: 80
Engines: Two 5,533kg (12,200lb) thrust Rolls-Royce Avon 531 turbojets
Cruising speed: 821km/h (510mph)
Ceiling: 10,000m (32,800ft)
Range: 2,419km (1,503 miles)
Weight: 48,000kg (105,822lb)
Span: 34·3m (112ft 6·25in)
Length: 32m (105ft)

RIGHT: *A Convair CV–880, a four-engined, medium-range jet airliner with a swept wing. Sixty-five were built, mostly for US domestic airlines, but the market had already been captured by the Douglas DC–8 and Boeing 707.*
BELOW: *A Douglas DC–8 of Braniff Airways. This was designed as a rival to the Boeing 707 and de Havilland Comet 4 jet airliners, and has been very successful, both in the passenger and freight versions.*
FAR RIGHT ABOVE: *A Boeing 707–400 of Air India. Originally intended purely for American domestic routes, the 707 was very soon used on routes throughout the world.*
FAR RIGHT CENTRE: *A Boeing 720B of Pakistan Airlines. This smaller version of the 707 uses Pratt & Whitney JT3D turbofans.*
FAR RIGHT BELOW: *One of BEA's de Havilland Trident 3Bs. The booster engine can be seen at the base of the rudder. They later served with British Airways.*

North Atlantic in October 1958.

This original Boeing 707 was followed by a whole family of 707 passenger and cargo variants, with different lengths and weights, and turbofan power. The passenger-carrying 707–320B and passenger/cargo 707–320C models are still in wide use. Also developed was the short-to-medium-range version known as the Boeing 720 or, with turbofans, 720B. Commercial 707 production ceased in 1982, but in 1988 airframes were still being made for AWACS and E–6A aircraft.

The second United States jet transport was the Douglas DC–8. It closely resembles the Boeing 707, and has likewise been produced in several versions. The first DC–8 flew on 30 May 1958, and the JT3C-powered DC–8–10 entered service with Delta Air Lines and United Air Lines on 18 September 1959. Then followed the –20 with more powerful JT4A engines, the –30 long-range aeroplane with the same engines, the –40 with Rolls-Royce Conway bypass engines and the –50 with JT3D turbofans. There were also freight, convertible, and mixed passenger/cargo versions. All had a span of 43·41m (142ft 5in) and a length of 45·87m (150ft 6in). Maximum seating was 177 and maximum weight varied between 120,200kg (265,000lb) and 147,415kg (325,000lb).

In 1966 the Series 60 DC–8 appeared. This was built in three main versions, the –61 with the same span as earlier models but a length of 57·09m (187ft 4in), seating for up to 257 and take-off weight of 147,415kg (325,000lb); the –62 with 45·23m (148ft 5in) span, 47·98m (157ft 5in) length, seats for 201 and take-off weight of 151,952kg (335,000lb); and the –63, combining the wing of the –62 with the fuselage of the –61. This had a maximum weight of 158,755kg (350,000lb) and seats for up to 269.

The DC–8–61 was put into service between Los Angeles and Honolulu by United Air Lines on 25 February 1967; the DC–8–62 entered service on the Copenhagen–Los Angeles route with SAS on 22 May 1967; and the DC–8 –63 was introduced by KLM on the Amsterdam–New York route in July 1967. There were cargo and convertible versions of the Series 60 aircraft, and a total of 556 DC–8s of all models was built before ending in 1972. In 1982–86 a total of 110 good DC–8–63s were upgraded to Super 70 standard with the quiet CFM56 turbofan engine giving 9,979kg (22,000lb) thrust.

The third United States constructor of large commercial jet transports was the Convair Division of General Dynamics, which produced two types. The first was the CV–880. It was smaller than the Boeing and Douglas types, its narrower fuselage having a maximum seating capacity of 130. It was powered by four 5,080/5,285kg (11,200/11,650lb) thrust General Electric CJ–805 engines, had a maximum weight of 87,770kg (193,500lb) and cruised at up to 989km/h (615mph). Only 65 were built.

The second Convair was the CV–990 (named Coronado by Swissair). It was designed for American Airlines and closely resembled the CV–880, but had 7,257kg (16,000lb) thrust General Electric CJ–805–23B turbofans, seating for up to 158 and a maximum cruising speed of 1,005km/h (625 mph). Distinctive external features were the four canoe-like shock bodies which extended aft of the wing. The CV–990 entered service with American Airlines and Swissair in March 1962.

Britain made an attempt to break into the so-called 'big' jet market, but with little success. The aircraft was the Vickers-Armstrongs VC10, which was in the same category as the Boeing 707 but had its four 9,525kg (21,000lb) thrust Rolls-Royce Conway bypass engines mounted in pairs on each side of the rear fuselage, and a high-mounted tailplane and elevators. The VC10 was designed for BOAC and was required to operate from restricted, hot and high runways which demanded specially good take-off and landing performance. Indeed, the VC10 could undertake some operations which were impossible for the early Boeings.

The standard VC10 could carry up to 139 passengers, had a maximum weight of 141,520kg (312,000lb), and a maximum cruising speed of 933km/h (580mph). The VC10 went into service on the London–Lagos route on 29 April 1964, and was followed by the Super VC10 which was 3·96m (13ft) longer, had more powerful Conways, seating for up to 163 and a maximum take-off weight of 151,952kg (335,000

and 304 economy-class seats, the maximum payload was 56,245kg (124,000lb) and the maximum cruising altitude 13,715m (45,000 ft). In spite of its size the aircraft was completely orthodox in appearance, except that it had four four-wheel main undercarriage units in order to spread the load on runways, taxiways and airport aprons. Maximum seating was originally quoted as 490.

The Model 747 underwent rapid development, with increased power and consequently higher permissible weights, and the 747–200B with 22,680kg (50,000lb) thrust Rolls-Royce RB.211s, 23,585kg (52,000lb) Pratt & Whitney JT9D–70s or 23,815kg (52,500lb) General Electric CF6–50Es has a brake-release weight of up to 371,943kg (820,000lb) and has taken off at about ten tons above this weight.

The Boeing 747 has also been built as the all-cargo 747F, with upward-swinging nose for front-end loading and a payload of up to 113 tons, and as the 747C, which can have all-cargo or all-passenger configuration or a combination of both.

In September 1973 Boeing flew the 747SR with structural reinforcement to allow high-frequency operation over short routes, which imposes greater stress on the structure – particularly the wings and undercarriage. This model can carry 500 passengers, 16 on the upper deck, but its take-off weight is limited to 272,154kg (600,000lb). Seven Boeing 747SRs were built for Japan Air Lines' domestic services, and they began operation on 10 October 1973.

All these Boeing 747s had the same external dimensions, but, on 4 July 1975, Boeing flew the first Model 747SP (Special Performance). This has a shortened fuselage, measuring 14·73m (48ft 4in) less in length than the other 747s, but its fin is 1·52m (5ft) higher.

ing the traffic growth without increasing aircraft movements was to build much bigger aircraft. At the same time these aircraft would reduce seat-mile costs and ease the noise problems in the vicinity of airports – an aspect of air transport which was beginning to reach generally intolerable levels.

The first of this new generation was made known in April 1966, when Boeing announced that it would build the Model 747 and that Pan American had placed an order for 25. In one step the Boeing was doubling the capacity, power and weight of the transport aeroplane. The media called it the Jumbo Jet.

In general appearance the Boeing 747 was similar to the Boeing 707, but was scaled up to have a wing span of 59·63m (195ft 8in), a length of 70·51m (231ft 4in) and a height from the ground to the top of the fin of 19·32m (63ft 5in). The maximum interior width is just over 6m (20ft), and the ceiling height 2·43m (8ft). Seating on the main deck can be nine- or ten-abreast (in tourist class) with two fore and aft aisles, and this feature has caused the term 'wide-body' to be applied to this category of airliner. The flight deck is on an upper level, and behind this is a passenger cabin which served as a first-class lounge in early 747s. A stairway connects the two levels. Three cargo and baggage holds have a total volume of 175·28m³ (6,190cu ft), which is about equal to the entire volume of a cargo Boeing 707.

The original engines in the Boeing 747 were four 18,597kg (41,000lb) thrust Pratt & Whitney JT9D–1 turbofans, with a diameter of just under 2·43m (8ft). The first quoted maximum take-off weight was 308,440kg (680,000lb), and landing weight 255,825kg (564,000lb). The first Boeing 747 flew on 9 February 1969, and when the type entered service with Pan American on the New York–London route on 22 January 1970, the brake-release weight had already risen to 322,956kg (712,000lb).

Pan American's aircraft had 58 first-class

Other differences include a much larger horizontal tail, a tandem pair of double-hinged rudders, modified landing gears and completely redesigned wing flaps. The latter are of the single-slotted type, requiring none of the giant track fairings which project behind the wings of other 747s. The SP can typically seat 299 passengers on the main deck plus 32 upstairs, and carry them 12,324km (7,658 miles). Maximum weight is up to 317,515kg (700,000lb).

In 1980 Boeing announced the 747SUD (stretched upper deck), which later was re-designated 747–300. Overall dimensions are unaltered, but the upper deck is extended 7.11m (23ft 4in) to the rear to increase its accommodation from 32 to 69 passengers, plus an optional crew rest area. Replacing the spiral staircase by a straight stairs enables seven more seats to be added on the main deck, and two new doors are added at upper-deck level. There are few other changes, but the revised body shape enables cruise Mach number to be raised from 0·84 to 0·85, adding some 11km/h (7mph) to a maximum of 996km/h (619mph). The –300 entered service in 1983, some of the deliveries being Combis (convertible passengers/cargo).

In 1985 Boeing announced an advanced long-range version of the –300, the 747–400. This is cleared to a weight of 394,625kg (870,000lb), has CF6–80C2, PW4256 or RB211–524G engines in the 26,300kg (58,000lb) class and many other changes including wings extended to 64.92m (213ft) with large winglets. The –400 carries 412 passengers 13,528km (8,406 miles).

The Boeing 747 was too big for some airlines which required large-capacity aircraft, and McDonnell Douglas and Lockheed both built very similar three-engined aircraft to meet this need. In layout both were wide-bodied aircraft with two wing-mounted engines and one tail-mounted, although the two companies adopted different methods of mounting the rear engine. In the Douglas DC–10 the rear engine is built into the fin structure, but in the Lockheed L.1011 Tri-Star it is within the fuselage with the air intake above the fuselage forward of the fin.

The first DC–10 flew on 29 August 1970.

This was the Series 10 version with 18,145kg (40,000lb) thrust General Electric CF6–6D turbofans, seating for 270 passengers in basic mixed-class configuration or a maximum of 345 in economy class, a maximum brake-release weight of 195,043kg (430,000lb) and a still-air range of 6,727km (4,180 miles). Span of the –10 is 47·34m (155ft 4in), and length 55·55m (182ft 3in). This version entered service on 5 August 1971, on American Airlines' Los Angeles–Chicago route, and on 14 August on United Air Lines' San Francisco–Washington route.

The most used DC–10 is the –30 model which first flew on 21 June 1972. This has the increased span of the –20/40, is powered by 23,133kg (51,000lb) thrust CF6–50C en-gines, has the same weight as the –20/40 but a range of 9,768km (6,070 miles). The DC–10–30 first went into service with Swiss-air on North Atlantic services on 15 December 1972. There are also cargo and conver-tible versions, and a P&W JT9D-engined Series 40 version. Excluding the USAF's KC–10 model (60 built), McDonnell Doug-las delivered 380 DC–10s.

In 1990 deliveries begin of the next-generation MD–11, a DC–10 stretched to a length of 61.2m (just over 200ft), to seat up to 405 passengers. The wings have large double winglets, the horizontal tail is much smaller and there are many other changes. With GE or P&W engines in the 27,220kg (60,000lb) class, the MD–11 will fly sectors up to 14,275km (8,870 miles) in length. Stretched versions seating 450 and a twin-engined ver-sion are being studied, but the design is slightly handicapped by using a wing origin-ally designed in 1966.

The Lockheed L.1011 TriStar, which first flew on 16 November 1970, has exactly the same span as the DC–10–10 but is slightly shorter at 54·43m (178ft 7½in). It was the first aircraft to be powered by the Rolls-Royce RB.211, which in its –22B form develops 19,050kg (42,000lb) of take-off thrust. The TriStar can carry 272 passengers in mixed-class configuration, 330 in coach class or up to 400 in economy class. Maxi-mum take-off weight is 195,043kg (430,000 lb), and the range with 272 passengers and

FAR LEFT ABOVE: *The Boeing 747SP–44 (Special Purpose) Matroosberg, of South African Airways. This is a very long-range version with a comparatively short body.*
FAR LEFT BELOW: *The latest wide-bodied airliner is the Airbus Industrie A300B.*
LEFT: *One of Middle East Airlines' Boeing 747–2B4Bs, with the Cedar of Lebanon crest on its tail. This airline now operates an extensive route network throughout the Middle East and to Africa, Asia and Europe. It suffered a setback in December 1968, when an Israeli attack on Beirut airport destroyed six of its airliners.*
BELOW: *Externally very like the A300 or A310, the Boeing 767-200ER has a bigger wing (without winglets) and narrower body. It typifies the new species of big twin-jet able to fly transoceanic routes.*
BOTTOM: *This dramatic view of Lufthansa's McDonnell Douglas DC–10–30 Düsseldorf landing at Hong Kong shows clearly the mounting of the rear engine.*

full fuel reserves is 6,275km (3,900 miles). Maximum payload is 38,782kg (85,500lb). The TriStar entered service on 26 April 1972, with Eastern Airlines.

At first Lockheed was unable to develop the L–1011 much because of the lack of a more powerful engine. The L–1011–100 offered a slight increase in fuel capacity, and the RB.211–254 of 21,773kg (48,000lb) thrust resulted in the L–1011–200 with slight-ly more range. The final model was the –500, with 22,680kg (50,000lb) engines, longer span but a shorter body, seating only 246–330 but carrying them up to 11,268km (6,090 miles). Altogether Lockheed built 249 L–1011s, ending in 1985. In 1987 the –250 was created by strengthening the original

L–1011–1, fitting new –524B engines, extra fuel and other changes. Among other things this almost doubles the range.

The fourth type of wide-bodied airliner is the Airbus Industrie A300B, which is slightly smaller than the DC–10 and TriStar, with a span of 44·84m (147ft 1in) and a length of 53·62m (175ft 11in). Unlike the other wide-bodied aircraft, the A300 is twin-engined, being powered by two wing-mounted 23,133kg (51,000lb) thrust General Electric CF6–50C turbofans. Maximum cabin width is 5·35m (17ft 7in) and seating ranges from 251 in mixed class to 336 in high-density nine-abreast layout.

The A300 first flew on 28 October 1972, and the B2 version entered service on the Paris–London route with Air France on 23 May 1974. It has a maximum take-off weight of 142,000kg (313,000lb), a cruising speed of 870km/h (541mph) and a range of up to 3,860km (2,400 miles). The B4 model has increased fuel capacity, a maximum take-off weight of 157,500kg (347,100lb) and a range of 6,000km (3,728 miles).

Sales were very slow to start, the airlines apparently finding it difficult to believe that a

RIGHT: *An Ilyushin Il–18 taxiing in at Heathrow Airport, 1959, one of the first aircraft to have a forward positioned entrance.*
BELOW RIGHT: *A Tupolev Tu–134 of Balkan Bulgarian Airlines. This aircraft is a development of the Tupolev Tu–124, with the engines moved from the wings to the tail and a 'T' tailplane fitted. It is a short-range airliner.*

windshear or microburst can cause it to hit the ground. Customers have a choice of the CFM56–5 or the IAE V.2500 engine, in the 10,400–11,340kg (23,000–25,000lb) class. The beautiful cabin seats up to 179, over ranges up to 5,837km (3,627 miles). The first flight took place in February 1987 and late in that year no fewer than 309 A320s had been ordered, plus 174 options, even though airline service was not to begin until spring 1988.

Airbus took a further giant gamble when in June 1987 it launched the A330/A340. The risk seemed less by the end of that year when 11 customers had signed for a combined total of 144 of these, the biggest commercial transports ever built in Europe. Both have the same extremely efficient wing of 58.65m (192ft 5in) span, with tip fences and unique computer-controlled flaps which adjust wing camber according to the burn-off of fuel so that cruising drag is always at a minimum. The A330 is a giant twin, with CF6–80C2 or PW4259 engines in the 27,216kg (60,000lb) class and seating up to 328. The A340 has four CFM56–5C engines, each of 13,880kg (30,600lb), carrying up to 295 passengers 12,695km (7,890 miles) or up to 14,270km

ABOVE FAR RIGHT: *No commercial transport can equal the development history of the McDonnell Douglas DC–9 which is in service with seven different lengths of fuselage. While the original version seated a maximum of 90 passengers, the MD–82 illustrated has a maximum capacity of 172.*
BELOW FAR RIGHT: *An early Tupolev Tu–104A, the first Soviet turbojet transport aircraft. This model carried 70 passengers and was powered by two wing-mounted 9,720kg (21,385lb) thrust Mikulin AM–3M–500 turbojets.*

in 1923 three air transport undertakings were formed. But before the formation of Soviet concerns, a Königsberg (now Kaliningrad)–Kowno–Smolensk–Moscow service had been opened, on 1 May 1922, by the joint Soviet-German airline Deruluft, using Fokker F.III single-engined monoplanes. This company, with various types of aircraft and with varying routes, was to operate between Germany and the USSR until 1937.

The three undertakings set up in 1923 were Dobrolet, Ukrvozdukhput in the Ukraine, and Zakavia. These organizations developed a number of routes, mainly with German aircraft, and in 1930 the first two were re-organized as Dobroflot, Zakavia already having been taken over by Ukrvozdukhput in 1925. In 1932 Aeroflot was created when all Soviet air services came under the Chief Administration of the Civil Air Fleet.

Until the German invasion of the Soviet Union in June 1941, Aeroflot gradually built up its network, and by 1940 had a route system measuring 146,300km (90,906 miles) and that year carried nearly 359,000 passengers and about 45,000 tons of cargo and mail.

During the war Aeroflot was engaged on

consortium linking together all the biggest aircraft companies in France, Germany, Britain, the Netherlands and Spain, and with engines and systems from the USA, might have created a good product. Gradually the excellence of the A300B became known, and once Eastern of the USA had bought 34 there was no looking back. In 1978 the smaller A310 was launched, seating around 218 and able to fly up to 9,175km (5,700 miles) with a full load. Features include an advanced two-crew cockpit, wingtip fences and a tailplane used as a trim tank for minimum cruising drag. In 1984 deliveries began of the big A300–600, with the rear fuselage of the A310 and many other advanced features, and in the –600R version able to carry 267 passengers and baggage 8,154km (5,066 miles), much more than double the range (with heavier payload) of the first version. By late 1987 sales of the A300B and 310 had reached 467 for 61 customers.

In 1984 Airbus decided to go ahead with a smaller single-aisle aircraft, the A320. Though it looks like a small A300, or even a 737, the A320 is actually the newest transport in the sky. It has futuristic avionics, manifest in the flight deck (with almost entirely colour electronic displays instead of 'instruments', and with small sidestick controllers leaving room for folding tables in front of the pilots). The pilots can do anything they like but can never put the A320 beyond design limits or in any kind of danger, and not even the fiercest

(8,870 miles) with 262 passengers. Airline service is due in 1992.

A fifth wide-body appeared in 1981. This was the Boeing 767, and it looked like a carbon copy of the European A300B, though in fact it has a bigger wing and narrower body (cabin width 4.72m, 15ft 6in, compared with 5.28m, 17ft 4in). Like the A300B and 310 only GE or P&W engines were fitted until in 1987 British Airways bought the 767 with Rolls-Royce engines. Boeing is marketing the –200 and –200ER with a length of 48.51m (159ft 2in) and the –300 and –300ER stretched to 54.94m (180ft 3in), the ER (Extended Range) models having extra fuel for ranges up to 12,500km (7,767 miles) with up to 210 passengers.

A few months later than the 767 Boeing also developed the 757. Similarly planned over many years, the 757 is the biggest of the narrow-body or single-aisle transports, with Rolls-Royce 535 or PW2037 engines of some 18,144kg (40,000lb) thrust and seating up to 239 passengers in a very long cabin the same width as the 707/727/737. The 757 entered service in 1983, and the basic 757–200 was later joined by the executive 77–52, the –200PF Package Freighter with a side cargo door and windowless cabin, and the convertible combi.

Soviet air transport

In 1921 or 1922 there were a few experimental air services in the Soviet Union, and

essential tasks, and when peace returned, faced the job of re-establishing and expanding its nationwide operations and building up an international route system. Its success is illustrated by the fact that Aeroflot now serves some 3,500 places in the USSR, operates to 65 other countries and, in 1975, carried more than 100 million passengers.

During the war the Soviet Union had received considerable numbers of military DC–3s from the United States, and in 1939 had already begun licence production of these under the designation PS–84, changed in September 1942 to Lisunov Li–2. In the early post-war years these Li–2s and ex-military DC–3s formed the backbone of the Aeroflot fleet, and some were supplied to neighbouring Communist countries.

But even while the war was being fought, Sergei Ilyushin's design bureau began work on the USSR's first postwar transport aeroplane – the Il–12. This was a low-wing 21/32-seat monoplane powered by two 1,650/1,775hp Shvetsov ASh–82FN engines. It had a nosewheel undercarriage, but was unpressurized. The type was introduced by Aeroflot on 22 August 1947, large numbers were built, and on the last day of November 1954 the Il–12s were joined by the improved but very similar Il–14.

In the mid-1950s the Il–12s were taking 33hr to fly from Moscow to Vladivostok, with nine intermediate stops, and on the Moscow – Sverdlovsk – Novosibirsk – Irkutsk route they took 17hr 50min with 14hr 35min flying time. These aeroplanes were obviously

not acceptable for the distances which had to be covered and in 1953, as part of a major modernization plan, the first Soviet jet transport was designed. This was the Tupolev Tu–104, which first flew on 17 June 1955, and went into service on the Moscow–Omsk–Irkutsk route on 15 September 1956, with a schedule of under seven hours.

The Tu–104 had a low-mounted swept-back wing, swept tail surfaces and two 6,750 kg (14,881lb) thrust Mikulin RD–3 or AM–3 turbojets close in to the fuselage. The pressurized cabins seated 50 passengers. The Tu–104 had a span of 34·54m (113ft 4in), a loaded weight of 71,000kg (156,528lb) and a cruising speed of 750/800km/h (466/497mph).

The Tu–104 transformed Soviet long-distance air services, but it was not particu-

larly economic and was therefore followed by the 70-passenger Tu–104A and 100-passenger Tu–104B. It is believed that between 200 and 250 examples were built and many are still in service after 20 years.

The Soviet aircraft industry next designed three types of propeller-turbine transport aircraft. Two of these were for heavy-traffic routes and the third was a long-range aircraft. Ilyushin produced the Il–18, which was to prove of major importance to Aeroflot, about 600 being put into service. This was a low-wing monoplane powered by four 4,000hp Ivchenko AI–20 engines, and initially having accommodation for 80 passengers. The Il–18 made its first flight on 4 July 1957, when it bore the name Moskva (Moscow), and it went into service on the Moscow–

Alma Ata and Moscow–Adler/Sochi routes on 20 April 1959. Several versions of Il–18 were built, including the Il–18V which has accommodation for up to 110, a maximum take-off weight of 61,200kg (134,922lb), a cruising speed of 625/650km/h (388/404mph) and a maximum-payload range of 2,500km (1,552 miles). About 100 Il–18s were sold to non-Soviet airlines.

The Antonov An–10 was of similar size to the Il–18, but was a high-wing monoplane designed to operate from poor aerodromes. Like the Il–18, the An–10 had four 4,000hp Ivchenko AI–20 engines, and when it entered service, on the Moscow–Simferopol route, on 22 July 1959, had 85 seats. The improved An–10A which followed had 100–110 seats.

The long-range aeroplane was unlike any other to be used in airline service. It was the Tupolev Tu–114, developed from the Tu–95 bomber. The 51·1m (167ft 7¾in) wing was low-mounted, swept-back 35 degrees and carried four massive Kuznetsov NK–12M turbines, each of 12,000hp and driving 5·6m (18ft 4½in) diameter eight-blade contra-rotating propellers. Later the 15,000hp NK–12MV was installed. Fully loaded, the Tu–114 had a take-off weight of 175,000kg (385,809lb), and for nearly a decade it was the biggest aeroplane in airline service. It was also the fastest propeller-driven airliner ever in service, having a maximum speed of 870km/h (470mph).

The Tu–114 entered service on the Moscow–Khabarovsk route on 24 April 1961;

went onto the Moscow–Delhi route in March 1963; opened the first Soviet transatlantic service, to Havana, on 7 January 1963; and the first to North America when it began operating to Montreal on 4 November 1966. Tu–114s began working joint Aeroflot-Japan Air Lines Moscow–Tokyo services on 17 April 1967, and bore JAL's name and badge as well as Aeroflot's. This large aircraft could carry 220 passengers, but standard seating was for 170.

Having re-equipped with medium- and long-range turbine-powered aircraft, Aeroflot set about the task of modernizing its short-haul fleet and introduced two new types in October 1962. One was the Tu–124, which was virtually a three-quarter-scale Tu–104, and the other was the Antonov An–24, which resembled the Fokker Friendship.

The Tu–124, with two 5,400kg (11,905lb) thrust Soloviev D–20P turbofans, had accommodation for 44 to 56 passengers and entered service on the Moscow–Tallinn route. About 100 were built, and it was replaced on the production lines by the similar, but rear-engined, 64/80-seat Tu–134 and TU–134A, which are now used in quite large numbers, some having been exported to East European airlines.

The An–24, with two 2,100hp Ivchenko AI–24 turboprops, is a high-wing monoplane. It has been built in several versions and in very large numbers, and it has accommodation for up to 50 passengers. There are cargo versions, including the An–24TV and An–26, and the An–30 survey aircraft. Numbers of An–24s have been exported.

TOP: *The British Aerospace 146 is quieter than any other jetliner and can also use much smaller airports. Most have been sold to airlines in the USA.*

ABOVE: *Unquestionably the most advanced civil aircraft yet built, the Airbus A320 entered service with Air France in April 1988. Later versions have winglets.*

As Aeroflot began developing its long-haul international and intercontinental routes, it required a jet transport to replace the Tu–114. To meet this requirement Ilyushin designed the Il–62, which closely resembles the British VC10. The Il–62 has four rear-mounted Kuznetsov or, in the Il–62M, Soloviev turbofans, and normal seating for 186. The Il–62 first flew in January 1963, but its development was prolonged and it did not enter service until March 1967, on domestic routes. On 15 September 1967, Il–62s began working the Moscow–Montreal services and on 15 July 1968 the extension to New York was opened.

Aeroflot has a vast number of local services, and a modern aeroplane was required to replace the Li–2s, Il–14s and smaller aircraft on these operations. The type chosen was the Yakovlev Yak–40, a unique tri-jet with the ability to operate from small, rough fields and carry up to 32 passengers. The Yak–40 flew in October 1966, went into service in September 1968 and has been exported to several countries including Afghanistan, Germany and Italy. The engines are 1,500kg (3,306lb) thrust Ivchenko AI–25 turbofans. Over 550 were built.

On 4 October 1968, a new Soviet airliner, also a tri-jet, made its first flight. This was the Tu–154, which has been designed to replace the large numbers of An–10s, Il–18s and Tu–104s. It has been reported that more than 600 were ordered for Aeroflot, and the type entered service on the Moscow–Simferopol and Moscow–Mineral'nye Vody routes in November 1971. Since then the Tu–154 has been developed through 154A, 154B and B–2 versions with greater power and various refinements; and the 154C freighter, well over 600 of these models being delivered. Large numbers were then made of the Tu–154M, with the NK–8–2U engines replaced by the Soloviev D–30KU–154–II and

with numerous other changes. The 154M seats up to 180, which it can carry 3,740km (2,324 miles).

In 1975 the first Yak–42 flew, but this trijet did not enter service until 1980, and an accident kept it off the timetable from 1982 until October 1984. Powered by three 6,500kg (14,330lb) Lotarev D–36 turbofans, the Yak–42 seats up to 120, with which it has a range of 1,900km (1,180 miles).

The first Il–86 flew in 1976. Looking like an inflated 707, this giant wide-body has four 13,000kg (28,660lb) NK–86 turbofans and seats up to 350 passengers, who can climb aboard from stairs to the lower deck, leave their coats and baggage and then climb up to the vast main deck. By 1992 this may be replaced in production by the Il–96–300, with a new wing of much greater span (57.66m, 189ft 2in) with winglets, new D–90A engines of 16,000kg (35,275lb) thrust, a new tail and many other changes. The fuselage is actually shorter, and will accommodate 235–300 passengers over ranges of up to 9,000km (5,590 miles).

In 1988 the Tu–204 should begin flight trials. Designed to replace the Tu–154, this looks like a 757, with two D–90A engines (as used on the Il–96–300) hung under an efficient tip-fenced wing of 42m (137ft 9.5in) span, a single-aisle fuselage seating 170–214 passengers, and a range with maximum payload of 2,400km (1,490 miles). The Tupolev bureau hope the 204 to be a thoroughly modern and competitive aircraft.

Supersonic transport

In Britain and France the decision was taken to produce jointly an airliner capable of cruising at more than Mach 2, about 2,143 km/h (1,332mph). This was the enormously costly Concorde project, with Aérospatiale and the British Aircraft Corporation being responsible for the airframe and Rolls-Royce

and SNECMA for the engines. The Concorde is a slim delta-winged aircraft with four 17,260kg (38,050lb) thrust Olympus 593 turbojets, a span of 25·37m (83ft 10in), a length of 61·94m (203ft 9in) and a maximum take-off weight of 185,065kg (408,000lb).

The first Concorde flew on 2 March 1969, and on 21 January 1976, the first supersonic passenger services were inaugurated – Air France flying the Paris–Dakar–Rio de Janeiro route and British Airways the London–Bahrein route. Air France also put Concordes on the Paris–Caracas route, and on 24 May 1976, Concordes of Air France and British Airways made simultaneous arrivals at Dulles Airport, Washington, inaugurating services from Paris and London respectively. Only nine Concordes were ordered for airline service, five for British Airways and four for Air France.

The Soviet Union was actually first to fly a supersonic transport, its Tupolev Tu–144, much like Concorde in appearance, making its first flight on the last day of 1968.

The United States held a design competition for a supersonic airliner, and selected the 298-passenger variable-geometry Boeing 2707 with four General Electric engines. This aircraft would have been 96·92m (318ft) long, spanning 53·08m (174ft 2in) with the wing swept forward and 32·23m (105ft 9in) with the wing swept back 72 degrees for supersonic flight. This design was then replaced by a pure-delta type, but finally the whole supersonic project was considered to be uneconomic and, after a long series of design hold-ups, it was cancelled.

TOP: *The Soviet Union's Tu–144 was the first supersonic transport to fly.*
ABOVE: *A full-scale mock-up of the Boeing 2707 supersonic transport project.*

Airlines

Following are general notes on the world's largest airlines offering international passenger and freight services, together with information on the main aircraft operated by them.

Adria Airways
Formerly Inex Adria, this Yugoslav airline flies A320s, DC–9s, MD–82s and Dash–7s on scheduled and charter services.

Aer Lingus/Aerlinte (Ireland)
Founded 1936. Ireland's national airline flies 747s, 737s, One-Elevens and Shorts 360s on scheduled passenger and cargo services.

Aeroflot
The Soviet Union's vast civil air operation is the biggest single 'airline', but it also handles agricultural, aeromedical, firefighting, research and many other duties, with some 8,000 aeroplanes and 3,000 helicopters.

Aerolineas Argentinas
Founded 1949. Argentina's national airline flies 727s, 737s, 747s and F28s on scheduled operations.

Aerolineas Centrales de Colombia
ACES has 18 Twin Otters and three jets on services radiating from Bogota.

Aeromexico
Mexico's flag carrier has DC–8s, 9s, 10s and MD–80s on scheduled services.

Aero Peru
The national Peruvian airline flies DC–8s, 727s and F28s on scheduled services.

Aero Trasporti Italiani
ATI, wholly owned by Alitalia, flies MD–80s, DC–9s and ATR.42s on intensive services in southern Italy.

Aerovias Nacionales de Colombia
Avianca has a history going back to 1919 and flies 707s, 727s, 747s and 767s on a major network.

Aerovias Venezolanas
Avensa has a fleet of 727s on domestic routes based on Caracas.

Air Afrique
Founded 1961. This carrier is run by 12 African countries with a French heritage, and flies A300s, DC–8s, DC–10s and a 727.
Passengers 426, 197
Fleet 12

Air Algérie
Founded 1953. This national carrier has a varied fleet including A310s, 727s, 737s, F27s, L–100s and 14 Ag–Cats (among others).

AirCal
Part of American Airlines, AirCal retains its identity on its fleets of 737s and 146s on intensive services in the western USA, Canada and Alaska.

Air Canada
Canada's flag carrier has DC–8s and 9s, 727s, 737s, 747s and 767s, and TriStars on scheduled passenger and cargo routes.

Air Espana
Trading as Air Europa, this Spanish/UK owned charter line flies 737s and 757s based on Mallorca.

Air Europe
This Gatwick-based tour and scheduled operator has a fleet of 737s and 757s.

Air France
Founded 1933. France's flag airline flies A300s, 310s, 320s and (soon) 340s, 727s, 737s and 747s, and seven Concordes on a global network.

Air India
Founded as Air-India International 1948. A national airline, AI has A300s and 310s, 747s and leased aircraft on scheduled international services.

Air Inter
France's main domestic airline flies A300s and 320s, Mercuries, Caravelles and F27s to 30 French cities.

Air Jamaica
This national carrier has A300Bs and 727s flying on international routes.

Air Lanka
Sri Lanka's airline flies 747, 737 and TriStar equipment on international services.

Air Madagascar
This airline flies domestic and long-haul routes with aircraft ranging from 747s and 737s, through 748s and Twin Otters to a selection of light planes.

Air Malta
Since 1974 this operator has used a fleet of 720Bs and 737s and is buying A320s.

Air Midwest
Based at Wichita, Kansas, this airline has a massive fleet of Metro, Brasilia and SF340 regional turboprops.

Air New Zealand
The NZ national carrier flies 747s, 767s, 737s and F27s on domestic and international routes.

Air Niugini (Papua New Guinea)
Founded 1973. The PNG airline flies A300s, F28s and Dash–7s.

Air NSW
Based at Sydney, the New South Wales airline has a fleet of F27s and F28s.

Air Ontario
This regional airline is replacing its big fleet of CV–580s with Dash–8s (20 for a start).

Air Portugal
This national carrier flies A310s, TriStars, 707s, 727s, 737s and Twin Otters.

Air UK
Extensive local and international services are flown with F27s, Shorts 360s and a One–Eleven.

Air Wisconsin
Known as United Express, reflecting a link with United, services radiate from Chicago and Washington with BAe 146s, F27s and Shorts 360s.

Air Zaïre
Founded as Air Congo 1961. This flag carrier relies on a DC–10, DC–8s, 737s and F27s.

Air Zimbabwe
Domestic and international routes are served with 707s and 737s.

Alaska Airlines
This airline flies a big network throughout the western USA and Alaska with 727s, 737s and MD–83s.

Alitalia
Founded 1946. Italy's flag airline flies 747s, MD–11s, A300Bs, MD–80s, DC–9s and various smaller machines.

All-Nippon Airways
ANA has a massive fleet of 747s, 767s, 727s, 737s, TriStars and YS–11s and is receiving 20 A320s.

American Airlines
Founded 1934. One of the giants, AA flies domestic and international services with over 400 jets including 727s, 747s, 767s, DC–10s, A300B–600Rs and MD–80s.

American Trans Air
This Indianapolis-based carrier has a fleet of TriStars and 727s.

America West Airlines
Based at Phoenix, Arizona, this regional airline flies 757s and some 60 737s.

Ansett Airlines
Founded 1936. One of the two giants of Australia, this carrier has 727s, 737s, 767s, F27s and Fokker 50s and is receiving A320s.

Atlantic Southeast Airlines
ASA is linked with Delta and flies Dash–7s, Shorts 360s, Brasilias and Bandeirantes.

Australian Airlines
One of the two giants of Australia, this carrier flies A300Bs, 727s, 737s, DC–9s, F27s and Jetstreams, and is receiving A320s.

Austral Lineas Aéreas
ALA radiates from Buenos Aires with One-Elevens and MD–81s.

Austrian Airlines
Founded 1958. This flag carrier has A310s, MD–81s, 87s and DC–9s, and Fokker 50s.

Aviaco
This Spanish domestic and charter airline flies DC–9s, F27s and Airtech CN–235s.

Balkan Bulgarian Airlines
Flies agricultural and GA services as well as scheduled services with Tu–154s and 134s, Il–18s, An–12s and 24s, Yak–40s, and Mi–8s.

Bangladesh Biman
This flag airline operates scheduled routes with DC–10s, 707s, F27s and F28s.

Braathens SAFE
This Norwegian carrier has a big fleet of 737s.

Braniff Airways
Much smaller than before a temporary shut-down in 1982, this famous line has 20 727s.

Britannia Airways
Despite its name this holiday operator has an all-American fleet of 737s and 767s.

British Airways
Founded 1972 by merger of BEA and BOAC. Despite its name this national flag carrier is trying to build up an all-American fleet of 737s, 747s, 757s, 767s and TriStars, plus seven under-used Concordes. It flies secondary routes with One-Elevens and 748s.

British Midland
BM's aggressive management has built services flown by the DC–9, ATP, F27, V. 800 and Shorts 360.

CAAC
The vast Civil Aviation Administration of China consists of many regional carriers, who have individually purchased 747s, 767s, 707s, 737s, A310s, Tu–154s, MD–80s (being built in China), Il–18s, BAe 146s, Tridents, IL–62s, An–12s, 24s and 26s, Dash–7s, Shorts 360s and many other types.

Canadian Pacific
CP flies fleets of DC–10s and 737s.

Cathay Pacific
Hong Kong's major airline flies non-stop to London and many other places with 747s and TriStars.

Ceskoslovenské Aerolinie
Founded 1923. CSA radiates from Prague with Il–62s, Tu–134s, Yak–40s and Il–18s.

China Airlines
Taiwan's flag carrier has 737s, 747s and 767s and A300Bs.

Comair
Radiating from Cincinnati, this Delta-linked carrier has busy fleets of SF.340s, Shorts 330s, Metros and Bandeirantes.

Condor Flugdienst
This holiday subsidiary of Lufthansa flies A300Bs and 310s, DC–10s, 727s and 737s.

Continental Airlines
Part of the Texas Air conglomerate, this vastly expanded carrier has almost 300 727s, 737s, 747s, DC–10s, DC–9s and MD–80s.

Cruzeiro do Sul
Founded 1942. This scheduled Brazilian carrier flies A300Bs, 737s and 767s.

Cubana
Founded 1930. Cuba's airline relies on the Tu–154, Il–14, 18, 62 and 76, Au–24 and 26 and Yak–40.

Dan-Air
This airline flies scheduled, contract and holiday services with the A300B, 727, 737, One-Eleven, BAe 146 and 748.

Delta Air Lines
Founded 1929. Delta is one of the giants, with over 450 jets including 727s, 737s, 757s and 767s, DC–8s, 9s, and 10s, TriStars and MD–88s.

Eastern Air Lines
Founded 1938. Another giant, this Miami-based operator has fleets of DC–10s, TriStars, A300Bs, A320s, 727s, 757s and DC–9s.

Egyptair
Founded as Misr-Airwork 1931. This national carrier flies 707s, 737s, 747s, 767s, and A300Bs.

El Al
Founded 1948. Israel's airline is all-Boeing, with 707s, 737s, 747s, 757s and 767s.

Emery Worldwide
This overnight package carrier has a big fleet of DC–8s and 727s.

Ethiopian Airlines
Founded 1945. Torn by war, this national airline flies 707s, 720s, 727s and 767s, ATR.42s and DH Canada STOLs.

Federal Express
This door-to-door package giant serves 85 countries with colossal fleets of MD–11s, DC–10s, 727s and well over 100 Caravans.

Finnair
Founded as Aero O/Y 1923. Finland's airline flies DC–9s, MD–80s and DC–10s, A300Bs, ATR42s and 72s and F27s.

Florida Express
This Orlando-based operator has a fleet of One-Elevens.

Flying Tiger Line
A famed cargo carrier, Tiger flies 727s and 747s.

Garuda Indonesia
Founded 1950. State-owned Garuda has 747s, A300Bs, A320s, DC–9s and 10s and F28s.

Gulf Air
Bahrein's airline flies 737s, 747s and TriStars.

Hapag-Lloyd
This charter/holiday airline is based at Hanover and has fleets of A300Bs and Cs, A310s, 727s and 737s.

Hawaiian Airlines
This local and long-haul operator flies TriStars, DC–8s and 9s, and STOL Dash–7s.

Horizon Air
Based at Seattle, this scheduled airline has fleets of F27s and F28s, One-Elevens, Dash–8s and Metros.

Iberia
Founded 1940. Spain's national carrier flies 747s, DC–9s and 10s, A300Bs and a big fleet of 727s.

Indian Airlines
Founded 1953. This airline serves 73 cities in and around India with A300s and 320s, 737s, F27s and 748s.

Interflug
The DDR (E. Germany) has about 300 agricultural and utility machines, plus Il–18s and 62s, Tu–134s and LET–410s.

Iran Air
Founded 1962. This flag operator flies 707s, 727s, 737s and 747s and A300Bs.

Iraqi Airways
Founded 1945. This national line has a motley fleet of 707s, 727s, 737s and 747s, Il–76s, An–12s and 24s, Falcons, JetStars and P.166s.

Japan Air Lines
Founded 1951. This privatized giant has the world's biggest fleet of 747s, as well as DC–10s, 767s, DC–8s and even two 727s.

JAT (Jugoslovenski Aerotransport)
Founded 1947. Yugoslavia's JAT has DC–9s, 10s and MD–11s, 707s, 727s and 737.

KLM Royal Dutch Airlines
Founded 1919. The famed line of the Netherlands has an unbroken 70-year history and flies 737s and 747s, DC–9s and 10s, A310s and Fokker 100s.

Korean Air
South Korea's mighty carrier has fleets of 707s, 727s and 747s, A300Bs, Cs and 600s, DC–10s and MD–11s, F27s and 28s and various smaller aircraft.

Kuwait Airways
Founded as Kuwait National Airways 1953. This national operator uses the A300 and 310, 727, 747 and 767 and two BAe. 125s.

Libyan Arab Airlines
Founded as Kingdom of Libya Airlines 1964. Services are flown with A310s, 707s and 727s, F27s and F28s.

Linjeflyg
Sweden's domestic line relies entirely on the F28.

LOT-Polish Airlines
Founded 1929. The national airline of Poland has a world network served by Il–62Ms, Tu–154Ms and 134As, Il–18s and Au–24s.

LTU
West Germany's Lufttransport-Unternehmen carries holiday and charter traffic with four versions of TriStars.

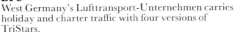

Lufthansa
Founded (original company) 1926. West Germany's giant flag airline flies the 727, 737 and 747, DC–10 and A300, 310, 320 and (soon) 340.

Malaysian
This growing national airline is equipped with 737s and 747s, DC–10s, A300Bs, F27s and Twin Otters.

Malev
Hungary's airline flies the Tu–134 and 154, and Il–18.

Martinair
This Dutch operator flies scheduled, charter and taxi services using a 747, DC–10s, MD–82s and jet and turboprop Cessnas.

Merpati Nusantara
Indonesia's second airline is a subsidiary of the first (Garuda); it is all-turboprop with a Vanguard, Viscounts, F27s, 748s, CN–235s, C.212s and Twin Otters.

Metro Airlines
Based at Dallas/Fort Worth, this regional carrier flies Jetstreams, CV–580s, Dash–8s and Twin Otters.

Mexicana
Founded 1924. This long-established airline has DC–10s and 40 727s.

Middle East Airlines
Founded 1945. MEA has been crippled by the Lebanon's civil war, but struggles on with 707s, 720Bs and 747s.

Midway Airlines
This carrier operates DC–9s and 737s on intensive routes from Chicago's old Midway airport.

Monarch Airlines
Charter and holiday services are flown from Luton and Gatwick with 737s and 757s.

Nigeria Airways
Founded 1958. A national carrier, Nigeria uses 707s, 737s, 747s, DC–10s and A310s.

Northwest Airlines
Based at St Paul, Minnesota, this giant flies 727s, 747s and 757s, DC–9s and 10s, MD–80s, A320s (100 on order), and is retiring the CV–580.

Olympic Airways
Founded 1957. Greece's flag airline uses the 707, 727, 737 and 747, A300B and Shorts 330.

Orion Air
A big US cargo line, Orion flies the 727 and 747, and DC–8 and 9.

Orion Airways
This British holiday line has A300Bs and 737s.

Ozark Air Lines
Merged with TWA, Ozark retains its identity and uses a fleet of many kinds of DC–9/MD–80.

Pacific Southwest Airlines
Based at San Diego, PSA has large fleets of the BAe 146 and DC–9/MD–80.

Pacific Western Airlines
Based at Calgary, Alberta, PWA is a one-type airline, with 27 737s.

Pakistan International Airlines
Founded 1955. PIA's fleet comprises 707s, 737s and 747s, A300Bs, F27s and Twin Otters.

PanAm
Founded in 1927. Perhaps the world's premier airline, Pan American World Airways flies 727s, 737s and 747s, and A300Bs, 310s and 320s.

Pelita Air Service
Part of a vast Indonesian oil company, Pelita flies L–100 Hercules, Transalls, BAe 146s, Dash–7s and F28s, an Albatross, CN–235s, C.212s, Skyvans, and such helicopters as the Puma, S–76, BO 105C, Bell 212, JetRanger and Alouette.

Philippine Air Lines
Founded as Philippine Aerial Taxi Co 1932. PAL operates the 747, DC–10, A300B, One-Eleven, 748 and Shorts 360.

Piedmont
Based in North Carolina, this swiftly growing line covers the USA and also serves London, England, using big fleets of 727s, 737s and 767s, and F28s.

Presidential Airways
Based at Washington DC, Presidential has fleets of 737s and BAe 146s.

Qantas Airways
Founded 1920. Australia's overseas operator, and also flying major internal routes, Qantas has 747s and 767s.

Royal Air Maroc
The airline of Morocco uses the 707, 727, 737, 747 and 757.

Royal Jordanian
Now privatized, this flag airline has 707s, 727s and 747s, TriStars, and A310s and 320s.

Sabena
Since 1923 this has been Belgium's national airline; it uses 737s and 747s, DC-10s, A310s and (soon) 340s, Xingus and SF.260s.

Saudia
Founded 1945. Saudi Arabia's airline flies the 707, 737 and 747, TriStar, A300–600, DC–8, F28, Gulfstream II and III and Beech A–100.

Scandinavian Airlines System
Founded 1946. SAS, flag line of Denmark, Norway and Sweden, uses the 747, DC–8, 9, MD–80 and MD–11, and F27 (plus hovercraft)

Simmons Airlines
Chicago-based Simmons serves local routes with 30 Shorts 360s and the bigger ATR 42.

Singapore Airlines
Tiny Singapore's airline is a major carrier, with 29 747s, plus fleets of A300s and 310s, and 757s.

South African Airways
Founded 1934. SAA flies 737s and 747s, and A300Bs.

Southern Air Transport
This Miami-based operator flies scheduled and charter cargo worldwide with 707s and L–100 Hercules.

Southwest Airlines
This Dallas-based regional line flies 83 737s.

Swissair
Founded 1931. The national Swiss line operates DC–9s, 10s, MD–81s and MD–11s, 747s, A310s and Fokker 100s.

Syrian Arab Airlines
Founded 1961. Syria has a mix of 727s and 747s, Il–76s, Tu–154s and 134s, Caravelles, An–26s, Yak–40s and Falcon 20s.

Tarom
Romania's airline mixes the Il–18 and 62, 707, Tu–154, One-Eleven, and An–24 and 26.

Thai Airways
This regional carrier flies A310s, 737s and Shorts 330s and 360s.

Thai Airways International
A global airline, Thai International is equipped with 747s, DC–10s and A300Bs and –600s.

Toa Domestic Airlines
Japan's third-biggest operator, TDA has large fleets of A300Bs, DC–9s and 10s, MD–80s and 87s, YS–11s and eight kinds of helicopter.

Transavia
This Dutch regional line flies 13 737s.

Transbrasil (Brazil)
Founded as Sadia 1955. This privately owned operator uses the 707, 727, 737 and 767.

Transport Aérien Transrégional
TAT is a French regional carrier flying the ATR 42, FH–227B, F28, Twin Otter and Beech 99.

Tunis Air
This flag carrier has the A300B, 727 and 737.

TWA
Founded as Transcontinental and Western Air 1930. One of the giants, NY-based Trans World operates 727s, 747s and 767s, TriStars, and MD–80s.

Turk Hava Yallari
Turkey's airline has DC–9s and 10s, 707s and 727s, A310s and F28s.

United Airlines
Founded 1931. Biggest airline outside the Soviet Union, UAL has 153 727s, 184 737s, 35 747s, 19 767s, 29 DC–8s, 55 DC–10s, and six ex-PanAm TriStars.

United Parcel Service
UPS offers 'next day delivery' through the USA and Europe with 42 DC–8s, 36 727s, six 747s, 20 757s and 11 turboprops.

US Air
This big regional carrier, which expects to acquire Piedmont and PSA, has nearly 200 jets: 727s and 737s, DC–9s, One-Elevens and Fokker 100s.

Varig
Since 1929 this Brazilian operator has expanded to run a global network with 707s, 727s, 737s, 747s and 767s, DC–10s and MD–11s, A300Bs and Electras.

VASP
Since 1933 this Brazilian operator has concentrated on the domestic market, with A300Bs, 727s and 737s.

Wardair
Based at Edmonton, Alberta, Wardair flies domestic and long-haul services with A310–300s (replacing DC–10s), as well as 747s.

Wilderöe
This line serves 36 Norwegian cities with STOL Dash–7s and Twin Otters.

Zantop International
This Detroit-based cargo carrier uses the DC–8, Electra and CV–640.

General Aviation

General Aviation (GA) is defined as all flying other than military and airline. It embraces private and business flying, all air sport, air taxi, agricultural, photographic survey and bush flying. The aircraft which GA uses are generally the smaller types, but not always. The King of Saudi Arabia's private Boeing 747, for example, strictly counts as a GA aeroplane as do the range of several dozen other airliners which are in private hands and are used as personal transports. There are more than 200,000 GA aircraft in the world, three-quarters of them in the United States. For comparison, there are about 10,000 airliners in the world. GA accounts for about 40 million flying hours a year, more than three times the airlines' total. Of this, 16 per cent is 'aerial work' flown by specialized commercial operators (agricultural, air survey, construction, photographic, etc.), and the remaining 84 per cent is flight instruction, transport and recreation. More than three-quarters of all GA flying is transport.

About 88 per cent of the GA fleet consists of single-engined lightplanes, the majority of them used for pilot training and personal flying. Some ten per cent are twin-engined piston types used for personal or business transport. There are slightly fewer than 2,000 business jets flying, which represents less than 1 per cent of the total in numbers but much more than that in terms of cost. There are about 6,000 GA helicopters. GA is easily the fastest-growing segment of aviation, with a growth rate which varies markedly with the world economy but which averages about six per cent a year. One reason for this growth is that the airlines tend to serve fewer and fewer airports as their fleets change from relatively large numbers of smaller jets to smaller numbers of the new wide-body types. In the USA only 425 airports out of a total of some 12,700 are now served by the airlines, leaving 12,275 served only by GA.

ABOVE: *The Beagle Pup was a British trainer that was manufactured briefly in the 1960s. It had a cruising speed of 200km/h (125mph).*
BELOW: *The 1930 de Havilland Puss Moth was one of the first monoplanes for the private owner. 250 were built.*

BOTTOM: *The very successful Britten-Norman Islander is an inexpensive STOL light twin that sells all over the world.*
CENTRE RIGHT: *The unique Polish PZL Mielec M-15 Belphegor jet biplane is intended as the replacement in the agricultural role of the ubiquitous Soviet-designed An-2.*

Of the approximately one million licensed pilots in the 'free' world, three-quarters are also in the United States. (All these figures exclude the USSR and China, and many of the world's smaller states who do not file returns with the International Civil Aviation Organization.) Second in the league is Canada, with 50,000 pilots and 16,000 aircraft. France is the European country where flying is most popular, with some 37,000 pilots and 6,000 GA aircraft. Australia has some 27,000 pilots and 4,000 aircraft. In South America both Argentina and Brazil have more than 20,000 pilots and 3,000 or 4,000 aircraft. As might be expected, these are all countries where distances are large and personal incomes relatively high. Britain has about 20,000 GA pilots and 3,000 GA aircraft.

GA manufacturers

The United States also completely dominates the manufacture of GA aircraft: it has often produced 15,000 or more GA aircraft in each year, though the total fluctuates. Britain pioneered the private aeroplane in 1925 with the de Havilland Moth, and enjoyed a thriving lightplane industry up until 1939. But the UK now manufactures only the Islander/Trislander utility aircraft and the British Aerospace 125 business jet in the general aviation field. Some GA aircraft are built in Europe and South America, but in general the world flies American types. Cessna, Piper and Beechcraft are the Big Three of GA manufacturing.

In the early 1980s damages inflicted by US courts for supposed negligence in manufac-

TOP RIGHT: *The 6-seat Cessna Centurion is as fast as many twin-engined designs. The turbo-charged version reaches 8,700m (28,500ft).*
ABOVE: *The Beech Baron is a luxurious twin-engined plane for the private owner. One version has a pressurized cabin.*

Type: Cessna 172
Accommodation: Four
Engines: One 150hp four-cylinder Lycoming
Cruising speed: 217km/h (135mph)
Ceiling: 3,993m (13,100ft)
Range: 1,022km (635 miles)
Loaded Weight: 1,043kg (2,300lb)
Span: 10·92m (35ft 10in)
Length: 8·21m (26ft 11in)

ture reached such proportions that the insurance needed to continue in business as an aircraft manufacturer became crippling in the GA field. Cessna, which used to build 8,000 aircraft a year, almost shut down except for the Citation business jet and the Caravan utility transport, and was sold to General Dynamics. Mighty Piper, sold to Lear Siegler, suspended production of everything except the pressurized Malibu, but in 1987 a new owner, Stuart Miller, announced some resumptions. Only Beech has been little affected, though output has fallen to below 300 per year. Total US output of new GA aircraft fell to a mere 1,000 in 1987.

Three-quarters of the aircraft which the American GA industry builds are single-engined types; some 14 per cent have more than one piston engine; eight per cent are agricultural designs; two per cent are turbo-prop-powered; and one per cent are jets.

In France the state-owned company Aérospatiale, a partner in the international enterprises that built Concorde and builds the entire range of Airbus types, also manufactures a range of smaller types including the ATR.42 and 72 commuter airliners, a range of helicopters, and the Rallye light aircraft. The Dassault-Breguet combine builds the Mystère/Falcon business jets as well as the renowned Mirage 2000 jet fighters. Reims Aviation builds and markets American Cessna lightplanes under licence from the American Cessna company. Avions Pierre Robin in Dijon builds high-performance touring lightplanes of its own design.

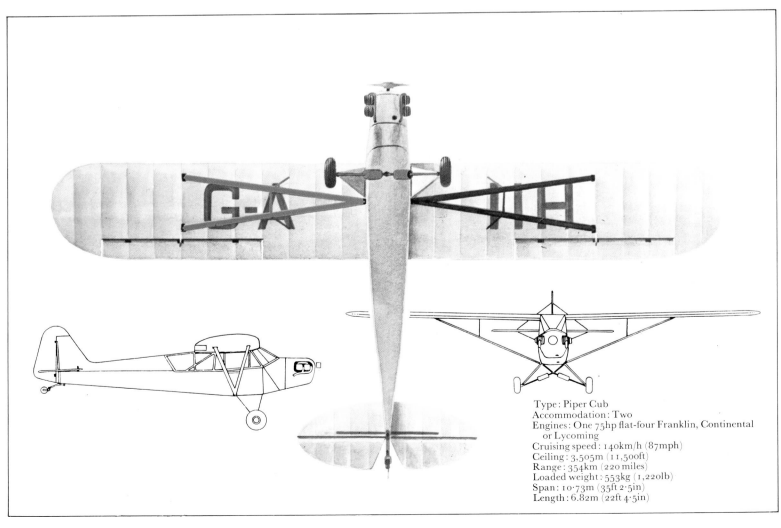

Type: Piper Cub
Accommodation: Two
Engines: One 75hp flat-four Franklin, Continental or Lycoming
Cruising speed: 140km/h (87mph)
Ceiling: 3,505m (11,500ft)
Range: 354km (220 miles)
Loaded weight: 553kg (1,220lb)
Span: 10·73m (35ft 2·5in)
Length: 6·82m (22ft 4·5in)

TOP LEFT: *This Rockwell Shrike Commander is used by the British Airports Authority for checking radio navigation aids.*

FAR LEFT: *The Grumman American Cheetah is a fast tourer for the private owner. It can carry four people at 237km/h (147mph).*

CENTRE LEFT: *The Rallye, built by the French company, Aérospatiale, is noted for its safe flying characteristics.*

LEFT: *Also French is the Dassault Falcon business jet. Like most modern business jets, it has turbofan rather than turbojet engines.*

ABOVE: *The Cessna Skymaster; twin booms behind the rear propeller carry the tail.*

BELOW: *The Rockwell Quail Commander is an American agricultural aircraft built in Mexico. It can spray liquids or chemical dusts.*

In Australia the state-owned Government aircraft factory built the twin-turboprop Nomad utility aircraft. In Canada the Canadair company has built the CL–215, a unique twin-engined amphibious water-bomber used for fighting forest fires in many countries. Fire bombers, able to dump huge quantities of water or special retardant chemical, are a major sector of GA. Several thousand of them are converted bombers and naval aircraft. A rare British example is the MacAvia conversion of the BAe 748, with a capacity of 7,571 litres (2,000 US gal).

Continental and Lycoming, both now divisions of large industrial groups (respectively, Teledyne and Avco) are the principal manufacturers of piston engines for GA aircraft. General Electric, Garrett Air Research, Canadian Pratt & Whitney and Rolls-Royce are the leading manufacturers of turbine engines in this field. Some half-dozen companies manufacture the navigation and communication equipment for GA aircraft. These systems, collectively called avionics (*avi*ation electr*onics*), may well account for a third of the total cost of the aircraft, and it is not unusual for even simple family aircraft to be as well equipped with radios as an airliner. The GA fleet also supports many businesses in the sales, maintenance and flying-instruction fields.

The modern lightplane

The earliest aircraft for the private owner were made of wood or metal structures with fabric covering. Almost all are now built of light alloys or fibre-reinforced (glass or carbon) composites. These structures need to be light to allow a good load of passengers, baggage and fuel to be carried.

Modern piston aero engines are almost invariably four- or six-cylinder with the cylinders lying flat, horizontally opposed in pairs, and air-cooled. Dual ignition systems are usual. With the necessity for these engines to be extremely reliable, manufactured cost is unfortunately relatively high. Wing flaps and dual controls for the pilots are standard on practically all types. The smaller aircraft have fixed landing gear and fixed-pitch propellers.

Even the least expensive training types have quite comprehensive instrumentation: airspeed indicator, altimeter, two gyro-driven attitude instruments and customary engine instruments such as an rpm gauge. They also generally have radios for navigation and communication. The faster single-engined types and almost all twins have retractable landing gear, constant-speed propellers, duplicated radios and, often, duplicated flight instruments. The more powerful engines have fuel injection rather than carburettors, and cowl flaps to vary the flow of cooling air. All but the smaller types now have autopilots. The basic radios – VHF communication, VHF Omni-Range (VOR) and Airborne Direction-Finding (ADF) navigation sets – are supplemented in the more expensive aircraft by radar transponders, Distance-Measuring Equipment (DME), computer-based area-navigation systems and weather radar. Turbine-powered GA aircraft are generally as well equipped as most airliners and are

operated in flight in much the same way.

The increasing complexity of systems and equipment, particularly in the larger and newer aircraft, makes correspondingly increased demands on the skill and training of the pilots. The required levels of these qualities may not always be in evidence, particularly in non-professional GA pilots. Largely for this reason, GA's accident record is consistently worse than that of the airlines and the military. The manufacturers do a fine job in building inherently safe aircraft, but the abilities of those who fly them do not always match the capabilities of their machinery. Most GA accidents are ascribed fairly to 'pilot error'; and the largest proportion of these accidents are the result of pilots attempting to fly in weather worse than they can safely cope with. Yet travel by GA aircraft is still generally held to be safer than travel by automobile.

Pilot training

One of the most useful and important GA activities is pilot training. Even those flying schools which exist to train airline pilots are effectively GA operations using GA aircraft types; and most military pilot trainees in most countries undergo at least some basic instruction or ability-grading in GA aircraft. These training aeroplanes are usually low-powered two-seaters of modest performance but correspondingly small fuel consumption.

Training for a private pilot's license varies around the world but almost invariably consists of ground instruction in such subjects as principles of flight, navigation, aviation law, flight planning and basic airframe, engine and instrument construction, all usually proceeding in parallel with airborne instruction. At least half of the flying is 'dual', i.e. with an instructor. Solo flying begins at the instructor's discretion, usually after 12 to 15 hours of dual flying, though often less. Gaining a commercial pilot's license takes longer, perhaps one or two years, and naturally involves more elaborate training.

The basic pilot's license generally qualifies the holder to fly single-engined aircraft in good weather and to carry non-paying passengers; further training and tests are required to fly multi-engined aircraft, or floatplanes, or in cloud using only instruments, or to give flying instruction. For the heavier piston-engined and all the turbine-powered types, a 'rating' on the license is required for that particular aircraft type as well. But the majority of amateur pilots never proceed beyond the first license. A Commercial Pilot's License (CPL) is required before a pilot may fly for hire or reward, whatever the aircraft type; there are hardly any exceptions to this rule.

Some airlines and governments sponsor young citizens for commercial pilot training. Third-world countries often have a state airline but no commercial-pilot training schools, or even a reservoir of ex-military pilots on which to draw. These countries must send their trainees abroad, usually to schools in the United States or Britain. One factor here is that English is the worldwide language of

Ballooning

Ballooning is the oldest kind of flying, almost 200 years old as against the powered aeroplane's 75 years. Except for some war applications – such as the use of tethered balloons as observation posts, and of free balloons to get messages and important people out of besieged Paris in 1870 – ballooning has always been an essentially sporting form of flight.

The very first balloons were hot-air filled; hydrogen balloons soon followed, and this gas largely supplanted hot air as a lifting substance until recent years and the invention of propane-gas burner systems. These offer a light, safe, simple and quite inexpensive source of hot air and allow the balloon to be airborne as little as 20 minutes after arrival at the launch site, whereas a hydrogen gas balloon takes hours to fill at a high cost.

Balloons, like aeroplanes, require airworthiness certificates, and the pilot-in-command needs a private pilot's licence (balloons). For this, a candidate needs a minimum of 12 hours' flying in balloons and to pass a simple medical, written papers and a flight test. Tuition is generally available from balloon manufacturers, balloon schools and already licensed pilots.

Ballooning was popular until World War I, and then declined until the 1960s when a great resurgence began which still continues. Today there are over 2,000 balloons flying in the USA and even more in the rest of the world – almost all propane-fuelled. The first world championships were held in New Mexico in 1972, with tests of the balloonists' abilities to ascend, maintain a set height and descend in timed steps measured by a barograph.

Calm summer evenings are the most popular time for a balloon flight, which begins with the big envelope laid out on the grass, downwind of its basket (usually still made of wicker, though some are of metal or glass-fibre) and the burner assembly. After a careful inspection of the whole balloon, the pilot light is lit and the main gas valve opened to generate a flame of some 1,093.3°C (2,000°F) and several feet long. The open neck of the balloon must be held open by hand at first; soon the envelope begins to swell with warm air. Crew members tug and heave at shroud lines, and may even run inside the envelope to help distend it properly.

Eventually the envelope contains enough hot air to lift itself and pull the basket upright. All the flight crew climb aboard, and after another long burn of 30 seconds or so the balloon lifts off. It is now a two-ton bubble of air heated to about 37.8°C (100°F), its buoyancy derived from the lower density of the hot air it contains. Further burns are required from time to time to maintain this buoyancy as the hot air cools.

Besides the burner valve, the balloonist also has a vent valve to release a small amount of hot air if he is climbing too fast; a rip panel to release a large amount quickly after landing; and a trail rope to slow his descent and forward speed before landing. He will also carry a welder's spark in case he has to relight the pilot burner – though even in the case of total burner failure the balloon should still hold its shape and descend slowly as a parachute.

Hang gliding

The current renaissance of hang gliding derives from work by the US National Aeronautics and Space Administration (NASA) on developing a foldable wing for spacecraft re-entering the atmosphere. NASA scientist Francis Rogallo invented an A-shaped fabric wing on a rigid frame, which has proved excellent for hang gliding. The present renewed interest in the sport began, like so many others, in Southern California but it has now spread throughout the world. Some of the more advanced gliders resemble conventional aeroplanes, but most are still variations of the original Rogallo delta wing.

A typical hang glider consists of a tough Terylene sail attached to a strong tubular A-frame, from which the pilot hangs in a harness. In front of him is a small triangular sub-frame which is rigidly attached to the main frame; the pilot holds this with both hands and steers the glider by swinging his

LEFT: *Hot air ballooning is growing rapidly in popularity, with almost 5,000 such balloons flying throughout the world.*
INSET: *A propane burner is used to heat the air in the balloon, generating a flame up into the balloon's envelope.*
BELOW: *Hang gliding is the least expensive way to go flying, and closest to man's dream of being a bird. The gliders derive their shape from space research.*

body about under it, thus moving the whole aircraft's centre of gravity. Some larger gliders have two harnesses and thus allow the dual instruction of novice pilots.

Hang gliders have poor glide ratios in comparison with conventional sailplanes, and thus require strong updrafts for continued flight. Favoured sites are those where a moderate to strong wind blows up and across a ridge of high ground. Most hang gliding is simple slope-soaring at low altitudes, though some experts have managed to climb sufficiently in slope lift to find thermals at high altitudes.

There are world and national contests for hang gliding, as with other aerial sports. The great appeal of hang-gliding is that it is the

cheapest form of flying; and, of all forms and ways of flying, it is the closest to Man's dream of 'being a bird'. It is also the only form of air sport which is good exercise: the glider, weighing about 17kg (35–40lb) has to be carried back up the hill after every flight. Neither the craft nor the pilot is required to be licensed.

Development work in hang glider design is concentrated on improving the efficiency of the basic Rogallo shape, and on improving its glide angle without impairing its natural docility and controllability. Whereas at first the pilot always hung vertically upright in his harness, the more advanced types now allow him to hang prone, thus reducing the drag of his body shape.

ABOVE: *A member of the British Red Devils parachute display team. Parachuting is judged on style while in free fall, and accuracy on touch-down.*
TOP RIGHT: *Another display parachutist. The sport has grown very rapidly since the 1950s, when it began in France – before this the uses of the parachute were almost wholly military.*
FACING PAGE
TOP: *The Glasflugel Hornet glider, a sailplane made of glass fibre. The use of this flexible material cuts down on problems of aerolasticity. See the chapter on airframes, page 86.*
BOTTOM: *Typical of modern sailplanes (competition gliders), the German Schleicher ASW 22B has a span of no less than 25 metres (marginally over 82ft).*

Parachuting

The first successful parachute descent was made by the Frenchman André Jacques Garnerin, jumping from a balloon at about 1,000m (3,280ft) near Paris on 22 October 1797. Captain Albert Berry made the first descent from an aeroplane, in America on 1 March 1912, when he jumped from a Benoist aircraft at about 500m (1,640ft) over St Louis.

Parachutes have saved the lives of possibly some 100,000 aviators in stricken aircraft. A handful of lucky men have even survived falls from high altitude *without* parachutes. One, Lt Col I. M. Chissov, a Russian, fell 7,000m (23,000ft) and survived, though he was badly injured. A British bomber crew-

man, Flt Sgt Nicholas Alkemade, jumped from his blazing aircraft without a parachute at about 5,500m (18,000ft) and survived without a single broken bone. The first emergency parachute descent was by a Polish balloonist, Jodaki Kuparento, after his balloon caught fire on 24 July 1808. One American in modern times has made more than 7,000 jumps; two Americans once made 81 jumps in 8hr 22min; and five RAF parachuting instructors once jumped together from 12,614m (41,383ft), falling free for more than 11,900m (39,000ft).

Perhaps the most astonishing parachute descent ever was that made by Capt Joseph Kittinger of the United States Air Force, who jumped from the gondola of a huge helium balloon at a height of 31,150m (102,200ft) over New Mexico on 16 August 1960. He fell free for 4min 38sec, through 25,817m (84,700ft), reaching a peak speed at about 18,300m (60,000ft) of almost 1,000 km/h (620mph), representing a Mach number of 0·93. The step by his balloon gondola's door was inscribed 'the highest step in the world', and his jump was the highest ever made, with the longest free fall. His parachute, with a barometric operating device, opened automatically.

Sport parachuting is a fast-growing sport, part of its appeal being that, unlike most other forms of air sport, it does not require a long and expensive training period. Early jumps are made with a static line attached to the aircraft to open the jumper's parachute automatically. Free fall comes later.

There are local, national and world contests in parachuting, when accuracy of touch-down is measured and style while man-oeuvring in free fall is judged.

Jumpers must wear two parachutes, one a reserve in case of a malfunction. Touch-down speeds are often low enough to allow experts to perform a stand-up landing in low winds. Landing is not like jumping from a particular height of, say, 3m (10ft) because the upper body, still sustained by the parachute, slows after the feet have touched. The important point to master in landing is not so much the vertical speed but rather the horizontal component caused by wind.

Soaring

Soaring – gliding, using several sorts of up-current in the air to stay aloft – was first developed after World War I. The peace treaty forbade defeated Germany from having powered aircraft, but did not ban gliders. In 1921 a German glider soared for 13 minutes, finally beating the Wright Brothers' 1911 time of 9 minutes 45 seconds. In 1922 a German pilot made the first soaring flight of more than an hour; in 1926 another was drawn up into a thunderstorm and made the first soaring flight away from his launch site. By 1935 the sport had advanced so that four contestants in a competition in Southern Germany flew more than 500km (300 miles) into Czechoslovakia. World records today reflect the extraordinary efficiency of modern sailplanes and the skills of their pilots. The record for distance in a straight line is 1,461 km (908 miles). The absolute altitude record is 14,938m (49,009ft).

Modern high-performance gliders are easily the most efficient aircraft of all, employing some of the most advanced aerodynamics and modern structures. Many are built largely of glass-fibre and/or carbon-fibre plastics. German designers and manufacturers are still in the forefront of the craft. Minimum still-air sink rates may be less than 0·6m/sec (2ft/sec), and glide ratios can approach 50:1. Such sailplanes have very long, narrow wings (to reduce induced drag) and require the pilot to lie back in an almost-prone position to reduce the fuselage frontal area. Most have wings and tailplanes which are quickly detachable so that they can be fitted in a long, narrow trailer for towing behind a car.

Aero-towing behind powered aircraft to a release height of about 700m (2,000ft) or winch-launching to about 250m (800ft) are the commonest methods of launching today, with car towing used to a lesser extent. A fairly recent development is the powered sailplane, which uses a small motor (usually a modified Volkswagen car engine) to climb to a useful height, when the engine is shut down and the aircraft is flown as a glider.

As with powered aircraft, all glider training is by dual instruction in two-seaters. Gliding clubs offer low-cost accommodation and catering, and encourage members to stay the weekend. Costs generally are kept low by the requirement for considerable voluntary assistance from members in all aspects of club life. The most basic qualifications are national certificates, after which there are international silver, gold and diamond badges for advanced performance. A Silver C, for example, requires a flight of at least five hours, a height gain of 1,000m (3,281ft), and a straight-line distance flight of at least 50km (31 miles). A Gold C requires five hours, 3,000m (9,842ft) height gain and 300km (186 miles) distance, which may be an out-and-return or a triangular flight. Diamonds may be added to a Gold C badge for a height gain (recorded on a barograph) of 5,000m (16,404ft); or a triangular or out-and-return flight of 300km (186 miles); or a distance flight of 500km (311 miles).

Gliding contests usually involve flying set tasks, typically out-and-return flights with photographs required of the turning points, these tasks being chosen by the contest judges having regard to the prevailing weather.

Gliders soar by using different kinds of rising air (it occasionally being possible to employ all kinds in a single flight). One is slope-soaring, riding the updraft of air over a hill ridge on a breezy day. The commonest 'lift' is that found in thermals – large bubbles of relatively warm air rising from near the surface of the earth as the sun heats it. (An extreme form of thermal lift is that under and inside a building thunderstorm.) Wave lift is found in mountainous regions where a strong, unstable air stream flows across the mountains; 'standing' waves form in the air, and these remain stationary with respect to the mountain ranges. Such lift is often very strong and smooth, and it can persist well into the stratosphere.

Flying Boats

The development of water-borne aircraft began very soon after the first successful landplane flights: after all, two-thirds of the earth's surface is water, and there are an enormous number of lakes, rivers and sheltered inlets which provide potential natural water aerodromes. Maritime aircraft seemed for a long time to have a big future: they were employed extensively for oceanic reconnaissance duties in both world wars, against submarines and for rescue work in World War II, and as long-range civil transports in the 1930s and 1940s. Yet today hardly any flying boats still exist; and the only water-borne aircraft currently in use in any numbers are a few light floatplanes operating in thinly-populated parts of the world, where land airports are few and lakes numerous.

Several factors combined to bring about the decline of the flying boat, most of them related to economics. Helicopters proved able to carry out air-sea rescue and anti-submarine roles in all weathers and, sometimes, more cheaply. All the world's cities, even those that were seaports, proved only too willing to build land airports with vast runways to attract landplane airliner services. Flying boats, with their big stepped hulls, could not be made to fly as fast as the slender airliners, and this became more marked as speeds rose. Because of their inherently higher drag they were more costly in fuel, and the boats also remained susceptible to damage from collision with floating debris, from waves when operating out at sea, and to the corrosive effects of salt water on aluminium alloys.

Yet flying boats had their attractions. Their big hulls allowed passengers a degree of room and comfort unknown in most landplanes. They were beautiful to watch, during take-off and landing, and still invoke affection among passengers and aviation enthusiasts.

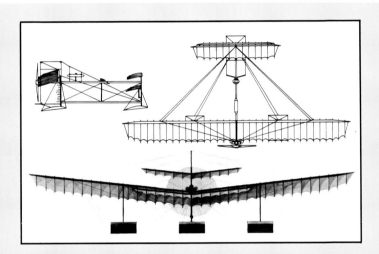

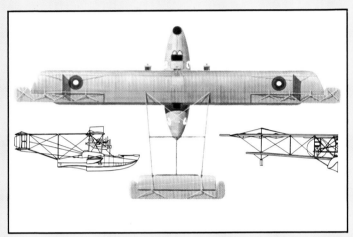

ABOVE: *Henri Fabre's* Hydravion *was the first successful floatplane. Built in 1910, it was powered by one 50hp seven-cylinder Gnome rotary engine, driving a pusher propeller, giving a maximum speed of 90km/h (55mph).*
RIGHT: *The Short Sunderland, which first flew with the RAF in 1937. It carried seven to ten Browning and Vickers machine guns and a bomb load of 700kg (2,000lb).*
BELOW: *The Dornier DO X, the world's largest aircraft when it was built in 1929. It had a maximum speed of 215km/h (134mph), and a range of 1,700km (1,056 miles). It could carry up to 150 passengers.*

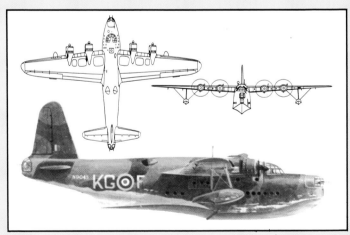

ABOVE: *The Navy Curtiss NC-4 the first aircraft to cross the Atlantic, did so in several stages. It was powered by four 400hp Liberty V-12 engines, one pusher, three tractor, giving a maximum speed of 145km/h (90mph) and a range of 2,360km (1,470 miles).*
BELOW: *The Saunders-Roe (Saro) Princess, a very large flying boat intended as a high capacity transatlantic transport aircraft. It first flew in August 1952, powered by ten Bristol Proteus turboprops, eight of which were coupled in pairs to contrarotating propellers. These engines gave a cruising speed of 580km/h (360mph) and a range of 8,480km (5,270 miles).*

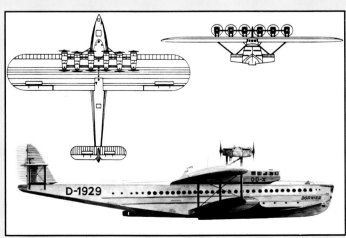

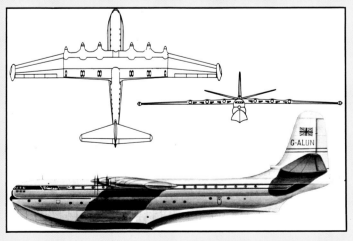

Early history

Henri Fabre, member of a family of Marseilles shipowners, developed the first aeroplane to take off from water. His very first Hydravion, built in 1909, had three engines coupled to drive a single propeller, and could not be coaxed into flight at all. His second design was also freakish in layout. It was of canard form, with the tail surfaces at the front and the wing and engine (driving a pusher propeller) at the back. The fuselage consisted simply of two wooden girders, astride the top one of which Fabre sat, his feet dangling down to a pair of pedals which warped the wings for lateral control, and his hands grasping a stick which moved the front elevator and rudder. The craft sat on three tiny floats which were almost wholly submerged when at rest. The flying surfaces resembled sails in that their fabric was taughtened to the end of each rib by spring hooks, and could be detached and reefed up. Only one surface of the wing was covered.

In this odd device, Fabre became the first man to take off from and land back on water, on 28 March 1910. His first flight covered some 500m (1,640ft) over the calm water of the harbour at La Mède near Marseilles. Not only had Fabre been first to succeed where many better-known aerial experimenters completely failed; but it was also his first flight. He succeeded because his exquisitely-built craft was extremely light in weight, and powered by one of the very first Gnome rotary engines, which was relatively light, but gave a powerful 50hp.

Fabre apparently lacked the funds to continue experimenting, though he continued to make floats for other experimenters. In 1911 he made some floats for a Voisin biplane landplane, which was thus transformed into the world's first amphibious aeroplane.

Glenn Curtiss

The great American pioneer of water-borne aircraft was Glenn Curtiss. He began his aviation experiments with wheeled landplanes, evolving a simple but efficient pusher biplane with hinged surfaces – ailerons – for roll control. As early as November 1908 he had tried (unsuccessfully) to turn his pioneering aeroplane, *June Bug*, into a seaplane by mounting it on floats and renaming it the *Loon*, after a species of water bird.

On 31 May 1910, he flew non-stop from Albany to New York City in a landplane fitted with a cork-filled canvas tube, a nose hydro-ski and wing-tip balance floats for flo-

BELOW: *Henri Fabre's 1910 50hp* Hydravion *was the first aircraft in the world to take off from water. It flew tail first.*

BELOW CENTRE: *This twin-engined Curtiss* America *flying boat was intended to fly the Atlantic in 1914, but war intervened.*

BELOW CENTRE: *Henry Ford (right) visited the Curtiss plant in 1912, and is here shown an 'aerial yacht' by Curtiss.*

BOTTOM: *Eugene Ely, flying a Curtiss pusher biplane, flew from San Francisco and landed near the battleship* Pennsylvania.

tation should he be forced to land on the Hudson river on the way. He foresaw the value to the US Navy of a true hydro-aeroplane, and tested a variety of floats on his biplane. With none of these could he achieve speeds sufficient for take-off from water.

The next winter he moved from the bitter cold of upstate New York to San Diego in California; and there, on 26 January 1911, he finally made his first take-off from water, in a pusher biplane employing a single central float with stabilizing outriggers.

On 14 November 1910, a (wheeled) Curtiss pusher flown by Eugene Ely became the first aeroplane ever to take off from a ship – the USS *Birmingham* moored in Hampton Roads, Virginia. On 18 January 1911, Ely completed the achievement by landing on the deck of the USS *Pennsylvania* in San Francisco Bay.

Curtiss flew the first aircraft with parallel twin floats, and demonstrated their natural lateral stability. He flew with a passenger; developed an amphibious wheel-and-float aeroplane; and had his seaplane lowered to the surface from a warship, flown, and then winched back aboard afterwards. On 12 January 1912, he flew the first successful boat-hulled aeroplane.

It was Curtiss who stumbled on the idea of a stepped hull to break the water's suction during take-off – a device which has been used on every successful seaplane and speedboat ever since. By 1912 he was manufacturing a two-seat Aeroyacht, and in the same year a boat-hulled Curtiss Model F hydroaeroplane became the first aeroplane ever to fly with the newly-invented Sperry gyroscopic autopilot. At least 150 Model Fs were eventually ordered by the US Navy.

World War I

During the war maritime aircraft were extensively used by both sides for reconnaissance duties and to drop bombs and launch torpedoes. Seaplane carriers were also used by both sides. Britain's first was a 7,500 ton converted collier which was the second Royal Navy ship to bear the name HMS *Ark Royal*. Others were converted cross-Channel steamers fitted with cranes and carrying about four aircraft each. The German merchant raider *Wolf* sank 28 Allied vessels in the Indian and Pacific Oceans, helped by the Friedrichshafen seaplane that she carried. This was the most famous of all the Friedrichshafen *FF* seaplanes, the *FF* 33e *Wolfchen*.

By the end of the war some very large flying boats were being built; by Zeppelin-Lindau in Germany whose boats were designed by Claude Dornier; Curtiss in America; and derivatives of Curtiss designs in Britain. The wing spans of some of these craft approached that of a modern Boeing 707; the Navy Curtiss NC flying boats, for example, spanned 38·40m (126ft). They were most literally flying 'boats', with a speedboat-style hull instead of a fuselage; the biplane wings, engines and tail surfaces were held above this hull by a web of struts and bracing wires.

Some 43 per cent of their all-up weight of 12,700kg (28,000lb) was useful load (a remarkable ratio for aircraft of any age), and could include some 4,500kg (10,000lb) of fuel. The prototype had three 400hp Liberty engines, but subsequent aircraft had four engines, giving a speed of between 107 and

137km/h (66 to 85mph). The engines had electric starters, for the propellers were too big to be turned by hand.

They were ready too late to take part in the war, but they achieved fame as participants in the US Navy's great 1919 project to make the first aerial crossing of the Atlantic. The attempt involved enormous surface vessel support; there were 68 navy destroyers, one stationed every 80km (50 miles) along the route to assist the flying boats by radio, smoke signal and pyrotechnic guidance.

Three of these 'Nancy boats' set off from Newfoundland on 16 May 1919, and one of them, the NC–4 commanded by Lieutenant Commander Albert C. Read, arrived safely in Lisbon on 27 May, via refuelling stops on the Azores, reaching Plymouth, England, on 31 May. The other two 'Nancies' came down in the open Atlantic; NC–1 sank after her crew had been rescued, but NC–3 taxied 322km (200 miles) to the Azores.

The Dornier Do X

An even slower Atlantic crossing was made by the huge Dornier Do X 12-engined flying boat, which took ten months to travel from Lake Constance to New York, where it arrived on 27 August 1931.

The Do X was the biggest aeroplane in the world in its time, spanning 48m (157ft 6in) and having a maximum weight of 56,000kg (123,459lb), and carrying aloft on one test flight 169 people – then a world record – comprising of 10 crew, 150 passengers and no less than 9 stowaways. (However, the aircraft's intended usual complement was no more than approximately 70 passengers). The Do X's hull had three decks, and contained a bar, smoking and writing rooms, a bathroom, a lounge and sleeping quarters, and a kitchen and dining room.

At first she was powered by twelve 525hp Siemens-built copies of the British Bristol Jupiter air-cooled engine. These were mounted in six back-to-back pairs on an auxiliary extra wing above the main wing. It was quickly found that the rear engines, mounted in the disturbed airflow from the front engines and the stub wing, tended to overheat, and there was a loss of lift and excessive drag. In an attempt to cure this, 12 American Curtiss Conqueror liquid-cooled engines of 600hp each were installed to replace the Jupiters, and the auxiliary wing was deleted. The Do X burned 1,828·57 litres (400 gallons) of petrol an hour to cruise at 175km/h (110mph), her maximum speed was 210km/h (130mph), and her ceiling when loaded was only 500m (1,640ft).

Dornier Wal

Dornier Wals were much more successful. They served with several early airlines, and two were used by Roald Amundsen on his 1925 attempt to reach the North Pole, one reaching 87° 43′ North – the farthest north any aeroplane had been – on 21 May 1925. Ramon Franco, brother of General Franco, was the first to fly a Dornier Wal across the South Atlantic in 1926, and another was flown across the North Atlantic to Chicago in August 1930 by Wolfgang von Grunau, the first east-west Atlantic crossing by a flying boat.

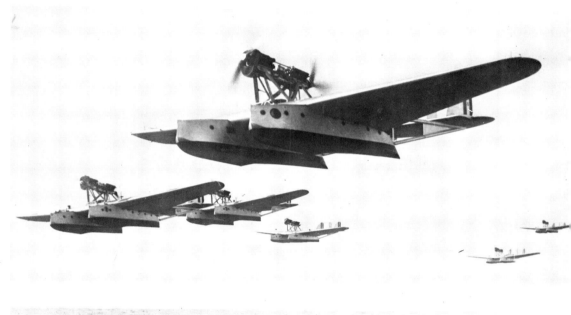

Other long-distance flights

Many epic long-distance flights were made in lighter aircraft fitted with floats. Perhaps the most remarkable were those of Francis (later Sir Francis) Chichester, who made the first solo air crossing of the Tasman Sea from New Zealand to Australia, between 28 March and 6 June 1931, in a little de Havilland Gipsy Moth seaplane, which he rebuilt *en route* after it had sunk at Lord Howe Island. Between 3 July and 14 August he made the first-ever solo flight between Australia and Japan, again in his seaplane.

An Italian design, remarkable because of its unorthodox configuration, was the Savoia-Marchetti S.55, a twin-hulled flying boat with an open cockpit, very thick wings, twin engines installed back-to-back and canted at a striking angle, and a tailplane carried on wire-braced booms. In 1926 S.55s set up 14 world records for speed, altitude, and distance with payload. In 1927 and 1928 they made long flights across the Atlantic (both north and south). In January 1931 12 S.55s led by the Italian Air Minister Marshal Italo Balbo made a flight in formation (though three aircraft crashed during the voyage) from Rome to Rio. In 1933 Balbo led a formation of 24 S.55s to Chicago for the World's Fair, making the journey of almost 10,000km (6,000 miles) in just under 49 flight hours.

Undoubtedly some of the greatest waterborne aircraft ever built were those designed to compete in the Schneider Trophy Races, described in Chapter 21. One of these machines was probably one of the greatest seaplanes ever built, although it was not completed in time to compete in the last contest, held in 1931. This was Italy's Macchi-Castoldi M.C. 72, which had a pair of V-12 Fiat engines mounted back-to-back in tandem, driving a pair of contra-rotating propellers. This installation solved the outstand-

ing problem which had plagued other Schneider designers: the effects of torque when their throttles were opened for take-off. The controllable pitch propeller had not been perfected in those days, and the Schneider racers had enormous fixed-pitch propellers of exceedingly coarse pitch. When the throttle was first opened and before the craft had gathered forward speed one float would dig in, and try to turn the machine in a circle. This was also the moment when the spray was worst, preventing the pilot from seeing where he was going. But there was no such problems with the Macchi M.C. 72, which

could make a straight take-off run under full control the whole time, because the contra-rotating propellers were free from torque. Four M.C. 72s were built, three crashing and killing their pilots; but the fourth, piloted by Warrant Officer Francesco Agello, set several remarkable speed records, including one of 709km/h (440mph), established on 23 October 1934, (by which time the aeroplane's double engine was developing more than 3,000hp). This remained an outright world speed record until 1939, and a seaplane record until it was beaten by a Russian turbo-jet-powered Beriev Be-10 in August 1961.

Transport Flying Boats

LEFT: *The 1934 Sikorsky S–42 was a 32-passenger flying boat designed for Pan American.*
FAR LEFT INSET: *Short Solents were used by BOAC in the late 1940s. Twelve were built.*
CENTRE LEFT INSET: *The Short S.26 G-class were enlarged developments of the Empire boats.*
LEFT INSET: *Two Breguet 530 Saigons served with Air France in the mid-1930s.*

Amphibious Aircraft

BELOW LEFT: *The Grumman Mallard amphibian was a 1946 design; only 59 were built.*
BELOW FAR LEFT INSET: *Pan American used Sikorsky S–38 ten-passenger, twin-engined amphibians to expand its services through Latin America.*
BELOW CENTRE LEFT INSET: *A few Douglas C–47s were experimentally fitted with amphibious floats, used as fuel tanks, in World War II.*
BELOW LEFT INSET: *More than 700 Supermarine Walruses served in World War II, mostly rescuing British pilots from coastal waters.*

World War II

Maritime aviation played a huge role in World War II – particularly in the form of carrier-borne landplanes, and most particularly in the Pacific Ocean war with Japan. Carrier flying, however, is described elsewhere in this book. Waterborne aircraft also operated in the war in large numbers; one type, the Consolidated PBY Catalina originally a 1935 design, was built in three countries (the USA, Canada, and the USSR) in a number totalling about 4,000 – the largest production run of any flying boat in history.

ABOVE: *The Martin PBM Mariner served as a transport, reconnaissance, air-sea rescue and anti-submarine aircraft with the US Navy in World War II.*

Some PBYs are still flying today as water bombers.

Flying boats and floatplanes were employed for oceanic patrol duties, in particular reconnaissance and anti-submarine work, and for rescue duties at sea.

Germany operated several Dornier derivatives of the Wal, as well as the Heinkel He 59 twin-engined floatplane and Blohm und Voss flying boats, including the large six-engined BV222 Wiking – the largest flying boat in service with any nation in World War II. They also built a prototype of an even

ABOVE CENTRE: *More than 4,000 Consolidated PBY Catalinas were built in the USA, Canada and the USSR; the type was the most prominent maritime aircraft of World War II.*

bigger boat, the 60·17m (197ft 4¾in) span BV238, with six 1,750hp engines.

A very large US boat was the Martin Mars, with a maximum take-off weight of 65,770kg (145,000lb). Originally a long-range patrol aircraft ordered by the US Navy in 1938, the Mars served principally as a cargo transport. Four were retired by the US Navy in 1950, and operated in Canada for many years thereafter as civil forest-fire water bombers. In naval service Mars flying boats set several records, including carrying 9,300kg (20,500 lb) of payload over 7,564km (4,700 miles) to

TOP: *Three Blohm und Voss Ha 139 seaplanes were used by Lufthansa for transatlantic mail flights in the late 1930s. This shows one on the German depot ship Schwabenland.*

Hawaii and back in 1944, and in carrying the largest number of passengers then flown, 301 plus a crew of 10, on a flight over California on 19 May 1946.

The principal British type was the Short Sunderland, a development of the pre-war Empire civil boats. This large machine had a span of 34.38m (112ft 9½in), a take-off weight (in the Mk III version) of 26,000kg (58,000lb), a maximum endurance of 13½ hours and maximum speed of 330km/h (205 mph) on four Bristol Pegasus 1,050hp engines.

Hughes Hercules

Even more grandiose, and still the largest aeroplane ever built, was the wooden Hughes Hercules, a 97.54m (320ft) span colossus powered by eight 3,000hp Pratt & Whitney piston engines, and designed to carry 700 passengers or freight and to weigh 177 tonnes (180 tons) at take-off. The Hercules was intended for production should a shortage of metal alloys develop; in the event there was no such shortage, and the war was over before the craft was ready. The prototype made only one flight, a hop of about a mile over Los Angeles harbour on 2 November 1947, when it was flown by the aircraft's sponsor, the billionaire Howard Hughes. It is on view today at Long Beach.

Saro Princess

Three prototypes of a gigantic new flying boat intended for commercial service were constructed in Britain between 1943 and 1953. The Saunders-Roe Princess was almost as large as the Hughes H.2, with a 66.75m (215ft) span and 149,685kg (330,000lb) take-off weight. It had the largest pressurized hull ever built, fully electric ('fly-by-wire') signalling to the controls, and was designed to carry a 18,000kg (40,000lb) payload at a cruising speed of 579km/h (360mph) at 11,278m (37,000ft). Power plants were ten 3,780hp Proteus turboprops, eight coupled in pairs and driving contra-rotating propellers.

There were also two other fascinating Saunders-Roe projects: the Duchess, a four-jet passenger flying boat; and the P.192, a fantastic dream of a 1,000 passenger five-deck 24-jet-engined flying boat, to fly between England and Australia.

Jet flying boats

The Saunders-Roe Company experimented in the early post-war years with a jet fighter flying boat, the SR/A1: the aircraft proved too slow (805km/h: 500mph) to be competitive with landplane fighters, which were already some 160km/h (100mph) faster. A later design for a waterborne jet fighter was the Convair XF2Y-1 Sea Dart which did

FAR LEFT: *The largest flying boat in the USA in its day, the Martin Mars served as a cargo transport with the US Navy in the 1940s. In the mid-1950s four were modified to serve as fire-fighting water bombers in Canada.*
CENTRE LEFT: *Short Sunderlands served as patrol and anti-submarine craft in World War II. They were heavily armed, with up to ten machine guns.*
LEFT: *The ten-engined Saunders-Roe Princess was intended as a transatlantic airliner for BOAC; but engine problems and the development of landplane airliners doomed the project.*

BELOW LEFT: *The Saunders Roe SR/A1 was the world's first jet fighter flying boat. It was planned for service in the Pacific in World War II, but did not fly until 1947. Three were built and test-flown; with a maximum speed of only 800km/h (500mph) they were too slow to succeed as fighters.*
BOTTOM LEFT: *The 1952 Convair XF2Y-1 Sea Dart was a revolutionary attempt to build a seaborne combat jet fighter, floating on a boat hull but rising on extendable hydroskis for take-off and landing. It is the only waterborne aircraft to have flown faster than sound. Only five were built, and one crashed.*

BELOW: *The Japanese Shin Meiwa PS-1 is one of the only two large flying boats of recent design in service today. It can operate in very rough water and has a notably wide range of operating speeds.*
BELOW CENTRE: *Six Martin XP6M SeaMasters were built in the late 1950s as prototypes of a bomber and mine layer for the US Navy; two crashed, and although three production SeaMasters were built, the programme was abandoned.*
BOTTOM: *The twin-turboprop Beriev Be-12 amphibian is in service with the Soviet Navy and is comprehensively equipped with radar and other detection devices.*

not have a conventional boat hull but instead rose up for take-off on hydro-skis. This project, too, was abandoned after only four examples had been built. The type remains the only supersonic seaplane.

The US Navy remained interested in flying boats for some years; their 1952 design competition led to the Martin SeaMaster four-jet patrol flying boat; however two prototypes crashed during flight testing, and only four other prototypes and three production aircraft were built, the programme being abandoned in 1960.

Japanese PS-1

Two major powers have, however, continued with military flying boat programmes. The Japanese Maritime Self-Defence Force operates the Shin Meiwa PS-1 four-turboprop anti-submarine and search-and-rescue flying boat and US-1 utility amphibian. Shin Meiwa continues to evaluate passenger and water bomber versions of the aircraft. The PS-1 employs a boundary layer control system, driven by a fifth engine, using blown flaps and control surfaces and has an automatic flight control system as well. Its long narrow hull serves both to reduce the high aerodynamic drag inherent in a flying boat design, and also enables it to operate in very rough water, in wind speeds of 47km/h (29mph) and wave heights of up to 4m (13ft). An unusual spray suppression device is a long groove built into the chine of the hull forebody. The aircraft has a high maximum speed of 547km/h (340mph); yet its touchdown speed is only 76km/h (47mph) and it can take off in 305m (1,000ft) to clear 15m (50ft).

In the search-and-rescue role the amphibian version can take off from a land base, fly 1,100km (684 miles) out to sea at a cruising altitude of 3,050m (10,000ft) at a speed of 425km/h (264mph) and then search at low level for five hours. The aircraft is then able to land on the water and spend an hour picking up survivors before returning to base.

The water-bomber version could carry as much as 16,257kg (35,840lb) of water, picking this up through hull scoops while contacting the water surface for only 20 seconds at a speed of 55 knots (34mph).

Russian flying boats

The Soviet Naval Air Force also continues to operate twin-turboprop Beriev Be-12 amphibians; the same design bureau earlier produced the piston engined Be-6 and the twin-jet Be-10 flying boats. The Be-12 holds a number of world records, including an altitude record in its class of 12,185m (39,977ft).

Canadair CL–215

A recent civil amphibious design is the Canadair CL–215, powered by two 2,100hp Pratt & Whitney R–2800 piston engines. This Canadian design is intended primarily as a water bomber for use against forest fires, and is in service with government agencies all over the world. Spanish aircraft are additionally equipped for search and rescue duties. The CL–215 can attack brush or forest fires with a variety of fire retardants: these include pre-mixed chemicals ground-loaded at a land aerodrome; short-term retardants mixed automatically with water, scoop-loaded while taxiing at speed across a lake; plain water scooped from any lake or the ocean; and pre-loaded foams for attacking oil fires. The aircraft can carry 5,455 litres (1,200 gallons) of water and can fill its tanks in under 20 seconds while skimming the water at 111km/h (69mph). It needs only a 1,220m (4,000ft) stretch of water for this purpose; on several occasions CL–215s have made more than 100 drops totalling almost 545,500 litres (120,000 gallons) of water in a single day. While fighting forest fires in Provence, full loads of water have been scooped from the Mediterranean in 2m (6ft) swells. Other flying boats have also been used for airborne fire-fighting and forestry protection. The Ontario Provincial Air Service pioneered these operations in 1924 with 13 HS2L flying boats, and now has about 50 aircraft, making it the largest service of its kind in the world.

Operating and piloting techniques

Operating and piloting techniques for water aircraft are interesting, and notably different from those of landplanes.

On the water with the motor cut or idling a seaplane always tends to weathercock into wind; power is always needed to turn it out of wind. At low power it tends either to creep into wind or drift back, depending on the relative strengths of wind, current and propeller thrust. The pilot should hold up-elevator at all times on the water, except towards the end of the take-off run. Taxiing is always done at low speeds, except where there is a considerable distance to cover and when wind and water allow fast taxiing 'on the step'. To get on the step the throttle is opened with the stick held hard back. At first the bows of the floats generate a considerable amount of spray – this reaches a peak at about 24km/h (15mph). As speed is gained the bows of the floats begin to rise out of the water and the spray moves back behind and below the propeller and the cabin. The aeroplane may then be taxied at speed. It is important not to stay at the in-between speeds where spray goes through the propeller, since in time it may cause damage: consequently seaplanes should be taxied either fast or very slowly.

When starting a turn out of the wind, wind and centrifugal force oppose each other, and turning is therefore easy. Turns through downwind require more care, as wind and centrifugal force are now combined to try to upset the craft. Furthermore, the downwind float may tend to dig in as the ship lists, causing more drag and requiring more rud-der and more power to overcome it. If the wind is strong enough the downwind wing-tip may even dig into the water, making capsizing a real possibility. Under such conditions it is best to close the throttle completely, whereupon the seaplane will slow down and then weathercock into wind by itself.

To beach the ship cross-wind it is desirable to come at the shore at a modest speed; if this is attempted too slowly the aircraft may weathercock around out of control just when it should be beaching. Furthermore, the bow wave from the floats serves to cushion the shock of contact with the shore if this is done at constant power. If power is reduced at the last minute the bows tend to drop with a bump just as contact is made.

It is easiest to come up to a boat, buoy or raft into wind; where possible manoeuvres on the water should be done downwind of one's objective, which is then approached straight and into wind at minimum speed. A straight-in approach also enables the pilot to feel out the effect of wind and water on the aircraft's heading while still out in the open, and to find the slowest (and therefore safest) speed at which to maintain positive control.

Modern floatplanes have retractable water rudders connected to the main rudder circuits, which make manoeuvring appreciably easier. They hang down below the floats, and can be retracted during the take-off and landing runs to avoid any risk of damage to them from floating objects and vibration.

On breezy days and in confined spaces it is often easiest to manoeuvre a seaplane downwind by sailing it much like a boat. Here again the water rudders should be retracted, since they will tend to oppose the turning moment of the air rudder as the ship drifts backwards. Cutting one magneto will further reduce the propeller thrust at idle and help the plane drift backwards, as will lowering the flaps and opening the cabin door. The plane can be steered when sailing backwards by slightly increasing power and turning the air rudder.

In open water on a windy day 'step turns' are a useful if tricky means of turning downwind. From weathercocked into wind, the pilot slams open the throttle and pulls hard back on the stick to get the craft partially planing. Power is then reduced, but the heels of the floats are kept well dug in for stability. The ship may then be turned out of wind and all the way round to downwind, steering with the rudder and taking great care to keep the tilting force of the wind and centrifugal force neatly in balance.

Take-off and landing, performed into wind, are, by comparison with manoeuvring on the water, comparatively easy. For take-off the water rudders are raised, throttle opened wide and the stick pulled all the way back. As speed is gained the bows of the floats rise to a certain height and then flatten out. By now the aircraft is starting to plane, and the stick should be eased forward to a neutral position so the float heels do not dig in. Soon the seaplane is running on the central part of the float bottoms alone – this is the minimum drag condition allowing fastest acceleration to take-off speed. Once this stage is reached a slight back-pressure on the stick will allow

the plane to leave the water. If too much back pressure is applied the float heels may dig in again and slow the aircraft down. Similarly, attempting the take-off run with the nose too low wets more of the floats and delays the actual take-off. One quickly learns to feel the angle at which the aircraft will be best able to plane.

Water landings are similar to wheel landings, except that there is a broader range of safe attitudes at touchdown. On rough water the landing should be full-stall with maximum flap.

Special techniques are employed for landing on very calm water, particularly on misty days, for then water and sky merge imperceptibly so that it is quite impossible to judge one's height above the surface. In such conditions (called 'glassy water'), one tries to land alongside a shoreline, a boat or some other floating object to give one depth perception. Otherwise one simply sets up a very shallow descent on instruments with power, at low airspeed and waits until the floats make (hopefully) gentle contact with the surface – which may well be the first indication of just where it is.

As part of his preflight inspection, the seaplane pilot checks the bilges of each water-tight compartment to be sure none is leaking. However, if he is mooring the aeroplane he

may deliberately flood some compartments to make the craft heavy and less likely to lift in a wind. If a float is ever accidentally holed (by floating timber or a submerged rock) the craft must be beached at once.

Operation from seawater brings special problems, since salt is highly corrosive to aluminium. Seawater-based craft are usually hauled up the ramp at the day's end and hosed down with fresh water; land-based

time goes by. Fibreglass floats are now available which do not corrode; however, they are not yet as tough and resistant to damage as those made from light alloy.

The weight and drag of float assemblies mean floatplanes are inherently inferior in performance to landplanes. Yet in a few regions such as northern Canada and Alaska, which are inaccessible to other aircraft, the floatplane is still supreme.

ABOVE: *The Canadair CL–215 is a piston-engined amphibian widely employed as a water bomber for attacking forest fires.*
BELOW LEFT: *Pioneer bush fliers in Canada had little in the way of airport facilities and so operated on floats in summer and skis in winter. Here a Fairchild 71 of Canadian Pacific Airways is seen with a variety of stores for one of the remote areas.*
BELOW: *This Cessna floatplane is a typical private commuter aircraft.*

amphibians should be landed in fresh water after sea operations to wash the salt water from them. Seaplane structures are carefully corrosion-proofed during manufacture, yet this proofing tends to wear away in spots as

The construction of water-based aircraft presents
unusual technological problems. They must be able
to float as well as to move along the water surface
at near-flying speeds. Basically, two approaches
have been developed: float construction and hull
construction. In the former, floats or pontoons are
substituted for wheels; in the latter, the aircraft is
built with a boat-like structure.

FAR LEFT: The He 45 was a general purpose military
aircraft used for training and reconnaissance.

BELOW FAR LEFT: This Tiger Moth is used for training
British Seaplane Club members in floatplane handling
techniques.

BELOW LEFT: The Macchi M.12 was a World War I
Italian type. A three-seat coastal patrol flying boat,
it had fore and aft gun positions connected by a tunnel
in the fuselage so that both could be manned by one
gunner. Side or wing tip floats were necessary to keep
this single-hulled aircraft upright.

BELOW: The Short Shrimp was typical of seaplanes of
the 1919 period. Of twin-float construction,
it also had a tail float attached directly to the
underside of the rear fuselage. The tail float carried
a hollow water rudder containing a spring-loaded
drop-plate to improve steering while taxiing.

LEFT: Sixty-three Supermarine Stranraers were built. They
served with the RAF and RCAF from 1936 to 1940.

ABOVE: This twin-float Gloster III.A biplane came
second in the 1925 Schneider Trophy race.

Lighter-than-Air

Balloons

It can be argued that the balloon is the most significant aircraft ever invented by man. Its original importance lay in the fact that for the very first time it enabled him to leave the surface of the Earth and travel freely in the air above, albeit at the whim of the breeze.

First demonstration of balloon flight

It has long been believed that the French brothers Etienne and Joseph Montgolfier were the originators of the hot-air balloon, the type of lighter-than-air craft with which man first achieved flight. But recent research has shown that in 1709 the Brazilian priest Bartolomeu de Gusmão demonstrated a model hot-air balloon at the court of King John V of Portugal. According to one Salvador Ferreira, the model balloon was constructed of thick paper and inflated by hot air. This came from 'fire material contained in an earthen bowl' which, as shown in a contemporary painting, was suspended beneath the open neck of the envelope.

On 8 August 1709, Gusmão presented his model balloon for examination by a distinguished and unimpeachable gathering which included the King, Queen Maria Anna, the Papal Nuncio and Cardinal Conti (later Pope Innocent III), together with princes, nobles, diplomats and other members of the court. It is recorded that the balloon rose to a height of 3·66m (12ft) before two valets, who feared it might set the curtains alight, ter-

PREVIOUS PAGES:
Hot-air balloons.

ABOVE: *The balloon built by the Montgolfier brothers, made of paper and filled with hot air. Its first ascent was made on 5 June 1783 at Annonay near Lyons.*
BELOW LEFT: *A drawing of the Montgolfier balloon, showing the hot air rising into it and (below) a plan of the balloon and its burner.*
BELOW RIGHT: *Lunardi's second balloon in which he, Mr Biggin and Mrs Sage (the first English lady aeronaut) ascended from St George's Fields on 29 June 1783. He had previously made an ascent on 13 May.*

minated its flight by knocking it to the ground. Thus, almost three-quarters of a century before the Montgolfiers, the principle of the hot-air balloon had already been demonstrated.

The Montgolfier brothers

Many stories, doubtless fanciful, have been told to explain the way in which Etienne and Joseph Montgolfier discovered that hot air rises, and how they concluded that if contained in an envelope of sufficient size it would have enough 'lift' to raise passengers into the air.

The choice of paper for the envelope of their first balloon must have seemed natural to the brothers, who were paper-makers by trade. The balloon, fairly rigid, stood over a pit containing wool and straw which when ignited filled the balloon's envelope with hot air. On 25 April 1783, the first successful Montgolfière took to the air at Annonay near Lyons in France. It is reported to have risen to a height of about 305m (1,000ft) and travelled about 91m (3,000ft) horizontally before it fell to the ground as the air in the envelope cooled.

The brothers gave a public demonstration at Annonay on 4 June 1783, when a new balloon rose to about 1,830m (6,000ft). This success resulted in a summons to the capital, so that King Louis XVI could see the Montgolfier's invention for himself. A balloon some 13m (42ft) in diameter was constructed especially for the event, and a basket hung

beneath it to carry the world's first aerial voyagers: a cock, a duck and a sheep.

The balloon was launched at the Court of Versailles on 19 September 1783, climbing to approximately 550m (1,800ft) before the astonished gaze of King Louis, Marie Antoinette and their court. It landed about 3·2km (2 miles) away, and there was some concern when the cock was discovered to be a little the worse for his adventure. Had he been weakened by the great altitude at which the balloon had flown? Further investigation suggested that he was probably suffering from the effects of being trampled or sat on by the sheep.

The moment was fast approaching for manned flight, for which the Montgolfiers created a magnificent new balloon 15m (49 ft) in diameter. Superbly decorated in a blue-and-gold colour scheme, it was emblazoned with the royal cipher, signs of the zodiac, eagles and smiling suns. Around its open neck was attached a wicker gallery capable of accommodating one or two men. In this vehicle the 26-year-old François Pilâtre de Rozier made a tethered flight to 26m (85ft) on 15 October 1783 and remained airborne for about 4½ minutes.

Just over a month later, on 21 November 1783, de Rozier and his passenger, the Marquis d'Arlandes, became the first men in the world to be carried in free flight by a balloon. They rose from the garden of the Chateau La Muette in the Bois de Boulogne, Paris, and were airborne for 25 minutes before landing

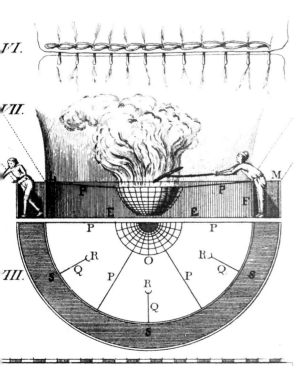

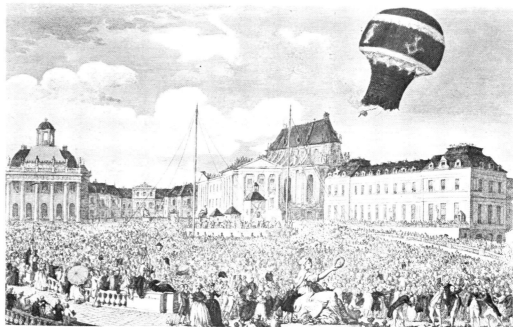

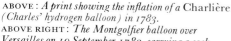

ABOVE: *A print showing the inflation of a* Charlière *(Charles' hydrogen balloon) in 1783.*
ABOVE RIGHT: *The Montgolfier balloon over Versailles on 19 September 1783, carrying a cock, a duck and a sheep.*
RIGHT: *The* Charlière *hydrogen balloon which came down at Gonesse near Paris on 27 August 1783, and was attacked and destroyed by the local inhabitants.*
BELOW RIGHT: *The hot air balloon in which Pilâtre de Rozier and the Marquis d'Arlandes made the first free balloon flight by man.*

about 8·5km (5·3 miles) from their departure point.

The first hydrogen balloon

A couple of decades before the Montgolfier flights, in 1766, the British scientist Henry Cavendish had isolated a gas which he called Phlogiston. Weighing only 2·4kg (5·3lb) per 28·3m³ (1,000cu ft) by comparison with 34·5 kg (76lb) for the same volume of air at standard temperature and pressure, it seemed likely to prove invaluable for lighter-than-air flight.

By 1790 this gas had been named hydrogen by the French chemist Lavoisier. Seven years earlier on 27 August 1783, Professor Jacques A. C. Charles had launched successfully a small, unmanned Phlogiston-filled balloon. Flown from the Champs-de-Mars, Paris, it was airborne for about 45 minutes before coming to earth some 25km (15·5 miles) away at Gonesse. There it was attacked with pitchforks by panic-stricken villagers, who believed it to be some strange device of the devil and who were not satisfied until it had been torn to shreds.

To construct the envelope of the balloon, Professor Charles had sought the assistance of two brothers named Robert. They had devised a method of rubberizing silk to make it leak-proof. With their help, following the successful flight of the model, Charles designed and built a man-carrying balloon. This balloon of 1783 was essentially the same as a modern gas-filled sporting design, and in

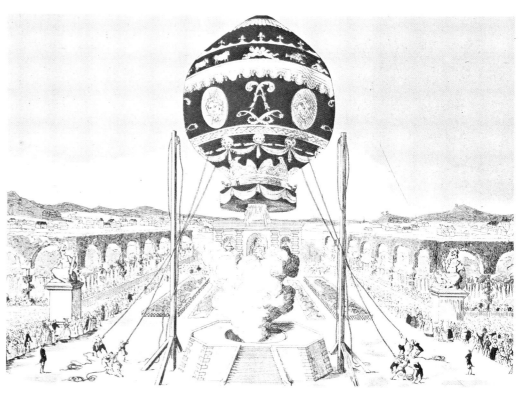

it Professor Charles and one of the Robert brothers, Marie-Noel, became the first men in the world to fly in a hydrogen-filled balloon. Ascending in front of an estimated 400,000 people from the Tuileries Gardens in Paris on 1 December 1783, they completed successfully a flight of 43·5km (27 miles) in two hours. The balloon was 8·6m (28·2ft) in diameter and had a volume of about 325·5m³ (11,500cu ft).

The balloon goes to war
The success of the *Charlière*, as Charles's craft was named, generated tremendous interest in the balloon as a sporting vehicle, a money-spinner for showmen and daredevils, a possible means of exploration and, of course, as a weapon of war. Within days of de Rozier's first flight in the Montgolfier's hot-air balloon, he had taken up one André Giraud de Vilette as a passenger. Afterwards, in a letter to the *Journal de Paris* de Vilette commented on the ease with which he had viewed the environs of Paris. 'From this moment,' he wrote, 'I was convinced that this apparatus, costing but little, could be made very useful to an army for discovering the positions of its enemy, his movements, his advances, and his dispositions . . .'

Not surprisingly, the French were the first to take the balloon to war. Four military balloons – the *Entreprenant*, *Celeste*, *Hercule* and *Intrepide* – were constructed and a Company of Aérostiers formed as a component of the Artillery Service under the command of Captain Coutelle. On 26 June 1794, Coutelle

ascended in the balloon *Entreprenant* during the battle of Fleurus in Belgium. Information signalled by Coutelle is believed to have played a significant part in the defeat of the Austrians. This flight represented the first operational use of an aircraft in war.

The spread of ballooning
Before the balloon's military debut, the first ascent by a woman had been recorded, as had first flights in Britain, Italy and America, and the first aeronaut, François de Rozier and a companion, Jules Romain, had been killed in a ballooning accident on 15 June 1785, while attempting a crossing of the English Channel from Boulogne in a combination hydrogen and hot-air balloon. The cause of the accident was unknown, but as the envelope was destroyed by flame it was believed that hydrogen had ignited during the venting of gas to control altitude.

The Channel had already been crossed, however, on 7 January 1785, by Jean-Pierre Blanchard of France and the American Dr John Jeffries. They took off from Dover, Kent, on a bitterly cold day and in a balloon which had only a small margin of lift when laden with the two men and their equipment. To avoid a ducking they had to throw overboard everything possible, including most of their clothes, but the two half-frozen gentlemen landed safely in the Forêt de Felmores, France, some two and a half hours later.

Balloons in America
Jean-Pierre Blanchard was well known as a

LEFT: *M. Charles and M. Robert descending in the presence of the Duc de Charle at Nesle on 1 December 1783, after making the first aerial voyage in a hydrogen balloon. It rose again with M. Charles alone in it and then descended after a trip of three miles.*
BELOW LEFT: *A military observation balloon in use at the siege of Mayence, 1794.*
BELOW: *Blanchard and Jeffries' historic cross-Channel flight on 7 January 1785, from Dover to the forest around Guines, near Calais, took two and a half hours to complete.*

ABOVE: *A tethered hydrogen balloon used by the French for observation purposes at the battle of Fleurus in Belgium on 26 June 1794, when they defeated the Austrians.*
RIGHT: *Jean-Pierre Blanchard making a flight over Berlin on 27 September 1788. He was one of the best known early balloonists after his cross-Channel flight.*
BOTTOM: *Another of Blanchard's pioneering balloon flights. In this picture he is shown at Lille, France, during a parade. A large crowd stand admiring.*

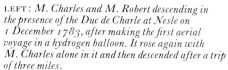

balloonist in Europe before his crossing of the English Channel. Dr Jeffries must have then suggested that he visit America, because eight years later Blanchard arrived in Philadelphia, where arrangements were made for a flight from the yard of the old Walnut Street Prison. At the appointed time, 10AM on 9 January 1793, a large part of the city's population turned out to witness the historic event. They numbered among them President George Washington, as well as John Adams, Thomas Jefferson, James Madison and James Monroe, all four of whom were subsequently to become presidents of the United States.

The president handed Blanchard a letter asking 'all citizens of the United States, and others, that in his passage, descent or journeyings elsewhere, they oppose no hindrance or molestation to the said Mr Blanchard.' After wishing him a safe journey, the president joined the assembled thousands to watch the balloon rise easily into the sky. Some 46 minutes later Blanchard landed safely about 24km (15 miles) away.

Ballooning in America expanded rapidly after this first demonstration. But unlike the French government, those in authority in America could visualize no military employment for the balloon. One significant flight caused ballooning in the United States to be regarded in an entirely new light. On 2 July 1859, John Wise, John La Montain and O. A. Gager flew 1,770km (1,100 miles) from St Louis, Missouri, to Henderson, New York.

This achievement brought the realization that long-distance travel by balloon was possible, and a number of people discussed seriously and planned a transatlantic flight. Among them was a flamboyant showman named Thaddeus Sobieski Constantine Lowe. He made a balloon 39·6m (130ft) in diameter with a reported gross lift of 20 tons. Named the *Great Western*, this balloon was weakened during a trial flight on 28 June 1860, and finally burst when the gas pressure proved too much. Finding that the *Great Western* was beyond repair, Lowe, still intent on flying across the Atlantic, built a second balloon, which was named the *Enterprise*.

Balloons in the American Civil War

In April 1861 the first shots of the American Civil War were fired. On 15 April President Abraham Lincoln called for troops and four days later balloonist James Allen and a friend were among those who volunteered, taking along two balloons. They were the spiritual fathers of today's enormous US Air Force, making their first military ascent on 9 June 1861. Other balloonists to join the Union Army included John Wise, John La Montain and Thaddeus Lowe.

On 18 June 1861, the latter demonstrated the potential of captive balloons for military reconnaissance. Climbing to a height of 152m (500ft) in the *Enterprise*, Lowe, a telegraph operator and an official of the telegraph com-

Balloon Corps of the Army of the Potomac.

But this first 'air force' was to last only 18 months. Though Lowe was a powerful driving force, the majority of his men were apathetic. When he resigned in May 1863 Lowe had made more than 3,000 ascents. His departure marked the end of the Balloon Corps, and almost 30 years were to pass before the US Army renewed its interest in military aviation.

Balloons in the Siege of Paris

When Paris was encircled by the Prussian army in September 1870 Chancellor Bismarck believed that the beleagured city would surrender quickly, being completely isolated from the rest of the world. But he had

pany sent the first air telegraph through a wire trailing to the ground. In it he told President Lincoln: 'This point of observation commands an area nearly 50 miles (80km) in diameter. The city, with its girdle of encampments, presents a superb scene...'

The president was suitably impressed, but it was not until early August 1861 that Lowe received his first official instructions. Ordered to construct a balloon of 708m³ (25,000cu ft) capacity, he flew the *Union* within three weeks. On 24 September Lowe made military history. He observed the fall of artillery fire from the *Union* and transmitted corrections by telegraph. The results were so good that Lowe was almost immediately tasked with providing four more balloons and crews to form the

not counted on the balloon as a means of carrying important passengers and dispatches out of Paris.

Under the leadership of Gaspard-Felix Tournachon the balloonists of Paris formed the *Compagnie d'Aérostiers Militaires*, intending to use captive balloons to observe Prussian army movements. On 23 September 1870, the balloon *Neptune* took off from the Place St Pierre. Piloted by Jules Duruof, *Neptune* landed three hours later at Evreux, some 97 km (60 miles) away. The 125kg (275lb) of mail and dispatches carried out of the city by Duruof was then sent on its way by more conventional means. The world's first airmail service had been inaugurated.

At the beginning of the siege there were

only five or six balloons in the city, so arrangements were made to mass-produce further examples. 'Production' lines were started at the Gare d'Orléans and at a derelict music hall in the Elysées-Montmartre. New pilots were also needed, and circus acrobats and sailors – selected because they were likely to have a good head for heights – were trained to fly the balloons out of Paris.

The balloons offered only a one-way service, leaving Paris in the direction of the wind. Messages from outside were brought into Paris by carrier pigeons flown out in the balloons. The number of communications a single pigeon could carry was very limited until a Monsieur Barreswil hit on the idea of photographing and reducing the letters and

ABOVE FAR LEFT: *A photograph taken in 1871
during the siege of Paris, showing the balloon
in which Gambetta escaped from the city, tethered
at the Place St Pierre.*
FAR LEFT: *Thaddeus Lowe's captive balloon at
Yorktown, used for observation purposes. It proved
extremely successful for military reconnaissance
during the Civil War.*
LEFT: *Another view of the balloon in which Gambetta
escaped at night from Paris. Balloons were also used
extensively to carry mail out of the city.*
ABOVE: *US military balloon detachment.*

dispatches so that one bird could transport
many more messages. Once delivered, these
'microfilms' could be enlarged to readability.
During World War II this idea was to be
developed into the Airgraph system to cater
for the large volume of forces (and later
civilian) airmail.

By the end of the siege, when a final sortie
flown on 28 January 1871, carried news of the
armistice, some 64 balloons had taken 155
passengers and crew to safety and 2½ to 3
million letters had reached destinations outside Paris. In the other direction, some 60,000
messages had been carried by 60 pigeons.

The siege gave rise to one other invention:
the anti-aircraft gun. Not surprisingly, the
Prussians were infuriated at the sight of important people and vital information floating
to safety over their heads. Krupps, the famous
armament manufacturers, responded by producing a special anti-balloon gun which
could be elevated to fire at balloons passing
over at altitude.

Military balloons in Britain
Although the first balloon flights in Britain
took place in 1784, it was not until 1878 that
the British Army showed any interest, allo-
cating £150 for the construction of a balloon.
Captains J. L. B. Templer and H. P. Lee were
responsible for developing the *Pioneer*, which
in 1879 became the first balloon to enter the
Army's inventory. The first balloon to be
actually used by the Army was Captain Tem-

pler's own *Crusader*, which entered service
shortly afterwards.

On 24 June 1880, a balloon detachment
was involved in military manoeuvres at Aldershot, Hampshire. Four years later the first
British balloon detachment ordered to fly in
support of military operations was sent to
Bechuanaland. Arriving in Cape Town on
19 December 1884, this unit comprised 2
officers, 15 other ranks and 3 balloons.
But order was restored in the colony without
a shot having to be fired, and they never saw
action.

In 1885 a detachment was sent to the
Sudan, and in 1899 a balloon was used to
keep an eye on the movements of the Boers
besieging Ladysmith.

LEFT: *Auguste Piccard designed this airtight, pressurized and air-conditioned gondola to be suspended from a balloon, in which he set an altitude record of 16,201m (53,153ft) in 1932. The previous year he had made the first flight into the stratosphere.*
BELOW LEFT: *The Italian Avorio-Prassone.*
BELOW RIGHT: *Not far from the British detachment in Flanders with their spherical observation balloon (late 1914) was a Belgian detachment complete with a German Drachen kite balloon.*

ABOVE: *The balloon detachment of the Prussian Army in 1884 at Tempelhof Field Berlin, with their partly inflated balloon.*
RIGHT: *A tethered kite balloon used for observation purposes by the German Army, seen here partially inflated.*
BELOW FAR RIGHT: *A row of barrage balloons in 1940. They were used by the British at possible targets such as airfields to present a hazard to enemy aircraft, and to deter them from mounting low-level attacks.*

Development of military balloons

The conventional spherical balloon is quite at home when travelling freely with the wind. But when restrained by a rope anchored to the ground it can spin and buck so badly that the observation crew can soon lose interest in everything but a rapid return to the ground.

To overcome this problem the kite balloon was developed and brought close to perfection by the Germans August von Parseval and H. B. von Sigsfeld. They took the partly controllable sausage-shaped observation balloon of the time and added a tail fin to keep it pointing steadily into the wind. The French army officer Captain Caquot later developed this design by improving the streamlining of the basic shape and introducing three inflated tail fins disposed at 120° intervals, with one pointing vertically downward. Like the single fin of the von Parseval Drachen, they were inflated by an airscoop facing into the wind. The Caquot type was used extensively for aerial observation both on land and at sea during World War I. By the beginning of World War II the balloon had long since been superseded by heavier-than-air craft in this role. Instead, barrage balloons were used in large numbers to present a hazard to enemy aircraft and deter low-level attacks.

One unusual military application of the balloon came during World War II, when the Japanese tried to attack North America from the home islands. Launched into the jetstream winds prevailing at a fixed altitude, bomb-carrying balloons proved capable of crossing the Pacific. They were equipped

with an ingenious barometric device which worked in conjunction with an automatic gas-discharge valve to keep the balloon at the desired altitude of between 9,144m (30,000ft) and 11,582m (38,000ft). The higher daytime temperature would have caused the balloon, if unchecked, to ascend until it burst as the gas expanded. The valve discharged gas, and so reduced buoyancy and caused the balloon to descend. When the required altitude was reached the barometric device energized an electrical ballast-discharge circuit, causing the balloon to rise again. Enough ballast was carried for the 9,978km (6,200 miles) Pacific crossing. When the last of this had been jettisoned, the weapon load of two incendiary bombs and one 15kg (33lb) anti-personnel device was released during the next low-level cycle. A small explosive charge was detonated to destroy the balloon.

It was calculated that weapon-release would occur over the not insignificant target represented by the North American continent. Of about 9,000 balloons launched, between 11 and 12 per cent are believed to have completed the crossing. Despite this comparatively large number of weapons over the target, the damage was confined largely to fires in open areas and proved of little consequence.

Only six people were killed in the campaign, and the bomb-carrying balloons were finally countered by a complete ban on radio and newspaper reports. As a result, the Japanese were unaware of the operation's technical success and called it off. Had they known that one in eight of these weapons were reaching the target they might have launched many more, equipping them with a far more potent payload.

Scientific use of balloons

Balloons have also played a significant part in scientific research from the earliest days. J. A. C. Charles made recordings of air temperature and changes in barometric pressure during his first ascent in a hydrogen balloon on 1 December 1783. In 1784 Jean-Pierre Blanchard and Dr Jeffries made similar recordings.

In 1804 the French scientist Gay-Lussac made two flights to investigate the way in which the Earth's magnetic field varied with increasing altitude. In Britain James Glaisher made 28 ascents for scientific purposes during 1862 to 1866. On one occasion Glaisher became unconscious with the balloon still ascending at an altitude which he estimated at 8,839m (29,000ft). He survived, but the companions of Gaston Tissandier were not so lucky; they died during an ascent in 1875. (See page 139). In Germany, Professor Assman continued the work begun by Glaisher, achieving more significant results with his more highly developed instruments. By 1932 the Swiss physicist Auguste Piccard had climbed 16,201m (53,153ft) in a sealed capsule suspended beneath his free balloon, having made the first flight into the stratosphere during the previous year. Some great heights have been reached, notably 34,668m (113,740ft) by US Navy reservist Commander Malcolm D. Ross on 4 May 1961. Unmanned balloons also have meteorological uses; some small ones indicate wind direction and speed, others carry instrument packages for transmitting data to earth.

Airships

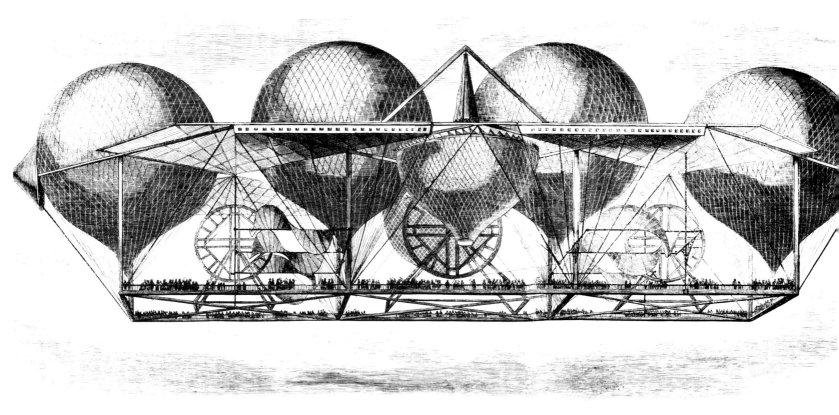

In the 18th and 19th centuries the balloon had made many things possible, but there were some serious problems to be overcome if men were to travel easily by air from point to point. These shortcomings can be summarized briefly: the balloon could travel only in the direction of the wind; and at the end of each flight all the costly lifting gas had to be valved off to allow the balloon to be packed and transported back to its home base. Such disadvantages were acceptable to sportsmen, but they made the balloon completely impractical for most military or commercial tasks. What was needed was an independent source of power to render balloons navigable.

Steerable balloons: the first ideas

Once airborne the balloon is wedded to the wind, moving in the same direction and at the same speed. It has no independent motion within the airstream which would allow the rudders or sails proposed by many inventors to affect its direction of movement. So despite some ingenious plans to utilize muscle power, the evolution of a steerable balloon had to await the development of a compact power unit. By propelling the vehicle independently of the wind, such an engine would make it possible for rudders and elevators to control direction.

While balloonists waited for a suitable powerplant they came to realize that the spherical balloon was hardly ideal in shape for steering anywhere. They therefore began to streamline it into a spindle shape not so different from more modern solutions. But

ABOVE: *Pétain's giant multiple airship built in 1850. It was constructed of several balloons linked together by a single undercarriage. It was a design very much ahead of its time, but in fact never flew.*
BELOW: *A drawing of an airship suspended from copper spheres in which a vacuum has been created, designed by Francis Lana, a Jesuit physicist, in 1670. It had a sail and oars for propulsion.*

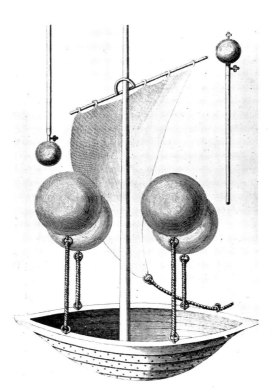

FAR RIGHT CENTRE: *Details for a proposed 'aerostatic machine' or airship designed by Meusnier in 1785. This model never got beyond the drawing board.*
FAR RIGHT BELOW: *The first successful airship constructed by Henri Giffard in 1852. It was powered by a small steam-engine also designed by Giffard, which developed approximately 3hp. He flew in it from Paris to Trappes on 24 September 1852.*

the early experimenters were disappointed to find that despite this 'ideal' shape, muscle-powered oars or propellers were incapable of steering their prototype airships.

The first practical airships

The autumn of 1852 saw French engineer, Henri Giffard, preparing for flight a vehicle which has the distinction of being the first powered and manned airship. He had constructed an envelope 43·9m (144ft) in length, pointed at each end and with a maximum diameter of 11·9m (39ft). Capacity of the envelope was 2,492m³ (88,000cuft), and from it were slung a car containing the power-plant and pilot.

The power source was a steam engine designed by Giffard which weighed about 45kg (100lb) and developed approximately 3hp. The engine, boiler, three-blade propeller and an hour's supply of water and fuel weighed 210kg (463lb). The propeller, 3·35m (11ft) in diameter, was driven at a maximum speed of 110rpm. The airship was to be steered by what Giffard called a keel, a triangular vertical sail mounted at the aft end of the long

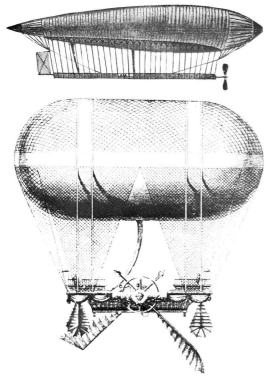

TOP: *The airship* La France *designed by Renard and Krebs in 1884, which made the first fully controlled flight by an airship.*
ABOVE: *Masse's design for an airship propelled and controlled by oars.*
RIGHT: *The nacelle and engines of the airship built by the French Tissandier brothers in 1883. It was driven by a 1·5hp electric motor. This illustration (engraved from a photograph) shows the brothers in it.*

horizontal pole to which the gondola was attached. It was effectively a rudimentary rudder.

On 24 September 1852, Henri Giffard piloted his airship from the Paris Hippodrome to Trappes some 27km (16·8 miles) away at an average speed of 8km/h (5mph). This flight must have been made in virtually still air if it was in fact steered: control must have been marginal at such a low speed, and the airship would have been carried away if there had been any wind at all. Giffard's craft was nonetheless the first true airship.

The next important development came from the German Paul Hänlein, who in December 1872 flew a small airship at Brünn in Central Europe. Its 6hp gas engine was fuelled by the coal gas which filled the envelope. Designed by Etienne Lenoir, this powerplant was a development of his basic single-cylinder gas engine, first demonstrated in 1860. Hänlein's airship was not a success, but it is remembered for its powerplant, the first internal-combustion engine to take to the air. In 1883 the Tissandier brothers of France built and flew an airship powered by a 1·5hp electric motor. It was driven by primitive batteries which together with the motor weighed about 200kg (440lb), giving an improbably low power: weight ratio of 1hp:293lb. Despite this, the Tissandier airship was at that time marginally the most successful.

The following year saw the most important of the early airships, *La France*, designed by Captains Charles Renard and A. C. Krebs

of the French Corps of Engineers. With an envelope of 1,869m³ (66,000cuft) capacity, it was 50·3m (165ft) long and had a maximum diameter of 8·23m (27ft). Beneath the envelope was suspended a lightweight gondola which extended for about two-thirds of the length of the vessel. It comprised a framework of bamboo covered by silk, and in it was mounted the powerplant, a 9hp electric motor drawing on specially developed batteries. This drove a four-blade tractor propeller of 7m (23ft) diameter. A total powerplant weight of 857kg (1,890lb) gave a power: weight ratio of 1:210. Though still impractically low, this figure nevertheless represented a considerable advance on the Tissandier engine.

On 9 August 1884, Renard and Krebs lifted off in *La France* from Chalais-Meudon, France. They completed a circular course of about 8km (5 miles), reaching a maximum speed of 23·5km/h (14·6mph) en route. This is regarded as the first fully controlled powered flight by a manned airship anywhere in the world. *La France* made a further six flights later that year and in the following one.

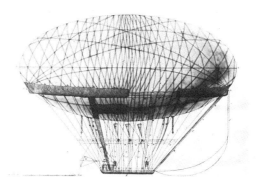

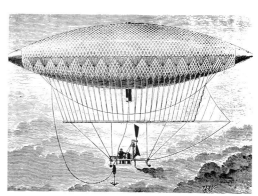

Development of the Airship

In spite of these achievements, electric motors powered by storage batteries were clearly impractical. Like the pioneers of heavier-than-air flight, those who believed that the airship represented the best method of air transport had to await the availability of a light and compact power source. Their patience was rewarded when an airship powered by a single-cylinder Daimler petrol engine was flown successfully on 12 August 1888. This was designed by the German Dr Karl Wëlfert, who was later to die in an airship. A flame from the petrol engine's exhaust is believed to have ignited gas being vented from the envelope.

Next on the scene with an airship powered by an internal-combustion engine was the Brazilian Alberto Santos-Dumont. He achieved a first successful flight in Paris on 20 September 1898, in a craft powered by a small motor-cycle engine. This little power-plant drove his *No 1* along quickly enough in relation to the day's light breeze that he could steer the craft in the chosen direction.

But it was his *No 6* of 1901 that brought Santos-Dumont real fame. In it he won the 100,000 franc prize offered by M Deutsch de la Meurthe, a wealthy member of the Aéro Club, for a flight of about 11·3km (7 miles) from the *Parc d'Aérostation* at St Cloud, around the Eiffel Tower and back. *No 6* was pointed at both ends and had an overall length of 33m (108ft 3in), a maximum diameter of 6m (19ft 8in) and a hydrogen capacity of 622m³ (21,966cu ft). Suspended from

the envelope was a lightweight girder keel supporting the small control basket which accommodated the pilot, engine, tanks, pusher propeller and triangular rudder. Power-plant was a Buchet petrol engine based on the Daimler-Benz design and fitted with a water-cooling system. Developing about 15hp, it drove a crude two-blade pusher propeller.

The first Zeppelin

The airships discussed so far belonged to the classes known as non-rigids and semi-rigids. In the former the shape of the envelope was maintained by gas pressure alone; in the latter a rigid keel permitted a lower gas pressure. One of the drawbacks of these two types was that diminishing gas pressure brought a

change in the shape of the vessel, and a serious pressure reduction could lead to the collapse of the forward end of the envelope. The resulting drag usually proved too much for the feeble powerplants of the day and was likely to thwart any attempts to steer the vehicle.

Though it was appreciated that a rigid airship would be far more practical, the weight penalty imposed by a rigid frame or envelope was unacceptable in the early years of airship development. The Austrian David Schwarz nevertheless designed an airship with a rigid aluminium-sheet envelope – reported to be 0·2mm (0·008in) thick – attached to a lightweight tubular framework braced by internal steel tension wires. Overall length was 47·55

FAR LEFT: *In 1901 Alberto Santos-Dumont, an adventurous Brazilian living in France, made a startling aerial tour of the Eiffel Tower in a dirigible of his own design powered by a petrol engine. He made a round trip from St Cloud in under 30 minutes.*

LEFT: *One of the dozen non-rigid airships built in France by Santos-Dumont during his pioneering years of 1898 to 1907. Suspended from the envelope is a keel supporting the basket, accommodating the pilot and the engine which drove a rudimentary propeller.*

ABOVE: *Count Ferdinand von Zeppelin's LZ3 (Z1) on its flight over Munich in 1909. This was one of the series of dirigibles he built during the early years of the 20th century, when he experimented with rigid airships, gradually increasing their size and power.*

LEFT: *Zeppelin's first airship, the LZ1, on its maiden flight on 2 July 1900. It made an 18 minute flight at Friedrichshafen on the Bodensee, with Zeppelin himself at the head of the five-man crew. It was under-powered and eventually scrapped in 1901.*

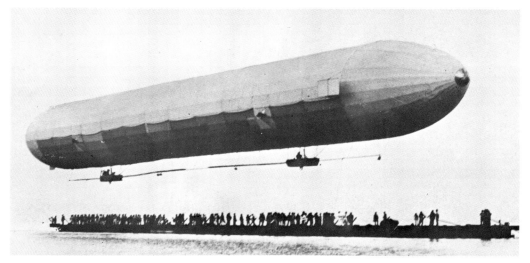

m (156ft) and gas capacity 3,681m³ (130,000 cu ft). Power was provided by a 12hp Daimler petrol engine driving four propellers, two of which were for propulsion and two for directional control. When the airship flew for the first time, on 3 November 1897, the belt drives to the propellers were found to be useless. The ship drifted for about 6·5km (4 miles) before landing heavily and breaking up in the wind.

Meanwhile, a significant figure had taken the stage in Germany, Count Ferdinand von Zeppelin. Visting America during the Civil War, he had taken to the air for the first time in one of the Army's balloons at St Paul, Minnesota. Impressed by this flight, influenced by a lecture on airship travel by the German Postmaster-General, and concerned at the comparative success of Renard and Krebs' *La France*, Count Zeppelin was convinced that it was his duty to provide Germany with a fleet of military airships.

In 1899 he began the construction of his first rigid airship, *Luftschiff Zeppelin 1*, (LZ1) in a floating shed on Lake Constance. Like many pioneers before and after him, the Count seems to have believed erroneously that in the event of a heavy landing the surface of the lake would prove more forgiving than the ground.

Zeppelin's airship was truly a giant for its day, measuring 128m (420ft) in length and having a maximum diameter of 11·73m (38 ft 6in). Its 11,327m³ (400,000cu ft) of gas was contained in 17 separate cells, an entirely new feature. Only its engines were less than huge: the two Daimlers together produced a mere 30hp for a weight of 771kg (1,700lb), giving a power:weight ratio of 1:57. Although this represented a great improvement on the airship *La France*, the enormous LZ1 needed large reserves of power if it was to be controllable in anything but still air. When the Zeppelin flew for the first time, on 2 July 1900, it was apparent immediately that it was grossly underpowered. Instead of flying at the expected 45km/h (28mph), the LZ1 had a maximum speed of only 26km/h (16 mph). This would have given marginal controllability if the vessel had been equipped with efficient aerodynamic controls. Miniature rudders at bow and stern were expected to provide directional control, and a 250kg (551lb) weight could be slid fore and aft on a track between the two gondolas to take care of vertical control.

Not surprisingly, after two more test flights giving a total airborne time of two hours, LZ1 was scrapped in 1901. It was not until four years later than von Zeppelin began construction of the LZ2, but by then an important airship had been built and flown in France.

BELOW: *A Lebaudy dirigible. This particular one was built for the British Army and is here seen leaving its hangar. In 1903 the first Lebaudy airship made the first controlled air journey by a practical dirigible, from Moisson to the Champ-de-Mars, Paris, a distance of 61km (38 miles). The airships were designed by Paul and Pierre Lebaudy.*

BELOW CENTRE: *One of the five airships designed and constructed by Ernest Willows, seen on 20 September 1913. They were all comparatively small dirigibles.*
BOTTOM: *The Spencer airship ascending at Ranelagh in Great Britain in 1903, watched by a crowd of admiring onlookers.*

is considered to be the first-ever controlled air journey by a practical dirigible. The Lebaudy airship subsequently suffered serious damage at Chalais Meudon. It was repaired and handed over in 1906 to the French government. Other examples were built for the French Army, Britain and Russia each acquired one, and a final example was built under licence in Austria.

Airship pioneers in America

The first true powered dirigible to be built in the United States was Captain Thomas S. Baldwin's *California Arrow*. With an envelope 15·85m (52ft) in length and powered by a two-cylinder motor-cycle engine built by Glenn H. Curtiss, it was flown for the first time on 3 August 1904, at Oakland, California, piloted by Roy Knabenshue. When Brigadier General J. Allen of the US Army subsequently saw it in flight at St Louis, Missouri, he was so impressed that he per-

The first controlled air journey

Under the guidance of an engineer, Henri Julliot, the brothers Paul and Pierre Lebaudy designed and completed an airship at Moisson, Seine-et-Oise, in 1902. It was strange in appearance, with an envelope 57·9 m (190ft) in length and pointed sharply at each end. Beneath the envelope – made of two-ply rubberized material and with a capacity of 2,000m³ (70,629cu ft) – was a steel-tube structure carrying a basketwork gondola to contain the powerplant and crew. The German-manufactured engine developed about 35hp, driving two pusher propellers.

First flown early in 1903, it made a number of successful local flights and, on 12 November 1903, was flown 61km (38 miles) from Moisson to the Champs-de-Mars, Paris. This

suaded the War Department to request tenders for the supply of a dirigible.

Baldwin's tender proved to be the lowest and on 20 July 1908, a 29·26m (96ft) dirigible powered by a 20hp Curtiss engine was delivered to Fort Myer, Virginia. When accepted on 28 August 1908, it was designated Dirigible No 1. The pilots trained to fly it were Lieutenants Frank P. Lahm, Benjamin D. Foulois and Thomas E. Selfridge. The Baldwin ship remained in service for three years before being scrapped, and was for all practical purposes the one and only US Army dirigible. Its engine and propeller can be seen to this day in the Smithsonian Institution.

At about this time the wealthy and eccentric journalist Walter Wellman hired Melvin Vaniman and Louis Godard to design and build a dirigible for a flight to the North Pole. The resulting *America* was 69·49m (228ft) long, had a capacity of 9,911m³ (350,000cu ft) and was powered by a single 80hp engine. A canvas-covered hangar for the airship was completed at Dane's Island, Spitzbergen, in September 1906. A year later, a first attempt to reach the Pole ended when the *America* was forced down in a snowstorm. The airship was recovered and fitted with a second engine, and another polar flight was tried two years later, but unfortunately this also proved a failure.

A transatlantic attempt followed in 1910, with the *America* lifting off from Atlantic City, New Jersey, on 15 October. Three days later the five-man crew was rescued off New England by a British ship, the RMS *Trent*.

Airship developments in Britain

The earliest practical British airship was the first of five designed and constructed by Ernest Willows between 1905 and 1914. Willows' *No 1* was only 22.56m (74ft) in length and was powered by a 7hp Peugeot engine. *No 3* achieved the first airship flight across the English Channel, travelling from London to Paris. The largest Willows ship, *No 5*, had only 1,416m³ (50,000cuft) of gas capacity, and they were distinguished (with the exception of *No 1*) by their swivelling engines, which assisted in manoeuvring the vessels laterally and vertically.

Britain's first military airship, British Army Dirigible No 1, was built by the Balloon Fac-

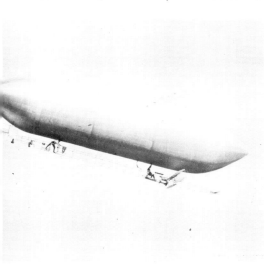

tory at Farnborough, Hants. Known unofficially as *Nulli Secundus*, this non-rigid had an envelope 37.19m (122ft) in length, 7.92m (26ft) in diameter, and with a capacity of 1,557m³ (55,000cuft). No suitable engine was available, so Balloon Factory commanding officer Colonel J. E. Capper obtained an Antoinette engine of about 50hp from Paris. This was installed along with the crew positions in a small car suspended beneath the envelope. *Nulli Secundus* flew for the first time on 10 September 1907, piloted by Colonel Capper and with the colourful Samuel F. Cody as engineer. A month later this airship flew from Farnborough to London, causing great excitement when it circled St Paul's Cathedral. Its maximum speed of about 26 km/h (16mph) ruled out a return journey that day, however, and a landing was made in the grounds of the Crystal Palace.

Redesigned in the following year, *Nulli Secundus* acquired a triangular-section fabric-covered keel which made her a semi-rigid, and the addition of pointed ends to the envelope provided another 28.3m³ (1,000cuft) of capacity. These changes improved maximum speed to 35km/h (22mph).

She was followed by the *Beta* (unofficially *Baby*), which measured 25.6m (84ft) in length and 7.52m (24ft 8in) in diameter, and had a capacity of 623m³ (22,000cuft). The original pair of 8hp Buchet engines were later re-

placed by a single 25hp radial engine driving two propellers. *Beta* was also rebuilt, and when flown in May 1910 had a gas capacity of 934m³ (33,000cuft) and a British-built Green in-line engine developing 35hp. In this form *Beta* could carry a crew of three and had a powered endurance of about five hours. The first practical airship to serve with any of Britain's armed services, *Beta* was flown at night in June 1910 and in a summer of activity was reported to have flown some 1,609km (1,000 miles) without any serious problems.

Later military airships included *Gamma* of 1912, which demonstrated a maximum speed of 51km/h (32mph); the 1912 *Delta*, capable of 71km/h (44mph); and *Eta* of 1913, reported to have a speed of 67.5km/h (42mph) and an endurance of up to ten hours.

ABOVE: *The British airship* HMA Gamma, *built in 1912, seen here leaving its hangar. This was a non-rigid military dirigible in service with the British Army, which attained a maximum speed of 51km/h (32mph).*
LEFT: *Baldwin's Dirigible no.1, constructed for the United States Army Signal Corps in 1908, and here seen in flight at Fort Myer, Virginia. It remained in service until 1911 and was the only practical non-rigid airship used by the US Army.*
BELOW: *The first British Naval airship* Mayfly *outside its shed on 23 September 1911. The structure was not sufficiently strong, and while being manoeuvred on the water by means of winches, it broke in two. It was returned to the shed and later dismantled without ever having flown; however, it did provide the builders with valuable experience.*

Airship development in Italy

Italy, the only other European nation actively interested in the development of airships, started by building a number of non-rigids and semi-rigids for the army's Corps of Engineers. The most important of these was the Forlanini semi-rigid of some 3,680m³ (129,995cu ft) capacity, which first flew in 1909. Airships of this type were to serve with the Italian Army for some years, although they could hardly be classed as highly successful experiments.

Flight and control of an airship

The endurance of an unpowered hydrogen balloon of any size depends largely on the amount of ballast it can carry. With its envelope completely full of gas, such a balloon is ballasted so that the weight of the crew, equipment and ballast are just enough to keep it on the ground. The discharge of a small amount of ballast will cause the balloon to rise, and as atmospheric pressure decreases the gas will expand and the volume of the envelope increase. Gas is able to vent off to prevent undue stress on the envelope, and eventually a height will be reached at which the balloon is in equilibrium. Further ballast must then be dropped if the balloon is to climb. If it is to descend, more gas must be valved off, and a second climb will call for the dropping of more ballast. Finally, when all the ballast has been discharged, the flight must be terminated.

A dirigible – a navigable airship – is affected by very different problems arising from the powerplant which gives it airspeed and directional control. Like the balloon, the dirigible can ascend vertically from the ground. Then, at a suitable height, the engines can be started and the ship can be steered in the required direction. The climb from ground level to the height at which the engines are started is made possible by the static lift provided by the gas within the envelope. Once the engines are started and the vessel is moving fast enough for its aerodynamic controls to be effective, the nose can be pitched up and dynamic lift created. In the same way, a negative dynamic force can be created by pitching the nose down.

USS Akron, 1931

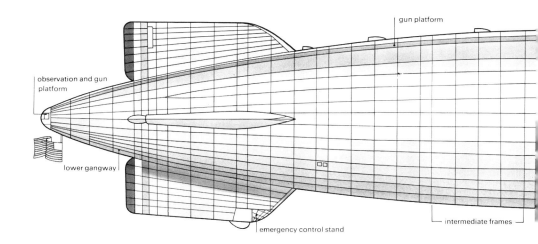

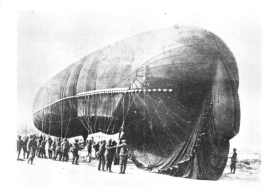

An airship usually carries water ballast, as opposed to the sand carried normally by balloons. But just as with balloons, airships have to discharge ballast if they are to rise from the ground under static lift.

If the airship is of the non-rigid or semi-rigid type, the envelope will contain a number of ballonets which are filled with air by a pump or via scoops placed in the slipstream of the propellers. They maintain pressure within the envelope so that it retains its aerodynamic shape. An envelope pressurized with gas alone would vent it off throughout the ascent. Then, on descent, the remaining gas would be compressed under the increasing atmospheric pressure. The envelope's internal pressure would fall, possibly causing a drag-inducing change in its aerodynamic form.

To prevent this, about 95 per cent of the envelope's volume is taken up by lifting gas, with the air-filled ballonets occupying the remaining 5 per cent. Air is vented off instead of gas as the airship climbs, and gas plus air maintain the aerodynamic shape. At optimum altitude the lifting gas will fill the envelope at the desired pressure, and the ballonets will be completely empty. The height to which an airship can climb without losing any lifting gas is known as its pressure altitude.

This simple arrangement becomes complicated when it is necessary to fly higher than the pressure altitude, and by the fact that the engines consume fuel. The airship will climb if ballast is discharged. Lifting gas will however have to be vented off as atmospheric pressure decreases. Secondly, the

as fuel is consumed. Thus Paul Hänlein's airship, with an engine fuelled by lifting gas from the envelope, was not quite as short-sighted a design as it first appears.

It was also proposed that as the airship's engines consumed liquid fuel, reducing gross weight, the hydrogen that had to be valved off to maintain equilibrium should be mixed with liquid fuel in the engines' carburation system. The vented gas would not therefore be a total loss, being added instead to the fuel stock and so helping to increase range. One of the features of the German *Graf Zeppelin* of 1928 was the use of a fuel known as *blaugas*, a petroleum vapour some 20 per cent heavier than air. This gas was stored in separate cells occupying some $30,000m^3$ ($1,059,434$cu ft), and the vessel's gross weight changed little as fuel was consumed. The later *Hindenburg* was equipped to take in water when flying through heavy rain to increase her ballast reserves.

Airships in World War I

By building the airships which operated with the pioneering airline, Delag, the designers of the Zeppelin Company learned a great deal that was to prove of enormous value in the development of military airships for the German Army and Navy.

While Delag was developing its airship services, a new design had appeared in Germany. A rigid airship designed by D. Johann Schütte and Heinrich Lanz, the Schütte-Lanz SL.1 made its first flight on 17 October 1911. Measuring 128m (420ft) in length and 17·98m (59ft) in diameter, the SL.1 had a capacity of about $16,672m^3$ (700,000cu ft). Its pair of 270hp engines gave a maximum speed of about 56km/h (35mph). One of the SL.1's most advanced features was an improved streamline shape, which had an important influence on the development of the Zeppelin airships when Schütte-Lanz became a part of the Zeppelin Company. But a second revolutionary idea, the use of wood for the SL.1's rigid frame, did not survive the amalgamation of the companies.

There was one other significant airship company in Germany: the concern established to build the non-rigid airships designed by Major von Parseval. These non-rigids had

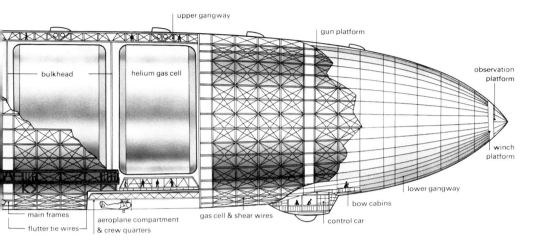

upper gangway · gun platform · observation platform · winch platform · lower gangway · bow cabins · control car · gas cell & shear wires · aeroplane compartment & crew quarters · main frames · flutter tie wires · bulkhead · helium gas cell

LEFT: *A partially inflated airship belonging to the Italian Army, used in the war against Turkey of 1911–12. This type rendered valuable service in exploration missions over unknown territory.*
BELOW LEFT: *One of the German Schütte-Lanz airships in trouble on the Belgian coast in 1915. It had to make a forced landing after suffering heavy damage from anti-aircraft fire during raids on the south coast of England. The first airship raids against London were launched in January 1915. The German Army ceased offensive airship operations in September 1916 after suffering serious losses in a bombing raid. The Navy, however, continued their raids on London until 1918.*

ABOVE: *A cutaway diagram of the Goodyear airship* Akron, *built for the United States Navy in 1931. It was 239m (785ft) in length and 41·4m (133ft) in diameter, and was powered by eight Maybach VL–II engines, giving a total of 4,480hp. It could attain a speed of 128km/h (79mph). On 3 April 1933 it crashed at sea off Barnegat, killing 73 people out of the 76 on board.*
BELOW: *The airship* Parseval III *at Munich in 1909. This was one of a number of non-rigid dirigibles constructed by Major August von Parseval, during the first years of the 20th century when his designs vied with Zeppelin's rigid airships for the attention of the German Navy.*

gross weight of the vessel will decrease as the engines consume fuel, and the airship will climb steadily unless gas is vented off. Changes in air temperature and temperature inversion present further control problems to the airship designer.

Here, very briefly, are some of the ways in which these problems can be overcome. If the lifting gas is heated on the ground, a volume of gas smaller than that normally required will expand to fill the envelope. The heating system can then be switched off as the airship starts its climb, and the falling gas temperature and external air pressure more or less cancel one another out. With such a system, lifting-gas pressure falls at about the same rate as external atmospheric pressure, making it unnecessary to vent gas during the climb above pressure altitude. An efficient means of condensing the water in the exhaust gas from the engines would provide additional ballast to keep the ship in equilibrium

pressure envelopes and were considered unsuitable for operational military use, but a small number were used for training.

Germany's principal military airships during World War I were the Zeppelins of the German Navy, which earned a fearsome reputation with their bombing raids on Britain. Though this was due more to the inadequacy of Britain's home defences than the accuracy and weight of the attacks, the Zeppelins were intensively developed, from the L3, with which Germany began the war, to the L70 and L71.

The L3 was 158m (518ft 4in) long, 14·78m (48ft 6in) in diameter and had a volume of 22,500m³ (794,575cu ft). Powered by three engines giving a total of 630hp, the L3 had a maximum speed of about 75km/h (46·5mph).

Four years later, Zeppelin attacks on Britain ended after the loss of the L70 and *Fregatten Kapitän* Peter Strasser, Chief of the German Naval Airship Division. L70 fell to the de Havilland D.H.4 patrol seaplane of Major Egbert Cadbury and Captain Robert Leckie on 5 August 1918. The L70 was 211m (692ft 3in) long and 23·9m (78ft 5in) in diameter, and had a volume of 62,178m³ (2,195,800cu ft). Its engines developed a total of 1,715hp and were reported to have speeded the airship through the air at 130km/h (81mph) during a trial flight.

Military airship development in Britain and France followed a different pattern, for neither nation envisaged those craft as war-winning strategic weapons. The two Allies proposed to use the airship mainly for naval

patrol, a role in which it was to give valuable service, especially for Britain.

France, pioneer of the dirigible, used about 25 airships – 21 non-rigids and 4 semi-rigids – during World War I. The former were mainly of the Astra-Torres type, a design in which the three lobes of the envelope ran from stem to stem. Designed by a Spaniard, Torres Quevedo, they were built and developed by the French *Astra Société des Constructions Aéronautiques*. The internal rigging of the Astra-Torres passed within the folds of the lower lobes, considerably reducing the parasite drag generated by the rigging. The other French manufacturer worthy of mention was Clément-Bayard, which built a small number of non-rigid airships of the Lebaudy type. The majority of the French airships were used

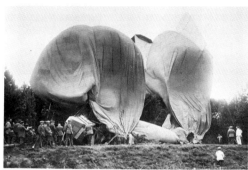

FAR LEFT: *The Gondola of a Zeppelin airship used in attacks against England.*
LEFT: *The French airship* Alsace *in 1915 after a forced landing.*
BELOW: *One of the US Navy's airships built by the Goodyear company.*
BELOW RIGHT: *An airship moored at Barrow-in-Furness in 1918. It is secured by the nose and free to swing round with the wind.*
ABOVE RIGHT: *The British Naval Airship C.23a starting off on submarine patrol.*
CENTRE RIGHT: *The rigid airship R.23, put into commission in 1916 as a training ship.*
BELOW FAR RIGHT: *Messrs Armstrong's airship shed at Barlow, Yorkshire, about 1917.*

in which many lives were lost, including that of Roald Amundsen.

The United States Army and Navy have at differing times been the world's major operators of military airships. The US Army's heyday lasted from 1921 to 1935, with the Navy taking over during World War II. In 1921 an agreement between the Army and Navy resulted in the former concentrating on the use of non-rigid and semi-rigid types for coastal and inland patrol, a division which lasted until the termination of US Army lighter-than-air operations in 1936.

As well as some notable rigid airships, the US Navy acquired a total of 241 non-rigids between 1917 and 1958. Some of these were excellent, fast vessels: the final examples of the N series, delivered between 1958 and 1960, had lengths of 122·83m (403ft), volumes of 42,929m³ (1,516,000cu ft), and maximum speeds of 145km/h (90mph). Of the Navy's rigid airships, the most successful was the German-built ZR–3 USS *Los Angeles*. Built subsequently in America by the Goodyear-Zeppelin Corporation were the 239·27m (785

ft) ZRS–4 *Akron* and ZRS–5 *Macon*. Fighter aircraft were carried inside and could be launched and recovered on a trapeze. Both of these airships were lost at sea: the *Akron* on 4 April 1933, with the loss of all but three of her crew; the *Macon* on 11 February 1935, with the loss of only two lives. This brought and end to the Navy's rigid airship programme.

Airships survive in the jet era

Although the USA no longer uses airships for military purposes, Goodyear has retained a fleet of four non-rigids for publicity purposes. Goodyear can claim to have built more airships than any other company in the world. Its 301st, the *Columbia III*, was completed in 1975. In 1986 this was dismantled and replaced by another ship of the same name, with volume of 5,740m³ (202,700cu ft) and cruising at 64km/h (40mph) on two 210hp engines.

By the late 1980s numerous companies in many countries were building airships. The vast majority are non-rigid and a high proportion use hot air rather than gas. Today's airship systems include several novel ideas. The helicopter pioneer Frank Piasecki built a demonstrator for his Heli-Stat, intended mainly for logging in forests. An airship envelope provided lift while a group of linked helicopters furnished vertical power and control. Sadly, this suffered a fatal accident. Another US firm, AeroLift, has built a small Cyclo-Crane comprising a rotating envelope carrying large aerodynamic blades, before attempting a 45-ton version for the US Army. In Canada Magnus Aerospace is developing giant spherical ships full of pressurized helium and gaining additional lift from spinning, creating the aerodynamic Magnus effect. The Magnus LTA–20 is to lift 54,430kg (120,000lb) of bulky cargo a distance of 80km (500 miles) in ten hours. In Britain the Wren RS.1 will be a pressurized metal-clad lifter with four 1,645hp pivoting turboprops giving the high cruising speed of up to 218km/h (136mph).

Most important of current airship programmes is the Sentinel 5000, by Airship Industries (UK) and Westinghouse (USA) for the US Navy. With a length of 129m (423ft 3in) and gross volume of 66,500m³ (2,348,430cu ft) this will be the biggest-ever non-rigid. A three-deck pressurized car will be the control centre for diesel and turboprop propulsion and a complete airborne early-warning radar and control system. Mission endurance will be 30 days.

R.100

In an attempt to resolve the question of whether the large, rigid airship was the ideal long-range commercial transport vehicle for the mid-1920s, Britain's first Labour government under Ramsay MacDonald promoted a programme of airship development. A sum of £1·3 million was allocated over a three-year period to cover the building of new sheds and mooring masts in Canada, Egypt, England and India; a limited amount of research into airship structures and aerodynamics; and the construction of two large intercontinental-range airships. Designated R.100 and R.101, these were to be designed and built respectively by the Airship Guarantee Company and the Air Ministry's Royal Airship Works at Cardington, Bedfordshire.

The Airship Guarantee Company, a subsidiary of Vickers Aircraft Company, represented the private-enterprise sector of the industry. Its design team – headed by the great inventor and engineer Barnes Wallis and including N. S. Norway (also known as the author Nevil Shute) as chief calculator, and a skilled metallurgist, Major P. L. Teed – found themselves based at the disused RNAS station at Howden, Yorkshire. Howden possessed a shed just large enough to accommodate the giant airship which Barnes Wallis was beginning to create on his drawing board.

The contract for the construction of R.100 was placed on 22 October 1924, and required the company to design and build the airship for a fixed price of £350,000; to demonstrate its capability with a return flight to India (a destination altered later to Canada); and to achieve a maximum speed of 113km/h (70 mph). A penalty clause provided for the deduction of £1,000 for each 0·8 km (0·5mph) that demonstrated maximum speed fell short of the specified maximum.

Duralumin – a light alloy composed of aluminium, copper, manganese, magnesium and silicon – had been developed in Germany by this time. Appreciating its potential as an aircraft-construction material, Vickers obtained world rights for the manufacture of duralumin outside Germany. Easily worked in its normalized state, duralumin hardens with age until it has almost as much tensile strength as an equal mass of steel, while weighing only one-third as much.

Already widely experienced in the design and building of airships, Barnes Wallis realized that a rigid duralumin frame would be both strong and significantly lighter than existing structures. At that time, however, there was no method of producing extruded tubes of duralumin with a thin section and large diameter. Typically, Wallis set about designing and building a machine to produce such tubes. They were fabricated in the end by helically winding thin-gauge flat strip and riveting it at close centres along the continuous overlap.

The giant craft slowly took shape, hanging from the roof of the Howden shed as the metallic jig-saw puzzle was pieced together. The hydrogen lifting gas was to be contained in 15 gas bags, the 14th and 15th of which were interconnected. The three-ply bags, made of cotton lined with two plies of gold-beaters' skin, were made by the Zeppelin Company at Friedrichshafen.

To ensure that the bags would not be chafed by their containing frames, Wallis devised a network of flexible wires which both restrained the bags and allowed them to expand and contract as the conditions of flight changed. From this arrangement evolved the geodetic method of construction, which he was later to develop and apply successfully to the production of military aeroplanes.

Three decks for the passengers and crew were located in the lower part of the hull between the fifth and sixth structural frames. The lower deck housed the crew; the upper deck provided a gallery lounge and the majority of the double and four-berth cabins; the remainder of the cabins were situated on the centre deck. In all, a hundred passengers could be carried in great comfort. The rest of the centre deck was occupied by a dining room that could seat 56 persons at a time. Having eaten, they could then take the air on the 12·2m × 4·6m (40ft × 15ft) viewing promenades on either side of the dining room.

The R.100 was powered by six 650hp Rolls-Royce Condor IIIA engines mounted in tandem pairs in three cars. One car was set centrally on the undersurface of the hull some 12·2m (40ft) ahead of the lower vertical fin, with the other two placed low on each side of the hull about 39·6m (130ft) aft of the passenger accommodation. In each car the forward engine drove a tractor propeller, the aft engine a pusher, and in addition each rear engine was fitted with reversing gear to facilitate

mooring operations.

On 16 December 1929, the R.100 flew for the first time, from Howden to Cardington under the command of Major G. H. Scott. Flight trials were completed at Cardington, and between 29 July and 16 August 1930, the airship, with a crew of 44, was flown to Canada and back. The outward flight took 78 hours against prevailing headwinds, the homeward crossing only 58 hours.

A little less than two months later, on 4 October 1930, the government-built R.101 crashed at Beauvais, France, on its maiden intercontinental flight, bringing British airship development to an end. The highly successful R.100 was taken out of commission and scrapped without making another flight.

These figures are taken from the original specification drawn up by Mr B. N. (now Sir Barnes) Wallis and from the hand-written reports of N. S. Norway during the airship's trials.

Length overall 219·4m (719·8ft)

Length of hull 216·2m (709·29ft)

Height overall 21·7m (136·8ft)

Maximum diameter 41·55m (136·33 ft)

Gross volume 141,585m³ (5,000,000 cu ft)

Fineness ratio 5·3

Fin area (total) 1,059m² (11,400 sq ft)

Hull weight 53,668kg (118,318lb)

Gross lift 159,052kg (350,650lb)

Fuel and oil 29,075kg (64,100lb)

Water ballast 19,586kg (43,180lb)

Maximum trial speed 131·16km/h (81·5mph)

Rotorcraft

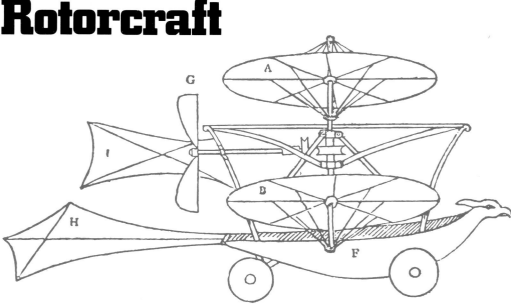

The development of rotorcraft, which term includes both helicopters and autogyros, or gyroplanes as they are sometimes called, has always been in a world of its own. Apart from ballooning, it was mainly with moving wings designed to emulate the flapping of bird flight that man made his first attempts to fly. From the legendary experiments of Icarus and the sketches of Leonardo da Vinci, right through to the notable contributions of Sir George Cayley in the 19th century, most of the proposed flying machines had some means of flapping or rotating the wings to produce lift.

Initially, the practical problems of mechanical complexity were quite insuperable, and it is only within the past 100 years that any significant attempts have been made to depart from the moving-wing concept. Then, as it came to be more clearly discerned that so many of the finest natural fliers – such as the eagle and the albatross – seldom flapped their wings, the fixed-wing philosophy began to make great strides. But there were still some adherents of rotating wings who were not deterred in their search for vertical take-off and landing. To them, the aeroplane was a highly dangerous form of flight and, although they were habitually ridiculed, there was a great deal of evidence to support their contentions. Most of the frequent aeroplane crashes of the early 20th century occurred during the critical manoeuvres of take-off and landing at comparatively high forward speeds.

The group of rotorcraft pioneers contracted as many gave up the complexities and turned to the much simpler fixed-wing aeroplane, but during the first 20 years of this century a handful of designers did meet with some limited success. Among the helicopters to fly successfully in that period, the Breguet-Richet No 1 was credited, on 29 September 1907, with being the world's first man-carrying helicopter to become airborne, albeit under restraint from tethering ropes controlled by the ground crew. This machine had a gross weight of 578kg (1,274lb) and was lifted by four biplane rotors of 8m (26ft 3in) diameter each and driven by a single Antoinette engine of 45bhp. It remained airborne for only about one minute outside the Breguet aircraft works at Douai, France, and reached a height of only about 0·6m (2ft).

On 13 November of that same year and also in France, at Lisieux, the Cornu tandem-rotor helicopter became the world's first rotating-wing aircraft to achieve unrestrained free flight, with the designer himself at the controls. This maiden flight deserves its place in history even though, at less than 30 seconds, it was of even shorter duration than Breguet's. The machine weighed 260kg

ABOVE: *This sketch of Sir George Cayley's 'Aerial Carriage', of 1843 vintage, reveals that he had a remarkable insight into the problems of rotating wing flight almost a century before the first practical helicopters began to emerge. The first Focke and Weir helicopters followed this configuration.*

BELOW: *Credited with achieving the world's first free flight by a man-carrying helicopter, this Cornu design remained airborne for less than a minute with the designer at the controls. The short flight was not long enough to give him time to become accustomed to the handling characteristics, even if control provision had been adequate. The fragile machine crashed and broke up on landing.*

ABOVE RIGHT: *A 1910 attempt to build a contra-rotating helicopter by Vuitton. Like other attempts of its day, it was not successful.*

BELOW RIGHT: *The Cierva C–19 Mk IV was the world's first autogyro to be put into quantity production. It was flown successfully by private pilots in a number of flying schools during the early 1930s and was the first type to demonstrate that rotating wing flight could be a practical proposition.*

BELOW FAR RIGHT: *The Cierva C–40 was the last autogyro design to be produced by the company. Detail design was by Dr James Bennett, based on Cierva's final research work before his untimely death. The rotorhead mechanism embodies means to provide for vertical 'jump' take-off.*

(573lb) and was powered by a 24hp Antoinette engine which drove the two twin-bladed 'paddle-wheel' rotors through a belt-and-pulley transmission system. To outward appearances, it gave the impression of having been built from parts of several bicycles and its short flight ended, regrettably, in the collapse of its frail tubular framework following a heavy landing. Still, it was a world first.

World War I brought a temporary halt to experiments but in the period that followed there were other limited successes. In France, the quadruple-rotor Oemichen helicopter,

which was additionally fitted with five small variable-pitch propellers for control purposes, actually achieved a world helicopter distance record with a flight of just over 500m (1,640ft) in 1924. Another quadruple-rotor machine, the de Bothezat helicopter, was also being flight-tested in the USA by the US Army Air Force and achieved a maximum duration of 1min 40sec. Both were incredibly complicated mechanical contraptions. Hardly less complicated was the Pescara helicopter which was lifted by twin co-axial, contra-rotating biplane rotors. This machine, built in France by a Spanish designer, proved itself

superior to the Oemichen by increasing the world helicopter distance record to just about 800m (2,640ft). It was no wonder that the helicopter proponents were ridiculed by fixed-wing aeroplane pioneers. The Atlantic had already been flown non-stop by Alcock and Brown in a Vickers Vimy bomber.

The Cierva autogiro

The breakthrough in rotating-wing design came with the invention by the Spaniard Juan de la Cierva of the autogyro. By 1925, he had proved the validity of his concept beyond all doubt and was invited to England to con-

Cierva's invention was based on a quite novel design philosophy which emerged from the mathematical analysis he made of contemporary helicopter designers' problems. They were then all using their various forms of rotor system with the blades set at a high positive pitch angle, rotated by the application of power to the drive shafts on which they were mounted. In effect, they were large propellers which screwed themselves upward against the resistance of the air. There were two principal problems. On the one hand, the application of engine power to turn the rotor drive shaft produced an equal and

work in a simple rotor system satisfied him that, if the blades were to be set at a *low* positive pitch angle and started in rotational motion, the rotating system could be towed through the air in such a way that the blades would maintain a constant rotational speed in a state of equilibrium, without any application of power.

Autorotation

He termed this phenomenon autorotation and applied it in practice by building a single rotor and mounting it on a pylon above a conventional aeroplane fuselage. Instead of a complex transmission system there was simply one bearing at the head of the pylon which allowed the rotor to turn freely. There were a few teething troubles initially but he soon developed an arrangement that worked satisfactorily. Take-off in the early machines was achieved by hand-starting the rotation of the blades and then accelerating across the aerodrome, pulled by a conventional engine-driven propeller. As forward speed increased on the take-off run, rotor rpm also increased until the state of equilibrium was reached. At this stage, which was normally after a run of a few hundred metres to reach a speed of some 48km/h (30mph), sufficient lift was being generated by the spinning blades for the machine to become airborne.

It was an elegant solution to the problem of flight with rotating wings. A simple, single rotor system could be used because, as it was freely rotating, there was no torque to counteract. The extended flight-test and develop-

tinue his research and development work with British sponsorship. During the next ten years, the English company he formed produced a succession of autogyro prototypes which established all the design theory and data on which the helicopter industries of the world were subsequently founded. Tragically, Cierva himself was killed at Croydon in 1936, at the premature age of 41, without seeing the extent to which his invention would be developed. Ironically, he was a passenger in a fixed-wing airliner when he lost his life; it was just such a crash as he had devoted his whole working life to eliminate.

opposite reaction which would turn the fuselage in the opposite direction when airborne. Known as torque reaction, this is a phenomenon familiar to all engineers. To avoid it, rotors on early helicopters were used in pairs, turning in opposite directions to counteract the torque.

Following from this need to use pairs of rotors came the requirement for complex mechanical transmission systems to transmit the engine power and then elaborate structures to support the systems. Cierva's discovery was that none of this was necessary. His calculations of the aerodynamic forces at

ment work which the first practical autogyro made possible yielded, in due course, a vast accumulation of technical data on which future improved designs could be based. Subsequent refinements to the system included mechanical means to start the rotor before take-off and means of precise control of its tip-path plane for safety in landing.

The latest forms of autogyro under development just before Cierva's untimely death had progressed still further. Methods were developed to enable the mechanical drive system to overspeed the rotor on the ground before take-off so that a sudden increase of blade pitch under the pilot's control would cause the machine to jump vertically without any forward run. The vertical jump was sustained for just long enough to allow the machine to gain normal forward flying speed under the influence of propeller thrust.

The autogyro was never able to hover but it could fly much more slowly than a fixed-wing aircraft, down to less than 40km/h (25

BELOW: *These Cierva C–30s, lined up outside the manufacturers' flying school, were the most popular autogyros produced. Several hundred were built in England and Europe. Like the C–40, they have no wings but this type was not able to take off vertically. It needed a short forward run.*

BOTTOM LEFT: *A historic picture of Igor Sikorsky with one of his early attempts at helicopter design, in 1910. Hardly surprising, it never flew.*
BOTTOM RIGHT: *The world's first helicopter to demonstrate extended precision hovering flight was the Focke Achgelis–Fa 61.*

BELOW FAR RIGHT: *Igor Sikorsky at the controls of his VS–300, the world's first practical single rotor helicopter. This photograph was taken in 1939. The tethering ropes were to prevent the machine overturning on its initial hovering trials.*

mph) without losing height, and could land with virtually no forward speed. If forward speed were reduced below this value in the air, the machine would begin to sink gently towards the ground. Rotor speed remained constant during the gliding descent, akin to the spinning descent of a sycamore seed. Since the control system governed the rotor's angle of tilt, and not that of the fuselage, the pilot was able to maintain full and precise control right to the point of touchdown.

Cierva C–30

The most effective of the many autogyros developed during the 1930s was the Cierva C–30, of which several hundred were produced by manufacturing licensees of the Cierva Autogiro Company in the UK and Europe. Modified versions of the type were also produced in the USA. The C–30 was a two-seat machine of some 862kg (1,900lb) gross weight and was powered by a 140bhp Armstrong Siddeley Genet Major radial engine. A power take-off from the rear of the engine crankshaft was used as a mechanical starting system for the three-bladed, 11·28m (37ft) diameter rotor. The two seats were arranged in tandem as open cockpits, with the pilot at the rear. The forward cockpit was located immediately below the rotorhead with its supporting pylon straddling the cockpit coaming. In this way, the weight of the passenger was exactly on the centre of gravity so that flying trim did not alter when the machine was flown solo.

The fuselage was a fabric-covered, tubular-steel structure, substantially the same as that of the Avro Cadet biplane, a successor to the famous 504K. The C–30 was, in fact, designed around this fuselage as Avro was Cierva's UK licensed manufacturer. From a production viewpoint it was advantageous to use an existing fuselage, although it was of course the rotor system rather than the fuselage which led to the C–30's success.

The fully articulated rotor was mounted universally on the pylon so that it could be tilted in any direction in response to movement of the pilot's hanging control column. Rotor blades were of aerofoil section and their construction was based on a tubular-

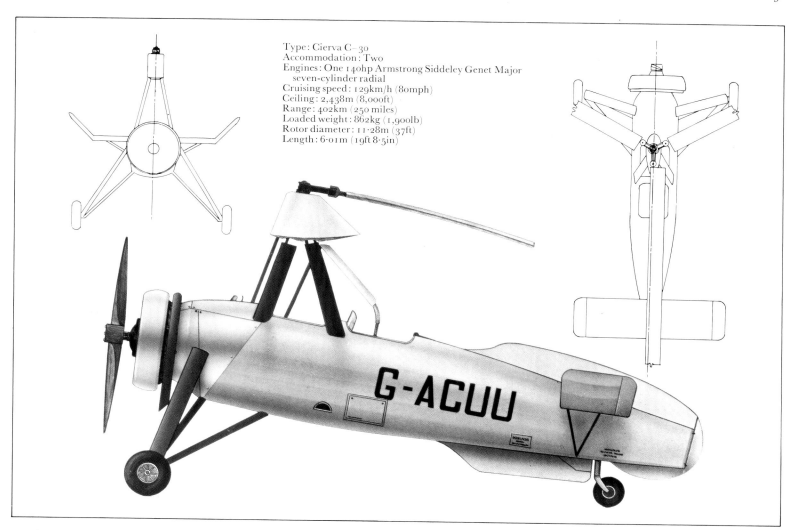

Type: Cierva C–30
Accommodation: Two
Engines: One 140hp Armstrong Siddeley Genet Major
seven-cylinder radial
Cruising speed: 129km/h (80mph)
Ceiling: 2,438m (8,000ft)
Range: 402km (250 miles)
Loaded weight: 862kg (1,900lb)
Rotor diameter: 11·28m (37ft)
Length: 6·01m (19ft 8·5in)

steel spar with spruce ribs and plywood covering. Blade chord was 25cm (10in) and the pitch angle was fixed at 2° positive. Normal rotational speed in the autorotative condition was 200rpm.

Design modifications to the standard fuselage included the fitting of a long-travel, soft-oleo landing gear, to accommodate the unique autogyro landing characteristics, and a steerable tailwheel operated by the pilot's rudder pedals. The tailplane was also of unique design. Fitted with upswept tips, the aerofoil camber was positive on the starboard side and negative on the port side to provide a movement about the longitudinal axis (anticlockwise when viewed from the rear) and thus counteract propeller torque in forward flight.

Design maximum speed was over 160km/h (100mph). It was not always possible to achieve this, particularly at full load, but the type had a useful cruising speed in the order of 130km/h (80mph) and carried fuel for just over two hours' endurance. The Cierva C–30 was sold widely for private and club flying and a number were also bought by the military agencies of several countries, mainly for army reconnaissance evaluation. Another military role, for which their slow-flying capabilities were found to be singularly well suited during World War II, was the calibration of ground radar stations.

If the war had not erupted when it did, it is more than likely that the pattern of progress might have been very different. In the years immediately preceding the outbreak

of war, however, there was a small handful of designers who, having derived benefit from the accumulated knowledge of autogyro rotor design, were beginning to show promising results with new helicopter projects. Prominent among them were Heinrich Focke in Germany, James Weir in Scotland, who had earlier been instrumental in sponsoring Cierva's developments in England, and Igor Sikorsky in the USA. Sikorsky had, in fact, built two unsuccessful helicopters in Europe in 1909, at the tender age of 19 years, and had then changed to fixed-wing aeroplane design after emigrating to America. He was prompted to look at the possibilities of helicopter design again by Cierva's autogyro successes which he enthusiastically acknowledged.

Pioneer helicopters

Three projects – the Focke Achgelis–Fa 61, the Weir W–5 and W–6, and the Sikorsky VS–300 – began to attract the interest of their respective military authorities in the late 1930s. They were all more complex than the autogyros, which by then were flying in quite large numbers, but the ability to sustain hovering flight was of special value in a variety of military roles. Even the vertical-jump take-off facility of the latest autogyros then flying was not considered by the military to be a suitable substitute for a hovering capability. Nor was the subsequent wartime use of autogyros on a scale sufficient to deflect their interest. So it was that the helicopter, when it finally began to emerge as a practical flying machine, was initially developed largely under military sponsorship and specifically as a military vehicle. This factor has influenced all its subsequent progress.

The demands made by the military upon the pioneer constructors were unbridled. There was a war to be won and the fledgling helicopter was seen to have enormous potential as a reconnaissance vehicle for swiftly moving ground forces. Initially, the interest came mainly from the US Army and the British Royal Navy. Government money was poured into development contracts with the result that by 1945 Sikorsky's Main and Tail Rotor (MTR) design had become firmly established as the classic configuration. Following the VS–300 prototype, three new designs were developed and put into limited production by the same company to meet military orders. These were the Sikorsky R–4, R–5 and R–6. The first and last were two-seaters, supplied to special units for pilot training and operational evaluation, while the R–5 was a larger helicopter with a lifting capacity of about 508kg (1,120lb). It was the world's first helicopter designed for a specific military role, having been ordered off the drawing board by the US Army in 1944, then adopted by the US Navy and then built in Britain as the Westland Dragonfly.

During World War II German companies led the world in helicopters. The first to go into production and service use was the Flettner F1 282 Kolibri, used from 1942 for shipboard observation and convoy escort. By far the world's most powerful helicopter, the Focke Achgelis Fa 223 Drache had side-by-side rotors driven by a 1,000hp BMW radial engine. Weighing 5,000kg (11,010lb), the Fa 223 could fly many service missions.

Post-war developments

After World War II, the established helicopter constructors turned their attention towards a possible commercial market. Bell was first to be awarded a commercial helicopter certificate of airworthiness, in 1946. This was for the Model 47, which remained in production for 27 years in many two- and three-seat variants. Sikorsky was not far behind, with a four-seat civil version of the R–5 designated S–51. It quickly became apparent, though, that the helicopter was anything but a motor car for the man in the street. Such was the cost structure established by the initial pressure of military procurement that manufacturers found the civil versions they were producing could be operated economically only in a limited number of highly specialized roles. Commercial sales were thus few and, with military interest becoming less intense, the crucial question of the hour was: 'Is the helicopter here to stay?'

In agricultural roles such as crop-spraying, for example, the high cost of the helicopter put it at a considerable disadvantage compared with fixed-wing aircraft. Commercial helicopter operators found themselves unable to penetrate more than about 20 per cent of the market, which comprised mainly those areas in which it was too difficult for the fixed-wing crop-sprayers to operate. The same applied in other aerial-work roles, par-

BELOW: *The Sikorsky S–58T is a quite recent twin turbine version of the original S–58, which first appeared in 1954. Over 1,000 S–58s were built.*
BOTTOM: *The first two-seat helicopter trainer was the Sikorsky R–4, of which several hundred were built in 1943/1944.*

RIGHT: *The Bell 47G helicopter is one version of a series that has been in production for more than 30 years.*
INSET: *The Bell 47J–2A was one of the first adaptations of a utility helicopter for aerial taxi work.*

ticularly in survey and construction engineering support. In some applications, however, mainly those concerned with operations in remote areas or mountainous terrain, the helicopter was able to perform time-saving miracles. In such uses its apparently high cost was immaterial since to do the job any other way would have cost even more.

Unfortunately, the opportunities to engage in such specialized work were few, and progress in developing commercial uses for the helicopter was painfully slow for the first few years after World War II. There was, nevertheless, still something of a helicopter euphoria during this period, with new projects making their appearance in countries all over the world. In Europe, the pre-war French and British pioneers applied themselves to further development, seeking to make up for the time lost during the war years.

In France, during this period, three groups in the nationalized aircraft industry were engaged in rival helicopter projects. SNCA du Nord began in 1947 but its interest sur-

vived only a few years. SNCA du Sud Ouest had acquired as part of German war reparations the services, for a few years from 1946, of members of a wartime Austrian design team. They had successfully built and flown the word's first jet-rotor helicopter, designed by Friedrich von Doblhoff in 1943. The Sud Ouest projects were thus all in the Single Jet-driven Rotor (SJR) category and led to the development of the SO–1221 Djinn helicopter, the only jet-rotor helicopter to go beyond the prototype stage into quantity production.

Other members of von Doblhoff's team went to Fairey Aviation in Britain, which led to the development of a Gyrodyne derivative with a jet-driven rotor and, from this, to the Rotodyne. Von Doblhoff himself went to the USA and joined McDonnell Aircraft Corporation to develop a ram-jet helicopter. No jet-rotor helicopter, however, has ever achieved any marked degree of commercial operating success.

Even the SO 1221 Djinn helicopter, the agricultural version of which is pictured here, was not a great success commercially. Production was discontinued after just over 100 had been built. A lightweight two-seater, the type was used mainly for pilot training and crop spraying. Its rotor was driven by what is known as 'cold' jets at the blade tips. This was, simply, compressed air bled from the compression chamber of the Turbomeca Palouste turbine engine. The air was ducted through a rotating seal at the rotorhead and thence through hollow spars to the blade tip nozzles.

Other forms of jet driven rotor have made use of pulse jets, pressure jets and ram jets. These all involve the ducting of fuel through the head and blades, to be burnt in small combustion chambers at the blade tips. Hence the contrasting term, 'hot' jets. Many such experimental helicopters have been built and some have flown extremely well; but the fuel consumption has always been prohibitive.

The third French company, SNCA du Sud Est, followed more conventional lines in the development of MTR designs, and produced the series of helicopters which have dominated French involvement in the field. The SNCA du Sud Est and Sud Ouest were later amalgamated in Sud Aviation, which itself was subsequently regrouped under what is now the single French nationalized aircraft constructor, Aérospatiale. Among the wide range of helicopters Aérospatiale produce, three, the Puma, Gazelle and Lynx, are manufactured jointly with Westland Aircraft Ltd under the Anglo-French collaboration agreement.

In Britain after World War II, the original Cierva Autogiro Company re-formed with many of its former key engineering staff and began the development of two new designs. The first to fly, the Cierva W–9, was a two-seat single-rotor helicopter but, instead of following exactly the classic MTR configuration, it used a laterally deflected jet at the tail in place of the tail rotor for torque compensation. Concurrently, the company was also building a much larger machine, the triple-rotor Cierva W–11 Air Horse, which flew

RIGHT: *The Fairey Rotodyne was an advanced design concept to attain high cruising speeds but the prototype was never developed to the production stage, largely for financial reasons.*
BELOW: *The French Djinn helicopter was the only jet-rotor helicopter to be produced in quantity. The duct from the turbo-compressor to the rotor head is clearly visible. This one has a spray boom fitted.*
BOTTOM: *Turbine powered, five-seat Gazelle helicopter, formerly in production by Aérospatiale and Westland Aircraft. Main rotor torque, in this type, is counteracted by a ducted fan in the tail fin named a fenestron.*

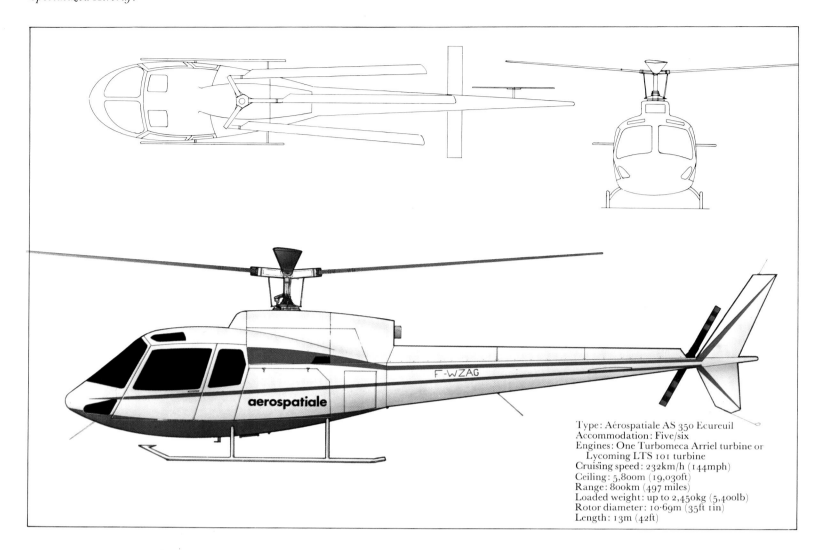

Type: Aérospatiale AS 350 Ecureuil
Accommodation: Five/six
Engines: One Turbomeca Arriel turbine or
        Lycoming LTS 101 turbine
Cruising speed: 232km/h (144mph)
Ceiling: 5,800m (19,030ft)
Range: 800km (497 miles)
Loaded weight: up to 2,450kg (5,400lb)
Rotor diameter: 10·69m (35ft 1in)
Length: 13m (42ft)

successfully at its design gross weight of 7,938kg (17,500lb).

This, in its day, was the largest helicopter in the world, with a cargo compartment 5·79m (19ft) in length capable of carrying wheeled vehicles. Volumetric capacity was in the region of 22·65m³ (800cu ft) and entry was by means of a ramp through clam-shell doors at the tail. The Air Horse was powered by a single water-cooled Rolls-Royce Merlin engine of 1,620bhp, the same engine that powered the famous Spitfire fighter. That the Air Horse did not mature to a successful conclusion was due mainly to lack of appreciation by the sponsoring British government of the need for a much higher level of funding to support so sophisticated a design.

Three other British aircraft constructors also entered the rotating wing field at the end of World War II: Westland Aircraft negotiated a manufacturing licence with Sikorsky to build the S–51 in Britain; Bristol Aeroplane Company began the development of a new MTR design with a cabin for five passengers; and Fairey Aviation Company developed a novel Compound Helicopter (CMP) project, also able to carry five passengers, named the Gyrodyne. All three were powered by the same type of engine, the Alvis Leonides nine-cylinder radial of 525 bhp. The Fairey Gyrodyne, with its novel compound design for superior cruising speed, was the first rotorcraft to take the world helicopter speed record above the 200km/h (124mph) mark. Over a 3km (1·86 miles) course, it attained 201km/h (124·9mph).

American initiative

In spite of inevitable setbacks in the comparatively early stages, helicopter projects grew both in number and variety, and nowhere was the profusion of new ideas so great as in the United States. At one time, just before 1950, there were more than 70 different active helicopter projects in America. Many were being built by small engineering companies or by individuals in private garages. Of these, only a few ever left the ground; many never progressed beyond the stage of being a gleam in their hopeful inventors' eyes.

One notable exception was the project of a young Californian graduate, Stanley Hiller Jr, who built his own back-yard Co-axial, Contra-rotating Rotor (CXR) helicopter in 1944. He was fortunate in having links with a major industrial corporation which helped him to surmount the initial hurdles. He was later to develop his own servo-paddle rotor control system. Hiller Helicopters Inc, which he formed, produced more than 1,000 helicopters based on this design feature during the ensuing two decades. Among other American companies, the two which had been predominant in pre-war autogyro development were also involved with new helicopter projects. Pitcairn Autogiro Company, which had been taken over by the giant Firestone Tyre & Rubber Company, produced a conventional MTR helicopter, but with a novel rotorspeed governor, while Kellett Aircraft Corporation concentrated its studies on what was then the unusual Twin Intermeshing-Rotor (TIR) configuration. The same design was also favoured by Kaman Aircraft Corporation, which developed it to build several hundred helicopters for the USAF.

The tandem-rotor helicopter was introduced because it was thought that limitations in feasible rotor diameters necessitated multiple rotors to lift heavier payloads. Piasecki Helicopter Corporation was the first to produce a practical Twin Tandem-Rotor (TTR) design, and derivatives of its first tandem-rotor machine, the XHRP–1 (jocularly known as the 'Flying Banana'), were still in production by Boeing Vertol in the mid-1970s. Among other configurations developed in the USA during the 1945 to 1950 period, SJR designs with rotors driven variously by pulse jets, pressure jets and ram jets all made their appearance – and, as often as not, their disappearance shortly thereafter.

By 1950, most of the weaker brethren had disappeared, leaving about 10 or 12 companies to form the nucleus of what was then a Cinderella industry struggling for recognition in a highly competitive aviation market.

Korean War

It was the outbreak of war in Korea, in the mid-summer of 1950, that provided the next major impetus to transform this small group of manufacturers into the thriving helicopter industry which now exists. The US Air Force and US Navy units equipped with the few hundred helicopters which had been delivered for evaluation were despatched to Korea for trials under active service conditions. Their performance was far beyond

the wildest expectations of the most optimistic military strategists.

The series of helicopter types principally involved were, initially, the Sikorsky H–5, derived from the S–51, Bell H–13 and Hiller H–23. Later, these were supplemented by the Sikorsky H–19 series, a larger machine capable of lifting ten men or a load of almost a ton, and the Piasecki HUP–1 series, a TTR design based aboard US Navy aircraft carriers and used for ship-to-shore work. These helicopters, in their performance of otherwise impossible rescue missions, confirmed beyond any doubt that rotating-wing aircraft were here to stay. With their unique versatility and independence of prepared landing strips, they were able to save the lives of thousands of soldiers, wounded or stranded behind enemy positions. This in itself was enough to ensure their unreserved acceptance as a new ancillary to fighting armies. Perhaps even more significant was the potential that the success of such rescue operations revealed.

Military tacticians were quick to realize that if the helicopter could quite easily infiltrate behind enemy lines to rescue wounded soldiers, it could equally penetrate with offensive personnel, weapons and supplies to mount attacks from almost any unexpected quarter. By the time the Korean war had ended, American military strategy had been completely re-orientated in line with this new philosophy. The new US Army was to be composed largely of highly mobile task forces mounted, supplied and supported entirely from the air by fleets of helicopters designed specifically for the variety of operational tasks involved. At the same time, the helicopter's potential in marine warfare, as an anti-submarine weapon and to support marine commando raids ashore, was realized.

The new philosophy was pioneered by the American armed forces and quickly taken up by those of other nations throughout the world, including the Soviet Union where the history of rotating-wing development had had early origins but not the same degree of practical development as elsewhere.

With military procurement pressure once again calling the tune, the helicopter industry made rapid strides in the decade following the Korean war. Military agencies were then in a much better position to write operational specifications for helicopters to fulfil special duties for which active-service experience had shown the need. Thus, with specific targets to achieve and the money to spend in terms of development contracts, helicopter construction was suddenly transformed into a boom industry and the various new designs began to roll off the production lines in their hundreds.

TOP LEFT: *The Sikorsky H–5 and its civil equivalent, the S–51, was the first American helicopter to be made in large numbers*
TOP RIGHT: *The Cierva Air Horse was the world's largest helicopter in its day but it was not a successful design and only two were built.*
ABOVE CENTRE: *The Piasecki HRP–1 was the first practical tandem rotor helicopter.*
ABOVE: *One of Germany's successful wartime helicopters, the Flettner Fl 282 was the first to go into production, in 1942.*
LEFT: *One of the small servo rotor 'paddles' is visible above the rotor head of this Hiller H–23B.*

Rotorcraft Configurations

Configuration Code & Category

RTG   **Running Take-off Autogyro**
STG   **Short Take-off Autogyro**
VTG   **Vertical Take-off Autogyro**
Freely rotating rotor, with forward propulsion by conventional propeller thrust. The rotor is tilted backward with respect to the flight path at all times. No rotor torque and no hovering capability.

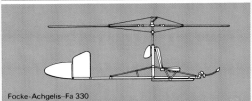

Cierva Autogyro

AGG   **Autogyro Glider**
Freely rotating rotor, usually started by hand. The equivalent of propeller thrust is supplied by a tow rope. In high winds, the AGG can be made to take off as a kite with an anchored tow rope.

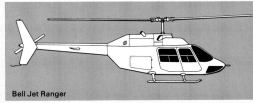

Focke-Achgelis-Fa 330

MTR   **Main and Tail Rotor helicopter**
Both lift and propulsion are provided by a single main rotor. The functions of the small tail rotor are to counteract main-rotor torque and to provide control in yaw. Both rotors are engine-driven through a gearbox and shaft system at a constant speed ratio, tail-rotor rpm being approximately five times main-rotor rpm.

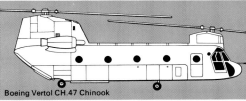

Bell Jet Ranger

TTR   **Twin Tandem-Rotor helicopter**
The twin rotors are contra-rotating to counteract torque and in some types they may overlap slightly. In such cases a synchronizing transmission shaft between the two rotor heads ensures that the two rotors keep a uniform speed so that blades cannot foul each other.

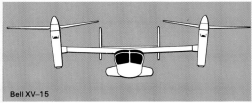

Boeing Vertol CH.47 Chinook

TSR   **Twin Side-by-side Rotor helicopter**
The twin rotors are contra-rotating to counteract torque and are mounted on lateral outriggers which may, but need not, be lifting surfaces. The rpm of the two rotors are synchronized but the blades do not normally overlap. A conventional tail unit often supplements rotor control.

Bell XV-15

TIR   **Twin Intermeshing-Rotor helicopter**
This is a cross between the TSR and CXR configurations. It avoids the mechanical complexity of co-axial gear systems but needs precise synchronization of the contra-rotating rotors. At the same time, the greater swept area increases rotor efficiency.

Kaman H-43 Huskie

CXR   **Co-axial contra-rotating Rotor helicopter**
The twin rotors are contra-rotating to counteract torque. Cyclic and collective pitch is applied to the two rotors in unison. Control in yaw is achieved by differential collective-pitch variation between the rotors so that the resulting difference in torque produces a turning moment.

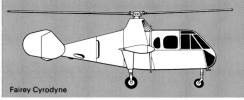

Cierva

CMP   **Compound helicopter**
A single lifting rotor provides the means of sustentation alone, while separate means of propulsive thrust is used to combine the propulsion and anti-torque functions. The plane of the rotor remains substantially parallel to the flight path in normal cruising flight to reduce aerodynamic drag for higher speeds.

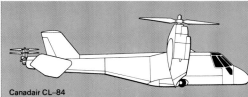

Fairey Cyrodyne

CVT   **Convertiplane**
The thrust produced is used in the vertical sense for take-off and to reach a forward speed at which the conventional fixed wing can take over the lift function. The vertical thrust is then converted, usually by mechanical means, into horizontal thrust for high-speed propulsion. For landing it is converted back again.

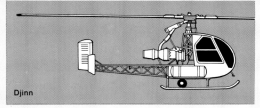

Canadair CL-84

SJR   **Single Jet-driven Rotor helicopter**
This may be similar in appearance to the MTR, but without the tail rotor as there is no torque when the rotor is rotated by jet propulsion at the blade tips. If a separate means of propulsive thrust is embodied in the design, the SJR can also be a VTG or a CMP.

Djinn

Typical Features

Most autogyros come within these three categories. Different types are obviously different in appearance but there is little to distinguish the categories from each other as variations in take-off performance are normally achieved through refinements to the rotorhead design. The distinctions become apparent only in a detailed inspection of the rotorhead mechanism.

Introduced by Focke Achgelis during World War II. The Fa-330 single-seater was built in quantity for use by German submarines on aerial spotting duties. Post-war versions, designed for towing behind a motor car or boat, became popular for pleasure flying in the 1950s and could be converted to ultra-light RTGs (*q.v.*) by the addition of an engine and propeller.

The classic helicopter configuration followed by the great majority of all helicopter constructors. This configuration is now equally applicable to the entire size range, from ultra-light single-seaters to the giant 56,000kg (123,450lb) Mi-26 helicopter.

Developed originally to provide greater lifting capacity when it was thought that single rotors would not be practicable in the sizes calculated to be necessary. This early hypothesis has since been disproved by progressive experience but the TTR configuration is still retained by Boeing Vertol types and there have been many other examples.

This was the original configuration adopted by Focke and Weir but it has hardly been used since, except in Russia. The TSR configuration, however, is suitable for some convertible helicopters, such as the tilt-rotor projects. One example of these, the V-22 Osprey, is under development in the USA. If convertible, they are categorized as CVTs (*q.v.*).

A variation on the TSR configuration to reduce the overall width and eliminate the structural weight of the outriggers. First used by Flettner in Germany and followed by Kellett and Kaman in the USA. There are still a few of the last-mentioned types in current operation. TIRs are known jocularly as 'egg-beaters', but the configuration has not been used for new projects since the 1950s.

This was the original Bréguet-Richet configuration. It has the advantage of compactness but aerodynamic interference between the downwash from the two rotors can cause problems. The possibility of mechanical interference between the blades is overcome in the latest example, by Cierva Rotorcraft, with an inter-rotor transfer link system.

In this category, the means of sustentation and the means of propulsion are designed to retain their separate functions throughout the flight envelope. The category was pioneered by the Fairey Gyrodyne and can be said to include any helicopter fitted with supplementary means of increasing horizontal propulsive thrust.

Tilt-wing and tilt-rotor aircraft come within this category. In the Canadair CL-84 the whole wing, engines and propellers are power-tilted to convert from one flight regime to another. In the Bell XV-15, the rotors and gearboxes on the wing tips are tilted while the wing itself remains fixed. Both fly as conventional aircraft in normal cruising flight.

Pioneered by Frederich von Doblhoff in Austria, the jet-driven rotor system has been applied many times in experimental helicopter projects. Only one, however, the French Djinn helicopter, has ever been put into quantity production and even this was for only a comparatively short period.

The turbine engine

Another coincident factor, which had a particularly important influence on the rapid rate of development from the mid-1950s onward, was the advent of the turbine engine and its application to helicopters. In fixed-wing aircraft, the shaft turbine as a powerplant to drive a conventional propeller – known as the turboprop – was not the sweeping success for which its designers had hoped. This was largely because it was so quickly superseded by the turbojet, which could produce much higher speeds and thereby justify its much higher costs.

For helicopters, though, the shaft turbine proved ideal. It had a much lower installed weight than the piston engine, which more than offset the greater fuel loads which had to be carried. Consequently, helicopter designers found for the first time that they had abundant power available. Moreover, the normal operating characteristics of the turbine, which permitted extended running without fear of damage at up to about 85 per cent power, was particularly well suited to a helicopter's requirement for long periods of continuous hovering in certain military roles. The high initial cost and high fuel consumption were no drawbacks to the military, while the much shorter warm-up period before take-off and smoother operation in flight were considerable advantages.

So, with the power available, much bigger helicopters were soon found to be practicable. Earlier fears of limitations to the size of rotor systems proved groundless in practice, and

Aerospatiale's boat-hulled Super Frelon seen in an air/sea rescue exercise.

the helicopter grew up. Before 1950, apart from the German Fa 223 and the British Air Horse prototypes, the biggest production helicopters powered by piston engines were in the region of 2,500kg (5,510lb) gross weight, with a lifting capability of some 680kg (1,500lb). Ten years later, at the outset of the Vietnam conflict in the early 1960s, transport helicopters like the Boeing Vertol CH–47A Chinook were in the air at gross weights of some 15,000kg (33,070 lb). The Chinook, with twin Lycoming turbines providing 5,300shp total installed power, had a useful load capability of more than 7 tons.

Speeds too had risen. Before 1950, the world helicopter speed record stood at 201 km/h (124·9mph). By 1963, a French MTR helicopter, the Sud Aviation Super Frelon, had taken it to 341km/h (211·9mph).

In addition to the technological progress, production quantities were also increasing substantially. Four major American constructors, Bell, Boeing Vertol, Hiller and Sikorsky, were all in the big league, each with more than 1,000 helicopters produced. Bell, in fact, was well in the lead with more than 3,000. Large numbers were also being built in Europe and the Soviet Union. Some indication of the extent to which the USA had by then developed the new strategic philosophy of helicopter mobility is that in the US Army alone there were some 5,000 qualified helicopter pilots. Their largest training school was equipped with more than 200 light helicopters, mainly of Bell and Hiller manufacture, and staffed by nearly 150 flying instructors.

Looking to the future at that time, the US Army sponsored a Light Observation Helicopter (LOH) design competition among the already extended industry, the prize for which was to be a production contract for 3,000 turbine-powered helicopters. With military contracts of such proportions in progress and in prospect, it was not surprising that helicopter constructors could spare but scant effort to meet the admittedly

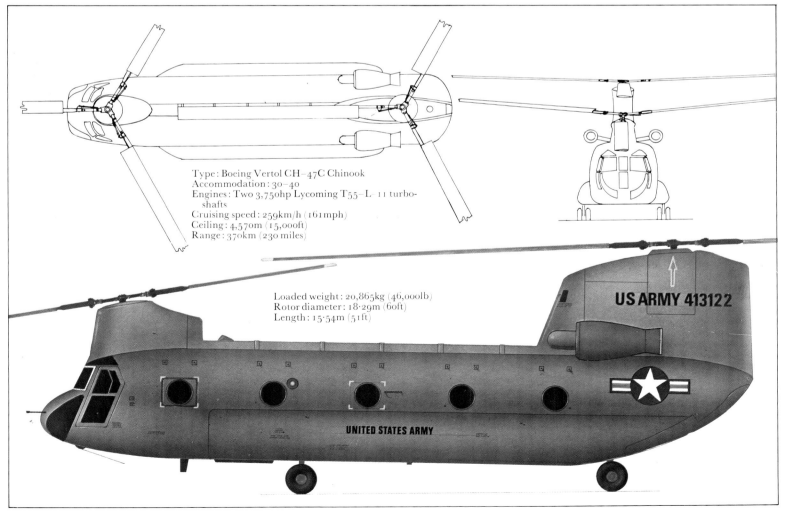

Type: Boeing Vertol CH–47C Chinook
Accommodation: 30–40
Engines: Two 3,750hp Lycoming T55–L-11 turboshafts
Cruising speed: 259km/h (161mph)
Ceiling: 4,570m (15,000ft)
Range: 370km (230 miles)

Loaded weight: 20,865kg (46,000lb)
Rotor diameter: 18·29m (60ft)
Length: 15·54m (51ft)

US ARMY 413122

UNITED STATES ARMY

small needs of commercial operators. Progress was nevertheless made in commercial applications and a few constructors did set up small sales organizations to supply civil adaptations of the military helicopters which comprised their main production. Their selling price was high, but for specialized tasks they could be operated on a cost-effective basis. After some 15 years of consistent, if slow, growth, there were by 1960 about 1,000 helicopters operated commercially by some 300 companies throughout the world. More than half the companies, though, may have been operating only one, two or possibly three small helicopters.

Specialized tasks

The multiplicity of tasks undertaken by these civil operators impinged upon almost every sphere of industrial and commercial activity. The thousands of kilometres flown on such tasks as overhead power-line construction and patrol, or the innumerable hours flown on crop-spraying and other agricultural work, would rarely be noticed except by those directly concerned. Similarly, the constant aerial support work done for offshore oil prospectors became commonplace.

By its very nature, the greatest proportion of all such operations was performed in remote areas and received little publicity. Only when a helicopter was used for spectacular work in a populous area did its unique attributes become more widely known. Typical of such a single feat was the placing of the cross on the summit of the rebuilt Coventry Cathedral by an RAF Belvedere. Following this remarkable demonstration of precision lifting, aerial crane work became more widely accepted as part of the commercial helicopter's repertoire. In particular, larger machines were used to lift air-conditioning equipment on to the flat roofs of 'hi-rise' buildings, a job which helicopters still are not infrequently called upon to do.

The growth of this particular application, combined with military needs for heavy-lift helicopters, led to the development of aircraft designed specifically for aerial crane work. In the Sikorsky S–60 and S–64 crane helicopters, provision is made in the aft of the cabin for a third set of pilot's controls, with the seat facing rearwards. When the actual lifting is to be done, the helicopter is flown with the normal controls into the hover approximately above the load. The captain then leaves his usual seat and takes over the hover with the third, aft-facing, set of controls. In this position he is looking directly downward on to the load for the precision manoeuvre and can instruct the second pilot, over the intercomm, to take over and resume normal flight as soon as the load has been secured or released, as the case may be.

Mainly because of their high cost, these specialized helicopters have not yet come into general use with commercial operators, whose fleets are made up mostly of small and medium-sized machines with gross weights ranging between some 1,360kg (3,000lb) and 6,350kg (14,000lb). Some of the larger operators, with big oil-rig support contracts, may exceed this upper figure if they have Sikorsky S–61 series helicopters engaged on this work.

BELOW: *Largest commercial helicopter in the West, the Boeing Vertol 234 Commercial Chinook normally seats 44 passengers. Alternatively it can carry a slung load of 12·7 t (28,000 lb).*
ABOVE RIGHT: *The Hughes OH–6A, winner of the original LOH design competition in America, is one of the fastest light helicopters in service. One reason for its success is the compact Allison T63 turbine which drives a fully articulated, four-blade main rotor.*
BELOW RIGHT: *The Kaman Huskie followed the Flettner intermeshing rotor concept. The type was in service with the USAF for rescue work.*

CENTRE FAR RIGHT, FAR RIGHT AND BELOW CENTRE FAR RIGHT: *Alternative methods of load carrying by the Sikorsky S–64 Sky-crane. A cargo pod can be secured directly below the centre section (below) or unwieldly loads, like the prefabricated building shown (above right), can be underslung. The close-up (above left) shows the pilot seated at the aft-facing controls with which he can hover the machine while securing or releasing the load.*
FAR RIGHT BELOW: *The Westland Sea King.*
BOTTOM RIGHT: *The Sikorsky S–61L fitted for helicopter airline service.*

Passenger transport

The one area in which the civil helicopter had not made any significant headway up till the early 1960s was the carrying of passengers. There were a few meritorious attempts, mainly in the USA, to set up helicopter airline services and these met with some, if qualified, success. The routes flown provided direct links between the principal airports of the area served – Los Angeles, San Francisco, Chicago and New York – and the outlying suburbs. Their success was qualified in that, although the adapted military transport helicopters themselves were technically capable of providing the services, their operating economics were such that the companies needed some measure of government subsidy. There have been a few isolated exceptions, but most were forced eventually to close down for this reason. In Europe, a helicopter airline service centred on Brussels, operated by the Belgian airline SABENA, was discontinued but a British Airways service between Cornwall and the Scilly Isles has continued.

In its smaller sizes, too, the helicopter was as far from becoming an effective means of personal transport for the man in the street as it had ever been. In fact, by 1960 the dream of a back-garden flying machine for the masses had been completely abandoned. It was still hoped, though, that the unique vertical take-off and landing ability would enable executive transport and aerial taxi work to become an important addition to the helicopter's extensive range of applications.

One of the first helicopters to be developed specifically for this latter role was the Bell Model 47J, named the Ranger. This was not a new design but an adaptation of the earlier Model 47G, of which it used all the dynamic components. In other words, it was a 47G in

all its rotating mechanisms but had a newly styled four-seat cabin in place of the earlier type's transparent plastic bubble which enclosed a bench seat for three, including pilot, sitting side-by-side. The bench was retained in the 47J for the three passengers, but the cabin was extended forward to provide space for a fourth, separate, seat for the pilot. The fuselage was more streamlined, with a stressed-skin monocoque structure instead of the open tubular-steel framework of the 47G.

With these improvements, the 47J Ranger was able to attain slightly higher speeds, up to a maximum of some 169km/h (105mph), and the type was used quite widely for passenger transport. In contrast, however, contemporary light fixed-wing aircraft of equivalent size and power could offer something like twice the speed at about one-third of the cost, so the helicopter was again operating in a strictly limited field. At that time, few more than 100 helicopters of this kind were in regular operation for company transport use and, in most cases, they were for a highly specialized requirement which virtually precluded the use of fixed-wing aeroplanes.

The position changed dramatically with the introduction of the small turbine engine, a by-product of the American LOH design competition previously mentioned. The turbine engines first used as helicopter powerplants produced some 1,000shp. Two types were used initially, the General Electric T58, with a power range between 900shp and 1,100shp, and the Lycoming T53 of some 1,000shp to 1,100shp. Both were free-shaft turbines, more details of which can be found in the engine section of this volume. These two power units gave rise to the development of helicopters such as the Bell UH–1A Iroquois series, a ten-seat MTR helicopter of 3,856kg (8,500lb) gross weight; the Kaman

H–43B, a TIR aircraft of similar size; and, with a twin turbine installation, to the 25-seat Vertol 107 TTR transport helicopter and the 28-seat Sikorsky S–61 series, the latter still in production in Italy in 1987. All these types were flying by 1960.

Although the 1,000shp turbines were regarded as small by contemporary fixed-wing standards, they were far too large for use in the four- or five-seat aircraft in the 1,300 to 1,400kg (2,850 to 3,085lb) gross-weight category. The French company Turboméca was developing a range of smaller fixed-shaft turbines producing some 400shp, but it was the Allison T63 free-shaft unit, designed specifically for the US Army LOH project, which provided the major breakthrough. It had a power-turbine spool no bigger than a two-litre oil can but produced 250shp at its best operating speed.

All three finalists in the LOH competition, the Bell OH–4A, Hiller OH–5A and Hughes OH–6A, used this Allison T63 turbine. The

declared winner was the Hughes OH–6A, a four-seater with a gross weight of 1,089kg (2,400lb) and a maximum speed of some 241 km/h (150mph), and mass-production of the type was set in motion. Of the other two, Bell developed its OH–4A prototype, which had been built to full civil airworthiness standards as one element of the design competition conditions, into what was to become the Model 206A JetRanger. A refined version of this type now constitutes the mainstay of light executive helicopter fleets. The additional speed conferred by the turbine engine was enough to make helicopter transport worthwhile to a much wider field of prospective users. Against a stiff headwind, the earlier 160km/h (100mph) helicopters were often no faster than a good motor car for conventional journeys. The extra 80km/h (50mph) of the turbine helicopter placed it well ahead in any wind conditions. Later, Hughes was also to produce a civil version, known as the Hughes 500, of the company's own turbine-engined competition winner.

Vietnam and its aftermath

When the Vietnam War began, the American armed forces were fully reorientated to the new military strategy of helicopter mobility. They were also partially equipped with a variety of light, medium and heavy transport helicopters, all turbine-powered, to put the new tactics into effect. The strategy met with virtually instant success, to the extent that some historians have dubbed the war in Vietnam 'The Helicopter War'. That the final outcome did not bring with it a conventional military victory for the new tactics was no fault of the helicopters.

The effect on the American helicopter industry of the initial successes in active service was extremely rapid expansion. Helicopters of all types were ordered in thousands; at one stage, Bell had one of the Beechcraft aeroplane factories almost exclusively engaged in producing fuselage and other components under sub-contract to meet the military demands. Experience in Vietnam also had a corresponding effect in other countries as

military strategists began to follow the American lead. The idea of an army going into action without helicopter support became totally outmoded.

The advances made during the ensuing decade and into the 1970s were largely a matter of degree. For example, the UH–1A Iroquois, designed by Bell as the Model 204, originally entered service just before 1960 as a ten-seat utility helicopter powered by a 1,100shp Lycoming T53–L–11 turbine. Its gross weight was 3,856kg (8,500lb), rotor diameter 13·41m (44ft) and disposable load 1,805kg (3,980lb). Through the years, 15 variants of the design have been produced, amounting to some 10,000 helicopters of this one series. The latest derivative, Bell's Model 214, has a maximum gross weight of 7,258kg (16,000lb) and a lifting capacity for external loads of up to 3,629kg (8,000lb). The rotor diameter has gone up to 15·24m (50ft) and the powerplant is a pair of 1,725hp General Electric CT7–2A turboshaft engines. This leaves a substantial power reserve available for high-altitude operation, maintaining sea-level performance up to about 6,000m (19,685ft).

Another current variant of the series, the 15-seat Bell 212, has a twin turbine installation for greater safety. The powerplant in this variant is the Canadian Pratt & Whitney PT6T–3, rated at 1,800shp total, in two coupled turbines either of which will sustain level flight if the other fails. The Model 212 rotor diameter is slightly smaller than that of the 214 at 14·69m (48ft 2½in) and the gross weight is 5,080kg (11,200lb).

Some indication of these advances' full extent can be gauged from the fact that the Lycoming T55 turbine in the model 214 is the same basic power unit used in a twin installation to power the 15-ton Boeing Vertol CH–47 Chinook helicopter. The Chinook itself is another typical example of the degree of advance made during the decade of the Vietnam war. The first prototype made its début in 1961 as a 14,969kg (33,000lb) grossweight TTR military transport helicopter. It was powered by twin Lycoming T55–L–7 turbines of 2,650shp each and had a disposable load capacity of 6,804kg (15,000lb). It was the first big helicopter in which the rear loading door opened downwards to form a drive-on ramp for loading wheeled or tracked vehicles into the cabin, which was 9·14m (30ft) long. Diameter of the two tandem rotors was 17·98m (59ft).

In the latest Chinook, the CH–47C Model 234, the maximum gross weight has grown to 22,680kg (50,000lb) and the disposable load can now be more than 11,340kg (25,000lb). The rotor diameter has remained substantially the same but the systems and transmission have all been modified and strengthened to take the considerably greater power output from the twin Lycoming T55–L–11C turbines. These are rated at a maximum of 3,750shp each, with emergency reserve up to 4,500shp each. Altogether, Boeing Vertol and the Piasecki Helicopter Corporation (which it absorbed) have produced more than 2,500 tandem-rotor helicopters.

Sikorsky Aircraft, which has also graduated mainly into the construction of large

helicopters, has produced even more, though not of tandem-rotor design. This company pioneered the boat-hulled helicopter with its S–61 series and has consequently been concerned mainly with production of aircraft for maritime applications. One of the latest versions of the S–61, known as the SH–3D Sea King, is in service with many navies for anti-submarine and air-sea rescue operations. One of their much publicized applications has been the retrieval of American astronauts on their splash-down return from space missions in the Atlantic or Pacific oceans. The Sea King is built under licence in several countries, in England by Westland Aircraft. Their version has a gross weight of 9,752kg (21,500lb). Disposable load is about 4,000kg (9,000lb). The machine is powered by twin Rolls-Royce Gnome H.1400–I turbines, rated at 1,660hp each, driving an 18·9m (62ft) diameter 5-bladed rotor. The Gnome turbines are derived from the General Electric T58 turbines which power the American-built version. In addition to all-weather navigation systems, the Sea King's equipment can include full sonar detection apparatus plus four torpedoes and four depth charges. It is operated by a crew of four.

At the other end of the size range, the helicopter war in Vietnam brought out the light observation helicopters in their thousands. First the Hughes OH–6A Cayuse and, later, the Bell OH–58A Kiowa, derived from the original OH–4A which was itself the forerunner of the five-seat civil JetRanger. Bell's latest JetRanger variant, known as the Long-Ranger, has a passenger cabin 0·61m (2ft) longer than that of the JetRanger to provide space for two extra seats. The rotor is of 1·12 m (3ft 8in) greater diameter and the power unit, an Allison 250–C20B turbine, has a maximum continuous rating of 370shp as opposed to the 270shp of its predecessor. Gross weight of the seven-seat LongRanger is up to 1,814kg (4,000lb) whereas the Jet-Ranger's gross weight is 1,452kg (3,200lb). Both types are still in full production.

Helicopter gunship

One more significant concept emerged from the implementation of the new military strategy in Vietnam. This concerned a much faster, heavily armed helicopter which could give close support and protection to troop-transport and supply armadas moving into forward battle zones. US Army commanders in the field found that aerial support available from conventional fixed-wing fighter squadrons was sometimes too remote and inflexible to be sufficiently effective for their specialized requirements. So the idea of the helicopter

recent attempts to revive the autogyro, commonly in the form of ultra-light single-seaters designed principally for amateur flying. There has been some progress in this field, though not without its problems, and the popularity of these diminutive rotorcraft has been spasmodic.

There is no reason why the autogyro could not be developed to a standard comparable in its own way with that of the present-day helicopter. It would never, however, be capable of sustained hovering flight, and it has always been precisely this facility from which the helicopter has derived its operating superiority.

Further design refinements, mainly to increase reliability, component life and comfort can be expected. Hingeless rotor systems such as in the new Westland Lynx and MBB BO105 are examples, as is the new nodal-

gunship was born, the best known example of which is probably the Bell AH–1 Huey-Cobra. This design uses all the dynamic components of the latest Bell UH–1 utility helicopter in a slender, streamlined fuselage to give it a speed of up to 354km/h (220mph). It carries a crew of two, pilot and air gunner/observer, and can be armed with a range of missiles, rockets, rapid-firing machine-guns and other weapons. The type was in quantity production before the end of the Vietnam war and proved extremely effective on active service.

After Vietnam, attention was once more turned towards civil applications, the most significant current trend being the design of light and medium-sized twin-turbine helicopters able to operate in full Instrument Flight Rules (IFR) conditions. This has entailed the development of special instrumentation, autostabilizers and navigation aids and has resulted in the production of such sophisticated executive transport helicopters as the West German five-seat Messerschmitt-Bölkow-Blohm (MBB) BO105, the Italian eight-seat Agusta A 109 or the American ten-seat Bell 222. The twin-engine reliability common to all three is essential to meet the requirements of IFR operation in air traffic control zones and also to fly into heliports located at the centre of populous areas. All three follow the classic MTR configuration and all are capable of speeds of about 240 km/h (150mph) or better.

In 1957 Mikhail Mil's bureau in the Soviet Union produced a pioneer twin-turbine helicopter, the Mi–6, which was far bigger than any predecessor. With a 35m (114ft 10in) five-blade rotor driven by two 5,500hp D–25V engines, it could carry 90 people or 12,000kg (26,455lb) internally. Mil then used two Mi–6 engine/rotor groups in the monster V–12, but this was clumsy. Instead his bureau produced the Mi–26, today's most capable helicopter. A conventional MTR design, this has an eight-blade main rotor of 32m (105ft) diameter, driven by two 11,240hp Lotarev D–136 engines. This great machine has a flight deck arranged for two pilots, navigator, engineer and loadmaster. In 1982 an Mi–26 lifted a cargo load of 25,000kg (55,115lb) to 4,100m (13,451ft). Normal gross weight is 56,000kg (123,450lb) and maximum speed 295km/h (183mph).

The Soviet Union has made great efforts with helicopters, partly to open up virgin territory and partly for combat missions. The Mil bureau has produced well over 10,000 of the Mi–8 and –17 family, each much bigger and more powerful than the Sea King, and several thousand Mi–24 ('Hind') battle helicopters bristling with sensors and weapons. The new Mi–28 ('Havoc') is in production, and the rival Kamov bureau has followed its naval Ka–25 and Ka–27 CXR families with the fast 'Hokum' gunship.

Right at the other end of the size range, there should also be mentioned the many

beam suspension system being introduced by Bell to reduce the effects of unavoidable vibration. Some marginal increases in cruising speeds may also be introduced to advantage. The use of 'exotic' materials will certainly increase. In the early days, rotor blades were manufactured on a tubular spar with wooden ribs and stringers, and covered in fabric. Today, extruded alloy spars are used, with stainless steel and glass-reinforced plastics also embodied. Titanium is beginning to replace steel for certain rotorhead components, and many more developments along these lines are on the way.

Today's helicopter speed record stands at 400.87km/h (249.09mph), set by a Westland Lynx. Yet the helicopter remains a fundamentally slow vehicle, inefficient in cruising flight. Perhaps half the helicopter market will eventually be taken by tilt-rotor machines, as described in the next section. A possible alternative is the stopped-rotor, being tested by the Sikorsky S–72X1, which by blasting compressed air from slits along each blade enables each blade to 'fly backwards' and thus be brought to a halt in flight. With the rotor stopped the S–72 can become a 582km/h (361mph) jet aeroplane, supported by a small wing.

TOP LEFT: *EH 101 three-engined helicopter, being developed by Britain and Italy.*
ABOVE LEFT: *The Soviet Mil Mi-26 is the biggest helicopter in the world.*
ABOVE: *The Bo-105 hingeless rotorhead system.*

Rotorhead Design

The heart of all rotorcraft, whatever the configuration, lies at the centre of the rotorhead. Through the geometry of this rotating mechanical assembly, all the power and all the major control functions are passed to the rotor system while back through it, as appropriate to the design, comes the principal response to aerodynamic effects.

All the basic principles of aerodynamics explained in the first chapter of this volume apply equally to the individual blades of the rotor system; but there is added complexity of effect due to the more complex motions of the blades with respect to the fuselage. In hovering flight, each blade is generating a constant lift proportional to the square of its own rotational speed throughout its 360° cycle of rotation. As soon as the machine begins to move forward, however, each blade produces lift proportional to the square of its rotational speed plus the forward speed on one side of the rotor and minus the forward speed as it passes round on the opposite side. The two sides are termed the advancing and retreating sides of the rotor disc.

Lift asymmetry between the advancing and retreating sides, which would obviously cause lateral instability, is avoided in most rotorhead designs by the inclusion of a flapping hinge on the hub, inside the blade root attachment. This allows a blade to flap upwards in response to its increasing lift as it comes into the influence of the higher-velocity airflow on the advancing side. This upward movement reduces the blade's angle of attack and, consequently, its lift. The reverse occurs as it passes round to the retreating side and flaps downward again in response to the lower-velocity air, increasing angle of attack and lift. Blade lift is thus automatically equalized between the two sides of the disc. In Cierva's original autogyros, plain steel bearings were used for the flapping hinges. Later, roller bearings took their place while, in the latest helicopter designs, elastomeric bearings needing no lubrication are used. The equivalent effect in so-called rigid-rotor helicopters is achieved through the medium of a flexible section at the root end of the blade spar itself.

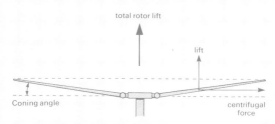

Equalization of lift over the swept area is also influenced by the rotor control system. In flight, the blades are held in position between the effects of lift and centrifugal force, the latter being approximately five or six times greater than the former. The overall rotor lift can be taken to act through a point at the hub centre and control of flight attitude is effected by means of tilting the rotor so that its overall lift vector moves away from the vertical, remaining perpendicular always to the plane of the spinning blades. The helicopter thus tilts with the rotor and then begins to move into the direction of tilt.

Tilting the rotor for control purposes is achieved either by mounting the hub on a universal joint and tilting it through a lever mechanism connected to the pilot's controls, or by fitting a universally mounted swashplate below the hub which is likewise connected to the pilot's controls. In this case, pitch-change bearings (to allow the blades torsional freedom about their longitudinal axes) are fitted in the blade-spar roots and connecting rods link the tiltable swashplate to pitch-control arms on the blades.

In the case of the tilting-hub method, as used in the early autogyros, a forward tilt of the hub introduced by pilot control reduces the pitch angle of the advancing blade with respect to the fuselage because it inclines the flapping hinge axis. At the same time, the same flapping hinge inclination increases the pitch angle of the retreating blade. A forward rotor tilt thus follows the forward hub tilt because the blades are made to fly low in the forward sector of the disc and high at the rear.

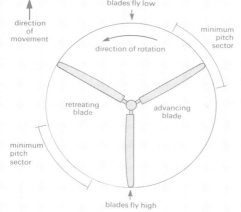

In addition, asymmetry of lift is influenced by the input of such a control tilt because the magnitude of blade lift is dependent both on the pitch angle and on the air velocity. Since the reduction of pitch resulting from the forward hub tilt takes effect in the advancing sector of the disc, and vice versa on the retreating side, the advancing blade loses lift while the retreating blade gains lift from the effect of the hub inclination. The flapping effect and the forward tilt effect are therefore working in harmony to equalize blade lift on the advancing and retreating sides of the rotor. Although the blade pitch change introduced by the control tilt does minimize the flapping motions of the blades, the flapping hinges are nonetheless necessary as some residual flapping still occurs which would otherwise lead to undesirable blade-root stresses and vibration.

The same applies to the alternative method using the tilting swashplate, which is known as

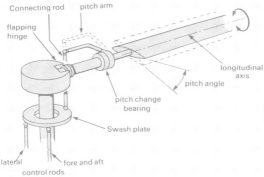

cyclic pitch control. A forward tilt of the swashplate causes the blade pitch arm connecting rods to be pulled down on the advancing side while, at the same time, they are being pushed up on the retreating side. This reduces the pitch angle of each blade with respect to the hub when advancing and increases it again when retreating, by twisting it about its pitch-change bearing. In this case, the effect on rotor tilt is accentuated. The cyclic variation of blade pitch is in the order of 10°, from a datum setting plus 5° on the retreating side to datum minus 5° in the advancing sector. Control movements to tilt either the swashplate or the universally mounted hub are made through a bearing assembly, since they are rotating with the rotorhead whereas the control-lever mechanism is mounted on the fuselage structure.

Use of the tilting-swashplate method also allows the introduction of what is known as collective pitch control. Its primary function is to provide for control of movement in the vertical sense at take-off and landing. Superimposed on the rotorhead cyclic pitch-control mechanism is another assembly which permits the swashplate itself to be raised or lowered, irrespective of its angle of tilt. Thus the connecting rods to the blade-pitch

arms can also be pushed up or down in unison, either to increase or decrease the pitch angle of all blades simultaneously. Associated engine-control linkages ensure that as the connecting rods move the blade-pitch arms upwards, increased power is applied to the rotor drive-shaft, and vice versa. This power control is often applied through the medium of an engine-speed governor, particularly with turbine-powered helicopters.

The range of collective pitch control available on a typical helicopter would be from 2° positive (the autorotative pitch) to about 14° positive. For take-off, the collective pitch lever would normally be raised to give a setting of some 10° on the blades, at which point the machine would lift off. To sustain forward flight, a continuous forward tilt is applied to the rotor through the controls; higher speeds are attained by increasing collective pitch and power. In this condition, the individual blades will be changing their pitch continuously at a rate of something like eight times per second, as they pass through their cycle of rotation. In the event of engine failure, the pilot action is to reduce collective pitch to the lowest value, whereupon the helicopter becomes an autogyro and glides gently downwards with the rotor in autorotation.

In the autogyro itself, blade pitch is normally fixed at the 2° setting and the rotor is not tilted forwards to provide propulsive thrust. This comes from the conventional propeller. In both the helicopter and the autogyro, lateral tilting of the rotor is used for sideways and turning control.

Reference to the section on aerodynamics will show that the reduction of blade lift resulting from its upward flapping motion is governed by the reduction in the blade's effective angle of attack. The reduced lift following a reduction of blade pitch angle represents a different means of achieving the same objective. The difference between angle of pitch and angle of attack is explained in the glossary.

There remains one further degree of freedom which may be allowed to the blades in some rotorhead designs. This is achieved by including drag hinges, also known as lead-lag hinges, which permit the blades limited motion with respect to each other in the plane of rotation. For a combination of aerodynamic and mechanical reasons, the blades tend to speed up very slightly as they pass through the advancing sector of the disc and slow down again as they traverse the retreating sector. These minor accelerations and decelerations with-

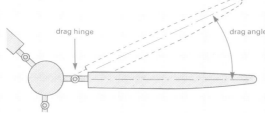

in the spinning rotor would give rise to vibrations if not suitably suppressed. Drag hinges are normally located between the flapping hinges and pitch-change bearings, but in some cases the flapping and drag hinges are coincident. Some designs dispense with drag hinges in favour of providing suitable stiffness in the blade spars themselves.

Rotors fitted with both flapping and drag hinges are known as fully articulated since the combination of the two constitutes the equivalent of a universal joint. Rotors with flapping hinges only are termed semi-articulated. The addition of hydraulically operated servo controls, in the larger helicopters, does not necessarily affect the geometry of the rotorhead, but other design refinements may do so. For example, if the flapping-hinge axis is designed to be set at an angle of less than 90° to the blade's longitudinal axis, when the blade flaps upwards the motion will also have in it an element of pitch reduction. The so-called rigid-rotor helicopters, which at the time of writing comprise only a small proportion of all rotorcraft, would be better described as hingeless-rotor machines. They function on much the same principles, using flexible members in place of hinges in order to reduce complexity.

V/STOL

The introduction of the turbojet engine, towards the end of World War II, came at a time when every available effort was being made by both sides to channel this new invention into a fighter aircraft. As events transpired, neither side was able to bring the first jet fighters into service until the war was virtually over.

After hostilities ceased, engine designers had more time to think of other future possibilities and vertical take-off, for both military and civil aircraft, was one obvious application. Even the early jet engines could develop a thrust of higher value than their own weight. By the time the Rolls-Royce Nene turbojet came into service, in the early 1950s, its maximum thrust of some 1,814kg (4,000 lb) was roughly twice its own weight. There was thus ample margin for the additional weight of a supporting framework.

To evaluate the feasibility of controlling such a concept, Rolls-Royce mounted two Nenes horizontally in a tubular-steel engine test-bed and modified the engine tailpipes to direct the jet efflux vertically downwards. When it first flew, in 1953, the four-legged framework created a worldwide sensation and was promptly dubbed the 'Flying Bedstead'.

Gross take-off weight of the machine was 3,264kg (7,196lb). Maximum vertical thrust from the two Nenes was about 3,629kg (8,000 lb) so there was no doubt about its vertical take-off capability. The main experimental purpose of the test rig was to evaluate the system devised for attitude control in hovering flight. Compressed air was bled from the two engine compressors, at 3·2kg (7lb) per second, into a common collector box and then ducted into four downward-facing nozzles, one positioned forward, one aft and one at either side.

The fore and aft nozzles each had a 10cm (4in) diameter orifice and produced some 132kg (290lb) average thrust at a 4·27m (14 ft) arm from the centre of gravity. To control pitch attitude, their thrust could be varied differentially by diverter valves connected to the pilot's control column. The nozzles could also be swivelled differentially from side to side by the rudder pedals, to provide control in yaw. The lateral nozzles were approximately half the diameter and half the distance from the centre of gravity, producing some 15·9kg (35lb) thrust each. This, too, could be varied differentially by lateral movement of the pilot's control column. The aggregate of all the control thrust remained constant and continuous during hovering flight and so it contributed slightly to vertical lift. After a few inevitable teething troubles had been overcome, the system worked reasonably well. About 380 tethered flights and 120 free

ABOVE: *The Rolls-Royce Flying Bedstead. The tubular steel pylon above the pilot formed no part of the mechanical design but was added solely for his protection in the event of a crash landing and subsequent overturn.*
BELOW: *The world's first multi-jet VTOL aircraft was the Short SC.1 delta wing research project. It provided valuable knowledge of automatic control systems.*
TOP RIGHT: *First of the vectored thrust VTOL aircraft was this Hawker Siddeley P.1127 prototype. From this design, the Kestrel was later developed.*
ABOVE RIGHT: *Hawker Siddeley Kestrels in formation. A batch of nine were produced for evaluation by a specially formed tripartite squadron of British, American and German pilots.*
RIGHT: *This French VTOL research aircraft was a modified version of the Dassault Mirage fighter. It was designated Mirage III–V. It remains the only VTOL aircraft to have exceeded Mach 2.*

flights were made during the comprehensive tests that followed.

The success of the 'Flying Bedstead' conjured up futuristic visions of Vertical Take-Off and Landing (VTOL) airliners operating direct between the centres of the world's capital cities, without the need for aerodromes and all the travelling delays associated with them. It was the fixed-wing aircraft designers' answer to the helicopter, then still in the throes of demonstrating its basic practicability. The vision spurred Rolls-Royce into the development of a special series of lightweight turbojet engines with a remarkably high power:weight ratio; in the first of the series, the RB.108, it was 8:1. The RB.108 was rated at 1,002kg (2,210lb) maximum take-off thrust for a basic dry weight of only 122kg (269lb). It was designed specifically for a vertical-lift application with provision for deflecting the angle of the jet efflux by a few degrees to aid control when hovering in flight.

The first use made of this power unit was by Short Brothers & Harland in the SC–1 delta-wing research aircraft. Powered by five RB.108s, four mounted vertically in two pairs for lift and one mounted horizontally for propulsive thrust, the SC–1 became the world's first VTOL aircraft with separate lift and cruise engines to achieve vertical take-off, in 1958. The first full transition – from the hover to fully wingborne forward flight and then back to the hover sustained only by the lift jets – was made in 1960.

Provision was made in the SC–1 for the

two pairs of lift engines to be swivelled fore and aft of the vertical. After take-off, they were swivelled aft a few degrees so that their thrust supplemented that of the propulsive engine in the transition to forward flight. On the approach to land, the lift engines could be swivelled forward a few degrees to provide a braking effect in the transition back to hovering flight. All five engines were fitted with a compressor bleed from which high-pressure air was fed into a common duct for hovering control. The duct system terminated in four small ejector nozzles, one at each wing tip, one at the nose and one at the tail, similar to the system used on the 'Flying Bedstead'.

The SC–1 was intended as the forerunner of a single-seat VTOL fighter which would use a more powerful version of the RB.108 engine for propulsive thrust. It was also intended as the scaled-down prototype of a much larger VTOL airliner. There was, however, another jet-lift system under concurrent development by Hawker and this, known as the vectored-thrust system, eventually proved superior.

Based on the Bristol Siddeley BS.53 Pegasus turbofan engine (see Chapter 4 for details), the vectored-thrust system was first tried in the Hawker P.1127, a developed version of which prototype was later named the Kestrel when it joined a trinational evaluation squadron of British, American and West German composition. Whereas in the Short SC–1 the greater part of thrust available was shut down during wing-borne flight, in the

vectored-thrust system of the Kestrel it could all be used to attain much higher forward speeds. The efflux from the turbofan engine was ejected through four swivelling nozzles, two forward and two aft on either side of the fuselage under the wing. For vertical take-off and landing, the nozzles were swivelled to direct the jets vertically downward, while for forward flight the nozzles were swivelled to the rear to give horizontal thrust.

First hovering trials of the P.1127 began in 1960 and the first full transition from vertical take-off to conventional forward flight and then back to vertical landing was achieved in the following year. By 1964, the aircraft had attained supersonic speed in a shallow dive and more powerful, truly supersonic, versions were under development.

Meanwhile, in Europe, the French company Marcel Dassault was developing a VTOL adaptation of one of its Mirage fighters. By replacing the SNECMA Atar turbojet with a smaller Bristol Siddeley Orpheus turbojet, space was made in the fuselage for installing eight Rolls-Royce RB.108 lift jets, mounted vertically in four pairs. This aircraft, called the *Balzac*, began flight trials in 1962 and was the forerunner of the Mirage III–V prototypes, which in turn were de-designed to lead on to a Mach 2 VTOL fighter-bomber. The production version of the Mirage III–V was planned to have a SNECMA-developed afterburning TF–306 (based on the Pratt & Whitney TF30) as its much more powerful main propulsion unit, and eight Rolls-Royce RB.162 jets for verti-

cal lift. The RB.162 was a developed version of the RB.108 and produced just over twice the thrust, 2,109kg (4,650lb). Development of the Mirage III–V was however discontinued after an unfortunate series of accidents, despite the prototypes having made successful transitions from hovering to forward flight and vice versa, and having achieved supersonic speed.

The more powerful Rolls-Royce lift jets were also chosen by the German company Dornier, which produced an ambitious project for a VTOL transport aircraft. It was designed as a high-wing monoplane with a cruising speed of 750km/h (466mph). Collaborators in the project were Vereinigte Flugtechnische Werke (VFW). The prototype, designated Do31E, had two banks each of four RB.162s to provide a total lift thrust of some 16,000kg (35,274lb), mounted in wing-tip pods. This was supplemented at take-off by the vectored thrust of twin Bristol Siddeley Pegasus 5 turbofans which combined to add some 10,433kg (23,000lb) to

LEFT: *The Ryan XV–5A fan-in-wing research aircraft was one of the concepts developed in the American attempts at VTOL design.*
ABOVE RIGHT: *The Lockheed Hummingbird, designated XV–4A by the US Army, employed an induced flow principle to achieve VTOL characteristics.*
RIGHT: *The Ryan X–13 Vertijet had no landing gear but took off from and landed on this adjustable hydraulic ramp. It proved impracticable.*

the static lift thrust. The production version was planned to have still higher take-off thrust, with ten RB.162 lift jets, and its gross weight was to be about 23,500kg (51,808lb).

Control in hovering flight was by a combination of differential thrust on the lift jets, for lateral control, and small 'puffer' nozzles at the tail for fore-and-aft control. Two of the tail nozzles were directed downwards and two upwards, all four being fed by ducted high-pressure air bled from the lift engines. Control in yaw was by differential inclination of the lift-engine nozzles.

Another German project, developed by the Entwicklungsring Sud research group formed by Bölkow, Heinkel and Messerschmitt in 1960, adopted yet another design configuration. In this project designated VJ–101C, a small high-wing monoplane of 6,010 kg (13,250lb) gross weight had six Rolls-Royce RB.108 lift jets installed in three pairs. One pair was mounted vertically in the fuselage, immediately abaft the pilot's cockpit, while the second and third pairs were in swivelling wing tip pods. For vertical take-off, all six lift jets were used, with the wing tip pods swivelled into the vertical position. To make the transition into forward flight, the wing tip pods were swivelled forwards through a 90° arc until their efflux provided horizontal thrust. As forward speed increased, with the wing taking over the lift function, the forward pair of lift engines was shut down. The prototype VJ–101C started flight trials in 1963 but the project was abandoned in the following year.

Concurrently, a variety of different projects was in the course of development in the USA. In 1963, Lockheed produced and flew its XV–4A jet-lift VTOL fighter prototype, powered by twin Pratt & Whitney JT12A–3 turbojets rated at 1,497kg (3,300lb) static thrust each. Named Hummingbird, the XV–4A was a mid-wing monoplane. The two engines, mounted horizontally in nacelles alongside the centre fuselage, were arranged to provide either horizontal or vertical thrust. For vertical take-off, the efflux from both engines was diverted through ducting into rows of downward-facing nozzles in the centre-fuselage compartment between the engine nacelles known as the nozzle chamber. Above and below the chamber, long doors in the upper and lower fuselage skin could be opened to allow free downward flow of the efflux, coupled with induced flow of ambient air through the upper doors to mix with and augment the jet efflux. The angle of the fixed nozzles was such that the aircraft hovered in a nose-up attitude. As in a helicopter, the application of forward control lowered the nose and thus introduced a rearward component of vertical thrust which imparted forward acceleration.

At approximately 145km/h (90mph), sufficient lift was being produced by the wing to allow the efflux from one of the turbojets to be diverted back to the direct propulsive function; conventional wing-borne flight was then established. This would be followed by diverting the second engine to direct horizontal thrust and closing the nozzle chamber doors. Design maximum speed was 837km/h (520mph). The system functioned reason-

ably well, but performance was not sufficient to warrant continuation of the project.

Ryan Aeronautical conceived a different way of achieving vertical take-off. Using what was known as the fan-in-wing system, the XV–5A prototype obtained its vertical thrust from two 1·59m (5ft 2½in) diameter, 36-blade lift fans mounted horizontally in the wings. A third, smaller, lift fan was mounted in the nose for control purposes. Around the periphery of each fan were fitted small turbine blades, or scoops, on to which the efflux from the twin General Electric J85–5 turbojets impinged to produce a fan speed of 2,640rpm at the full rated power of 1,206kg (2,658lb) static thrust on each engine.

Hinged semi-circular doors in the upper and lower wing surfaces above and below the fans were opened during the take-off. Below each fan there was also a series of transverse louvres, adjustable under pilot control. A thumb-wheel on the control column was used to rotate these louvres and so deflect the fan thrust rearwards to gain forward speed from

the hovering position. Differential rotation of the louvres could be demanded through the pilot's normal controls to provide roll and yaw control in the hovering and slow-speed phases. Conventional wing-borne flight was established in this aircraft at approximately 225km/h (140mph), whereupon the turbojet efflux was transferred from the peripheral scoops on the fans to normal tail-pipe ejection and the fan doors were closed.

Flight trials of the Ryan XV–5A began in 1964 but, although the concept proved practicable, it too was later abandoned. Ryan was also involved in the experimental development of a 'tail-sitter' jet aircraft, the X–13, which was designed to take off vertically from a gantry. This, however, proved impracticable for a number of reasons. More successful was the company's contribution, with Hiller Helicopters, to the Ling-Temco-Vought XC–142A tilt-wing project for a military transport. Powered by four 2,850shp General Electric T64–1 shaft turbines, driving 4·72m (15 ft 6in) diameter variable-pitch propellers, the XC–142A was designed to carry up to 7,112kg (15,680lb) of wheeled vehicles and other cargo in its 9·14m (30ft) long cabin. Its maiden flight was in 1964, but it too never entered production.

Other North American companies concerned with tilt-wing projects have been Boeing Vertol and Canadair, the latter company's twin-engined CL–84 having been longest in development. The use of ducted fans or ducted propellers has been yet another concept in the USA's search for a vertical take-off

LEFT: *Ling-Temco-Vought XC–142A military transport tilt-wing aircraft. The project was designed by a consortium of American companies, including Hiller and Ryan, for the US armed forces. It flew successfully on test but has not been built in quantity.*
BELOW: *Canadair's CL–84 tilt wing aircraft has a horizontal co-axial rotor system designed by Cierva Rotorcraft mounted abaft the rudder. It provides fore-and-aft control in the hover and is stopped and locked in cruising flight.*
BOTTOM: *US Marine Corps AV–8A Harrier. US manufacturing licencees for the type are McDonnell Douglas and the USMC designation is AV–8A.*
RIGHT: *Apart from the Harrier, the Yakovlev Yak-38 is the only jet-lift combat aircraft in the world. Single- and two-seat versions are used by the Soviet navy.*

capability. It has been proved feasible by Hiller, Piasecki and others, but is hardly practical for general service. Bell Aerosystems, an associate of the helicopter constructor, developed a project for an aircraft in which four ducted propellers mounted on two stub wings were capable of being tilted through 90° to provide either vertical or horizontal thrust. Powered by twin General Electric T58 shaft turbines of 1,250shp each, the machine was known as the X–22A. A similar aircraft, using four tilting propellers without the ducts, was built by Curtiss Wright and designated X–19A.

Between 1950 and 1970, the US military authorities sponsored the development of almost any apparently feasible VTOL system for practical evaluation. None of the methods, however, was as simple as the British vectored-thrust principle, so it has transpired that a developed version of the Kestrel became the first jet-lift aircraft to enter service. In fact it came about because of the abrupt cancellation in 1965 of the Hawker P.1145, an impressive supersonic fighter/attack aircraft powered by a vectored BS.100 engine of 15,000kg (33,000lb) thrust, which had been planned for the RAF and Royal Navy. In its place the government substituted a much smaller 'P.1127RAF', later named Harrier. This flew in 1966.

Despite its designation the Harrier was a totally new design, with a small wing with anhedral and 34° sweep. Thanks to development of the Pegasus to 9,752kg (21,500lb) thrust the Harrier could fly useful missions, always operating in the STOVL (short takeoff vertical landing) mode, and on rough surfaces making short rolling landings to avoid ingesting debris. It took years for people to comprehend that the greatest attribute of the Harrier was its survivability; in the event of war it did not need to be parked

Rockets and Guided Missiles

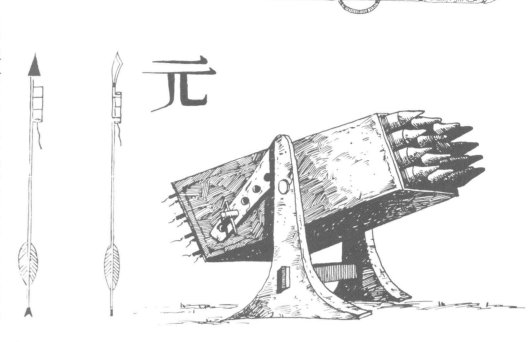

The Chinese are believed to have been the first to use missiles which fly by the reactive effect of their hot exhaust gases. Mongols besieging the town of Kai-fung-fu, north of the Yellow River, in 1232 felt the impact of 'arrows of flying fire' which the townsfolk launched in large numbers. The arrows were described as making a noise like thunder and travelling a great distance, spreading fire and destruction where they fell. There is no mention in the early literature of these fire-arrows being launched from a bow, and it seems likely that the Chinese stumbled upon the reactive effect after incendiary arrows (which had tubes of burning compounds bound to the shaft) were observed to fly faster and farther than conventional arrows.

Accounts exist of rocket-propelled fire-arrows being used in the siege of Siang-yang-fu in 1271, and the Mongols themselves seem to have introduced them into Japan. They were launched from Mongol ships during the

battle of Tsu Shima in 1274 and were used again during the land attack on Iki Shima.

The next five centuries saw rockets being used as fireworks and sometimes as weapons. One can trace the history of such devices – particularly fireworks – from China through the Mongolian intervention in Japan to Korea, Java and India.

Although mention of the use of saltpetre in explosive mixtures had been made in Arab literature of the 13th century, it was Roger Bacon, an English Franciscan monk, who established the proper mixture ratios of sulphur, saltpetre and charcoal which put 'gunpowder' on the map. Such refinements led, in about 1300, to the first crude firearms which used the principle of a tube, closed at one end, from which a ball was propelled by

the sudden explosive ignition of gunpowder.

For a long time the war rocket was relatively neglected. It was not until the latter part of the 18th century, when British troops in India came under fire from Indian-built rockets in the two battles of Seringapatam of 1792 and 1799, that impetus was given to further development.

The Indian rockets used by the troops of Haidar Ali, prince of Mysore, had iron tubes bound by leather thongs to a bamboo stabilizing stick. They weighed between 2·7 and 5·4kg (6 and 12lb) and travelled over 915m (3,000ft). The rockets bounced and skittered over the ground and were particularly damaging to the British cavalry.

Congreve rockets

It was this encounter with rockets in India that led to more serious work in England through the efforts of Col William (later Sir William) Congreve, who began with the aim of doubling the range of the Indian rockets. His 14·5-kg (32-lb) weapon was iron-cased with a conical nose cap, measured 107cm (42in) in length and had a diameter of 10cm (4in). It contained a 3·17kg (7lb) charge. The 4·57m (15ft) stabilizing stick was detachable. It slipped into metal straps on the rocket head.

Under Congreve experiments were begun at the Royal Laboratory at Woolwich to develop incendiary rockets for use against the French during the Napoleonic Wars. The first major engagement took place in 1806 when some 2,000 Congreve rockets were dis-

charged against the town of Boulogne. They were fired from 18 boats which had been quietly launched from parent ships and rowed into the bay.

The following year a still more spectacular rocket attack was launched on Copenhagen which left much of the town ablaze. Congreve rockets were in action again in 1813 at the battle of Leipzig, and later that year Danzig fell after a series of rocket attacks.

In the war of 1812 between Britain and the United States, Congreve rockets repelled American troops in the battle of Bladensburg (1814) and were effective against infantry, cavalry and ships. The sight of the rockets bombarding Fort MacHenry inspired Francis Scott Key to write 'The Star-Spangled Banner' which later became the American national anthem. Warheads were of two types, those which discharged carbine balls with a shrapnel effect, and those with an incendiary mixture which started fires. In the end a whole family of Congreve rockets had been produced ranging from 8·2, 10·9, 14·5 and 19kg (18; 24, 32 and 42lb) to 'blockbusters' of 45·4 and 136·2kg (100 and 300lb). Rockets were referred to as 'pounders' – the 'pounder' designation actually referred to the weight of a lead ball that would fit into the internal diameter of a rocket.

Although Congreve made considerable advances in manufacture, rockets were still relatively inefficient and lacking in accuracy. A major disadvantage was the long balancing stick which inevitably accounted for a

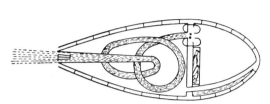

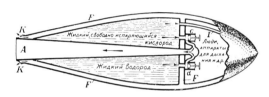

large part of the rocket's total dead weight.

William Hale, another British subject, eliminated the balancing stick by placing exhaust deflectors to impart spin at the nozzle end. These spin-stabilized rockets were used in various campaigns, notably by the Austrian rocket corps and the Dutch colonial service.

Some 2,000 were made in the US for use in the Mexican War. In the US Ordnance Manual of 1862 the 7·26kg (16lb) Hale rocket was credited with a range of 2km (1·25 miles).

Congreve also made spin-stabilized rockets later in his career. By the turn of the century, however, the black powder rockets were being outclassed by new developments in field artillery.

The rocket pioneers

The next big advances were of a more theoretical nature. In 1881 a Russian explosives expert, Nikolai Ivanovich Kilbalchich, was arrested and sentenced to death for his part in an assassination plot against Tsar Alexander II. While awaiting execution he sketched a design for a man-carrying platform propelled by gunpowder cartridges fed continuously to a rocket chamber. The chamber, mounted in a gimbal frame, could be swung manually on its axis to change the platform's path of flight as it rose into the air.

Kilbalchich's sketches were not discovered until after the Russian Revolution of 1917. In the meantime, a Russian genius of humble birth – Konstantin Tsiolkovsky – had greatly advanced rocket theory. By 1883 he was convinced that a rocket would work in the vacuum of space by the recoil effect of its exhaust gases. In 1903 he published his first treatise on space travel, in which he boldly advocated the use of liquid propellants.

During the early years of the 20th century Tsiolkovsky filled his notebooks with advanced ideas for rockets and with the arithmetic of space travel. He considered a whole range of liquid propellants including liquid oxygen and liquid hydrogen, liquid oxygen and gasolene or kerosene. He recommended controlling the thrust by regulating the flow of liquids by valves and proposed cooling the

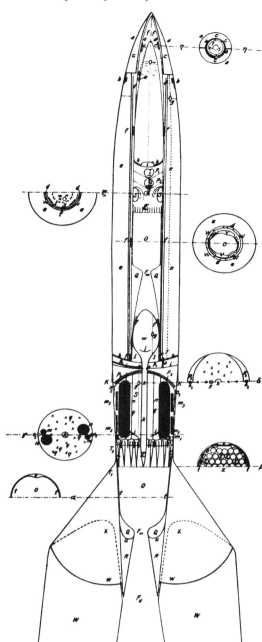

combustion chamber in which they burned by passing one of them through a double wall or jacket.

He proposed stabilizing rockets by the gyroscopic effect of a rotating flywheel and controlling their flight outside the atmosphere by means of rudders, or jet-vanes, working in the rocket's exhaust or by swivelling the nozzle. In this the Russian pioneer was anticipating *thrust vector control* in which a rocket is steered by changing the direction of its exhaust. In modern rockets this is either done by moving vanes in the exhaust gases, swivelling the rocket nozzle, or injecting fluid into a fixed nozzle from the side.

Tsiolkovsky also drew attention to the

ABOVE: *Robert H. Goddard was the great American rocket pioneer. The picture shows him loading a solid-fuel rocket of bazooka type in 1918. In 1926 he launched the world's first liquid fuel rocket.*
LEFT: *Modell B rocket to explore the upper atmosphere proposed by Hermann Oberth in 1923. Although never built, it inspired the rocket experimenters of the VfR ('German Society for Space Travel').*

ABOVE: *From this tower on 17 July 1929 Goddard launched a 'weather rocket' with a barometer, thermometer and camera. Instruments were recovered by parachute.*
BELOW: *By the mid-thirties Robert H. Goddard was already launching liquid fuel rockets at Roswell, New Mexico. This four-chamber L–7 rocket climbed 61 metres (200ft) on 7 November 1936.*

potential value of rockets built on the 'step' principle. The first step, or stage, would propel the top stages until its propellants were used up. The next stage would fire when the first had been separated, and so on until the final stage and its cargo (payload) would reach the speed required to complete the journey.

If speed is great enough and the rocket is flying parallel with the Earth above the atmosphere, Tsiolkovsky calculated, 'the centrifugal force cancels gravity and after a flight which lasts as long as oxygen and food supplies suffice, the rocket spirals back to Earth, braking itself against the air and gliding back to Earth. . . .'

Though he never built a rocket, Tsiolkovsky combined in these few ideas the basic requirements for large rocket missiles and space travel. His theories were soon to be corroborated by rocket pioneers in other countries, notably Robert Esnault-Pelterie in France (1913), Robert H. Goddard in the United States (1919) and Hermann Oberth in Germany (1923).

It was Goddard who launched the world's first liquid-propellant rocket at Auburn, Massachusetts, on 16 March 1926. Fuelled by gasolene and liquid oxygen, it flew for just 2·5 seconds and landed 56·08m (184ft) from the launch stand. Its average speed was 103km/h (64mph).

Next to succeed – but only just – was a German, Johannes Winkler, whose rocket made a brief hop of just over 3m (10ft) near Dessau on 21 February 1931. Three weeks later, however, the same rocket ascended 305m (1,000ft). It was fuelled with liquid methane and liquid oxygen.

Wernher von Braun

The year 1927 had seen the foundation of the *Verein für Raumschiffarte.V.(VfR)*, the German Society for Space Travel. For the rocket enthusiasts it was a time when small amateur groups could still make important contributions by building and firing rockets. The *VfR* concentrated on the improvement of liquid-fuel motors, launching its Mirak and Repulsor rockets from a converted ammunition depot in Reinickendorf – a Berlin suburb – which they called the *Raketenflugplatz* (rocket flying field).

Leading members of the *VfR* included Hermann Oberth, Walter Hohmann, Guido von Pirquet, Max Valier, Rudolf Nebel, Klaus Riedel, Kurt Hainish and Willy Ley. A youth of 18 who joined in 1930 was to become a legend in his own time: Wernher von Braun.

By now the liquid-fuel rocket had been firmly established and in August 1931 the *VfR* launched a Repulsor rocket to an altitude of 1,006m (3,300ft) which floated back to the ground by parachute. The experiments at the *Raketenflugplatz* began to attract international attention, but Germany was in the grips of an economic depression and membership of the *VfR* fell significantly. Money ran short and the Berlin authorities – quite understandably – objected to rockets being fired within the city limits.

It was clear that further progress would have to depend on gaining some kind of government support. Accordingly, in the summer of 1932, Nebel and von Braun set up a demonstration for the German Army near Kummersdorf. It was an appropriate moment. Hitler was a rising star (he came to power in 1933) and the Army recognized that rockets were outside the scope of the Versailles Treaty which forebade the manufacture of aircraft in Germany. By the end of the year work at the *Raketenflugplatz* had ended and von Braun had received an invitation to conduct experimental work for his doctor's thesis on rocket combustion at the Army's proving ground.

Other early experiments

By that time the *VfR* had fired the enthusiasm of people in other countries and March 1930 saw the foundation of the American Interplanetary Society (later the American *Rocket* Society), which carried out a wide range of experimental work with liquid-fuel rockets before World War II. The British Interplanetary Society, barred from making rocket experiments under the Explosives Act of 1875, nevertheless made a number of significant theoretical contributions to rocket flight, including the first engineering concept of a vehicle for landing on the Moon. The Society, founded in October 1933, is still active today.

Small groups of rocket enthusiasts (unknown to the Germans, Americans or British) also were at work in the Soviet Union and on 17 August 1933 a small rocket fuelled by liquid oxygen and solidified gasolene (gasolene and colophony, a dark-coloured resin obtained from turpentine) placed in the combustion chamber reached a height of 400m (1,312ft). This rocket, known as the GIRD 09, was designed by M. K. Tikhonravov and built by a team which included Sergei P. Korolev, an engineer who was later to achieve fame as the chief designer of Soviet spaceships.

TOP: *Mirak 2 rocket built by the* VfR *('German Society for Space Travel') in 1931. The rocket motor was inside the liquid oxygen tank, left, and the fuel in the tubular extension.*
ABOVE: *The automobile manufacturer Fritz von Opel experimented with rocket propulsion and flew this rocket glider at 153km/h (95mph) on 30 September 1929. It had a cluster of 16 solid fuel Sander rockets.*

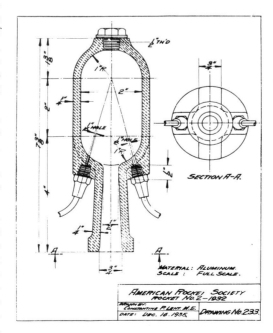

ABOVE: *Second rocket motor built by the American Interplanetary Society.*
RIGHT: *GIRD X was the first Soviet rocket to use a liquid oxidant and a liquid fuel. On 25 November 1933 it soared nearly 4,900m (16,076ft).*

How a Rocket Works

It is essential to appreciate from the beginning that a rocket is, quite simply, another form of heat engine. It burns fuel to produce heat and can, in its most simple form and without any moving parts, transform this heat energy into motion. In so doing, the rocket conforms to the principle of reaction, expressed by the scientist Sir Isaac Newton as his Third Law of Motion: every action produces an equal and opposite reaction.

This may, at first, seem difficult to understand, but one of the easiest ways to learn how a rocket works is to think about a rubber balloon. If it has been blown up, and the neck through which it was inflated tied to prevent air from escaping, then

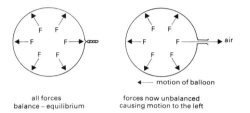

| all forces | forces now unbalanced |
| balance – equilibrium | causing motion to the left |

the balloon is full of air at an inner surface pressure equal to that of the atmospheric pressure on the outer surface.

So that the balloon can maintain its shape, without being distorted in any way by the atmospheric pressure, the air within the rubber envelope exerts an equal pressure on every point of the inner surface. It is thus in equilibrium:

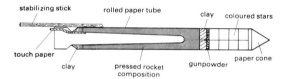

provided there is no leak, change of temperature or atmospheric pressure, it will maintain its form and volume.

If we now release the tie securing the neck of the balloon, we open a channel to atmosphere and the air rushes out. At all other points within the balloon air continues to press against the inner surface – only at the orifice of the neck has the air nothing to push against. The balloon is no longer in equilibrium and, if released, will fly off in a direction diametrically opposite to the point at which the air is escaping.

Perhaps you think that the escaping air is pushing against the atmospheric air that surrounds it. It is fairly easy to prove that this is not the case. At sea level under conditions known as the International Standard Atmosphere (ISA), the air has a pressure of $71\cdot8$kg/m^2 ($14\cdot7$lb/sq in) and a relative density of $1\cdot0$. At a height of 6,705m (22,000ft) the pressure has already fallen to only $30\cdot3$kg/m^2 ($6\cdot21$lb/sq in) and its relative density to less than $0\cdot5$. At 18,590m (61,000ft) the pressure is less than 5kg/m^2 (1lb/sq in) and the relative density is only $\cdot0897$.

As we climb higher into space, both pressure and density cease to have any significance, and this is a good point at which to carry out our balloon experiment yet again. This time it will need very little pressure to inflate it to the same size as at sea level, and when we release the tie at the neck we shall find that, despite the much reduced inner pressure, it will travel away from the air release point at an even greater speed. This is because, firstly, the air flowing from the balloon has no atmospheric pressure to slow its egress, which means that the 'opposite reaction' will be greater. Secondly, the bulky volume of the

balloon travelling through the air meets with no resistance from a dense atmosphere and can, therefore, move more easily.

All ordinary firework rockets work in precisely the same way: instead of air for propulsion, they burn within their casing combustible material which produces a large volume of gas. The rocket continues to react to the expulsion of this gas until all the gas producing materials are consumed. And because the rocket can carry its own oxidant (oxygen source) and fuel to burn with the oxygen, it is quite independent of atmospheric oxygen and can operate in the vacuum of space. In this respect it differs com-

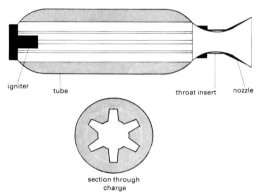

section through charge

pletely from the aero-engine which is an air-breathing engine, able to operate only within Earth's atmosphere, and only to altitudes at which it can obtain adequate supplies of atmospheric oxygen to permit combustion of its hydrocarbon fuel.

Weapon-carrying ballistic rockets usually burn solid propellants. Because it is more difficult to control the thrust of solid-propellant motors, liquid oxidants and fuels have proved better for space vehicles, since liquid-propellant motors can be shut down and re-ignited to permit positioning and control.

ABOVE LEFT: *Forces acting in a balloon.*
LEFT: *Firework rocket*
ABOVE: *Solid rocket motor with case-bonded charge.*
BELOW: *Pressure-fed rocket with regenerative cooling.*
BELOW RIGHT: *Pump-fed rocket with regenerative cooling.*

ABOVE: *The US Airforce Titan rocket.*

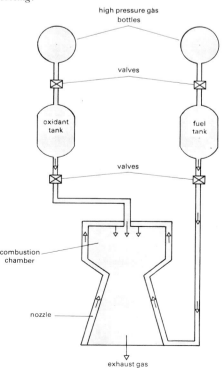

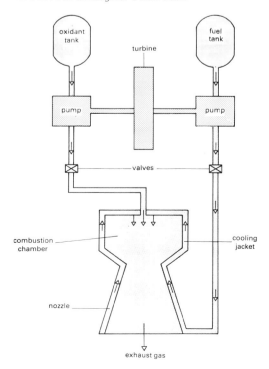

The V-2

The major advances in the 1930s, however, were taking place in Germany. In a remarkably short space of time the Army was making significant experiments in a bid to raise the efficiency of rocket motors. The work was under the direction of Captain Walter Dornberger (a qualified engineer) and Wernher von Braun. The first rocket Aggregate One, or A–1, was used for static tests only. It was just over 1·5m (4·9ft) long and had a diameter of 30·5cm (12in). The motor, fuelled by alcohol and liquid oxygen, was regeneratively cooled and had a design thrust of 300kg (660 lb). When ignited, the rocket blew up because an explosive mixture had accumulated in the combustion chamber, a problem which afflicted many of the early liquid-fuel rockets.

Success, however, was not to be long delayed. In December 1934, from the island of Borkum in the North Sea, two A–2 rockets fuelled by liquid oxygen and alcohol climbed to altitudes of about 2,400m (7,900 ft). Stability was maintained by the action of a 'brute-force' gyro set between the propellant tanks.

This double success led to more money becoming available to the researchers, who already had plans for a larger A–3 rocket with several important technical innovations. A three-axis gyro-control system worked exhaust vanes and fin-mounted rudders; a liquid-nitrogen pressure system replaced a thick-walled nitrogen bottle as a means of pressurizing the propellant tanks. New alcohol and oxygen flow valves were designed to eliminate ignition explosions. The rocket also had the virtue of not needing a launching ramp. Because of the new stabilization system, it could be launched vertically, 'standing on its fins'.

Although firings carried out from the Baltic island of Griefswalder Oie were disappointing – in the first, the recovery parachute opened prematurely only five seconds after lift-off – they were successful enough to show the merit of the basic design.

While this work was going on in Germany, Goddard was making similar progress near Roswell, New Mexico. On 28 March 1935, he fired a gyro-stabilized rocket to an altitude of 1,463m (4,800ft) and over a distance of 3,962m (13,000ft). The average speed was 885km/h (550mph). The flight path of the rocket was corrected by the gyro working steerable vanes in the exhaust. The rocket was 4·51m (14·8ft) long and had an empty weight of 35·61kg (78·5lb).

With the resurgence of German militarism, huge sums of money became available to von Braun's research team, from both the German Army and the Luftwaffe. This allowed work to begin in 1935 on a large experimental rocket establishment near the village of Peenemünde on the Baltic coast.

The major project was to be a large artillery rocket for which the designation A–4 had already been chosen. To achieve it, features of the design would be tested on a small-scale A–5 rocket. Launched at Griefswalder Oie in mid-1938, without the guidance system, the rockets reached heights of 12,875m (42,240ft). Over the next two years different control techniques were tried and

some A–5s, launched on inclined trajectories to achieve maximum range, were made to fly along radio guide beams to improve their accuracy.

Thus was born the concept of the A–4 rocket, designed to carry a 1,000kg (2,205lb) warhead over a distance of 275km (170 miles). This involved major advances in almost every department of rocket engineering. The A–4, which weighed more than 12,500kg (27,557lb), did not, however, achieve immediate success at Peenemünde. The first example, set upright on its fins on a small launch table, failed to lift off as the thrust of its engine gradually faded; it toppled over and exploded. The second A–4, held stable by its gyro-controlled exhaust vanes, flew perfectly for more than 45 seconds, then began to oscillate from side to side and finally broke apart in the air.

After necessary strengthening modifications had been made, the third A–4 made a perfect ballistic flight, reaching a maximum height of 85km (53 miles) before splashing into the Baltic some 190km (118 miles) from the launch pad. The date was 3 October 1942 – World War II had already been in progress for three years. In the eyes of the German High Command, the rocket had now become something far more important than a mere artillery weapon. Before adequate trials could be carried out, Hitler ordered the A–4 into large-scale production as a means of bombarding London and Britain's Home Counties.

Mobile A–4 batteries were organized and

TOP LEFT: *Forerunner of the German A–5 and A–4 (V–2) rockets was the A–3 static-tested at Kummersdorf in 1936. When launched in 1937 at Griefswalder Oie, a tiny island in the Baltic, three rockets went out of control and crashed soon after lift-off due to troubles with the gyro-control system. However, they gave the Germans experience with three-axis gyros and exhaust rudder control.*

TOP RIGHT: *First A–4 (V–2) rocket to perform successfully is prepared for launching at Peenemünde. On 3 October 1942 it travelled 190km (118 miles) reaching a peak altitude of some 85km (53 miles). Two others had misbehaved. In the spring of 1942, the first toppled over on the launch pad and exploded. Four weeks later the second made a smooth take-off, penetrated the so-called 'sound barrier' but then went out of control.*

ABOVE: *A–4 rocket being raised into the vertical firing position on the Meillerwagon transporter-erector. After a hurried test programme at Peenemünde, the first production models appeared early in 1943. In June 1944, an A–4 fired from Peenemünde went off course and crashed in Sweden. The debris, brought to England, gave vital clues to the rocket's capability.*

rockets were deployed at coastal sites. The A–4 attack on London began on 8 September 1944, from a site near the Hague in Holland; and at last the Goebbels propaganda ministry was able to reveal the existence of the V–2–Retaliation Weapon No 2 – against which there was no available defence.

Some 4,320 V–2s were launched between 6 September 1944 and 27 March 1945. Of these some 1,120 were directed against London and the London Civil Defence area, killing 2,511 people and seriously injuring nearly 6,000 more.

When the war in Europe ended, the full extent of the V–2 operation was at last open to inspection, including the extensive underground factory in the Kohnstein Hills, near Nordhausen, where V–1 flying bombs and V–2 rockets had been made on a production line basis. The principals at Peenemünde, in the face of the Russian advance, had fled westward, preferring to place themselves in the hands of the Western Allies. General Dornberger and von Braun surrendered to the welcoming and open arms of the US 7th Army on 2 May 1945.

Wernher von Braun's interrogation report showed the depth of progress that had been achieved by the Peenemünde team, and that it had kept alive the spaceflight ideal despite the pressures of the military establishment. In order to improve the range of the V–2, test

rockets had been fitted with swept-back wings to see if they would glide in the upper atmosphere. One had flown with some degree of success in 1944, and drawings existed for a version in which a pressure cabin replaced the warhead so that a man could fly in space. Even more ambitious was the A–9/A–10 project in which an improved A–4, with dart-like wings, was to be launched in flight from the nose of a huge liquid-fuel booster to fly some 5,000km (3,100 miles).

Although the scheme never got beyond the design stage in 1942, it was a clear indication of the shape of things to come. In his interrogation report, von Braun described a coming era when multi-stage rockets would launch artificial satellites, allow the construction in orbit of giant space platforms and put men on the Moon. Within a quarter of a century, he would be making his own prophecies come true.

Other war-time developments

The V–2 was only part of the story, for the arts of war had taken steps in many other directions. Much of the rocket work in Germany had been devoted to countering Allied air raids on the homeland though, in practice, few of the new weapons saw service by the war's end.

From Peenemünde came the ground-to-air missile Wasserfall (Waterfall), which outwardly resembled a scaled-down V–2 with

stubby cruciform wings. The need to keep the missile fully fuelled and ready for operations, however, led to the use of Visol, a by-product of petroleum, as the fuel which ignited in the combustion chamber on contact with the oxidant, nitric acid. Like the V–2, Wasserfall took off vertically from a simple platform.

This supersonic missile was expected to intercept bombers at a slant range of 48km (30 miles) and at an altitude of 9,650m (31,650ft), steering being effected by graphite vanes acting in the exhaust in combination with aerodynamic rudders. Guidance depended on the missile centering itself in a radar beam directed at the target. Early trials, however, were restricted to visual sighting and radio control by 'joystick' from the ground. It was proposed to fragment the entire rocket body from the detonation of 90kg (200lb) of high explosive to achieve a devastating shrapnel effect.

Another promising German ground-to-air missile of the period was the Hs–117 Schmetterling (Butterfly), which had the appearance of a small swept-wing monoplane. Ramp-launched, it was accelerated into flight by solid-fuel rocket boosters mounted above and below the fuselage. One version used a BMW liquid-propellant rocket motor of 159 kg (350lb) thrust. It flew at subsonic speed and was expected to make its kill at heights up to 8,850m (29,050ft), again under joy-

BELOW: *A requirement to increase the range of the V–2 led to the A–4b. This was basically a strengthened V–2 fitted with sweptback wings and enlarged aerodynamic rudders. From this von Braun believed a spaceplane could be developed to carry a pilot 644km (400 miles) in 17 minutes.*

BELOW RIGHT: *The first successful launch of an A–4b took place at Peenemünde on 24 January 1945. This was probably the first winged guided missile to fly faster than sound. However, on the downward leg of the trajectory, a wing broke off and the extended-range experiment was frustrated.*

BOTTOM: *Also developed at Peenemünde was the Wasserfall surface-to-air missile. Based on the aerodynamics of the A–5 and its A–4 big brother, it used storable propellants which ignited on contact (nitric acid and Visol, a petroleum by-product). It came too late to be used in the war.*

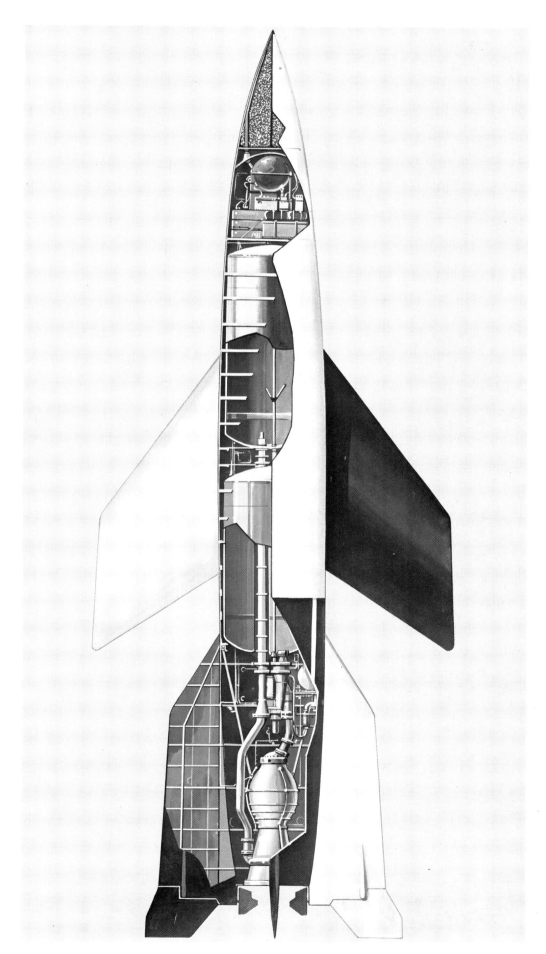

stick control by the ground operator.

The techniques of control were highly advanced for their day. Not only did Schmetterling have a kind of Machmeter to control the flight velocity but self-homing systems were under development. One was an infrared device with a 20° cone of sensitivity; a gyro-stabilized iconoscope ('electric eye') was designed to operate in conjunction with a proximity fuze using the radar principle.

The Russians, meanwhile, had made important advances of their own. Before World War II, Korolev had worked on a rocket-powered flying bomb (Project 212) following experiments with a rocket-powered glider. Ramp-launched from a rocket-propelled carriage, the small monoplane was designed to carry a 30kg (66lb) warhead a distance of some 50km (31 miles). Its ORM–65 rocket engine, designed by Valentin Glushko, burned kerosene and nitric acid.

Although test launchings got underway in 1939, the flying bomb was not put into production. Korolev had been working on a much improved cruise missile (Project 212A) which embodied some very advanced features of rocket propulsion and guidance, but the war caused such projects to be abandoned in favour of less sophisticated weapons such as the Katyusha barrage rockets and air-launched rocket bombs.

Initial post-war developments

At the end of the war, Germany stood head and shoulders above all others in the technology of rockets and guided weapons, and major efforts were made by America and Russia to round up as many rocket specialists as could be found. Many of the principals, including von Braun and a large part of his Peenemünde team, were encouraged to continue their work in the United States. Others were taken to Russia and put to work in collectives which worked in isolation from mainstream Soviet technology so that secrets could be preserved.

In the meantime, work in America, spurred by the pioneering efforts of Goddard (who died in August 1945), had led to a research group at the California Institute of Technology to developing in 1944 to 1945 a small rocket called Wac-Corporal. It was powered by a nitric acid/aniline engine of 680kg (1,500lb) thrust. Launched from a tower by a solid-fuel booster, the rocket sent a payload of instruments to an altitude of 70km (43·56 miles).

Like some of the German anti-aircraft missiles and the rocket motors developed in Russia the propellants were self-igniting and of a type which could be stored in the tanks of a rocket. This was to lead to a series of missiles for the US Army which could be held at instant readiness. It was also to result in experiments in which Wac-Corporals mounted as second stages of modified V–2s were to show the prospect for reaching the greater speeds and altitudes required for long-range ballistic rockets and space travel. On 24 February one such vehicle, blasting off from White Sands, New Mexico, sent the smaller Wac to a record height of 393km (244 miles). The rocket's maximum speed was 8,286km/h (5,150mph). Others of the series

were launched from Cape Canaveral by the US Army Ordnance to obtain maximum horizontal range.

In the Cold War era the rocket age was exploited with a new frenzy. Guided missiles of many kinds were developed in both offensive and defensive roles, with the major powers striving to match blow for blow each new technical advance.

Surface-to-air missiles

One of the first operational missiles to employ storable liquids was America's MIM–3A Nike-Ajax, the ramp-launched anti-aircraft missile which took off with the assistance of a three-finned solid-fuel booster. The missile itself had triangular cruciform wings and steerable nose vanes and was powered by a nitric acid/aniline motor. Guidance was by radio command. One radar tracked the target aircraft and another tracked the missile. A computer continuously integrated the two plots and caused steering signals to be sent to the missile, vectoring it towards the

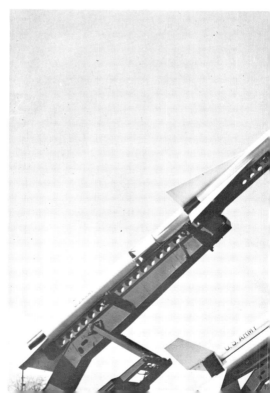

TOP ROW

FAR LEFT: *Captured German V–2 carries an American Wac-Corporal rocket as second stage: White Sands, New Mexico, February 1949.*

THIRD LEFT: *Early version of the US Army's Nike-Ajax surface-to-air missile on test at White Sands, New Mexico.*

SECOND LEFT: *US Navy Talos surface-to-air missiles on shipboard launcher.*

LEFT: *Bomarc surface-to-air missile of USAF Air Defense Command which could attain Mach 4 speeds and altitudes above 30,480m (100,000ft).*

MIDDLE ROW

FAR LEFT TOP: *Members of a Soviet SA–4 Ganef surface-to-air missile battery train in conditions of mock gas attack.*

FAR LEFT BOTTOM: *Soviet training exercise with SA–2 Guideline surface-to-air missile.*

CENTRE: *The Russian SA–5 Gammon high-altitude air defence weapon.*

SECOND RIGHT TOP: *Seacat, Britain's close-range anti-aircraft missile is launched from the Swedish ship* Sodermanland.

SECOND RIGHT BOTTOM: *Masurca, an early French surface-to-air missile, was first deployed on the frigates* Suffren *and* Duquesne.

BELOW: *Bloodhounds of No 30 Surface-to-Air Missile (SAM) Squadron at RAAF Base, Williamtown, New South Wales, Australia.*

BOTTOM ROW CENTRE: *US Army's earliest family of air-defence missiles: left to right, Nike Ajax, Nike Hercules and Nike Zeus.*

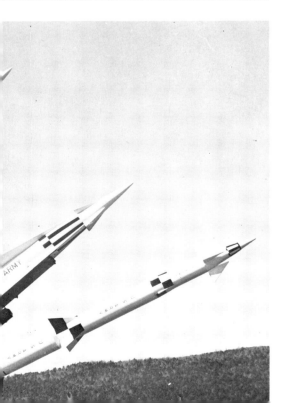

target. When fairly close, the missile became self-homing, responding to ground-radar reflections from the target.

Its Russian counterpart, (SA–2, NATO code-name Guideline – Russian missiles are identified by their NATO code names throughout) burned nitric acid and a hydrocarbon fuel and was also guided by automatic radio command. Evidence of the effectiveness of this weapon came on 1 May 1960, when it brought down Lt Francis Gary Powers in his high-flying U–2 reconnaissance aircraft.

Other anti-aircraft weapons followed in quick succession. In America the MIM–14A Nike-Hercules gave more height and range, as did the LIM–49A Nike-Zeus, which had a small nuclear warhead. Bomarc, an aircraft-like missile with cropped delta wings, had twin underslung ramjets and a rocket booster. It could reach a speed twice that of sound and heights of 18,288m (60,000ft) and had a range comfortably exceeding 161km (100 miles).

Russia's response to Nike-Zeus was the SA–5 Gammon, claimed to be effective against both high-flying aircraft and missiles. Anti-aircraft missiles for the Soviet Fleet were the SA–N–1 Goa and SA–N–2 Guideline. US naval weapons in the same class were Talos, Tartar and Terrier.

Britain, too, gave priority to anti-aircraft defence. Early developments were the solid-fuel Thunderbird and the ramjet-powered Bloodhound, both of which responded to semi-active radar homing. For air defence at sea there were the close-range Seacat, the longer-range Seaslug and, later, Sea Dart. From France came Masurca, a two-stage solid-fuel missile which first appeared as a beam-rider and then with self-homing capability. A missile which depends on semi-active radar homing incorporates a seeker which locks onto radar reflections from a target which is being illuminated from another source, e.g. a ground radar or, in the case of an air-to-air missile, the fire-control radar of the launch aircraft. *Active radar* hom-

BELOW: *A requirement to defend small infantry units in the field from low-flying aircraft led to the Blowpipe, a British man-portable supersonic weapon. When Blowpipe is fired a sliding ring tail fin assembly locks into place at the rear of the missile as it leaves the launcher. It is steered from radio command signals by thumbstick control.*
BOTTOM: *The SA–7 Grail is Russia's equivalent of the US Army's Redeye. It is used to defend small units against attack from low-flying aircraft and helicopters. It is also used by communist-supported guerrilla forces.*

BELOW: *Crotale anti-aircraft missiles. This highly compact French all-weather anti-aircraft system was the outcome of a South African requirement under the name Cactus. A similar system has been adopted by the French Air Force for the defence of airbases against fast low-flying strike aircraft. A full Crotale system comprises up to three wheeled vehicles. Two are combined launch and command vehicles and the other a surveillance radar vehicle. Missiles have an effective range of 500m to 8·5km (0·5 to 5 miles) against aircraft diving supersonically at ground targets.*

BOTTOM: *A particularly effective Soviet low-altitude anti-aircraft system is the SA–3 Goa. Operating with X-band fire control radar NATO code name Low Blow, it is associated with Flat Face radar which picks up approaching targets. With the SA–2 Guideline the weapon has been widely used in South-East Asia and the Middle East. SA–N–1, the naval version, has been widely deployed by the Soviet Fleet on Kashkin, Kotlin, Kresta 1 and Kynda-class ships using different target acquisition and fire control radars. It is being superseded by the SA–N–3 Goblet.*

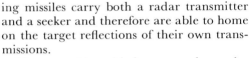

ing missiles carry both a radar transmitter and a seeker and therefore are able to home on the target reflections of their own transmissions.

Man-portable guided weapons have also made their mark for the self-defence of Army units against low-level air attack. America's tube-mounted Redeye, launched from the shoulder, was designed to home automatically on the engine heat of an attacking aircraft. Russia's equivalent – the SA–7 Grail – is believed to have a slant range of 2·9 to 4·0km (1·8 to 2·5 miles). Britain has produced the Blowpipe, a similar weapon but which the operator steers along his line of sight by radio signals.

The post-war years also saw a major growth in arms exports as the leading powers manoeuvred on the chessboard of world influence. The Russian surface-to-air missile SA–2 Guideline, as well as becoming established in the Warsaw Pact countries, was supplied to Cuba, Indonesia, North Vietnam and certain Middle East countries. A Mk 2 version supplied to Egypt was in action against the Israeli Air Force during the 1967 Six-Day War.

In the Yom Kippur War of 1973, Israeli pilots faced Soviet anti-aircraft weapons deployed in depth. While batteries of SA–2 Guideline missiles engaged their aircraft at long range, SA–3 Goa took on low-flying targets. A relatively new anti-aircraft missile, the SA–6, was fired from triple mounts on tracked vehicles which advanced with the Egyptian infantry, giving them effective cover. Of 104 Israeli aircraft lost in 1973, 100 are said to have fallen to missiles.

Significant Missiles of the Post-War Era

Abbreviations: H.e., high explosive (chemical); KT, kiloton, the explosive power, or yield, of 1,000 tons of TNT; LOS, line-of-sight; MT, megation, equivalent to the explosive power of 1,000,000 tons of TNT; MIRV, multiple, independently targetable re-entry vehicle; NA, not available; RV, re-entry vehicle; SAC, semi-active command; SAR, semi-active radar; SD, snap down; SU, snap up; TTR, target-tracking radar; TVC, thrust vector control. Data shown in *italics* provisional.

Surface-to-air missiles

Designation/Country	Length m/f	Diameter cm/in	Span cm/in	Weight kg/lb	Range km/miles	P, Propulsion; G, Guidance; C, Control; W, Warhead
Crotale (France) Tube-launched from tracked AFV	2·89 (9·4)	15 (5·9)	54 (21)	80–85 (176–187)	0·5–8·5 (0·3–5·3)	**P:** Solid rocket. **G:** Radio command. **C:** Cruciform canard vanes. **W:** H.e., fragmentation, IR proximity fuze.
Roland I (Germany) * Radar, Roland II.	2·40 (7·9)	16 (6·3)	50 (20)	65 (143)	0·2–6·3 (0·1–3·7)	**P:** Solid rocketboost/sustainer. **G:** Infra-red autogathering, then SAC to LOS*. **C:** Deflector in motor exhaust. **W:** H.e., with 65 projectile charges, proximity fuze.
RBS 70 (Sweden) Tube-launched by infantryman from tripod stand. Altitude: 3km (1·86 miles)	1·32 (4·33)	10·6 (4·17)	NA	23·5 (51·8)	5 (3·1)	**P:** Solid rocket boost (jettisonable), solid rocket sustainer. **G:** Laser beam rider. Target tracking by stabilized optical system. **C:** Cruciform rear fins. **W:** H.e., prefragmented, impact or laser proximity fuze.
Bloodhound Mk 2 (UK) Speed: M2+. Altitude: NA	7·67 (25·2)	54·6 (21·5)	282 (111)	NA	80+ (50+)	**P:** Four solid rocket boosters, 2 Thor ramjets. **G:** Constant wave semi-active radar homing. **C:** Twist and steer, pivoting wings. **W:** H.e., proximity fuze.
Blowpipe (UK) Speed: 1·5. Altitude 1,500m (4,920ft). Tube launched from shoulder.	1·34 (4·4)	7·6 (3)	27·4 (10·75)	NA	3 (1·9)	**P:** Solid rocket boost/solid sustainer. **G:** Radio command line-of-sight. **C:** Cruciform canard fins. **W:** H.e., fragmentation or shaped charge, impact fuzes.
Rapier (UK) Speed: M2+. Altitude: 3,000m (9,84oft).	2·24 (7·3)	12·7 (5)	38 (15)	42·5 (94)	0·5–7 (0·3–4·3)	**P:** Dual-thrust solid rocket. **G:** SAC to LOS, optical or with Blindfire radar. **C:** Cruciform rear fins. **W:** H.e., hollow charge, impact fuze.
Seacat/Tigercat (UK) Seacat on wide variety of ships including cruisers, destroyers, frigates. Tigercat is land based version.	1·47 (4·8)	19 (7·5)	64 (25·2)	63–68 (139–150)	5+ (3·1+)	**P:** Dual thrust solid. **G:** Radio-command to line-of-sight (TV autogathering). **C:** Cruciform wings. **W:** H.e., proximity fuze.
Sea Dart (CF–299) (UK) Speed: M3·5. Altitude: 25,000m (82,020ft). RN type 42 destroyers. Type 82 destroyer.	4·36 (14·3)	42 (17)	91 (36)	550 (1,213)	80+ (50+)	**P:** Solid rocket boost, Odin ramjet. **G:** Semi-active radar. **C:** Cruciform rear fins. **W:** H.e., fragmentation, proximity fuze.
Seaslug Mk 2 (UK) Altitude: 15,000m+ (49,213ft). Last four RN County-class destroyers.	6·1 (20·0)	41 (16·1)	144 (57)	NA	45+ (28+)	**P:** Four solid rocket boosts, solid sustainer. **G:** Beamrider, infra-red auto-gathering/command line-of-sight. **C:** Cruciform rear fins. **W:** H.e., proximity fuze.
Seawolf (PX 430) Speed: 800m/sec+ (2,625ft/sec+). Shipborne with six-barrelled auto-controlled launcher. RN County-class GM destroyers Type 22 frigates etc.	2·0 (6·56)	19 (7·5)	56 22	80–84 (176–185)	5 (3·1)	**P:** Dual thrust solid rocket. **G:** Semi-active command to line-of-sight/TV plus TTR. **C:** Cruciform rear fins. **W:** H.e., impact + proximity fuzes.
Thunderbird Mk 2 (UK) Speed: M2+. Altitude NA.	6·34 (20·8)	53 (21)	162 (64)	NA	50–80 (31–50)	**P:** Four solid-rocket boosts, solid sustainer. **G:** Semi-active radar homing. **C:** Cruciform rear fins. **W:** H.e., proximity fuze.
Improved Hawk MIM–23B (US) Speed: M2·5. Altitude: 30/18,000m (98/59,000ft)	5·08 (16·6)	37 (14·5)	120 (47)	625 (1,378)	40 (25)	**P:** Dual thrust solid. **G:** Semi-active constant wave radar. **C:** Elevons on long chord cropped delta wings. **W:** H.e., blast fragmentation.
Nike Hercules MIM–14B (US)	12·70 (41·6)	88 (34·6)	*228* (90)	4,800 (10,584)	30/150 (19/93·2)	**P:** Four × solid rocket boosts, solid sustainer. **G:** Radar command. **C:** Elevons on long chord delta wings. **W:** H.e., fragmentation or nuclear, command detonation.
Red Eye FIM–43A (US) Altitude: 2,500m (8,200ft).	1·2 (3·6)	7·0 (2·75)	14·0 (5·51)	8·2 (18)	3·4 (2·1)	**P:** Dual thrust rocket. **G:** Infra-red homing, optical aiming. **C:** Cruciform flip-out canard vanes. **W:** H.e., fragmentation, smooth casing.
Patriot MIM–104A (US) Replacement for Hawk and Nike-Hercules. Speed: M3.	5·18 (17·0)	40·6 (16·0)	92 (36·2)	NA	NA	**P:** Solid rocket. **G:** Semi-active radar homing. **C:** Cruciform rear fins. **W:** H.e., or nuclear.
Standard ER RIM–67A (US) Speed: M2·5+. Altitude: 20,000m+ (65,62oft). Four missiles on M730 tracked vehicle.	8·23 (27·0)	35 (13·8)	*157* (61·8)	*1,360* (3,000)	56 (35)	**P:** Solid rocket boost, solid sustainer. **G:** Semi-active radar. **C:** Cruciform rear fins. **W:** H.e., proximity fuze.

Missile						Propulsion / Guidance / Control / Warhead
Stinger **FIM-92A** (US)	1·52 (5·0)	7·0 (2·75)	9·0 (3·5)	15·1 (33·3)	4·8 (3·0)	**P:** Dual thrust solid. **G:** Infra-red homing proportional navigation. **C:** Cruciform flip-out vanes. **W:** H.e., fragmentation, smooth casing.
Guideline **SA-2** (USSR) Altitude: 18,288m+ (60,000ft+)	10·6 (34·7)	51 (20)	170 (67)	2,300 (5,070)	40–50 (25–30)	**P:** Solid rocket booster, liquid sustainer. **G:** Automatic radio command, radar tracking of target. **C:** Cruciform steerable fins. **W:** H.e., 130kg (286lb).
Goa **SA-3** (USSR) Altitude: 12,192m (40,000ft+).	6·7 (22·0)	45·7 (18)	122 (48)	NA	25–30 (15–18)	**P:** Two-stage solid rocket. **G:** Radio command, radar terminal homing. **C:** Cruciform steerable foreplanes. **W:** H.e.
Ganef **SA-4** (USSR) Altitude: 18,000m (59,000ft). Two mounted on tracked launcher.	9·1 (30·0)	80 (32)	228·6 (90)	1,000 (2,200+)	70 (42)	**P:** Integral ramjet, four wrap-round solid rocket boosters. **G:** Radio command. **C:** Cruciform steerable wings. **W:** H.e.
Gammon **SA-5** (USSR) Altitude: 30,000m (98,425ft).	16·4 (54·0)	86 (34)	365 (144)	NA	160+ (100+)	**P:** Solid boost, solid sustainer. **G:** Radio command, semi-active radar homing. **C:** Cruciform steerable fins plus wing-mounted ailerons. **W:** H.e.
Gainful **SA-6** (USSR) Speed: M2·8. Altitude: 100/18,000m (328/59,055ft). Triple launcher on modified PT 76 tank.	6·2 (20·3)	33·5 (13·2)	124 (48·8)	550 (1,212)	59·5 (37)	**P:** Integral solid fuel ramjet, solid rocket boost. **G:** Radio command, semi-active radar homing. **C:** Cruciform steerable wings? **W:** H.e., 80kg (176lb).
Grail **SA-7** (USSR) Speed: M1·5. Altitude: 45/1,500m (148/4,920ft)	1·35 (4·4)	7 (2·75)	NA	NA	2·9–4·0 (1·8–2·5)	**P:** Solid rocket boost/sustainer. **G:** Infra-red homing, optically aimed. **C:** Cruciform flip-out steerable(?) wings and tail fins. **W:** H.e., 2·5kg (5·5lb), probably fragmentation.
Gecko **SA-8** (USSR) Altitude: 50–6,000m (164–4,920ft). On six-wheeled amphibious vehicle.	3·2 (10·5)	20·2 (8·25)	64 (25)	NA	12·0 (7·5)	**P:** Dual thrust solid rocket. **G:** Radar command. **C:** Cruciform canard fins and foreplanes. **W:** H.e.
Gaskin **SA-9** (USSR)	NA	NA	NA	NA	5 (3·1)	**P:** Solid rocket motor. **G:** Infra-red homing. **C:** Cruciform wings and tail fins. **W:** H.e., Twin-quad canister on modified BRDM Scoutcar.

Air-to-air missiles

Missile						Propulsion / Guidance / Control / Warhead
Rb 72 (Sweden)	260 (102·3)	17·5 (6·9)	60 (24)	110 (242)	8 (5)	**P:** Solid rocket. **G:** Infra-red. **C:** Cruciform rear fins. **W:** H.e.
Magic **R.550** (France) Dog-fight missile, high-g manoeuvres.	274 (107·8)	15·7 (6)	66 (26)	90 (198)	0·3–7·0 (0·2–4·3)	**P:** Solid rocket. **G:** Infra-red (fast reacting, all-aspect). **C:** Cruciform canard fins. **W:** H.e., impact + IR proximity fuze.
Super R.530 (France) Altitude: 21,300m (70,000ft). Speed: M4·5+.	354 (139)	26 (10)	90 (35·4)	200 (441)	29–35 (18–22)	**P:** Dual thrust solid rocket. **G:** Semi-active radar. **C:** Cruciform rear fins. **W:** H.e. + proximity fuse.
Red Top (UK) Speed: M3·0.	327·7 (129)	22·1 (8·75)	91 (35·8)	150 (330)	11 (7)	**P:** Solid rocket. **G:** Infra-red, (lead collision). **C:** Cruciform rear fins. **W:** H.e.
Skyflash (UK) Based on Sparrow air-frame.	366 (144)	20·3 (8·0)	102·3 (40·3)	192·8 (425)	25·5 (15·8)	**P:** Solid rocket. **G:** Advanced semi-active radar seeker (SU and SD capability). **C:** Cruciform wings. **W:** H.e. + radar proximity fuze.
Falcon **AIM 26B, HM 55, R627** (US) Speed: M2·0.	213 (84)	27·9 (11·0)	61 (24)	115 (253)	8–10 (5–6)	**P:** Solid rocket. **G:** Semi-active radar homing. **C:** Cruciform rear fins. **W:** H.e. (AIM–26B), nuclear (AIM–26A).
Phoenix **AIM-54A** (US) Speed: M5·0+. On F-14A.	396 (156)	38 (15)	91·4 (36)	380 (838)	200+ (124+)	**P:** Solid rocket. **G:** Cruise: semi-active radar, active radar final 16km (10 miles). **C:** Cruciform rear fins. **W:** H.e., proximity + contact fuzes.
Sidewinder IC **AIM-9C/D** (US) Speed: M2·5.	292 (114)	12·7 (5)	63·5 (25)	84 (185)	10–18 (6–11·2)	**P:** Solid rocket. **G:** Semi-active radar (AIM–9C), infra-red (AIM–9D). **C:** Cruciform triangular canard fins. **W:** H.e. + contact fuzes.
Sparrow **AIM-7E** (US) Speed: M4·0. Also surface-to-air application.	365 (144)	20 (8)	102 (40)	205 (452)	25–50 (15–31)	**P:** Solid rocket. **G:** Semi-active, constant-wave radar. **C:** Cruciform wings. **W:** H.e., proximity plus contact fuzes.
Acrid **AA-6** (USSR) Speed: M2·2.	590 (232) IR 628 (247) SAR	36 (14·2)	225 (88·5)	650–850 (1,433– 1,874)	37 (10)	**P:** Solid rocket. **G:** Infra-red or radar homing. **C:** Cruciform canard fins and wing-mounted surfaces. **W:** H.e., 100kg (220lb).

Name						Characteristics
Apex **AA-7** (USSR)	430 (169)	24 (9.45)	105 (41.3)	NA	28 (17)	**P:** Solid rocket. **G:** Infra-red or radar homing. **C:** Cruciform rear fins. **W:** H.e.
Aphid **AA-8** (USSR)	210 (83.7)	NA	NA	NA	5–8 (3.1–5)	**P:** Solid rocket. **G:** Infra-red or radar homing. **C:** Cruciform canard fins (?). **W:** H.e.

Air-to-ground missiles

Name						Characteristics
AS-30 (France) Also 'fire-and-forget' AS 30 laser version. *When launched at M0.9 at low level.	3.78 (12.4)	34.2 (13.5)	100 (39.4)	520 (1,146)	2–3 (1.2–1.86) 12 (7.4)*	**P:** Solid rocket boost, solid rocket sustainer. **G:** Radio command, late models SAC to LOS. **C:** Deflectors in motor efflux. **W:** H.e., 230kg (507lb) general purpose or semi-armour piercing fuze.
Martel **AS-37** (France) Martel AJ163 (length 3.9m [12.8ft]), is radio command/TV guided.	4.16 (13.7)	40 (15.7)	120 (47)	530 (1,168)	30 (18.6)	**P:** Solid rocket boost, solid rocket sustainer. **G:** Passive radar homing. **C:** Cruciform rear fins. **W:** H.e., 150kg (330lb), proximity fuze.
Bullpup A **(AGM-12B)** (US)	3.20 (10.5)	30.5 (12)	101 (40)	258 (571)	11 (6.8)	**P:** Storable liquid rocket (pre-packaged). **G:** Radio command. **C:** Cruciform canard fins. **W:** H.e., 113kg (250lb).
Maverick **(AGM-65)** (US)	2.46 (8.07)	30.5 (12)	71 (28)	210 (463)	25–30 (15.5–18.6)	**P:** Solid rocket. **G:** Automatic TV homing or IR or semi-active laser. **C:** Cruciform rear fins. **W:** H.e., 59kg (130lb).
Shrike **(AGM-45A)** (US)	3.05 (10)	20.3 (8)	91.4 (36)	177 (390)	16 (10)	**P:** Solid rocket. **G:** Passive radar homing. **C:** Cruciform wings. **W:** H.e., fragmentation.
Standard Arm **(AGM-78)** (US)	4.57 (15)	30.5 (7.7)	109 (240)	816 (1,800)	25 (15.5)	**P:** Dual-thrust solid rocket. **G:** Passive radar homing. **C:** Cruciform rear fins. **W:** H.e.
Kipper **AS-2** (USSR) Speed: M1.2. Belly-mounted on TU-16. Badger C. Also anti-shipping role.	9.5 (31)	NA	4.6 (15)	NA	180–210 (112–130)	**P:** Underslung turbojet. **G:** Radio command, mid-course; radar homing. **C:** Aircraft type. **W:** H.e.
Kitchen **AS-4** (USSR) Speed: M2+. Belly-mounted on Tu-22 Blinder B.	11.3 (37)	NA	245 (96)	6,000+ (13,228+)	300 (186)	**P:** Storable liquid rocket. **G:** Inertial plus mid-course (?). **C:** Aircraft type. **W:** H.e.

Ballistic missiles, tactical

Name						Characteristics
Frog 7 (USSR) On ZIL-135 erector-launcher.	9.1 (29.8)	55 (21.6)	NA	NA	60 (37)	**P:** Dual thrust solid. **G:** Spin-stabilized, (unguided). **C:** Cruciform fixed fins. **W:** H.e., chemical or nuclear.
Lance **MGM-52C** (US)	6.10 (20.0)	56 (22)	198 (78)	1,530 (3,373)	70–120 (43–74.5)	**P:** Single-stage storable liquid rocket. **G:** Simplified inertial. **C:** Fluid injection into single nozzle. **W:** Nuclear 10 KT or h.e. cluster.
Pershing **MGM-31A** (US) Deployed on erector-launcher. Being replaced by terminally guided Pershing 2.	10.5 (34.5)	101.6 (40)	NA	4,535 (10,000)	160–640 (160–460)	**P:** Two-stage solid, Thiokol XM105, Thiokol XM106. **G:** Simplified inertial. **C:** Efflux deflectors, stage 1; steerable fins, stage 2. **W:** Nuclear 60–400KT.

Anti-tank missiles

Name						Characteristics
Harpon (France)	121.5 (47.8)	16.4 (6.4)	50 (19.7)	30.4 (67)	500–3,000 (1,640–9,842)	**P:** Solid rocket boost/sustainer. **G:** Wire, automatic command to line-of-sight. **C:** Deflectors in sustainer exhaust. **W:** H.e., shaped charge.
Swingfire (UK)	107 (42.0)	17 (6.7)	37 (14.5)	34 (75)	140–4,000 (459–13,123)	**P:** Solid rocket boost/sustainer. **G:** Wire command to line-of-sight. **C:** Jetavator. **W:** H.e., 6.8kg (15lb) hollow charge. Container launched.
Tow **BGM-71** (US)	117 (46.0)	15.2 (6)	34 (13.4)	24.5 (54)	65–3,750 (213–12,303)	**P:** Solid rocket boost/sustainer. **G:** Wire command to line-of-sight, automatic optical tracking. **C:** Flip-out fins. **W:** H.e., shaped charge. Tube launched.
Sagger **AT-3** (USSR)	87 (34.2)	12 (4.7)	46 (18.1)	11 (24.2)	500–3,000 (1,640–9,842)	**P:** Solid rocket. **G:** Wire command to line-of-sight. **C:** Not available. **W:** H.e., shaped charge.

Anti-ship missiles

Exocet **MM–38** (France) Speed: Mo·93–0·95.	5·20 (17)	35·0 (13·8)	104 (41)	735 (1,620)	5–40+ (3·1–25+)	**P:** Solid rocket, boost and sustainer. **G:** Cruise: inertial plus radio altimeter; attack: active radar seeker. **C:** Cruciform rear fins. **W:** H.e., 165kg (364lb), hexolite in steel block, delay fuze.
Kormoran (W. Germany)	4·40 (14·4)	34·4 (13·5)	100 (39·3)	600 (1,323)	35+ (22+)	**P:** Solid rocket, boost and sustainer. **G:** Cruise: autopilot plus radio altimeter; attack: active radar seeker. **C:** Cruciform rear fins. **W:** H.e., 160kg (353lb), clustered projectile charges, delay fuze.
Gabriel (Israel)	3·35 (11)	32·5 (12·8)	138·5 (54·5)	400 (882)	20+ (12+) 40+ (25+)	**P:** Solid rocket, boost and sustainer. **G:** Cruise: autopilot; radio altimeter; attack: radar/electro-optical. **C:** Cruciform rear fins. **W:** H.e., 150kg (331lb).
Sea Skua **(CL 834)** (UK)	*2·8* *(9·2)*	*22·2* *(8·7)*	*67* *(26·3)*	*200* *(441)*	*14* *(9)*	**P:** Solid rocket. **G:** Semi-active homing responsive to target illumination by radar. **C:** Cruciform canard fins. **W:** H.e., 35kg (77lb).
Harpoon **AGM–84A/RGM–84A** (US)	4·58 (15·0) 3·84 (12·6) air launch	34·3 (13·5)	91·4 (36)	667 (1,470) 522 (1,151) air launch	110 (68)	**P:** Solid rocket, boost turbojet cruise 300kg (661lb) thrust. **G:** Cruise: advanced autopilot; attack: active radar seeker. **C:** Cruciform fins. **W:** H.e., 227kg (500lb), shaped charge, contact, delay fuze.
Scrubber **SS–N–1** (USSR) Speed: Mo·9. Operational 1958–59.	*7·6* *(25·0)*	*100* *(39·3)*	*460* *(180)*	*4,080* *(9,000)*	*110–185* *(68–115)*	**P:** Solid rocket, boost turbojet cruise. **G:** Cruise: autopilot/radio command; attack radar or IR homing. **C:** Aircraft-type. **W:** H.e. On Kildin and Krupny class destroyers.
Styx **SS–N–2** (USSR) Speed: Mo·9. Operational 1960. On Komar and Osa fast patrol boats.	*6·25* *(20·5)*	*75* *(29·5)*	*275* *(108·2)*	*2,500–3,000* *(5,513–* *6,615)*	*40+* *(25+)*	**P:** Solid rocket, boost turbojet cruise. **G:** Cruise: autopilot/radio command; attack: command active radar. **C:** Aircraft-type control (ailerons + rudders). **W:** H.e.
Shaddock **SSC–1/** **SS–N–3** (USSR)	*10·9–13·8* *(35·7–42)*	*100* *(39·3)*	*210+* *(82·6+)*	*11,790* *(26,000)*	*450+* *(280+)*	**P:** Two × solid rocket boost; turbojet cruise. **G:** Cruise: radio command; attack: radar or IR homing. **C:** Aircraft-type. **W:** H.e., or nuclear KT range. M2·0–2·5.

Cruise or Stand-off missiles, strategic

Hound Dog **AGM–28B** (US) Speed: M2+. Withdrawn 1965.	13·0 (42·5)	71·1 (28)	370 (146)	4,603 (10,104)	1,207 (750)	**P:** Pratt & Whitney J52–P–3, turbojet 3,400kg (7,496lb) thrust. **G:** Inertial. **C:** Aircraft-type. **W:** Single thermonuclear IMT.
SRAM **AGM–69A** (US)	4·27 (14)	45 (17·7)	NA	*1,000* *(2,205)*	64–161 (40–100)	**P:** Two-pulse solid. **G:** Inertial plus terrain avoidance. **C:** Cruciform rear fins. **W:** Single thermonuclear 200KT.
ALCM **AGM–86A** (US) Speed: Mo·55 cruise. Mo·7 attack.	4·3 (14·0)	61 (24)	NA	860 (1,896)	1,207+ (750+)	**P:** Single Williams Research F 107 turbofan, 300kg (661lb) thrust. **G:** Inertial and terrain comparison. **C:** Aircraft-type. **W:** Single thermonuclear.
Tomahawk YBGM 109 (US) Speed: Mo·7 attack. Modified 391cm (154in) Tomahawk adapted for launch from ships, submarines or land vehicles.	5·56 (18·3)	52 (20·5)	228 (90)	NA	2,414+ (1,500+) 3,600* (2,237)*	**P:** Williams F 107 turbofan, 300kg (661lb) thrust + solid booster. **G:** Inertial and terrain comparison. **C:** Cruciform rear fins. **M:** Thermonuclear or H.e.

Ballistic missiles, strategic, land based

Scarp **SS–9** (USSR) Silo hot launch. No. in service 1976: 200+.	*35·0* *(115)*	*300* *(118)*	NA	*12,000* *(7,456)*		**P:** Two-stage storable liquid. **G:** Inertial. **C:** Steerable vernier motors. **W:** Mod: 3 FOBs. Mod: 4 3 MRVs or test MIRVs.
Sego **SS–11** (USSR) Service entry 1966. Silo hot launch. Total over 1,000.	*19·5* *(64)*	*240* *(94)*	NA	*10,500* *(7,456)*		**P:** Two-stage storable liquid. **G:** Inertial. **C:** Not available. **W:** Single RV 1–2 MT (Mod. 1), three MRVs 300–500 KT each. (Mod. 3).
Scaleboard **SS–12** (USSR) Service entry 1969. Mobile on MAZ 543 carrier.	*11·0* *(36)*	NA	NA	*700–800* *(435–497)*		**P:** Single-stage rocket. **G:** Inertial. **C:** Steerable aerodynamic rudders (?). **W:** Nuclear or h.e.
Scapegoat **SS–14** (USSR) Mobile on Scamp transporter-erector (modified JS III tank chassis).	*10·6* *(35)*	*140* *(55)*	NA	*4,000* *(2,485)*		**P:** Two-stage solid. **G:** Inertial. **C:** Not available. **W:** Thermonuclear.
SS–16 (USSR) Silo hot launch. Also mobile role.	*20* *(66)*	*200* *(79)*	NA	*9,000* *(5,592)*		**P:** Three-stage solid. **G:** Inertial. **C:** Not available. **W:** Single RV IMT + or 3 MIRVs and penetration aids.
SS–17 (USSR) Silo cold launch. Number in service mid-1976: 30 (in modified SS–11 silos).	*24* *(79)*	*260* *(102)*	NA	*10,000* *(6,214)*		**P:** Two-stage storable liquid. **G:** Inertial. **C:** Not available. **W:** Four? IMT KT MIRVs.

Missile						Details
SS–18 (USSR) Silo cold launch. No. in service late 1976: 50. Mod 2 entered service late 1976.	37 (121)	340 (132)	NA		Mod 1 10,500+ (6,525+) Mod 2 9,250+ (5,748+)	**P:** Two-stage storable liquid. **G:** Inertial. **C:** Not available. **W:** Mod. 1. Single RV 40–50 MT; Mod. 2. Eight to ten MIRVs 1–2 MT each.
SS–19 (USSR) Silo cold launch. No. in service late 1976: 140.	24 (79)	260 (102)	NA	10,000 (6,214)		**P:** Two-stage storable liquid. **G:** Inertial. **C:** Not available. **W:** Six MIRVs 400–500 KT.
SS–20 (USSR) Employs first two stages of SS–16. Max. range lightweight thermo-nuclear warhead. About 600 on mobile launchers.	16 (55)	200 (79)	NA	2,400–7,000 (1,491–4,350)		**P:** Two-stage solid. **G:** Inertial. **C:** Not available. **W:** Single IMT RV or 3 MIRVs.
Minuteman II LGM–30F (US) Silo-launched.	18·2 (60)	183 (72)	31,750 (70,000)	11,000+ (6,835+)		**P:** Three-stage solid; lift-off thrust 91,000kg (200,621lb) × 60sec. **G:** Inertial. **C:** 4 moveable nozzles (stages 1 and 3); fluid inertion into single fixed nozzle (stage 2). **W:** Single RV1–2MT plus penetration aids.
Minuteman III LGM–30G (US) Silo-launched.	18·2 (60)	183 (72)	34,475 (76,000)	13,000+ (8,078+)		**P:** Three-stage, solid (improved stage 3). **G:** Inertial. **C:** As Minuteman II. **W:** Three MIRVs of 200 KT each plus penetration aids.
Titan II LGM 25C (US) No. in service: 54. Silo launched. All withdrawn by 1984.	31·4 (103)	305 (120)	149,690 (330,000)	15,000 (9,320)		**P:** Two-stage storable liquid. **G:** Inertial. **C:** Gimballed engines, both stages. **W:** Single RV 10MT plus penetration aids.
SSBS S3 (France) Silo launched.	13·7 (45)	150 (59)	25,800+ (56,880+)	3,000+ (1,864+)		**P:** Two-stage solid; lift-off thrust 55,000kg (121,254lb) × 76sec. **G:** Inertial. **C:** 4 Gimballed nozzles (stage 1); Freon injection into single fixed nozzle. **W:** Single RV IMT plus penetration aids.

Ballistic missiles, strategic, submarine based

Missile						Details
Polaris A–3 UGM–27C (US/UK) In service 1960–79.	9·75 (32)	137 (54)	15,850 (35,000)	4,626 (2,875)		**P:** Two-stage solid. Lift off thrust 36,287kg (80,000lb). **G:** Inertial. **C:** 4 rotating nozzles (stage 1), fluid injection into single fixed nozzle (stage 2). **W:** Three MRVs, 200 KT each.
Poseidon C–3 UGM–73A (US) Still in Royal Navy service.	10·36 (34)	188 (74)	29,484 (65,000)	4,626+ (2,875+)		**P:** Two-stage solid. **G:** Inertial. **C:** One gimballed nozzle each stage. **W:** 10 MIRVs 50 KT each at 5,200km (3,231 miles), or 14 50 KT at 4,000km.
Trident I C–4 UGM–93A (US)	10·36 (34)	188 (74)	32,000 (70,560)	7,000 (4,350)		**P:** Three-stage solid. **G:** Inertial + stellar navigation. **C:** Not available. **W:** Eight MIRVs 100 KT.
Sawfly SS–N–6 (USSR) In Yankee-class nuclear submarines; 16 per vessel.	12·8 (42)	175 (68·9)	18,000–20,000 (39,690–44,100)	2,400 (1,491) 3,000 (1,864)		**P:** Two-stage storable liquid rocket. **G:** Inertial. **C:** Not available. **W:** Thermonuclear IMT.
SS–N–8 (USSR) In Delta-class nuclear submarines; 12 in Delta 1, 16 in Delta 2.	14+ (45·9+)	180–200 (70·8–78·7)	20,000+ (44,100)	7,800 (4,847)		**P:** Two-stage storable liquid rocket. **G:** Inertial + stellar navigation. **C:** Not available. **W:** Thermonuclear 1–2 MT.
SS–NX–17	NA	NA	NA	NA		**P:** Two-stage solid rocket. **G:** Inertial + stellar navigation (?). **C:** Not available. **W:** Thermonuclear, single RV or MIRVs.
SS–NX–18 Replaces SS–N–8 in Delta 1 and 2 class submarines.	NA	NA	NA	8,046+ (5,000+)		**P:** Two-stage storable liquid rocket. **G:** Inertial + stellar navigation (?). **C:** Not available. **W:** Thermonuclear, MIRVs.
MSBS M2/M20 (France) In French ballistic submarines, 16 per vessel. Replaced by M4.	10·4 (34·1)	150 (59)	20,000 (44,100)	1,864 (3,000)		**P:** Two-stage solid rocket. **G:** Inertial. **C:** Four gimballed nozzles, stage 1; Freon injection into single fixed nozzle, stage 2. **W:** Thermonuclear, 500KT in M2 and 1 MT in M20.

Anti-ballistic missiles

Missile						Details
Galosh ABM–1 (USSR) 64 launchers at four sites defend Moscow.	18–19 (59–62)	240–270 (945–106)	NA	NA	300 (186)	**P:** Multi-stage rocket, 4 first stage nozzles. **G:** Radar command. **C:** Not available. **W:** Probably nuclear.
Spartan XLM–49A (US) 'Safeguard' system high-altitude, area defence. Never used.	16·8 (55·2)	107 (42)	3·0 (9·8)	13,000 (28,660)	460 (750)	**P:** Three-stage solid rocket, Thiokol TX 500, TX 454, TX 239. **G:** Radar command. **C:** Aerodynamic cruciform fins/fins and gas-jets on final stage. **W:** Nuclear.
Sprint 1 (US) 'Safeguard' system high-acceleration terminal defence. Never used.	8·3 (27)	137 (54)	NA	3,402 (7,500)	40+ (25+)	**P:** Two-stage solid rocket. **G:** Radar command. **C:** Thrust vector control and aerodynamic cruciform fins, stage 2. **W:** Nuclear.

LEFT: *US Army Lance was designed to replace the tactical missiles Honest John and Sergeant. It employs pre-packaged liquid propellants. Maximum range is about 120km (75 miles).*
BELOW: *MGR–3 Little John, the compact, solid-fuel, air-transportable, artillery rocket, could have either a nuclear or conventional h.e. warhead.*
BELOW RIGHT: *The highly successful French SS–11 wire-guided anti-tank missile. More than 160,000 were produced for many countries.*

CENTRE LEFT: *Soviet FROG tactical missiles during manoeuvres in Minsk. It is operated by Warsaw Pact countries, Egypt and Syria.*
CENTRE RIGHT: *Soviet troops photographed on manoeuvres with SCUD B tactical missiles. Missiles are elevated for vertical firing from a small launch table at the back on the lines of the German V–2.*
BOTTOM: *The Vigilant anti-tank missile was a British development. It was easily carried and with it one man could destroy a heavy tank single-handed.*

Tactical missiles

Battlefield weapons include large, unguided and spin-stabilized rockets with conventional high explosive, chemical or nuclear warheads launched from tracked or wheeled chassis. American examples are the MGR–1B Honest John, MGR–3A Little John and MGM–52B/C Lance. Russia has produced a whole family of such weapons known to the West as Free Rocket Over Ground (FROG). More complex missiles of medium range – some able to attack near battlefield areas – also quickly made their mark.

From the beginning Russia concentrated on giving tactical missiles high mobility by mounting them on cross-country vehicles which could be driven off the road and concealed in woods and forests. The missiles, raised into a vertical firing position by the carrier vehicle's hydraulic erector, are placed on a small launch table and fired from the vehicle by remote control. Typical examples bear the NATO code-name Scud A and Scud B; Scaleboard has much greater range. America produced Redstone, Corporal, MGM–29A Sergeant and MGM–31A Pershing.

Early missiles were beam-riders, centring themselves on radio beams aligned with the target; others were guided by radio com-

mand. Such systems, however, are vulnerable to electronic counter-measures and they were rapidly superseded by inertial guidance systems with pre-set target instructions which could not be 'jammed'. The MGM–13B Mace – an American battlefield missile in the form of a ramp-launched pilotless aircraft – had a Goodyear Atran terrain/map-comparison guidance system, a forerunner of the highly advanced terrain-comparison systems used in America's latest hedge-

hopping cruise missiles.

France too has developed its own surface-to-surface battlefield missile, Pluton. This highly effective solid-fuel weapon has inertial guidance and operates from the AMX–30 tracked launcher.

Nor has the infantryman been neglected. To meet the challenge of the new battle tanks, a whole range of anti-tank missiles have become operational with the world's armies. The MGM–51A Shillelagh – developed for

BELOW: *The US Army Sergeant. This medium-range artillery missile became operational in 1962, was air-transportable and could be rapidly emplaced and fired by a six-man crew. The faster reaction time achieved over earlier missiles reflected the use of a solid propellant, reduced ground handling equipment and the use of digital electronics. The project originated at the Jet Propulsion Laboratory which had previously developed the liquid-propellant Corporal surface-to-surface missile.*

BELOW: *Mace, a modernized 'flying-bomb' of aircraft scale, was towed to the launch site on a vehicle which served the double purpose of transporter and launcher. The powerplant was an Allison J33–41 turbojet. Launch was assisted by an under-tail solid rocket booster. Now superseded by medium-to-long range tactical ballistic missiles and specialized strike aircraft, Mace was deployed on Okinawa in the Far East and in West Germany. It could carry a nuclear warhead more than 1,900km (1,200 miles).*

BOTTOM: *The French tactical missile Pluton is operated from a tracked carrier known as the AMX–30. The vehicle also embodies the associated data processing and control equipment. Range is 10 to 120km (6·25 to 74 miles). Warhead can be either nuclear or conventional high explosive.*
INSET: *A French Harpon missile being fired from its carrier. The wire-guided missile can be used against tanks and fortifications and has an automatic system of guidance which confers deadly accuracy.*

the US Army – was launched from a combination gun/launcher carried by a variety of vehicles, including the General Sheridan lightweight, armoured reconnaissance vehicle and the M60A1E2 medium tank. The missile itself had flip-out fins and was steered by deflecting the exhaust of the sustainer rocket according to signals from a microwave guidance system. The operator kept his optical sight fixed on the target and the missile responded automatically. The same gun/launcher could fire conventional 152mm (6in) ammunition.

Another significant US anti-tank weapon is the MGM–71A Tube-launched, Optically sighted, Wire-guided (TOW). The missile and its tripod-mounted launcher can be operated by two men. Its effective range is 1,830m (6,000ft). In Vietnam, TOW was mounted in launch pods carried by Huey helicopters. All the 'gunner' had to do was hold a target such as a supply truck in the cross-hairs of his stabilized sight. The pilot could do limited evasive manoeuvres after missile launch and the weapon would still score a bullseye at a range of 3,000m (9,840ft). In four weeks, two helicopters registered 47 kills, including the destruction of 24 Communist tanks.

Equally effective is the British anti-tank missile Swingfire. Powered by a two-stage rocket motor, it covers 4,000m (13,120ft) in 25 seconds, though even this is not as fast as either TOW or the Euromissile HOT. Command signals generated by a thumb control stick are transmitted over fine wires (breaking strain 10·4kg [23lb]) and the missile is programmed automatically to the operator's line of sight. Path changes are effected by a swivelling tail-end 'jetavator'. The High-Explosive Anti-Tank (HEAT) warhead consists of a copper cone backed by a shaped charge which can punch a hole through 56cm (22in) of armour.

When Egyptian troops stormed across the Suez Canal on 6 October 1973, special units engaged Israeli tanks with Russian-built Sagger missiles, which they unpacked from suitcase-like containers. With this weapon, the infantryman attaches the warhead to the missile and unfolds the guidance fins. A launch rail is set up on the container's lid and Sagger mounted on it. Wire connections can then be run out to a firing unit which the operator uses to launch and steer the missile in conjunction with an optical sight and a joystick. With this and other weapons some 200 Israeli tanks were destroyed in one day.

Air-to-air missiles
During World War II, Germany expended much effort on air-to-air guided weapons. The 90·7kg (200lb) Henschel Hs 298 was actually put into production in 1945 just as the war was ending. It took the form of a small swept-wing monoplane powered by a

TOP LEFT: *Shillelagh was launched from the same 152mm gun as used to fire conventional ammunition.*
TOP: *Probably the world's longest-range air-to-air missile is the AIM–54A Phoenix which arms the US Navy's F–14A Tomcat high performance interceptors.*
ABOVE: *TOW anti-tank missile being fired from a Bell HueyCobra helicopter.*
LEFT: *One of the most effective early-generation anti-tank weapons is Swingfire which became operational with the British Army in 1969.*

dual-thrust solid-fuel motor, with fast- and slow-burning propellants for boost acceleration and sustained flight, one inside the other. Its maximum speed was about Mach 0·8. Launched at a distance of up to 1·6km (1 mile) from the target, it was radio-controlled and had a warhead which was detonated by a proximity fuze.

A less advanced weapon was the Ruhrstahl X–4, which had a crude form of self-homing and a maximum range of some 3·2 km (2 miles). Like some other German guided weapons, it was impervious to radio jamming since control signals were transmitted through fine, 0·2mm (0·008in) insulated wires which unwound from bobbins on the wings and connected it to the launch aircraft.

The missile, which had cruciform wings, was spin-rotating and controlled by solenoid-operated spoilers on the tail fins. In normal flight these vibrated at 5 cycles per second

but, in order to change the flight path, were made to operate asymmetrically by control signals from the launch aircraft. The smaller X–7 variant was designed by the same team for air-launching against ground targets.

The lessons learnt in Germany were quickly applied by the United States, the USSR, Britain and France. Early examples of American air-to-air missiles were the AIM–4E/F Falcon, which armed the F–101, F–102 and F–106 interceptor/fighters, and the AIM–7E Sparrow, which became standard armament on the F–4 Phantom. Both employ semi-active radar homing and depend on continuous-wave radar beams being directed from the launch aircraft and they respond to the reflections from the target. Sidewinder, on the other hand, has an Infra-Red (IR) seeker which steers the missile towards any heat source of a certain intensity, such as an engine or its exhaust.

The Soviet Union replied with a large

ABOVE : *Another formidable anti-tank missile is the Soviet AT–3 Sagger which Egyptian forces used with devastating effect against Israeli tanks during the Yom Kippur War. Apart from two-man infantry deployment, it is also mounted on a number of vehicles.*
BELOW : *Saab 37 Viggen attack aircraft with some alternative war-loads : on the aircraft three 600kg (1,320lb) Saab 04E air-to-surface missiles; in front two 300kg (660lb) Saab 05A ASMs, four AAMs (Falcons and Sidewinders), and four pods with 36 135mm air-to-surface rockets and other external stores.*

TOP : *Sparrow III on a modified UH–2C Seasprite helicopter for tests at the Naval Air Station, Point Mugu, California. This highly successful US air-to-air missile, which became operational in 1958, has appeared on a small number of the Western world's radar-equipped interceptors and strike aircraft. It has since appeared in many variants for air and ship-launched applications.*
ABOVE : *HOT – the high-speed, long-range anti-tank missile – launched from a 'Jagdpanzer Rakete' tracked combat vehicle. Developed jointly by France and West Germany, it can be fired by the infantry from ground positions and from vehicles, ships and helicopters.*

RIGHT: *Soviet AA–2 Atoll air-to-air missile on MiG–21PFM interceptor of the Indian Air Force. The weapon closely resembles the US Sidewinder and has similar infra-red guidance. Typical mounts are the MiG–21 Fishbed, MiG–23 Flogger and MiG–25 Foxbat. It has been produced in quantity for the Warsaw Pact countries and has been widely exported – to North Korea, North Vietnam, Egypt, Syria, Cuba, Iraq and Finland, and built in India under licence.*
BELOW: *Buccaneer strike aircraft with four Martel air-to-surface missiles, developed jointly by France and Britain. One version (AS.37) homes automatically into enemy radars, the other (AJ.168) is guided from the launch aircraft by television. Aircraft operating Martel include the Jaguar and Mirage III–E. Nimrod, Atlantique, and Sea King helicopter.*

solid-fuel missile (NATO code-name Alkali) which armed all-weather versions of the MiG–19. Like Sparrow and Falcon, it homed on the reflections from the target of the interceptor's radar signals.

Russian designers have generally tended to develop two versions of their air-to-air missiles, one with semi-active radar homing and the other with an IR homing head. Both versions are then flown on the same interceptor to enhance the probability of a kill. Two versions of Anab, for example, were carried by the Yak–28P, Su–9 and Su–11 all-weather interceptors. The Tu–28P could accommodate four of the larger AA–5 Ash missiles, two with radar homing and two with IR.

Britain, too, was well to the fore with missile armament during the 1950s and 1960s. Firestreak, an IR-homing pursuit course weapon, became operational on Lightning and Sea Vixen interceptors. It was followed by the IR-guided Red Top, which could attack from virtually any direction by virtue of a lead-collision IR guidance system.

France's first-generation air-to-air missile, the Matra R.511, depended on fairly crude semi-active radar homing and was carried by Vautour and Mirage IIIC interceptors. It was superseded by the Matra 530, a much more effective weapon not limited to pursuit-course attack. Like the Soviet Anab and Ash, it was produced in both IR and semi-active radar homing versions.

One of the most promising of the new breed of air-to-air weapons in Sky Flash, which was developed in Britain from the well tried Sparrow airframe. Using a new and advanced semi-active radar seeker, it is designed to engage targets from sea level to high altitudes and can 'snap down' to inter-

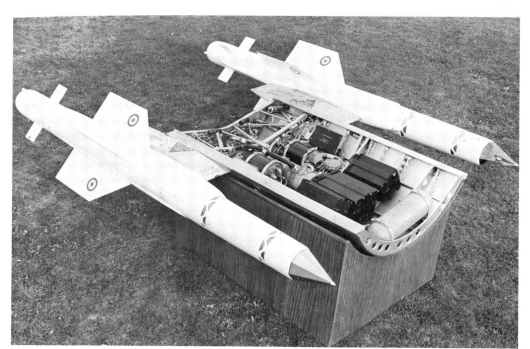

ABOVE CENTRE: *The French describe their R550 Magic as a 'dog-fighter'. At the tail are rectangular fins and aft of the nose two sets of vanes one fixed, one steerable. In close-range air combat Magic is capable of making high-g turns exceeding those of the most manoeuvrable piloted aircraft. It is in service with the French forces on several types of aircraft.*

ABOVE: *Firestreaks mounted on the missile pack of a Lightning interceptor. The first-generation British infra-red guided missile was limited to pursuit-course engagements but regularly demonstrated a success rate of over 85 per cent in trials. It was superseded by Red Top which was faster, had greater range and could make its attack from almost any angle.*

cept low-flying aircraft. It is claimed to be particularly effective in electronic-warfare environments where the enemy is operating 'jammers' and other devices.

The Matra Magic 'dogfight' missile, which can manoeuvre at very high g levels, is another highly promising example.

Air-to-surface missiles
Germany played a leading role in developing anti-shipping guided missiles in World War II when glide-bombs were used against Allied merchant shipping. The Hs 293, which took the form of a small monoplane, had an underslung rocket motor which gave a thrust of 590kg (1,300lb) for 12 seconds. The maximum speed was about 600km/h (375mph).

Released at 8 to 9·7km (5 to 6 miles) from the target, the Hs 293 was radio-controlled from the launch aircraft, where the pilot visually tracked a flare in the tail of the missile. Experiments were also made with a crude television system to give the crew of the launch aircraft a missile-eye view of the approaching target. There was even a scheme to control such relatively long-range glide missiles by wire transmission to prevent the Allies jamming the control signals. An enlarged version, developed in 1944, was the Hs 295.

The Hs 294 was a more advanced weapon, designed to attack warships in a spectacular manner. After being visually guided by radio from the launch aircraft, the missile entered the sea within some 45·7m (150ft) of the target ship, shedding the wings and rocket motor to complete the attack under water. It was then exploded below the waterline either by the action of a proximity fuze or automatically at a predetermined depth.

Air-to-surface (tactical)
The need to pick off point targets in Vietnam – especially bridges defended by anti-aircraft missiles, and single vehicles carrying supplies from the North – led the Americans to introduce the so-called 'smart' bomb. Paveway was a conventional bomb fitted with a laser seeker and extra control surfaces. A laser beam aimed at the target by a supporting aircraft meant the bomb, released from an attack aircraft such as a Phantom, could steer itself towards the bright spot of laser energy reflected from the target. This allowed it to score a direct hit. From such devices have come a whole range of precision guided missiles which allow launch aircraft to leave the scene immediately the weapon has been released.

Another example of precision guidance is found in the AGM–65A Maverick. It has a small television camera which focuses on the target, steering the missile on to it with lethal effect. Many other versions followed.

A European example of a precision tactical missile is Martel, which exists in two versions. One, built in Britain, is TV-guided; the other, built in France, homes on to radar emissions.

So effective are such techniques that TV-guided Mavericks have been launched against ground targets from radio-controlled pilotless drones, removing the human element entirely from the attack zone.

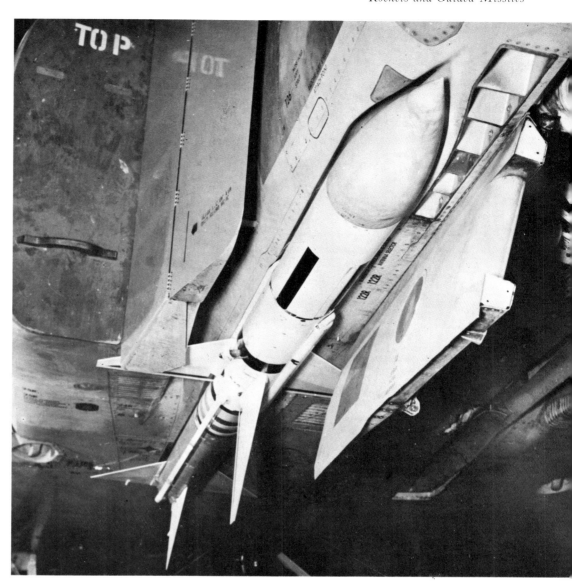

TOP : *The Sky Flash air-to-air missile, installed on an RAF McDonnell Douglas F-4 Phantom. The all-weather missile, based on the US AIM–7E Sparrow, has totally new British electronic equipment and can be launched at subsonic or supersonic speeds. It is designed to intercept a target flying faster and at higher or lower altitudes than the launch aircraft.*

ABOVE : *Falcon air-to-air missiles in production for the US Air Force. Since the weapon first entered service in 1955 there has been a whole family of missiles with variations in guidance system, body design, propulsion and warheads. The two basic types have infra-red and semi-active guidance respectively. Versions of Falcon were licence-built in Sweden.*

TOP: *AGM–28A Hound Dog strategic missile was designed to be released from the B–52 bomber some 965km (600 miles) from a target. It carried a thermonuclear warhead.*

CENTRE: *Soviet SS–4 Sandal was the medium-range ballistic missile with which Premier Khruschev confronted President Kennedy during the Cuban crisis of October 1962.*

ABOVE LEFT: *Subsonic cruise missiles are carried by long range bombers of the Soviet Naval Air Force for attacking ships and coastal targets. Photograph shows an Egyptian Tu–16 Badger with two AS–5 Kelt missiles on underwing pylons. It is rocket powered and has a maximum range of about 150km (93 miles). In the nose is a large radar.*

ABOVE RIGHT: *Line-up of Blue Steel missiles at R.AF Scampton. The large stand-off missile was carried by RAF Mk 2 Vulcan and Victor aircraft of RAF Strike Command as part of the UK nuclear deterrent. Towards the end of its operational life crew training was directed towards penetrating hostile airspace at low level. Blue Steel was superseded in the deterrent role by the RN Polaris submarines.*

Air-to-surface (strategic)

Efforts to enhance the ability of strategic bombers to reach their targets led to air-launched 'stand-off' missiles which fly on as the bomber turns away. Among the first of such weapons was the AGM–28B Hound Dog carried by Boeing B–52Gs and Hs of the US Air Force's Strategic Air Command. Powered by an underslung turbojet, this 4,355kg (9,600lb) winged missile had a range of some 1,110km (690 miles). It carried a thermonuclear warhead and flew under inertial control.

Blue Steel, the British equivalent, was carried by Victor B.2s and Vulcan B.2s of RAF Strike Command. It was powered by a Stentor two-chamber rocket engine fuelled by kerosene and High Test Peroxide. One chamber was for cruising flight, the other for a high-speed dash. The missile was 10·7m (35ft) long, had a maximum diameter of 127 cm (50in) and a wing span of 3·96m (13ft). It was particularly effective in allowing bom-

bers to 'hedge-hop' and attack at low level.

Russia is also prominent in this league. As early as 1961 a stand-off missile, about 15m (50ft) long and resembling a swept-wing turbojet fighter, made its appearance beneath the fuselage of a Tu–95 Bear B. It was given the NATO code-name Kangaroo. Designed to attack peripheral targets within the NATO area, it was guided by radio from the launch aircraft.

Smaller winged missiles were developed mainly for anti-shipping duties. Two turbojet missiles which bore the NATO code-name Kennel were carried by the Tu–16 Badger B on underwing pylons. A small radar mounted above the engine air intake in the nose allows the weapon to home on to its target automatically. A land-based version for coastal defence is called Samlet.

A more recent air-launched weapon is the rocket-powered Kelt, two of which can be carried under the wings of the Tu–16 Badger G. Yet another winged missile for anti-shipping use is Kipper, which is belly-mounted on the Tu–16 Badger C. It resembles a swept-wing aircraft with an under-slung turbojet.

New Soviet cruise missiles are now being encountered by the West, which expects that they will profit from the latest advances in turbojet and rocket/ramjet technology. The Tupolev Backfire B bomber, for example, can accommodate two large stand-off missiles beneath the fixed portions of its variable-geometry wings and under the fuselage.

Intercontinental ballistic missiles

Marshal Zhigarev, Commander-in-Chief of the Soviet Air Forces, had pressed the importance of inter-continental rockets as early as 1946. Before such ambitions could be realized, much research and development was needed to improve rocket technology in four major areas: propulsion, structures, guidance and warheads.

The Russians began by improving the German V–2. Starting in 1945, the Gas Dynamics Laboratory/Experimental Design Group (GDL–OKB) sought to expand the know-how embodied in the RD–1/RD–3 family of rocket engines, variants of which had also been used for the assisted take-off of aircraft. Performance of the original V–2 was improved by modifying the engine cooling system, reducing the water content of the fuel, increasing the chamber pressure and temperature, and lengthening the nozzle.

By the end of 1947 the first launchings were being made of improved V–2 rockets built by Soviet engineers on Soviet soil. A secret launch base was used at Kapustin Yar (literally 'cabbage crag'), south-east of Stalingrad (now Volgograd).

The Russian version of the German missile had an RD–101 rocket engine. The same basic vehicle was later to appear as a geophysical rocket, the V–2–A. Experience with this vehicle led to the first Soviet short-range ballistic missile, known to the West as the SS–3 Shyster and to the Russians as Pobeda (Victory). The missile first appeared in a Red Square military parade in 1957, but US radar stations in Turkey had long since detected its early trials and listened in to its

telemetry, radio signals which carry coded data.

The engine, developed between 1952 and 1953, was the RD–103 which burned kerosene and liquid oxygen to produce a vacuum thrust of some 50,000kg (110,230lb). America's equivalent was the US Army's Redstone, developed by the von Braun team at Redstone Arsenal, Alabama.

Next to appear in Russia was the SS–4 Sandal, powered by a four-chamber RD–214 engine of 74,000kg (163,140lb) vacuum thrust. Developed between 1955 and 1957, this engine had variable thrust in flight, achieved by adjusting the supply of propellants to the gas generator of the turbopumps. It was this missile which featured in the Cuban missile crisis of 1962, when Nikita Khruschev tried to put pressure on the United States by stationing offensive missiles on its 'doorstep'. Reconnaissance photographs of Cuba showed that concrete had been poured for launch pads with support vehicles and missiles drawn up around the sites. The United States was within range of a missile capable of carrying a nuclear warhead over 1,770km (1,100 miles). Operational flexibility had been greatly improved compared with earlier missiles because Sandal employed storable liquids – nitric acid and kerosene – and could thus be held ready, fully fuelled, for long periods.

Although the Cuban sites were dismantled at the end of the confrontation, and the missiles returned to the Soviet Union, the episode proved to be the final justification for the

TOP LEFT: *The US Army's Redstone surface-to-surface missile was based on experience gained from the German V–2. It had steerable refractory vanes operating in the rocket efflux.*
TOP RIGHT: *Soviet geophysical rockets in the 'Cosmos' pavilion at the USSR Exhibition of Economic Achievement, Moscow. In the foreground is a Soviet version of the German V–2 (NATO code name*

Scunner) modified with a jettisonable instrument-carrying nosecone and ejectable instrument canisters on the body. At left is the later geophysical rocket based on the early ballistic rocket Shyster. The nosecone is open to reveal a spherical instrument container.
ABOVE: *While the United States and the Soviet Union were negotiating the basis of a SALT 2 agreement to limit the number of strategic delivery systems, the USSR*

was building bigger missiles to achieve a much greater throw-weight of nuclear explosives within agreed numerical ceilings. This chart shows the scale of the build-up compared with contemporary US weapons. From left: Titan II, SS–7, SS–5, SS–8, SS–4, Minuteman II, SS–11, SS–9, Minuteman III, SS–13, SS–16, SS–19, SS–17, SS–18. For details see missile table, pages 379 to 383.

BELOW: *Thor intermediate-range ballistic missiles were stationed in eastern England between 1958 and 1963. They were operated by the Royal Air Force but the US Air Force had authority over the thermonuclear warheads. When Atlas and Titan ICBMs brought the USSR within range from bases in the United States, Thor was withdrawn from England. The British Government planned to replace it with Blue Streak installed in underground silos but this was cancelled in 1960.*

build-up of missile forces in the United States. Already, Thor intermediate-range missiles stood ready at many RAF bases in eastern England pending the development of the big Intercontinental Ballistic Missiles (ICBMs) which could reach Soviet targets from launch sites in the United States. These rockets, fuelled with liquid oxygen and kerosene, had a range of some 2,775km (1,725 miles), putting Moscow into the front line of any nuclear exchange. Thor, instead of employing exhaust vanes to steer, had an inertial guidance system which issued commands to the rocket motor itself; the engine was gimbal-mounted and thus the thrust direction was changed for course corrections.

In the meantime the Russians had built a second major test centre north of Tyuratam in Kazakhstan. The big intercontinental rocket was rolled ponderously out to the launch pad for its first firing. In 1957 came the triumphant (if terse) announcement from the *Tass* news agency: . . . a super long-distance international ballistic multi-stage rocket flew at . . . unprecedented altitude . . . and landed in the target area.

It was many years before the West was to be acquainted with the design, although the broad features were known to Western intelligence. At lift-off, no fewer than 20 main thrust chambers fired, as well as 12 smaller swivel-mounted engines for control. Sixteen of the main chambers were in four strap-on boosters which were thrown off sideways as the vehicle left the atmosphere. The central core of the rocket, powered by four main chambers, drove the vehicle to its final cut-off velocity as four Vernier motors responded to commands from the inertial guidance system. The warhead separated successfully and continued to its target on a ballistic trajectory.

The engine of the central core was the RD-107 of 96,000kg (211,650lb) vacuum thrust. Each of the four boosters had an RD-108 engine of 102,000kg (224,870lb) vacuum thrust. All of them burned liquid oxygen and kerosene. Launch pads for these huge rockets were elaborate and extremely vulnerable to nuclear attack. At the test centre they comprised a large reinforced-concrete platform linked with the preparation building by a standard-gauge railway. Rockets assembled horizontally on a transporter-erecter were transferred to the launch pad by diesel locomotive. After being elevated into a vertical position on the pad, they were fuelled from rail-mounted tankers moved up on adjacent track.

America's reply to Korolev's ICBM was the 1½-stage Atlas, which broke new ground by having thin but pressurized stainless-steel tanks which withstood the thrust loads of the rocket engines without heavy reinforcement. Fuel (kerosene) and oxidant (liquid oxygen) were kept in separate compartments.

The original Atlas tank, 18·29m (60ft) long and 3·05m (10ft) diameter had no internal framework and comprised the entire airframe from propulsion bay to nosecone. At the base was a single gimbal-mounted rocket engine with two small swivel-mounted engines outboard for roll control and Vernier adjustment of velocity. In a jettisonable skirt – into which the main sustainer engine projected – were two thrust chambers drawing their propellants from the same tanks.

All engines burned at lift-off, producing a total thrust exceeding 163,295kg (360,000lb). The outboard engines were jettisoned with the skirt after 145 seconds of flight. The sustainer accelerated Atlas to a velocity of about 24,140km/h (15,000mph), when it was shut off by the guidance system. Launch tests began at Cape Canaveral, Florida, as the Russians were preparing to launch their first ICBM at Tyuratam – but early success was to be denied. In June 1957, a test missile blew up soon after take-off and the following September another went off course and dis-

LEFT: *Titan II – the largest and most powerful ICBM in the US inventory – is launched from an operational type of silo at Vandenberg AFB, California. Top left, twin jets of smoke roar from vents on either side of the 44.5m (146ft) silo as the first stage ignites. Top right, the nosecone appears above ground level. Left to right, bottom row, the missile rises quickly out of the silo until the full 31.4m (103ft) length appears. The total number of missiles of this type deployed in the United States at present was 54 in 1963–83.*

BELOW: *The 18.3m (60ft) tank for an Atlas intercontinental ballistic missile is moved to the final assembly area of the manufacturing plant in San Diego, Calidornia. The thin-skinned stainless steel tank is pressurized to maintain its shape. During final assembly the tank was mounted to the propulsion section and other systems – guidance and control, power supply, telemetry, etc – were added. Although no longer an operational weapon, Atlas has been used widely in US space programmes.*

BELOW: *An Atlas ICBM blasts off from the US Air Force Missile Test Centre at Patrick Air Force Base, Florida. The weapon is long retired from operational service, its role being taken over by the Titan II and missiles of the Minuteman family. Towards the end of its life, Atlas provided targets for US anti-ballistic missiles launched on test from Kwajalein Atoll in the Pacific. Atlas also performed flawlessly in all the Mercury manned orbital flights, and was used in the launch vehicles Atlas-Agena and Atlas-Centaur.*

integrated completely in the atmosphere.

Only the booster engines were live in these initial tests. But when Atlas was launched with all engines firing, on 2 August 1958, success was complete; the rocket splashed into the Atlantic some 4,000km (2,500 miles) downrange. Although testing was far from complete, the US Defence Department lost no time in ordering the missile into production and Atlas D became operational on 'soft' sites in the United States in 1959.

In the meantime, a two-stage ICBM of more conventional design had been tested in January with a dummy top stage. The rocket – Titan I – employed the same propellants as Atlas and the engine gave a lift-off thrust of some 136,080kg (300,000lb). The vulnerability of missiles on open sites to nuclear attack, however, had already made clear the need for protective measures. Efforts to improve the reaction time of Atlas D led to above-ground 'coffin sites' at Vandenberg Air Force Base (AFB), California, and at Warren AFB, Wyoming. These offered a degree of protection from blast effects. The missiles were stored horizontally and fuelled behind protective walls before being elevated into a vertical launch position.

Titan I was stored in a deep underground silo. When the launch command was given,

the silo doors opened and the rocket was raised to the surface by an elevator. Like Atlas, however, Titan I suffered the disadvantage of having to have liquid oxygen loaded on board immediately before launch; this added precious minutes to its reaction time, which was about 15 minutes. An improved version of Atlas, the F model, was tested in 1961. This too was designed for protection in an underground silo.

In the Soviet Union, the two-stage ICBM known to the West as SS–7 Saddler, became operational in 1961. Unlike some other missiles of the period, it was omitted from Red Square military parades. Evidence from reconnaissance satellites showed it to be a two-stage missile over 30·5m (100ft) tall and some 3m (9·84ft) in diameter. It was thought to employ storable liquid propellants.

The Russians were then discovered to be installing missiles in underground silos from which they could be launched direct, the exhaust flame being diverted through vents on two sides of the silo. This rocket was thought capable of delivering a 5-megaton (MT) warhead over a range of more than 10,460km (6,500 miles). Another two-stage rocket of similar range, the SS–8 Sasin, became operational in 1963.

Strenuous efforts were being made in the

United States to develop ICBMs which could be launched from inside their silos. By 1963, the two-stage Titan II was in service. This used fuel and oxidant which (as in some contemporary Russian missiles) ignited on contact. This big ICBM stands 31·39m (103 ft) high, has a range of more than 8,047km (5,000 miles) and carries a warhead of over 5-MT yield. With propellants stored in the missile, Titan II has a reaction time of one minute from its fully hardened underground silo. Fifty-four of them are still operational at air force bases in Arizona, Kansas and Arkansas.

Russia replied to the new Titan with another two-stage rocket, also employing storable propellants and silo-launched. This is the ICBM known in the West as the SS–9 Scarp, which entered service in 1965 and has since appeared in four different versions. The so-called Mod 1 version carries a single re-entry vehicle of about 20-MT yield; Mod 3 has been used to test Fractional Orbit Bombardment System (FOBS) vehicles.

The missile build-up of the 1960s, however, was only beginning. The need to achieve faster-reacting missiles which could be held at instant readiness underground, or deployed in mobile launch systems, led to solid-fuel rockets of high performance. At the same time, advances in inertial guidance and reductions in the size and weight of thermonuclear warheads allowed ICBMs of high destructive power to be designed very much smaller than Atlas and Titan.

In the United States, the Minuteman I ICBM was officially declared operational in December 1962, only four years after the programme began. Eleven months later, Minuteman Wing I at Malmstrom AFB, Montana, and Wing II at Ellsworth AFB, South Dakota, were fully operational and missiles were being installed at Minot AFB, North Dakota. No fewer than 1,000 silos were built for Minuteman ICBMs, which lie hidden below ground in the wheat and cattle country of North and South Dakota, Wyoming, Montana and Missouri. Each flight of ten missiles is controlled from a metal capsule located 15m (50ft) underground and which is manned 24 hours a day, on a rota basis, by two officers of Strategic Air Command.

The original three-stage missiles have now been replaced by new models of increased range, accuracy and warhead power. The latest version, Minuteman III, carries three Multiple, Independently targetable Re-entry Vehicles (MIRVs). The MIRVs are carried in a post-boost 'bus' and are discharged against different targets according to instructions from a computer in the missile.

The Soviet counterpart of Minuteman I is the silo-launched SS–13 Savage, which entered service in 1968. It is believed to carry a single warhead of between 1- and 2-MT yield or three Multiple Re-entry Vehicles (MRVs) which explode in the same general area as each other but which are not individually guided towards specific targets.

Soviet designers developed the SS–13 on a modular basis. The second and third stages were used in a tracked transporter-erecter (Scamp) to achieve a mobile system which can be driven to different parts of the country and operated from points of concealment, among trees or in mountain passes for example. Some have been deployed near the border with China. A longer-range development of the same launch technique is Scrooge, which carries a larger missile (possibly also derived from the SS–13) in a long tubular container. The SS–11 Sego liquid-propellant rocket, however, although in the same general class as the SS–13, had greater performance and was produced in greater numbers.

France joined the 'ballistic club' in August 1971, when the first squadron of nine two-stage, solid-fuel *Sol-Sol Balistique Stratégique* (SSBS) missiles became operational in silos on the Plateau d'Albion in Haute Provence, Southern France. These S1 missiles could launch a 150-kT warhead a distance of 3,000 km (1,864 miles). Two years later a second squadron of nine S2 missiles became operational. These have a range of 3,150km (1,957 miles). In 1979 to 1980, these early missiles are to be replaced by the slightly smaller but more effective S3, which is designed to carry a 1·2-MT thermonuclear warhead. Manufactured at the French Atomic Centre at Pierrelatte, warheads of this type have been tested amid environmentalists' protests at

the *Centre d'Expérimentation du Pacifique*, Fangataufa Atoll, near Mururoa.

China, too, began to build ballistic missiles in the 1960s. The first examples appear to have been based on the Soviet Shyster/Sandal-class medium-range missiles. They have a range of over 1,700km (1,056 miles), can carry either h.e. or nuclear warheads and have been operational, according to intelligence reports, since about 1966.

By the end of the decade it was clear that missiles of longer range were on the way. First to be tested was a single-stage rocket known in the West as the CSS–2 which employed storable liquid propellants and had a range of some 4,000km (2,485 miles). A multi-stage variant of this rocket is thought to have launched China's first artificial satellite on 24 April 1970.

The mid-70s saw the introduction of multi-stage ICBMs of limited range – the CSS–3. Installed in underground launch cells in north-west China they are capable of reaching Moscow and the European part of the Soviet Union. Subsequently the Chinese deployed a complete spectrum of silo-based and mobile missiles, air-launched cruise missiles and submarine-launched missiles. Today their designers in this field are completely self-sufficient.

A later development, the CSSX–4 – a true ICBM with a range of some 11,000km (6,835 miles) – brought the United States within range of thermonuclear warheads. It was expected to be operational about 1980 in limited numbers.

Tests of nuclear warheads are made at Lop Nor in Sinkiang Province. The first atomic test took place there on 16 October 1964, the first thermonuclear (hydrogen-bomb) test on 17 June 1967. Subsequent tests have indicated that the warheads of China's ICBMs will have an estimated explosive yield of some 3 MT.

The main missile test centre – from which the Chinese also launch their satellites – is at Chuang Ch'eng-tze in the sparsely-populated, rolling wastes of Inner Mongolia.

Submarine-launched Ballistic Missiles

The increasing vulnerability of land-based ICBMs to thermonuclear attack led to the super-powers installing ballistic missiles in submarines. The US Navy began with Polaris A–1, which became operational aboard the Fleet Ballistic Missile (FBM) submarine USS *George Washington* on 15 November 1960, with a full complement of 16 missiles. By 4 October 1967, the forty-first FBM submarine, USS *Will Rogers*, was on patrol.

Thirty-three of these vessels were assigned to the Atlantic Fleet and eight to the Pacific, target selection and assignment being under the control of the Joint Chiefs of Staff. The first five vessels, which originally carried Polaris A–1, were re-fitted to carry the longer-range Polaris A–3; all the 616-class vessels have since been converted to accommodate the larger-diameter Poseidon C–3 missile.

Broadly, the development history of these two-stage sea-going missiles reflects the in-

TOP FAR LEFT: *A Minuteman ICBM installed in its silo. The total force of 1,000 missiles deployed in the United States comprises 450 LGM–30F (II) and 550 LGM–30G (III). The original Minuteman I is no longer in service.*
CENTRE FAR LEFT: *The Soviet equivalent of the USAF's Titan II was the SS–9 Scarp. It could carry a single 20–25 megaton warhead or three 5 megaton multiple re-entry vehicles (MRVs).*
BOTTOM FAR LEFT: *The three-stage SS–13 Savage which employed solid propellants was an answer to Minuteman I but only about 60 missiles were ever deployed – at Plesetsk south of Archangel.*
ABOVE LEFT: *The French constructed underground launch silos on the Albion Plateau in Haute Provence. This is one of their Sol-Sol Balistique Stratégique (SSBS) missiles on a fully enclosed transporter-erector.*
ABOVE: *The increasing vulnerability of land-based ICBMs to nuclear attack led to sea-based missiles. Here a Polaris is being loaded aboard the nuclear submarine USS George Washington.*
LEFT: *This Soviet tracked carrier bears the NATO code name Scamp; it contains the two-stage missile Scapegoat. Weapons of this type were operational on the Chinese border.*

creasing demand for more range, payload and accuracy. The Polaris A–2, slightly longer than the original, led to the A–3 which was a big advance on both earlier models. The Navy-industry team described it as 'an 85 per cent new missile', with 60 per cent more range and only a minimal increase in overall size. In fact, it had the same diameter of 137 cm (54in) and was just 30·5cm (12in) longer. The maximum effective range was about 4,635km (2,880 miles).

Britain abandoned the Blue Streak Inter-mediate-Range Ballistic Missile (IRBM) in April 1960 in favour of the American Skybolt air-launched ballistic missile. Skybolt, how-ever, was subsequently cancelled by the Americans, and by the Nassau agreement of December 1962 Britain was to buy Polaris missiles, without warheads. By the end of 1969 the base at Faslane on the Clyde in Scotland had been built and all four of the British-produced Polaris submarines had been commissioned.

They have been named HMS *Resolution* (launched September 1966), HMS *Renown* (February 1967), HMS *Repulse* (November 1967) and HMS *Revenge* (March 1968). Each carried 16 Polaris A–3s with warheads of British design and manufacture. Like their American contemporaries, each submarine was later converted to fire Poseidon, while much bigger vessels are built to carry Trident.

Russia, too, was in the race and, after gaining experience with large and primitive solid-propellant missiles carried in both diesel- and nuclear-powered submarines (in two and three vertical launch tubes extend-ing from the base of the hull through an extended bridge fin), much improved mis-siles began to appear.

Russian Yankee-class nuclear submarines have 16 SS–N–6 Sawfly missiles apiece in the pressure hull. The two-stage Sawfly has appeared in three versions: Mod 1 has a single re-entry vehicle of about 1–MT yield; Mod 2 has improved performance; and Mod 3 has three MRVs. Late models have a range of some 3,000km (1,865 miles).

Greater surprises were in store when high-resolution photographs of Soviet submarine yards were secured by US Air Force recon-naissance satellites. At Severdovinsk they uncovered the new Delta-class submarines which carry the SS–N–8 ballistic missile. This can achieve ranges of 7,885km (4,900 miles) and 12 are carried in Delta 1 vessels, 16 in Delta 2s.

For a submarine-launched missile, SS–N–8 performance is outstanding. Test firings carried out from a submarine in the Barents Sea in October 1974 resulted in high-accuracy impacts in the central Pacific, sug-gesting that the United States could be attacked from as far away as the North Sea.

The SS–N–8 is believed to obtain a mid-course navigational fix from a stellar-inertial guidance system. It is thought to employ a storable liquid propellant of the type which the Americans dismissed as too dangerous in their own FBM submarines in the face of counter-launched missiles, torpedoes and nuclear depth charges.

America's counterpart of the Delta 2 sub-

FAR LEFT: *Polaris A–3 starts a ballistic test down the Altantic Missile Range. The weapon could be distinguished from its predecessors by the bullet-shaped nosecone; it had new rocket motors, control systems and guidance package. Range exceeded 4,020km (2,500 miles).*
CENTRE LEFT: *Armed with multiple, separately targeted warheads, Poseidon C3 has greater ability to penetrate enemy defences. It was carried by 31 of the US Navy's FBM submarines which had 16 launch tubes each.*
FAR LEFT BELOW: *Britain's Blue Streak intermediate range ballistic missile was intended to be launched from silos in Britain and the Middle East. The project was cancelled in 1960 due to soaring costs and the increasing vulnerability of land-based missiles.*
LEFT: *Skybolt ballistic missiles on a B–52 bomber. Britain was preparing to operate the weapon from V-bombers when America cancelled the project. Instead, Polaris missiles were offered for use in British-built nuclear submarines.*
BELOW: *This Soviet two-stage silo-launched IRBM is code-named Sasin by NATO. It was being phased out of service in 1976 to make way for newer land-based and sea-based missiles.*
BOTTOM: *Test model of the French MSBS ballistic missile bursts from the sea at the Toulon test centre. A small fleet of French nuclear-powered submarines has since put to sea with 16 missiles apiece. The M–1 missile has a range of 2,590km (1,400 miles).*

marine is the Trident, which is planned to carry no fewer than 24 missiles with ranges of between 7,400–7,800km (4,600–4,850 miles). Each UGM–93A Trident I missile can carry eight MIRV warheads of 100-kT yield. Associated with Trident is the world's most advanced re-entry vehicle, the General Electric Mk 500 Evader. This is the first Manoeuvrable Re-entry Vehicle (MARV), which can make abrupt directional changes in flight in order to confuse anti-ballistic missile defences.

The Chinese also have begun to develop their own ballistic missiles for launching from nuclear powered submarines.

France, too, has a fleet of FBM nuclear-powered submarines, each of which carries 16 two-stage solid-fuel *Mer-Sol Balistique Stratégique* (MSBS) missiles of French design and manufacture. The M1 and M2 versions carry a 500-kT warhead with ranges of 2,593km (1,610 miles) and 3,148km (1,955 miles) respectively. The M1 was installed in the submarines *Le Redoutable* and *Le Terrible*, and the M2 in *Le Foudroyant*. M20 missiles with 1–MT thermonuclear warheads are carried in *L'Indomptable*. The fifth and sixth French missile submarines, *Le Tonnant* and *L'Inflexible*, to be commissioned in 1979 and 1983, will carry the larger M4 missile with six or seven multiple re-entry vehicles, each having a 150-kT yield.

Anti-ballistic missiles

The 'missile race' caused the super-Powers to seek ways of protecting vital targets from the worst effects of a nuclear attack. Since most ICBMs have multiple warheads and/or eject decoys, this is no easy task and a determined attack would surely saturate any practical defence system.

In the United States the Safeguard Anti-Ballistic Missile (ABM) system was developed

LEFT: *America's fast-reacting anti-missile missile is Sprint. The photograph shows a prototype on test. It is designed to destroy incoming warheads which have already succeeded in penetrating the first line of defence supplied by the long-range Spartan ABM.*
RIGHT: *Spartan is the long-range component of the US anti-ballistic missile system, Safeguard, established at Nekoma, North Dakota in November 1974. Its short range counterpart is Sprint. The first operational test was made in August 1970 when a Spartan, launched from Kwejalein Atoll, scored a direct hit on a Minuteman I. Other successes followed, but the Nekoma site was deactivated in 1975 as an economy measure.*

LEFT: *Inside this casing is Russia's long-range anti-missile missile which bears the NATO code name Galosh. 64 missiles were installed at four sites on the perimeter of Moscow in the late 1960's. Modifications have since been made to associated radars to make the system effective against China's long-range missiles which have the Soviet capital within range from sites in north-west Sinkiang.*

largely as a means of protecting Minuteman ICBM bases. The first sites were earmarked for Malstrom AFB, Montana, and Grand Forks AFB, North Dakota, each complex having 100 missiles – 30 long-range LIM–49A Spartan and 70 short-range Sprint. (However, see below).

Controlled by a Missile Site Radar (MSR) the three-stage Spartan is silo-launched to intercept missile re-entry vehicles above the atmosphere. The third stage is ignited by ground command at a time which depends on the precise path of the incoming warhead as worked out by fast computers on the ground. Interception, carried out automatically by radar command, ends with the detonation of a thermonuclear warhead.

The two-stage Sprint was meant to catch ICBM warheads which had eluded Spartan *after* they had entered the atmosphere. In consequence it has a very high acceleration and is launched from its underground cell by a separate charge, the first stage igniting in mid-air. Rapid course changes after launch are brought about by a thrust vector control system by fluid injection into a single propulsion nozzle, and then during second stage flight by aerodynamic controls.

Associated with Safeguard are Perimeter Acquisition Radar (PAR) located at strategic points around the United States which provide the MSRs with preliminary tracking data. Supporting data comes from radars external to the United States and US Air Force early warning satellites which can detect the launch of ICBMs and Submarine-Launched Ballistic Missiles (SLBMs) from geo-stationary orbit by sensing energy emitted in the rocket's exhaust.

Similar developments have, of course, taken place in the Soviet Union. Four sites of 16 ABM–1 Galosh anti-ballistic missiles defend Moscow. They operate in conjunction with two Try Add radar sites which contain target tracking and guidance/interception radars. Advance warning of missile attack depends on large Hen House phased-array radars in peripheral areas of the country and Dog House acquisition radars near the capital.

A new SH–4 exo-atmosphere ABM and a Sprint-type high-acceleration ABM have been tested together with more advanced radars. Advance warning satellites have also reached the test stage.

SALT negotiations

The enormous build-up of thermonuclear weapons by the major Powers in the 1960s led to the Strategic Arms Limitation Talks (SALT). A five-year agreement signed by the United States and the Soviet Union on 26 May 1972, allowed the USA 1,054 land-based ICBMs, 44 FBM submarines and 710 SLBMs. The USSR could deploy 1,618 land-based ICBMs, 62 FBM submarines and 950 SLBMs. There was no restriction, however, on the number of warheads a missile could carry. Strategic bombers were also left out of the equation, and no limits were placed on medium-range missiles of the type which could easily threaten to destroy Western Europe.

Curbs were also placed on anti-missile missiles. Under Salt 1, each super Power was allowed two ABM complexes with 100 missiles apiece. This allowed the Russians to preserve their Moscow ABM defences and, if desired, to place ABM's at one other location. America elected not to set up an ABM system near Washington but continued with the installations near the Minuteman base at Grand Forks, North Dakota, other sites under construction at this time being aban-

Fu
Poc

Magnetic Compass

BELOW: *In 1977 the US Air Force was directed to deploy new precision-guided cruise missiles. The winged missiles (next page) follow the contour of the earth at low level to strike their targets with deadly accuracy. They are a far cry from their predecessor – the Fieseler Fi 103, or V–1 flying bomb, used by the Germans during World War II for the indiscriminate bombardment of London and south-east England. It was powered by an Argus As 014 pulse-jet engine. At a pre-set range the V–1 was put into a dive cutting off the supply of fuel to the engine. The cut-out of the engine gave people in the area a few seconds In which to take cover before the missile hit the ground and exploded. Data : length 7·3m (24ft 5in) ; wing span 5·4m (17ft 8in) ; launch weight with 568 litres (150 gal) fuel 2,203kg (4,959lb) ; thrust 300kgf (660lbf) ; warhead 848kg (1,870lb) ; speed 563km/h (350mph) ; range 260km (161 miles).*

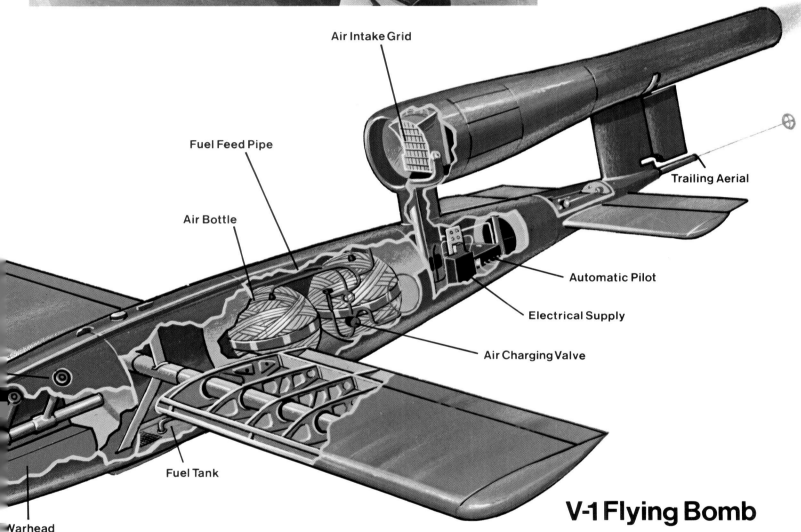

Air Intake Grid

Fuel Feed Pipe

Air Bottle

Trailing Aerial

Automatic Pilot

Electrical Supply

Air Charging Valve

Fuel Tank

Warhead

V-1 Flying Bomb

doned despite heated public debate.

The *status quo* was recognized in 1974 when a rider was added to the 1972 Treaty which limited each country to just one ABM complex. In 1975, however, the Grand Forks ABM sites were mothballed as an economy measure – just as they were becoming operational, leaving Moscow with the only 'alert' ABM complex.

Efforts to embrace a wider concept of strategic arms limitations in the interests of détente led to a second round of negotiations (SALT 2), which resulted in President Ford and Soviet Communist Party chairman Brezhnev meeting in Vladivostok on 23 to 24 November 1974. This meeting ended with

the so-called Vladivostok Accord, in anticipation of a ten-year agreement under which both countries would place a ceiling of 2,400 on all strategic nuclear-delivery systems– land-based, airborne and submarines. At most, 1,320 could have MIRV warheads.

Though the SALT 2 agreement was never ratified, both the USA and Soviet Union at least paid lip-service to it. The biggest argument concerned the Soviet SS–25 missile. Each side was allowed one new 'light' ICBM, and the Soviet choice was SS–23. Widespread deployment of SS–25 violated the treaty, unless one accepted the Soviet explanation that this totally different weapon was only a modification of the 'legal' SS–13 depicted on

p.394. The SS–20 escaped any limitation as it was held not to be strategic, despite its range of up to 5,000km (3,100 miles). SS–20 has caused much alarm because of its multiple high-yield warheads, great accuracy and the fact that, as it is fired from a mobile launcher, it is impossible to counter or even to count (1988 estimates were all in excess of 500 launchers deployed, most with quick reload capability).

When President Carter cancelled the B–1 bomber on 30 June 1977 he gave the go-ahead to two types of cruise missile, the USAF AGM–86B for launch from the B–52 and the versatile Tomahawk which is launched by the Navy from surface ships and

submarines and by the USAF from road vehicles deployed in Europe. Anti-nuclear groups in Western Europe were given special orders to campaign violently against 'cruise', because the difficulty of intercepting these small and agile weapons posed the Soviet defences with major problems. Subsequently AGM–86B became a major weapon for the B–1B and ATB (Advanced Technology Bomber). Like Tomahawk, it is powered by a Williams F107 turbofan in the 272-kg (600-lb) thrust class and carries a variety of warheads (usually a W–80 nuclear) up to 2,500km (1,550 miles). The Soviet outcry

was swelled by the decision to deploy 108 launchers in West Germany for the high-precision Pershing II battlefield support missile, despite the fact that these replace, on a one-for-one basis, the earlier Pershing Ia with a 400-kT warhead, far more destructive than the small W–85 warhead of the radar-map-guided Pershing II.

Fortunately, most of these problems appeared to begin to evaporate by the welcome signing of an INF (Intermediate-range Nuclear Force) treaty between President Reagan and Secretary Gorbachev in December 1987. This should progressively eliminate all

'theatre' nuclear weapons, with full opportunity for inspection by both sides. Even more significantly, it is generally hoped that this historic treaty will lead to a further agreement eliminating ICBMs and strategic SLBMs. By 1988 the Soviet strength in such weapons was prodigious, and even the USAF had begun to deploy significant numbers of its only new missile, the MGM–118 Peacekeeper. Initially this 88,450-kg (195,000-lb) ICBM was loaded into former Titan II silos, but a mobile railway-based version has also been developed. This appears now unlikely to be deployed.

ABOVE LEFT: *The US Navy Tomahawk. Three versions were planned for strategic, anti-ship and land-based roles. First flight with the terrain-comparison (Tercom) guidance system was made on 27 January 1977.*
ABOVE RIGHT: *The Soviet navy was well to the fore with cruise missiles in the early Sixties. Here a Styx surface-to-surface missile is put aboard a fast patrol boat of the Soviet Navy.*

BELOW: *Another Soviet anti-shipping cruise missile bears the NATO code name Shaddock. The coastal defence version is carried on an eight-wheeled vehicle which resembles a road tanker. Other versions of Shaddock are installed in similar weather-proof containers aboard ships of the Soviet Navy.*
INSET BELOW: *For launch the tank-like container is elevated by hydraulic jacks and the missile is fired from above the cab with the help of attached rocket boosters.*

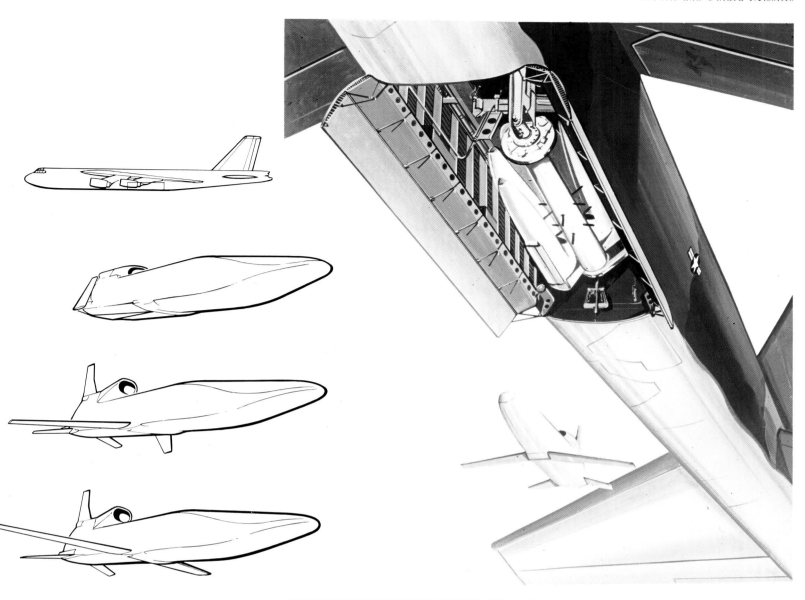

ABOVE: *B–52 bomb bay showing rotary launcher containing eight packaged ALCM's. Alternative arrangements are possible with extended range missiles.*
RIGHT: *The new-generation American flying-bomb which disturbed the SALT 2 negotiations with the USSR. Called the Air-Launched Cruise Missile (ALCM) it can fly a precise course to its target at tree-top level that avoids radar.*

Space Exploration

RIGHT: *Sputnik ('Fellow Traveller') 1, the world's first artificial satellite. It was launched on 4 October 1957 from the Tyuratam-Baikonur cosmodrome by Russia's newly-developed intercontinental ballistic missile designed by Dr S P Korolyov. It was little more than an orbiting radio transmitter with long 'whip' aerials, and weighed 83·6kg (184lb). The casing had a diameter of 58cm (22·8in). The orbit ranged between 215 and 939km (134 and 588 miles) inclined at 65·1 deg to the equator, and this spherical satellite reached a velocity of over 28,400kph (17,750mph). Sputnik 1 burned up in the atmosphere after 92 days on 4 January 1958. The same basic rocket that launched this satellite was later fitted with a top stage to launch Russia's cosmonauts, see page 403.*

FAR RIGHT: *A month later came an even bigger surprise for the new 'Space Age' world with the orbiting of the dog Laika in a small pressure cabin. Laika was fed and watered automatically and sensors attached to the body gave basic medical data which was relayed to Earth by radio. Sputnik 2, which remained attached to the orbiting carrier rocket, circled the Earth for 5½ months, then burned up as it came into contact with the Earth's atmosphere. The Russians had made no plans for recovery but Laika had long since lost consciousness and died as the air supply ran out. The photograph shows a replica of the spacecraft as it appeared on the nose of the launch vehicle. Laika's pressure cabin is the cylinder at the bottom; the radio transmitter is housed in the smaller section above.*

Russia shook the world on 4 October 1957, by launching the world's first artificial satellite. This signal event in world history had been made possible by the big intercontinental ballistic missile developed by the team under Sergei P. Korolev. The satellite itself, called *Sputnik 1*, weighed 83·6kg (184lb). It took the form of a polished aluminium sphere of 0·58m (1·9ft) diameter with long whip antennae. The distinctive 'bleeps' of its radio transmitter, picked up by radio stations around the globe, signified the dawn of a new age.

America's position in the field of rocketry was seriously challenged, and President Eisenhower felt the need to re-assure the American public, on television and radio, that the nation's defences were sound. In Moscow there was unsuppressed glee. Enthusiastically, *Pravda* commented that 'The launching of the first satellite proved that . . . the possibility of flying to the Moon is reasonable and justified.'

In the United States efforts were redoubled to bring to the launch pad at Cape Canaveral the three-stage Vanguard rocket which had previously been announced as part of America's contribution to the International Geophysical Year (1957–58). The objective was the launching of a 9kg (19·8lb) scientific satellite. The first rocket with all stages 'live' was ready on the launch pad on 6 December 1957. Within its nose cap was a test satellite of 13·3cm (6·4in) diameter, which weighed just 1.47kg (3·25lb).

However, when Vanguard's engine was ignited it failed to develop sufficient thrust, and instead of blasting off into space the rocket toppled over and erupted in a sea of flames. The tiny satellite, thrown clear of the inferno was still bleeping.

Almost immediately the Eisenhower administration agreed to allow the US Army Ballistic Missile Agency to develop their Jupiter C rocket as a satellite launcher. The concept had already been worked out by rocket scientist Dr Wernher von Braun and his associates, and it took a remarkably short time to bring it to the pad.

On 1 February 1958, the rocket – Juno I – lifted off from Cape Canaveral to put into orbit America's first artificial satellite, *Explorer 1*. The satellite swung round the Earth in a wide, elliptical orbit and the data it sent back led to the discovery that the Earth is girdled by electrically-charged particles from the Sun, trapped by the Earth's magnetic field (ie, the Van Allen radiation belts, named after the scientist who devised the experiment).

Soon afterwards a Vanguard rocket succeeded in placing a 1·47kg (3·25lb) test satellite into orbit, which Nikita Khruschev promptly dubbed America's 'grapefruit' satellite. Further salt was rubbed into the wound when, two months later, Russia launched *Sputnik 3*, a big cone-shaped geophysical laboratory nearly 1,000 times as heavy.

By then a major re-think of space policy was taking place in America, leading to the National Aeronautics and Space Act which was passed on 29 July 1958. This opened the way for the formation, from the nucleus of the existing National Advisory Committee for Aeronautics (NACA), of an important new body, the National Aeronautics and Space Administration (NASA), inaugurated on 1 October 1958. All space-related functions were to be transferred to NASA over the next four years, and the civilian space agency was thus able to acquire the services of the Army Ballistic Missile Agency and von Braun's concepts for large (Saturn) launch vehicles. In March 1960 von Braun was appointed director of NASA's Marshall Space Flight Center at Huntsville, and additional funds were given to support the Saturn rocket programme.

In the meantime, the Russians had not been idle. On 12 April 1961, they pulled off the supreme achievement of putting a man into orbit – using a modification of the same type of rocket that had launched the first sputniks. That man was Yuri Gagarin. He travelled in a ball-like re-entry capsule, attached to a service module which kept the craft supplied with air (oxygen and nitrogen at about 1 atmosphere pressure). The cone-shaped service module had small thrusters for changing the craft's attitude in space and a retro-rocket which, fired at a backward angle, braked it for re-entry. Only the cosmonaut's capsule returned to Earth; the service module, which separated before re-entry, was burned up by friction in the atmosphere.

Gagarin made one circuit of the Earth in 108 minutes, proving that a human being

BELOW: *The lifting power of the big Russian ICBM received still greater emphasis in May 1958 with the launching of Sputnik 3, a cone-shaped geophysical laboratory. It weighed 1,327kg (2,926lb), but otherwise failed to set any new records.*
RIGHT: *America's Vanguard 2 satellite in place on the launch vehicle before fitment of the jettisonable nosecone. Launched on 17 February 1959 into a high elliptical orbit it should circle the Earth for more than 100 years before it eventually burns itself out in contact with the Earth's denser atmosphere.*

BELOW: *Before America launched a man into Earth orbit, sub-orbital tests were made in Mercury capsules. This Redstone rocket launched astronaut Alan Shepard on a 15-minute flight over the Atlantic on 5 May 1961.*
BOTTOM: *The Vostok launcher. Russia kept the big rocket that launched Yuri Gagarin into orbit on 12 April 1961 secret for six years. A replica was eventually shown to the world at the Paris Air Show in 1967.*

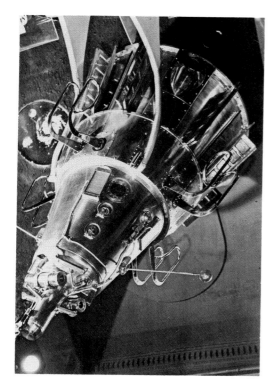

could survive in space at least for relatively short periods, despite the high accelerations experienced at take-off and re-entry, a period of weightlessness and exposure to radiation while in orbit.

America was still some way from a similar achievement, having restricted initial tests to ballistic lobs over the Atlantic. On 5 May 1961, Alan Shepard, in the Mercury capsule *Freedom 7*, was launched by a Redstone rocket to reach a height of 187km (116 miles) and travel a maximum distance of 478km (297 miles). Just over two months later Virgil 'Gus' Grissom made a similar flight in the capsule *Liberty Bell* splashing down 488km (303 miles) downrange. Although his capsule became waterlogged and sank, he was rescued in good shape by helicopter. (Grissom was later to lose his life with Ed White and Roger Chaffee in the Apollo spacecraft fire at Cape Canaveral on 27 January 1967.)

These early US lobs had one advantage over those carried out by Russia. The astronauts themselves used hand controls for changing the attitude of their spacecraft in flight, providing valuable experience for America's first attempt to place a man into orbit. The opportunity fell to John Glenn, who lifted off from Cape Canaveral on 20 February 1962, aboard the capsule *Friendship 7* on the nose of a modified Atlas ICBM. Glenn completed three orbits of the Earth in a flight which lasted 4 hours 55 minutes. Although the Russian Gherman Titov had already made a spaceflight lasting over 25 hours, Glenn's flight was a milestone

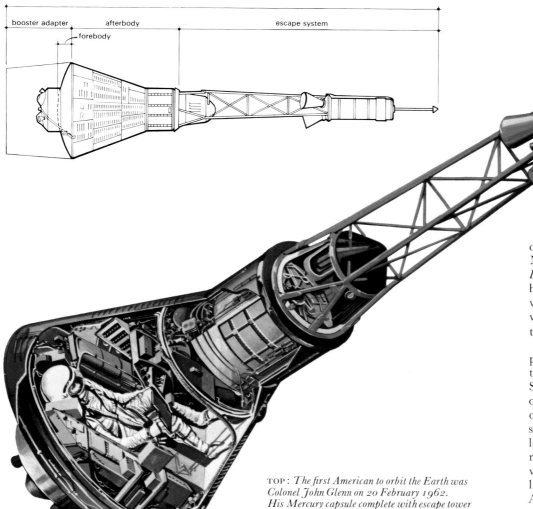

booster adapter | afterbody | escape system
forebody

in American space exploration history.

More American and Russian manned flights followed in swift succession. Valentina Tereshkova, the first spacewoman, made a spaceflight of nearly 71 hours in June 1963. Russia launched a three-man spacecraft, *Voskhod 1*, in October 1964, and less than five months later Russia's Alexei Leonov 'space-walked' from the orbiting *Voskhod 2*.

America followed this with the two-man Gemini spacecraft, from which Ed White spacewalked in June 1965, manoeuvring in space with the help of a hand-held 'gas-gun'. In *Gemini 7* Frank Borman and James Lovell remained aloft for nearly two weeks, during which they flew close to *Gemini 6*, manned by Walter Schirra and Tom Stafford. Manned spacecraft docked with Agena 'targets' in orbit, and astronauts perfected the art of spacewalking.

Events were now moving swiftly towards a major American challenge. On 25 May 1961, President Kennedy told a joint session of Congress, 'I believe that this nation should commit itself to achieving the goal, before the decade is out, of landing a man on the Moon and returning him safely to Earth . . .' Thus was born Project Apollo.

Unmanned Moon probes

Before men could land on the Moon it was necessary to send unmanned probes to explore its surface. Russia succeeded first in

TOP: *The first American to orbit the Earth was Colonel John Glenn on 20 February 1962. His Mercury capsule complete with escape tower weighed 1,935kg (4,265lb) and was just under 8m (26ft) long. Glenn made three orbits of the Earth before landing in the Atlantic some 338km (210 miles) north-west of San Juan, Puerto Rico. When he re-entered the atmosphere there was concern that the capsule's heat shield may have become detached. Flaming chunks of debris, glowing bright orange, streaked past his window. In fact, it was the retro-rocket pack which had failed to jettison and was instead burning up – not the heat shield.*

BELOW: *Although Soviet cosmonauts normally land on Russian soil they have to be ready to make an emergency landing on water. For example, the Soyuz 23 cosmonauts inadvertently splashed down in a Russian lake. This photograph shows Alexei Leonov during a training session in a Soyuz capsule.*

crash-landing an object (*Lunik 2*) on the Moon, in 1959. In the same year they sent *Lunik 3* around the Moon to photograph the hidden side. Another major achievement was the landing of *Luna 9* in February 1966, which sent the first television pictures from the Moon's surface.

America sent Ranger spacecraft, which photographed the Moon continuously until they crash-landed on it. The three-legged Surveyor craft examined the soil by means of a mechanical scoop operated under remote control from Earth. Another American spacecraft, Lunar Orbiter, photographed large areas of the Moon from orbit. These robot forays proved that the Moon's surface was safe for a manned landing, and allowed landing sites to be selected for the forthcoming Apollo missions.

Men around the Moon

At last – in the face of circumlunar flights by Russian unmanned Zond spacecraft – NASA gave the go-ahead to launch three men into lunar orbit aboard *Apollo 8*. During Christmas 1968 Frank Borman, James Lovell and William Anders gave television viewers on Earth their first sight of the Moon's bleak surface from a distance of 112·6km (70 miles).

This was the turning point in the 'Moon Race'. After a final dress rehearsal by the *Apollo 10* astronauts, a manned landing was at last in sight.

After separating from the *Apollo 11* command ship in lunar orbit, Neil Armstrong and Edwin Aldrin began their descent onto the Moon's Sea of Tranquillity, skirting a boulder field to put their Lunar Module safely on to the surface. The date was 20 July 1969.

A large part of the world, linked by television, watched the misty image of Neil Armstrong slowly descend the ladder to jump lightly on to the virgin moondust. He was soon joined by Aldrin. The two men, quickly adjusted to low gravity conditions. They set up scientific instruments and obtained 21·8kg (48lb) of moon samples.

At the end of their nearly 22-hour stay, they blasted off in the ascent stage of the Lunar Module, leaving the descent stage behind. A metal plaque records their visit: 'Here, men from planet Earth first set foot upon the Moon, July 1969 AD. We came in peace for all mankind.'

By the end of 1972, five more Apollo craft

had landed and 12 men had left footprints in the Moon.

More than 380kg (837·7lb) of lunar rock and soil were brought to Earth by the six successful missions, ranging from loose soil and surface rocks to core samples from *mare* ('lunar seas') and highlands. The mare rocks turned out to be relatively rich in titanium and iron, and the highland rocks had a greater abundance of aluminium. A typical *Apollo 11* soil sample contained: oxygen (40 per cent), silicon (19·2 per cent), iron (14·3 per cent), calcium (8·0 per cent), titanium (5·9 per cent), aluminium (5·6 per cent) and magnesium (4·5 per cent).

Of the rocks taken from the first four landing sites, none was older than 4,200 million years nor younger than 3,100 million years. The soils were older than the rocks, and dated back 4,500 or 4,600 million years. The age of the crater Copernicus, dated by material found in its 'rays', appeared to be only about 850 million years.

A total of 166 man-hours had been spent in surface exploration at six different sites, and the astronauts had travelled more than 96km (59·6 miles). Five Apollo Lunar Surface Experiment Packages – science stations powered by nuclear energy – continued to send data to Earth years after the astronauts had departed. A total of 60 major scientific experiments in geology, geophysics, geochemistry and astrophysics were left behind on the Moon. Another 34 were conducted in lunar orbit from the command ships.

In the light of this tremendous achievement, one must inevitably ask what had happened to the Russian challenge. In 1968–69 cosmonauts seemed to be on the brink of making circumlunar flights. By linking up separate spacecraft modules in orbit, they might even – eventually – have landed on the Moon. The immense success of the Apollo programme, however, put Soviet plans for manned moonflight into abeyance, and effort was concentrated on perfecting automatic Moon samplers, able to land, collect soil and rock specimens, and bring them back to Earth. Even as the *Apollo 11* astronauts were preparing to land, a robot spacecraft, *Luna 15*, was being manoeuvred in lunar orbit under control from a Russian ground station. There was speculation that this was an attempt to obtain a Moon sample and fly it to Earth in a final bid to up-stage Apollo. In the event, the probe crash-landed.

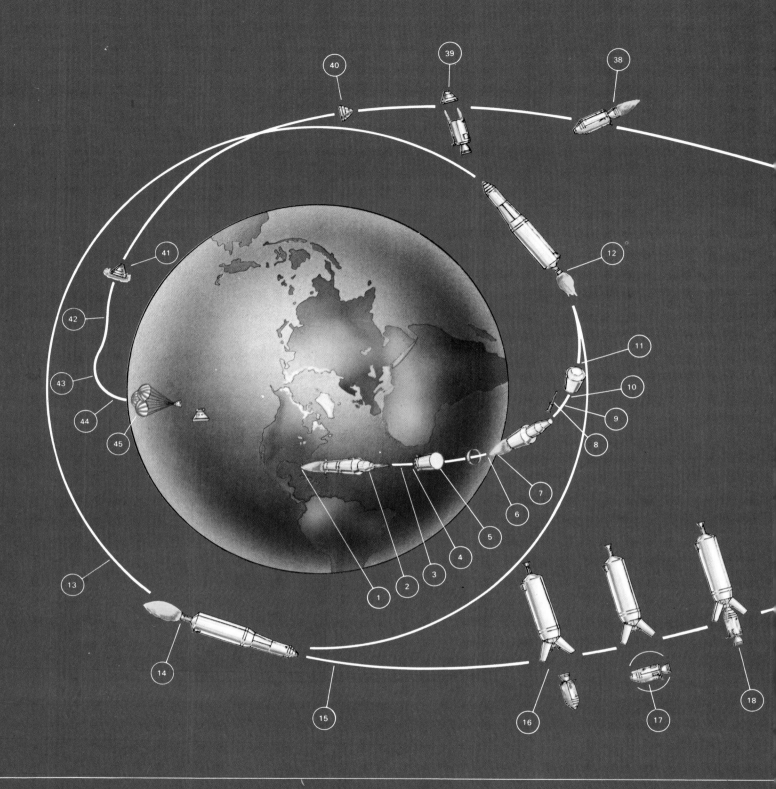

Apollo Space Mission

1	S-1C engines ignition
2	Lift-off
3	S-1C powered flight
4	S-1C engines cutoff
5	S-1C/S-11 separation, S-1C retro, S-11 ullage
6	S-11 engines ignition
7	S-1C/S-11 interstage jettison
8	Launch escape tower jettison
9	S-11 powered flight
10	S-11 engines cutoff
11	S-11/S-1VB separation, S-11 retro, S-1VB ullage
12	S-1VB engine ignition
13	Translunar injection 'GO' decision
14	S-1VB engine ignition
15	Translunar injection
16	CSM separation from LM adapter
17	CSM 180° turnaround
18	CSM docking with LM/S-1VB
19	CSM/LM separation from S-1VB, S-1VB jettison
20	SM engine ignition
21	SM engine ignition
22	SM engine firing lunar orbit insertion
23	Lunar orbit insertion
24	Begin lunar orbit evaluation
25	Pilot transfer to LM
26	CSM/LM separation
27	LM descent engine ignition
28	LM descent
29	Touchdown
30	Lift-off
31	LM—RCS ignition
32	Rendezvous
33	CSM/LM initial docking
34	Transfer crew and equipment from LM to CSM
35	CSM/LM separation and LM jettison
36	Transearth injection
37	SM engine ignition
38	SM engine ignition
39	CM/SM separation
40	Orient CM for re-entry
41	122,000m (400,000ft) altitude penetration
42	Communication blackout period
43	61,000m (200,000ft) altitude
44	76,200m (250,000ft) altitude
45	Deploy main chute at 3,000m (10,000ft)

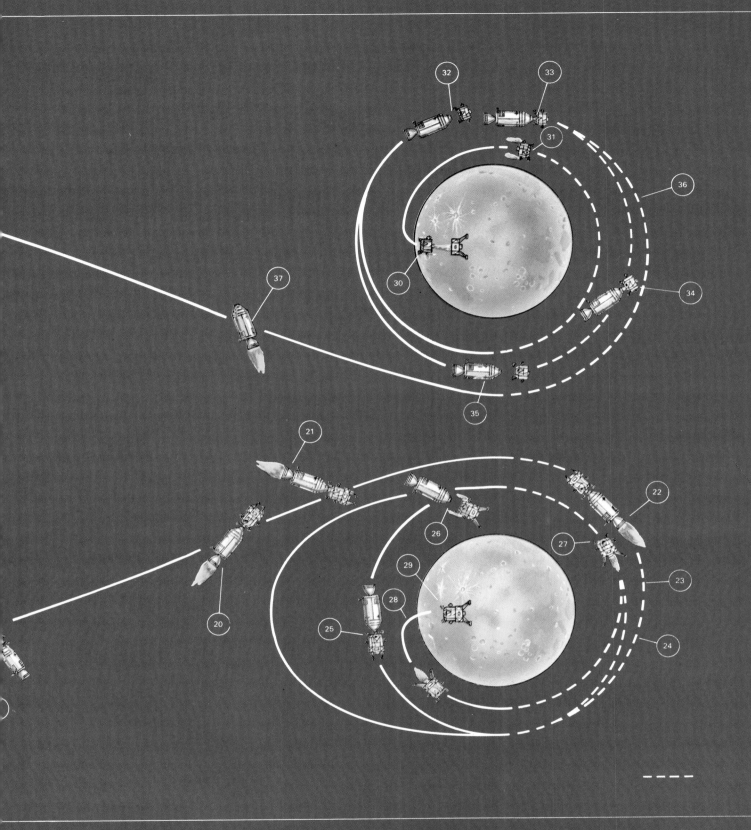

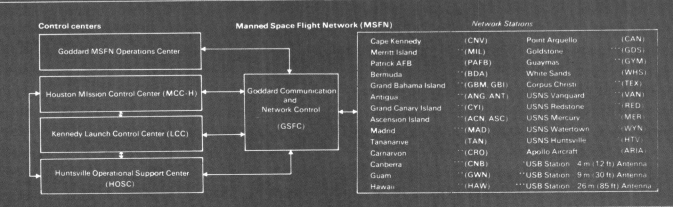

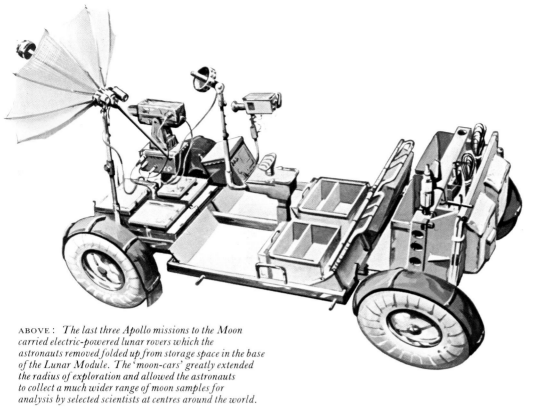

ABOVE: *The last three Apollo missions to the Moon carried electric-powered lunar rovers which the astronauts removed folded up from storage space in the base of the Lunar Module. The 'moon-cars' greatly extended the radius of exploration and allowed the astronauts to collect a much wider range of moon samples for analysis by selected scientists at centres around the world.*

In September 1970 *Luna 16* soft-landed, drilled 35cm (13·8in) into the Moon's surface, transferred a 100 gramme (3·5oz) core sample to a capsule and blasted it back to the USSR. Two months later *Luna 17* succeeded in landing a Lunokhod roving vehicle, which the Russians steered over the lunar surface by remote radio control.

Interplanetary probes

Similar means of robot exploration were developed in the United States. With the 'Moon Race' won, Congress looked with disfavour on more manned missions as new priorities and the unpopular war in Vietnam began to engage its attention. Any long-term prospects of astronauts pressing on to Mars were abandoned.

Instead, research was directed to developing an increasing number of ingenious automatic vehicles to range deeper into the Solar System. Built at a fraction of the cost of manned spacecraft, they could be sent to places where men dare not go: deep into the crushing atmosphere of Venus, for example, and through the intense radiation belts of Jupiter.

In 1962 America had already flown *Mariner 2* past Venus and probed the planet with instruments which indicated surface temperatures above the melting point of lead. Russia then parachuted instrument capsules through the planet's thick carbon dioxide atmosphere to find that its surface pressure was 90 to 100 times that on Earth. There was strong competition between Russia and America to probe the secrets of Mars, with fly-by and orbital missions. The US *Mariner 9* spacecraft became the first artificial satellite of the Red Planet on 13 November 1971, as a dust storm raged on Mar's surface. When the dust storm cleared *Mariner 9*, scanning from orbit, discovered huge volcanoes, canyons and features like ancient dried-up river beds.

At the same time Russia was attempting to land instrument capsules. The first was destroyed on impact; the second came down in the region between Electris and Phaethontis (about 48 deg S, 158 deg W), but failed just as it began to send a picture.

In 1974 *Mariner 10* flew past Venus on its way to Mercury, showing for the first time that the tiny planet nearest the Sun had a moon-like surface of craters, mountains and valleys.

Pioneers 10 and *11* swung around the giant planet Jupiter before departing on different paths. The first is due to leave the Solar System in 1987; the second is headed for the ringed planet Saturn, which it should reach in 1979. Calculations show that *Pioneer 10* will reach the environs of the star Aldebaran in the constellation of Taurus after some 1,700,000 years. Both craft carry pictorial greetings to any extra-terrestrial intelligence that may find them in the remote future, giving information on the senders, their home planet known to them as Earth, and its position in the Galaxy.

In 1973 Russia again launched a spacecraft to Mars, using the powerful Proton D–I–e rocket and spacecraft of a new generation – *Mars 4, 5, 6* and *7*. The results were disappointing. One arrived in orbit around Mars; the others flew past, one doing so unintentionally when its braking engine failed to fire. However, the biggest setback was the loss of the two landing capsules released by the *Mars 6* and *7* mothercraft as they approached the planet.

Despite these costly setbacks, Russia pressed on and, in November 1975 two large automatic spacecraft succeeded both in landing capsules on Venus and going into orbit. The capsule of *Venera 9* plunged into the planet's atmosphere at 10·7km/sec (6.6 miles/sec) and at an angle of 20 degrees. When speed had dropped to 250m/sec (820ft/sec) the probe threw off clamshell protective covers and deployed a parachute. After it had floated down to a height of 50 km (31 miles) from the surface, the parachute jettisoned and the capsule was slowed in the thickening atmosphere by a disc-like drag brake attached to the body. It hit the surface at 7–8m/sec (23–26ft/sec), the landing shock being absorbed by a ring of crushable metal. The capsule transmitted data (via its orbiting mothercraft) for 53 minutes, including the first picture to show surface conditions on Venus.

Clearly visible was a level surface strewn with sharp-edged rocks 30 to 40 cm (12 to 16in) across, which might have come from volcanoes. Atmospheric pressure at the surface was about 90 atmospheres, and the temperature 485°C (905°F).

The capsule released by *Venera 10* came down some 2,200km (1,367 miles) from the first landing site, transmitting for 65 minutes and sending a picture showing outcrops of rock among rock debris. There the atmosphere was 92 atmospheres, which suggested that the craft had dropped into a lower lying region.

Two years later America sent two Viking spacecraft into orbit around Mars, on 19 June and 7 August 1976 respectively. The first released its landing craft onto the dry plains of Chryse Planitia on 20 July, seven years to the day since men first walked on the Moon. Photographs received on Earth showed a scattering of rocks; the surface had a red-brown colouring and the sky was a pinkish hue. The thin carbon dioxide atmosphere had a pressure of 7·69 millibars.

On 3 September 1976, the Lander of Viking *2* made its touchdown in Utopia Planitia some 200km (125 miles) west of the crater *Mie*, which has a diameter of about 100km (62 miles). It is believed that some of the rocks in the pictures came from this impact crater.

In a highly successful series of experiments, both robot Landers scooped up Martian soil and transferred samples to an onboard biology laboratory for automatic analysis. The results revealed that the main constituents of Mars' soil were iron, calcium, silicon, titanium and aluminium. But whether the reactions of treated soil samples are evidence of chemical or biological activity remain an open question. The results indicated a surprising amount of water and oxygen in the soil, but there was no certain evidence of organic material.

Meanwhile, the Viking Orbiters had made extensive photographic surveys of the sur-

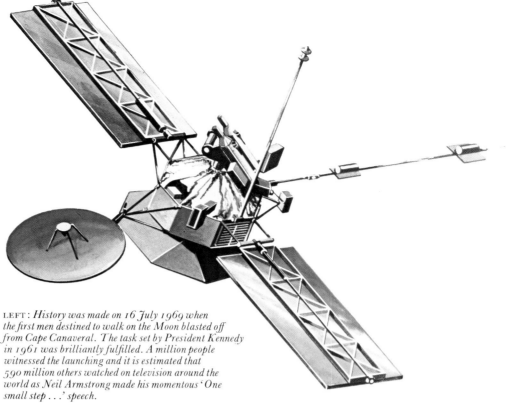

LEFT: *History was made on 16 July 1969 when the first men destined to walk on the Moon blasted off from Cape Canaveral. The task set by President Kennedy in 1961 was brilliantly fulfilled. A million people witnessed the launching and it is estimated that 590 million others watched on television around the world as Neil Armstrong made his momentous 'One small step . . .' speech.*

LEFT INSET: *The first man to leave footprints in the moondust was Neil Armstrong who later took this picture of his companion Edwin Aldrin. The two men landed on the Sea of Tranquillity in the Lunar Module Eagle on 20 July 1969 while the third member of the team, Michael Collins, remained aboard the orbiting command ship* Columbia *taking charge of Earth Communications.*
TOP: *America's Mariner 10 made a tour of the inner planets in 1973–75. After taking photographs of the Earth and the Moon, it swung past Venus and on to Mercury, the tiny planet nearest the Sun. Mercury was found to have a moon-like landscape with craters, mountains and valleys – also a distinctive magnetic field. Previous launches of the successful Mariner series had already probed the secrets of Mars.*

INSET: *Lunokhod looked like something out of a Jules Verne fantasy. In fact, it was a highly functional electric-powered robot with TV 'eyes' which sent pictures to a Russian control station. When its batteries needed re-charging the lid opened to expose solar cells which generated electricity from sunlight. Throughout its long lunar life this vehicle was to be directed successfully from Earth.*
ABOVE: *Lunokhod 1 was landed on the Sea of Rains by the Luna 17 carrier in November 1970. After being driven off ramps under remote control from Earth, it was steered on a series of journeys lasting 322 Earth days. The Russians claim it travelled a distance of 10,542m (34,588ft), sending back TV pictures and scientific data, and travelling on gradients up to 30 deg in the Mare Ibrium.*

Clarke wrote: 'No communications development which can be imagined will render the chain obsolete, and since it fills what will eventually be an urgent need, its economic value will be enormous.'

Although Clarke and the BIS did all they could to promote the scheme, it was left to America to put the 24-hour satellite into commercial practice with *Early Bird* built by the Hughes Aircraft Company in 1965. Today's communications explosion is the result. Matching the Earth's rotation above the Atlantic, Pacific and Indian Oceans are big, drum-shaped satellites of the Intelsat network, any one of which can relay up to 6,000 telephone calls, or 12 colour television programmes or a combination of the two.

By 1975 the chain linked North and South America, Europe and Africa, and girdled the Far East via 111 antennae at 88 Earth stations in 64 countries. (By the use of land-lines from the existing Earth stations, the system actually reaches more than 100 countries on a full time basis).

Since 1965 transoceanic telephone traffic has grown from an estimated three million calls a year to more than 50 million in 1974, and it is predicted that calls will increase to 200 million by 1980. Indeed, because of satellites, a three-minute telephone call between New York and London is 55 per cent cheaper today than when *Early Bird* entered service. And satellites are the only way to send television around the globe.

The ripples of Clarke's idea are spreading continually. Indonesia, for example, ordered from Hughes a domestic satellite system of the kind pioneered in Canada and America. Unlike satellites which span oceans and link continents, a domestic satellite focuses its power into a beam that concentrates coverage onto a specific nation. The Indonesian satellite *Palapa* (a name which signifies national unity) opened its service to 130 million people in the world's largest archipelago on 17 August 1976, in celebration of the Republic's 31st year of independence. It provides telephone, television, radio, telegraph and data services to the populated areas of 5,000 of the nation's 13,000 islands, which stretch over 5,000km (3,107 miles). Plans were also set in hand to use the satellite to broadcast educational television programmes throughout Indonesia.

The big test of the social applications of space technology began in 1974 in a programme co-ordinated by the US Department of Education and Welfare. The powerful ATS–6 'umbrella' satellite was placed in geostationary orbit just west of the Galapagos Islands in the Pacific to transmit educational and medical services directly to low-cost television receivers in scores of isolated communities in the United States.

face, using two high-resolution cameras which provided spectacular new details of the ancient river beds, the equatorial canyon and the polar caps. Temperature measurements revealed that, contrary to expectations, the latter consist mainly of water ice and not dry ice (frozen CO_2).

Earth satellites
The results of deep space astronomy conducted from unmanned Earth satellites and manned space stations have been no less exciting. Orbiting above the Earth's hazy and turbulent atmosphere, they have opened up a 'new' universe to which we were practically blind – a universe of fantastic spectral range extending far beyond visible light and radio waves to X-rays, ultraviolet and gamma rays. Astronomers have begun to make fundamental discoveries about supernovae (exploding stars), pulsars (pulsating stars), neutrons (superdense stars), 'black holes' and quasars (quasi-stellar objects). The effect of changing conditions of the Sun upon the Earth's environment has also been under constant study.

Not all the work is in the interests of pure science. Meteorological satellites keep us informed from day to day of the world's changing weather conditions and give advance warning of destructive storms like hurricanes and typhoons. Early examples of such satellites were Tiros and Nimbus (US) and Meteor (USSR). They send television pictures of cloud, ice and snow cover on the day and night sides of the Earth, and also provide data on the thermal energy reflected and emitted by the Earth and its atmosphere, important for the greater understanding of meteorological conditions.

At the same time the super-powers lost no time in exploiting the military potential of space – especially by the use of photo-reconnaissance satellites and satellites able to provide advance warning of missile attack.

Earth satellites have also become 'radio stars', to assist the precision navigation of ships and aircraft in all weathers, and networks of military satellites have been established to maintain reliable, fade-free communications with naval vessels. *Elint* (electronic intelligence) missions are also carried out from orbit, with satellites 'listening in' to defence radars and communications, logging their codes and frequencies.

After the USA, in 1962 and then in 1975, had shelved development of an anti-satellite rocket (Saint), the Russians began testing 'killer' satellites within the Cosmos programme which fragmented close to orbiting targets. Other military test satellites were observed to eject re-entry vehicles at the end of a single orbit. These were the so-called Fractional Orbit Bombardment System (FOBS) vehicles which, launched in anger, could attack the United States through the 'back door' of the South Pole.

Nonetheless, enormous efforts have been made by the super-powers to apply the fruits of space technology to human need. More and more people are becoming involved in Space in their ordinary work – geologists, oilmen, doctors, farmers, ecologists, teachers and others. And although few of them are likely ever to see a rocket blast off from a launch centre, they will be called upon to use their skills, via space, to help solve some of mankind's most terrifying problems – poverty and famine, over-population, illiteracy, pollution, and perhaps even the energy crisis.

It all began with a space age communications explosion sparked by a technical memorandum which the 'space prophet' Arthur C. Clarke placed before the Council of the British Interplanetary Society (BIS) in 1945.

LEFT: *America triumphed by landing two Viking robots on Mars in 1976. The first came down on* Chryse Planitia *on 20 July, the second on* Utopia Planitia *on 3 September. Each took samples of the reddish Martian soil with a mechanical scoop and analysed them. Although there were strong chemical reactions, there was no positive sign of any animal or vegetable life.*

BELOW: *Technicians of the British Aircraft Corporation (now British Aerospace) check the antennae group of Intelsat 4, one of the world's most advanced communications satellites.*

CENTRE BELOW: *GEOS 1 was to have been Europe's first geo-stationary satellite for research into the Earth's magnetosphere. A launch mishap left it stranded in an unplanned orbit.*

ABOVE: *Tiros 3, an American meteorological satellite, used to transmit pictures of cloud cover and the Earth's terrain to assist in meteorological research and the understanding of weather patterns.*

BELOW: *Comsat 1, a US communications satellite in geo-stationary orbit used to relay telephone calls and TV broadcasts.*

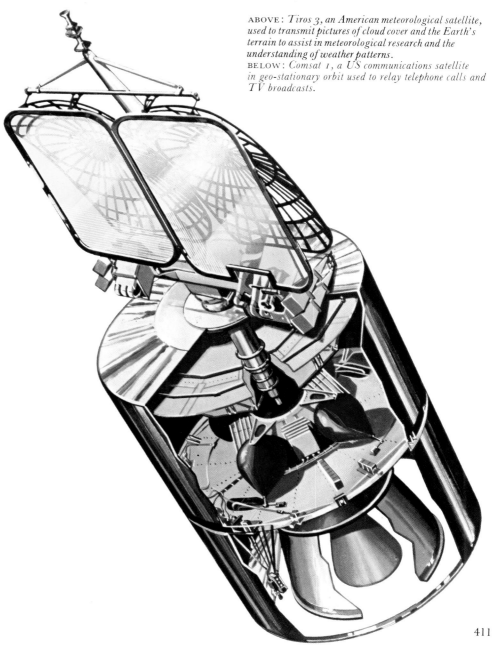

Teachers participating in the experiment were able to see instructors, pictures and charts 'bounced' to them from the satellite across hundreds of miles, and to ask questions as if they were in the same lecture hall. The satellite allowed doctors in city hospitals to 'visit' patients in remote regions of Alaska, monitor their medical condition and advise locums of specialized medical treatments. Medical data, including electro-cardiograms and X-rays, were sent to doctors over the space link.

The same versatile platform was later moved eastward, by operating a small rocket motor aboard the satellite, to take up a position some 35,880km (22,295 miles) above Lake Victoria in East Africa. From this lofty vantage point, it broadcast directly to children and adults in some 5,000 towns and villages in seven states of India.

Wernher von Braun one of the creators of that experiment claimed that 'ATS–6 could

turn out to be the most important advance since movable type as a means of reaching people now separated by vast geographical, economic and cultural barriers. Its successors may indeed eradicate illiteracy from the face of the Earth.'

All the time space investment is producing new opportunities for social advance. 'If I had to chose one spacecraft, one Space Age Development, to save the world,' said NASA Administrator Dr James C. Fletcher, 'I'd pick ERTS and the satellites which I believe will be evolved from it later in this decade.' He was speaking about the first of two observation satellites – later re-named Landsat – which take special photos of the Earth. Rocketed into polar orbits some 925km (575 miles) high, they circle the globe every 103 minutes. Every 18 days they view the same spot anywhere in the world at the same local time of day.

The 952kg (2,100lb) Landsats carry television cameras as well as radiometric scanners, to obtain data in various spectral ranges of visible light (red, blue and green) and infrared, which show up hidden features on the ground. The data obtained relates to crop species, crop quality, moisture content of the soil, and soil fertility over wide areas. This in turn helps to find the best places to develop new land for farming and the best time to plant and harvest crops for maximum yield.

Satellite observation also promises to achieve an early prediction of the total harvest, so that food supplies are known in advance and can be set against future need. An early discovery was that crops and forest areas damaged by blight or insects showed up blue-black in the satellite photos and healthy vegetation pink or red.

Landsat observations have also shown how controlled grazing of cattle in the drought and famine-stricken Sahel region of Africa can lead to the reclamation of desert land for productive use. And in cases of severe flooding – along the Mississippi and in devastated areas of Bangladesh, for example – the satellite images have given vital information for the resettlement of whole communities.

India is now moving into the field of space technology with determination. Having had a scientific satellite launched by the Russians in April 1975, India is now building her own satellites and launch vehicles for meteorological study and storm warning, Earth resource observation and help with agriculture. If, for instance, the onset of the monsoon rains can be predicted by satellites – information which is crucial to the transplantation of rice – millions of dollars a year could be saved. And in these regions the saving of a rice crop is not just a financial gain; it could mean the

ultimate difference between life and death.

Other developing countries are catching on fast. Kenya, Zaïre, Iran and Venezuela were among the first to set up ground stations so that resource data supplied by America's Landsats could be applied without delay in their own particular territories. Already the Landsats have made important discoveries in many parts of the world in the search for deposits of minerals and fossil fuels.

An early revelation was the 'geological inference' that the oil and gas deposits of the Alaskan North Slope are much larger than had been previously believed. The satellites obtained multi-spectral images of previously invisible 'large lineaments and structural breaks' bearing no relationship to surface geology. It is such hidden features, discernible only by satellite, that are providing fresh hope of discovering oil and mineral deposits throughout the world.

In retrospect, it is that sort of data, streaming back to us from spacecraft sent to explore the Moon, Mars and other planets, that has been most valuable. It has prepared the way for a spectacular and much needed rediscovery of our own planet. And, what is even more remarkable, these benefits are coming from only a small share of the one cent of the tax dollar which currently supports the space programme in the United States.

Space stations

The desire to gather information on Earth's natural resources and expand other useful areas of science and technology, led to the first manned space stations. After teething troubles with test models, the Russians launched a series of 18·5-tonne (18·2-ton) Salyut orbital laboratories. Cosmonauts docked with the stations at intervals in Soyuz ferries to carry out various research tasks, which ranged from photographing the Earth's resources to making experiments in biology, medicine and industrial processing under conditions of weightlessness. Telescopes were carried to observe the Sun and the stars from above the atmosphere.

The much larger Skylab space station launched by America on 14 May 1973, carried out a wide range of similar experiments which included use of the Apollo Telescope Mount to increase knowledge of the Sun and its influence on Earth's environment. Metals were melted in a small electric furnace, and experiments made in the growth of crystals of a kind which could be of value to the electronics industry. Photographs of the comet Khoutek were taken by astronauts.

Skylab was made from the S–IVB stage of a Saturn rocket. Fitted out with two-storey accommodation, it included a large

workshop area for experiments and, on the 'lower floor', a wardroom and living area including a combined washing and toilet compartment and separate sleeping compartments.

Although damaged during the launching – accomplished by a two-stage Saturn V – the first two boarding parties of three men each were able to carry out running repairs. The workshop's external meteoroid shield and one of the large solar 'wings' had been torn off by air pressure as the rocket passed through the atmosphere; the other 'wing' was held fast by a piece of torn metal. Not only was the first boarding party able to erect a temporary sunshade over part of the workshop to reduce the temperature, but the astronauts were able to climb out of the station to cut away the obstruction and release the jammed solar 'wing'. The second team brought a larger sunshield which they erected without great difficulty. In the end Skylab exceeded all expectations, with three highly successful astronaut missions lasting 28, 56 and 84 days respectively.

Not long after, Russia and America pooled their resources in a joint space experiment which entailed astronauts and cosmonauts training in each others' countries. This was the celebrated Apollo-Soyuz Test Project (ASTP).

The mission got underway in July 1975, when a three-man Apollo spacecraft docked with a two-man Soyuz 225km (140 miles) above the Earth. Both craft had been modified to make their life support and docking

BELOW: *Detached S-IVB stage of a Saturn rocket photographed from Apollo 7 spacecraft in Earth orbit. The Skylab space-station was based on a modification of the S-IVB. Within the big hydrogen fuel tank compartments were built to provide two-floor accomodation for three astronauts. An experiment chamber, which occupied the upper third of the tank, allowed the astronauts to float about freely doing various zero-g experiments. Below this was the crew work area and living quarters which featured every possible refinement to make this cramped space an endurable working environment.*

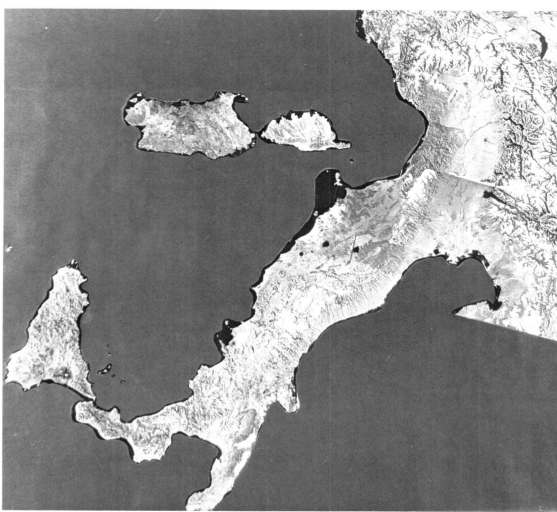

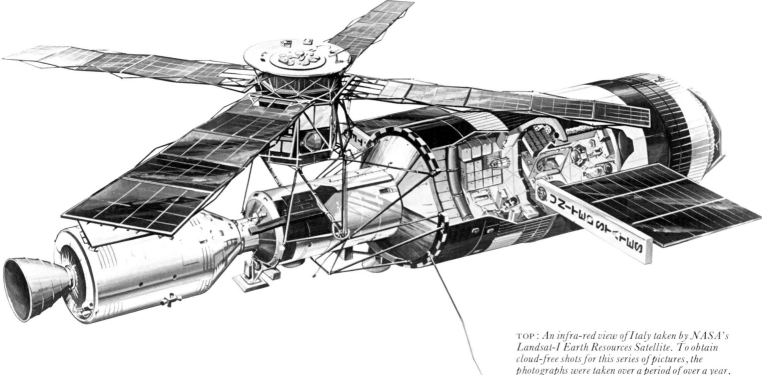

systems compatible. The crews visited one another and carried out joint experiments before separating their ships and landing in their respective recovery areas. The project proved the feasibility of space rescue missions and the ability of nations to co-operate in the joint construction and supply of space stations and in their peaceful use.

The Space Shuttle

Meanwhile, a revolutionary new space vehicle which promises to bring down the cost of space travel was taking shape in America. Unlike conventional rockets, which fall to destruction each time a spacecraft is launched, this winged craft – blasting off like a rocket and flying home like an aeroplane –

TOP: *An infra-red view of Italy taken by NASA's Landsat-1 Earth Resources Satellite. To obtain cloud-free shots for this series of pictures, the photographs were taken over a period of over a year, between 1972 and 1973 as the satellite cruised in polar orbit.*

ABOVE: *Skylab was badly damaged when it was launched on 14 May 1973. It arrived in orbit minus a solar 'wing' with the meteoroid shield ripped off and the other 'wing' jammed shut by a piece of torn metal. The first boarding party had to carry out emergency repairs. Astronauts spacewalked to extend the remaining solar 'wing'. They also fixed a sunshade to prevent Skylab from overheating. In the end near-disaster was turned to triumph in a series of spectacular experiments. At the time of this flight, Skylab took the record for the heaviest object ever to have orbited the Earth.*

can be re-used 100 times or more. Called the Space Shuttle, it consists of the winged Orbiter, about the size of a DC–9 airliner, a large external tank and two powerful solid rocket boosters. The Orbiter has a crew of three – pilot, co-pilot and mission specialist.

Launched vertically from the pad like a normal rocket, it rides on the large external tank which supplies the ascent propellants (liquid oxygen and liquid hydrogen). The two solid fuel rockets, mounted on the sides of the tank, assist during the lift-off and jettison when the craft has climbed some 45km (28 miles). They are intended to be recovered after parachuting into the sea. Some of them will be used again after being cleaned and re-filled with propellant.

The big tank is jettisoned just before the 'spaceplane' arrives in orbit under the thrust of small manoeuvre engines. This is the only part which does not return to Earth, as it burns up in the atmosphere during re-entry.

The first Orbiter, OV–101 *Enterprise*, underwent captive and then free flight tests in 1977, carried aloft on a modified Boeing 747 to make unpowered landings at NASA Dryden (Edwards) in California. OV–102 *Columbia* made the first actual mission from Cape Canaveral in April 1981, flying a 36-orbit mission with John Young as commander and Capt Robert Crippen as pilot. After STS–5 (the fifth mission) it was modified to carry Spacelab and a full crew of six. After STS–9 it was returned to Rockwell at Palmdale to be brought to the latest operational standard, returned to service in 1986. OV–099 *Challenger* made its first flight in April 1983. It became the first operational Orbiter, cleared for 100 missions between major overhauls, but was destroyed by failure of a solid boost motor on mission 51L on 28 January 1986. This delayed the programme while the boost motors and various other items were modified. OV–103 *Discovery* first flew on 30 August 1984, and OV–104 *Atlantis* first flew on 3 October 1985. These last two Orbiters are lighter than their predecessors, partly because insulation blankets replace most of the thick tiles (which tended to become detached in flight). A sixth Orbiter was being built in 1988 to replace *Challenger*, and the whole programme was expected to pick up again in that year.

Meanwhile, the European Ariane, a traditional expendable launch vehicle made by a French-led European group, has won greatly expanded commercial business through non-availability of Shuttle launches. Ariane launches are from Kourou, in Guiana. Japan and other countries are also selling payload space aboard expendable space rockets. Yet the economic attraction of reusable launchers is obvious. In 1985 France began full-scale development of the Hermes, similar to a smaller version of the NASA Shuttle Orbiter, able to carry up to six crew and 3,000kg (6,614lb) of cargo. British Aerospace and Rolls-Royce have proposed a much more advanced and economic vehicle in which nothing is thrown away and the takeoff is like any other aeroplane. Called Hotol (Horizontal takeoff and landing), this would be a large (60m, 197ft long; 230,000kg, 507,065lb) vehicle looking slightly like a Concorde. Its RB.545 engines would burn oxygen from the air for the first nine minutes of flight, thereafter switching to stored liquid oxygen. On paper it renders Hermes obsolete, but as this is written the British have no international support and are being criticised for not joining wholeheartedly in the Hermes programme!

LEFT AND BELOW: *The Space Shuttle is intended to reduce costs by making flight into orbit more like normal air travel. It takes off vertically like a rocket with crew and cargo and flies home like an aeroplane. When it is not carrying Spacelab, it can launch satellites or retrieve them from orbit for repair or servicing. The scale drawing shows the Shuttle Orbiter mounted on the External Tank with the two Solid Rocket Boosters alongside. Overall length is 56.1m (184ft). At over twice the cost of the Apollo project, America's Space Shuttle will – by 1990 – lead the way to 'space colonies'.*

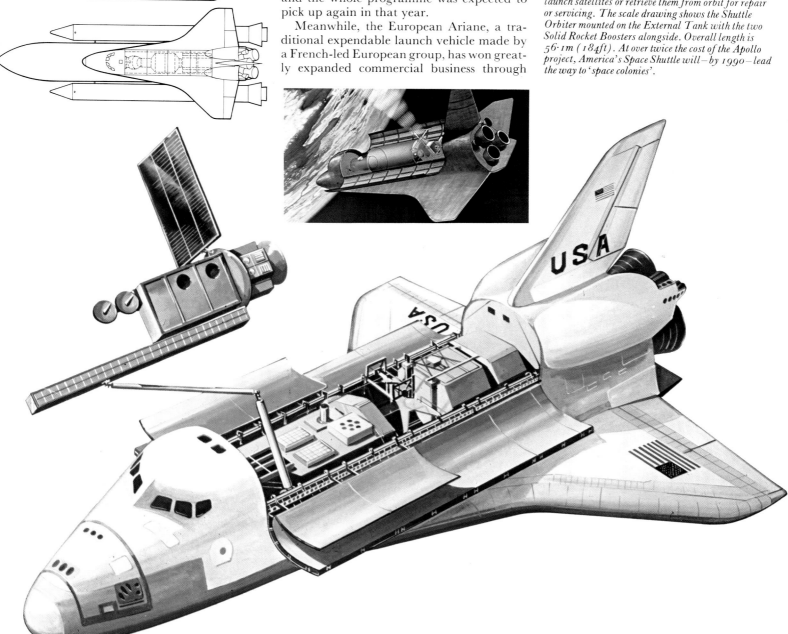

Significant Space Missions

Manned spacecraft

Spacecraft/ country	Date	Crew	Weight kg (lb)	Earth orbit parameters Perigee km (miles)	Apogee km (miles)	Incl. to equator deg.	Notes
Vostok 1 (USSR)	12 April, 1961	Yuri Gagarin	Sphere-cylinder 4,725 (10,419)	169 (105)	315 (196)	64·95	First man in space. One Earth orbit. Flight time 108min.
Freedom 7 (US)	5 May, 1961	Alan Shepard, Jr	Cone-frustum 1,829 (4,032)	Sub-orbital trajectory reaching 187km (116 miles) altitude, 478km (297 miles) range.			First American in space. Flight time 15min 22secs. Mercury MR–3.
Liberty Bell 7 (US)	21 July, 1961	Virgil Grissom	Cone-frustrum 1,829 (4,032)	Sub-orbital trajectory reaching 190km (118 miles) altitude, 487km (303 miles) range.			Craft sank after splashdown. Flight time 15min 37secs. Mercury MR–4.
Vostok 2 (USSR)	6–7 August 1961	Gherman Titov	Sphere-cylinder 4,730 (10,430)	166 (103)	232 (144)	64·93	First flight to exceed one day. 17 orbits. Flight time 25·3hr.
Friendship 7 (US)	20 February, 1962	John Glenn, Jr	Cone-frustum 1,352 (2,981)	159 (99)	265 (165)	32·54	First American in orbit. Three orbits. Flight time 4hr 55min 23sec. Mercury MA–6.
Aurora 7 (US)	24 May 1962	Scott Carpenter	Cone-frustum 1,349 (2,975)	154 (96)	260 (162)	32·5	Repeated Glenn's flight. Flight time 4hr 56min 5sec. Mercury MA–7.
Vostok 3 (USSR)	11–15 August 1962	Andrian Nikolayev	Sphere-cylinder 4,730 (10,430)	166 (103)	218 (135)	64·98	First television from manned spacecraft. Flight time 94hr 22min.
Vostok 4 (USSR)	12–15 August 1962	Pavel Popovich	Sphere-cylinder 4,730 (10,430)	169 (105)	222 (138)	64·95	First time two manned craft in orbit simultaneously. Approached to about 5km (3 miles) of Vostok 3. Flight time 70hr 57min.
Sigma 7 (US)	3 October 1962	Walter Schirra, Jr	Cone-frustum 1,370 (3,201)	153 (96)	285 (177)	32·55	Almost doubled previous US flight time. Flight time 9hr 13min 11sec. Mercury MA–8.
Faith 7 (US)	15–16 May 1963	Gordon Cooper, Jr	Cone-frustum 1,370 (3,020)	161 (100)	267 (166)	32·54	First extended US flight. Flight time 34hr 19min 49 sec. Mercury MA–9.
Vostok 5 (USSR)	14–19 June 1963	Valery Bykovsky	Sphere-cylinder 4,730 (10,430)	162 (101)	209 (130)	64·97	Longest flight to date. Flight time 119hr 6min.
Vostok 6 (USSR)	16–19 June 1963	Valentina Tereshkova	Sphere-cylinder 4,713 (10,392)	168 (104)	218 (135)	65·09	First woman in space. Flight time 70hr 50min.
Voshkod 1 (USSR)	12–13 October 1964	Konstantin Feoktlstov Vladimir Komarov Dr Boris Yegorov	Sphere-cylinder 5,320 (11,731)	177 (110)	377 (234)	64·90	First three-man crew in space. Flight time 24hr 17min.
Voskhod 2 (USSR)	18–19 March 1965	Pavel Belyayev Aleksei Leonov	Sphere-cylinder 5,682 (12,529)	167 (104)	475 (295)	64·79	Leonov took first space walk (about 10min). Flight time 26hr 2min.
Gemini 3 (US)	23 March 1965	Virgil Grissom John Young	Cone-frustum 3,220 (7,100)	160 (99)	240 (149)	33·00	First US two-man flight. Flight time 4hr 53min.
Gemini 4 (US)	3–7 June 1965	James McDivitt Edward White, II	Cone-frustum 3,540 (7,800)	162 (101)	281 (175)	32·53	White walked in space for 21min. Flight time 97hr 56min 11sec.
Gemini 5 (US)	21–29 August 1965	Gordon Cooper, Jr Charles Conrad, Jr	Cone-frustum 3,180 (7,011)	197 (122)	303 (188)	32·61	Tested man in prolonged weightlessness. Flight time 190hr 55min 14sec.
Gemini 7 (US)	4–18 December 1965	Frank Borman James Lovell, Jr	Cone-frustum 3,200 (7,056)	215 (134)	321 (199)	28·87	Endurance flight of nearly two weeks – a record that stood for 4½ years. Flight time 330hr 35min 31sec.
Gemini 6 (US)	15–16 December 1965	Walter Schirra, Jr Thomas Stafford	Cone-frustum 3,800 (8,379)	258 (160)	271 (168)	28·89	Approached to nearly 30cm (1ft) of Gemini 7, thus achieving first 'rendezvous' (close formation flight) of manned orbiting spacecraft. Flight time 25hr 51min 24sec.
Gemini 8 (US)	16 March 1966	Neil Armstrong David Scott	Cone-frustum 3,789 (8,355)	159 (99)	265 (165)	28·91	First docking (joining) of manned–unmanned craft in orbit; ended early because of malfunction. Flight time 10hr 41min 26sec.
Gemini 9 (US)	3–6 June 1966	Thomas Stafford Eugene Cernan	Cone-frustum 3,680 (8,114)	270 (168)	272 (169)	28·86	Demonstration of extended walk, rendezvous. Flight time 72hr 21min.

Mission	Date	Crew	Shape / weight kg (lb)	Perigee km (miles)	Apogee km (miles)	Inclination	Notes
Gemini 10 (US)	18–21 July 1966	John Young Michael Collins	Cone-frustum 3,630 (8,004)	160 (99)	268 (167)	28·85	Retrieved scientific experiment package during rendezvous with unmanned orbiting Agena craft. Flight time 70hr 46min 39sec.
Gemini 11 (US)	12–15 September 1966	Charles Conrad, Jr Richard Gordon, Jr	Cone-frustum 3,630 (8,004)	161 (100)	280 (174)	28·83	Docking tests. Flight time 71hr 17min 8sec.
Gemini 12 (US)	11–15 November 1966	James Lovell, Jr Edwin Aldrin, Jr	Cone-frustum 3,630 (8,004)	243 (151)	310 (193)	28·78	Aldrin walked in space 129min. Flight time 94hr 34min 31sec.
Soyuz 1 (USSR)	23–24 April 1967	Vladimir Komarov	Sphere-cylinder + 2 'wings' 6,000 ? (13,230?)	198 (123)	211 (131)	51·64	Komarov, killed in re-entry crash, becomes history's first space flight fatality. Flight time 26hr 40min.
Apollo 7 (US)	11–22 October 1968	Walter Schirra, Jr Donn Eisele Walter Cunningham	Cone-cylinder 14,690 (32,391)	231 (144)	297 (185)	31·63	First US three-man flight. Flight time 260hr 9min 3sec.
Soyuz 3 (USSR)	26–30 October 1968	Georgiy Beregovoi	Sphere-cylinder + 2 'wings' 6,000? (13,230?)	177 (110)	203 (126)	51·66	Beregovoi, at 47, oldest man in space, manoeuvred near unmanned Soyuz 2. Flight time 94hr 51min.
Apollo 8 (US)	21–27 December 1968	Frank Borman James Lovell, Jr William Anders	Cone-cylinder 28,400 (62,622)	191 (119)	191 (119)	32·60	First men to visit vicinity of a celestial body by coming to within 112·6km (70 miles) of the Moon. First circumnavigation of Moon by men. Flight time 147hr 0min 42sec.
Soyuz 4 (USSR)	14–17 January 1969	Vladimir Shatalov Aleksei Yeliseyev Yevgeniy Khrunov	Sphere-cylinder + 2 'wings' 6,625 (14,608)	161 (100)	215 (134)	51·73	First docking of two manned craft; first crew transfer between orbiting craft. Flight time 71hr 22min.
Soyuz 5 (USSR)	15–18 January 1969	Boris Volynov Aleksei Yeliseyev Yevgeniy Khrunov	Sphere-cylinder + 2 'wings' 6,585 (14,520)	210 (130)	233 (145)	51·69	Soyuz 4 and 5 docked and transferred two cosmonauts by EVA from Soyuz 5 to Soyuz 4. Flight time 72hr 46min.
Apollo 9 (US)	3–13 March 1969	James McDivitt David Scott Russell Schweickart	Cone-cylinder 22,030 (48,576)	203 (126)	229 (142)	32·57	First in-space test of Lunar Module (Moon landing craft); first crew transfer between craft through interior connection. Flight time 241hr 0min 54sec.
Apollo 10 (US)	18–26 May 1969	Thomas Stafford John Young Eugene Cernan	Cone-cylinder 28,870 (63,658)	183 (114)	184 (114)	32·56	Stafford and Cernan flew in separated Lunar Module to within 15km (9·4 miles) of Lunar surface, in final rehearsal for Lunar landing. Flight time 192hr 3min 23sec.
Apollo 11 (US)	16–24 July 1969	Neil Armstrong Edwin Aldrin, Jr Michael Collins	Cone-cylinder 28,800 (63,504)	183 (114)	184 (114)	32·51	First Lunar landing by men, Armstrong and Aldrin on Sea of Tranquillity. Stay time 21hr 36min 21sec. Flight time 195hr 18min 35sec.
Soyuz 6 (USSR)	11–16 October 1969	Georgiy Shonin Valeriy Kubasov	Sphere-cylinder + 2 'wings' 6,577 (14,502)	192 (119)	231 (144)	51·68	Soyuz 6, 7 and 8 operated as a group flight without actually docking. Flight time 118hr 21min.
Soyuz 7 (USSR)	12–17 October 1969	Viktor Gorbatko Anatoliy Filipchenko Vladislav Volkov	Sphere-cylinder + 2 'wings' 6,570 (14,487)	210 (130)	223 (139)	51·65	See above. Flight time 118hr 43min.
Soyuz 8 (USSR)	13–18 October 1969	Vladimir Shatalov Aleksei Yeliseyev	Sphere-cylinder + 2 'wings' 6,646 (14,654)	201 (125)	227 (141)	51·65	See above. Flight time 118hr 51min.
Apollo 12 (US)	14–24 November 1969	Charles Conrad, Jr Alan Bean Richard Gordon, Jr	Cone-cylinder 28,790 (63,482)	183 (114)	199 (124)	32·56	Second manned landing on Moon. Conrad and Bean on Ocean of Storms. Stay time nearly 31hr 31min. Flight time 244hr 36min 25sec.
Apollo 13 (US)	11–17 April 1970	James Lovell, Jr John Swigert Fred Haise	Cone-cylinder 28,890 (63,702)	186 (116)	186 (116)	32·56	Explosion of oxygen tank in Service Module prevented planned Moon landing. Astronauts safely recovered after circumnavigation of Moon. Flight time 142hr 54min 41sec.
Soyuz 9 (USSR)	1–19 June 1970	Andrian Nikolayev Vitaliy Sevastyanov	Sphere-cylinder + 2 'wings' 6,500 (14,332)	176 (109)	227 (141)	51·64	Longest-duration manned spaceflight to date. Flight time 424hr 59min.
Apollo 14 (US)	31 January–9 February 1971	Alan Shepard, Jr Edgar Mitchell Stuart Rossa	Cone-cylinder 29,229 (64,450)	186 (116)	186 (116)	32·56	Third manned landing on Moon, Shepard and Mitchell on Fra Mauro. Stay time 33hr 31min. Flight time 216hr 1min 57sec.
Soyuz 10 (USSR)	22–24 April 1971	Vladimir Shatalov Aleksei Yeliseyev Nikolai Rukavishnikov	Sphere-cylinder + 2 'wings' 6,575? (14,497?)	209 (130)	258 (160)	51·60	Docked with Salyut 1 space station, launched 19 April 1971, but crew did not board. Flight time 47hr 46min.

Mission	Dates	Crew	Shape / weight				Notes
Soyuz 11 (USSR)	6–29 June 1971	Georgiy Dobrovolsky Vladislav Volkov Victor Patsayev	Sphere-cylinder + 2 'wings' 6,565? (14,476?)	189 (111)	209 (130)	51·57	Docked with Salyut 1 for 22 days. Crew died as they returned to Earth when faulty valve allowed air to escape from cabin. Flight time 570hr 22min.
Apollo 15 (US)	26 July–7 August 1971	David Scott James Irwin Alfred Worden	Cone-cylinder 30,340 (66,900)	169 (105)	173 (108)	32·56	Fourth manned landing on Moon, Scott and Irwin on Hadley-Apennine region. First use of Lunar Roving Vehicle. Stay time 66hr 55min. Flight time 295hr 11min 53sec.
Apollo 16 (US)	16–27 April 1972	John Young Charles Duke, Jr Thomas Mattingly, II	Cone-cylinder 30,358 (66,939)	169 (105)	178 (111)	32·56	Fifth manned landing on Moon, Young and Duke on Descartes region. Set up first astronomical observatory. Used Lunar Roving Vehicle. Stay time 71hr 2min. Flight time 265hr 51min 5sec.
Apollo 17 (US)	7–19 December 1972	Eugene Cernan Harrison Schmitt Ronald Evans	Cone-cylinder 30,340 (66,900)	169 (105)	178 (111)	32·56	Sixth manned landing on Moon, Cernan and Schmitt on Taurus Littrow region. Used Lunar Roving Vehicle. Stay time 75hr. Flight time 301hr 51min 59sec.
Skylab 2 (US)	25 May–22 June 1973	Charles Conrad Joseph Kerwin Paul Weitz	Cone-cylinder 13,780 (30,385)	156 (97)	359 (223)	50·04	First astronauts to board Skylab Space Station, carried out repairs. Flight time 672hr 49min 49sec.
Skylab 3 (US)	28 July–25 September 1973	Alan Bean Jack Lousma Owen Garriott	Cone-cylinder 13,860 (30,561)	159 (99)	230 (143)	50·03	Second crew to board Skylab Space Station. Flight time 1427hr 9min 4sec.
Soyuz 12 (USSR)	27–29 September 1973	Vasily Lazarev Oleg Makarov	Sphere-cylinder 6,570? (14,487?)	181 (112)	229 (142)	51·58	Test of improved Soyuz for Salyut ferry missions, with chemical batteries replacing solar 'wings'. Flight time 47hr 16min.
Skylab 4 (US)	16 November 1973–8 February 1974	Gerald Carr Edward Gibson William Pogue	Cone-cylinder 13,980? (30,826?)	154 (96)	224 (139)	50·04	Third and final crew to board Skylab Space Station. Man's longest space-flight to date. Flight time 2017hr 15min 32sec.
Soyuz 13 (USSR)	18–26 December 1973	Pyotr Klimuk Valentin Lebedev	Sphere-cylinder + 2 'wings' 6,570? (14,487?)	188 (117)	247 (153)	51·57	Conducted experiments in biology, astrophysics and Earth resource observation. Flight time 188hr 55min.
Soyuz 14 (USSR)	3–19 July 1974	Pavel Popovich Yuri Artyukhin	Sphere-cylinder 6,570? (14,487?)	195 (121)	217 (135)	51·58	Docked with Salyut 3 for 353hr 33min. Flight time exceeded 370hr.
Soyuz 15 (USSR)	26–28 August 1974	Gennardy Sarafanov Lev Demin	Sphere-cylinder 6,570? (14,487?)	173 (108)	236 (147)	51·62	Failed to dock with Salyut 3 on 27 August because of fault in automatic control system. Flight time 48hr 12min.
Soyuz 16 (USSR)	2–8 December 1974	Anatoly Filipchenko Nikolai Rukavishnikov	Sphere-cylinder + 2 'wings' 6,680? (14,729?)	184 (114)	291 (181)	51·80	ASTP flight test. Tested docking system of Soyuz and reduction of cabin pressure as required for joint docking experiment. Flight time 143hr 4min.
Soyuz 17 (USSR)	11 January–9 February 1975	Alexei Gubarev Georgi Grechko	Sphere-cylinder 6,570? (14,487?)	185 (115)	249 (155)	51·63	Docked with Salyut 4 for 677hr 8min. Flight time 709hr 20min.
Soyuz 'anomaly' (USSR)	5 April 1975	Vasily Lazarev Oleg Makarov			Sub-orbital		Attempt to reach Salyut 4 failed, due to launch vehicle third stage malfunction. Cosmonauts recovered safely in Western Siberia.
Soyuz 18 (USSR)	24 May–26 July 1975	Pyotr Klimuk Vitaly Sevastyanov	Sphere-cylinder 6,570? (14,487?)	186 (116)	230 (143)	51·69	Docked with Salyut 4. Flight time 1,511hr 20min.
Soyuz 19 (USSR)	15–21 July 1975	Alexei Leonov Valery Kubasov	Sphere-cylinder + 2 'wings' 6,680 (14,729)	186 (116)	220 (137)	51·78	ASTP mission. Docked with Apollo CSM/DM for about 2 days for crew exchanges and joint experiments. Flight time 142hr 31min.
Apollo 18 ASTP (US)	15–24 July 1975	Thomas Stafford Vance Brand Donald Slayton	Cone-cylinder 13,860? (30,561?)	170 (106)	228 (142)	51·76	See above. Flight time 217hr 28min.
Soyuz 21 (USSR)	6 July–24 August 1976	Boris Volynov Vitaly Zhdobov	Sphere-cylinder 6,570? (14,487?)	246 (153)	274 (170)	51·59	Docked with Salyut 5 on 7 July Flight time 1,182hr 24min.
Soyuz 22 (USSR)	15–23 September 1976	Valery Bykovsky Vladimir Aksyonov	Sphere-cylinder + 2 'wings' 6,570? (14,487?)	185 (115)	296 (184)	64·75	Intercosmos mission. Carried MKF–6 multi-spectal cameras made in East Germany. Flight time 189hr 54min.
Soyuz 23 (USSR)	14–16 October 1976	Vyacheslav Zudov Valery Rozhdestvensky	Sphere-cylinder 6,570? (14,487?)	188 (117)	224 (139)	51·61	Failed to dock with Salyut 5 on 15 October because of fault in automatic control system. Flight time 48hr 6min.

Soyuz 24 (USSR)	7–25 February 1977	Viktor Gorbatko Yuri Glazkov	Sphere-cylinder 6,570? (14,487?)	218 (135)	281 (175)	51·60	Docked with Salyut 5 on 8 February. Flight time 424hr 48min.
Soyuz 25	9–11 October 1977	Vladimir Kovalyonok Valery Ryumin	Sphere-cylinder 6560 (14,450)	120	159	52	Attempted docking with Salyut 6 space station; soft docking accomplished, but emergency return made after repeated hard-docking failure. Duration of flight 48hr 44min 45sec.
Soyuz 26	10 December 1977–14 March	Yuri Romanenko Georgi Grechko	Sphere-cylinder 6560 (14,450)	120	159	52	Romanenko made manual docking with Salyut 6, both crew made long EVA and then returned in Soyuz 27 after 96 days 10hr. Duration of flight 96 days 10hr 0min 7sec.
Soyuz 27	10–16 January 1978	Vladimir Dzhanibekov Oleg Makarov	Sphere-cylinder 6560 (14,450)	120	159	52	First manned resupply mission, docked with Salyut 6 and later exchanged crews and departed to leave rear docking port free for unmanned Progress 1 tanker. Duration of flight 142hr 59min.
Soyuz 28	2–10 March 1978	Alexei Gubarev Vladimir Remek (Czechoslovakia)	Sphere-cylinder 6560 (14,450)	120	159	52	Docked with Salyut 6; long celebration followed by many (mainly Czech) tasks. Duration of flight 190hr 16min.
Soyuz 29	15 June–2 November 1978	Vladimir Kovalyonok Alexander Ivanchenkov	Sphere-cylinder 6560 (14,450)	120	159	52	Docked with Salyut 6 on 17 June after Romanenko/Grechko had left; over 200 experiments, EVA and Earth photography, and visited by five other craft including Progress 2, 3 and 4 tankers. Duration of flight 139 days 14hr 48min.
Soyuz 30	27 June–5 July 1978	Piotr Klimuk Miroslav Hermaszewski (Poland)	Sphere-cylinder 6560 (14,450)	120	159	52	Docked with Salyut 6; numerous tasks, mainly Polish. Duration of flight 190hr 3min.
Soyuz 31	26 August–3 September 1978	Valery Bykovsky Sigmund Jähn (E Germany)	Sphere-cylinder 6560 (14,450)	120	159	52	Third Intercosmos (multi-national) mission to Salyut 6; E. German tasks. Returned in Soyuz 29 capsule. Duration of flight 188hr 49min.
Soyuz 32	25 February–19 August 1979	Vladimir Lyakhov Valery Ryumin	Sphere-cylinder 6560 (14,450)	120	159	52	Reactivated Salyut 6 and completed many tasks. Soyuz 33 failure resulted in cancellation of planned manned visits, but Progress 6 and 7 brought supplies including giant radio telescope (and changed orbit). Duration of flight 175 days 35min 37sec.
Soyuz 33	10–12 April 1979	Nikolai Rukavishnikov (first Soviet civilian) Georgi Ivanov (Bulgaria)	Sphere-cylinder 6006 (13,230)	120·5	162	52	Major propulsion failure aborted Salyut 6 docking. Highly stressed re-entry. Duration of flight 47hr 1min.
Soyuz 35	9 April–11 October 1980	Leonid Popov Valery Ryumin	Sphere-cylinder 6560 (14,450)	120·5	162	52	Reactivated Salyut 6 (Soyuz 34 was changed to unmanned mission); many tasks, and hosted three manned missions plus Progress 10 and 11. Returned via Soyuz 37. Duration of flight 184 days 20hr 12min.
Soyuz 36	26 May–3 June 1980	Valery Kubasov Bertalan Farkas (Hungary)	Sphere-cylinder 6560 (14,450)	120·5	162	52	Docked with Salyut 6, many tasks (most Hungarian). Returned via Soyuz 35. Duration of flight 188hr 46min.
Soyuz T–2	5–9 June 1980	Yuri Malyshev Vladimir Aksyonov	Sphere-cylinder 7006 (15,432)	120·5	162	52	Docked with Salyut 6 in bigger (3-man) Soyuz, with other changes, new crew spacesuits, left orbital module docked to enlarge Salyut capacity and save retro fuel. Duration of flight 94hr 19min 30sec.
Soyuz 37	23–31 July 1980	Viktor Gorbatko Pham Tuan (Vietnam)	Sphere-cylinder 6560 (14,450)	120·5	162	52	Docked with Salyut 6. 30 experiments. Returned via Soyuz 36. Duration of flight 188hr 42min.

Mission	Date	Crew	Type (mass)				Notes
Soyuz 38	18–26 September 1980	Yuri Romanenko Arnaldo Mendez (Cuba)	Sphere-cylinder 6560 (14,450)	120·5	162	52	Docked with Salyut 6, 15 experiments. Landed within 1.8 miles of target. Duration of flight 188hr 43min 24sec.
Soyuz T–3	27 November–10 December 1980	Leonid Kizim Oleg Makarov Gennardy Strekalov	Sphere-cylinder 7006 (15,432)	120·5	162	52	Docked with Salyut 6. Intensive repair schedule included numerous upgrades and replacements. Docked Progress 11 used to raise orbit. Duration of flight 307hr 7min 42sec.
Soyuz T–4	12 March–26 May 1981	Vladimir Kovalyonok Viktor Savinykh	Sphere-cylinder 7006 (15,432)				Docked with Salyut 6, 30 experiments with equipment replacing third seat. Duration of flight 74 days 17hr 37min.
Soyuz 39	22–30 March 1981	Vladimir Dzhanibekov Jugderdemidyin Gurragcha (Mongolia)	Sphere-cylinder 6560 (14,550)				Docked with Salyut 6, numerous experiments. Duration of flight 188hr 42min.
STS–1 **OV–102** *Columbia*	12–14 April 1981	John Young Robert Crippen	Winged orbiter 99,543 (219,258)				First Shuttle Orbiter flight. Vehicle assessed, photographed by KH-11. Duration of flight 30hr 20min 52sec.
Soyuz 40	15–22 May 1981	Leonid Popov Dumitru Prunariu (Romania)	Sphere-cylinder 6560 (14,550)				Docked with Salyut 6; last two-man Soyuz vehicle. Studied atmosphere and magnetic field. Duration of flight 188hr 41min 52sec.
STS–2 **OV–102** *Columbia*	12–14 November 1981	Joseph Engle Richard Truly	Winged orbiter 104,741 (230,708)				First mission by recycled spacecraft. Pallet and remote manipulator fitted. Earth surveillance tasks. Duration of flight 30hr 13min 11sec.
STS–3 **OV–102** *Columbia*	22–30 March 1982	Jack Lousma Gordon Fullerton	Winged orbiter 106,883 (235,425)	244 (130)	255 (130)	38	Third flight of 102, first with payload bay packed with equipment. Many faults did not affect mission. Landed at White Sands. Duration of flight 192hr 5min.
Soyuz T–5	13 May–10 December 1982	Anatoli Berezovoi Valentin Lebedev	Sphere-cylinder 7006 (15,432)	192 (343)	231 (356)	51·6	First mission to Salyut 7, supported by Progress 13–16. Returned in Soyuz T–7. Duration of flight 211 days 9hr 5min.
Soyuz T–6	24 June–2 July 1982	Vladimir Dzhanibekov Alexander Ivanchenkov Jean-Loup Chrétien (France)	Sphere-cylinder 7006 (15,432)	189 (343)	233 (356)	51·6	Visited Salyut 7 in intensive work programme. Returned in same craft. Duration of flight 189hr 50min 52sec.
STS–4 **OV–102** *Columbia*	27 June–4 July 1982	Thomas Mattingly II Hank Hartsfield	Winged orbiter 109,716 (241,664)	295	302	28·5	First US manned flight with military payloads, as well as first commercial load. Duration of flight 169hr 9min 31sec.
Soyuz T–7	19–27 August 1982	Leonid Popov Alexander Serebrov Svetlana Savitskaya	Sphere-cylinder 7006 (15,432)	228 (292)	280 (302)	51·6	Visited Salyut 7. Savitskaya managed life-science tasks. Returned in T–5. Duration of flight 189hr 52min 24sec.
STS–5 **OV–102** *Columbia*	11–16 November 1982	Vance Brand Robert Overmyer William Lenoir Joseph Allen	Winged orbiter 109,505 (241,200)	294	317	28·5	First commercial mission, launching SBS–C and Anik C–3; first US crew with no spacesuits and no means of escape. Duration of flight 122hr 14min 26sec.
STS–6 **OV–99** *Challenger*	4–9 April 1983	Paul Weitz Karol Bobko Story Musgrave Don Peterson	Winged orbiter 117,372 (258,529)	280	290	28·5	First flight of 99, satellite launched and EVA work in payload bay by Musgrave and Peterson. Duration of flight 120hr 23min 42sec.
Soyuz T–8	20–22 April 1983	Vladimir Titov Gennardy Strekalov Alexander Serebrov	Sphere-cylinder 7006 (15,432)	180	228	51·6	Rendezvous radar failed and docking with Salyut 7 was aborted. Duration of flight 48hr 17min 48sec.
STS–7 **OV–99** *Challenger*	18–24 June 1983	Robert Crippen Rick Hauck John Fabian Sally Ride Norman Thagard	Winged orbiter 112,219 (247,178)	295	302	28·5	Four-man one-woman crew; deployed Palapa, Anik and SPAS. Duration of flight 146hr 23min 59sec.
Soyuz T–9	27 June–23 November 1983	Vladimir Lyakhov Alexander Alexandrov	Sphere-cylinder 7006 (15,432)	197 (324)	227 (337)	51·6	Docked with Salyut 7, with Star Module at other end; supplied by Progress 17 and 18. Duration of flight 149 days 10hr 46min.
STS–8 **OV–99** *Challenger*	30 August–5 September 1983	Richard Truly Daniel Brandenstein Dale Gardner Guion Bluford William Thornton	Winged orbiter 110,200 (242,732)	295	302	28·5	Night launch, Insat 1A placed in orbit and other tasks, followed by night landing. Duration of flight 145hr 8min 40sec.

Mission	Date	Crew	Vehicle	Perigee	Apogee	Inclination	Notes
STS–9 **OV–102** *Columbia*	28 November–7 December 1983	John Young Brewster Shaw Jr Owen Garriott Robert Parker Ulf Merbold Byron Lichtenberg	Winged orbiter 112,419 (247,619)	239	251	57·0	Spacelab placed in orbit and crew of two three-man teams (Merbold a West German) worked 12-hr shifts on 72 experiments. Duration of flight 10 days 7hr 47min 23sec.
STS–41B **OV–99** *Challenger*	3–11 February 1984	Vance Brand Robert Gibson Bruce McCandless Ronald McNair Robert Stewart	Winged orbiter 113,705 (250,452)	306	322	28·5	Palapa and Westar lost in unusable orbits, but McCandless tested manned manoeuvring unit in untethered EVA up to 100m distant. Return to Kennedy. Duration of flight 191hr 15min 54sec.
Soyuz T–10	8 February–1 October 1984	Leonid Kizim Vladimir Solovyov Oleg Atkov	Sphere-cylinder 7006 (15,432)	198 (289)	219 (297)	51·6	Long and busy stay in Salyut 7, supplied by Progress 19 to 23, visited by Soyuz T–11 and –12. Returned in T–11. Duration of flight 236 days 22 hr 49min.
Soyuz T–11	3–11 April 1984	Yuri Malyshev Gennardy Strekalov Rakesh Sharma (India)	Sphere-cylinder 7006 (15,432)	196 (284)	223 (296)	51·6	Visited Salyut 7, landed in Soyuz T–10. Duration of flight 189hr 40min.
STS–41C **OV–99** *Challenger*	6–13 April 1984	Robert Crippen Dick Scobee Terry Hart George Nelson James van Hoften	Winged orbiter 115,431 (254,254)	218 (493)	465 (498)	28·5	Faulty Solar Max (launched 1980) recovered, repaired and returned to orbit, and Long Duration Exposure Facility activated. Duration of flight 167hr 40min.
Soyuz T–12	17–28 July 1984	Vladimir Dzhanibekov Svetlana Savitskaya Igor Volk	Sphere-cylinder 7006 (15,432)	198 (334)	225 (354)	51·6	Stayed 11 days at Salyut 7, EVAs with versatile electron-beam tool. Duration of flight 11 days 19hr 14min 36sec.
STS–41D **OV–103** *Discovery*	30 August–4 September 1984	Hank Hartsfield Michael Coats Steven Hawley Judy Resnik Mike Mullane Charles Walker	Winged orbiter 98,417 (216,778)	297	314	28·5	Walker (McDonnell Douglas) from industry; SBS-4, Leasat 1 and Telstar 3C placed in orbit and giant solar array erected. Landed at Edwards. Duration of flight 144hr 56min.
STS–41G **OV–99** *Challenger*	5–13 October 1984	Robert Crippen Jon McBride Sally Ride Kathy Sullivan David Leestma Paul Scully-Power Marc Garneau	Winged orbiter 109,768 (241,780)	346	359	57·0	Deployed Earth radiation budget satellite, extensive EVA. Landed at Kennedy. Scully-Power Australian and Garneau Canadian. Duration of flight 197hr 23min 33sec.
STS–51A **OV–103** *Discovery*	8–16 November 1984	Rick Hauck David Walker Dale Gardner Joseph Allen Anna Fisher	Winged orbiter 118,802 (261,679)	294 (342)	298 (362)	28·5	Leasat 2 and Anik D2 placed in orbit, and Palapa and Westar (on which Lloyds had paid out $180 million) retrieved for resale. Duration of flight 191hr 45min 54sec.
STS–51C **OV–103** *Discovery*	24–27 January 1985	Thomas Mattingly II Loren Shriver Ellison Onizuka James Buchli Gary Payton	Winged orbiter 113,905 (250,891)	300	300	28·5	Classified military mission to place elint payload in geostationary 22,300-mile orbit using Inertial Upper Stage. Duration of flight 73hr 33min 13sec.
STS–51D **OV–103** *Discovery*	12–19 April 1985	Karol Bobko Don Williams David Griggs Jeffrey Hoffman Rhea Seddon Jake Garn Charles Walker	Winged orbiter 113,013 (248,927)	315	460	28·5	Garn, Senator and head of NASA budget committee. Anik C1 and Leasat 3 placed in orbit, but latter failed to activate despite EVA. Severe crosswind caused decision to land future missions at Edwards. Duration of flight 167hr 55min.
STS–51B **OV–99** *Challenger*	29 April–5 May 1985	Robert Overmyer Frederick Gregory Norman Thagard William Thornton Don Lind Lodewijk van den Berg Taylor Wang	Winged orbiter 112,084 (246,880)	346	358	57·0	Elderly crew (41 to 56 years) but enough data to fill 44,000 200-page books and 3 million frames of video. Monkeys and rats caused problems. Duration of flight 168hr 8min 50sec.
Soyuz T–13	6 June–26 September 1985 (Dzhanibekov)	Vladimir Dzhanibekov Viktor Savinykh	Sphere-cylinder 7006 (15,432)	198 (356)	222 (359)	51·6	Reactivated 'dead' Salyut 7 in very severe conditions. Duration of flight (Dzhanibekov, Savinykh returning in T–14) 112 days.
STS–51G **OV–103** *Discovery* *Discovery*	17–24 June 1985	Daniel Brandenstein John Creighton John Fabien Shannon Lucid Steven Nagel Patrick Baudry Abdul Aziz Al-Saud	Winged orbiter 116 462 (256,524)	352	356	28·5	Baudry from France and Prince Aziz Al-Saud from Saudi Arabia. Morelos A, Arabsat and Telstar 3D placed in orbit and Spartan deployed and retrieved; also SDI hi-power laser test. Duration of flight 169hr 38min 58sec.

Mission	Date	Crew	Craft / weight kg (lb)				Remarks
STA–51F **OV–99** *Challenger*	29 July– 5 August 1985	Gordon Fullerton Roy Bridges Jr Tony England Karl Heinze Story Musgrave Loren Acton Joh-David Bartoe	Winged orbiter 114,796 (252,855)	311	319	49·5	Spurious warning shut down a main engine (Fullerton over-rode another shutdown warning), resulting in low orbit, but outstanding scientific results. Duration of flight 190hr 45min 27sec.
STS–51I **OV–103** *Discovery*	27 August– 2 September 1985	Joseph Engle Richard Covey William Fisher Michael Lounge James van Hoften	Winged orbiter 107,898 (237,661)	351 (314)	380 (449)	28·6	Aussat, ASC–1 and Leasat 4 placed in respective orbits; then huge Leasat 3 captured and put through complex electronic repair. Duration of flight 170hr 14min 42sec.
Soyuz T–14	17 September– 23 November 1985 (not Grechko)	Vladimir Vasyutin Georgi Grechko Alexander Volkov	Sphere-cylinder 7006 (15,432)	196 (344)	223 (345)	51·6	After eight days aboard Salyut 7 Grechko returned with Dzhanibekov in T–13; Vasyutin became ill and returned with Volkov and T–13's Savinykh. Duration of flight 64 days 21hr 52min.
STS–51J **OV–104** *Atlantis*	3–7 October 1985	Karol Bobko Ronald Grabe David Hilmers Robert Stewart William Pailes	Winged orbiter c.113,500 (250,000)	476 (320)	515	28·5	First flight of 104; military mission including two DSCS–3 comsats, plus NASA Bios experiment. Duration of flight 97hr 45min 30sec.
STS–61A **OV–99** *Challenger*	30 October– 6 November 1985	Hank Hartsfield Steven Nagel Guion Bluford James Buchli Bonnie Dunbar Reinhard Furrer Ernst Messerschmid Wubbo Ockels	Winged orbiter 96,734 (213,070)	322	333	57·0	First leased mission (W Germany, for Spacelab D1 and many experiments); two German and one Dutch payload specialists and mission controlled from Oberpfaffenhofen. Duration of flight 168hr 44min 51sec.
STS–61B **OV–104** *Atlantis*	27 November– 4 December 1985	Brewster Shaw Jr Bryan O'Connor Mary Cleave Woody Spring Jerry Ross Rodolfo Vela Charles Walker	Winged orbiter 118,700 (261,455)	323	383	28·5	Night launch; Ku–2, on up-rated PAM–D2 upper stage, and Mexican Morelos B (overseen by Mexican Vela) placed in their orbits, and 45-ft tower assembled in Space.
STS–61C **OV–102** *Columbia*	12–18 January 1986	Robert Gibson Charles Bolden Jr Steven Hawley George Nelson Franklin Chang-Diaz Robert Cenker Bill Nelson	Winged orbiter 115,985 (255,471)	324	386	28·5	After many delays (mission planned for 18 December) OV–102 flew first time in over two years. Ku–1 placed in orbit. Kennedy landing needed, but diverted to night landing at Edwards.
STS–51L **OV–99** *Challenger*	28 January 1986	Dick Scobee Michael Smith Judy Resnik Ronald McNair Ellison Onizuka Gregory Jarvis Christa McAuliffe	Winged orbiter c.113,500 (250,000)	—	—	—	Solid rocket booster joint ring failed 73s after lift off, blowing vehicle apart. Result was a 2.5-year hiatus in Shuttle programme.
Soyuz T–15	13 March– 16 July 1986	Leonid Kizim Vladimir Solovyov	Sphere-cylinder c.7006 (15,432)	193 (332)	238 (339)	51·6	Docked with new Mir (crew quarters of a permanent space station), with six docking ports; also (6 May) visited Salyut 7 and supplied by first new TM and Progress 26 returning to Mir 26 June in T–15. Duration of flight 125 days 1min.
Soyuz TM–2	5 February 1987 (see remarks)	Yuri Romanenko Alexander Laveikin	Sphere-cylinder c.7006 (15,432)	263	301	51·6	Docked with Mir. Returned 29 July with Laveikin and two Cosmonauts from TM–3. Duration of flight 174 days.
Soyuz TM–3	22 July 1987 (see Remarks)	Alexander Viktorenko Alexander Alexandrov Mohammad Faris (Syria)	Sphere-cylinder c.7006 (15,432)	236	300	51·6	Docked with Mir. TM–2 returned on 29 July with Viktorenko, Faris and Laveikin (TM–2). Duration of flight (TM–3) 159 days.
Soyuz TM–4	20 December 1987 (see remarks)	Vladimir Titov Musakhi Manarov Anatoly Levchyenko	Sphere-cylinder c.7006 (15,432)	260	298	51·6	Docked with Mir. Levchyenko returned in TM–3 on 28 December with Romanenko (who had been in space 326 days) and Alexandrov. Duration of flight (TM–4) not known at press time.

Facts, Feats
and Records

Great Flying Feats

On 13 January 1908 at Issy Les Moulineaux in France Henry Farman flew the first circular flight in Europe, thus winning the 'Grand Prix d'Aviation' donated by two French benefactors. He covered a distance of 1km (·625 miles) in a modified Voisin biplane and remained airborne for 1 minute 28 seconds. Though a mere hop by modern standards this was the most impressive flight that had been made in Europe to that date. Yet in America the Wright Brothers, before they temporarily gave up flying in 1905, had already covered 40·25km (25 miles) in one flight and stayed up more than half an hour at a time; and they could turn and fly figure

PREVIOUS PAGES:
Gnat fighters of the Red Arrows aerobatic team.

eights with ease. In Europe no one knew or believed this until, in 1908, the Wrights returned to flying, and gave their first demonstrations in public. Wilbur took a Flyer to France, while Orville stayed home to fly their new two-seat Flyer for the US Army.

The Wrights' achievement
Wilbur's first flight in Europe was a two-minute circuit of Le Mans race-course on 8 August 1908. He astonished spectators with the degree of control he had over his machine. 'One of the most exciting spectacles in the history of applied science,' wrote one. 'Good grief!' exclaimed another, 'We are beaten . . . we don't exist.' When he finished his French demonstrations at Auvours on the last day of 1908 Wilbur Wright had made

more than 100 flights, with flying-time of more than 25 hours; he had taken up passengers on more than 60 occasions; had attained an altitude of 110m (360ft); and had made one flight of more than 2 hours.

Orville's flights in America began impressively, but ended tragically on 17 September with structural failure of his Flyer while aloft.

Louis Blériot
One of the French experimenters most inspired by watching Wilbur's demonstrations was Louis Blériot. His 1909 Type XI monoplane was his first really successful design, and in it he made a cross-country flight of 42km (26 miles) from Etampes to Chevilly on 13 July. On 25 July he made the first aerial crossing of the English Channel. This flight

won him the £1,000 prize offered by the London *Daily Mail* newspaper to the first person to achieve this feat as well as attracting extraordinary publicity around the world. It brought home to people for the first time that the aeroplane might in time become a formidable weapon of war. Blériot's flight was also a feat of great courage: the 37 minutes it took was about as long as his frail 25hp 3-cylinder Anzani engine had ever run without stopping. He had no compass or real means of navigation for the misty crossing and he was still suffering the effects of an injury received in an earlier crash. (His arrival on the Cliffs of Dover was in truth another crash-landing.) The feat earned him a fortune. In addition to the £1,000 newspaper prize he soon had orders for more than 100 of his Type XI monoplanes. The design of the XI was interesting in that it incorporated many features that we have since come to take for granted in aeroplane design: monoplane wings, tractor propeller directly attached to the engine crankshaft, hinged tilting stick and rudder pedal controls, and a covered-in fuselage structure. (The Wright Flyers by contrast had none of these features.)

ABOVE: *Two important early aircraft which were seen at the Reims Air Show in 1909. Top, the Voisin biplane; bottom, the Blériot 12 monoplane.*
BELOW: *A reconstruction of Blériot's famous Type XI monoplane.*
BOTTOM LEFT: *Wilbur Wright repairing his launching rail at Le Mans, 1908.*
BOTTOM RIGHT: *Louis Blériot and his Type XI monoplane at Dover after his historic cross-Channel flight on 25 July 1909.*

Reims air show

The first flying meeting took place soon after Blériot's Channel crossing, between 22 and 29 August at Reims in France. It was sponsored by local champagne interests and 38 aircraft were entered. Of these, 23 actually got airborne to make 120 flights. Speed records were established by Blériot and the American Glenn Curtiss of between 75 and 77km/h (47 to 48mph); an Antoinette monoplane climbed to 155m (508ft); and the distance prize was won by Henry Farman, who covered 180km (112·5 miles) in 3 hours 5 minutes. Among the many impressed visitors was Lloyd George, who commented: 'Flying machines are no longer toys and dreams . . . the possibilities of this new system of locomotion are infinite.' However, as much as the many fine flights made at Reims, most visitors were impressed by the general appalling unreliability of the engines.

1909–1910

On 7 September 1909 the first pilot of a powered aeroplane was killed and on 22 September the second, and on 6 December,

the third. Since the beginning of the 19th century five men and one woman had been killed in parachuting accidents; two experimenters in glider crashes; and a considerable number in balloons. There were no less than 32 aviation fatalities in 1910. That year, however, also saw more than 20 aviation meetings held, in Europe, the USA and Egypt; the first night-flying accomplished, in the Argentine; and the first lady to qualify as a pilot, Baroness de Laroche. The Hon C. S. Rolls, the wealthy sportsman who was involved in the founding of Rolls-Royce cars, made the first out-and-return non-stop Channel crossing on 2 June 1910 and lost his life on 12 July during a meeting at Bournemouth when his French-built Wright Flyer broke up in the air. On 17 August a French-American pilot, J. B. Moisant, crossed the Channel carrying his mechanic as a passenger.

The Peruvian pilot, Georges Chavez, flying a Blériot monoplane made the first aerial crossing of the Alps, through the Simplon pass, although he lost his life at the end of the attempt while trying to land at Domodossola. Curtiss biplanes made several remarkable flights in the USA: from Albany to New York City; and New York to Philadelphia and back when radio was used for the first time. Eugene Ely took off in a Curtiss biplane from a platform built on the USS *Birmingham* – another first in aviation. Glenn Curtiss himself made the first bombing tests, dropping dummies onto flagged buoys which marked the outline of a warship.

FAR LEFT: *Farman, using the aeroplane as a sporting vehicle, here makes a graceful turn round a pylon in the type of biplane used by Grahame-White in 1910.*
BELOW FAR LEFT: *Not all early flights met with success. This is Baroness de la Roche's wrecked Voisin in July 1910.*
LEFT: *The Paulhan triplane, one of the new types of military aeroplanes exhibited at the Reims show in 1911.*
BELOW: *Tryggve Gran's aeroplane at Cruden Bay, Scotland before the start of its flight across the North Sea in July 1914.*
BOTTOM LEFT: *W. Newell with his parachute at Hendon. He was the first man to make a parachute drop from an aeroplane over Great Britain, on 9 May 1914.*
BOTTOM RIGHT: *Georges Chavez at the start of his flight over the Alps which ended in disaster when he crashed on landing.*

by William Randolph Hearst, but arrived outside the specified 30-day limit. Another American pilot flew a Curtiss 90 miles over water–from Key West to Havana, Cuba. Curtiss flew the first practical seaplane in 1911. And Orville Wright made a nine-minute flight in a glider which remained a soaring world record for ten years. Near the end of the year a *Concours Militaire* was held at Reims, for the display of aeroplanes that might be used in war. And one *was* used, for the first time, when an Italian pilot, in a Blériot, made an hour-long flight on 22 October 1912 to observe Turkish positions near Tripoli. The first air mail flights were made. The first parachutist to drop from an aeroplane was Captain Albert Berry in the USA, and American Harriet Quimby became the first woman to fly the Channel. In England Frank McClean flew a Short float-plane through Tower Bridge and was promptly in trouble with the police (as has been every other pilot to try it since) when he landed on the Thames by Westminster.

1912–1918

During 1912 and 1913 the world airspeed record was raised no less than ten times by two French pilots, Jules Vedrines and Maurice Prévost, flying very advanced De-perdussin monocoque racers. The first such record set was Vedrine's 145km/h (90mph) on 13 January 1912. By 29 September next year Prévost had achieved 203km/h (126mph).

Two outstanding flights of 1913 were the first air crossing of the Mediterranean, by Roland Garros in a Morane-Saulnier monoplane. He flew 729km (453 miles) from southern France to Tunisia in just under 8 hours. The first flight from France to Egypt was made, in stages, by Vedrines in a Blériot. The Norwegian pilot Tryggve Gran was the first man to fly the North Sea, in a Blériot, on 30 July 1914.

Civil flying generally ceased in Europe during World War I. But South America was far from the conflict and two remarkable record flights were made there. The Argentinean Army pilot Teniente Candelaria made the first crossing of the Andes from east to west on 13 April 1918. He flew a Morane and reached an altitude of about 3,960m (13,000 ft). A Chilean pilot, Teniente Godoy, crossed the Andes from west to east on 12 December in a Bristol monoplane.

1911

By 1911 aviation had become an enormously popular public spectacle, and the first air races were held that year. The first non-stop flight from London to Paris was made on 12 April in a Blériot by Pierre Prier, who flew the 402km (250 miles) in a little less than 4 hours. There were races (in stages) from Paris to Rome and Madrid and a return trip to Paris via London, Brussels and Amiens. In America Calbraith P. Rodgers made the first coast-to-coast flight in a Wright biplane. The 6,437km (4,000 miles) took him 82 flying hours in 82 stages over a period of 49 days. He had 19 crashes en route and was followed all the way by a special train containing spares as well as his wife and mother. He had hoped to win a $50,000 prize for the journey offered

1919

Substantial prize money had already been put up before the war for the first air crossing of the Atlantic Ocean. As soon as the war was over many began to plan for the attempt. Newfoundland, the easternmost province of Canada, had never seen an aeroplane, but, in the spring and summer of 1919, it became a familiar dateline for press reports of preparations for the transatlantic crossing. When the two British aviators John Alcock and Arthur Whitten Brown arrived as latecomers in the race they were hopefully asked to pay $25,000 for the temporary lease of one small meadow when land could be bought there for 35 cents an acre! The prize for the first successful crossing was £10,000 (then $50,000), put up by the *Daily Mail* in 1913 for the first flight between England and America to be completed within 72 hours.

Meanwhile the US Navy planned an ambitious Atlantic crossing with 3 huge Curtiss flying-boats aided by a string of destroyers stationed every 80·5km (50 miles) across the ocean. Two of the flying-boats were wrecked during the attempt, but the third, the NC–4, left Newfoundland on 8 May. She landed twice to refuel, in the Azores and in Portugal, and finally arrived in England on 31 May – the first Atlantic crossing by any aircraft. It was a formidable achievement, with 6,315km (3,925 miles) covered in 57 hours 16 minutes flying time at an average airspeed of 126·8km/h (78·8mph)

News of the US Navy's attempt stampeded two landplane crews also in Newfoundland into making ill-advised and hasty starts from there on 18 May. First off were Harry Hawker and MacKenzie Grieve in a Sopwith biplane named, appropriately, the *Atlantic*. An hour later followed Frederick Raynham and William Morgan (who had lost a leg in the war) in a Martinsyde. They crashed on take-off, but survived. Nothing was heard of Hawker and Grieve for a week; then a Danish tramp steamer which had no radio signalled to a semaphore station in the Hebrides that she had picked up the Sopwith's crew who had ditched alongside her after experiencing trouble with the aircraft's engine.

Alcock and Brown left Newfoundland on 14 June. They were flying a converted twin-engine Vickers Vimy bomber fitted with extra fuel tanks. It was so overloaded that spectators watching its take-off run were convinced it would crash. Even when airborne and headed out to sea they were still struggling for altitude. Later, flying higher but in cloud, they lost control of the aircraft and it fell into a spin from which it was only recovered just above the waves. They also became badly iced-up, with rime clogging the engine air intakes. Brown crawled out between the biplane wings to each engine in turn to clear them. Another hazard awaited them even after they had arrived safely over Ireland: the green field on which they chose to land was in fact a soft peat bog, and the Vimy went over on its nose in a crash-landing. Nevertheless, Alcock and Brown made the first ever non-stop Atlantic crossing, flying 3,041km (1,890 miles) in 16 hours 28 minutes. Subsequently, both were knighted by King George V for the feat. Sir John

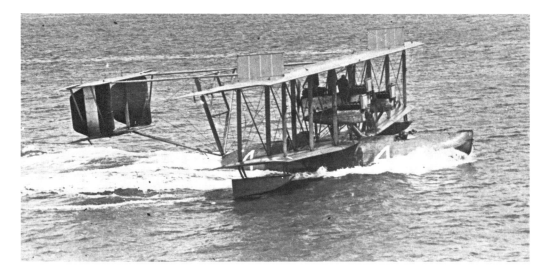

Alcock did not have long to enjoy his fame for he lost his life on 19 December that year while flying in bad weather over France.

In July 1919 the Atlantic was crossed for the first time by an airship, the British R.34. It set out from Scotland on 2 July, captained by Major G. H. Scott of the RAF and carrying Lieutenant Commander Zachary Lansdowne of the US Navy as an observer. The R.34 also carried the first transatlantic aerial stowaway, a rigger who, disgruntled at being left off the crew roster, had hidden among the gas-bags. The airship battled headwinds, and over Newfoundland met thunderstorms that threatened to wrest control of it from the crew, who radioed for help from the US Navy. The crew recovered control and made their way unaided to Long Island, where they landed at Montauk (with just two hours' fuel remaining) on 6 July. The R.34 was the first big airship – 202·69m (665ft) long – to be seen in the United States and attracted huge crowds during its short stay. Just before midnight on 9 July it took off, circled central Manhattan with its searchlights playing, and set off on the return journey. It arrived back in England in 75 hours 3 minutes, helped by good tailwinds.

Thus within two months the Atlantic was crossed four times, by relatively crude and frail machines and yet without a single life being lost on any of the attempts.

A Vimy was used later that year when two Australians, brothers Ross and Keith Smith, made the first flight from England to Australia, in just under 28 days, from 12 November to 10 December. Their flight from Hounslow to Darwin covered 18,175km (11,294 miles). It earned them knighthoods and a £10,000 prize donated by the Australian government. There is another and more tragic coincidence with Alcock and Brown's flight: Sir Ross Smith was killed in an accident not long afterwards, near Brooklands in England on 13 April 1924 while flying a Vickers Viking – the same type in which Sir John Alcock had been killed.

TOP LEFT: *The crew of the US Navy Curtiss NC–4 flying boat which made the first transatlantic flight during May 1919, making seven stops on the way. Three other aircraft set out at the same time, but were not successful.*

CENTRE LEFT: *The British airship R.34, shown here as it came in to land at Mineola, Long Island, on 6 July 1919. It was the first aircraft of any type to make a double crossing of the North Atlantic.*

LEFT: *Alcock and Brown standing by their Vickers Vimy a few days after their Atlantic flight on 14–15 June 1919. They made the crossing from St John's, Newfoundland, to Clifden, County Galway, Ireland.*

TOP: *The United States Navy Curtiss NC–4 flying boat that made the first transatlantic crossing, photographed here as it was entering the harbour at Lisbon, Portugal, where it stopped to refuel.*

ABOVE: *The Vickers Vimy converted bomber in which Ross and Keith Smith won the £10,000 prize offered by the Australian Government for the first flight from Great Britain to Australia, seen here with the crew.*

RIGHT: *The Vickers Vimy was a favourite plane with the first long distance fliers. Here the model in which Alcock and Brown made the first non-stop Atlantic crossing takes off from Newfoundland.*

1920-1925

A Vimy was also the chosen mount of Pierre van Ryneveld and Christopher Quintin Brand when they set off from Brooklands for South Africa on 4 February 1920. When they crashed at Wadi Halfa the South African government provided them with another Vimy in which to continue. This in turn they crashed at Bulawayo on 6 March and it was replaced by a DH-9. They arrived at Wynberg Aerodrome, Cape Town, on 20 March . . . the first men to fly from England to South Africa. They too received knighthoods and £5,000 prize-money.

In the United States, the first coast-to-coast crossing in a single day was accomplished by James H. Doolittle in a DH-4. He flew from the east coast of Florida to San Diego, California, on 4 September 1922 with one refuelling stop in Texas. His flying time was 21 hours 19 minutes over a distance of 3,480km (2,163 miles). (Doolittle, then an Army lieutenant, was to go on to greater fame as a racing pilot and later still organizer and leader of the first US bombing raid on Tokyo during World War II.)

On 2-3 May 1923 two other US Army aviators, Lieutenants O. G. Kelly and J. A. MacReady, flew a Fokker T.2 monoplane from Long Island to San Diego for the first non-stop coast-to-coast flight. They covered 4,050km (2,516 miles) in 26 hours 50 minutes. A little later they established a new world endurance record for aeroplanes by staying aloft in the Fokker for 36 hours 5 minutes over a 4,052km (2,518 miles) measured

ABOVE: *The de Havilland DH-4B flown by Smith and Richter being refuelled in mid-flight by another DH-4.*
ABOVE INSET: *The crew of the first mid-air refuelling flight in June 1923 at Rockwell Field, California.*
RIGHT: *One of the Douglas World Cruisers which made the first round-the-world flight in 1924.*
ABOVE FAR RIGHT: *Charles Lindbergh standing in front of his aeroplane,* Spirit of St Louis, *a modified version of a Ryan monoplane mail carrier.*
FAR RIGHT: *Another view of the* Spirit of St Louis, *in which Lindbergh made the first solo crossing of the North Atlantic on 20-21 May 1927.*

course. The record was beaten on 27-28 August the same year when two other US Army aviators, Captain L. H. Smith and Lieutenant J. P. Richter, kept a DH-4B aloft for 37 hours 15 minutes using in-flight refuelling – their DH-4 was flight-refuelled 15 times from another DH-4.

In 1924 the US Army mounted an even more elaborate exercise than that of the US Navy's transatlantic flight of 1919 – the first around-the-world flight. Four Douglas World Cruisers, open-cockpit single-engined biplanes, were used. Of these two finished the flight – flying 42,398km (26,345 miles) in 363 hours 7 minutes of flying time.

This American project was not the first attempt to fly around the world. Sir Ross Smith had been killed testing a Vickers

amphibian preparing for such an attempt in 1922. An Englishman, Norman Macmillan, using a DH-9 landplane and then a Fairey III-C seaplane, had got as far as the Bay of Bengal in 1922, before the Fairey was wrecked in a storm. They planned a second attempt in 1923, but abandoned it when their supply yacht was arrested in San Pedro, California, for breaches of the US prohibition laws. A French expedition was mounted in 1923 with 5 aircraft and 14 personnel, but they did not get far. There were three other attempts proceeding at the same time as the successful American one: that of Squadron Leader Stuart MacLaren of the RAF, who actually saw the Douglas World Cruisers fly overhead while he was on the water in Burma waiting out a storm; that of Lieutenant Antonio

Locatelli of the Italian Air Service, who flew with the Americans on one leg, but got no further than Iceland; and that of Major Pedro Zanni of Argentina, who crashed his first aeroplane in French Indochina in August and his spare in Japan in May 1925.

There is a point in all this that is insufficiently emphasized by most historians. Most great flying feats and record flights become achievable due to technical advances and that if whoever *did* first achieve them had *not* succeeded, some other aviator, not far behind him, surely would have.

Charles Lindbergh

It was thus, though it is none the less praiseworthy for that, with the flight that is perhaps the single greatest feat in flying history: Charles Lindbergh's first solo Atlantic crossing, from New York to Paris, on 20–21 May 1927. Technically, it was made possible by the development of the Wright Whirlwind engine which possessed a reliability, power: weight ratio and fuel efficiency of a new order. The financial spice for such a flight was a $25,000 prize put up in 1919 by Raymond Orteig. Before Lindbergh's flight six men had already died in the many attempts on the Orteig prize. France's two greatest surviving fighter aces of World War I, René Fonck (who survived) and Charles Nungesser (who vanished in mid-Atlantic with his navigator), were among the failed contenders for the prize.

Lindbergh chose to make his flight solo in a single-engined plane in contrast to several of

his rivals who had chosen trimotors or twins with multi-man crews. Lindbergh's decision was based on the consideration that the weight he would save by flying alone on one engine would be more useful as extra fuel capacity. He commissioned a modified version of a Ryan monoplane mail-carrier, named *Spirit of St Louis*. He designed and controlled the modifications (principally to increase its fuel capacity) himself. Throughout his preparations he doubted that he would be ready in time to cross the Atlantic before his competitors and kept an alternate set of charts of the Pacific.

His take-off roll, which began at 07.52 on 20 May 1927 was one of the great cliff-hangers of all time. The earth was soft with rain, and spectators followed in cars with fire

extinguishers convinced they would be needed since his plane was accelerating so slowly. Lindbergh just staggered over the wires at the end of the field, and set off at low altitude northwestward. Lindbergh battled ice, an involuntary spin in mid-ocean (as had Alcock and Brown) and the craving to sleep. When he arrived over Le Bourget airfield near Paris in the darkness of his second night aloft, at 22.24 on 21 May, traffic jammed every access road from all directions and such a wave of humanity poured over the field that he could hardly find space to land without injuring someone. He had been airborne 33 hours 39 minutes (after having hardly had any sleep the previous night) and had flown 5,810km (3,610 miles) non-stop, averaging 173km/h (107·5mph). His aircraft, which

hangs today in the National Air & Space Museum in Washington, had a performance that was extraordinary then and is still remarkable now. With a 14m (46ft) span, 975·5kg (2,150lb) empty weight and 237hp engine, it could carry 1,700 litres (450 US gallons) of fuel, for a range of about 7,000km (4,350 miles). Its useful load was thus considerably greater than its empty weight. It offered Lindbergh (the only pilot who ever flew it) no direct forward visibility for the huge fuel tank was where the windshield should have been. He could only peer obliquely forward through side panels.

Perhaps the major significance of Lindbergh's feat was that it enthused many ordinary people, who had never given aviation much attention, with flying's possibilities.

BELOW: *The Fokker F.VIIB–3M trimotor* Southern Cross, *in which Charles Kingsford Smith and a crew of three made the first full crossing of the Pacific by air, between 31 May and 9 June 1928, from Oakland to Brisbane.*

BELOW RIGHT: *Bert Hinkler working on his Puss Moth, in which he flew from Toronto via New York and South America to Africa, the first solo flight across the South Atlantic.*

BOTTOM: *Jim Mollison and Captain Broad, a test pilot, standing in front of Mollison's aeroplane, the* Heart's Content, *in which he made the first east-west crossing of the Atlantic during 18–19 August 1932.*

RIGHT: *The LZ 127* Graf Zeppelin *arriving at Hanworth in August 1931. This airship made the first and only airship flight round the world during August 1929, in a total time of 21 days, 7 hours and 34 minutes.*

BELOW FAR RIGHT: *Members of the crew of Byrd's Antarctic expedition grouped in front of the Curtiss-Wright plane which was used by the expedition for aerial surveys.*

BOTTOM RIGHT: *Amelia Earhart standing by her Lockheed Vega monoplane. She was the first woman to make a solo crossing of the North Atlantic, during 20–21 May 1932, from Newfoundland to Northern Ireland.*

1927–1929

In the same year as Lindbergh's flight, on 28–29 June, two US Army Lieutenants, Albert F. Hegenburger and Lester J. Maitland, in a Fokker C–2 trimotor, made the first non-stop flight from the USA westwards to Hawaii. They flew 3,874km (2,407 miles), from Oakland to Honolulu, in 25 hours 50 minutes.

The South Atlantic was also first flown non-stop that year, on 14–15 October, when Dieudonné Costes and Joseph Le Brix of France flew a Breguet XIX from Senegal to Brazil covering 3,420km (2,125 miles) in 19 hours 50 minutes.

The first solo flight between England and Australia was made by H. J. L. 'Bert' Hinkler in a prototype Avro 581 Avian (7–22 February 1928). An Avian III was used by Lady Heath to make the first solo flight by a woman from South Africa to London (12 February–17 May 1928). At about the same time another woman pilot was setting out in the opposite direction, from London for South Africa. Lady Bailey, flying a Cirrus II Moth, left London on 9 March 1928 and returned on 16 January 1929 having in that period flown round Africa, crashed her Moth and replaced it with another.

Charles Kingsford Smith with a crew of 3 made the first full crossing of the Pacific by air between 31 May and 9 June 1928 in a trimotor Fokker F.VIIB–3M, *Southern Cross*. Flying from Oakland to Brisbane via Honolulu and Fiji, they covered 11,890km (7,389 miles) in 83 hours 38 minutes flying time. Kingsford Smith captained the same Fokker

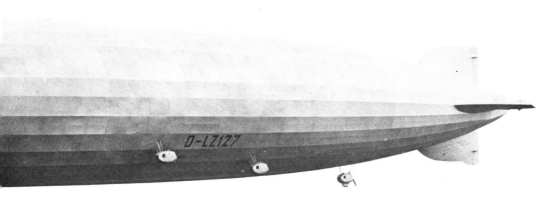

for the first air crossing of the Tasman Sea. He flew from Sydney to Christchurch in 14 hours 25 minutes on 10–11 September of that year.

Two RAF pilots, G. Jones Williams and N. H. Jenkins, made the first non-stop flight from England to India, in a Fairey Long-range Monoplane, between 24–26 April 1929. They flew 6,647km (4,130 miles) in 50 hours 37 minutes – more than 2 days and nights aloft without landing.

Graf Zeppelin

The first, and only, airship flight around the world, and one of the most romantic journeys ever, was made by the *Graf Zeppelin*, captained by Dr Hugo Eckener, between 8–29 August 1929. It was only the second world flight by any aircraft. German airships were noted for their comfort, good food and fine wines, and it is reported that, with the clocks on board being advanced 1 hour every 7, the approximately 20 passengers, some of whom had paid $9,000 for the trip, complained they were being fed too often!

Commander Byrd

The first flight over the South Pole took place on 28–29 November 1929 in a Ford 4–AT Trimotor commanded by Commander R. E. Byrd of the US Navy and flown by Bernt Balchen with two other crew members. The flight, from and returning to the expedition's base at 'Little America' on the edge of Antarctica, took 18 hours 59 minutes. It is interesting that no less than 10 aircraft were in Antarctica with various expeditions that summer. The first aerial expedition to the South Pole (a British one) had been planned as early as 1920.

Byrd, with Floyd Bennett as co-pilot, had earlier made the first-ever flight over the North Pole, out of and back to Spitsbergen, on 9 May 1926 in a Fokker F.VIIA–3M. He had been greatly assisted on this occasion by the Norwegian explorer Roald Amundsen, who had been the first man ever to reach the South Pole.

1930s

Long-distance and over-ocean records continued to be set throughout the 1930s. Clyde Pangborn and Hugh Herndon, in a Bellanca, made the first non-stop flight from Japan to the USA between 3–5 October 1931. The flight, from Tokyo to Washington state, lasted 41 hours 13 minutes. The great American woman pilot, Amelia Earhart, made the first female solo crossing of the North Atlantic from Newfoundland to Londonderry, in a Lockheed Vega monoplane during 20–21 May 1932. James Mollison made the first solo east–west Atlantic crossing, from Dublin to New Brunswick on 18–19 August 1932 in a de Havilland Puss Moth. Mollison had set an Australia-to-London speed record in 1931 in a Moth, and from London to Cape Town in 1932, both times using the same aeroplane. In 1933 Mollison made the first solo east–west crossing of the South Atlantic, in the course of a marathon flight which took him from Lympne, England to Natal in Brazil in a time of 3 days, 10 hours 8 minutes.

Mollison was married to the British aviation heroine Amy Johnson, who had made in 1930 an impressive solo flight – the first by a woman – from England to Australia. She flew a de Havilland D.H.6OG Gipsy Moth. Jim and Amy Mollison flew the North Atlantic together during 22–24 July 1933 in a twin-engined DH Dragon, from Pendine in Wales to a crash-landing at Bridgeport, Connecticut.

A world long-distance record, and the first-ever non-stop flight from England to South Africa, was made by Squadron Leader O. R. Gayford and Flight Lieutenant G. E. Nicholetts in a Fairey Long-range Monoplane between 6–8 February 1933. They flew 8,595km (5,431 miles) in 57 hours 25 minutes.

The first solo flight around the world was made by the one-eyed American pilot Wiley Post between 15–22 July 1933. Piloting a Lockheed Vega, *Winnie Mae*, he flew 25,099km (15,596 miles) in 7 days 18 hours 49 minutes.

Sir Charles Kingsford Smith accompanied by P. G. Taylor made the first flight from Australia to the USA between 22 October–4 November 1934 flying a Lockheed Altair. Amelia Earhart used a Lockheed Vega monoplane to make the first solo flight from

LEFT: *Amy Johnson (Mrs Jim Mollison) in the cabin of her aircraft* Desert Cloud *in November 1932.*
LEFT INSET: *Amy Johnson setting out in her small biplane for the lone flight from Croydon to Australia on 5 May 1930. It was the first such flight by a woman.*

Significant Air Races

Race	Date	Place/Course	Winner	Aircraft	Speed/time
First Gordon Bennett aeroplane race	29 October 1910	Belmont Park, N.Y./100km (62.2 miles)	Claude Grahame-White	100hp Gnome-Blériot	1hr 1min 4.74sec
Circuit of Europe	18 June–7 July 1911	Paris–Reims–Liège–Spa–Liège–Verloo–Utrecht–Breda–Brussels–Roubaix–Calais–Dover–Shoreham–London–Shoreham–Dover–Calais–Amiens–Paris	Lieutenant de Vaisseau Conneau	Blériot XI	
Daily Mail Circuit of Britain	22–27 July 1911	Brooklands–Hendon–Harrogate–Newcastle–Edinburgh–Stirling–Glasgow–Carlisle–Manchester–Bristol–Exeter–Salisbury Plain–Brighton–Brooklands/1,625km (1,010 miles)	Lieutenant de Vaisseau Conneau	Blériot XI	22hr 28min
Daily Mail First Aerial Derby	8 June 1912	Hendon–Kempton Park–Esher–Purley–Purfleet–Epping–High Barnet–Hendon/130km (81 miles)	T. O. M. Sopwith	Blériot	Av. 94km/h (85.5mph) 1hr 23min 8.4sec
First Kings Cup Air Race	8–9 September 1922	Croydon–Glasgow–Croydon	Captain F. L. Barnard	DH.4A	198.9km/h (123.6mph) 6hr 32min 50sec
First Pulitzer Trophy Race	1920	Mitchell Field, Long Island, N.Y./closed-circuit	Lieutenant Corliss C. Moseley	638hp Verville-Packard	251.87km/h (156.5mph)
Last Pulitzer Trophy Race	1925	Mitchell Field, Long Island, N.Y./closed-circuit	Lieutenant Cyrus K. Bettis	620hp Curtiss R3C–1	400.72km/h (248.99mph)
First Thompson Trophy Race	August 1930	Chicago/closed-circuit	Speed Holman	450hp Laird Solution	324.9km/h (201.9mph)
Last Thompson Trophy Race	September 1949	Cleveland, Ohio	Cook Cleland	4,000hp F2G–1 Corsair	639.04km/h (397.07mph)
First Bendix Trophy Race	4 September 1931	Burbank–Cleveland	James Doolittle	535hp Laird Super Solution	358.94km/h (223.03mph)/11hr 16min 10sec
Last Bendix Trophy Race	September 1949	Rosamond Dry Lake, California–Cleveland	Joe DeBona	2,270hp F6C Mustang	756.63km/h (470.136mph)/4hr 16min 17sec
MacRobertson Race	20–23 October 1934	England–Australia/18,329km (11,333 miles)	Charles W. A. Scott/Tom Campbell Black	DH.88 Comet	255.7km/h (158.9mph)/70hr 54min 18sec
Schlesinger Air Race	29 September–1 October 1936	Portsmouth–Johannesburg/9,897km (6,150 miles)	C. W. A. Scott/Giles Guthrie (only finishers)	Vega Gull	186.7km/h (116mph)/2 days 4hr 56min
Circuit of the Alps	25 July 1937	Zurich International Flying Meeting/125 miles	Oberst Karle Franke	Bf 109V13	387km/h (240.9mph)
London–Christchurch (NZ) Air Race	8–9 October 1953	London–Christchurch, New Zealand/19,623km (12,193 miles)	Flight Lieutenants R. L. E. Burton and D. H. Gannon	English Electric Canberra	795.887km/h (494.48mph)/23hr 51min 10sec
Daily Mail Transatlantic Air Race	4–11 May 1969	London–New York	Lieutenant Commanders Brian Davies and Peter Goddard, RN	McDonnell-Douglas Phantom FG.1	4hr 36min 30.4sec flying time

The Schneider Trophy

When the French armaments industrialist Jacques Schneider first put up his trophy for seaplane races in 1912 he can hardly have envisaged that it would become a tournament of enormous world-wide interest and that the competing aircraft, despite the extra drag from their huge floats, the size of which was fixed by the principles of buoyancy, would become the fastest machines in the world at the time. Their designers extended the frontiers of high-speed flight as they wrestled with the problems of minimum-drag airframes, control-surface flutter, propeller torque and cooling the big engines. The early Schneider races were all biplanes, strut- and wire-braced. By the time the series ended in 1931 the superiority of the low-drag monoplane for high-speed flight had been proved and solutions provided for the design of thin-winged mono-

planes, necessary for faster aircraft.

Britain had won the last pre-war contest in Monaco in 1914. The first post-war race, held at Bournemouth in 1919, was declared null and void. Italy won the 1920 and 1921 contests at Venice; but Britain won the contest again at Naples in 1922. In 1923, at Cowes, it was won by a US Navy-sponsored team of 465hp Curtiss CR–3 racers. The Americans declined to hold a race in 1924 after all European entries had been withdrawn; but they won the 1925 contest with a 600hp Curtiss R3C–2 flown by Lieutenant James Doolittle at 374km/h (232mph). The 1926 contest was won by the 882hp Italian Macchi M.39. Britain won the 1926 race at Venice with a Supermarine S.5 of the RAF's High Speed Flight, flown by Flight Lieutenant S. N. Webster, who during the contest set a world seaplane speed record of 455

km/h (283mph).

Thereafter it was agreed by the competing nations, in view of the costs and work involved, to hold the contest every other year. Flying Officer H. R. D. Waghorn won the 1929 contest at Spithead, in a Supermarine S.6 at 528km/h (328mph). The S.6s were powered by a special Rolls-Royce V–12 engine that developed almost 2,000hp, and was a forerunner of the Merlin. The British government, in view of the poor economic state of Britain, declined to sponsor a team for the 1931 contest. A private individual, Lady Houston, donated £100,000 to allow Britain to host the race. However, no other nation was ready to take part and unlike the USA in 1924, Britain chose to win the contest, an uncontested 'walkover', with an S.6B powered by a 2,350hp Rolls-Royce racing engine. Under the original

terms of the contest Britain thereby retained the trophy, having won it three times running. This same S.6B, flown by Flight Lieutenant G. H. Stainforth on 29 September 1931, set up a world absolute speed record of 658km/h (490mph).

The aircraft that competed for the Schneider Trophy were somewhat freakish craft, and seaplanes rather than landplanes only because the contest rules required that they be so. Their handling was tricky, they were dangerous to fly, and many of them crashed. Yet despite all that they made a great contribution to the development of aviation. One example will suffice. The Supermarine designer R. J. Mitchell later used the experience he gained from his Schneider racers to help in the design of the Spitfire, a decisive factor in the outcome of the Battle of Britain – an event which changed history.

LEFT: *Lieutenant James H. Doolittle standing on the float of the Curtiss R3C–2 seaplane in which he won the 1925 Schneider Trophy at Baltimore, USA. He won with an average speed of 374.3km/h (232.5mph), which pushed up the winning speed by over 88km/h (55mph) more than in the previous contest.*

BELOW: *The Supermarine S.6 N247 in which Flying Officer H. R. Waghorn won the Schneider Trophy in 1929 at Cowes. The S.6's average speed was an astonishing 528.9km/h (328.6mph), and its victory was the second consecutive one for Britain, leaving only one more to be gained to win the trophy outright, which Britain did in 1931.*

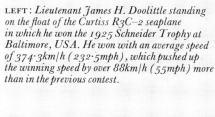

Date	Place	Winner	Country	Aircraft	Engine	Speed km/h (mph)
16 April 1913	Monaco	Maurice Prévost	France	Deperdussin	160hp Gnome	73·63 (45·75)
20 April 1914	Monaco	Howard Pixton	Britain	Sopwith Schneider	100hp Gnome Mono	139·66 (86·78)
10 September 1919 (Contest later declared null and void)	Bournemouth	Guido Janello	Italy	Savoia S.13	250hp Isotta-Fraschini	176·66 (109·77)
21 September 1920	Venice	Luigi Bologna	Italy	Savoia S.12	550hp Ansaldo	172·55 (107·22)
7 August 1921	Venice	Giovanni di Briganti	Italy	Macchi M.7	250hp Isotta-Fraschini	189·74 (117·90)
12 August 1922	Naples	Henri Biard	Britain	Supermarine Sea Lion II	450hp Napier Lion	234·48 (145·70)
28 September 1923	Cowes	David Rittenhouse	USA	Curtiss CR–3	465hp Curtiss D–12	285·46 (177·38)
26 October 1925	Baltimore	James Doolittle	USA	Curtiss R3C–2	600hp Curtiss V–1400	374·28 (232·57)
13 November 1926	Hampton Roads	Mario de Bernardi	Italy	Macchi M.39	882hp Fiat AS.2	396·70 (246·50)
26 September 1927	Venice	S. N. Webster	Britain	Supermarine S.5	900hp Napier Lion VIIA	453·28 (281·66)
7 September 1929	Calshot	H. R. Waghorn	Britain	Supermarine S.6	1900hp Rolls-Royce 'R'	528·87 (328·63)
13 September 1931	Calshot	J. N. Boothman	Britain	Supermarine S.6B	2350hp Rolls-Royce 'R'	547·31 (340·08)

Hawaii to the US mainland on 11–12 January 1935. Amelia Earhart's solo flight lasted 18 hours 16 minutes.

Jean Batten of New Zealand made the first solo South Atlantic crossing by a woman on 13 November 1935 flying a Percival Gull. She used the same aircraft to make the first flight from Britain to New Zealand by a woman, in the fastest time ever of 11 days 45 minutes (5–16 October 1936).

Other events of that time foretold the way aviation was headed. The prototype Bf 109 first flew on 28 May 1935; the first Hurricane on 6 November; and the first Spitfire on 5 March 1936. The specially designed Me 209 racer, flown by Fritz Wendel, set a world speed record of 755·14km/h (469·22mph) on 26 April 1939. This remained the world record for piston-engined aircraft for 30 years. The first successful helicopter, the Focke Achgelis Fa 61 V1, made its first flight on 26 June 1936. Frank Whittle ran the world's first turbojet engine on 12 April 1937. The first jet aircraft to fly was the Heinkel He 178, on 27 August 1939. The first British jet, the Gloster E.28/39, did not make its maiden flight until 15 May 1941. Germany also flew the first rocket-propelled aircraft, a modified glider named the *Ente* piloted by F. Stamer. The flight lasted a minute, on 11 June 1928.

Post-war achievements

A rocket-powered aircraft, air-launched from under the wing of a bomber, made the first manned supersonic flight. The Bell XS–1 was flown by Major Charles Yeager on 14 October 1947. Rocket aircraft remain those which have flown fastest and highest, even within the atmosphere. A North American X–15A–2 research aircraft, air-launched from under the wing of a B–52 jet bomber and piloted by W. J. Knight, reached a speed of Mach 6·72 (7,297km/h/4,534mph) on 3 October 1967. The highest altitude achieved during the X–15 programme was 107,860m (354,200ft), reached by J. A. Walker on 23 August 1963.

The first non-stop flight around the world was made by a piston-engined Boeing B–50 Superfortress in March 1949, captained by James Gallagher and manned by a crew of 18. The *Lucky Lady II* flew about 36,225km (22,500 miles) at an average speed of 378·35km/h (235mph), and was refuelled in the air four times: over the Azores, Saudi Arabia, the Philippines and Hawaii. Three 8-jet Boeing B–52 bombers flew around the world in formation and non-stop in 1957, covering 39,147km (24,325 miles) in 45 hours 19 minutes – less than half the time the B–50 had taken only 8 years earlier. The B–52s were refuelled four times, and followed a route that took them over the Atlantic to Morocco, to Dharan, across India and Ceylon, and home to California via Guam. Each aircraft carried a crew of 9 instead of the usual 6, and one tail gunner set a record of a kind as the first man to circumnavigate the world facing backwards.

The world has been circumnavigated since World War II by innumerable pilots, including at least one in a homebuilt aero-

plane. Among the women who have flown around the world is Miss Sheila Scott who achieved her feat in 33 days 3 hours, between 18 May–20 June 1966, in a single-engined 260hp Comanche. Her other records include flying from London to New York in 24 hours 48 minutes on 4–5 May 1969; and from London to South Africa in 3 days 2 hours 14 minutes, between 6–9 July 1967. The world has been circumnavigated over the North and South Poles by, among others, Max Conrad flying solo in a light twin. This pilot has made a number of remarkable long distance solo flights. He holds a world class record for distance, at 11,211·83km (6,967 miles) in a single-engined Comanche. At the age of 62 he flew solo from Cape Town to St Petersburg, Florida, in a Twin Comanche.

The more important world records since World War II have generally been competed for by air forces and governments, with private individuals lacking the financial resources to embark on such ventures. A number of memorable world records have, however, been set in recent years. The Lockheed test pilot Darryl Greenamyer, in a privately sponsored project, flying his much-modified Grumman Bearcat Special *Conquest One*, took the world airspeed record for piston-engined aircraft, achieving 776·449 km/h (483·041mph) on 16 August 1969. On 30 July 1983 Frank Taylor, again of the USA, took the record to 832·12km/h (517·06mph) with a modified P–51 Mustang.

James R. Bede, flying a sailplane which he had converted to motor power, gained the world record for distance around a closed circuit for piston-powered aeroplanes, when he covered 17,083km (10,615 miles) between 7–10 November 1969 in 70 hours 15 minutes. This, however, was totally eclipsed by one of the greatest achievements in the story of aviation: a flight which simultaneously gained the World Records both for distance in a straight line and distance in a closed circuit! This is because the flight was non-stop around the world, *without* air refuelling.

The credit goes to Dick (brother of Burt) Rutan and Jeana Yeager, who built and crewed a remarkable aeroplane named *Voyager*. Made entirely of carbon fibre and other fibre-reinforced composites, it has a wing of 33.77m (110ft 9.5in) span, with a central nacelle carrying a canard foreplane joined at its tips to the front of twin booms carrying rear fins. The engine used throughout the flight was a highly economical TCM Voyager 200, a water-cooled piston engine of 110hp. For the takeoff only, a 130hp TCM 0–240 was mounted on the nose, with feathering propeller. Empty weight of this amazing machine was a mere 1,217kg (2,683lb), whereas the weight of fuel was 3.181kg (7,012lb). The flight of 42,934km (26,678 miles) took 9 days 3 min 44 sec, starting and finishing at Edwards, California.

In September 1975 a Lockheed SR–71A aircraft, en route to be parked at the Farnborough Air Show, set a record of 1 hour, 54 minutes 56·4 seconds for the New York to London sector of its flight – a distance of 5,584km (3,470 miles). It returned to California, 9,085km (5,645 miles) in just 3 hours, 47 minutes 35·8 seconds. The same type of aircraft holds the world airspeed record at 3,529·56km/h (2,193·17mph), and the (sustained) altitude record of 25,929m (85,069ft). Yet it is less than 80 years since Louis Blériot made that first crossing of the 30·6-km (19-mile) wide English Channel and when the world speed and altitude records were a mere 77·3km/h (48mph), and just over 152·4m (500ft) in altitude.

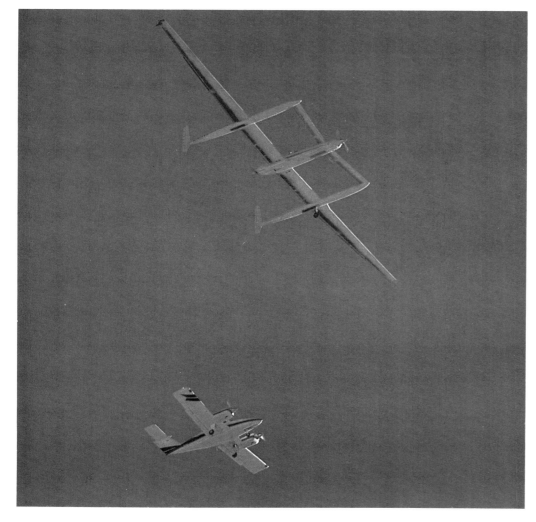

OPPOSITE PAGE
FAR LEFT: *The New Zealand girl Jean Batten arriving at Croydon after her solo flight from Australia to Britain which ended after 15 days on 29 April 1935.*
CENTRE LEFT ABOVE: *Sheila Scott in the cockpit of her single-engined Piper Comanche, in which she later flew round the world, between 18 May and 20 June 1966.*
CENTRE LEFT BELOW: *The veteran American pilot Max Conrad checking up on his engine after landing safely in Rome at the end of a non-stop flight from Chicago.*

THIS PAGE
LEFT ABOVE: *The Bell XS–1, a rocket-powered aircraft which made the first manned supersonic flight on 14 October 1947, air-launched from under the fuselage of a bomber.*
LEFT: *On 14–23 December 1986 the* Voyager, *crewed by Dick Rutan and Jeana Yeager, flew non-stop around the world on internal fuel. It took off on two small piston engines (113/130hp) at a weight of 9,700lb (4,400kg).*

Air Crimes

Crimes connected with aviation fall into four main categories: theft, smuggling, sabotage and hijacking. They may be categorized as follows:

Theft

Unless the villain is a pickpocket, or one prepared to snatch hand-luggage from the overhead rack, theft rarely occurs *within* an aircraft. It is in the terminal and its environs that theft flourishes. In the terminal itself the loss of luggage is commonplace, either before departure, after arrival, or at some transit airport. There have been enormous losses by bullion robberies from airport warehouses, as well as pilfering from stacks of cargo awaiting shipment, which offer tempting opportunities for amateur and professional alike.

Terminal car parks – where automobiles are left for extended periods – are also targets for thieves, who know that in many instances the theft may not be reported for two or more weeks after the event. The only safeguard against theft, whatever form it takes, lies in constant and increased vigilance by police, security patrols and by the public generally.

Smuggling

Smuggling, in some form or other, whether it be watches, drugs, diamonds, currency, or any other dutiable or forbidden items, has flourished since the beginning of international air travel, and customs officers at airports throughout the world remain constantly alert for such attempts. Hiding places for would-be smuggled items frequently display imagination – inside hats and babies' nappies, hidden in hollow heels and wooden legs, and concealed around the figures of stout men and women, who become surprisingly thin when examined by customs officials.

Some smugglers are prepared to gamble with their lives. For example, some attempt to smuggle drugs by swallowing liquid cannabis contained in rubber contraceptives. Although smuggling by swallowing is an old practice, this idea is comparatively new – only about two years old, in fact. Although this idea originated in America, customs investigators in Europe are concerned because the idea is fast spreading on their side of the Atlantic. Some cannabis smugglers give themselves away by refusing a cup of tea or drink of water when being quizzed by investigators. The drugs are usually swallowed in India just before a flight, and nothing else must be swallowed until they are passed naturally – usually in the United States. Liquid swallowed could bring about an embarrassing situation, or worse still, cause death by perforation of the contraceptive. Light aircraft have also been used

frequently for smuggling cocaine and marijuana from South America, Mexico, Jamaica and elsewhere into the United States and Canada.

Another comparatively new form of smuggling is that of human beings – mainly illegal immigrants – into the United States and into Britain. Many succeed with the connivance of staff already employed at international airports. Many attempts have been very ambitious. On one occasion a chartered airliner arrived at Heathrow, London, from the East; its passengers, all carrying musical instruments, were ushered through customs, gaining entry as an orchestra. Two weeks later the airliner returned to take them back, but not one member of the 'orchestra' was seen again.

Sabotage

This may be regarded as the most cowardly of all crimes, especially when innocent men, women and children lose their lives in the explosion or subsequent crash, as so often happens. Saboteurs fall into three categories: those with strong political or ideological motives; those who threaten sabotage in the hope of gain – either in cash or the release of persons from custody; and demented, compulsive murderers.

From their very nature, aircraft are particularly vulnerable to sabotage, and almost any concessions can be wrung from the operating companies – at least when the lives of passengers and crew are in jeopardy.

The important subject of sabotage, and how to combat it, was discussed at a conference organized by the International Civil Aviation Organization (ICAO) in the summer of 1970. This conference, held in Montreal, revealed that there had been 14 sabotage or armed-attack cases during the 18 months prior to the end of June 1970. A six-point manifesto was issued, in which one item called for 'concerted action' in suppressing such criminal acts. Today, although acts of sabotage appear to be less frequent, it is possible that some aircraft whose disappearance is marked down as 'cause unknown', could be traced to an act of sabotage.

Hijacking

Of all modern aviation crimes, 'hijacking' is the most recent. Isolated incidents of aerial piracy have been brought to the notice of the public since 1930, but the first official statement concerning this crime came from the Tokyo Convention of September 1963, held to discuss 'Offences Committed on Board Aircraft'. The Convention stated that '. . . if a person on board unlawfully or by force or threat of force seized an aircraft, or wrongly interfered with the control of it in

put down passengers, mail and cargo originating from the territory of the state of registry of the aircraft, the privilege to take on passengers, etc. destined for such territory and, finally, to take on or put down passengers, etc. coming from or destined to the territory of any contracting state). Only 12 states are parties to this agreement, and states prefer to negotiate the granting of the traffic rights on a bilateral basis.

The Chicago Convention deals in detail with many technical and economic aspects of international civil aviation, and in Part II establishes the International Civil Aviation Organization (ICAO) and sets its constitutional rules.

The International Civil Aviation Organization (ICAO), created by the Chicago Convention, came into being on 4 April 1947. It is a specialized agency in the United Nations system and its membership – 135 states – makes it practically universal. The Headquarters of ICAO is in Montreal, Canada, and Regional Offices are located in Paris (European Office), Dakar (African Office), Bangkok (Far East and Pacific Office), Cairo (Middle East and Eastern African Office), Mexico City (North American and Caribbean Office) and Lima (South American Office).

The aims and objectives of the Organization are to develop the principles and techniques of international air navigation, and to foster the planning and development of international air transport to insure the safe and orderly growth of international civil aviation throughout the world, encourage the arts of aircraft design and operation for peaceful purposes, promote safety of flight in international air navigation and to promote generally the development of all aspects of international civil aeronautics.

The supreme body of ICAO is the Assembly, which meets every three years. The executive body is the Council, a permanent body composed of 30 states elected by the Assembly. The Air Navigation Commission is a permanent body of 15 technical experts appointed by the Council from among persons nominated by contracting states. The Air Transport Committee dealing with economic matters is composed of Representatives on the Council. The Legal Committee of ICAO was established by the Assembly, and is open to participation by all states. The President of the ICAO Council is elected by the Council, and is an international civil servant; he has no vote in the Council. The chief executive officer is the Secretary General who is appointed by the Council.

From the legal point of view, the essential function of ICAO is the formulation and approval of international standards and recommended practices in the field of civil aviation. These standards are approved by the Council and designated as Annexes to the Chicago Convention. All contracting states undertake to collaborate in securing the highest practicable degree of uniformity in regulations, standards, procedures, etc., but may file a difference if they find it impracticable to comply in all respects with any international standard or procedure. So far the Coun-

cil of ICAO has approved 17 Annexes to the Chicago Convention containing international standards and recommended practices.

The International Air Transport Association (IATA) is a non-governmental organization of airlines, and was created in 1945 to succeed the International Air Traffic Association of 1919. It was incorporated in Canada by Act of Parliament and has the status of a corporation, to which Canadian law attributes an international character. IATA is sometimes compared with an international cartel of airlines – particularly because of its rate-making functions. However, since in many states the airlines are subject to anti-trust legislation, and since the tariffs and rates are subject to governmental approval, IATA should not legally be considered as a business cartel.

Apart from technical and economic regulation of the airline industry, IATA performs important functions in the legal field by establishing uniform air transport documents and general conditions of carriage which are accepted uniformly in the tariffs of a great majority of world airlines.

The international character of civil aviation necessitates a high degree of uniformity of legal regulations. The unification of law is best achieved by multilateral international agreements, the rules of which are transformed into the national legislation of contracting states. Since the establishment of ICAO, all agreements in this field have been prepared by the ICAO Legal Committee and adop-

ted by Diplomatic Conferences convened and held under the auspices of ICAO. The following air law conventions have been adopted so far.

Warsaw Convention for the Unification of Certain Rules Relating to International Carriage by Air (1929). This Convention, to which 108 states are parties, is the most widely accepted unification of private law. It unifies the rules on documentation in the carriage of passengers, baggage and cargo (passenger ticket, baggage check and air waybill). Secondly, it unifies the regime of liability of the carrier. The carrier is liable on the basis of fault, but there is a presumption of fault and the burden of proof is reversed. The extent of liability of the carrier is limited to 125,000 Poincaré francs (less than US $10,000) in respect of passengers, and to 250 francs (about US $20) per kilogramme of baggage or cargo. Finally, the Convention unifies the question of jurisdiction by defining the courts before which an action may be brought.

The Convention was amended in 1955 by **The Hague Protocol**, which simplified the documentation and doubled the limit of liability in respect of passengers. In 1961 a Supplementary Convention was adopted at *Guadalajara*, which extends the applicability of the Warsaw Convention to the 'actual' (as distinct from 'contractual') carrier. In 1966 the Warsaw Convention was *de facto* amended by a private agreement of air carriers ('**Montreal Agreement**, 1966') with respect to carriage to, from or via the territory of the USA.

For such carriage the carrier is absolutely liable with respect to passengers, and the limit of liability is $75,000.

The **Guatemala City Protocol** of 1971 (not yet in force) further simplified the documentation requirements with respect to passengers, introduced the principle of absolute liability and increased the limit of liability to about US $100,000.

The **Montreal Protocols** of 1975 (not yet in force) would furthermore replace the gold value clause (Poincaré franc) in the system of the Warsaw Convention by the Special Drawing Rights of the International Monetary Fund.

Geneva Convention on the International Recognition of Rights in Aircraft (1948). The purpose of this Convention, to which 38 states are parties, is to ensure protection of rights in aircraft (such as the right of creditors), when the aircraft and its equipment are situated in another country or where the registration of the aircraft is changed from one country to another.

Rome Convention on Damage Caused by Foreign Aircraft to Third Parties on the Surface (1952). There are 28 states parties to this Convention, which introduces the regime of absolute liability for damage caused by aircraft on the surface. The liability is limited and the amount varies with the weight of the aircraft. The Legal Committee of ICAO is now studying a possible amendment of the Rome Convention in order to make it clear that it would deal only with 'physical impact' damage and not with damage caused by noise, vibrations or sonic boom. The system of the limits of liability will also be reviewed. The preparation of a new separate instrument on liability for damage caused by noise and sonic boom is under study in ICAO.

Under the auspices of ICAO three Conventions in the field of criminal law have been prepared, the last two of them dealing with criminal acts against the safety of civil aviation such as 'hijackings'. They are the **Tokyo Convention on Offences and Certain Other Acts Committed on Board Aircraft** (1963); the **Hague Convention for the Suppression of Unlawful Seizure of Aircraft** (1970) and the **Montreal Convention for the Suppression of Unlawful Acts Against the Safety of Civil Aviation** (1971). For further details see the Chapter on Air Crimes on pages 436–439.

LEFT: *Some of the passengers from an airliner hijacked to Damascus on 29 August 1969. It was a TWA Boeing 707 on a flight to Lydda, and was forced to land at Damascus by two Arab terrorists, who then destroyed the aircraft there after allowing the passengers to be evacuated. This attack was another example of hijacking for political reasons — the terrorists declared that it was a reprisal for the recent sale by the United States of 50 Phantom jet fighters to the Israeli armed forces.*

Famous Names

Ader, Clément (1841–1925)
French pioneer, whose batwinged *Éole* monoplane became, on 9 October 1890, the first man-carrying aeroplane after Mozhaiski to achieve a powered 'hop' into the air.

Alcock, Sir John (1892–1919)
Who, with Lieutenant Arthur Whitten Brown (later Sir Arthur) as navigator, achieved the first non-stop crossing of the North Atlantic in a Vickers Vimy bomber, 14 to 15 June 1919.

Aldrin, Edwin (1930–)
Astronaut and crew member of the *Apollo XI* spacecraft which carried the first men to land on the Moon's surface, on 21 July 1969.

Amundsen, Roald (1872–1928)
Renowned explorer, a member of the crew of 16 of the Italian airship *Norge* which flew over the North Pole on 12 May 1926.

Antonov, Oleg K. (1906–1984)
Antonov began with the design of gliders and light transports, and his An-2 of 1947, despite being a biplane, outsold every other post-1945 aircraft with over 20,000 built. His design bureau at Kiev continues to produce turboprop and jet transports, including the world's largest aeroplane, the An-124.

Archdeacon, Ernest (1863–1957)
Whose prize awards did much to foster European aviation, was the first passenger to be carried in an aeroplane in Europe, on 29 May 1908.

Arlandes Marquis d'
First aerial passenger in history, who accompanied Pilâtre de Rozier in a Montgolfier hot-air balloon on 21 November 1783.

Armstrong, Neil (1930–)
Commander of the *Apollo XI* spacecraft which carried the first men to the Moon, and the first man to step on the lunar surface on 21 July 1969.

Bacon, Roger (1214–1292)
English scientist-monk who, so far as is known, was the first person to write in a scientific manner about the possibility of manned flight. His *Secrets of Art and Nature*, in which he first put forward such ideas, was written about 1250.

Bader, Sir Douglas (1910–1987)
Famous British fighter ace of World War II, who achieved his 22 credited victories while flying with two artificial legs. His legs were terribly injured and amputated following a pre-war flying accident.

Balbo, General Italo (1896–1940)
Italian Air Minister, who led two mass flights of Savoia-Marchetti flying-boats across the South and North Atlantic in 1931 and 1933 respectively.

Balchen, Bernt (1899–1973)
Pilot of the Ford Trimotor *Floyd Bennett*, which recorded the first flight over the South Pole on 29 November 1929.

Baldwin, F. W. (Casey)
First British subject (a Canadian) to fly a heavier-than-air machine. This flight was made at Hammondsport, N.Y., 12 March 1908, in an AEA Red Wing.

Ball, Albert (1896–1917)
First great British fighter ace of World War I, was credited with his 47th victory on 6 May 1917. He died on the following day when his aircraft crashed after following a German single-seater scout aircraft into dense cloud.

Banks, F. Rodwell (1898–1986)
Naval officer in World War I, Banks devoted a lifetime to aero engines, becoming famed for his special fuel 'cocktails' that won the 1929 and 1931 Schneider Trophies. In World War II Director of Engine Production, later director of Bristol, Blackburn, de Havilland and Hawker Siddeley.

Barber, Horatio (1875–)
Pilot of a Valkyrie monoplane which, on 4 July 1911, carried the first recorded consignment of air-freight between Shoreham and Hove in Sussex.

Barnard, F. L.
Commodore of Britain's early Instone Airline, who won the first King's Cup Air Race on 9 September 1922. Captain Barnard was flying Sir Samuel Instone's de Havilland D.H.4A *City of York*.

Batten, Jean (1909–c. 1980)
Pioneer woman pilot from New Zealand, whose exploits included the first solo air crossing of the South Atlantic by a woman, 11–13 November 1935.

Beachey, Lincoln
Daring pioneer pilot in America, first became famous for an airship flight around the Washington Monument on 13 June 1906.

Béchereau, Louis
Responsible primarily for the design of the streamlined, circular, monocoque wooden fuselage of the highly successful French Deperdussin monoplanes of 1912/1913.

Bell, Alexander Graham (1847–1912)
Famous inventor born in Edinburgh, was associated with Glenn Curtiss and Thomas E. Selfridge in Bell's Aerial Experiment Association which, in 1903 to 1905, carried out successful aeroplane experiments in America, following the work of the Wright brothers.

Bennett, D. C. T. (1910–1986)
New Zealander Don Bennett was a famed commercial pilot in the 1930s. In World War II he formed and commanded RAF Bomber Command's Pathfinder Force. After the war he formed British South American Airways and other organizations.

Birkigt, Marc (1878–1953)
Swiss engineer who in 1904 founded with Damien Mateu the Hispano-Suiza Motor Company. During World War I the company turned to production of aero-engines. Another famous product of the company was the 20mm Hispano gun.

Bishop, R. E. (1903–)
Designer of the de Havilland D.H.106 Comet 1, which on 2 May 1952, operated the world's first passenger service by a turbojet-powered aircraft.

Bishop, William A. ('Billy') (1894–1956)
Canadian ace pilot of World War I, credited with 72 confirmed victories. Awarded Britain's highest military honour, the Victoria Cross, he rose subsequently to the rank of Honorary Air Marshal in the Royal Canadian Air Force.

Black, Tom Campbell
Co-pilot of the de Havilland D.H.88 Comet which won the 1934 MacRobertson race from Mildenhall, England, to Melbourne, Australia, in a time of 71 hours.

Blériot, Louis (1872–1936)
French aviator pioneer who, flying a Blériot XI monoplane of his own design, made the first crossing of the English Channel in an aeroplane, on 25 July 1909.

Boeing, William (1881–1956)
Co-pilot with Eddie Hubbard of the Boeing C–700 seaplane which made the first international air mail flight from Canada to the USA, on 3 March 1919, he was the founder of the now famous Boeing Company.

Boothman, John N. (1901–1957)
A member of the Royal Air Force High Speed Flight of 1931, and pilot of the Supermarine S.6B racing seaplane which won the Schneider Trophy outright for Great Britain on 13 September 1931.

Brabazon of Tara, Lord (Moore-Brabazon, J. T. C.) (1884–1964)
First Englishman to make an officially recognized aeroplane flight in England, on 30 April 1909, and holder of the first Aviator's Certificate in Great Britain, awarded by the Aero Club on 8 March 1910.

Braun, Dr Wernher von (1912–1977)
German rocket pioneer, under whose supervision was developed the world's first ballistic rocket missile, the V–2 (A–4), first launched successfully on 3 October 1942. He was one of the leading American space scientists.

Breguet, Louis (1880–1955)
French pioneer of rotary-wing flight. In association with Professor Richet he designed and built a helicopter which, on 29 September 1907, was the first to lift a man from the ground. This was not a controlled flight, the aircraft being stabilized by four men with poles.

LeBris, J. M. (1808–1872)
A French sea captain who studied bird flight on his ocean voyages. In 1856 he completed construction of a glider which was launched from a horse-drawn cart. A second aircraft, built in 1868, was launched frequently, with ballast instead of a human pilot.

Butler, Frank Hedges (1856–1928)
Aware of the encouragement which early aviation activities in Europe were receiving from the Aéro-Club de France, was the first to suggest the formation of an Aero Club in Britain. This was duly registered on 29 October 1901.

Byrd, Richard E. (1888–1957)
American explorer, first man to fly over both North and South Poles, on 9 May 1926, and 29 November 1929, respectively.

Cabral, Sacadura
Who, together with Gago Coutinho, achieved the first aeroplane crossing of the South Atlantic in 1922. Their arrival in Brazil, on 16 June, marked the end of an epic three-month journey involving three aircraft.

Camm, Sir Sydney (1893–1966)
Famous British aircraft designer, who joined the H. G. Hawker Engineering Company in the early 1920s. Knighted in 1953 for his services to aviation, his name will always be associated with the Hawker series of fighters.

Capper, J. E. (1861–1955)
Officer Commanding the Balloon Section of the British Army in 1903 and pilot of the first British Army airship, *Nulli Secundus*, which flew for the first time on 10 September 1907.

Castoldi, Dr Ing Mario
Famous aircraft designer, whose work is associated with the Italian company Aeronautica Macchi. He is known universally for the superb racing seaplanes which he designed to compete in the Schneider Trophy Contests.
Section of the British Army in 1903

Cayley, Sir George (1773–1857)
British inventor, now regarded universally as the 'Father of Aerial Navigation'. As early as 1799 he engraved on one side of a silver disc – now in London's Science Museum – a diagram showing the forces of lift, drag and thrust.

Cessna, Clyde V. (1880–1954)
In 1911 a Kansas farmer went to a visiting 'air circus'. He was so impressed he bought a Blériot. After many vicissitudes he designed the Model A, a cantilever high-wing monoplane, and went into production at a factory in Wichita in 1927. Cessna weathered the Depression and in 1935 handed over a prosperous company to nephew Dwane Wallace. Cessna went back to till the soil, and the company that bears his name has delivered 178,000 aircraft.

Chadwick, Roy (1893–1947)
Famous aircraft designer of the British A. V. Roe Company, renowned for the design of the Avro Lancaster bomber which played a significant role in World War II.

Chanute, Octave (1832–1910)
American railway engineer, who became one of three great pioneer designers of gliders. He began to build hang gliders in 1896.

Charles, Professor J. A. C. (1746–1823)
Designed and built the first hydrogen-filled balloon. This made the first ascent, unmanned, of a hydrogen balloon on 27 August 1783. He was also one of the first men to be carried aloft in such a balloon on 1 December 1783.

Chavez, Georges (1887–1910)
Peruvian pioneer pilot, made the first aeroplane flight over the Alps, in a Blériot monoplane, on 23 September 1910. Success turned to disaster, for Chavez was killed when his aircraft crashed when landing at Domodossola, Italy.

Chichester, Sir Francis (1901–1971)
British aviator, sailor and navigator extraordinary, achieved the first solo flight in a seaplane between New Zealand and Japan in 1931.

Cierva, Juan de la (1886–1936)
Spanish designer of rotary-wing aircraft, whose C–4 *Autogiro* flew successfully for the first time at Getafe, Spain, on 9 January 1923.

Coanda, Henri (1885–1972)
Romanian aviator engineer who, in 1910, built the world's first jet-propelled aeroplane. This was not powered by a turbojet, but by what today would be called a ducted-fan. He subsequently became a significant aircraft designer with the British & Colonial Aeroplane Company (later Bristol).

Cobham, Sir Alan (1894–1973)
Gained early renown for a number of long distance flights, including a 8,047m (5,000 mile) circuit of Europe in 1921, London to Cape Town and return in 1926 and a 37,015m (23,000 mile) flight around Africa in 1927.

Cochran, Jacqueline
Who, flying a North American F–86 Sabre fighter aircraft on 18 May 1953, became the first woman in the world to pilot an aeroplane at a speed faster than sound.

Cody, Samuel Franklin (1861–1913)
Born an American, made the first officially recognized aeroplane flight

OPPOSITE PAGE
TOP LEFT : *Douglas Bader (left)*.
TOP RIGHT : *Floyd Bennett (left)*.
CENTRE : *Jean Batten*.
BOTTOM LEFT : *Armstrong, Collins and Aldrin*.
THIS PAGE : FAR LEFT : *Dr Wernher von Braun*
BELOW : *Octave Chanute*.
TOP CENTRE : *Richard E Byrd*.
BOTTOM : *Samuel F Cody*.

in Britain, in the *British Army Aeroplane No. 1*, which he designed and built. This flight, of 424m (1,390ft) was made on 16 October 1908. Cody later became a British citizen.

Conrad, Max
American pilot who made numerous record long-distance flights in light aircraft, including Casablanca–Los Angeles, non-stop in 58.5 hours. Between 1928 and 1974 he logged 52,929 hours' flying time.

Cornu, Paul
Pioneer designer of rotary-winged aircraft, achieved the first free flight of a man-carrying helicopter on 13 November 1907. This was in a twin-rotor aircraft of his own design, flown at Lisieux, France.

Costes, Dieudonné
With Joseph le Brix, achieved the first non-stop crossing of the South Atlantic flying the Breguet XIX *Nungesser-Coli* on 14 to 15 October 1927.

Coutelle, Capitaine J.-M.-J. (1748–1835)
Commander of the French Company of Aérostiers in the Artillery Service who, on 26 June 1794, carried out

military reconnaissance from the tethered observation balloon *Entreprenant* at the battle of Fleurus – the first operational use of an aircraft in warfare.

Curtiss, Glenn H. (1878–1930)
American pioneer designer and aircraft builder. He made the first official public flight in the USA of more than 1km (0·62 miles) on 4 July 1908. His Curtiss Aeroplane and Motor Company was to become famous for the design and construction of naval aircraft.

Dassault, Marcel
French aircraft designer who, under the original name of Marcel Bloch, was responsible for the design of a great variety of aircraft. During World War II his brother adopted the Resistance code-name of d'Assault. After the war, in 1946, both brothers adopted the family name Dassault, and Marcel rebuilt his aircraft company under this famous name.

Delagrange, Léon (1873–1910)
French pioneer pilot who accomplished the first aeroplane flight in Italy in May 1908, and in which country he flew the world's first woman aeroplane passenger on 8 July 1908. On 17 September 1908, he flew non-stop for 30min 27sec at Issy-les Moulineaux, then the longest duration recorded in Europe.

Doolittle, James H. (1896–   )
Renowned American pilot, who has contributed much to aviation. He made the first one-day transconti-

nental crossing of the United States in 1922, won the 1925 Schneider Contest at Baltimore, made the first instrument landing in fog in 1929, set a World Speed Record in 1932 and became a national hero when he led a raid on Tokyo in April 1942.

Dornier, Dr Claude (1884–1969)
German pioneer designer and aircraft builder, first employed in the development of large all-metal flying-boats for Count von Zeppelin. This early work culminated in the evolution of some superb flying-boats in the between-wars period, some of which remained in service until 1970.

Douglas, Donald Wills (1892–1981)
Fully qualified at MIT, Douglas helped Martin build bombers in 1917, but formed his own company in 1920 and soon received orders from the US Navy and Army (including the first aircraft to fly round the world). His fame increased with the DC-1 of 1933, from which stemmed all the Douglas Commercials. By 1966 the now vast firm was deluged with orders and at the same time

going broke, and it was forced to find a financially strong partner, the result being McDonnell Douglas.

Douhet, Giulio (1869–1930)
Italian exponent of air power whose outspoken criticism of Italy's conduct of World War I earned him a court-martial. Subsequently restored to command, his book *Command of the Air* is a classic argument for strategic bombing.

Dunne, J. W. (1875–1949)
British pioneer designer of tailless aircraft. His then-advanced designs were aimed at producing an auto-

matically-stable aircraft. His first really successful aircraft – the Dunne D.5 – flew in 1910.

Durouf, Jules
French balloonist, pilot of the first balloon to ascend from besieged Paris, on 23 September 1870, to initiate the world's first 'airline' and 'airmail' services.

Earhart, Amelia (1898–1937)
American pioneer pilot, the first woman to achieve a solo North Atlantic flight in May 1932. Subsequently she made the first solo flight from Hawaii to California in 1935. She disappeared over the Pacific during a round-the-world flight attempt in 1937.

Eckener, Dr Hugo (1868–1954)
Early associate of Count von Zeppelin. After the death of the latter he headed the Zeppelin Company and was responsible for development and construction of the LZ.126 later *Los Angeles, Graf Zeppelin* and *Hindenburg*.

Ellehammer, Jacob C. H. (1871–1946)
Danish inventor who designed and built an early aeroplane with which he achieved 'hopping' flight in 1906.

Ely, Eugene
American pilot with the Curtiss Aeroplane & Motor Company who achieved the first aeroplane take-off from a ship, the *USS Birmingham*, on 14 November 1910. On 18 January 1911 he piloted the first aeroplane to land on a ship, the *USS Pennsylvania*, anchored in San Francisco Bay.

Esnault-Pelterie, Robert (1881–1957)
French aircraft designer and founder of the REP Aircraft Company, his developed REP 2 *bis* monoplane

flying successfully on 15 February 1909. He is better remembered as a pioneer researcher into the possibility of space flight.

Fabre, Henri (b. 1883)
French pioneer of the seaplane, whose Hydravion became the first aircraft to take off from water on 28 March 1910. Fabre was to become well-known as a designer and constructor of seaplane floats.

Fairey, Sir Richard (1887–1956)
An electrical engineer, Fairey achieved such success with model aeroplanes – made in large numbers for retail sale – that he threw up his job, joined the Blaire Atholl Syndicate, became chief engineer for the Short brothers, and in 1915 formed Fairey Aviation Ltd. Few companies have produced such a diversity of aircraft, even though Fairey himself ran the British Air Commission in the USA in 1942–45. Pressures were such that he was a sick man from 1945.

Farman, Henry (1874–1958)
British-born pilot and aircraft builder who became a naturalized French citizen in 1937, and was the first Briton to establish an internationally ratified world distance record. A distance of 771m (2,530ft), it was achieved in a Voisin aircraft on 26 October 1907.

Flynn, Reverend J. (1880–1951)
Founder of the Australian Flying Doctor Service which was inaugurated on 15 May 1928.

Fokker, Anthony H. G. (1890–1939)
Aviation pioneer, pilot and founder of the company which today bears his name in the Netherlands.

Gagarin, Yuri (1934–1968)
Russian cosmonaut, first man to orbit the Earth in spaceflight, in the *Vostok 1* spacecraft, on 12 April 1961.

Garnerin, André Jacques (1769–1823)
A Frenchman, recorded the first successful parachute descent from a vehicle in flight on 22 October 1797. He descended from a balloon which was at a height of about 915m (3,002ft).

Garros, Roland (1888–1918)
Made the first air crossing of the Mediterranean on 13 September 1913 flying a Morane-Saulnier monoplane. This French pilot has another niche in aviation history, recorded in World War I, when he used a crude method of firing a machine-gun through the propeller disc to make head-on attack possible.

Giffard, Henri (1825–1882)
French pioneer airship builder, whose steam-engine powered vessel, in which he flew from Paris to Trappes on 24 September 1852, is regarded as the world's first successful airship.

Glenn, John (1921–   )
American astronaut who, on 20 February 1962, became the first American to orbit the Earth, in the Mercury capsule *Friendship 7*.

Goddard, Robert H. (1882–1945)
American pioneer of rocket propulsion for space flight. His research and development of high energy liquids led to the launch of a liquid-propel-

lant rocket of his own design, the world's first, on 16 March 1926.

Grahame-White, Claude (1879–1959)
Pioneer British pilot and competitor in many of the early aviation races; developer of the aerodrome at Hendon, north London, which became the venue of many important aviation meetings, and is now the site of the Royal Air Force Museum.

Green, Charles (1785–1870)
British balloonist and showman who, by 1854, had made well over 500 ascents. His *Royal Vauxhall* balloon, first flown in August 1836, could carry 12 persons when the envelope was filled with coal gas, and as many as 28 when pure hydrogen was used for lift.

Grumman, Leroy R. (1895–1982)
A highly qualified engineer, Grumman worked for Grover Loening and became chief designer. In 1929 Loening sold out, but backed Grumman in forming his own company which went into operation in January 1930. The first job was to repair a Loening amphibian, but soon Grumman invented a way of completely retracting landing gear into a seaplane float strut, or into the flanks of a fuselage. This was the start of a swiftly growing company. Grumman retired as Chairman in 1966.

Guidoni, A.
One of the significant names in Italian aviation history, he pioneered the naval torpedo-carrying aeroplane. His first successful launch of a 340kg (750lb) torpedo was made in February 1914.

Gusmão, Bartolomeu de (1686–1727)
Luso-Brazilian priest who, on 8 August 1709, demonstrated before King John V of Portugal and his court a practical model of a hot-air balloon which rose to a height of about 3·6m (11ft).

Hamel, Gustav (1889–1914)
Pioneer aviator, and pilot of the Blériot monoplane which, on 9 September 1911, carried the first official air mail in Britain, between Hendon and Windsor.

Handley Page, Sir Frederick (1885–1962)
British aviation pioneer, founder of Handley Page Ltd in 1909 and builder of Britain's first strategic bomber aircraft in World War I. His name will be perpetuated in aviation

history for the Handley Page slot, an aerodynamic feature contributing to aircraft safety since 1920.

Hargrave, Lawrence (1850–1915)
Australian pioneer of kite design, whose development and perfection of the box-kite in 1893 was an important contribution to the construction of early European aircraft.

Havilland, Sir Geoffrey de (1882–1965)
One of the best-known British aviation pioneers, whose career extended from the very beginning of powered flight to the advent of the jet age. His name is linked in aviation history with the D.H.60 Moth that his company produced in 1925, and which fostered the private flying movement around the world.

Hawker, Harry G. (1889–1921)
Australian pioneer pilot, contender and recipient of a £1,000 consolation prize in the *Daily Mail* Hydro-Aeroplane Trial of August 1913. He was a founder of the H. G. Hawker Engineering Company at Kingston-upon-Thames, Surrey, and his name is perpetuated in various Hawker companies.

Heinkel, Professor Ernst (1888–1958)
German pioneer aircraft designer. Amongst a whole series of superb aircraft, he is likely to be best remembered for the He 178, the world's first turbojet-powered aircraft, which flew for the first time on 27 August 1939.

Henson, William Samuel (1805–1888)
Early British disciple of Sir George Cayley, he designed and patented a heavier-than-air craft, the *Aerial Steam Carriage*, in 1843. Although never built in full-size form, it incorporated many advanced ideas.

Hill, G. T. R. (1895–   )
British designer of tailless aircraft who, like J. W. Dunne, sought to evolve an extremely stable aircraft. Most advanced was the Pterodactyl Mk V, built by the Westland Aircraft Company, which flew in 1932.

Hinkler, H. J. L. ('Bert') (1892–1933)
Australian pilot who made the first solo flight from England to Australia, between 7 and 22 February 1928. Later, between 27 October and 7 December 1931, he made the first solo flight in a light aircraft, a de Havilland Puss Moth, from New York to London.

Hughes, Howard (1905–1976)
American billionaire and movie magnate, in 1935 built an advanced racing aircraft with which, on 13 September 1935, he established a landplane speed record of 567·026km/h (352·388mph). He was later to build the *Hercules*, a 180-ton giant flying-boat which he flew on 2 November 1947.

Ilyushin, Sergei Vladimirovich (1894–1977)
Ilyushin was a mechanic and then pilot in the Czarist bomber forces in 1917. Despite this he rapidly rose in Soviet organizations, becoming chief of NTK-UVVS (the air force technical committee) in 1926, and a lead designer at the central design bureau. In 1938, when head of GUAP (administration of aviation industry) he commuted in a single-seat Yak; a mechanic forgot the oil and the forced landing scarred 'Il' for life. On his 80th birthday he received his third Hero medal and Order of Lenin.

Immelmann, Max (1890–1916)
Renowned German pilot of World War I, whose name is always linked with those of Oswald Boelcke and Manfred von Richthofen. This legendary trio developed German fighter tactics. Immelmann died after achieving 15 victories in air combat.

Jatho, Karl (1873–1933)
German pioneer designer and constructor, whose powered aeroplane made short 'hops' in 1903.

Johnson, Amy (1903–1941)
British airwoman who became famous by achieving the first female solo flight from England to Australia between 5 to 24 May 1930 flying a de Havilland Gipsy Moth. She vanished flying in 1941.

Junkers, Professor Hugo (1859–1935)
German aircraft designer and pioneer of all-metal aircraft construction, whose Junkers J.I, the world's first all-metal cantilever monoplane, flew for the first time on 12 December 1915.

Kingsford Smith, Sir Charles
Pacific crossing, between 31 May and 9 June 1928; the first crossing of the Tasman Sea, 10/11 September 1928; and the first flight from Australia to the United States between 22 October and 4 November 1934.

Knight, W. J.
Pilot of the North American X-15A-2 rocket-powered research aircraft which, on 3 October 1967, he flew at a speed of 7,297km/h (4,534mph), making the X-15 the fastest aeroplane ever flown.

Krebs, A. C.
Associate of Charles Renard, whose electric-motor-powered airship, *La France*, flew for the first time on 9 August 1884.

Langley, Samuel Pierpont
(1834-1906)
American railway surveyor and civil engineer, who demonstrated successful steam-powered model aeroplanes in 1896. A full-size man-carrying aircraft was developed from these in 1903, but two attempts to launch the *Aerodrome*, as it was called, proved unsuccessful.

Latécoère, Pierre (1883-1943)
French aircraft manufacturer and founder of the Lignes Aériennes Latécoère, which he established to forge an air route between France and South America. On 12 May 1930 a Latécoère 28 floatplane *Comte de la Vaulx* first completed the trans-Atlantic route which had been Pierre Latécoère's dream.

Latham, Hubert (1883-1912)
French pioneer pilot, renowned for his two unsuccessful attempts to beat Louis Blériot in achieving the first crossing of the English Channel in a heavier-than-air craft.

Levavasseur, Léon (1863-1922)
French pioneer aviation engineer and designer of the graceful Antoinette monoplanes and engines which contributed importantly to the early development of aviation.

Lilienthal, Otto (1848-1896)
German pioneer glider-builder and pilot, one of the most important names in aviation history. Before his death on 10 August 1896 – the result of a flying accident the previous day – he had made thousands of flights and had recorded meticulously the results of his theoretical and practical research work.

Lindbergh, Charles (1902-1974)
American airmail pilot who, on 21 May 1927, landed at Paris after a 33½-hour flight non-stop from New York. This historic flight was the first solo non-stop flight across the North Atlantic, and brought everlasting fame to Lindbergh.

Link, Edward Albert
American inventor of a ground-based flight trainer which could reproduce the behaviour and response

to controls of an aeroplane in flight. His first trainer, sold in 1929, has led to today's simulators.

Lippisch, Professor Alexander
(1894-   )
German aircraft designer responsible for much research into advanced aerofoils. He designed the DFS 194 which, powered by a 272kg (600lb) thrust Walter rocket, was the first successful liquid-rocket-powered aircraft, flown in 1940 by Heini Dittmar, and he contributed to the development of tailless and delta aircraft.

McCurdy, John A. D.
Canadian engineer who made the first successful controlled flight in a heavier-than-air machine (AEA *Silver Dart*) in the British Common-

wealth by a British subject. Baddeck, N.S., 23 February 1909.

McDonnell, James S. (1899-1980)
'Mr Mac' traced his Scottish ancestry back to the 13th century, but he grew up in Arkansas. He flew with the US Army, got his Master's degree from MIT, and designed for Huff-Daland, Stout and Ford, before building his first aircraft in 1929. After designing for Martin he formed McDonnell Aircraft in St Louis in July 1939. One

Manly, Charles M. (1876-1927)
Pilot for American Samuel Pierpont Langley, and designer of the remarkably advanced five-cylinder radial air-cooled petrol engine which powered Langley's *Aerodrome*.

of its early jobs, gratefully received, was to make some parts for Douglas transports and bombers. Mr Mac never thought that in 1966 his company would take over mighty Douglas.

Maxim, Hiram Stevens
(1840-1916)
American who, while resident in England, designed and built a giant flying-machine of 31·70m (104ft) wing span. Powered by two 180hp steam-engines of his own design, and launched along a special railway track in 1894, it lifted at a speed of 64km/h (40mph) but broke its restraining guard rails and came to a halt.

Mermoz, Jean (-1936)
Pilot of the French Lignes Aériennes Latécoère and later Cie Générale Aéropostale, renowned for his pioneering flights across the South Atlantic. He disappeared without a trace on 7 December 1936 during his 24th trans-Atlantic flight.

Messerschmitt, Willy
(1898-1985)
Famous German aircraft designer who as early as 1921 collaborated in the design of a successful glider. It was followed by other gliders and a series of lightplanes before his creation of the Bf 108 Taifun in 1934, Messerschmitt's first all-metal aircraft. Then came what must be regarded as his masterpiece, the Bf 109 fighter, often claimed as the most famous German aircraft of all time.

Mitchell, R. J. (1895-1937)
British aircraft designer of the Supermarine Aviation Works, Southampton, Hants, responsible for the design of the S.4, S.5, S.6, S.6A and S.6B racing seaplanes which gained the Schneider Trophy for Britain. These aircraft had an important influence on Mitchell's Spitfire design.

Mitchell, William E. ('Billy')
(1880-1936)
Controversial exponent of air power, and commander of all US Army squadrons at the Western Front in

THIS PAGE
TOP LEFT: *Sir Charles Kingsford Smith.*
ABOVE CENTRE: *Charles Lindbergh.*
ABOVE: *Hiram Maxim (centre front).*
LEFT: *Otto Lilienthal.*
OPPOSITE PAGE
TOP LEFT: *Jim Mollison.*
TOP CENTRE: *Hermann Oberth.*
TOP RIGHT: *Umberto Nobile*
CENTRE LEFT: *Billy Mitchell (right).*

World War I. Court-martialled in 1926 for his outspoken views, he resigned his commission. In 1946, ten years after his death, he was posthumously awarded the Congressional Medal of Honour.

Mollison, J. A. (1905–1959)
British long-distance pilot, first to make a solo east–west crossing of the North Atlantic, on 18/19 August 1932. He was also the first man to fly solo east–west across the South Atlantic and the first man to cross both North and South Atlantic.

Montgolfier, Etienne (1745–1799) and **Joseph** (1740–1810)
French makers of the world's first hot-air balloon capable of lifting a man, and whose balloon, launched on 21 November 1783, was the first to carry men in free flight.

Mozhaiski, Alexander F. (1825–1890)
Russian designer of an odd-looking monoplane which, powered by two steam-engines, achieved a short 'hop' of 20 to 30m (66 to 98ft) in 1884 after take-off down an inclined ramp.

Nesterov, Lieutenant
Pilot of the Imperial Russian Army who, on 27 August 1913, was the first pilot in the world to execute the manoeuvre known as a loop, flying a Nieuport Type IV monoplane.

Nobile, Umberto (1885–   )
Italian designer of the semi-rigid airship *Norge* in which he, together with Lincoln Ellsworth and Roald Amundsen, flew over the North Pole on 12 May 1926. Nobile was later to build the *Italia*, which was lost after flying over the Pole on 24 May 1928.

Northrop, John Knudsen (1895–1981)
Small and quiet-spoken, Northrop designed the first aircraft for Loughead (later spelt Lockheed), then produced the brilliant Lockheed Vega, and from 1929 designed some of the first cantilever stressed-skin aircraft for his own company and for Douglas (of which Northrop was for a time a subsidiary). In 1953 he retired to a life of consultancy. The future USAF 'stealth' bomber, built by Northrop, owes much to his pioneer work on flying-wing aircraft.

Oberth, Hermann (1894–   )
Austrian pioneer of the theory of rocket power and the problems of interplanetary navigation. His thesis *The Way to Space Travel* is regarded as a basic classic of the theory of astronautics.

Ohain, Pabst von (1911–   )
German pioneer of jet propulsion who, financed by Ernst Heinkel, demonstrated his first turbojet engine of about 250kg (551lb) thrust in September 1937.

Opel, Fritz von (1899–1971)
German tycoon of the motor industry, who financed development of several rocket-powered aircraft. Most successful was the Opek-Sander-Hatry Rak. 1, first flown by von Opel on 30 September 1929.

Parke, Wilfred
British naval pilot, was the first to demonstrate the method of recovery from a spin on 25 August 1912. Prior to that time spinning aircraft had almost invariably crashed.

Paulhan, Louis (1883–1963)
French pioneer pilot and winner of the *Daily Mail* £10,000 prize for the first flight from London to Manchester.

Pégoud, Adolphe (1889–1915)
The first pilot in the world to demonstrate an aircraft in sustained inverted flight, at Buc, France, on 21 September 1913.

Pénaud, Alphonse (1856–1880)
French inventor who built and flew a successful rubber-driven model aeroplane in 1871. He was the first to establish some basic aerodynamics, including the suggested use of control surfaces on wings and tail.

Phillips, Horatio (1845–1924)
Pioneer British designer who patented, in 1884, a wing section which contributed to the design of modern aerofoil sections. His multiplane of 1907 had some 200 narrow-chord wings of venetian-blind form.

Pilcher, Percy (1866–1899)
British pioneer glider builder and pilot, inspired and given practical help by the German Otto Lilienthal. Pilcher, like his mentor, died when one of his gliders crashed, on 2 October 1899, the first Englishman to lose his life in an heavier-than-air craft.

Piper, William T. (1881–1970)
In 1929 Piper, a successful construction engineer in the oil industry, was thinking about retiring. He invested in Taylor Brothers, a local Pennsylvania builder of lightplanes. The Depression killed the company, Piper bought it for $761 and got it going again, but did not think of changing the name until 1937 when it became Piper Aircraft. The plant was destroyed by fire in the same year, but Piper – assisted by three sons who became vice-presidents – soon made it a giant.

Platz, Reinhold
German aircraft designer for Fokker, responsible for the basic design of the DrI triplane and D VII biplane fighter aircraft built during World War I, and post-war airliners.

Porte, John C.
A member of the British Royal Naval Air Service, who developed from the Curtiss *America* flying boat – which he had assisted to design – a superb series of RNAS flying boats during World War I, known as Felixstowe 'boats.

Post, Wiley (1899–1935)
Renowned American long-distance pilot who, flying the Lockheed Vega monoplane, *Winnie Mae*, made the

first solo round-the-world flight between 15 and 22 July 1933.

Prévost, Maurice
French pioneer pilot and winner of the first Schneider Trophy Contest, held at Monaco on 16 April 1913. His low average speed of 73·63km/h (45·75mph) was due to his having to complete an extra lap to comply with the rules of the contest.

Reitsch, Hanna (1912–1984)
Famous German airwoman, and the first woman in the world to pilot a helicopter. She demonstrated the Focke Achgelis Fa 61 twin-rotor helicopter in the Deutschlandhalle in Berlin in 1938.

Renard, Charles (1847–1905)
French airship pioneer who, with the assistance of A. C. Krebs, designed and built the first practical airship. Powered by an electric motor, it reached a speed of 19km/h (12mph) when first flown on 9 August 1884.

Roe, Sir Alliott Verdon (1877–1958)
British pioneer aircraft designer, constructor and pilot, and founder of the famous Avro company. On 13 July 1909 he flew his Roe I triplane to become the first Briton to fly an all-British aeroplane.

Rolls, The Hon. Charles S. (1877–1910)
Associate of Henry Royce in the founding of Rolls-Royce Limited, accomplished the first double crossing of the English Channel in a French-built Wright biplane on 2 June 1910. He was killed on 12 July 1910 when his aircraft crashed at Bournemouth.

Royce, Sir Henry (1863–1933)
Co-founder with C. S. Rolls of Rolls-Royce Limited, he masterminded the development of all the company's aero engines up to and including Rolls-Royce Merlin.

Rozier, Francois Pilâtre de (1754–1785)
First man in the world to be carried aloft in a tethered Montgolfier balloon on 15 October 1783; first to make a free flight, in company with Marquis d'Arlandes, on 21 November 1783; and the first to die in a ballooning accident on 15 June 1785.

Santos-Dumont, Alberto
(1873–1932)
Brazilian aviation pioneer resident in Paris, was first to fly round the Eiffel Tower in his No. 6 dirigible, and the first to achieve a manned, powered and sustained aeroplane flight in Europe in his 14 *bis* tail-first biplane on 12 November 1906.

Schneider, Jacques (1879–1928)
Son of a French armaments manufacturer, in 1912 he offered a trophy for international competition to speed the development of waterborne aircraft. The resulting Schneider Trophy Contests had significant influence on the design of high-speed aircraft.

Scott, Charles W. A.
(1903–   )
Pilot of the de Havilland D.H.88 Comet who, together with Tom Campbell Black, won the 1934 Mac-Robertson England–Australia air race. Charles Scott had, in 1931, made two record-breaking solo flights, England to Australia and Australia to England, in a de Havilland Gipsy Moth.

Scott, Sheila
British record-breaking woman pilot of the post World War II years. Between 18 May and 20 June 1966 she completed a solo round-the-world flight in a record time – the longest solo flight by any pilot of any nationality, and the first round-the-world flight by a British pilot.

Seguin, Laurent and Louis
French designers of the Gnome rotary engine which, from 1908, was the most important of early aircraft engines. The performance of the Gnome rotaries was surpassed by

new designs evolved during World War I.

Shepard, Alan B. (1923–   )
First American to travel in space, was launched in a sub-orbital trajectory in the Mercury capsule *Freedom 7* on 5 May 1961. In his 486km (302 mile) flight he attained a height of 185km (115 miles).

Short, Eustace (1875–1932), **Horace** (1872–1917) **and Oswald** (1883–1970)
British pioneers who founded the nation's first aircraft factory. By early 1909 they were building their first six Short-Wright biplanes under licence. The company exists to this day, under the title of Short Brothers & Harland, and is still designing and building aeroplanes.

Sikorsky, Igor A. (1881–1972)
Russian pioneer aircraft designer and builder, creator of the world's first four-engined aircraft, the *Bolshoi*, which flew on 13 May 1913. Later becoming a naturalized American citizen, he concentrated on the development of rotary-winged aircraft, his VS–300, which was first flown on 14 September 1939, being

developed into what is now regarded as the world's first practical single-rotor helicopter.

Smith, Sir Keith (1890–1955) **and Sir Ross** (1892–1922)
Australian brothers who, between 12 November and 10 December 1919, achieved the first flight from England to Australia in a Vickers Vimy bomber. The brothers won a prize of £10,000 from the Australian Government, and both were knighted.

Smith-Barry, R. R.
Founder of the School of Special Flying at Gosport, Hampshire, in July

1917, marking the beginning of the superb flying training which became a feature of the Central Flying School of the Royal Air Force.

Sohn, Clem (d. 1937)
American 'bird-man' who devised webbed 'wing' and 'tail' which enabled him to manoeuvre in semi-gliding 'flight' after jumping from an aircraft, landing finally by parachute. He was killed on 25 April 1937 when his parachute failed to open fully.

Sopwith, Sir Thomas
(1888–   )
British pioneer pilot and founder of the Sopwith Aviation Company at Kingston-upon-Thames, Surrey; later a director and subsequently Chairman of the Hawker Siddeley Group.

Sperry, Lawrence B. (1892–1923)
American designer of gyro-stabilizing mechanisms to control an aircraft in flight. His earliest device was fitted to a Curtiss flying boat in 1912.

Stringfellow, John (1799–1883)
British associate of W. S. Henson in the construction of the 6.10m (20ft)

span, Aerial Steam Carriage. Stringfellow was responsible in particular for its steam-engine power plant.

Sukhoi, Pavel Osipovich
(1895–1975)
Contemporary of Andrei Tupolev at the TsAGI (Central Aero and Hydrodynamic Institute), he has been concerned primarily with the design of fighter aircraft. The greatest success of his design bureau has come in the years since 1953, with the development of a family of jet fighters.

Tank, Kurt
Contemporary with Germany's

Willy Messerschmitt, Kurt Tank achieved similar fame as an aircraft designer with the Focke-Wulf company. His first significant design was the Fw 56 Stösser trainer of 1933, but he is remembered especially for the Fw 190, regarded as the most advanced fighter aircraft in the world when it entered service in 1941.

Temple, Félix du (1823–1890)
French naval officer who designed and built a powered monoplane aircraft which, piloted by an unknown young sailor in 1894, gained the distinction of being the first manned aircraft to achieve a brief 'hop' after launch down a ramp.

Tereshkova, Valentina (b. 1937)
Russian cosmonaut, the first woman to travel in space, in *Vostok 6*, 16 to 18 June 1963, completing 48 orbits of the Earth before landing.

Thomas, George Holt
(1869–1929)
British founder of Aircraft Transport and Travel Ltd., first British airline to be registered (on 5 October 1916). On 25 August 1919 this company inaugurated the world's first daily scheduled international commercial airline service.

Trenchard, Hugh (1873–1956)
British soldier turned aviator, regarded as the 'Father of the Royal Air Force', who fought strenuously to keep it an independent service. He was to become Lord Trenchard, Marshal of the Royal Air Force.

Trippe, Juan T. (1899–   )
Airline pioneer and founder of Pan American World Airways System in 1927. Before then Trippe had organized Long Island Airways with seven World War I aeroplanes. For many years Pan American dominated American civil aviation.

Tsiolkovsky, Konstantin
(1857–1935)
Russia's 'father of astronautics' who, as early as 1898, suggested the use of liquid propellants to power space rockets. He first discussed the principle of reaction, proved theoretically that a rocket would work in a vacuum, and made the first serious calculations relating to interplanetary travel.

Tupolev, Andrei Nikolaevich
(1888–1972)
Most famous internationally of all Soviet aircraft designers, he helped to found the TsAGI (Central Aero and Hydrodynamic Institute) in 1918, becoming the head of its aircraft design department in 1920. His design and development of the Tu–104 jet-powered airliner earned him the Lenin Prize and the Gold Medal of the FAI, and in 1970 he was made an Honorary Fellow of the Royal Aeronautical Society. Apart from the creation of some important bomber and civil transport aircraft, the Tupolev design bureau's Tu–144 was the world's first supersonic airliner to fly.

Turnbull, W. R.
Canadian engineer who tested aerofoils in a wind tunnel he built in 1902. He may have been the first to recommend dihedral to increase stability. He developed the first controllable pitch propeller between 1916 and 1927.

Twiss, L. P. (1921–   )
Pilot of the Fairey Delta 2 research aircraft which, on 10 March 1956, set the first world speed record established at over 1,609km/h (1,000mph). His average speed for the two runs over the selected course was 1,822km/h (1,132mph).

which, powered by a 22hp Anzani engine, made its first flight on 9 December 1909.

Wallis, Sir Barnes N. (b. 1887)
British aviation engineer and designer, designer of the rigid airship R.100, originator of geodetic construction, inventor of the skip bomb which breached the German Moehne, Eder and Sorpe dams, and of the earthquake bombs which helped to bring Allied victory in Europe. Post World War II he developed the variable-geometry wing.

Watt, Sir Robert Watson
(1892–1973)
Leader of the British research team which, in the three years prior to the outbreak of World War II, designed and created a radar system for the detection and location of enemy aircraft.

Wright, Orville (1871–1948) and Wilbur (1867–1912)
The proud names of the two American brothers renowned in aviation history for designing, building and flying the first man-carrying aeroplane in the world to achieve powered, controlled and sustained flight, on 17 December 1903.

Yakovlev, Alexander (b. 1906)
Helped build gliders 1923, labourer at VVA (Soviet AF academy) 1924, cadet 1927, where built first aircraft (AIR–1). Many successful trainers and sporting types, but fame came with I–26 (Yak–1) which led to over 37,000 World War II fighters. Subsequently most senior of surviving Soviet designers, still at work in giant design bureau named after him.

Yeager, Charles (1923–   )
United States Air Force pilot who, on 14 October 1947, piloted the Bell XS–1 rocket-powered research aircraft at a speed of 1,078km/h (670mph) (an equivalent of Mach 1·015)

at a height of 12,802m (42,000ft). His was the first supersonic flight in history.

Zeppelin, Ferdinand von
(1838–1917)
Founder of the German Zeppelin Company at Friedrichshafen on Lake Constance. His first rigid airship, the LZ.I, flew for the first time on 2 July 1900.

Vinci, Leonardo da (1452–1579)
Italian genius born in 1452 who is renowned in aviation history for drawings which show his early designs for a rotary-wing aircraft and a parachute. He believed from his observations of bird flight that man would be able to achieve flight by his own muscle power.

Voisin, Charles (1882–1912) and Gabriel (1886–   )
French pioneer aircraft builders. Flying a boxkite glider towed by a launch, Gabriel recorded the first flight from water by a manned aircraft on 6 June 1905.

Walden, Henry W.
Designer and builder of the first successful monoplane aircraft to fly in America. This was the Walden III

Wenham, F. H. (1824–1908
His classic paper *Aerial Locomotion* was read to the Aeronautical Society in 1866. One of the leading theorists of his day, he also built the world's first wind-tunnel in conjunction with John Browning.

Whittle, Sir Frank (1907–   )
British designer of the aircraft gas turbine engine, who ran the first aircraft turbojet engine in the world on 12 April 1937. It later powered the Gloster Meteor jet fighter.

Willows, E. T. (–1926)
British designer of some small, but successful, semi-rigid airships, the first of which flew in 1905. He patented the design of swivelling propellers to simplify control of an airship in flight.

Chronology

1709-1914

1709
8 August
Bartolomeu de Gusmão demonstrated the world's first model hot-air balloon.

1783
25 April
First flight of a Montgolfier hot-air balloon.
4 June
First public demonstration by the Montgolfiers of a small hot-air balloon.
19 September
A Montgolfier balloon carried a sheep, a duck and a cockerel into the air, the first animals to fly.
15 October
First man carried in tethered flight in a Montgolfier hot-air balloon.
21 November
First men carried in free flight in a Montgolfier hot-air balloon.

1 December
First men carried in free flight in a hydrogen balloon.

1784
4 June
First woman carried in free flight in a Montgolfier hot-air balloon.

1785
7 January
First crossing of the English Channel by balloon.

1793
9 January
First free flight of a manned hydrogen balloon in America.

1794
26 June
Observation balloon used at battle of Fleurus, the first operational use of an aircraft in war.

1797
22 October
First successful parachute descent from a balloon made by Frenchman A. J. Garnerin.

1849
22 August
First air raid by balloons: the Austrians launched pilotless balloons against Venice, armed with time-fuze-controlled bombs.

1852
24 September
First flight of the world's first powered and manned airship.

1861
1 October
American Army's Balloon Corps was formed.

1866
12 January
Foundation of the Aeronautical Society of Great Britain (later the Royal Aeronautical Society).

1870
23 September
First balloon ascent from besieged Paris at the height of the Franco-Prussian War.

1884
9 August
The dirigible *La France* made the world's first flight of a powered, manned and controlled airship.

1890
9 October
Clément Ader's bat-winged *Éole* monoplane made a brief hop.

1896
10 August
Death of Otto Lilienthal as the result of a gliding accident.

1898
In this year the *Aéro Club de France* was founded.

1899
2 October
Death of Percy Pilcher as the result of a gliding accident.

1900
2 July
Count von Zeppelin's first rigid airship made its first flight.

1901
19 October
Alberto Santos-Dumont flew his No. 6 dirigible round the Eiffel Tower, Paris.
29 October
Foundation of the British (later Royal) Aero Club.

1903
17 December
Wright brothers achieve world's first powered, controlled and sustained flights by a man-carrying aeroplane.

1905
12 October
Foundation of the *Fédération Aéronautique Internationale* (FAI).

THIS PAGE
TOP LEFT: *The siege of Fleurus, 1794.*
TOP RIGHT: *Santos-Dumont's No 6 airship.*
FAR LEFT: *The first (unmanned) Montgolfière.*
ABOVE LEFT: *Garnerin and parachute.*
FACING PAGE
TOP LEFT: Nulli Secundus II.
TOP RIGHT: *Chavez over the Alps, 1910.*
CENTRE: *S. F. Cody in his No 3, 1909.*
BELOW: *Cody and the 1912 biplane.*

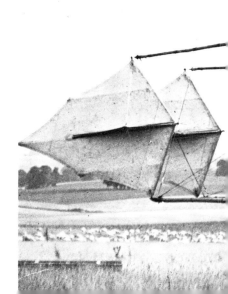

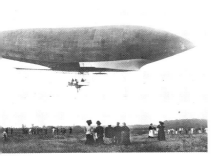

1906
12 November
Alberto Santos-Dumont makes first powered and sustained flight by a man-carrying aeroplane in Europe.

1907
10 September
First flight of *Nulli Secundus*, first British army airship.
13 November
Paul Cornu's helicopter made world's first flight by a manned rotary-wing aircraft.

1908
16 October
First officially recognized aeroplane flight in Britain made by S. F. Cody.

1909
14 May
First flight of over 1·6km (1 mile) in Britain recorded by S. F. Cody.
25 July
First aeroplane flight across the English Channel made by Louis Blériot in his Type XI monoplane.
25 July
First official aeroplane flight in Russia, made by Van den Schkrouff flying a Voisin biplane at Odessa.
29 July
First official aeroplane flight in Sweden, made by French pilot Legagneux flying a Voisin biplane.

2 August
The US Army bought its first aeroplane from the Wright brothers.
22–29 August
World's first international aviation meeting held at Reims, France.
30 October
First official aeroplane flight in Romania made by Louis Blériot flying a Blériot monoplane.
9 December
First official aeroplane flight in Australia. Colin Defries flew an imported Wright biplane over the Victoria Park racecourse at Sydney.

1910
8 March
Frenchwoman Mme la Baronne de Laroche became the first certificated woman pilot in the world.
10 March
First night flights in an aeroplane made by Frenchman Emil Aubrun.
28 March
First successful powered seaplane, built and flown by Henri Fabre, takes off from water.
2 June
First 'there and back' double crossing of the English Channel by an aeroplane, flown by the Hon C. S. Rolls.
27 August
First transmission of radio messages between an aeroplane and a ground station made by James McCurdy in America.
11 September
First crossing of the Irish Sea, from Wales to Ireland (Eire) made by Robert Loraine in a Farman biplane.
23 September
Georges Chavez made the first aeroplane flight over the Alps in a Blériot monoplane.
14 November
A Curtiss biplane flown by American pilot Eugene Ely was the first aircraft to take off from a ship.

1911
18 January
Eugene Ely, flying a Curtiss biplane, recorded the first landing of an aeroplane on a ship.
18 February
World's first official air mail flight at Allahabad, India, flown by Frenchman Henri Pequet.
12 April
Pierre Prier, flying a Blériot monoplane, made the first non-stop flight from London to Paris.
18 June
The Circuit of Europe, the first real international air race, started from Paris.
1 July
First flight of the US Navy's first aeroplane, a Curtiss A–1 Hydroaeroplane.
22 July
First 'Round Britain' air race started from Brooklands, Surrey.

1911
9 September
First official air mail flight in Great Britain, made by Gustav Hamel in a Blériot monoplane, which carried air mail between Hendon and Windsor.
17 September–5 November
First coast-to-coast flight across the United States made by Calbraith P. Rodgers in a Wright biplane.
23 September
First official air mail flight in the United States, made by Earl L. Ovington flying a Blériot monoplane.

22 October
First use of an aeroplane in war, Italian Capitano Piazza making a reconnaissance of Turkish positions in his Blériot monoplane.

1912
13 April
The British Royal Flying Corps was constituted by Royal Warrant.
12 November
A Curtiss A–1 Hydroaeroplane, flown by Lt Ellyson, was launched by a compressed-air catapult.

1913
16 April
First Schneider Trophy Contest was won at Monaco by Maurice Prévost.
13 May
First flight of a four-engine aeroplane, the *Bolshoi* biplane designed by Igor Sikorsky, at St Petersburg, Russia.
23 September
First flight across the Mediterranean Sea made by Roland Garros in a Morane-Saulnier monoplane.

1914
1 January
World's first scheduled transport service by an aeroplane, the Benoist Company's St Petersburg to Tampa service in Florida, USA.
4 August
Beginning of World War I.
30 August
Bombs were dropped on Paris, the first to be dropped from an aeroplane on a city.

1915-1933

1915
19 January
First air raid on Britain by Zeppelin airships.
3 March
NACA, National Advisory Committee for Aeronautics, predecessor of NASA, established in the USA.
31 May
Zeppelin LZ.38, an airship of the German Army, made the first Zeppelin attack on London.
12 December
The Junkers J.1, the world's first all-metal cantilever monoplane, flew for the first time.

1916
15 March
First use of US Army aircraft in military operations, in support of a punitive expedition into Mexico.
31 May
A British naval seaplane spotted for the British fleet at the Battle of Jutland, the first use of an aeroplane in a major fleet battle.
12 September
The Hewitt-Sperry biplane, in effect the world's first radio-guided flying-bomb, was tested in the USA.
5 October
First British airline company, Aircraft Transport and Travel Ltd, registered by George Holt Thomas.

1917
20 May
First German submarine (U–36) sunk as the direct result of action by an aircraft.
2 August
A Sopwith Pup, flown by Sqdn Cdr E. H. Dunning, made the first landing of an aeroplane on a ship under way.

1918
11 March
The first regular and scheduled international air mail service in the world began, between Vienna and Kiev.
1 April
British Royal Air Force formed.
13 April
First flight across the Andes, from Argentina to Chile, made by Lt Luis C. Candelaria.
24 June
First official air mail flight in Canada, from Montreal to Toronto, by a Curtiss JN–4, flown by Capt B. A. Peck, RAF.
12 August
First regular air mail service inaugurated in the USA, with a New York to Washington route.

1919
3 March
First American international air mail service established between Seattle and Victoria, British Columbia.
8–31 May
First crossing of the North Atlantic by air achieved by Lt Cdr A. C. Read and US Navy crew, flying a Navy/Curtiss NC–4 flying-boat.
14–15 June
First non-stop crossing of the North Atlantic by air, achieved by Capt John Alcock and Lt Arthur Whitten Brown in a Vickers Vimy bomber.
2–6 July
First crossing of the North Atlantic by an airship, the British R.34 made a return flight between 9–13 July.
7 August
First flight across the Canadian Rocky Mountains, from Vancouver to Calgary, made by Capt E. C. Hoy.
25 August
World's first scheduled daily international commercial airline service began, Aircraft Transport and Travel's London to Paris service.
1 November
America's first scheduled international commercial air service began, between Key West, Florida, and Havana, Cuba, by Aeromarine West Indies Airways.
12 November–10 December
First flight from England to Australia completed by Australian brothers Keith and Ross Smith.

1920
4 February–20 March
First flight from England to South Africa made by Lt-Col P. van Ryneveld and Sqdn Ldr C. Q. Brand, using three aircraft.
7–17 October
First trans-Canada flight, from Halifax, N.S., to Vancouver, B.C., by aircraft and airmen of the Canadian Air Force.
16 November
QANTAS (Queensland and Northern Territory Aerial Services), Australia's famous international airline, was first registered.

1921
22 February
Inaugural flight of the USA's first coast-to-coast air mail service, between San Francisco and New York.
5 December
Australia's first regular scheduled air services inaugurated by West Australian Airways.

1922
4 September
First coast-to-coast crossing of the US in one day achieved by Lt J. H. Doolittle flying a de Havilland DH-4B.
2 November
First scheduled service operated by Australia's QANTAS, between Charleville and Cloncurry.

1923
9 January
Spaniard Juan de la Cierva's first successful rotating-wing aircraft, the C-4 *Autogiro*, made its first flight.
2–3 May
First non-stop air crossing of the US accomplished by Lts O. G. Kelly and J. A. MacReady flying a Fokker T–2 monoplane.

US Army Air Service, made the first round-the-world flight.
1 July
Inauguration of the USA's first regular transcontinental air mail service.

1926
16 March
American R. H. Goddard launched the world's first liquid-propellant rocket.
9 May
First flight of an aeroplane over the North Pole, the Fokker F.VIIA/3m *Josephine Ford* piloted by Floyd Bennett.

27 June
First successful in-flight refuelling of an aeroplane, made during the setting of a world endurance record by the US Army Air Service.
4 September
The *Shenandoah*, the USA's first helium-filled rigid airship, made its first flight.

1924
1 April
Establishment of Imperial Airways, Britain's first national airline.
6 April–28 September
Two out of four Douglas DWC floatplanes, flown by pilots of the

28-29 November
First aeroplane flight over the South Pole, made in a Ford Trimotor, the *Floyd Bennett*, piloted by Bernt Balchen.

1930
5-14 May
The first solo flight from England to Australia was made by a woman pilot, England's renowned Amy Johnson flying a de Havilland Gipsy Moth.

15 May
An aircraft of Boeing Air Transport carried the first airline stewardess, Ellen Church, on a commercial flight.

1931
13 March
First successful modern firing of a solid-fuel research rocket in Europe, by Karl Poggensee near Berlin.

13 September
Schneider Trophy won outright for Britain, by a Supermarine S.6B seaplane flown by Flt-Lt J. N. Boothman.

3-5 October
First non-stop flight from Japan to the USA made by Clyde Pangborn and Hugh Herndon in a Bellanca aircraft.

1932
20-21 May

1927
20-21 May
First non-stop solo crossing of the North Atlantic, from New York to Paris, achieved by Charles Lindbergh in the Ryan monoplane *Spirit of St Louis*.

28-29 June
First non-stop flight between the USA and Hawaii, made by Lts A. F. Hegenberger and L. J. Maitland in the Fokker C–2 monoplane *Bird of Paradise*.

14-15 October
First non-stop aeroplane crossing of the South Atlantic, from Saint-Louis, Senegal, to Port Natal, Brazil, made by Capt D. Costes and Lt-Cdr J. Le Brix in the Breguet XIX *Nungesser-Coli*.

1928
7-22 February
Sqdn Ldr H. J. L. Hinkler made the first solo flight from England to Australia.

15 May
Inauguration date of the Australian Flying Doctor Service.

31 May-9 June
First flight across the Pacific Ocean, from San Francisco to Brisbane, made by Capt C. Kingsford Smith and C. T. P. Ulm, in the Fokker F.VIIB/3m *Southern Cross*.

11 June
First flight of the world's first rocket-powered aircraft in Germany.

8 October
The first cinema film was projected in an aircraft in flight.

1929
30 March
First commercial air route between London and Karachi, India, inaugurated by Imperial Airways.

24-26 April
First non-stop flight from England to India, made by Sqdn Ldr A. G. Jones Williams and Flt-Lt N. H. Jenkins in a Fairey Long-range Monoplane.

8-29 August
First airship flight around the world, made by the German *Graf Zeppelin* commanded by the famous Dr Hugo Eckener.

Amelia Earhart achieved the first solo crossing of the North Atlantic by a woman.

1933
6-8 February
First non-stop flight from England to South Africa, made by Sqdn Ldr O. R. Gayford and Flt-Lt G. E. Nicholetts flying a Fairey Long-range Monoplane.

3 April
First aeroplane flights over Mount Everest, a Westland P.V.3 flown by the Marquess of Clydesdale and a Westland Wallace flown by Flt-Lt D. F. McIntyre.

15-22 July
American pilot Wiley Post, flying the Lockheed Vega *Winnie Mae*, made the first solo round-the-world flight.

31 December
First flight of the Soviet Polikarpov I–16, first monoplane fighter with enclosed cockpit and retractable landing gear to enter squadron service.

FACING PAGE
TOP : *Airship R.34 at its moorings, 1919.*
CENTRE : *First in -flight refuelling.*
BELOW LEFT : *First American air mail service.*
BELOW RIGHT : *The first round-the-world flight.*
THIS PAGE
TOP : *De Havilland US mail carrier.*
ABOVE CENTRE : *Wiley Post and his Lockheed Vega,* Winnie Mae, *1933.*
ABOVE : *The* Graf Zeppelin *at Lakehurst, New Jersey, in 1929.*
LEFT : *The Australian Flying Doctor Service in action.*
BELOW : *The Russian Polikarpov I–16.*

1934-1987

1934
8–9 August
First non-stop flight from Canada to Britain by J. R. Ayling and L. G. Reid, from Wasaga Beach, Ontario, to Heston.
20 October
Start of the MacRobertson Air Race from England to Australia.
22 October–4 November
First flight from Australia to the United States.
8 December
Inauguration of the first regular weekly air mail service between England and Australia.

1935
11–12 January
First solo flight by a woman from Honolulu, Hawaii, to the USA.
13 April
Inauguration of the first through passenger service by air from England to Australia.
6 November
First flight of the Hawker Hurricane, Britain's first monoplane fighter with enclosed cockpit, retractable landing gear and eight machine-guns.
22 November
First scheduled air mail flight across the Pacific Ocean, San Francisco to Manila, Philippines.

1936
March
First Spitfire flight.
26 June
The Focke Achgelis–Fa 61, the first entirely successful helicopter in the world, made its first flight.

1937
12 April
Frank Whittle in Britain successfully ran the world's first aircraft turbojet engine.
6 May
The German airship *Hindenburg* destroyed by fire in the USA.

1938
26 June
The first through flying-boat service from England to Australia was operated by Short C Class 'boats of Imperial Airways.
20–21 July
First commercial aeroplane flight across the North Atlantic, flown by the *Mercury* upper component of the Short-Mayo composite.

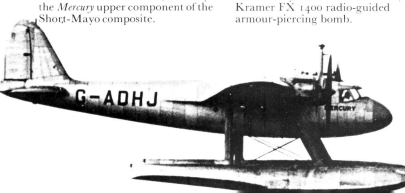

1939
28 June
Inauguration of Pan American's weekly New York to Southampton service, flown with Boeing Model 314 flying boats.
5 August
Imperial Airways began an experimental Southampton to Montreal and New York mail service with Short C Class flying boats, which were refuelled in flight.
27 August
The German Heinkel He 178, the world's first aircraft to fly solely on the power of a turbojet engine, made its first flight.
2 September
Beginning of World War II.

1941
5 April
First flight of the Heinkel He 280, the world's first aircraft designed specifically as a turbojet-powered fighter.

15 May
The first British aircraft powered by a turbojet, the Gloster Whittle E.28/39, flew for the first time at Cranwell, Lincolnshire.
20 May
The Luftwaffe's largest airborne assault of the war was launched against the island of Crete.
2 October
The German Messerschmitt Me 163, the world's first rocket-powered combat aircraft, was flown at a speed of 1,004km/h (624mph) during early tests.

1942
7–9 May
The Battle of the Coral Sea, the first naval battle in history which was fought entirely by carrier-based aircraft.
3 October
First successful launching of a German V–2 (A–4) ballistic rocket.

1943
9 September
The Italian battleship *Roma* was hit and sunk by a German Ruhrstahl/Kramer FX 1400 radio-guided armour-piercing bomb.

1944
12 July
Britain's first jet fighter, the Gloster Meteor Mk 1, began to enter service with the Royal Air Force.
3 October
Germany's first jet fighter, the Messerschmitt Me 262, began to enter operational service.

1945
14 March
First operational use of the British *Grand Slam* bomb, weighing 9,980kg (22,000lb).
6 August
First military use of an atomic bomb, when a nuclear device was exploded over the Japanese city of Hiroshima.
3 December
A de Havilland Vampire I, used for deck-landing trials aboard HMS *Ocean*, was the first jet fighter to operate from the deck of an aircraft carrier.

1946
24 July
First manned experimental use of an ejection seat: the 'guinea pig', Bernard Lynch, ejected at 515km/h (320mph).

1947
14 October
The Bell XS–1 rocket-powered research aircraft became the first aircraft in the world to exceed the speed of sound, piloted by Charles Yeager, USAF.

1948
26 June
Beginning of the Berlin Airlift.

1950
25 June
Beginning of the Korean War.
29 July
A Vickers Viscount made the world's first scheduled passenger service to be flown by a turboprop-powered airliner.

1951
21 February
An RAF English Electric Canberra bomber achieved the first non-stop unrefuelled crossing of the North Atlantic by a jet-powered aircraft.

1952
2 May
The world's first regular passenger service flown by a turbojet-powered airliner, the de Havilland Comet 1.

1953
18 May
American airwoman Jacqueline Cochran, flying a North American

F–86 Sabre, became the first woman in the world to fly at supersonic speed.

1954
15 July
First flight of the USA's first turbojet-powered commercial aircraft, the Boeing 707.

1956
21 May
The world's first airborne hydrogen bomb was dropped on Bikini Atoll in the Pacific Ocean.

1957
4 October
First man-made satellite, *Sputnik 1*, launched by the Soviet Union.
19 December
First transatlantic passenger service flown by a turboprop-powered airliner.

1958
4 October
First transatlantic passenger service flown by turbojet-powered airliner, the de Havilland Comet 4.

1974
14 August
First flight of a Panavia MRCA (now Tornado) multi-role combat aircraft prototype.
23 December
First flight of a Rockwell International B–1 strategic bomber prototype.

1976
21 January
Concorde SST introduced into international commercial airline service simultaneously by Air France and British Airways.
20 July
Touchdown on Mars of the first Viking Lander.

1977
25 February
Two-manned *Soyuz 24* docked with the *Salyut 5* spacelab during a flight lasting 424 hr 48 min.
23 August
Gossamer Condor man-powered aircraft wins long-offered £50,000 Kremer Prize for figure-eight flight round two pylons 0.5 mile apart.

12 June 1979
Same pilot, Bryan Allen, pedals Gossamer Albatross into headwind across Channel from Folkestone to Cap Gris Nez.

14–23 December 1986
Crewed by designer/builders Dick Rutan and Jeana Yeager, piston-engined *Voyager* flies non-stop around the world, 42,934km (26,678 miles), starting with 4,576 lit (1,006 gal) of fuel, in 9 days 3 min 44 sec.

12–14 June 1987
Gulfstream IV business jet flies around the world with four refuelling stops in total time of 45 hrs 25 min.

1961
12 April
Soviet spacecraft *Vostok 1* launched into Earth orbit carrying Russian cosmonaut Yuri Gagarin, the first man to travel in space.

1963
16 June
Russian cosmonaut Valentina Tereshkova, launched into Earth orbit in *Vostok 6*, became the first woman to travel in space.

1966
3 February
Soviet *Luna 9* spacecraft soft-landed on the Moon's surface and transmitted the first pictures of the lunar scene.
20 June
British airwoman Sheila Scott landed at Heathrow after setting a new woman's record of 33 days 3 mins for a round-the-world flight.

1967
3 October
North American X–15A–2 rocket-powered research aircraft was flown at a speed of 7,279km/h (4,534mph) by W. J. Knight.

1968
31 December
Soviet Tupolev Tu–144 prototype made its first flight, the first super-sonic commercial transport aircraft to fly.

1969
9 February
First flight of the world's first wide-body airliner, the Boeing Model 747.
2 March
The 001 prototype of the Anglo-French Concorde supersonic transport aircraft made its first flight at Toulouse, France.
1 April
The Hawker Siddeley Harrier, the world's first V/STOL close support aircraft, first entered squadron service.
21 July
Neil Armstrong, US astronaut, became the first man to step on the Moon's surface.

1970
12 September
Launch of Soviet unmanned spacecraft *Luna 16*, which landed on the Moon, collected soil samples,

FACING PAGE
TOP: *The Messerschmitt Me 262.*
ABOVE CENTRE: *The E. 28/29.*
CENTRE LEFT: *Hindenburg.*
CENTRE RIGHT: *Bell XS–1 under B–29*
BOTTOM: *Mercury part of Short-Mayo*
THIS PAGE
TOP: *The Hawker Kestrel.*
BELOW: *Vostok launcher at Le Bourget.*
BOTTOM LEFT: *North American X–15.*
BOTTOM RIGHT: *The Lunar Rover.*

took off from the Moon under remote control and was recovered success-fully on 24 September.

1971
31 July
A Boeing Lunar Roving Vehicle was driven on the Moon's surface for the first time.

Vought Corsair 02U–1 – biplane.

Glossary

Absolute ceiling
The maximum altitude above sea-level at which a heavier-than-air craft can maintain level flight.

Accumulator
A reservoir to supply a fluid under pressure.

ADF
Automatic direction finder. An automated system of radio direction finding (RDF), *(q.v.)*.

Aerobatics
Manoeuvres, other than those necessary for normal flight, carried out voluntarily by an aircraft's pilot.

Aerodrome
An area intended for the take-off and landing of aircraft, and including generally the associated buildings and equipment for the operation of aircraft.

Aerodynamics
The science dealing with air in motion and the reactions of a body moving in air.

Aerodyne
See aeroplane.

Aerofoil
A suitably shaped structure which, when propelled through the air, generates lift.

Aeronautics
Dealing with flight within the Earth's atmosphere.

Aeroplane
A heavier-than-air aircraft.

Aerostat
A lighter-than-air aircraft.

Afterburner
A thrust-augmenting feature of a gas turbine engine.

AI
Airborne interception. An airborne radar system evolved to assist in the location and interception of hostile aircraft.

Aileron
Movable control surface hinged to aft edge of each wing, usually adjacent to wingtip, to control rolling movements.

Airbrake
An aerodynamic surface which can be deployed in flight to increase drag.

Aircraft
Vehicles, both lighter- and heavier-than-air, which derive their flight capability from reactions of the air.

Airfield
See aerodrome.

Airfoil
See aerofoil.

Airframe
The name given to an aircraft's structure, excluding the power plant.

Airlift
The carriage of personnel or supplies by air.

Airplane
See aeroplane.

Airscrew
See propeller.

Airship
Popular name for a steerable lighter-than-air craft. More accurately a dirigible.

Airstrip
A natural surface adapted – often in an unimproved state – for the operation of aircraft. Located usually in a forward area, such strips are sometimes reinforced by steel netting.

Aldis lamp
Signalling device, projecting a narrow beam of light, visible only at the point to which it is aimed by a sight.

Altimeter
A height recording instrument.

Altitude
Height.

Amphibian
A heavier-than-air craft which can operate from land or water surfaces.

Analogue computer
A computing device, based on the principle of measuring as opposed to counting.

Angle of attack
The angle at which an aerofoil surface meets the oncoming airstream.

Angle of incidence
The angle at which an aerofoil surface is attached to a fuselage in relation to the longitudinal axis.

Anhedral
The angle which an aerofoil surface makes to the fuselage when the wing or tailplane tip is lower than the root attachment.

Area rule
A concept for shaping an aircraft for minimum drag in supersonic flight.

Artificial horizon
A gyro-stabilized flight instrument which shows graphically pitching and rolling movements of the aircraft, to aid the pilot to maintain level flight.

ASI
Air-speed indicator (see IAS).

Aspect ratio
The ratio of the span to the mean chord of an aerofoil surface.

Astrodome
Transparent dome on fuselage upper surface to permit celestial navigation.

Astronaut
Crew member of a spacecraft.

Astronautics
Dealing with flight in space.

ASW
Anti-submarine warfare.

Automatic pilot
A gyro-stabilized system to maintain an aircraft in steady flight.

Autopilot
See automatic pilot.

Autorotation
The continued automatic rotation of a rotary wing by means of air flowing through it due to forward or downward motion of a gyroplane.

APU
Auxiliary power unit carried within an aircraft for such tasks as main engine starting, ground air conditioning, and the provision of electric, hydraulic or pneumatic power in the air or on the ground.

Ballast
Carried in an aerostat to permit variation of lift.

Ballistic missile
One which becomes an unguided free-falling body in the unpowered stage of its flight, and subject to ballistic reactions.

Ballonnet
An air-filled compartment in the envelope of an airship, used to maintain a desired gas pressure at varying altitudes.

Balloon
An unsteerable lighter-than-air craft.

Biplane
A fixed-wing aircraft with two superimposed wings.

Blimp
Small non-rigid, occasionally semi-rigid, airship.

Mirage III – delta wing.

Convair F–102A – one of the first applications of the area rule.

Boundary layer
The thin stratum of air in contact with the skin surface of the airframe.

Buffet
Irregular oscillations imposed upon an aircraft's structure by turbulent airflow or conditions of compressibility.

Bungee
A tensioning device, usually of rubber cord. Used commonly for landing gear shock-absorption of simple, lightweight aircraft.

Cabin
An enclosed compartment for the crew and/or passengers of an aircraft.

Camber
The curvature of an aerofoil surface.

Canard
An aircraft which has its main wing or lifting surface at the aft end of the airframe.

Cantilever
Describing a beam or other structural member which is supported at one end only.

CAT
Clear air turbulence. Invisible, and sometimes violent whirlpools of air, created usually by the shearing action between jet streams or other high velocity winds, and a larger mass of slow moving air.

Ceiling
Operating height of an aircraft.

Centre of gravity
(CG), the point of a structure at which the combined total weight is considered to act.

Centre of pressure
The point at which the left of an aerofoil surface is considered to act.

Centre-section
The central portion of a wing.

Chaff
Narrow metallic foil strips dropped by an aircraft to confuse an enemy's radar system. Known also by its original British code name Window.

Chord
The distance from the leading- to trailing-edge of an aerofoil.

Cockpit
An open compartment in an aircraft's fuselage to accommodate a pilot, other crew member, or passenger.

Constant-speed propeller
One in which the pitch angle of the blades is changed constantly in flight to maintain engine revolutions at their optimum for maximum torque.

Cosmonaut
Term used by the Soviet Union to describe a crew member of a spacecraft.

Cowling
Usually the fairing which encloses a power plant.

Delta wing
One which, viewed in plan, has the appearance of an isosceles triangle: the apex leads, the wing trailing-edge forming the flat base of the triangle.

Digital computer
A computing device, based on the principle of counting as opposed to measuring.

Dihedral
The angle which an aerofoil surface makes to the fuselage when the wing or tailplane tip is higher than the root attachment.

Dirigible
See airship.

Ditch
To set down a landplane on water, due to emergency conditions.

Dive brake
A drag-inducing surface which can be deployed to reduce the speed of an aircraft in a dive, sometimes to prevent structural limitations being exceeded, and to improve controllability. A similar aerodynamic device used to steepen an aircraft's glide path is termed an air brake.

DME
Distance measuring equipment, a navigational aid.

Dope
A cellulose base paint used to waterproof and/or render taut the fabric covering an aircraft.

Drag
The resistance of an aerofoil to its passage through the air, creating a force parallel to the airstream.

Drag chute
A heavy-duty parachute attached to an aircraft's structure which can be deployed to reduce its landing run.

Drift
The lateral movement of an aircraft away from its true heading.

Drone
A pilotless aircraft controlled remotely from the ground, known also as an RPV (*q.v.*).

Drop tank
An auxiliary tank, generally containing fuel, which can be jettisoned if required.

ECM
Electronic countermeasures. Airborne electronic equipment designed to neutralize or reduce the effectiveness of an enemy's radar or other devices which employ or are affected by electromagnetic radiations.

Elevator
Movable control surface hinged to the aft edge of the tailplane (stabilizer) to permit control about the aircraft's lateral axis.

Elevons
Control surfaces which on a tailless aircraft combine the duties of ailerons and elevators.

Elint
Electronic intelligence, regarding the location, volume, direction and type of an enemy's electronic devices.

Envelope
The gas container of a lighter-than-air craft, or of air in a hot-air balloon or airship.

Exhaust-gas analyser
An instrument to determine and indicate the ratio of fuel to air fed into an engine.

FAA
Fleet Air Arm (UK).

FAC
Forward air controller. An officer operating in a forward position with a land force, and with suitable radio equipment to call in attack or bomber aircraft for close-support operations.

Fairing
An addition to a structure to reduce its drag coefficient.

Fillet
A faired surface to maintain a smooth airflow at the angular junction of two components for instance between wing and fuselage.

Fin
Fixed vertical surface of tail unit, to provide stability about the aircraft's vertical and longitudinal axes.

Flap
Usually a hinged wing trailing-edge surface which can be lowered partially to increase lift, or fully to increase drag.

Flaperon
Control surface which can serve the dual functions of a flap or aileron.

Flight deck
Term for a separate crew compartment of a cabin aircraft.

Flight envelope
An aircraft's performance range characteristics.

Flight plan
A written statement of the route and procedures to be adopted for other than local flights.

Flight simulator
Permits the practice of flight operations from a ground-based training device. Some are for general flight routines, but many are configured for training in the operation of a specific aircraft.

FLIR
Forward-looking infra-red, which allows a pilot to see land surface features under no-visibility conditions.

Floatplane
An alternative name for a seaplane, i.e. an aircraft which operates from water and is supported on the water's surface by floats.

Flutter
An unstable oscillation of an aerofoil surface.

Flying-boat
A heavier-than-air craft which operates from water, and in which the hull or body is the main flotation surface.

Flying wires
Bracing wires of a non-cantilever wing which take the load of the wing in flight.

Fully-feathering propeller
One in which, in the event of engine failure, the leading-edge of each blade can be turned to face the airstream. This prevents windmilling and minimizes drag.

Fuselage
The body of an aircraft.

GAM
Guided air missile. A guided and generally self-propelled weapon, designed to be air-launched from an aircraft.

Piper Super Cub – high wing.

GAR
Guided air rocket. A guided, rocket-propelled weapon, designed to be air-launched from an aircraft.

GCA
Ground controlled approach. Landing procedure when a pilot receives radio instructions for his approach and landing from a ground-based controller who monitors the operation by radar.

Glider
An unpowered heavier-than-air craft.

Gosport tube
Flexible speaking tube used for communication between instructor and pupil in early training aircraft.

Ground effect
The effect upon an aircraft of the cushion of air compressed against the ground by a helicopter's rotor or a low-flying fixed-wing aircraft.

Gyroplane (gyrocopter, 'Autogiro')
An aircraft with an unpowered rotary wing.

Harmonize
To align a fighter aircraft's guns and gunsight to produce a desired pattern of fire at a given range.

Helicopter
An aircraft with a powered rotary wing.

Helium
A valuable lifting gas for lighter-than-air craft which is not inflammable.

High-wing monoplane
One in which the wing is mounted high on the fuselage.

Horizontally-opposed engine
In which two banks of in-line cylinders are mounted horizontally opposite to each other on a common crankcase.

HUD
Head-up display, which projects performance and/or attack information on to an aircraft's windscreen for constant appraisal by the pilot.

Hull
The fuselage or body of a flying-boat.

HVAR
High velocity aircraft rocket.

Hydrogen
The lightest known lifting gas, for lighter-than-air craft; unfortunately highly inflammable.

IAS
Indicated air-speed, from an ASI, which progressively reads lower as altitude increases and air density decreases.

ICBM
Intercontinental ballistic missile (see ballistic missile).

Icing
The process of atmospheric moisture freezing upon the external surfaces of an aircraft.

IFF
Identification, friend or foe; an electronic interrogation device.

ILS
Instrument landing system, to permit safe landing in conditions of bad visibility.

ILS marker
A radio marker beacon of an instrument landing system.

Inertia starter
Starting device of many piston-engines, the inertia of a hand- or electrically-driven flywheel being connected to the engine by clutch.

In-line engine
In which cylinders are one behind another.

INS
Inertial navigation system. This relies upon the ability of three highly sensitive accelerometers to record, via a computer, the complex accelerations of an aircraft about its three axes. From this information the computer can continually integrate the aircraft's linear displacement from the beginning of the flight, pinpointing the position of the aircraft at all times.

Interdiction
The prevention of, or interference with, an enemy's movements and communication.

Interference drag
Drag induced by conflicting airstreams over adjacent streamlined structures.

ISA
International Standard Atmosphere. An agreed standard of sea-level pressure and temperature (1,013·2 millibars at 15°C) to facilitate the comparison of aircraft performance figures.

JATO
Jet-assisted take-off. The augmentation of an aircraft's normal thrust for take-off, by the use of a jet engine.

Kinetic heating
The heating of an aircraft's structure as a result of air friction.

Kite
A tethered heavier-than-air craft sustained in the air by lift resulting from airspeed in a horizontal wind.

Landing weight
The maximum weight at which an aircraft is permitted to land.

Landing wires
Bracing wires of a non-cantilever wing which support the wing when it is not in flight.

Landplane
A heavier-than-air craft which can operate from land surfaces only.

Leigh-light
An airborne searchlight used for ASW operations in World War II.

Lift
The force generated by an aerofoil, acting at right angles to the airstream which flows past it.

Longeron
A primary longitudinal member of a fuselage structure.

Loran
Long-range navigation, a navigational aid.

Low-wing monoplane
One with the wings attached at or near the bottom of the fuselage.

Mach number
The speed of a body as measured by the speed of sound in the medium in which that body is moving. Hence, Mach 0·75 represents three-quarters of the speed of sound. The speed of sound in dry air at 0°C (32°F) is about 331m (1,087ft) per second (1,193km/h; 741mph). Named after the Austrian physicist Ernst Mach.

Mean chord
The area of an aerofoil divided by its span.

Mid-wing monoplanes
One with the wings attached midway on the fuselage.

Minimum control speed
The lowest practicable speed at which an aircraft can be flown and controlled.

Monocoque
Aircraft structure in which the skin carries the primary stresses.

Monoplane
A fixed-wing aircraft with one set of wings.

NACA
National Advisory Committee of Aeronautics (US, predecessor of NASA).

NASA
National Aeronautics and Space Administration (US).

NATO
North Atlantic Treaty Organization.

Navigation lights
Mounted on the exterior of an aircraft so that its dimensions and position can be seen at night.

Omni
Omnidirectional radio range, a navigational device giving bearing from the transmitter, irrespective of the location or heading of the aircraft.

Ornithopter
Name given to a flapping-wing aircraft. Only model ornithopters have so far, proved capable of flight.

Parachute
A collapsible device which, when deployed, is used to retard the descent through the air of a falling body. Worn originally by aircrew for escape from a stricken aircraft, the parachute has since been adopted for paratroopers (parachute, equipped troops), and for such uses as the dropping of cargo and supplies, recovery of spacecraft, and to reduce the landing run of high-speed aircraft.

Parasite drag
Induced by structural components which make no contribution to lift.

Parasol monoplane
One in which the wing is mounted above the fuselage.

Pitch
The movement of an aircraft about its lateral axis, i.e. that which extends from wingtip to wingtip, resulting in up and down movements of the tail.

Pitot head
Comprising two tubes, one to measure impact pressure, the other static pressure. They are used in conjunction with an airspeed indicator which registers airspeed from the pressure difference between the two tubes.

Planing
The condition when a flying-boat's hull or seaplane's floats are supported by hydrodynamic forces.

Pressurization
A condition in an aircraft's cabin, cockpit or a compartment, where the air is maintained at a pressure higher than that of the surrounding air. This is to compensate for the lower pressure at high altitude, permitting the normal respiratory and circulatory functions of crew and passengers at heights abnormal to the body.

Port
To the left of, or on the left-hand side of an aircraft.

Propeller
Rotating blades of aerofoil section, engine driven, to create thrust for the propulsion of an aircraft. Sometimes called an airscrew.

Pulse jet engine
Essentially an aerodynamic duct with a series of spring-loaded inlet valves at the forward end. Works on intermittent combustion principle and will run in a static condition.

Pusher propeller
A propeller mounted aft of an engine.

RAAF
Royal Australian Air Force.

Radar
A method of using beamed and directed radio waves for location and detection of a variety of objects, and for navigational purposes.

Radial engine
In which the radially disposed cylinders are fixed and the crankshaft rotates.

RAE
Royal Aircraft Establishment (UK).

RAF
Royal Air Force.

Ramjet engine
Essentially an aerodynamic duct in which fuel is burned to produce a high-velocity propulsive jet. It requires acceleration to high speed before any thrust is produced.

RATO
Rocket-assisted take-off. Describes the augmentation of an aircraft's normal thrust for take-off, by the use of a rocket motor.

RCAF
Royal Canadian Air Force.

RDF
Radio direction finding. A method of obtaining by radio the bearing of an aircraft to two or three ground stations to determine its position.

Reversible-pitch propeller
One in which the blades can be turned to provide reverse thrust to reduce an aircraft's landing run.

RFC
Royal Flying Corps (UK, predecessor of RAF).

Rib
A component of an aerofoil structure, shaped to locate the covering or skin of such structure in the desired aerofoil section.

RNAS
Royal Naval Air Service (UK, predecessor of FAA).

RNZAF
Royal New Zealand Air Force.

Rocket engine
One which consists simply of an injector, combustion chambers and exhaust nozzle, and which carries with it a liquid or solid fuel and an oxidizer, allowing it to operate outside of the Earth's atmosphere.

Roll
The movement of an aircraft about its longitudinal axis, i.e. that which extends through the centre-line of the fuselage, representing a wing-over rolling movement.

Rotary engine
In which radially disposed cylinders rotate around a fixed crankshaft.

Rotor
A rotary-wing assembly, comprising the rotor blades and rotor hub, of an autogyro or helicopter.

RPV
Remotely piloted vehicle (see Drone).

Rudder
Movable control surface hinged at aft edge of fin to permit directional control.

SAAF
South African Air Force.

Sailplane
An unpowered heavier-than-air craft designed for soaring flight.

Scanner
The antenna of a radar installation which moves (scans) to pick up signal echoes.

Scarff ring
Movable mounting for a machine-gun.

Seaplane
A heavier-than-air craft which operates from water, and in which floats provide the flotation surface.

Semi-monocoque
Aircraft structure in which frames, formers and longerons reinforce the skin in carrying the load stresses.

Skin
The external covering of an aircraft's structure.

Skin friction
Drag induced by rough or pitted skin surfaces.

Slat
An auxiliary aerofoil surface, mounted forward of a main aerofoil.

Slot
The gap between a slat and the main aerofoil surface, which splits the airflow at the wing leading-edge to maintain smooth airflow at greater angles of attack.

Spacecraft
A vehicle to travel in space – that is outside the Earth's atmosphere – which can be manned or unmanned.

Spar
A principal structural member of an aerofoil surface.

Specific fuel consumption (sfc)
The amount of fuel required per hour by an engine per unit power.

Spoilers
Drag-inducing surfaces used for both lateral control and for lift dumping to improve braking and reduce landing run.

Stall
Describes the condition of an aerofoil surface when its angle of attack becomes too great and the smooth air flow over the upper surface breaks down, destroying lift.

Starboard
To the right of, or on the right-hand side of an aircraft.

Short SC–1 – VTOL.

Step
A break in the smooth undersurface of a flyingboat hull or seaplane float to resist suction and facilitate planing.

STOL
Short take-off and landing.

Streamline
To shape a structure or component so that it will cause minimum aerodynamic drag.

Stringer
An auxiliary, lightweight structural member, the primary use of which is to establish the desired external form of a structure before covering or skinning.

Strut
Usually a tube or rod in compression, frequently separating the wings of a biplane.

Subsonic
Flight at speeds below the speed of sound.

Supercharger
A form of compressor to force more air or fuel/air mixture into the cylinders of a piston-engine (to supercharge) than can be induced by the pistons at the prevailing atmospheric pressure.

Supersonic
Flight at speeds above the speed of sound.

SV-VS
Soviet Military Aviation Forces (*Sovietskaya Voenno-Vozdushnye Sily*).

Swept wing
In which the angle between the wing leading-edge and the fuselage centre-line is less than 90 degrees.

Tabs
Small auxiliary surfaces which are used to trim an aircraft so that the pilot does not have to offset manually loads induced by control surfaces.

Tailplane
(Stabilizer) horizontal aerofoil surface of tail unit, fixed or with small angular adjustment, to provide longitudinal stability.

Thickness/chord ratio
The thickness of an aerofoil expressed as a ratio of the wing chord.

Thrust
A force generated by propeller or jet efflux which propels an aircraft through the air. For stable forward flight the forces of thrust and drag must be equal.

Tractor propeller
The term for a propeller mounted forward of an engine.

Transonic
Relating to the phenomena arising when an aircraft passes from subsonic to supersonic speed.

Triplane
A fixed-wing aircraft with three superimposed wings.

Turbofan
A gas turbine in which air is ducted from the tip of the fan blades to bypass the turbine. This air, added to the jet efflux, gives high propulsive efficiency.

Turbojet
A gas turbine engine which produces a high velocity jet efflux.

Turboprop
A gas turbine which, through the medium of reduction gearing, also drives a conventional propeller.

Turboshaft
A gas turbine which, through the medium of a gearbox, drives a power take-off shaft. Used most commonly for rotary-wing aircraft.

USAAC
United States Army Air Corps (US, predecessor of USAAF).

USAAF
United States Army Air Forces (US predecessor of USAF).

USAAS
United States Army Air Service (US, predecessor of USAAC).

USAF
United States Air Force.

USMC
United States Marine Corps.

USN
United States Navy

Variable-geometry wing
One which can be swept or spread in flight to provide optimum performance for take-off/landing, cruising and high-speed flight.

Variable-pitch propeller
One in which the pitch angle of the blades can be changed in flight to give optimum performance in relation to engine load.

Vee-engine
In which two banks of in-line cylinders are mounted at an angle on a common crankcase.

V/STOL
Vertical and/or short take-off and landing.

VTOL
Vertical take-off and landing.

Wing warping
Early method of lateral control in which a flexible wing was twisted, or warped, to provide control similar to that of an aileron.

Yaw
The movement of an aircraft about its vertical axis, i.e. that which passes vertically through the junction of the lateral and longitudinal axes, representing movement of the tail unit to port or starboard.

Air Museums

The major international air shows of the world are those held in alternate years at Paris, France (1977 and 'odd' years) and Farnborough, England (1978 and 'even' years). The Paris Show is organized by the Union Syndicale des Industries Aeronautiques et Spatiales, and is normally held during the first week in June. This has been staged at Le Bourget Airport, Paris, for many years, although it may be transferred to a new site in a few years' time. The Farnborough Show is arranged by the Society of British Aerospace Companies, and has been held at its present site since shortly after World War II. The Farnborough display usually takes place during the first week in September.

Other important trade shows are held at Hannover in the Federal German Republic – this takes place during the first week in May in the same year as Farnborough shows – and at Iruma, near Tokyo in Japan, during October.

Surprisingly the USA does not have a regular showpiece to display the products of its aerospace industry, but there are two specialist shows held annually. These are the annual conventions of the National Business Aircraft Association – the location of this varies but it is held late in September – and of the Experimental Aircraft Association – which is held at Oshkosh, Wisconsin, during the first week in August. The latter is a gathering of amateur aircraft constructors from all over the United States and beyond. Other regular events in the USA are the National Air Races, held at Stead Field, Reno, Nevada, during September, and the Confederate Air Force annual air show, a unique gathering of surviving World War II aircraft types, held in October, at Harlingen, Texas.

There are many major aviation museums in countries all over the world. Much effort has been put into preserving and restoring significant historical aircraft, especially in the last ten years, a period which has seen the growth of many new museums in Europe, the USA and elsewhere. Although some of this work is comparatively recent, it is fortunate that much of the world's early aviation heritage had already been preserved for posterity – for example, 6 original Wright biplanes still exist in museum collections, together with over 30 Blériot monoplanes, and – from a more modern era – more than 70 Supermarine Spitfires.

The tabulation below includes all major museums that have aircraft on display and are open to the public, with the location of each, the number of aircraft actually on display, and some indication of the particular specialization of each.

ARGENTINA

Museo Nacional de Aeronautica, Aeroparque Airport, Buenos Aires.
Twenty aircraft, relating to the history of the Argentine Air Force and aircraft industry, from a Comper Swift of 1930 to an I.A.35 Huanquero of 1957.

AUSTRALIA

Camden Museum of Aviation, P.O. Box 72, Kogarah, New South Wales.
Twenty-two aircraft, almost all of military types, from a D.H. Moth of 1927 to two Sea Venom jet fighters of 1955.

Moorabbin Air Museum, P.O. Box 242, Mentone, Victoria. (At Moorabbin airfield.)
Ten aircraft of both military and civilian types, dating from a D.H. Moth of 1929 to a Gloster Meteor jet fighter of 1951.

Warbirds Aviation Museum, Mildura Airport, Mildura, Victoria.
Ten aircraft, all military types.

BELGIUM

Musée de l'Armée et d'Histoire Militaire, Palais du Cinquantenaire, Brussels.
Sixty aircraft relating to the history of Belgian military and civil aviation.

BRAZIL

Air Museum, Campo dos Afonsos, Rio de Janeiro.
Ten aircraft, connected with the history of Brazilian military aviation.

CANADA

Calgary Centennial Planetarium, Mewata Park, Calgary, Alberta.
Six relatively modern aircraft are displayed here.

Canadian National Aeronautical Collection, Rockcliffe Airfield, Ottawa, Ontario.
An excellent and varied collection with forty exhibits dating from 1918 to the present day, ranging from a Sopwith Camel to a Vickers Viscount. Products of the Canadian aircraft industry and aircraft which have served with Canadian airlines and the Armed Forces are comprehensively covered. About four aircraft each at Ottawa International Airport, the National War Museum, and the Museum of Science and Technology, all in Ottawa.

Canadian Warplane Heritage, Hamilton Civic Airport, Hamilton, Ont.
Fairey Firefly, Vought Corsair, B-25 Mitchell and about six other World War II aircraft, all in flying condition or being restored to it.

CZECHOSLOVAKIA

Letecka Expozice Vojenskeho Muzea, Prague-Kbely.
Forty aircraft, including gliders, light aircraft, fighters, bombers and helicopters, mostly dating from World War II or later.

National Technical Museum, Kostelni Str. 42, Prague 7.
Thirteen aircraft, dating from a 1910 Blériot, to a Mraz Sokol light aircraft of 1947.

DENMARK

Egeskov Museum, DK-5772, Kvaerndrup. (South of Odense.)
Thirteen aircraft, mostly light aircraft and gliders owned by the Royal Danish Aero Club.

FRANCE

Musée de l'Air, Aeroport du Bourget, Paris; *and* 8 rue des Vertugadins, Chalais Meudon, Hauts-de-Seine.
Two separate exhibitions, with a total of over sixty aircraft on display, together giving a comprehensive view of French aviation history, with extra exhibits from other countries.

GERMAN FEDERAL REPUBLIC

Deutschesmuseum, 800-München (Munich).
Fifteen aircraft, covering the history of German aviation, from an early Lilienthal glider to a Dornier DO 31 vertical take-off transport.

Hubschraubermuseum (Helicopter Museum), Bückeburg. (Near Minden.)
Seventeen helicopters of German, British, French and US manufacture mostly being modern designs.

Luftwaffen Museum, 2082-Uetersen, Holstein. (Near Hamburg.)
Twenty aircraft, mostly related to the post-1955 history of the Luftwaffe, but also including a Messerschmitt Bf109 and Heinkel He 111.

HOLLAND

Aviodome, National Aeronautical Museum, Schiphol Airport, Amsterdam.
Seventeen aircraft, covering the history of Dutch aviation, from a replica Fokker Spin (Spider) of 1911 to a Fokker S.14 jet trainer of 1951.

Luchtmacht Museum, Soesterburg Air Force Base. (Between Utrecht and Amersfoort.)
Twelve aircraft connected with the history of the Royal Netherlands Air Force, mostly of World War II or more modern design.

INDIA

Indian Air Museum, Palam Airfield, New Delhi.
A collection of seventeen aircraft relating to the history of the Indian Air Force, from a unique Westland Wapiti of 1935 to a Folland Gnat jet fighter.

ITALY

Museo Aeronautico Caproni di Taliedo, Via Durini 24, Milano.
Thirty-two aircraft, mostly of Italian (and many of Caproni) design. Oldest is a Caproni Ca.6 of 1910, newest a Bernardi Aeroscooter of 1961.

Museo Nazionale della Scienza e della Technica, Via San Vittore 21, Milano.
Twenty aircraft, mostly of Italian design, dating from 1910 to 1950.

NEW ZEALAND

Museum of Transport and Technology, Western Springs, Auckland 2.
Twenty aircraft, both military and civilian types, from the Pearse monoplane of 1904 to a de Havilland Vampire of 1952.

POLAND

Army Museum, Warsaw.
A display of nine military aircraft ranging from a Yak-9 of 1942 to a MiG–15 of 1954.

Museum of Aeronautics and Astronautics, Kracow.
Thirty-three aircraft, ranging from a replica Blériot XI to a Yakovlev Yak-23 jet fighter.

ROMANIA

Central Military Museum, Bucharest.
Eleven aircraft of Romanian, Italian, Russian and Canadian design, from a replica Vuia monoplane of 1906 to a MiG–15 jet fighter.

SOUTH AFRICA

South African National Museum of War History, Erlswold Way, Saxonwold, Johannesburg.
Nine aircraft are displayed, all being British or German types used in World War II.

SWEDEN

Air Force Flymuseer, F13M Malmslätt, near Linköping.
Over forty aircraft are held at this Royal Swedish Air Force base, including a wide variety of Swedish, British, German, Italian and American designs relating to the history of the Air Force.

SWITZERLAND

Verkehrshaus, Lidostrasse 5, CH–6006 Lucerne.
Thirty aircraft, dating from the Dufaux biplane of 1910 to modern jets such as a Swissair Convair 990 airliner, covering the complete spectrum of the history of the Swiss aircraft industry.

THAILAND

Royal Thai Air Force Museum, Don Muang Air Force Base, Bangkok.
Twenty-eight aircraft of types used by the Royal Thai Air Force, from a Boeing P–12 of 1931 to a North American F–86 Sabre of 1953.

UNITED KINGDOM

Cosford Aerospace Museum, Royal Air Force Cosford, Shropshire. (Near Wolverhampton.)
Twenty-five aircraft, most of which are World War II or modern types, ranging from a Spitfire to a TSR–2.

Acknowledgments

The publishers would like to thank the following individuals and organizations for their kind permission to reproduce the photographs in this book. In particular special thanks is given to John W. R. Taylor and Michael J. H. Taylor for the material used from their superb collection of prints and transparencies. Any photographs not listed below are from this collection. Aer Lingus 272 centre; The Aeroplane 238, 454–455 below; Aerospatiale 385 inset; Air Canada 136; Air France 19 below, 123 above right, 236 centre, 248 below, 249 above and centre, 273 above right; Air Ministry 168–169, 171 left, 343 below left; Air Portraits 193 above; Airship Industries 338 bottom; American Airlines 243 right, 256 above left; Basil Arkell 343 below right, 345 below, 346 above and below, 347 above and centre left, 351 above, 352, 353 above left, above centre, centre left and below centre, 354 above, 355 left inset and right inset and below inset, 356 right, 360 above left and below right, 362 centre, (Peter Shepherd) 459 centre; Associated Press 74 above centre, 85, 87 below, 154–155, 155 centre; Associated Newspapers 339 below; Australian News and Information Bureau 49 below, 132, 133 top right, 155 below; Avoco Lycoming 75 below; Ron Barnfield 125 above right; Bell Helicopter Co. 355 centre; Lewis Benjamin 290 below; Bildarchiv Preussischer Kulturbesitz 33 above, 34 inset, 34–35, 35 inset, 50 above right, 316 above, 317 above right, centre right and below right, 319 above right, centre left and below left, 320 above, 321 left, 324 above, 454 above left, 455 above left and centre; Boeing Company 104–105, 243 left, 247, 255 left, 268 above, 269 above, 275 centre, 276–277 above, 278 above left, above right, and below, 279 above left, 280 above left, 280–281 above, 391 below, 401 below; Douglas Botting 339 above; British Aircraft Corporation 272 left, 411 above left and below left; British Airways 264 below, 285 left, 141 below; C E Brown 235, 260 above, 263; David Calkin 314–315; Camera Press Ltd. 195 below, centre left, centre right, below left and below right, 446 above left, 449 centre right; J Charleville/SAF 96 centre left; CP Air 311 below left; D D L 240 above centre; Daily Telegraph 298 inset, (Mike Sheil) 298; Daily Telegraph Colour Library/The Photo Source 2–3, 366–7; Dassault-Breguet Aviation 74 below; William Gordon Davis 38 below, 183 above, 326 left, 427 below, 453 above centre right, 454 above right; Dornier-Werke GmbH 241 above right, 261 above; Douglas Aircraft Co 254–255 above, 257 above; E S A/Spacelab 414 below; Robert Estall 133 top left; Mary Evans Picture Library 40 above; Finnair 240 above right, 270; Flight International 63 above right, 142 left, 160 below; 176 below, 177 above right, 184 above, 189 above, 271 below right, 433 right; Fokker 234 above, 276 centre right; Ford Motor Company 242 centre; F P G/McFadden Air Photo 306 below; Kenneth Gatland 369 inset and below, 375 below, 376 above left and above right, 384 centre left and below, 385 above right, 391 above right, 392 left, 393 left and right, 396 above left, 397 below, 405 centre right, 409, 414 above, 447 above left, 459 above; General Dynamics Corporation 281 centre; James Gilbert front jacket, endpapers, 4–5, 8–9, 20, 21 left and right, 103 below, 104 left centre, 186 above, 187 above centre, 226–227, 290 above, 291 above left and below right, 292 above and centre, 293 centre left, 294 below, 295 above, 296 inset, 300 right, 301 above, 306 above centre inset and below centre inset, 311 below right, 312–313, 340, 341; Golden Ray 236 left, Goodyear Aerospace Co. 330–331 above, 338

centre; Bill Gunston 73 above right; Clive Hart 12 right, 14 left, 17 left, 23 below, 24 above left and above right, 26–27, 28 above left, above centre and above right, 29 above and below right, 30 above, 33 below, (Smithsonian Institution) 22 above and below, 23 above, 27 above, 29 below left, (Science Museum, Crown Copyright) 24 below, (British Museum) 28 below left and below right; F Haubner 282; Hawker Siddeley Aviation 95 below, 273 left, 362–363, 388 left, 389 above; Mike Hooks 120 below, 121 below, 122, 123 below left, below right, 124, 125 left and below right, 126 above right, above left and below, 127 above left, above centre, above right and below, 128, 129 left, centre and right, 130 above and below, 131 above centre and below, 133 below, 134 above and below, 134–135, 137 above and below; Angelo Hornak (From "Early Flying Machines 1799–1909" by Charles Gibb-Smith. Published by Eyre Methuen) 17 above right, 25 above left, centre left, 37 above, 39 above, (From "History of Rocketry and Space Travel" by Werner Von Braun and Frederick Ordway. Published by Nelson) 368 left, above, centre and below right, 369 above, above centre and below centre, 370 above left, above centre, above right and below, 371 above, below left, 373 left and above right, 375 above, 451 above centre; P Howard 139, 140 left, 141 above, 143 right, 144 above left and below; Hudson's Bay Company 245 above right; Leslie Hunt 167 centre, 190 below, 191 above and centre; Robert Hunt Library 39 centre and below; Imperial War Museum 65 above right, 119, 148–149 below, 171 below right, 172, 174 above, 181 below left and below right, 182 above left, 187 above and below, 189 below, 304 below left and above right, 321 below right, 322 below left and below right, 328 above and centre, 329 above, 333 above, centre, below left and below right, 334 above left, 373 below right; Italian Air Ministry 177 above left; Keystone Press Agency 121 above left, 133 centre, 154 above, 436 inset, 437 inset, 436–437, 438, 439, 440, 441, 442, 442–443, 443, 444, 445; K L M 232 below left; Howard Levy 103 above, 182 below, 186 centre and below, 295 right, 347 below, 355 above, 405 below; Library of Congress 48 below; Lockheed 267 above, 396 above right; Lufthansa 232 above left, 232–233 above, 248 above, 249 below, 281 below, 304 below right, 307 above; McDonnell Aviation Corporation 74 above, 164 right; Harry McDougall 239; Mansell Collection 12 left, 13, 53 below left, 318 below; Martin Marietta 410; Otto Menge 307 above centre; Ministry of Defence 163 above, 363 right; 386 below left; Minnesota Historical Society 322 above left; J G Moore Collection 91 above, 181 above centre; Kenneth Munson 63 below right; Musee de L'Air 160 above; N A M C 272 right; N A S A 412 left, 446 below left; Port Authority, New York and New Jersey 283 above; High Commissioner for New Zealand 121 above right; Norsk Teknisk Museum 336 above, centre left and centre right, 425 centre; Novosti Press Agency 145 below, 197 above left and above right, 395 below, 397 centre, 400 above right and inset, 402 left and right, 403 left, 404, 409 inset, 449 above centre left, 452 below right, 453 left, 457 below right; Panagra 252 below; Pan Am 264 above; Photri 110 above, 145 above right, 405 above centre, above right and centre left, 408 and 408 inset, 411 right, 413 right, 459 below right; Pilot Press 74 top, 155, 180 right, 180 below left, 190 below, 202, 204 below, 205 centre, 210, 211, 224 below, 279 281 centre, 284 below, 297 top right, 338 below, 356 top and below left, 363 top;

Popperfoto 104 above, 113 below, 148–149 above, 151, 178 below left, 325 above left, 334 below, 336 below right, 337 above, 339 below, 422–423, 422 inset, 426 centre, 430 below, 430–431 above, 432, 432 inset, 434 below right and above right, 446 below right, 448 above left and centre right, 452 centre and above right, 453 above right, 456 above; Pratt and Whitney 58 left; QANTAS Air Lines 41 below, 245 above left; Radio Times Hulton Picture Library 10–11, 15 left, 17 below right, 36 below, 42 above, centre left, below left and right, 43 above, below left and below right, 46 above and below, 47 below, 50 above left and below left, 51 above left, 50–51 centre, 51 centre right and below right, 52 centre and below, 53 above and below right, 54 above, 76–77 below, 101 above, 316 below right, 317 above left, 318 above, 324 below, 325 below left, above right, centre right and below right, 328 below, 329 below, 334 above right and inset, 342 above and below, 343 above, 424 below, 425 below left, 426 above and below, 427 centre, 448 below left; Pierre Regout 230 below right; John Rigby 6–7, 87 above, 146–147, 177 below, 280 below, 293 centre right, 297 above, 420–421; Herbert Rittmeyer 95 above, 99 centre, 197 centre, 291 above right, 353 above centre left; Rockwell International 180 below; Rolls Royce, Aero Division 84, 84 inset; Ronan Picture Library 454 below right; Ryan 242 left; SABENA 236 right, SAS-ABA 265 below; Satour 133 above; The Science Museum 25 right and below left, 30 below left, 32 above left, 32–33 below, 41 above, 48–49, 450–451; Peter Selinger 301 below; Short Bros 240 below; Smithsonian Institution 37 below, 47 above, 162 below, 320 below, 447 above right, 450 below; Spectrum 30 below right, 31, 71 centre left, 123 above left, 193 centre, 224 left, 296–297, 300 left, 403 below, 405 left, 413 left: Frank Spooner/Gamma 435 below; Tom Stack and Associates 335 above left; Tony Stone Associates Ltd 299; John Stroud 1, 78 below, 92 above, 228 above and below, 229, 230 left and above right, 231 above, 232–233 centre, 233 right, 234 below, 237 below, 240 above left, 241 above left and below, 246 above and below, 250 below, 252 above, 253 below, 254 centre and right, 256 below, 257 above, 258 above, 259 above left, below left and right, 260 below, 265 above, 266 above, 268 below right, 269 below left and below right, 271 below left, 283 below, 284 above left and above right, 306 above right inset, 456 below left, 458 below centre; Suddeutcher Verlag 40 below, 50 centre and below right, 51 above right, 316 below left, 319 above left, 321 above right, 322 above right, 323 above and below, 327 above and below, 330 below, 331, 332 above left and above right, 334–335, 335 above right and centre left, 337 below, 425 above and below right, 431 centre and below, 434 left, 446 centre and above right, 447 below left, above centre and below right, 448 below right, 449 above left and below right, 450 centre, 451 above left, 452 above left, 454 below left, 455 above right, 456 below right, 457 below centre; Swiss Air 237 above left and above right; Teledyne Ryan Aeronautical 67 centre, 167 below left and below right; TWA 243 centre, 254 left; United Airlines 232–233 below, 242 right; US Airforce 303 centre right, 329 centre, 372, 428 above and inset, 456 centre; US Army 398 above; US Navy 303 below, 332 below; Vought 448 below centre left; John Warwick 326 right; Gordon S. Williams 244–245, 253 centre, 307 below centre.

Illustrations drawn by Mike Badrock, The County Studio, Bill Hobson, Nigel Osborne and Tom Stimpson.